Birkhäuser

More information about this series at https://link.springer.com/bookseries/4927

Twenty Five Years of Modern Tsunami Science Following the 1992 Nicaragua and Flores Island Tsunamis. Volume II

Edited by
Utku Kânoğlu · Yuichiro Tanioka · Emile A. Okal ·
Maria Ana Baptista · Alexander B. Rabinovich

Previously published in *Pure and Applied Geophysics* (PAGEOPH),
Volume 177, No. 3, 2020

Birkhäuser

Editors
Utku Kânoğlu
Department of Engineering Sciences
Middle East Technical University
Ankara, Turkey

Emile A. Okal
Department of Earth and Planetary Sciences
Northwestern University
Evanston, IL, USA

Alexander B. Rabinovich
Department of Fisheries and Oceans
Institute of Ocean Sciences
Sidney, BC, Canada

Shirshov Institute of Oceanology
Russian Academy of Sciences
Moscow, Russia

Yuichiro Tanioka
Institute of Seismology and Volcanology
Hokkaido University
Sapporo, Japan

Maria Ana Baptista
Instituto Politécnico de Lisboa
Instituto Superior de Engenharia de Lisboa
Lisbon, Portugal

ISSN 2504-3625
Pageoph Topical Volumes
ISBN 978-3-030-86456-9

Cover illustration: Viacheslav K. Gusiakov, Pure and Applied Geophysics (Pageoph), Global Occurrence of Large Tsunamis and Tsunami-like Waves Within the Last 120 years (1900–2019), published 2020, Springer Nature Switzerland AG

This book is published under the imprint Birkhäuser, www.birkhauser-science.com by the registered company Springer Nature Switzerland AG
The registered company address is: Gewerbestrasse 11, 6330 Cham, Switzerland

Contents

Tsunami Warning

Tsunami Probability and Hazard Assessment

Tsunami Tide Interaction

Landslide Generated Tsunamis

Tsunami Hydrodynamics

Pure Appl. Geophys. 177 (2020), 1183–1191

https://doi.org/10.1007/s00024-020-02451-x

Pure and Applied Geophysics

Introduction to "Twenty Five Years of Modern Tsunami Science Following the 1992 Nicaragua and Flores Island Tsunamis, Volume II"

UTKU KÂNOĞLU,[1] YUICHIRO TANIOKA,[2] EMILE A. OKAL,[3] MARIA ANA BAPTISTA,[4,5] and ALEXANDER B. RABINOVICH[6,7]

Abstract—Following the first volume (PAGEOPH, 2019, 176, No. 7), twenty-four papers on tsunamis are included in the PAGEOPH topical issue "Twenty five years of modern tsunami science following the 1992 Nicaragua and Flores Island tsunamis: Volume II," reporting on the frontiers of tsunami science and research. The first two papers overview meteorological tsunamis, discussing progress since the 1992 Daytona event, and examining the March 2017 Persian Gulf destructive event. The next four papers review historical tsunami events, starting with a paper providing statistics for the last 120 years. The 2018 Kodiak event is investigated in the following two papers. A set of five papers discusses tsunami-warning methodologies specifically for the Australia and Nankai (Japan) regions, and general tsunami warning approaches. Probabilistic tsunami hazard assessment including case studies for two Australian coasts and the Pacific Coast of Central America, as well as discussion regarding the effect of shallow slip amplification uncertainty, and tsunami hazard assessment for the Port of Ensenada, Baja California, are presented in the next five papers. Two papers discuss tsunami tide interaction, and the following two investigate landslide-generated tsunamis, specifically a tsunami landslide scenario study for the Maltese Islands, and the 1694 Ambon, Indonesia tsunami. Tsunami hydrodynamics studies investigating shoaling on steep continental slopes and transmission of long surface, and tsunami-like waves are presented in the last two papers.

Keywords: Tsunami observations and detection, DARTs, tsunami modelling, tsunami earthquake, tsunami warning and hazard assessment, landslide generated tsunami, meteotsunamis, tsunami statistics and probability.

1. Introduction

The Nicaragua tsunami earthquake (Kanamori 1972) on 2 September 1992 and the Flores Island, Indonesia catastrophic tsunami on 12 December 1992 opened a 25-year period of numerous devastating earthquakes and tsunamis, and marked the beginning of "modern tsunami science era" (Okal 2019). This period included the 26 December 2004 Sumatra tsunami (Synolakis and Kong 2006), killing about 230,00 people, and the 11 March 2011 Tohoku (East Japan) tsunami killing almost 20,000 people and resulting in the Fukushima Dai-ichi nuclear power plant accident (Satake et al. 2013a).

In our previous issue (Kânoğlu et al. 2019), we briefly summarized significant developments in the last 25 years. Those are:

- The International Post-Tsunami Surveys became systematic after the 1992 Nicaragua and Flores Island tsunamis. These surveys turned out to be the leading resource of crucial scientific information about the events to estimate tsunami risk in specific coastal regions, but also to help tsunami scientists reconstruct the seismic sources of the events, in order to verify tsunami numerical models and identify scientific gaps in our knowledge.
- Further, the Tsunami Bulletin Board (TBB) network (tsunami_bb@infolist.nws.noaa.gov) was initiated after the 1992 events, linking all tsunami scientists practically throughout the entire world.

[1] Department of Aerospace Engineering, Middle East Technical University, 06800 Ankara, Turkey. E-mail: kanoglu@metu.edu.tr
[2] Institute of Seismology and Volcanology, Hokkaido University, Sapporo, Japan. E-mail: tanioka@mail.sci.hokudai.ac.jp
[3] Department of Earth and Planetary Sciences, Northwestern University, Evanston, IL 60208, USA. E-mail: e-okal@northwestern.edu
[4] Instituto Superior de Engenharia de Lisboa, Instituto Politécnico de Lisboa, Lisbon, Portugal. E-mail: mavbaptista@gmail.com
[5] Instituto Dom Luiz, Universidade de Lisboa, Lisbon, Portugal.
[6] Department of Fisheries and Oceans, Institute of Ocean Sciences, 9860 West Saanich Road, Sidney, BC V8L 4B2, Canada. E-mail: a.b.rabinovich@gmail.com
[7] P.P. Shirshov Institute of Oceanology, Russian Academy of Sciences, 36 Nakhimovsky Pr, Moscow 117997, Russia.

- Following the events of the last 25 years, and specifically, after the 2004 Indian Ocean tsunami, the worldwide sea-level network was upgraded. As a result, each new measurable tsunami event now provides high-quality sea level data from hundreds of instruments, which enables scientists to investigate tsunamis with higher accuracy, significantly improving their general understanding. In addition to coastal measurements, a worldwide network of DART[1] stations was deployed to obtain data during open ocean propagation. This network became the backbone of Tsunami Warning Systems (Titov 2009). Further, a seafloor network of geophysical observatories was installed at the shelf-continental slope of Vancouver Island in 2009, further improving tsunami measurements. All these measurements provide an enormous amount of data that not only allows tsunami scientists to verify their numerical models (Synolakis et al. 2008) but also the reconstruction of the source in real-time under operational tsunami warning conditions (Titov et al. 2016).
- Tsunami numerical models advanced substantially, as a combination of access to modern computers, and to high-resolution bathymetric/topographic data, of better characterizations of tsunami sources (e.g., Titov et al. 2005; Fujii and Satake 2007; Wei et al. 2008), of the explosion of data from the field, i.e., from deep ocean and near-shore measurements, and from field surveys, as well as of theoretical progress, involving for example the effects of Earth elasticity and seawater stratification on open-ocean tsunami propagation (Watada et al. 2014; Allgeyer and Cummins 2014). In this context application to real-time forecasting made substantial progress during past 25 years (Titov et al. 2016).
- Active development of paleotsunami studies enabled tsunami scientists to investigate historical events that occurred hundreds, even thousands of years ago, for example, reliably reconstructing, the Cascadia Subduction Zone earthquake and major tsunami of 26 January 1700 (Satake et al. 1996; Atwater et al. 2005). Paleotsunami findings fundamentally improve tsunami statistics, and result in much more reliable tsunami probabilistic hazard studies for tsunami-prone coastlines.

As a result, the catastrophic events of 1992–2018 and their intensive investigation resulted in tremendous progress in tsunami science, hazard mitigation, and warning. Moreover, these developments were reported in many papers in numerous journals including Pure and Applied Geophysics (PAGEOPH) topical issues (Satake and Imamura 1995; Satake et al. 2007, 2011a, b, 2013a, b; Cummins et al. 2008, 2009; Rabinovich et al. 2015a, b, 2018, 2019; Geist et al. 2016). Besides, the two catastrophic tsunamis of 2010 (Chile) and 2011 (Tohoku), as well as other strong events, generated so much attention and substantial new information that they motivated the publication of an extra, intersession volume (Rabinovich et al. 2014). In addition, high interest regarding the Illapel (Chile) earthquake and tsunami of 16 September 2015 resulted in a topical collection of regular PAGEOPH papers "Chile-2015" that were later published as a book (Braitenberg and Rabinovich 2017). Also, another PAGEOPH topical volume, related explicitly to landslide-generated tsunamis, was prepared by Bardet et al. (2003), in the wake of the destructive 1998 Aitape, Papua New Guinea landslide-generated tsunami. All these volumes document research at the frontiers of tsunami science and reflect progress made continuously in tsunami warning and hazard mitigation. Many papers published in these volumes have become classics and been highly cited.

During the 28th International Tsunami Symposium, held in Bali on 21–23 August 2017, and followed by a field trip to Babi Island, Flores on 24–25 August, the commission suggested preparation of the present topical issue. The first volume, "Twenty Five Years of Modern Tsunami Science Following the 1992 Nicaragua and Flores Island Tsunamis, Volume I," was published by PAGEOPH (2019, 176, No. 7, Kânoğlu et al. 2019). It included the overview papers by Okal (2019), Gusiakov et al. (2019) and Arcos et al. (2019), specially prepared for this volume totalling 22 papers.

[1] DART = Deep-ocean Assessment and Reporting of Tsunamis, is an effective network of deep-ocean stations elaborated for continuous monitoring of tsunami waves in the open ocean and early tsunami warning (Titov 2009; Mungov et al. 2013).

The current Volume II is the direct continuation of Volume I. Altogether, this volume comprises 24 papers that cover a broad spectrum of scientific questions and present frontiers of modern tsunami science.

2. *Meteorological Tsunamis*

A relatively new subject in tsunami science, which began to attract much attention in recent years, is meteorological tsunamis ("meteotsunamis") that are hazardous tsunami-like waves of atmospheric origin. Several destructive events of this kind occurred in the last years in various regions of the world oceans. In the opening paper of this volume, Rabinovich (2020) presents one of the first attempts to overview and classify the strongest events since 1992. A total of 51 selected events over the past 27 years were examined and described, that took place in the Mediterranean, the Black Sea, Japan, South Korea, Australia, the Great Lakes, South Africa, the USA, Canada, Brazil, the Netherlands, and other countries and regions. All meteotsunami events were separated into four groups: "Good-weather harbour," "Good-weather open-coast," "Bad-weather harbour (storm seiches)," and "Bad-weather open-coast." "Good-weather" meteotsunamis are most typical for the Mediterranean region, while "bad-weather" events mainly occur on the Atlantic coasts of the USA and Europe.

One of the most devastating meteotsunamis impacted the northeast coast of the Persian Gulf (PG) on 19 March 2017, inundating the shore in the area of Dayyer for a distance of ~ 1 km and resulting in the deaths of five people (Salaree et al. 2018). Heidarzadeh et al. (2020) examine this event in detail, using the data from 12 sea-level stations and 47 high-resolution air pressure records along the PG. The results show that the event was very local; the total length of the affected coast was ~ 80 km. The atmospheric processes during 18–22 March were found to be very active, and a number of tsunamigenic air pressure disturbances were observed propagating eastward over the PG with speeds of 21–38 m/s. The Froude number, Fr, estimated as the ratio of the air disturbance speed to the long-wave speed, on 19 March 2017 in the Dayyer region was close to resonance, $Fr \sim 0.9$–1.1, which is highly favourable for meteotsunami generation.

3. *Historical Tsunami Events*

The section on historical tsunami events includes four papers starting with Gusiakov (2020), which provides comprehensive statistics on tsunami wave heights for the last 120 years (from 1900 to 2019). All yearly-maximum tsunamis were divided into four groups: seismogenic (65%), landslide-generated (19%), volcanic (8%), and meteorological (8%). The temporal distribution of the collected extreme runup values shows that the annual occurrence of large tsunamis was relatively stable throughout the twentieth century, but with some increase during the last 27 years (since 1992).

On 8 November 1905, a large earthquake occurred offshore Mount Athos, North Aegean Sea in Greece. Triantafyllou et al. (2020) conclude that the occurrence of a tsunami of approximately 3 m runup is a consequence of the landslide triggered by the earthquake. The determination of the magnitude indicates a 7.2 event. The numerical simulation of the tsunami using two landslide scenarios concludes that a slide of ~ 2.7 million m^3 reproduces well the historical observations.

Melis et al. (2020) investigated the 23 July 1949 Chios Earthquake in the Central Aegean Sea, one of the largest in the region. The seismic moment of $M_0 = 7 \times 10^{26}$ dyn cm was obtained form mantle surface waves, ranking the Chios event second only to the 1956 Amorgos earthquake (Okal et al., 2009). Besides, they carried out a field survey collecting data from surviving witnesses and confirmed a small tsunami. They presented two possible fault planes, which can be distinguished using an extensive dataset of reported macroseismic intensities, but cannot be resolved from tsunami simulations. Also, no ancillary sources such as underwater landslides are required to explain the reported amplitudes.

The paper by Bocchini et al. (2020) addresses the tsunamigenic potential of moderate magnitude earthquakes and the challenge they pose for tsunami early warning. They present a comprehensive study of the 1 July 2009 event, an earthquake of magnitude ∼ 6.5, 80 km offshore Crete, Greece and the companion tsunami. The numerical simulation of the earthquake-induced tsunami compares well with the eyewitness reports, suggesting a high tsunamigenic potential for high-angle reverse faults in the upper-plate above the plate interface, significantly higher than that of interplate earthquakes of similar or even slightly larger moment magnitude.

4. *Tsunami Source Models and Case Studies*

Hossen et al. (2020) present a new algorithm to identify sea surface displacement and then earthquake slip distribution through tsunami waveform using time reverse imaging based on a Green's function approach. They apply their methodology to the 23 January 2018 Kodiak earthquake and model its source. They also discuss that while GPS and seismic data could provide relevant information, their methodology only needs tsunami data as input, and proves computationally efficient.

The 2018 large Kodiak earthquake generated a trans-Pacific tsunami. Wang et al. (2020) study tsunami amplification and transformation along its path using tsunami waveforms observed at 21 open-ocean sites and 27 coastal tide gauges. Wavelet analysis revealed the strong dispersive character of the propagated tsunami waves. The abrupt jump in water depth of about 4 cm detected at DART 46409, located nearby the epicenter of the 2018 Kodiak earthquake, appears to have been due to an earthquake-induced seafloor subsidence.

5. *Tsunami Warning*

The recent developments in the tsunami modelling components of the Indonesia Tsunami Early Warning System (InaTEWS) are presented in Harig et al. (2020). They consist of a dual system running in parallel, with a high-resolution scenario database precomputed with the finite element model TsunAWI, and the GPU-parallelized linear long-wave model easyWave, handling events outside the database coverage. They discussed tsunami warning products, i.e., tsunami wave heights and arrival times, estimated through TsunAWI and easyWave. In addition, they review the performance of the warning system for the past events pointing out to system updates.

Reliable and robust determination of the initial tsunami source is essential for tsunami early warnings. Even though near-field Global Navigation Satellite System (GNSS) observations have shown promise to constrain source inversion, they remain sparse along most active faults. Chen et al. (2020) present an automated earthquake finite source inversion methodology that combines near-field GNSS data and mid-range teleseismic P displacement waveforms. They implemented their algorithm successfully as part of the prototype of JPL's GPS-Aided Tsunami Early-Detection system, and tested it for many recent events. They present their real-time results and discuss future improvements.

The Joint Australian Tsunami Warning Center (JATWC) issues three levels of threat: Land Threat, Marine Threat or No Threat along the coast. Greenslade et al. (2020) evaluate the threshold for a Land Threat using tsunami inundation simulations. The result indicated that the threshold value was found to be about 50 cm.

Fauzi and Mizutani (2020) develop a real-time tsunami inundation forecasting method, using machine-learning algorithms. The method was applied to the hypothetical Nankai megathrust earthquake in Japan. The results show that the method can be fast (less than 1 s) and yield results comparable with nonlinear numerical inundation modelling. These constitute promising results for a real-time tsunami inundation forecasting.

Recently, offshore tsunami observations have become important for tsunami early warning. Meza et al. (2020) develop a methodology to optimize the design of an offshore tsunami observation network, which proved significantly useful for the implementation of a network in locations where the deployment of denser arrays is difficult.

6. *Tsunami Probability and Hazard Assessment*

Zamora and Babeyko (2020) study tsunamigenic sources from the Middle America subduction zone that affect the Central American coast. They performed a probabilistic tsunami hazard analysis (PTHA) that considers local and regional seismic sources, in order to estimate the probabilities of exceeding specific tsunami amplitudes along the Central American Pacific coast. The study shows that tsunami heights are likely to exceed 2 m at the 50–80% probability level for 500-year time exposure along the coasts of northwestern Costa Rica, El Salvador, and Nicaragua, southern Colombia, and northern Ecuador. They discuss the large dependence of PTHA on model assumptions, which require integration of all possible uncertainties to perform rigorous hazard models.

A large slip concentrated in a shallow plate interface was observed in several recent large tsunamigenic earthquakes. Scala et al. (2020) propose a method for a PTHA incorporating underthrust earthquakes with enhanced shallow slips. The mean hazard curves obtained using the proposed method show increased probabilities for larger inundation heights, as compared to the typical curves derived from depth-independent slip distributions.

Davis and Griiffin (2020) performed a PTHA sensitivity study for the Australian coastline for six far-field sources considering two rigidity models combined with three slip models. They found that the tsunami hazard offshore Australia is insensitive to the choice of rigidity model, yet significantly affected by the choice of slip model. The fixed-area-uniform-slip model produces lower hazard than both a variable-area-uniform-slip model and a spatially heterogeneous slip model.

Kain et al. (2020) investigate the potential impact at Hobart Airport and nearby coastal communities from a tsunami along the Puysegur subduction zone, off New Zealand's southwest coast. They simulate a maximum credible earthquake (M_w 8.7) and provide input to local hazard management plans. They identify severe inundation along the east coast, with a flow depth of more than 4 m. They suggest palaeotsunami studies, which would not only validate the results of their modelling, but also provide inundation extent and recurrence intervals of events having previously affected Tasmania.

Ortiz-Huerta (2020) investigate velocity field in and around the Port of Ensenada, Baja California, Mexico induced by the hypothetical tsunamis resulting from M_w 9.3 earthquakes around the Pacific Ocean. Tsunami induced currents were determined as 4–6 knots ($\sim$ 2–3 m/s) at the harbour entrance and 2–4 knots ($\sim$ 1–2 m/s) inside and outside the harbour. Currents are negligible for depths greater than 120 m, which is approximately 15 km offshore from the harbour. Ortiz-Huerta (2020) further suggests deploying an adequate array of pressure gauges and current-meters operating in real-time for marine traffic control.

7. *Tsunami Tide Interaction*

Didenkulova and Pelinovsky (2020) solve the nonlinear shallow water-wave equations analytically over a sloping beach in the presence of a tide in addition to incoming waves. They analyse the influence of the tide on tsunami runup characteristics. First, using an approximate theory, they consider a static nearshore tide, uniform in space during the tsunami travel to the shore. They then consider the case where the tide is not static, showing that the runup oscillations can be presented as a linear superposition of tsunami and tide, even for strongly nonlinear incoming tsunamis. They discuss that nonlinear effects are expressed as weak variations in shoreline velocity with runup and rundown velocities increasing during low tide and high tide, respectively.

Wilson and Power (2020) examine the effect of accounting for tides in numerical models used to estimate the potential impact of tsunamis on coastal regions. A series of numerical experiments were provided by the authors to demonstrate the differences between tsunami models using static or dynamic tide inputs; the simulations were done for two New South Wales estuaries, Sydney harbour, and port hacking, Australia. The effects of geomorphological constrictions were found to affect downriver maximum water levels, tsunami wave heights, upriver water accumulation, and inundation maxima.

8. *Landslide Generated Tsunamis*

Mueller et al. (2020) investigate the vulnerability of Malta against tsunami events considering four scenario events; two landslide sources from slides at outer Malta Plateau and Gela, and two earthquake sources mimicking the 365 A.D. western Hellenic Arc and 1693 south-east of Sicily events. All scenario events caused inundation in densely populated areas, with flow depth exceeding 10 m. Their results show that the Malta Escarpment and Sicily amplify and reflect waves resulting in hot spots along the eastern coastline. Their overall conclusion is the existence of a potential tsunami hazard for Malta, with further studies requiring a better understanding of source mechanisms, as well as probabilistic hazard and risk assessment.

Using detailed historical accounts Pranantyo and Cummins (2020) study the 1674 Ambon, Indonesia tsunami runup, ranking as one of the country's deadliest with 100 m runup and 2300 fatalities. They conclude that runup observed on the north coast of Ambon was generated by an earthquake-triggered coastal landslide; they consider as plausible seismic sources from the Seram Megathrust, the South Seram Thrust, and faults local to Ambon. They use a two-layer tsunami model with an approximate volume of 1 km^3 and explain the observed tsunami runup along the Ambon coast. They emphasize that landslide tsunamis are significant for tsunami hazards along the Indonesian coast, such as the 1992 Flores, 2018 Palu, and Sunda Strait tsunamis.

9. Tsunami Hydrodynamics

In the context of Synolakis' (1991) previous study of the evolution of long waves over a sloping beach under Green's law, George et al. (2020) investigate the amplification of long waves over the continental shelf. They show that Green's law is a good approximation for sufficiently gentle continental slopes, i.e., when the width of the slope is large compared to the wavelength. They discuss asymptotic values for the transmission and reflection coefficients for a sharp interface, showing that incoming waves reflect more at steeper slopes, resulting in less amplification. Further, they discuss that many realistic tsunamis fall into an intermediate region, an example being the 2011 Tohoku earthquake tsunami approaching the Oregon coast.

Ermakov and Stepanynts (2020) consider a transformation of long linear waves in an ocean with variable depth. The authors calculate the transmission and reflection coefficients as functions of frequency and total depth for three typical models of bottom profile variation: (1) piecewise-linear, (2) piecewise-quadratic, and (3) hyperbolic tangent profiles. For all these cases, exact solutions are obtained, analysed and graphically illustrated.

10. Conclusions

In 2018, while these last two volumes were progressing, the Palu, Sulawesi, Indonesia (e.g. Heidarzadeh et al. 2019; Omira et al. 2019; Putra et al. 2019) and Anak-Krakatau, Indonesia (e.g. Muhari et al. 2019; Paris et al., 2020) highly destructive tsunamis occurred. Even though we included papers presenting substantial developments over the various field of tsunami science in these two volumes, the 2018 Palu and Anak events showed that "our progress is, and will remain, a never-ending story," (Okal 2019). These two events alone left us with questions as to why the tsunami warning system failed to save lives in Indonesia, what their precise sources were, and how the social dynamics of a region can affect early warning systems (UNDRR and UNESCO-IOC 2019). We hope that papers in this volume will help resolve issues raised by the Palu and Anak tragedy events, as well as other challenging issues which might be raised in the future.

Acknowledgements

We acknowledge Professor Kenji Satake (University of Tokyo, Japan) and Dr. Vasily Titov (PMEL/NOAA, Seattle, WA, USA) for their support. We thank the authors who contributed papers to these topical volumes. Finally, we would like to especially thank the reviewers who shared their time, effort, and expertise to maintain the scientific rigour of this volume. AR's contribution was partially supported by the Russian State Assignment of IORAS 0149-2019-0005.

Publisher's Note Springer Nature remains neutral with regard to jurisdictional claims in published maps and institutional affiliations.

References

Allgeyer, S., & Cummins, P. (2014). Numerical tsunami simulation including elastic loading and seawater density stratification. *Geophysical Research Letters, 41*(7), 2368–2375. https://doi.org/10.1002/2014gl059348.

Arcos, N. P., Dunbar, P. K., Stroker, K. J., & Kong, L. S. L. (2019). The impact of post-tsunami surveys on the NCEI/WDS global historical tsunami database. *Pure and Applied Geophysics, 176*(7), 2809–2829. https://doi.org/10.1007/s00024-019-02191-7.

Atwater, B. F., Musumi-Rokkaku, S., Satake, K., Tsuji, Y., Ueda, K., & Yamaguchi, D. K. (2005). *The Orphan Tsunami of 1700: Japanese Clues to a Parent Earthquake in North America*. US Geological Survey Professional Paper No. 1707. https://doi.org/10.3133/pp1707.

Bardet, J.-P., Synolakis, C. E., Davies, H. L., Imamura, F., & Okal, E. A. (2003). Landslide tsunamis: Recent findings and research directions. *Pure and Applied Geophysics, 160*(10–11), 1793–1809. https://doi.org/10.1007/s00024-003-2406-0.

Bocchini, G. A., Novikova, T., Papadopoulos, G. A., Agalos, A., Mouzakiotis, E., Karastathis, V., et al. (2020). Tsunami potential of moderate earthquakes: The July 1, 2009 earthquake (Mw 6.45) and its associated local tsunami in the Hellenic Arc. *Pure and Applied Geophysics*. https://doi.org/10.1007/s00024-019-02246-9.

Braitenberg, C., & Rabinovich, A. B. (Eds.). (2017). *The Chile-2015 (Illapel) Earthquake and Tsunami*. Basel: Birkhäuser/Springer. https://doi.org/10.1007/978-3-319-57822-4.

Chen, K., Liu, Z., & Song, T. Y. (2020). Automated GNSS and teleseismic earthquake inversion (AutoQuake Inversion) for tsunami early warning: Retrospective and real-time results. *Pure and Applied Geophysics*. https://doi.org/10.1007/s00024-019-02252-x.

Cummins, P. R., Kong, L. S. L., & Satake, K. (2008). Introduction to "Tsunami science four years after the 2004 Indian Ocean tsunami. Part I: Modelling and hazard assessment". *Pure and Applied Geophysics, 165*(11–12), 1983–1989. https://doi.org/10.1007/s00024-008-0423-8

Cummins, P. R., Kong, L. S. L., & Satake, K. (2009). Introduction to "Tsunami science four years after the 2004 Indian Ocean tsunami, Part II: Observation and data analysis". *Pure and Applied Geophysics, 166*(1–2), 1–7. https://doi.org/10.1007/s00024-009-0442-0

Davis, G., & Griiffin, J. (2020). Sensitivity of probabilistic tsunami hazard assessment to far-field earthquake slip complexity and rigidity depth dependence: Case study of Australia. *Pure and Applied Geophysics*. https://doi.org/10.1007/s00024-019-02299-w.

Didenkulova, I., & Pelinovsky, E. (2020). Tsunami runup on a plane beach in tidal environment. *Pure and Applied Geophysics*. https://doi.org/10.1007/s00024-019-02332-y.

Ermakov, A. M., & Stepanynts, Y. A. (2020). Transformation of long surface and tsunami-like waves in the ocean with a variable bathymetry. *Pure and Applied Geophysics*. https://doi.org/10.1007/s00024-019-02259-4.

Fauzi, A., & Mizutani, N. (2020). Machine learning algorithms for real-time tsunami inundation forecasting: A case study in Nankai region. *Pure and Applied Geophysics*. https://doi.org/10.1007/s00024-019-02364-4.

Fujii, Y., & Satake, K. (2007). Tsunami source of the 2004 Sumatra-Andaman earthquake inferred from tide gauge and satellite data. *Bulletin of the Seismological Society of America, 97*(1A), S192–S207. https://doi.org/10.1785/0120050613.

Geist, E. L., Fritz, H. M., Rabinovich, A. B., & Tanioka, Y. (2016). Introduction to "Global tsunami science: Past and future, Volume I". *Pure and Applied Geophysics, 173*(12), 3663–3669. https://doi.org/10.1007/s00024-016-1427-4.

George, J., Ketcheson, D. I., & LeVeque, R. J. (2020). Shoaling on steep continental slopes: Relating transmission and reflection coefficients to Green's law. *Pure and Applied Geophysics*. https://doi.org/10.1007/s00024-019-02316-y.

Greenslade, D. J. M., Uslu, B., Allen, S. C. R., Kain, C. L., Wilson, K. M., & Power, H. E. (2020). Evaluation of Australian tsunami warning thresholds using inundation modelling. *Pure and Applied Geophysics*. https://doi.org/10.1007/s00024-019-02377-z.

Gusiakov, V. K. (2020). Global occurrence of large tsunamis and tsunami-like waves within the last 120 years (1900–2019). *Pure and Applied Geophysics*. https://doi.org/10.1007/s00024-020-02437-9.

Gusiakov, V. K., Dunbar, P. K., & Arcos, N. (2019). Twenty-five years (1992–2016) of global tsunamis: Statistical and analytical overview. *Pure and Applied Geophysics, 176*(7), 2795–2807. https://doi.org/10.1007/s00024-019-02113-7.

Harig, S., Immerz, A., Weniza, Griffin, J., Weber, B., Babeyko, A., et al. (2020). The tsunami scenario database for the Indonesia Tsunami early warning system (InaTEWS): Evolution of the coverage and the involved modeling approaches. *Pure and Applied Geophysics*. https://doi.org/10.1007/s00024-019-02305-1.

Heidarzadeh, M., Muhari, A., & Wijanarto, A. B. (2019). Insights on the source of the 28 September 2018 Sulawesi Tsunami, Indonesia based on spectral analyses and numerical simulations. *Pure and Applied Geophysics, 176*(1), 25–43. https://doi.org/10.1007/s00024-018-2065-9.

Heidarzadeh, M., Šepić, J., Rabinovich, A., Allahyar, M., Soltanpour, A., & Tavakoli, F. (2020). Meteorological tsunami of 19 March 2017 in the Persian Gulf: Observation and analyses. *Pure and Applied Geophysics*. https://doi.org/10.1007/s00024-019-02263-8.

Hossen, J. M., Sheehan, A. F., & Satake, K. (2020). A multi-fault model estimation from tsunami data: An application to the 2018 M7.9 Kodiak earthquake. *Pure and Applied Geophysics*. https://doi.org/10.1007/s00024-019-02433-z.

Kain, C. L., Lewarn, B., Rigby, E. H., & Mazengarb, C. (2020). Tsunami inundation and maritime hazard modelling for a maximum credible tsunami scenario in southeast Tasmania, Australia. *Pure and Applied Geophysics*. https://doi.org/10.1007/s00024-019-02384-0.

Kanamori, H. (1972). Mechanism of tsunami earthquakes. *Physics of the Earth and Planetary Interiors, 6*(5), 346–359. https://doi.org/10.1016/0031-9201(72)90058-1.

Kânoğlu, U., Tanioka, Y., Okal, E. A., Baptista, M. A., & Rabinovich, A. (2019). Introduction to "Twenty five years of modern tsunami science following the 1992 Nicaragua and Flores Island tsunamis, Volume I". *Pure and Applied Geophysics, 176*(7), 2757–2769. https://doi.org/10.1007/s00024-019-02266-5.

Melis, N., Okal, E. A., Synolakis, C. E., Kalogeras, I. S., & Kânoğlu, U. (2020). The Chios, Greece earthquake of 23 July 1949: Seismological reassessment and tsunami investigation. *Pure and Applied Geophysics*. https://doi.org/10.1007/s00024-019-02410-1.

Meza, J., Catalán, P. A., & Tsushima, H. (2020). A multiple parameter methodology for placement of tsunami sensor networks. *Pure and Applied Geophysics*. https://doi.org/10.1007/s00024-019-02381-3.

Mueller, C., Micallef, A., Spatola, D., & Wang, X. (2020). The tsunami inundation hazard of the Maltese Islands (central Mediterranean Sea): A submarine landslide and earthquake tsunami scenario study. *Pure and Applied Geophysics*. https://doi.org/10.1007/s00024-019-02388-w.

Muhari, A., Heidarzadeh, M., Susmoro, H., Nugroho, H. D., Kriswati, E., Supartoyo, et al. (2019). The December 2018 Anak Krakatau volcano tsunami as inferred from post-tsunami field surveys and spectral analysis. *Pure and Applied Geophysics, 176*, 5219–5233. https://doi.org/10.1007/s00024-019-02358-2.

Mungov, G., Eblé, M., & Bouchard, R. (2013). DART® tsunameter retrospective and real-time data: A reflection on 10 years of processing in support of tsunami research and operations. *Pure and Applied Geophysics, 170*(9–10), 1369–1384. https://doi.org/10.1007/s00024-012-0477-5.

Okal, E. A. (2019). Twenty-five years of progress in the science of "geological" tsunamis following the 1992 Nicaragua and Flores events. *Pure and Applied Geophysics, 176*(7), 2771–2793. https://doi.org/10.1007/s00024-019-02244-x.

Okal, E. A., Synolakis, C. E., Uslu, B., Kalligeris, N., & Voukouvalas, E. (2009). The 1956 earthquake and tsunami in Amorgos, Greece. *Geophysical Journal International, 178*, 1533–1554. https://doi.org/10.1111/j.1365-246X.2009.04237.x.

Omira, R., Dogan, G. G., Hidayat, R., et al. (2019). The September 28th, 2018, tsunami in Palu-Sulawesi, Indonesia: A post-event field survey. *Pure and Applied Geophysics, 176*, 1379–1395. https://doi.org/10.1007/s00024-019-02145-z.

Ortiz-Huerta, L. G. (2020). Far-field tsunami hazard assessment along the Pacific Coast of Mexico by historical records and numerical simulation. PART II: Tsunami-Induced currents in the Port of Ensenada, Baja California. *Pure and Applied Geophysics*. https://doi.org/10.1007/s00024-019-02319-9.

Paris, A., Heinrich, P., Paris, R., & Abadie, S. (2020). The December 22, 2018 Anak Krakatau, Indonesia, landslide and tsunami: Preliminary modeling results. *Pure and Applied Geophysics, 177*(2), 571–590. https://doi.org/10.1007/s00024-019-02394-y.

Pranantyo, I. R., & Cummins, P. R. (2020). The 1674 Ambon tsunami: Extreme run-up caused by an earthquake-triggered landslide. *Pure and Applied Geophysics*. https://doi.org/10.1007/s00024-019-02390-2.

Putra, P. S., Aswan, A., Maryunani, K. A., Yulianto, E., & Kongko, W. (2019). Field survey of the 2018 Sulawesi tsunami deposits. *Pure and Applied Geophysics, 176*(6), 2203–2213. https://doi.org/10.1007/s00024-019-02181-9.

Rabinovich, A. B. (2020). Twenty-seven years of progress in the science of meteorological tsunamis following the 1992 Daytona Beach event. *Pure and Applied Geophysics*. https://doi.org/10.1007/s00024-019-02349-3.

Rabinovich, A. B., Borrero, J. C., & Fritz, H. M. (2014). Introduction to "Tsunamis in the Pacific Ocean: 2011–2012". *Pure and Applied Geophysics, 171*(12), 3175–3182. https://doi.org/10.1007/s00024-014-0894-8.

Rabinovich, A. B., Fritz, H. M., Tanioka, Y., & Geist, E. L. (2017). Introduction to "Global tsunami science: Past and future, Volume II". *Pure and Applied Geophysics, 174*(8), 2883–2889. https://doi.org/10.1007/s00024-017-1638-3.

Rabinovich, A. B., Fritz, H. M., Tanioka, Y., & Geist, E. L. (2018). Introduction to "Global tsunami science: Past and future, Volume III".*Pure and Applied Geophysics, 175*(4), 1231–1237. https://doi.org/10.1007/s00024-018-1851-8.

Rabinovich, A. B., Geist, E. L., Fritz, H. M., & Borrero, J. C. (2015a). Introduction to "Tsunami science: Ten years after the 2004 Indian Ocean tsunami, Volume I". *Pure and Applied Geophysics, 172*(3–4), 615–619. https://doi.org/10.1007/s00024-015-1038-5.

Rabinovich, A. B., Geist, E. L., Fritz, H. M., & Borrero, J. C. (2015b). Introduction to "Tsunami science: Ten years after the 2004 Indian Ocean tsunami, Volume II". *Pure and Applied Geophysics, 172*(12), 3265–3270. https://doi.org/10.1007/s00024-015-1175-x.

Salaree, A., Mansouri, R., & Okal, E. A. (2018). The intriguing tsunami of 19 March 2017 at Bandar Dayyer, Iran: Field survey and simulations. *Natural Hazards, 90*(3), 1277–1307.

Satake, K., & Imamura, F. (1995). Introduction to "Tsunamis: 1992–94". *Pure and Applied Geophysics, 144*(3–4), 373–379. https://doi.org/10.1007/BF00874373.

Satake, K., Okal, E. A., & Borrero, J. C. (2007). Tsunami and its hazard in the Indian and Pacific Oceans: Introduction. *Pure and Applied Geophysics, 164*(2–3), 249–259. https://doi.org/10.1007/s00024-006-0172-5.

Satake, K., Rabinovich, A.B., Dominey-Howes, D., & Borrero, J. C. (2013a). Introduction to "Historical and recent catastrophic tsunamis in the world: Volume I. The 2011 Tohoku tsunami". *Pure and Applied Geophysics, 170*(6–8), 955–961. https://doi.org/10.1007/s00024-012-0615-0.

Satake, K., Rabinovich, A. B., Dominey-Howes, D., & Borrero, J. C. (2013b). Introduction to "Historical and recent catastrophic tsunamis in the world: Volume II. Tsunamis from 1755 to 2010". *Pure and Applied Geophysics, 170*(9–10), 1361–1367. https://doi.org/10.1007/s00024-012-0616-z.

Satake, K., Rabinovich, A., Kânoğlu, U., & Tinti, S. (2011a). Introduction to "Tsunamis in the World Ocean: Past, present, and future. Volume I". *Pure and Applied Geophysics, 168*(6–7), 963–968. https://doi.org/10.1007/s00024-011-0304-4.

Satake, K., Rabinovich, A., Kânoğlu, U., & Tinti, S. (2011b). Introduction to "Tsunamis in the World Ocean: Past, present, and future. Volume II". *Pure and Applied Geophysics, 168*(11), 1913–1917. https://doi.org/10.1007/s00024-011-0278-2.

Satake, K., Shimazaki, K., Tsuji, Y., & Ueda, K. (1996). Time and size of a giant earthquake in Cascadia inferred from Japanese tsunami records of January 1700. *Nature, 379*, 246–249.

Scala, A., Lorito, S., Romano, F., Murphy, S., Selva, J., Basili, R., et al. (2020). Effect of shallow slip amplification uncertainty on probabilistic tsunami hazard analysis in subduction zones: Use of long-term balanced stochastic slip models. *Pure and Applied Geophysics*. https://doi.org/10.1007/s00024-019-02260-x.

Synolakis, C. E. (1991). Green's law and the evolution of solitary waves. *Physics of Fluids A, 3*(3), 490–491. https://doi.org/10.1063/1.858107.

Synolakis, C. E., Bernard, E. N., Titov, V. V., Kânoğlu, U., & González, F. I. (2008). Validation and verification of tsunami numerical models. *Pure and Applied Geophysics*, *165*(11–12), 2197–2228. https://doi.org/10.1007/s00024-004-0427-y.

Synolakis, C. E., & Kong, L. (2006). Runup measurements of the December 2004 Indian Ocean tsunami. *Earthquake Spectra*, *22*(S3), S67–S91. https://doi.org/10.1193/1.2218371.

Titov, V. V. (2009). Tsunami forecasting. In E. N. Bernard & A. R. Robinson (Eds.), *The sea, ideas and observations on progress in the study of the seas* (Vol. 15. Chap. 12, pp. 371–400). Cambridge: Harvard University Press.

Titov, V. V., González, F. I., Bernard, E. N., Eble, M. C., Mofjeld, H. O., Newman, J. C., et al. (2005). Real-time tsunami forecasting: Challenges and solutions. *Natural Hazards, 35*(1), 35–41. https://doi.org/10.1007/1-4020-3607-8_3.

Titov, V., Kânoğlu, U., & Synolakis, C. (2016). Development of MOST for real-time tsunami forecasting. *Journal of Waterway Port, Coastal, and Ocean Engineering, 142*(6), 03116004. https://doi.org/10.1061/(asce)ww.1943-5460.0000357.

Triantafyllou, I., Zaniboni, F., Armigliato, A., Tinti, S., & Papadopoulos, G. A. (2020). The large earthquake (~ M7) and its associated tsunami of 8 November 1905 in Mt Athos, Northern Greece. *Pure and Applied Geophysics*. https://doi.org/10.1007/s00024-019-02363-5.

UNDRR and UNESCO-IOC (2019). Limitations and Challenges of Early Warning Systems: A Case Study of the 2018 Palu-Donggala Tsunami. United Nations Office for Disaster Risk Reduction (UNDRR), Regional Office for Asia and the Pacific, and the Intergovernmental Oceanographic Commission of United Nations Educational, Scientific and Cultural Organization. (IOC Technical Series No 150).

Wang, K., Thomson, R. E., Rabinovich, A. B., Fine, I. V., & Insua, T. L. (2020). The 2018 Alaska-Kodiak tsunami off the west coast of North America: A rare midplate tsunamigenic event. *Pure and Applied Geophysics*. https://doi.org/10.1007/s00024-019-02427-x.

Watada, S., Kusumoto, S., & Satake, K. (2014). Traveltime delay and initial phase reversal of distant tsunamis coupled with the self-gravitating elastic Earth. *Journal of Geophysical Research: Solid Earth, 119*(5), 4287–4310. https://doi.org/10.1002/2013jb010841.

Wei, Y., Bernard, E. N., Tang, L., Weiss, R., Titov, V. V., Moore, C., et al. (2008). Real-time experimental forecast of the Peruvian tsunami of August 2007 for US coastlines. *Geophysical Research Letters, 35*(4), 25. https://doi.org/10.1029/2007gl032250.

Wilson, K. M., & Power, H. E. (2020). Tsunami modelling with static and dynamic tides in drowned river valleys with morphological constrictions. *Pure and Applied Geophysics*. https://doi.org/10.1007/s00024-019-02411-0.

Zamora, N., & Babeyko, A. Y. (2020). Probabilistic tsunami hazard assessment for local and regional seismic sources along the Pacific coast of Central America with emphasis on the role of selected uncertainties. *Pure and Applied Geophysics*. https://doi.org/10.1007/s00024-019-02372-4.

(Published online March 5, 2020)

Pure Appl. Geophys.177 (2020), 1193–1230

https://doi.org/10.1007/s00024-019-02349-3

Pure and Applied Geophysics

Twenty-Seven Years of Progress in the Science of Meteorological Tsunamis Following the 1992 Daytona Beach Event

Alexander B. Rabinovich[1,2]

Abstract—Meteorological tsunamis ("meteotsunamis") are hazardous tsunami-like waves of atmospheric origin. They have typical periods from a few minutes to about 3 h and typical spatial scales from a few hundred meters to approximately 100–150 km. The waves have different local names in different regions of the world: "rissaga" in the Balearic Islands (Spain), "marrobbio" in Sicily (Italy), "šćiga" in the Adriatic Sea (Croatia), "milghuba" in Malta, and "abiki" in Japan. Meteotsunamis have markedly different generation mechanisms than storm surge or rogue waves, and are mainly produced by direct air pressure forcing. Several recent destructive meteotsunami events have attracted considerable attention to the phenomenon. The present paper is one of the first attempts to classify and overview the strongest events. A total of 51 selected events over the past 27 years are examined and described. Some of these events are well known (e.g. the events of 4 July 1992 Daytona Beach, Florida, 15 June 2006 Ciutadella Harbour, Spain, and 13 June 2013 East Coast of USA), while others have only been mentioned in the media and on the Internet. The list of events includes those that have occurred in the Mediterranean, the Black Sea, Japan, South Korea, Australia, the Great Lakes, South Africa, the USA, Canada, Brazil, the Netherlands and other countries and regions. All meteotsunami events are separated into four groups: "Good-weather harbour", "Good-weather open-coast", "Bad-weather harbour (storm seiches)" and "Bad-weather open-coast". "Good-weather" meteotsunamis are most typical for the Mediterranean region, while "bad-weather" events mainly occur on the Atlantic coasts of the USA and Europe.

Key words: Meteotsunami, tide gauge records, tsunami-like waves, rissaga, marrobbio, abiki, šćiga, seiches, harbour oscillations, Proudman resonance.

The topical tsunami issue: "Twenty Five Years of Modern Tsunami Science, Volume II" of *Pure and Applied Geophysics*.

[1] Institute of Ocean Sciences, Fisheries and Oceans Canada, 9860 West Saanich Road, Sidney, BC V8L 4B2, Canada. E-mail: A.B.Rabinovich@gmail.com

[2] P.P. Shirshov Institute of Oceanology, Russian Academy of Sciences, 36 Nakhimovsky Prosp, Moscow 117997, Russia.

1. Introduction

On the clear and calm evening of 3 July 1992 (local time, but 4 July 1992 UTC), when the sea was almost quiet, a huge, sudden tsunami-like solitary wave impacted Daytona Beach on the Atlantic coast of Central Florida. The wave injured about 75 people and severely damaged parked vehicles and beach infrastructure (Churchill et al. 1995). If the wave had occurred during the 4th of July Independence Day holiday on the following afternoon, thousands of vehicles and tens of thousands of people on the beach would have been at highest risk (Sallenger et al. 1995).

The natural assumption in the case of such unexpected devastating waves, is that they were wave caused either by a seismic event or by an offshore landslide. But at the time of the Daytona event, no seismic activity was detected, and no signature of submarine landslide was found (Sallenger et al. 1995). As a consequence, the two groups working on the event (Sallenger et al. 1995; Churchill et al. 1995), came to the same conclusion, that the wave was produced by a squall line and associated pressure jump that had been traveling along the coast of Florida at the speed of long ocean waves and resonantly generating an amplified "*squall-line surge*". The authors tried to locate other historical examples of destructive "non-seismic" tsunami-like waves, but the only similar event they could find was the catastrophic flood of 26 June 1954 in Lake Michigan, which also appeared to be induced by a resonantly moving squall-line (Ewing et al. 1954; Platzman 1965).

In fact, strong, sometimes catastrophic, high-frequency atmospherically-induced sea level oscillations had been known for a long time in certain regions of the World Ocean, including Ciutadella Inlet (Harbour), Menorca Island, Spain (Tintoré et al. 1988; Gomis et al. 1993); Vela Luka Bay, Korčula Island, Croatia (Orlić 1980); Mazara del Vallo, Sicily, Italy (Colucci and Michelato 1976); Nagasaki Bay, Japan (Honda et al. 1908; Hibiya and Kajiura 1982); and Longkou Harbour, China (Wang et al. 1987). For this type of oscillation, Nomitsu (1935) and Defant (1961) suggested the general term: "*meteorological tsunamis*". The Daytona event of 3 July 1992 was very close to a generic meteorological tsunami event, but at the same time it might be considered as unique in that:

1. It occurred in a region where such waves had never been reported before.
2. The wave height was extreme: according to Churchill et al. (1995) it was $\sim$ 3 m, while Sallenger et al. (1995) report a height of 6 m. It is evident that this was one of the highest atmospherically induced tsunami-like waves ever observed.
3. The wave injured 75 people, many more than any other known wave of this type.
4. All "hot" places listed above are harbours, bays or inlets; the destructive waves generated there had a character of *extreme seiches* (cf. Rabinovich 2009), i.e. multiple standing oscillations. The wave recorded at Daytona had the character of a *solitary wave*.

In many respects, the Daytona event was a milestone in the examination of such phenomenon. It was strong, unexpected, destructive and demonstrated that such waves can occur anywhere in the world's oceans. Until the beginning of 1990s, the term "*meteorological tsunami*" (or "*meteotsunami*"), had never been considered in the frame of general tsunami studies, despite the fact that "*tsunamis of unknown origin*" (that were likely meteotsunamis) were common in tsunami catalogues (cf. Soloviev and Go 1974, 1975; Lander and Lockridge 1989; Lander 1996; O'Loughlin and Lander 2003). However, a few years after the Daytona event, Rabinovich and Monserrat (1996, 1998) and then Monserrat et al. (2006), following Nomitsu (1935), showed the marked analogy between these anomalous atmospherically-induced waves and seismically generated tsunamis and introduced the term "*meteotsunami*" to the tsunami community. As a result, this phenomenon began to attract more and more attention. It became clear that devastating tsunami-like flood events, locally known as *rissaga* in the Balearic Islands, Spain (Tintoré et al. 1988; Gomis et al. 1993), *šćiga* in the Adriatic Sea (Orlić 1980), *marrobbio* (*marrubbio*) on the coast of Sicily (Colucci and Michelato 1976; Candela et al. 1999), *milghuba* in Malta (Drago 1999, 2009), *abiki* in Japan (Honda et al. 1908; Hibiya and Kajiura 1982) and *extreme seiches* in the Great Lakes (Ewing et al. 1954), are from the physical point of view the same type of processes; they have different source and generation mechanism from those of the ordinary tsunami waves, but hydrodynamically are almost identical, and have similar propagation and transformation properties. In tide gauge records, without knowing the source, it is virtually impossible to distinguish meteotsunamis and seismically generated tsunamis (Monserrat et al. 2006). What is also important, this new understanding and knowledge helped to explain some mysterious hazardous events of the past (cf. Okal et al. 2014; Gusiakov 2019).

The physical similarity and inability to distinguish meteotsunamis from seismically generated tsunamis appear to be the main reason why about 10 years ago meteorological tsunamis began to be actively included into modern tsunami catalogues and tsunami glossaries (compare Tsunami Glossary 2006 and 2016 versions). Nevertheless, as was indicated by Pattiaratchi and Wijeratne (2015), "*meteotsunamis are still an underrated hazard*". According to Gusiakov et al. (2019), during the period 1992–2016 altogether there were 290 tsunami events; 244 of them were of seismic origin ($\sim$ 84%), 22 landslide-generated (7.6%), 11 volcanic (3.8%), 8 meteotsunamis (2.8%) and 5 of unknown origin (1.7%). From these 290 events, 63 had maximum wave heights $H_m > 1$ m, i.e. they were *potentially damaging*. At the same time, according to the information of Romero et al. (2019) and Sebastian Monserrat (2019, Pers. Comm.), in Ciutadella Harbour, Menorca Island alone (one of the meteotsunami hottest spots) during the period of 1992–2018, there were 71 (!) meteotsunami events with $H_m > 1$ m. Thus, the total number of damaging meteotsunamis in Ciutadella alone was greater than the total number of potentially hazardous seismic tsunamis all over the world! Certainly, meteotsunamis are typically local events, they do not affect the entire ocean like the 2004 Sumatra (Indian Ocean) tsunami or the historical 1946 Aleutian and 1960 Chile tsunamis, and do not create 40-m runup like the 2011 Tohoku tsunami. Still, meteotsunamis are destructive and dangerous, and much more common than had been thought before.

Okal (2019) critically evaluated and overviewed 47 strong tsunamis of "geological origin" (i.e., seismic, landslide and volcanic), starting with the 1992 Nicaragua event. As a continuation of his study, we review here the 51 greatest meteotsunami events (Fig. 1). This number is conditional. In particular, for Ciutadella Harbour, we considered only one (strongest) event: the catastrophic flood of 15 June 2006 (Jansà et al. 2007; Vilibić et al. 2008). All other events between 1992 and June 2006 (35 events with $H_m > 1$ m) are combined into one event overview, while all subsequent events (35 events with $H_m > 1$ m that occurred during the period of July 2006–2018) into another. The destructive events of 25–27 June 2014 in the Adriatic Sea, on the western coast of Sicily and in the area of Odessa, the Black Sea, are described individually, but altogether are considered as one event. As part of the present study, we describe only the largest known events for the period 1992–2019, consisting of meteotsunamis with recorded heights $H_m > 1$ m, or those which created noticeable damage and/or attracted specific attention. It is unlikely that the list is complete, as strong events could have occurred in sparsely populated regions or coincided with a low tide and remain unnoticed. In any case, we believe that even this preliminary list will be helpful.

The paper is organized as follows. In Sect. 2, we present definitions of meteotsunamis and explain the difference between meteotsunamis and other long-wave phenomena. In Sect. 3, we describe the main types of meteorological tsunamis, while in Sect. 4 we use a concrete example to illustrate the origin of these waves and to demonstrate the high correlation between meteotsunamis and atmospheric disturbances. Section 5 is the major section of the paper that includes an overview of the 49 known events. Finally, Sect. 6 discusses some specific questions related to meteotsunami studies and ends with a conclusion of the main results.

2. *What are Meteorological Tsunamis?*

Japan is most affected by hazardous tsunami waves. Significant atmospherically induced seiches (known in Japan as "*second undulations of tides*") are also common in numerous bays and inlets of the Japanese coast and these waves are sometimes destructive (cf. Rabinovich 2009). For this reason, these oscillations have attracted much more attention in Japan than in any other country. The pioneering study of Honda et al. (1908) became a handbook for many generations of scientists working on this problem. The authors described and examined seiche oscillations in almost 100 bays and harbours of Japan. They indicated that atmospheric processes are the main source of non-seismic sea level harbour oscillations and found that the strongest oscillations (locally known as "*abiki*"), with wave heights of up to 1.5–2.0 m, are observed in Nagasaki Bay. Honda et al. (1908) also overviewed some other places around the world with known strong seiches and mentioned the Atlantic coast of Canada, Malta and "a similar phenomenon in Mazzara, Sicily, where it had long be termed **Mirabia** or **Marrobbio**".

The work by Honda et al. (1908) initiated active studies of these oscillations along the coast of Japan. Nomitsu (1935) was probably the first to pay attention to the pronounced similarity between seismic tsunamis and atmospherically generated seiches in certain bays of the Japanese coast and suggested the term "*meteorological tsunami*" to extreme, atmospherically induced, long ocean waves. Nomitsu described highly intense waves in Osaka Bay and Biwa Lake generated by a severe typhoon on 21 September 1934, which impacted the Kansai District and pointed out that: "…The meteorological tsunamis and seiches are not essentially different in nature and may be discussed together. Similarly, we should expect the corresponding fluctuations of the water level whenever the motive forces increase or decrease **suddenly**."

In his fundamental "Physical Oceanography", Defant (1961) generalized the previous studies and

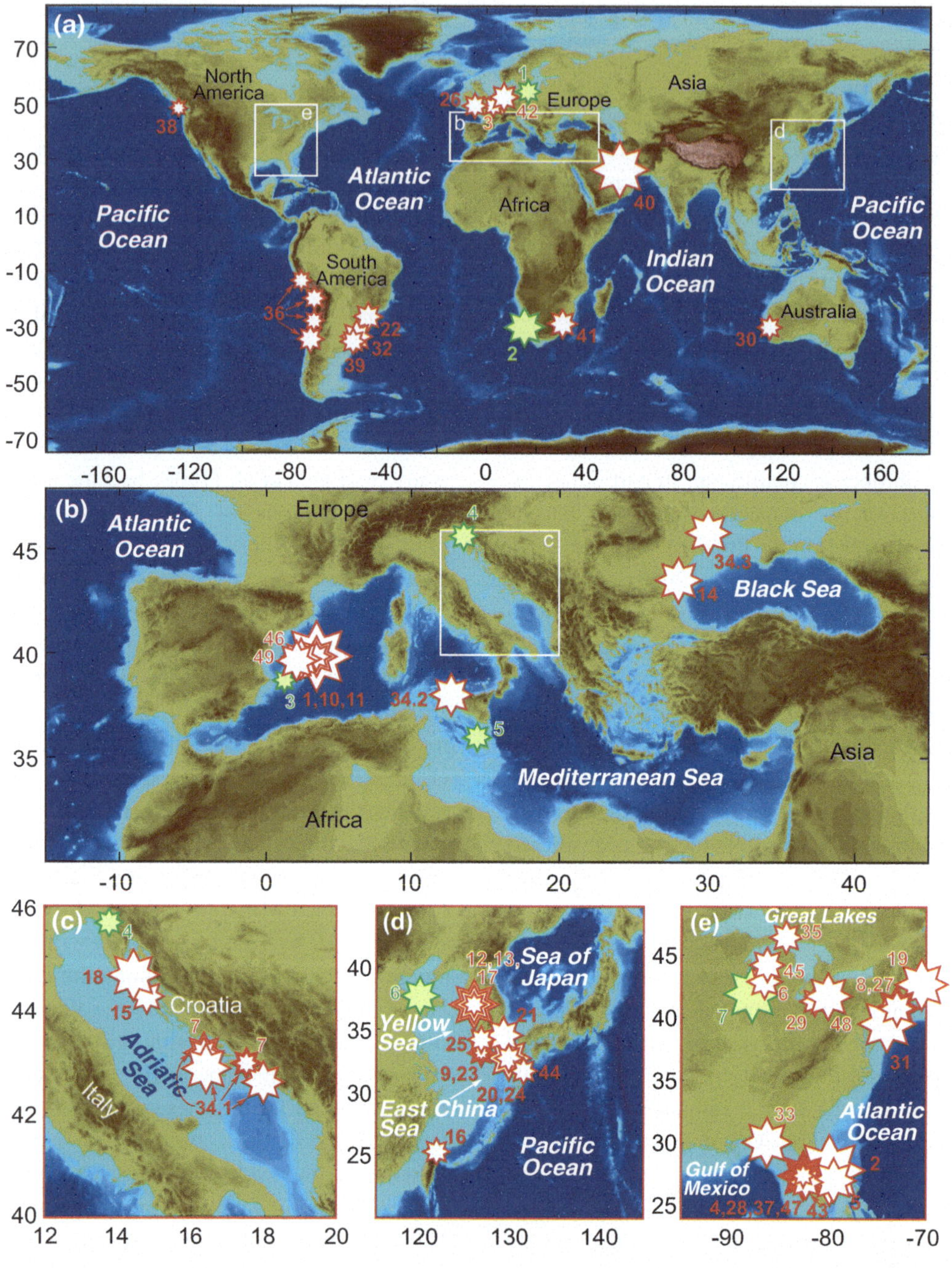

Figure 1

Maps showing the locations of the 51 most prominent meteotsunami events of 1992–2019 considered in the present study (denoted by white stars with red borders) and a few historical events (denoted by *yellow stars* with green borders). **a** World Ocean, showing locations of boxes "b", "d" and "e" enlarged in the lower panels; **b** Mediterranean and Black seas, with location of box "c" enlarged in the bottom left panel; **c** Adriatic Sea; **d** Southeast Asia and adjacent marginal seas; and **e** East Coast of the United States, the Gulf of Mexico and the Great Lakes. Red numbers in all maps indicate meteotsunami events listed in Sect. 5; green numbers are sites of selected historical meteotsunamis: "1"—Southwestern coast of the Baltic Sea (*Seebär* events), "2"—South Africa, "3"—Sant Antoni, Ibiza Island, Spain; "4"—Trieste, Italy; "5"—Malta (*milghuba*); "6"—Longkou Harbour, China, and "7"—southern Lake Michigan wrote: "The meteorological tsunami are nothing but seiches of bays and of the shelf, but they exceed in intensity the normally smaller amplitudes of the seiches… A phenomenon similar to that of tsunami occurs on the Baltic coast and is called "Seebär". The Seebär always sets in quite unexpectedly when the water is calm, often even smooth; out from the sea a high-towering wave rolls on to the breech, and floods it to a height of 1 or 2 m… The phenomenon of "Resaca" in the harbours of northern Spain might be related to the Seebär…The cause of such catastrophic seiches appears to be certainly meteorological, whereas a more precise synoptical explanation of the whole phenomenon is still lacking."

Remarkably, Defant (1961) describes quite precisely the nature and main properties of meteorological tsunamis and indicates a few places of the World Ocean where they are observed. "*Resaca*" is a Spanish equivalent of the Catalan term "*rissaga*", which means "drying"; significant seiches ("*resaca*") are common in harbours of the northern (Atlantic) coast of Spain (cf. Frère et al. 2014); however, they are not as strong as "*rissagas*" in bays and ports of the Balearic Islands, in the Western Mediterranean. Defant's description of "*Seebär*" is based on several old (1910s–1920s) publications reporting these waves on the German coast of the Baltic Sea. More recently, significant oscillations of a similar type have been observed on the coast of Finland (Pellikka et al. 2014).

According to Monserrat et al. (2006), *meteorological tsunamis* (*meteotsunamis*) are long ocean waves, which have approximately the same spatial and temporal scales as ordinary tsunamis, and can affect coastal regions in a similar destructive way, but which are generated not by underwater earthquakes, volcanic explosions or landslides, but by atmospheric disturbances such as hurricanes, frontal passages, squall lines, internal atmospheric waves and jumps of atmospheric pressure (cf. Rabinovich 2009; Vilibić et al. 2014a). Meteotsunamis can impact coastal regions that may not be vulnerable to earthquake-generated tsunamis, but like typical tsunamis they can produce severe damage to coastal infrastructure and threaten coastal communities (Heidarzadeh et al. 2019).

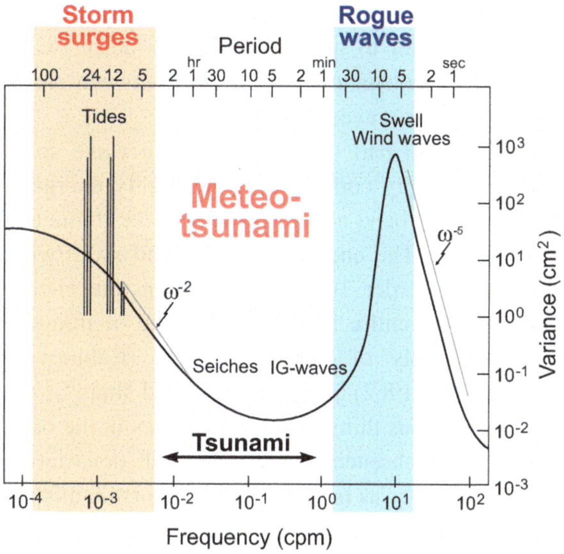

Figure 2
Spectrum of surface gravity waves in the ocean (modified from Rabinovich 2009). The frequency bands of storm surge, tsunami/meteotsunami and rogue waves are indicated. The solid thin black lines denote the theoretical spectral limits of ω^{-2} and ω^{-5} for open-ocean background spectra

Meteorological tsunamis have long been considered as a very rare and local phenomenon that occurs only in specific harbours. For scientists not directly involved into investigation of this phenomenon or in tsunami waves, in general, this term is new and not well understood. Two questions are most typical:

- What is the difference between a "meteotsunami" and a "storm surge"?
- What is the difference between a "meteotsunami" and a "rogue wave"?

As illustrated by Fig. 2, *Storm surge* is a *low-frequency* process, with typical periods from several hours to approximately 1 week (Pugh and Woodworth 2014). They occupy the *red* band of the ocean-wave spectrum. In contrast, *rogue waves*, also known as *freak waves* and *killer waves*, are *high-frequency* phenomena with characteristic periods from a few seconds to 30 s, corresponding to the *blue* part of the wave spectrum.[1] Meteotsunamis have exactly the

[1] This definition in spectral analysis of geophysical data is in analogy with light: a ''red spectrum'' means that the main energy is concentrated at low frequencies, while a ''blue spectrum'' has most energy at high frequencies (cf. Thomson and Emery 2014).

same periods as ordinary tsunami waves, ranging from 1 to 2 min to about 3 h. In the ocean wave spectrum, meteotsunamis are located between storm surge and rogue waves (Fig. 2).

The tsunami part of the open-ocean wave spectrum looks empty compared to the highly energetic tidal frequency band and swell/wind wave frequency band (Fig. 2). The energies of tides and swell/wind waves are of order 10^3–10^4 cm^2, while the energy throughout the entire intermediate range of frequencies is typically of order 1–10 cm^2 (Rabinovich 2009). Munk (1962) jokingly remarked that "…the most conspicuous thing about long waves in the open ocean is their absence." However, such devastating long ocean waves as tsunamis and meteotsunamis are clearly within this particular "empty" range of frequencies. Also, harbour oscillations and seiches in coastal regions are specifically associated with this frequency band and in certain bays, inlets and lakes can substantially increase the spectral level. In fact, both tsunamis and meteotsunamis most typically resemble the character of harbour oscillations.

Meteorological tsunamis differ from storm surge and rogue waves not only because they have different spatial and temporal scales, but also because they have considerably distinct generation mechanisms and affect different oceanic regions (Rabinovich and Šepić 2016). Storm surge is formed by the combined effect of low air pressure and wind, and wind generally playing the determining role in producing a destructive surge. That is why strong storm surges usually occurs in *extensive shallow-water regions*, in particular in the Bay of Bengal (catastrophic Bangladesh floods of 1864, 1876, 1970 and 1991; cf. Pugh and Woodworth 2014), in the northern part of the Gulf of Mexico (the 2005 hazardous "Katrina surge"; Pugh and Woodworth 2014), and the Gulf of Finland in the Baltic Sea (the prominent "St. Petersburg's floods"; cf. Kulikov and Medvedev 2013). The genesis of rogue waves ("killer waves") normally involves the superposition of intensive wind waves and/or swell. Consequently¸ these waves have approximately the same periods as wind waves (Fig. 2) and are observed in the *open ocean* during or after strong storms (Kharif et al. 2009).

In contrast to storm surge (mainly formed by wind forcing) and rogue waves, meteotsunamis are mainly produced by *direct air pressure forcing*. Typically, the initial water disturbance is formed on the *shelf*, but then the propagating wave strongly amplifies, first as a forced wave produced by the *Proudman resonant* effect (Proudman 1929; Vilibić 2008), and then, when the wave is approaching the coast as a free wave, by shoaling effects as described by Green's Law. Considerable additional amplification can also be caused as a result of harbour resonance in specific bays and inlets. There are several "hot spots" in the world oceans, where the resonant generation of meteotsunamis occurs on a regular basis. However, sometimes powerful meteotsunamis also take place in unexpected regions.

3. Types of Meteotsunamis

The term "*meteorological tsunami*" is now widely used for extreme tsunami-like waves of atmospheric origin (cf. Pattiaratchi and Wijeratne 2015; Rabinovich and Šepić 2016). These waves come in a variety of forms and the 1992 Daytona Beach event does not look similar to "*rissaga*" waves in Ciutadella Harbour or "*abiki*" in Nagasaki Bay. As shown in Fig. 3, there are several types of meteotsunamis, but without going into detail, all of them may be divided into two principal groups: "*bad-weather*" and "*good-weather*" meteotsunamis.

"*Bad-weather*" in this context is considered as a synonym for "*stormy weather*", i.e., as the weather commonly associated with large-scale intensive atmospheric processes, such as typhoons, hurricanes, deep cyclones, cold front passages and "*derechos*".[2] Most often such meteotsunamis are observed on the USA East Coast, in the Gulf of Mexico, on the Pacific coast of Japan and along the North Sea and Atlantic coasts of Europe (Pasquet et al. 2013; Dusek et al.

[2] A *derecho* is a widespread, long-lived convectively induced windstorm associated with a series of rapidly moving thunderstorms and/or squall lines. Most often *derecho*s are formed in the central part of the United States and then propagate eastward. *Derecho* is one of the most important tsunamigenic weather processes, responsible for meteotsunamis on the Atlantic coast of the USA and in the area of the Great Lakes (Šepić and Rabinovich 2014).

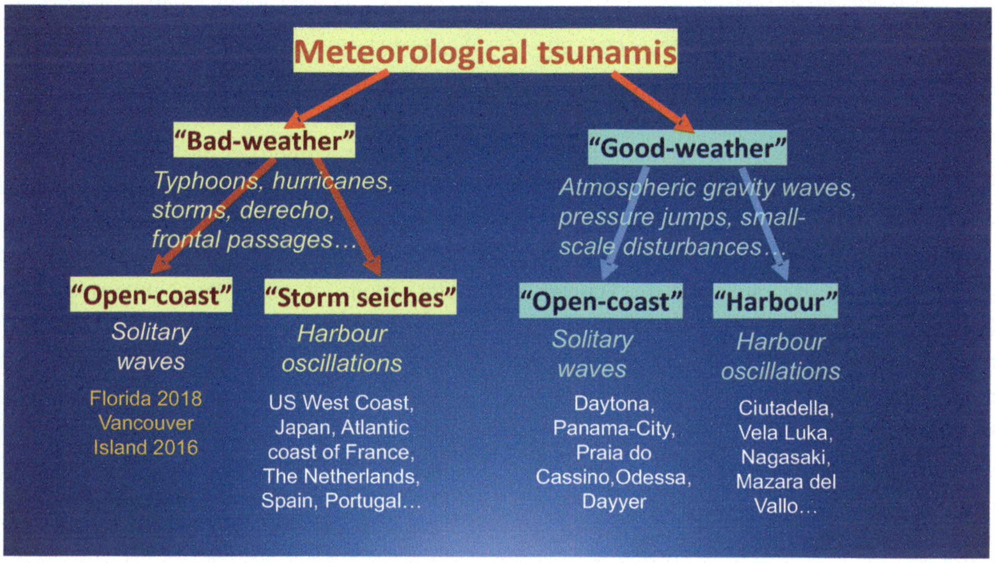

Figure 3
Classification of the meteorological tsunamis

2019; De Jong et al. 2003). There are two distinctive features characterizing this group of events:

1. They can affect large segments of the mainland coastline (up to several hundred kilometers; cf. Thomson et al. 2007; Bailey et al. 2014).
2. They often coincide with other dangerous types of long-wave oscillations, in particular, with storm surge and infragravity (IG) waves (cf. Thomson et al. 2007, 2019); the summation of waves from several hazardous phenomena strongly increases the damaging effect of these waves.

It is logical to expect that extreme weather events can give rise to extreme seiche waves through strong forcing. The waves can be threatening and destructive, but at least they can be effectively forecast.

Much more mysterious and surprising are "good-weather" meteotsunamis. For the public, and even for geophysical scientists, it is more difficult to imagine that a devastating wave (or waves) occurring on a calm and quiet day has an atmospheric origin. That is why other reasons for the event are invariably first proposed, including earthquakes, submarine slides, internal tides or even underwater explosions (cf. Chapman and Giese 2001; Ranguelov et al. 2008; Nikonov and Fleyfel 2015; Salaree et al. 2018; Šepić et al. 2018a).

It is important to emphasize that in the present definition, the term "*good weather*" means "*calm ocean weather*" and is used in contrast to "*bad (stormy) weather*". In fact, as was indicated by Šepić et al. (2015), tsunamigenic atmospheric systems have very specific properties: wave ducting, unstable mid-layers, jet-stream, etc. The corresponding atmospheric synoptic patterns, which can be described as "tumultuous atmosphere", support the generation of numerous intense small-scale atmospheric disturbances,[3] which subsequently resonantly trigger meteotsunamis in coastal areas. However, all these atmospheric processes take place at high altitudes and are not felt by people on the ground, who report that the weather during the corresponding event was "nice" and "quiet". In particular, the weather during the 1992 Daytona Beach event was described as "calm", "cloudless" and "sunny" (a squall line that was propagating southward from Georgia did not reach Daytona Beach; Churchill et al. 1995). This is

[3] It appears that such atmospheric systems also support formation of air-mass thunderstorms and squall lines, but in most cases they do not coincide exactly with the time and place of observed meteotsunamis.

the reason why beaches are often crowded during these events (Daytona Beach, Praia Do Cassino, Odessa, etc.) and that is why these events are described as "unexpected" (in strong contrast to extreme seiches caused by hurricanes, like "Irma" and "Harvey" in 2018).

The strongest and most harmful meteotsunamis—1978 Vela Luka, 1979 Nagasaki Bay, 1984 and 2006 Ciutadella, 1992 Daytona Beach and 2017 Dayyer, Persian Gulf—are specific examples of "good-weather events"! These devastating *good-weather* meteotsunamis occur not because of the extreme forcing, but because the system comes into *resonance*. In this regard, each strong meteotsunami is an analogue of the "Tacoma Bridge"—the classical civil engineering example of the destruction of a major bridge due to resonance effects (https://www.youtube.com/watch?v=j-zczJXSxnw). In fact, the fluid stream flowing back and forth within Vela Luka harbour (https://www.youtube.com/watch?v=WeTcoEuXEZo) resembles the swinging oscillations of the Tacoma Bridge in 1940.

The division into "bad-weather" and "good-weather" meteotsunamis is somewhat subjective. The principal difference is that intense seiches of the first type are mostly caused by *strong forcing* (but resonant effects can significantly amplify them), while those of the second type are *caused* by strong resonance. Both types of meteotsunamis have quite different wave properties when observed in embayments (bays, inlets or harbours) or on the open coast (along a straight beach). In the first case, they resemble *harbour oscillations*, i.e., multiple standing oscillations with specific natural (*eigen*) frequencies (cf. Rabinovich 2009), while meteotsunamis on a straight beach are normally just a solitary wave or a few such waves. This means that we can distinguish four main types of meteotsunamis (Fig. 3).

3.1. Good-Weather Harbour Meteotsunamis

Rissaga waves observed in Ciutadella Inlet, Menorca Island (Fig. 4) are typical examples of good-weather harbour meteotsunamis. Significant meteotsunamis in this inlet (see inset in Fig. 4b) appear to be observed more often than in any other location in the world (Monserrat et al. 1991, 2006; Marcos et al. 2009). Meteotsunamis (*rissaga*) with wave height > 1 m are recorded in this inlet 2–3 times each summer. Maximum wave heights $\sim$ 4–6 m were observed here in 1984 and 2006 (Monserrat et al. 2006; Romero et al. 2019).

Ciutadella Inlet has very consistent, almost monochromatic, seiche oscillations with a period of 10.5 min, corresponding to the fundamental (Helmholtz) mode of this inlet. Numerous meteotsunamis observed at this site had precisely this period. As an example, Fig. 4a (bottom plot) shows the *rissaga* of 25–26 May 2008. Strong sea level oscillations with wave heights of more than one meter continued at this site for about 10 h, causing powerful hazardous flows surging back and forth along the inlet.

Meteotsunamis in Ciutadella Inlet are normally accompanied by significant seiche oscillations on the neighbouring coasts of Mallorca and Ibiza islands (Rabinovich and Monserrat 1996, 1998). These oscillations commonly begin 1–1.5 h earlier than on the coast of Menorca Island and potentially can be used as a predictor for *rissaga* in Ciutadella Inlet (Marcos et al. 2009). Figure 4a (upper plot) shows seiches recorded at the time of the *rissaga* event at Cala Rajada, located on the northeastern coast of Mallorca Island (Fig. 4b); seiche oscillations at this site were highly correlated with those at Ciutadella but were approximately three times weaker.

The consistent periods of harbour meteotsunamis are a key property that can be used effectively for short-term (operative) and long-term (probability) forecast of these hazardous waves. It is well known that the main spectral peaks for different seismic tsunami events at the same site are almost identical, while the spectral peaks from the same tsunami event but at different sites are substantially different (cf. Honda et al. 1908; Rabinovich 1997). The same is true for meteotsunamis (Monserrat et al. 1998). The similarities in the spectra for various events demonstrate the dominant influence of local topography over the individual characteristics of the tsunami/meteotsunami source. To illustrate this property, we selected two records from Sant Antoni, Ibiza Island, Baleares, Spain (Fig. 5). The uniqueness of these records is that over one-month period in 2003, both a

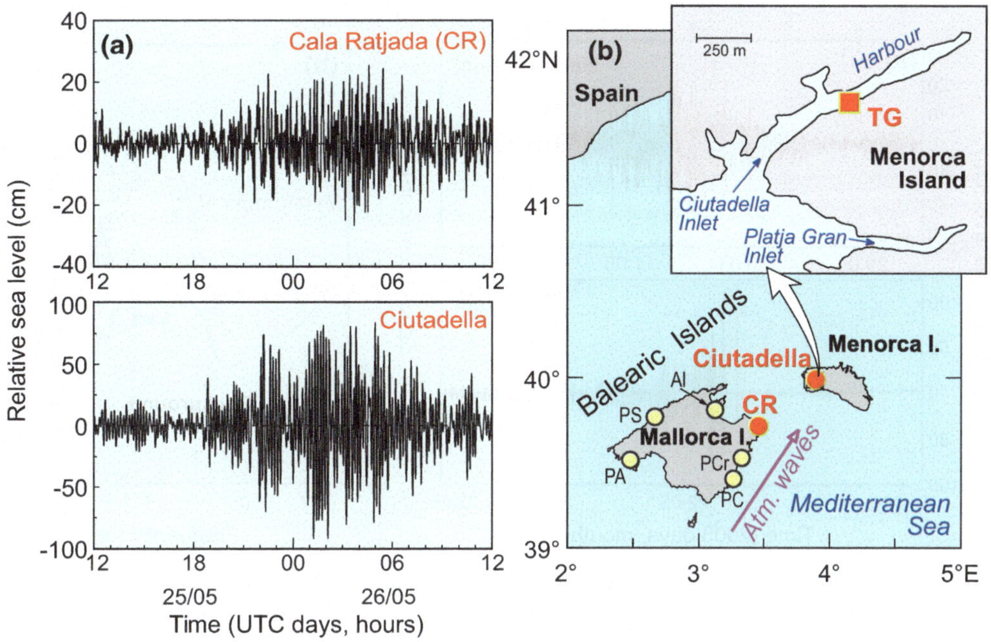

Figure 4

a The meteotsunami of 25–26 May 2008 recorded at two stations on the Balearic Islands, Spain: Cala Ratjada (Mallorca Island) and Ciutadella (Menorca Island); the records were de-tided and high-pass filtered with a 3-h Kaiser-Bessel window, the vertical scale in the two plots is different. **b** Map presenting the locations of the two reference stations denoted by red circles (CR = Cala Ratjada), and five sites on Mallorca Island denoted by yellow circles (Al = Alcudia, PS = Puerto Soller, PA = Puerto Andraitx, PC = Porto Colom and PCr = Porto Cristo) with common *rissaga* waves. The purple arrow indicates the approximate direction of atmospheric waves during the event; the inset shows the geometry of Ciutadella Inlet, Menorca Island, TG indicates the location of a tide gauge inside the inlet

meteotsunami and a seismically generated tsunami occurred at this station. More specifically:

1. Active atmospheric processes over the Western Mediterranean generated a meteotsunami on 30 April 2003, which was observed along the coasts of the Balearic Islands; at Ciutadella the maximum trough-to-crest wave height was ~ 90 cm (Romero et al. 2019). This meteotsunami with wave heights of up to 55 cm was clearly recorded at Sant Antoni (Fig. 5a, upper plot).
2. Three weeks later, at 18:44 UTC on 21 May 2003, a M_w 6.8 thrust-fault earthquake occurred off the northern coast of Algeria (Bouhadad et al. 2004). The earthquake caused a significant tsunami that spread throughout the Western Mediterranean and was recorded on the coasts of Spain, France and Italy (Alasset et al. 2006). The tsunami affected the Balearic Islands and was measured at several stations (Vich and Monserrat 2009), including Sant Antoni, Ibiza Island (Fig. 5a, bottom plot).

The meteotsunami and tsunami measurements at the same station enabled us to compare their spectral properties. Our spectral analysis procedure is similar to that described by Thomson and Emery (2014). The sampling interval of the records was 2 min; the examined time period for both records was 1024 min (~ 17 h), the quiet period between these two events (~ 12 days) was used to examine the properties of background oscillations. To improve the spectral estimates, we used a Kaiser–Bessel (KB) spectral window with half-window overlaps prior to the Fourier transform. The length of the window was $N = 256$ (512 min), yielding $\nu = 6$ degrees of freedom for the event spectra and $\nu = 124$ for the background spectrum. The results are shown in Fig. 5b.

The time series of the tsunami and meteotsunami look quite different (Fig. 5a), yet their spectra are almost identical and have nearly coincident prominent peaks at 20.5 and 15.8 min and a weaker peak at

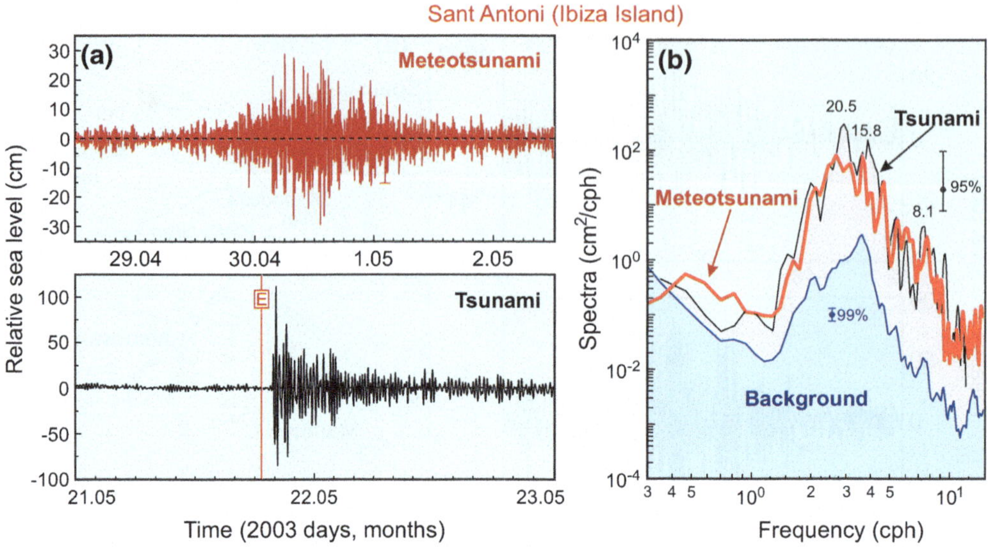

Figure 5
a The meteotsunami of 30 April 2003 and the Algerian (Boumerdès) tsunami of 21 May 2003 recorded at Sant Antoni, Ibiza Island, Spain; the records were de-tided and high-pass filtered with a 6-h Kaiser-Bessel window. The solid vertical red line labelled "E" denotes the origin time of the 2003 earthquake. **b** Spectra of the tsunami and meteotsunami records shown in **a** and background oscillations at the same station. Periods (in min) of the main spectral peaks are indicated. The 95% confidence level applies to the tsunami and meteotsunami spectra, the 99% confidence level to the background spectra. The location of Sant Antoni is indicated in Fig. 1b by a yellow star and a green label "3"

period of 8.1 min. The coincidence of these spectra demonstrates that (1) with respect to the spectra, there is no difference between the tsunami and meteotsunami waves; and (2) at the same station the two spectra are practically indistinguishable. The background spectrum repeats the shape of the event spectra but is about two orders of magnitude weaker. This similarity is further evidence for the dominant influence of local topography on arriving tsunami/meteotsunami waves. It also means that it is difficult to reconstruct the spectral characteristics of the source based solely on the data from coastal tide gauges. However, Rabinovich (1997) and Monserrat et al. (1998) suggested a method to bypass this problem and separate the influences of the topography and the source on the observed tsunami spectrum. The method is based on the assumption that the transfer function describing the linear topographically induced transformation of long waves approaching the coast, is the same for tsunami/meteotsunami waves and the ever-present background long waves.

The general character of meteotsunamis at other sites in the Mediterranean (in particular, on the coasts of Italy and Croatia where such events are common; cf. Vilibić and Šepić 2009) are typically similar to those recorded on the coasts of the Balearic Islands and can be described as "good-weather harbour meteotsunamis". Research studies done in various parts of the world indicate that such meteotsunamis are generated by small-scale intense atmospheric disturbances or by atmospheric gravity waves (cf. Thomson et al. 2009). These disturbances emerge in specific atmospheric systems associated with "tumultuous atmospheres" but, in general, are not associated with any extreme weather conditions (Belušić and Strelec Mahović 2009; Šepić et al. 2015).

3.2. *Good-Weather Open-Coast Meteotsunamis*

Figure 6 shows the sea level variations due to the meteotsunami of 28 March 2014 at Panama City, northern Florida, USA (Event No. 30 in Sect. 6). Similar to meteotsunamis in the Mediterranean, this event also occurred during relatively calm and quiet weather but had a totally different character. Whereas in Ciutadella Inlet and other Mediterranean "hot

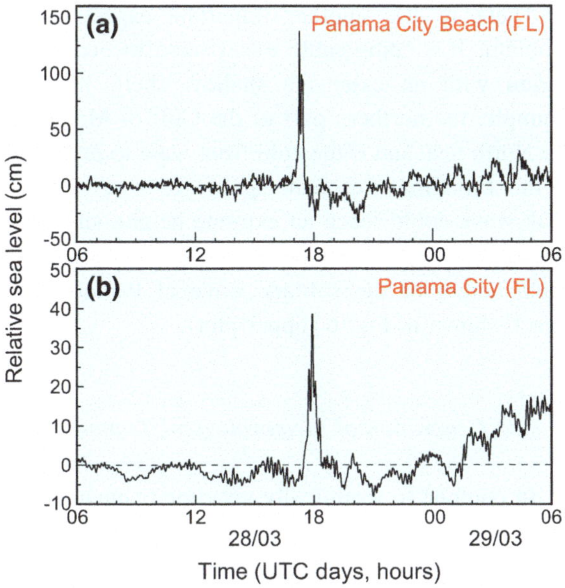

Figure 6
The meteotsunami of 28 March 2014 recorded at two tide gauges on the coast of Northwest Florida, USA: **a** Panama City Beach, Station code #8729210 and **b** Panama City, #8729108; the distance between stations is 18.4 km. Both records were de-tided. Note the different vertical scale in the plots

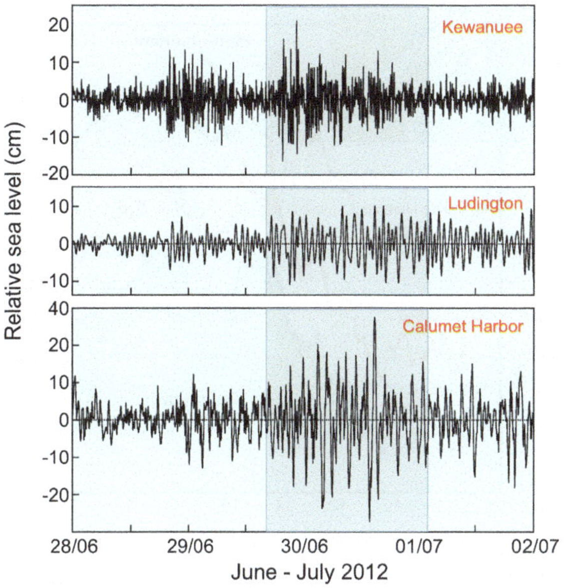

Figure 7
De-tided and high-pass filtered (with 6-h Kaiser-Bessel window) records at three stations in Lake Michigan from 28 June to 2 July 2012. Shaded area indicates intense sea level oscillations (storm seiches) associated with the propagation of *derecho* over the Great Lakes (based on data from Šepić et al. 2014)

spots" the recorded waves are generally long-ringing seiches, the meteotsunami that destructively impacted beaches in the area of Panama City was a solitary wave. There are no records of the waves that inundated Daytona Beach on 4 July 1992 (Event #2), or which dangerously impacted the beaches of Odessa on 27 June 2014 (Event #34-3), but eyewitness reports and existing videos (Churchill et al. 1995; Sallenger et al. 1995; Šepić et al. 2018b) show their close similarity to the wave recorded at Panama City Beach and Panama City (Fig. 6). These two stations are located approximately 18 km apart and both clearly recorded the impacting wave; however, at Panama City it was 3.5 times smaller and had a longer duration (about 70 and 30 min, respectively).

3.3. *Bad-Weather Harbour Meteotsunamis (Storm Seiches)*

Significant seiches are common on the Atlantic coast of the USA and in the Great Lakes. During strong storms, normally associated with incoming hurricanes, deep cyclones or *derechos*, these seiches are intensified considerably and can be identified as "meteotsunamis" (cf. Dusek et al. 2019). Figure 7 shows storm seiches recorded at three stations in Lake Michigan during the *derecho* of 29–30 June 2012 (cf. Šepić and Rabinovich 2014). Substantial strengthening of seiches during the storm activity is evident. Such seiches and associated reversing currents can lead to the breaking of mooring lines, fenders and piles and can seriously damage anchored boats (Rabinovich 2009). The negative effect of storm seiches is strongly amplified by their superposition with storm surge (cf. Thomson et al. 2007).

3.4. *Bad-Weather Open-Coast Meteotsunamis (Peak Wave)*

Severe storms can not only generate strong seiches but also produce a solitary "*peak wave*". Most often, such peak waves result from the propagation of a cold front associated with an intense cyclonic system. An example of this type, shown in Fig. 8 (Event #34), was caused by the interaction on 14–15 October 2016 of typhoon "Songda", which had crossed the Pacific Ocean, and a deep cyclone that was moving northward along the coast of British

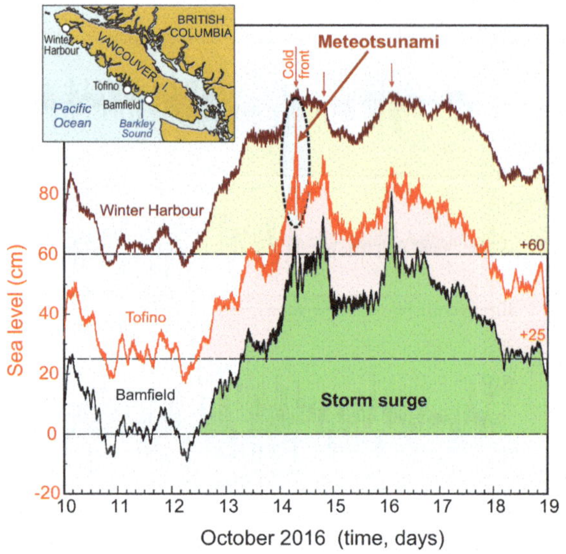

Figure 8
Residual (de-tided) sea level records at three Canadian Hydrographic (CHS) tide gauge stations located along the coast of Vancouver Island, British Columbia, for the period 10–19 October 2016. The records at Tofino and Winter Harbour are offset for the presentation purpose for + 25 and + 60 cm, respectively. The sea level rise associated with series of storm surges is marked by shading. Thin red arrows denote times of the main cold front (solid arrow) and two secondary fronts (dashed arrows). The dagger-like meteotsunami, recorded at Tofino on 14 October 2016, is indicated. The location of the stations is shown in the inset

Columbia. These two large-scale atmospheric processes together resulted in an extremely low atmospheric depression, intense storm and significant sea level oscillations, including strong storm surges, seiches and infragravity waves. A particularly interesting phenomenon was observed at Tofino, on the west side of Vancouver Island (see inset in Fig. 8), where a propagating cold front on 14 October 2016 produced a sharp, dagger-like 40-cm increase in sea level, i.e., a "peak-wave meteotsunami". A similar sharp sea level peak (but not as sharp and strong as at Tofino) was also observed at Bamfield, but not at Winter Harbour located in the northern part of Vancouver Island. Figure 8 clearly demonstrates a potentially hazardous combined effect ("double jeopardy"; Thomson et al. 2007) consisting of a superposition of a meteotsunami ($\sim$ 40 cm) and a storm surge ($\sim$ 55 cm). Separately, neither feature would have been dangerous, but when coincident they can be quite destructive.

There is also another important aspect of the problem. If a comparable event were to occur in a region with an extensive shallow shelf, like, for example, the northern part of the Gulf of Mexico or the North Sea, and if the cold front were to propagate at the Proudman resonant speed, $U \sim c$, then the peak-wave could reach an extreme height, similar to that of the 1992 "Daytona wave" or, at least, the wave height of the solitary wave at Panama City Beach shown in Fig. 6 (upper plot).

4. *Generation of Meteorological Tsunamis*

In contrast to seismically generated tsunamis, the generation of meteorological tsunamis is far from straightforward. The initial tsunami source formed by a major thrust fault earthquake is an extensive area of several hundred kilometres with a large vertical water displacement of several metres or more. Thus, it is logical that such huge displacements of water will produce great waves at a nearby coast and even distantly. However, the height of an initial meteotsunami source, induced by a small-scale atmospheric disturbance, is only a few centimetres (roughly, 1 cm of sea level rise for each 1 hPa fall in air pressure according to the "inverse barometer" effect). It seems counterintuitive that such a negligible initial source could create such highly destructive waves! Nevertheless, there is considerable evidence that this happens. The atmospheric origin of *rissaga* waves in the Baleares, *marrobbio* at Sicily, *šćiga* in the Adriatic Sea, *abiki* at Kyushu Island, Japan, extreme waves on the coast of Florida and in the Great Lakes, as well as in many other places in the world, is well established. The entire process of wave amplification from 3 cm in the open sea to $\sim$ 500 cm at the head of an inlet is explained by Monserrat et al. (2006) based on the study by Hibiya and Kajiura (1982) of the catastrophic meteotsunami at Nagasaki Bay (Japan) on 31 March 1979. The initial pressure jump of $\sim$ 3 hPa, formed over the western part of the East China Sea, moved eastward at the speed of $U \sim 31$ m/s. As the generated long waves propagated across the sea, they amplified from 3 cm to 16 cm as a result of the Proudman resonance (the long wave speed in the sea, c, was from 22 to 39 m/s). Between the outer shelf (depth 60 m) and the head of Nagasaki Bay, the arriving waves were amplified by a factor of 2.4 up to a height of 45 cm due to the combined effects of topographic funnelling and shoaling on the shelf and inside the bay. Finally, the resonant amplification in the bay of the incoming wave train formed catastrophic oscillations with a period of about 35 min and a maximum recorded wave height of 279 cm in the middle of Nagasaki Bay and an estimated wave height of about 480 cm at the head of the bay. Three people were killed by these waves. A similar *cascade amplification* process is likely responsible for producing extreme meteotsunamis in other regions of the world oceans.

Monserrat et al. (1998) demonstrated that microbarographic air pressure data can be efficiently used to precisely reconstruct sea level changes during a *rissaga* event. Also, modern numerical models allow us to reliably hindcast meteotsunami events based on air pressure data only (cf. Vilibić et al. 2004, 2008; Orlić et al. 2010). Moreover, the recently developed Adriatic Sea and Coast (AdriSC) Meteotsunami Forecast system, based on high-resolution sea level and air pressure observations and real time modelling, can realistically reproduce meteotsunami events and be used for operational meteotsunami warning (Denamiel et al. 2019). The close correlation between high-frequency variations in air pressure and sea level oscillations is clearly seen in Figs. 9 and 10. We used simultaneous 6-day microbarograph and tide gauge records from the same two stations, Ciutadella (Menorca Island) and Cala Ratjada (Mallorca Island), which are shown in Fig. 4b. Ciutadella Inlet (see inset in Fig. 4b) is the key basin for the meteotsunami study, probably the ideal site for a thorough examination of the generation mechanisms and physical properties of this phenomenon (cf. Tintoré et al. 1988; Gomis et al. 1993; Monserrat et al. 1998; Liu et al. 2003). The inlet is about 1 km long, 100 m wide and 5 m deep, with the harbour located in the upper part of the inlet. The fundamental period of the inlet (Helmholtz mode), estimated based on long-term observations (Rabinovich and Monserrat 1996, 1998; Monserrat et al. 1998; Marcos et al. 2009) is about 10.5 min. Numerical models yield the same numbers (Rabinovich et al. 1999; Liu et al. 2003). The particular geometry of Ciutadella Inlet and its high Q-factor[4] result in pronounced resonant amplification of long waves arriving from the open sea. Another important factor is the southeastern shelf of the Balearic Islands, which is conducive to the Proudman resonance of propagating atmospheric disturbances and long ocean waves. Thus, the *rissaga* waves in Ciutadella Inlet are produced by the combined effect of two resonant factors: (1) resonantly amplified long waves propagating over the shelf and arriving at the Ciutadella mouth; and (2) the resonant amplification of the incoming waves in the inlet itself (Monserrat et al. 2006). *Rissaga* waves with periods varying from 5 to 45 min are also observed at other bays along the coast of the Balearic Islands, in particular at Cala Ratjada (Figs. 4a, 9a), but normally they are much weaker than in Ciutadella Inlet (compare Fig. 9a, b, bottom plots).

High-frequency air pressure records at Cala Ratjada and Ciutadella (Fig. 9, upper plots) show four distinct trains of atmospheric internal gravity waves propagating over the region. The waves had typical trough-to-crest wave heights of 2–3 hPa and were precisely coincident with four trains of sea level oscillations. The redundant sea level oscillations began immediately after the arrival of the atmospheric waves, providing clear evidence that the atmospheric waves generated the sea level oscillations and were continuing to pump energy into them. After passage of the atmospheric wave train, the sea level oscillations gradually decayed, after having reached characteristic trough-to-crest heights of approximately 30 cm at Cala Ratjada and 100 cm at Ciutadella. This means that the recorded oscillations were 15–50 times greater than sea level changes that would result from the "inverted barometer" effect of

[4] The quality (Q) factor is a measure of the resonant increase of wave heights for waves arriving from the open ocean and of the energy decay in the system: the higher Q, the stronger the amplification of the incoming waves and slower the energy decay (Rabinovich, 2009). The Q-factor strongly depends on the shape of the basin: Miles and Munk (1961) concluded that narrowing a harbour entrance increases the Q-factor and, consequently, the amplification of the arriving waves. This means that the construction of dams, dikes, and walls to protect a harbour from wind waves constricts the entrance width and leads to strong resonant amplification of seiche oscillations inside the harbour. This effect became known as "*harbour paradox*". Intense meteotsunamis typically occur in harbours and inlets with high Q-factor.

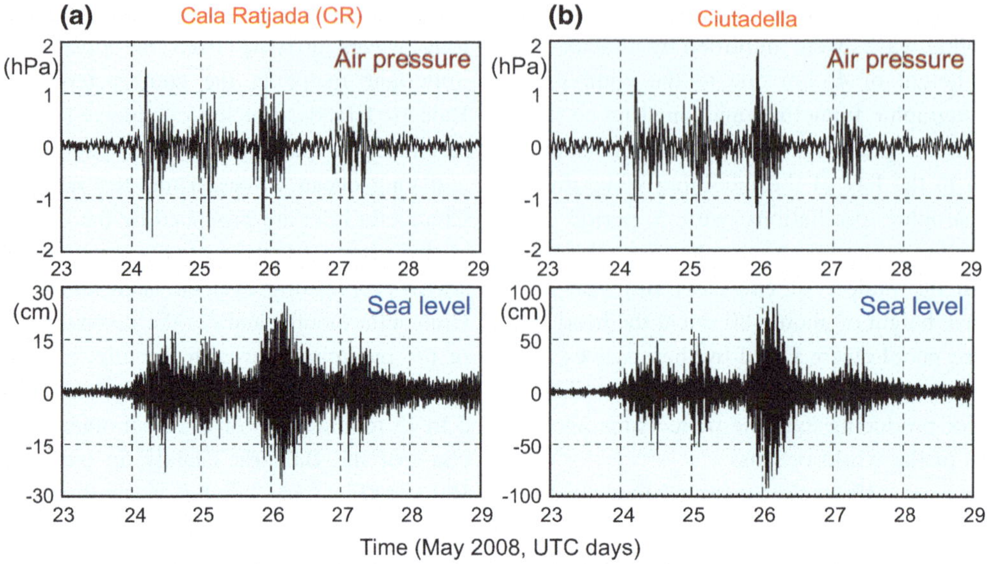

Figure 9
High-pass filtered with 3-hour Kaiser-Bessel window air pressure and de-tided sea level records at two stations on the Balearic Islands, Spain: **a** Cala Ratjada (Mallorca Island) and **b** Ciutadella (Menorca Island) for the period of 23–29 May 2008. Four trains of atmospheric waves and associated sea level oscillations are evident in the records. Note that sea level records at **a** Cala Ratjada and **b** Ciutadella have a different vertical scale

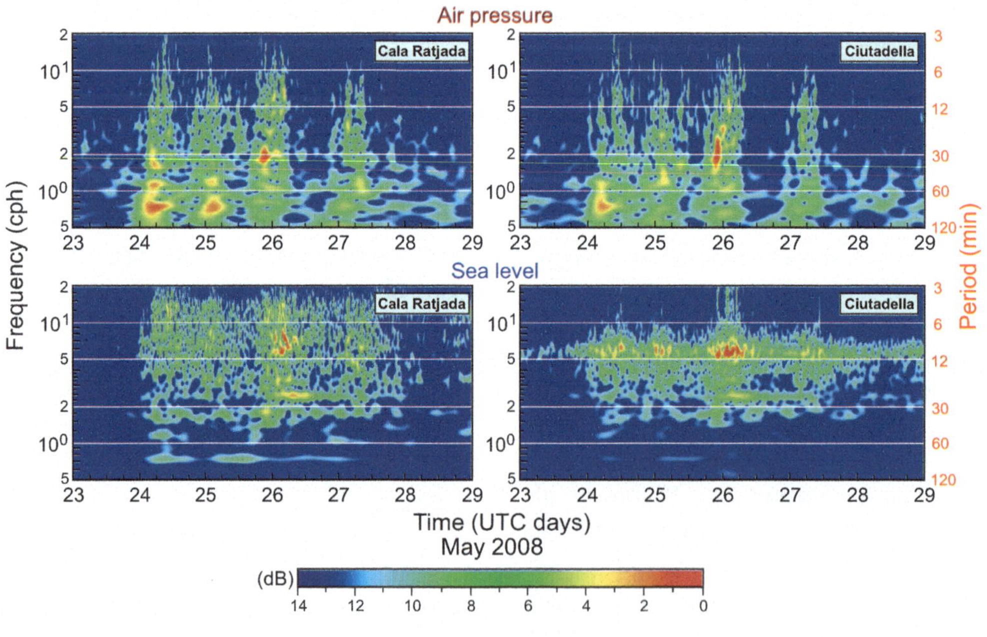

Figure 10
Frequency–time (f–t) diagrams of the air pressure and sea level records at Cala Rajada and Ciutadella (Balearic Islands, Spain), shown in Fig. 7. The fundamental period T_0 = 10.5 min of the inlet is evident in Ciutadella record

roughly 1 cm per 1 hPa. As indicated above, the main reasons for this major difference is the Proudman resonance, which occurred on the shelf of the Balearic Islands, and harbour resonance in Cala Ratjada and Ciutadella Inlet. It is important to emphasize that an atmospheric disturbance cannot directly generate seiches in such small basins as Cala Ratjada and Ciutadella because of the dissimilar scales of the corresponding bay and atmospheric disturbance; the waves are formed in an external basin and then enter the inner bays through the open entrance.

It is obvious that the strength of the atmospheric disturbance (pressure jump or wave heights in a train of internal atmospheric waves) and the air pressure gradient are the key factors in the formation of meteorological tsunamis. In general, hazardous meteotsunamis are caused by pronounced and sharp air pressure disturbances. However, understanding the correlation between the air disturbance heights and the wave heights of generated seiches is not a trivial matter; as even very strong atmospheric disturbances sometimes do not generate noticeable meteotsunamis, while moderate meteorological disturbances can produce a very strong sea level response (Rabinovich and Monserrat 1998). Šepić and Rabinovich (2014) introduced the ''*generation coefficient*'', R_{max}, determined as the ratio between the maximum heights of the induced sea level oscillations and of the respective atmospheric pressure disturbance. This coefficient characterizes the efficiency of a meteotsunami generation at various stations for a particular event. For the four events, shown in Fig. 9, R_{max} = 13.5, 11.1, 18.7 and 9.6 (for Cala Ratjada) and R_{max} = 38.5, 54.3, 50.1 and 59.9 (for Ciutadella). For comparison, for a meteotsunami generated by the *derecho* of 29–30 June 2012 on the East Coast of the United States and in the Great Lakes, this coefficient was from 5 to 24 (Šepić and Rabinovich 2014). The sites with high R_{max} can be identified as "hot spots", i.e., as sites with high risk of destructive meteotsunamis, and undoubtedly Ciutadella Inlet is one of the hottest in the world. The event of 26 June 2008, when the height of the air pressure disturbance was 2.7 hPa at Cala Ratjada and 3.4 hPa in the area of Ciutadella, while the atmospherically generated seiches (Fig. 4) were up to 52 and 172 cm, respectively, corresponded to some of the most prominent events recorded at these stations.

The frequency–time (*f*–*t*) diagrams, which are similar to wavelet plots (cf. Thomson and Emery 2014), give additional information about the records shown in Fig. 9 and specific features of the events at specific sites (Fig. 10). These plots enable us to examine the nonstationary properties of the signal and to evaluate the evolution of meteotsunamis as a function of frequency, *f*, and time, *t*, and to display the correlation in time between the external forcing (atmospheric pressure disturbances) and the resulting sea-level response.

Four trains of atmospheric waves passing over the region are evident in the *f*–*t* diagrams as burst-like increases in the atmospheric wave energy that look very similar for the two stations (Fig. 10, upper plots). All four events were polychromatic and occupied a broad frequency band of 0.5–10.0 cph (periods from 2 h to 6 min). The *f*–*t* diagrams for sea-level oscillations at Cala Ratjada and Ciutadella (Fig. 10, bottom plots) are significantly different from each other and from the air pressure diagrams. The individual sea level events are highly correlated with the air-pressure bursts, but they are not as separately distinct as those for the air pressure signals because of the relatively slow sea level energy decay after each event. At Cala Ratjada, the generated seiches were "wide-band" and spanned the entire range of frequencies from 1.5 to 15 cph (periods of 40–4 min). In contrast, at Ciutadella the motions were "narrow-band"; with the dominant period of 10.5 min associated with the fundamental mode of the Ciutadella Inlet, strongly prevailing. Another visible band of the increased energy, also noticeable at Cala Rajada, is found at a period of $\sim$ 25 min. which appears to be related to the shelf resonance (Monserrat et al. 1998; Liu et al. 2003). The marked resonant properties of Ciutadella Inlet and adjacent shelf are probably the main factors responsible for hazardous meteorological tsunamis (*rissaga*) at this site.

5. *Significant Meteotsunamis, 1992–2019: A Timeline*

In this section we describe, in chronological order, the strongest and most important meteotsunamis of 1992–2019. We have mainly included events with trough-to-crest wave heights of more than 1 m; however, there are certain exceptions, when recorded waves were lower than this threshold but attracted high scientific or public attention and/or caused noticeable destruction of coastal infrastructure. Some of these events were thoroughly examined and described in reports and scientific papers, while others were only stated in media, or mentioned in some publications, without any details. The list of events has a very preliminary character; further investigations and archive search can significantly supplement it. The types of the events are indicated: (1) GWHO = "good weather harbour oscillations" ("*harbour meteotsunami*"); (2) GWSW = "good weather solitary wave" ("*beach meteotsunami*"); (3) BWHO "bad weather harbour oscillations"; (2) BWSW = "bad weather solitary wave". The locations of the events are shown in Fig. 1.

Following the suggestions of Orlić (2015) and the principles used for cataloguing meteorological tsunamis in the Adriatic Sea (http://jadran.izor.hr/~sepic/meteotsunami_catalogue/), we assigned for each event a *quality indicator* (QI), selected from a five-grade scale as follows:

1. elementary description of sea-level variability;
2. advanced description of sea-level variability;
3. description of sea-level variability combined with an analysis of meteorological data;
4. analysis of oceanographic and meteorological data supplemented by oceanographic or meteorological model results;
5. analysis of oceanographic and meteorological data combined with both oceanographic and meteorological model results.

5.1. List of Events

1. *Ciutadella Harbour, Menorca Island, Baleares, Spain,* 21 *June* 1992–23 *May* 2006*; Type:* GWHO; QI = 2–3.
   Ciutadella Harbour is a narrow inlet on the western coast of Menorca Island (see inset to Fig. 4b) and the fundamental period of this bay is 10.5 min (cf. Rabinovich and Monserrat 1996, 1998). This is one of the "hottest" and best-known places in the world, where meteotsunamis, locally known here as *rissaga*, occur regularly. Such waves with trough-to-crest wave heights of more than 1 m occur here every summer (Tintoré et al. 1988; Gomis et al. 2003; Monserrat et al. 2006). According to Romero et al. (2019) and Monserrat (2019, Pers. Comm.) in 14 years (June 1992–May 2006, i.e. before the catastrophic event of 15 June 2006), *rissagas* with wave heights > 1.5 m had been recorded in Ciutadella 14 times. The strongest events (with wave heights $\geq$ 2.0 m) were observed on 21 June 1992 (2.01 m), 31 July 1996 (2.00 m), 31 July 1998 (2.30 m) and 23 May 2006 (2.20 m). Substantial tsunami-like oscillations (*rissagas*) also occur in several other bays and harbours of the Balearic Islands, in particular, in Platja Gran (Menorca), Porto Colom, Porto Cristo, and Port Andraitx (all Mallorca) (Fig. 4b), and in Sant Antoni (Ibiza) (indicated in Fig. 1b), but *rissagas* in Ciutadella normally feature the maximum height (Rabinovich and Monserrat 1996; Marcos et al. 2009).
2. *Daytona Beach, Pacific coast of Florida, USA,* 3–4 *July* 1992*; Type:* GWSW; QI = 3.
   On 3–4 July 1992 a devastating solitary ocean wave hit Daytona Beach, eastern Florida, injured 75 people and significantly damaged properties and beach infrastructure. The wave first arrived at the shore at the north end of Daytona Beach and progressed southward with a speed of ~ 13–14 m/s, affecting in total about 48 km of the shoreline and lasting at each location for about 40–60 s (Churchill et al. 1995). Observers reported a large wave approaching the beach like a wall of white water that had the "roar" of a breaking wave. At least 20 vehicles on the beach were lifted up by the wave resulting in numerous collisions (Sallenger et al. 1995). The drivers, whose vehicles were swept by the wave, reported either a single large crest wave or a large crest wave followed by two much smaller waves. According to Churchill et al. (1995), the maximum wave height was about 3 m above normal tide, while Sallenger et al. (1995) reported that the wave was 6 m high. Speculations that the wave was caused by an underwater seismic event or by a mudslide on the continental shelf were not supported by observations (Sallenger et al. 1995). As a result, both groups of authors working on this event (Sallenger et al. 1995;

Churchill et al. 1995) concluded that the wave was produced by the squall line, which moved parallel to the mainland coast from Georgia towards central Florida for several hundred kilometers. The satellite images indicate clear skies over the Florida Peninsula during the entire day of the event, however an extensive cloudiness associated with a mesoscale convective system was observed over Georgia and South Carolina (Churchill et al. 1995). The squall, identified based on radar measurements, originated in that region and then for several hours propagated southward with a speed of 13 m/s, which matched the shallow-water speed of long ocean waves over the local shelf, creating favourable conditions for the Proudman resonance (Proudman 1929). This appears to be the key factor promoting the generation of the anomalous wave. The 1992 Daytona Beach meteotsunami was probably the strongest ever recorded on a straight beach (not in a harbour) and is classified as a "*beach meteotsunami*".

3. *Rotterdam, The Netherlands,* 1 *January* 1995; *Type:* BWHO; QI = 3.
   The port of Rotterdam, located on the southern coast of the North Sea, is one of the "hottest spots" along the entire northern seacoast of Europe. Powerful harmful seiches occur in this port quite often. As was shown by De Jong et al. (2003) and De Jong and Battjes (2004), these seiches are mainly induced by the passage of intense low-pressure weather systems over the southern North Sea, in particular, by sharp cold fronts and trailing convection cells. The maximum wave heights (> 1.0 m) occur at the SE end of the Caland Canal near *Rosenburgse Sluis*. These oscillations, and especially intense associated currents, significantly affect the port activity and cause damage to boats. The dominant observed seiche period here, associated with the fundamental (Helmholtz) mode is 90 min, however, depending on the tidal height (high/low tide) it changes from 85 to 100 min. One of the strongest seiche events at this site occurred on 1 January 1995, when the maximum measured wave height was $\sim$ 1.55 m (De Jong et al. 2003).
4. *Western Florida, USA,* 25 *March* 1995; *Type:* GWSW; QI = 3.
   On 25 March 1995, a large solitary wave, of undetermined provenance, washed ashore along the Gulf Coast of Florida from Tampa Bay to south of Naples (Paxton and Sobien 1998). The wave was typically described by eyewitnesses as 3 m or greater, breaking between 0.5 and 3 km offshore, and taking 120–180 s to arrive at the shore. Just prior to the wave's arrival at the beach, witnesses reported a rapid water runout and then abrupt runup of 2–3-m vertical height. This wave was generated and amplified by a large-amplitude atmospheric gravity wave propagating southeastward over the eastern Gulf of Mexico. The wave appears to have had a resonant nature, associated with coincidence of the atmospheric wave and the water wave speeds (Paxton and Sobien 1998).
5. *Port Canaveral, Atlantic coast of Florida, USA,* 19 *June* 1996; *Type:* GWSW; QI = 3.
   A strong solitary wave with recorded trough-to-crest wave height of 1.19 m was measured at Port Canaveral, east coast of Florida (Dusek et al. 2019). The sampling interval at the corresponding tide gauge was 6 min; thus, the actual wave height may have been several centimetres larger. No noticeable waves were observed at nearby tide gauges. According to Dusek et al. (2019), Port Canaveral is the hottest site along the entire East Coast of the USA, with the wave recorded at this station on 19 June 1996 the strongest during the entire period of 1996–2017.
6. *Lake Michigan, USA,* 31 *March* 1998; *Type:* BWHO; QI = 3.
   The Great Lakes have a long-recorded history of destructive tsunami-like waves caused by convective storms (Anderson et al. 2015; Bechle et al. 2015, 2016). The most catastrophic event in Lake Michigan occurred on 26 June 1954, when seven people were killed (Ewing et al. 1954; Platzman 1965). A large meteotsunami hit the Lake Michigan shoreline at White Lake Harbor, 45 miles south of Ludington, MI, on 31 May

1998. The event was associated with a massive and severe weather event.

7. *Stari Grad and Mali Ston bays, Adriatic Sea, Croatia, 27 June* 2003*; Type:* GWHO; QI = 5.
Extreme sea level oscillations and strong current reversals were observed in several bays and harbours of the Eastern Adriatic on 27 June 2003. The most dramatic reports were from Stari Grad Bay on the Island of Hvar, and Mali Ston Bay located between the mainland Dalmatian coast and the Pelješac Peninsula (Vilibić et al. 2004). At Stari Grad, according to the meteotsunami catalogue, the maximum wave height was 3.5 m http://jadran.izor.hr/~sepic/meteotsunami_catalogue/. At Mali Ston the maximum observed height was up to 1.3 m, resulting in a propagating meteobore and severe damage to shellfish farms by swift currents. The event was found to be caused by the resonant matching of the atmospheric disturbance propagating with a speed of 22 m/s toward the funnel-shaped bays and the longwave speed over the 50-m deep sea shelf (Vilibić et al. 2004). According to Belušić et al. (2007), the source of the original perturbation is traced back to the Alpine region, where the potentially unstable air in the westerly current rises over the Alps resulting in convection and wave development. The wave and convective activity propagated southeastward down the eastern Adriatic coast. Due to the potentially unstable air above the Adriatic, the system additionally amplifies along its track. The maximum height of the "two-head" abrupt disturbance was ~ 5 hPa (Belušić et al. 2007).

8. *Mid-Atlantic coast, USA, 9 December* 2005*; Type:* BWHO; QI = 3.
A strong winter storm affected the East Coast of the United States on 9 December 2005. The storm was caused by the combination of low-pressure systems from the Midwest and the Gulf of Mexico; in 24 h the air pressure south of Newfoundland deepened by about 36 hPa! The storm was accompanied by a heavy snowfall over coastal New England and intense seiches. Maximum wave heights up to 1.04 m were recorded at Providence, Rhode Island (RI), which is one of the hottest sites along the East Coast (Dusek et al. 2019).

9. *Cheju (Jeju) Island, South Korea, 9 February* 2005*; Type:* Unknown; QI = 1.
This meteotsunami, the first reported for the coast of South Korea, caused verified damage in coastal areas (Kim et al. 2017).

10. *Ciutadella Harbour, Menorca Island, Baleares, Spain, 15 June* 2006*; Type:* GWHO; QI = 4.
On 15 June 2006 a destructive flood event (extreme "*rissaga*") affected the west coast of Menorca Island, Baleares. The main impacted area was Ciutadella Harbour (Monserrat et al. 2006; Jansà et al. 2007). This was one of the strongest meteotsunamis ever observed in the world oceans. A sudden first negative (ebb) wave of more than 4 m caused a catastrophic drying of the main harbour located in the upper part of the inlet (see inset in Fig. 4b). Then, a few minutes later, a huge flood wave came (Vilibić et al. 2008). The maximum wave height was 4–5 m (even 6 m according to some witness reports). The length of this harbour is only 1 km; the fundamental period is 10.5 min (Rabinovich and Monserrat 1996, 1998). Thus, an extreme water flow moved back and forth along the harbour, creating huge sea level differences (for 4–5 m) at the head of the harbour and crushing everything in its way. Around 100 boats were strongly damaged and 35 of them were sunk. The total damage cost was estimated to be of more than 30 million euros. A very abrupt pressure jump, accompanied by strong wind, caused by a squall line system, was found to be responsible for this extraordinary event. The atmospheric pressure dramatically increased in Palma de Mallorca by $\Delta P \approx 7$ hPa in only half an hour, and more impressively, the last 5 hPa occurred in less than 10 min (Vilibić et al. 2008). A similar convective disturbance 22 years earlier, on 21 June 1984, was identified to be the reason of another catastrophic *rissaga* in Ciutadella (Tintoré et al. 1988; Monserrat et al. 2006). The 2006 event was the subject of intensive study and numerical modelling (cf. Vilibić et al. 2008).

11. *Ciutadella Harbour, Menorca Island, Baleares, Spain,* 18 *June* 2006–16 *July* 2018*; Type:*

GWHO; QI = 2–3.

After the destructive event of 15 June 2006 strong meteotsunamis (*rissagas*) kept occurring regularly in Ciutadella Harbour. In 12 years (June 2006–July 2018) there were 35 events with maximum wave heights $\geq$ 1.0 m, six of them higher than 1.5 m. An example of one of these events is shown in Figs. 4 and 9. The strongest events (with wave heights $\geq$ 2.0 m) were observed on 18 June 2006 (2.00 m), 25 May 2008 (2.00 m) and 26 May 2008 (2.25 m)[5] (Romero et al. 2019; Sebastian Monserrat 2019, Pers. Comm.). The dominant period of all these oscillations was 10.5 min. Strong rissaga waves with a period of 5.5 min were also observed in Platja Gran Inlet, west coast of Menorca Island, adjacent to Ciutadella Harbour.

12. *Western coast of South Korea, Yellow Sea, 27 August* 2006*; Type:* Unknown; QI = 1.

A meteotsunami impacted the western coast of South Korea, the region which appears to be under the highest risk of destructive meteotsunamis (Fig. 1d). One fisherman went missing near the coast of Gunsan, in the central part of the western coast (Kim et al. 2017).

13. *Western coast of South Korea, Yellow Sea, 31 March* 2007*; Type:* GWHO; QI = 2.

One more event was reported on this coast; anomalous tsunami-like sea level oscillations were recorded along the west coast of Korea on March 29–31, 2007. These oscillations had maximum amplitudes up to 1.50–1.65 m at stations Yeonggwang and Wido (maximum trough-to-crest wave heights of > 2.5 m) and pronounced wave amplitudes at number of other stations (Cho et al. 2013; Choi et al. 2014). According to Eom et al. (2019), four people were killed, two injured and the total damage cost was $2 million, including flooding of houses, loss of fish farms, a ship overturning and inundation of nuclear plant coolant pumps. The recent studies of Kim et al. (2016, 2017) revealed that the meteotsunami originated in the region of Shandong, eastern China. Analysis of the observations demonstrated that these sea level oscillations coincided with an atmospheric pressure jump of 5.3 hPa moving eastward over the Yellow Sea with an estimated speed of 24 m/s. This speed was nearly equal to the propagation speed of the long ocean wave (22–30 m/s), indicating that sea level oscillations were enhanced via the Proudman resonance.

14. *Bulgarian coast, western Black Sea, 7 May* 2007*; Type:* GWHO; QI = 4.

Strong tsunami-like sea level oscillations of non-seismic origin were observed on 7 May 2007 on the Bulgarian coast of the Black Sea. These 2–3 m oscillations were first assumed to be produced by a submarine landslide (Ranguelov et al. 2008), however, further investigation of this event revealed that they were caused by an atmospheric disturbance traveling in the NNE direction (Vilibić et al. 2010) with an amplitude of 2–3 hPa and a propagation speed of ~ 16 m/s. Numerical modelling of the event showed that such a disturbance is capable of generating large tsunami waves and strong currents over the shallow regions adjacent to the sites where maximum sea level oscillations had actually been reported (Dalboka, Kavarna, Balchik, Shabla, and Bolata). Therefore, this event has the potential to be classified as a meteotsunami, the first one ever reported in the Black Sea.

15. *Široka Bay, Ist Island, the Adriatic Sea, Croatia,* 2007*; Type:* GWHO; QI = 4.

Ist Island is a small Croatian island situated northwest from the city of Zadar. It has an X-shape and consists of two hilly parts with two funnel-shaped bays between them: Kosirača Bay oriented to the NW and Široka Bay opened to the SE. On 22 August 2007 Široka Bay was strongly impacted by an unexpected flood that caused severe damage to boats and coastal infrastructure; according to the Croatian meteotsunami catalogue (http://jadran.izor.hr/~sepic/meteotsunami_catalogue/), the maximum wave height was up to 4 m. Witness reports and numerous videos show that the sea level rose about 1.5 m above the seafront and flooded neighboring streets and houses (Belušić and

[5] These estimates are related to the very head of the harbour, where all heights were significantly larger than at the tide gauge location, shown in Figs. 4 and 8.

Strelec Mahović 2009). Šepić et al. (2009) examined the event and found that the intense inundation of the bay was caused by a double resonance: (1) Open-sea wave amplification through the Proudman resonance (coincidence of the speed of the atmospheric disturbances, 21–24 m/s, and the longwave speed offshore of Ist Island) and (2) topographic amplification of incoming waves inside the affected bay. It appears that the air pressure disturbances had been formed over the Apennines and then propagated as ducted gravity waves to the eastern Adriatic.

16. *Northern Taiwan, East China Sea,* 29 *January* 2008*; Type:* GWHO; QI = 2.
    Significant seiches regularly occur on the northern (East China Sea) coast of Taiwan (Lin and Liang 2017). The strongest event, recorded on 29 January 2008, was found to be caused an abrupt drop of atmospheric pressure (about 1.5 hPa in 5 min) propagating eastward with a mean speed of ~ 17 m/s. Maximum wave heights of > 1.5 m were observed at two stations: Jhuwei and Taipei. Two boats were significantly damaged during this event. Significant seiches on the north coast lasted for about 1–2 h.
17. *Western coast of South Korea, Yellow Sea,* 4 *May* 2008*; Type:* GWHO; QI = 2.
    The event that occurred on 4 May 2008 on the western coast of South Korea was one of the most catastrophic meteotsunamis in human history. Unexpected, devastating tsunami-like waves in the area of Boryeong (northwestern coast of South Korea) washed 36 people away in the sea; 9 of them were killed and 15 injured (Yoo et al. 2010; Kim et al. 2017). An atmospheric pressure jump of 2–3 hPa was measured at the Boryeong observatory. According to CCTV video image analysis, the maximum height of the waves was 1.3 m, the dominant period was 3.1 min, and the current speed was nearly 7.3 m/s (Yoo et al. 2010). However, the wave height observed at the Boryeong station was only 10 cm; thus, the exact factors responsible for destructive amplification of this wave in the main affected area remain unclear (Cho et al. 2010).
18. *Mali Lošinj, Adriatic Sea, Croatia,* 15 *August* 2008*; Type:* GWHO; QI = 3.
    Mali Lošinj is a town on Lošinj Island in Western Croatia. On 15 August 2008 a tsunami-like sea wave flooded the town. According to the meteotsunami catalogue http://jadran.izor.hr/~sepic/meteotsunami_catalogue/, the trough-to-crest height of the maximum wave was 3.5 m. Belušić and Strelec Mahović (2009) examined the event and found that it was caused by a cold front propagating from the Apennines that initiated an abrupt jump of atmospheric pressure of about 4 hPa. Large incoming open-sea waves, combined with the sharp harbour resonance, were the main reasons for the strong effect at Mali Lošinj.
19. *Boothbay (Maine), East Coast of USA,* 28 *October* 2008*; Type:* BWSW; QI = 4.
    Unexpected tsunami-like waves strongly affected the East Coast of USA on 28 October 2008. The greatest impact was in Boothbay, where the zero-to-peak amplitude exceeded 2 m, while the maximum trough-to-crest wave height was ~ 3.6 m; several boats and marine structures were seriously damaged (Whitmore and Knight 2014). A cold front was found to be moving over the area at the time of the event, and the internal gravity waves (IGW), characterized by strong pressure changes (up to 2.5 hPa/3 min), were detected by radar in the rear of the front (Vilibić et al. 2014b). The intensity, speed and direction of the IGWs were favourable for generation of a meteotsunami wave on the shelf of the Gulf of Maine. The event initiated the first project to develop a meteotsunami forecasting capability for the USA, under a 2011 NOAA funded project focusing on meteotsunami forecasting for the US East Coast (Whitmore and Knight 2014).
20. *Soutwestern Kyushu Island, Japan,* 24–25 *February* 2009*; Type:* GWHO; QI = 3.
    Kyushu Island, the southernmost of the four main islands of Japan, is a region where large amplified seiches/meteotsunamis, locally known as "*abiki*", occur most frequently, and repeatedly reach destructive strength. According to Shiga et al. (2007), *abiki* occur about four–five times a year on average, mostly during the period from December to April. The initial open-ocean sea

level disturbances are formed in the area west of Kyushu, including Tsushima Strait, while the atmospheric disturbances, responsible for their generation, originate in the mountains of south-eastern China (Tanaka 2010). The tsunamigenic atmospheric disturbances are transferred across the East China Sea by a jet stream toward western Japan; two processes in the middle atmosphere support their longevity: wave-ducting (cf. Monserrat and Thorpe 1996) and wave-conditional instability of the second kind (CISK) (Lindzen 1974).

On 24–25 February 2009 the west and south coasts of Kyushu Island, and several small islands located nearby, were strongly impacted by a destructive meteotsunami (*abiki*). The area most affected was Nagasaki Bay on the western coast of Kyushu Island, which is one of the "hottest" meteotsunami places in the world, where powerful *abiki* events occur quite often. The strongest known event in this bay was on 31 March 1979, when the maximum instrumentally recorded wave in the middle of the upper harbour (located in the upper part of the bay) was 2.79 m, and the reconstructed height at the head of the harbour was 4.8 m (Hibiya and Kajiura 1982; Rabinovich 2009); that event was responsible for the death of three people. The meteotsunami that hit southwest Kyushu on 24 February 2009, had a recorded maximum height of 2.90 m (in Uraichi Bay), which was even higher than the maximum measured *abiki* wave during the 1979 Nagasaki event (Tanaka 2010).

Long-wave *abiki* oscillations, which started during the night of 24 February 2009, impacted the entire western and southern coasts of Kyushu Island. Maximum amplitudes, exceeding 1 m, were observed not only at Nagasaki Bay (1.57 m), but also in Makurazaki Bay (1.41 m) and Nakanoshima Bay (1.53 m) (Tanaka 2010). The ordinary seiche oscillations in these three bays rapidly intensified at 22:00 JST (13:00 UTC). Typical periods of the recorded oscillations were from 5 to 30 min.

In the morning hours of 25 February 2009, the oscillations along the entire southwestern coast of Kyushu and at numerous neighbouring islands began to amplify. The most affected island was Kamikoshiki, from the group of Koshiki Islands located 40 km west from Kyushu (Asano et al. 2012). This island was subjected to a destructive meteotsunami with a maximum sea level height of 3.1 m observed at the head of the fishing port Oshima, in the upper part of Urauchi Bay (Tanaka 2010). Urauchi Bay has a *T*-shape, with a length of the "foot" of $\sim$ 3.0 km and the width of the "bar" of $\sim$ 2.5 km; Oshima Harbour forms the right "arm" of this "bar". From 30 boats strongly damaged in western Kyushu, 24 were broken in the region of the Koshiki Islands, and 18 of them specifically in Oshima Harbour (Tanaka 2010; Asano 2012). A field survey in the area of Urauchi Bay showed that the local dike was strongly damaged and partly disintegrated. The trail of flooding showed that the amplitude between 08:30 and 08:40 JST was as high as 2.90 m, and the maximum high-water level was 68 cm higher than the benchmark level at the Oshima fishing port (Tanaka 2010). The recorded maximum height was larger than the previous record for western Kyushu (2.79 m in Nagasaki on 31 March 1979). The particular *T*-shape of Urauchi Bay appears to be responsible for such a strong effect.

On 25 February 2009 the *abiki* waves also caused serious damage in Sakitsu Bay, located inside of Yokaku Bay, on the southwestern coast of the Akamasu Islands (Tanaka et al. 2014). This group of islands is situated at the entrance to Nagasaki Bay, the "hottest" region for meteotsunamis.

21. *Tsushima Island, Tsushima Strait, Japan,* 15 *July* 2009*; Type:* GWHO; QI = 3.

On 15 July 2009, a meteotsunami was recorded in Tsushima Strait and flood damage occurred on the west coast of Tsushima Island. While most meteotsunamis in southern Japan occur in the winter/spring season (excluding those caused by typhoons), the 15 July 2009 event was exceptional. The meteotsunami was found to be related to the frontal system ("Baui front") propagating southeastward across the East China Sea. This system continuously generated trains of atmospheric gravity waves through the CISK mechanism (cf. Lindzen 1974). The ducted atmospheric gravity waves (cf. Monserrat and Thorpe 1996) formed 1–2 hPa air pressure disturbances, which produced long ocean waves resulting in meteotsunamis in coastal areas. The propagation speed of the atmospheric waves was estimated as 25–30 m/s, particularly favourable for the Proudman resonance in Tsushima Strait (Tanaka 2012), where depths are typically about 140 m.

Eight houses were flooded and three fishing boats were damaged by the meteotsunami in the Sasuna area on North Tsushima Island. According to the report of Nagasaki Marine Observatory (2009), local residents observed an extraordinary sea-level change at 14:45 JST. The sea surface rapidly dropped, and within 5 min the seabed was exposed in some inlets of the bay in the Sasuna area. The sea level rose again at 14:50 JST and it flooded over the quay wall at 14:55. The Nagasaki Marine Observatory estimated the maximum wave height of the oscillations as 2.4 m (Tanaka 2012).

22. *Florianópolis and Armação, Santa Catarina, Brazil,* 19 *November* 2009*; Type:* GWSW; QI = 1.
    A strong tsunami-like wave hit two coastal Brazilian towns, Florianópolis and Armação, in state of Santa Catarina. The beach infrastructure was partly damaged.
23. *Southern coast of South Korea, Cheju (Jeju) Strait,* 2 *January* 2010*; Type:* Unknown; QI = 1.
    On 2 January 2010 a meteotsunami affected the southern coast of Korea and islands located in Cheju Strait; one person in the area of Komundo Island went missing during this event (Kim et al. 2017).
24. *Western Kyushu Island, Japan,* 3 *March* 2010*; Type:* GWHO; QI = 4.
    One of the largest meteotsunamis (*abiki* waves) on the western coast of Kyushu Island occurred on 3 March 2010, when maximum wave amplitudes exceeding 1 m were recorded at several bays and harbours of this coast, including Nagasaki and Makurazaki bays (Fukuzawa and Hibiya 2019). Simultaneously, triangle-shaped atmospheric disturbances of $\sim$ 1 hPa were recorded at several weather stations along the coast of western Kyushu.
    An intensive numerical study indicated that the original ocean waves were formed in the East China Sea (ECS) and amplified through the Proudman resonance during their eastward propagation. After leaving the shallow continental shelf of the ECS they kept propagating to the east as free waves, being further amplified through reflection and refraction by the complicated topographic configuration of Kyushu Island. A secondary Proudman resonance occurred over the elongated shelf westward of Makurazaki Bay, promoting significant longterm *abiki* oscillations in this bay (Fukuzawa and Hibiya 2019).
25. *Southern coast of South Korea, Cheju (Jeju) Strait,* 26 *April* 2011*; Type:* GWHO; QI = 2.
    Daeheuksando is a small town located on the Heuksan Islands, a group of isolated islands located about 100 km southwest from the mainland coast of Korea. The Port of Daeheuksando forms a ring with a very narrow entrance and evident resonant properties. The combination of the external resonance of incoming waves on the Heuksan shelf and the strong internal harbour resonance caused great amplification of meteotsunamis (e.g. 31 March 2007). The combined effects of the shelf (external) resonance and the harbour (internal) resonance on 26 April 2011 induced a large meteotsunami in this port and caused significant property damage (Kim et al. 2017).
26. *Yealm Estuary, English Channel, southern coast of England,* 27 *June* 2011*; Type:* GWHO; QI = 3.
    On 27 June 2011 tsunami-like waves struck the Yealm Estuary, located on the southern coast of England, near Plymouth. This event was caused by a meteotsunami driven by convective cells extending from the Bay of Biscay into the English Channel (La Manche) (Tappin et al. 2013). The tsunami flooded the Yealm Estuary at about 9:30 UTC as a bore propagating upriver. At Marazion, the sea withdrew before the wave came in. Tidal anomalies of 20 cm were reported at Newlyn, around 30 cm at Plymouth and Weymouth, and 40 cm at Portsmouth.

Approximately 300 km of the coastline were affected by this event (Tappin et al. 2013). The event was explained by convective atmospheric activity in the Bay of Biscay and English Channel (Tappin et al. 2013).

27. *Mid-Atlantic coast, USA,* 8 *December* 2011*; Type:* BWHO; QI = 3.
A powerful storm affected the mid-Atlantic coast of the United States on 8 December 2011 and produced significant seiches at a great number of sites. According to Dusek et al. (2019), intense seiches (> 20 cm) were recorded at 24 tide gauges located along the mainland coastline of the USA: from Cutler Farris Wharf, Maine, in the north to Myrtle Beach, South Carolina, in the south. The entire affected coast was ~ 1500 km. The maximum wave height of 80 cm was measured at Providence, Rhode Island.
28. *Naples, Southwestern Florida, USA,* 11 *January* 2012*; Type:* GWSW; QI = 3.
Paxton (2016) examined meteotsunamis observed on the southwestern (Gulf) coast of Florida for the period 2007–2015 and found that the largest meteotsunami during this period was recorded on 11 January 2012, with a maximum trough-to-crest wave height of 0.92 m. The event began at 14:00 UTC, with an initial water level of -0.05 m and reached its peak at 15:00 UTC, with a final water level of 0.87 m. The barometric pressure decreased from 1012.7 hPa to 1011.0 hPa ($\Delta = -1.7$ hPa) over 12 min. The rapid decrease in the air pressure indicates a probable resonant effect: the radar imagery showed a frontal system moving south at a speed of 13 m/s, which is close to the long-wave speed, $c \approx 14$ m/s, on the shelf of Naples that has a typical depth of ~ 20 m..
29. *Lake Erie, USA,* 27 *May* 2012*; Type:* BWSW; QI = 4.
On 27 May 2012 two atmospheric convective systems generated a series of tsunami-like waves (meteotsunami) in Lake Erie (Anderson et al. 2015). The resulting waves swept three swimmers a distance of 0.5 miles offshore, inundated a marina, and capsized a boat along the southern shoreline. The modelling results of Anderson et al. (2015) revealed that the convective systems generated a series of waves in the lake incident to the southern shore that then reflected toward the northern shoreline and reflected again to the southern shore, resulting in spatial wave focusing and edge wave formation that combined to impact recreational users near Cleveland, OH.
30. *Southwestern Australia,* 10 *June* 2012*; Type:* GWHO; QI = 3.
Southwestern Australia is one of the global hot-spots for meteotsunamis with the annual occurrence of about 25 events (Pattiaratchi and Wijeratne 2014, 2015). The main reasons of these meteotsunamis are squalls, tornadoes, thunderstorms, and frontal passages. Although meteotsunamis there are not catastrophic, for many sites these events are responsible for the maximum sea level elevations ever observed along the coast. The strongest meteotsunami event occurred on 10 June 2012 and was caused by a low-pressure system moving almost parallel to the southwestern Australian coast. The maximum trough-to-crest wave heights were recorded at Busselton (1.10 m) and Bunbury (1.03 m) (Pattiaratchi and Wijeratne 2014).
31. *Mid-Atlantic coast, USA,* 13 *June* 2013*; Type:* BWSW/HO; QI = 4.
On 13 June 2013 at approximately 15:00 UTC a low-end "*derecho*" reached the US Mid-Atlantic coast. A typical *derecho* consists of numerous downburst clusters that are, in turn, comprised of many smaller downbursts. *Derechos* are responsible for extensive devastating effects and frequent casualties (Šepić and Rabinovich 2014). The *derecho* of 13 June 2013 was produced by a mesoscale convective system (MCS) over the Midwestern United States that destructively propagated eastward with a speed of ~ 21 m/s (Bailey et al. 2014).
A few hours later, around 19:30 UTC, tsunami-like waves crashed upon Barnegat Inlet (New Jersey) and Falmouth Harbor (Massachusetts), despite clear skies and calm weather (Bailey et al. 2014). At about 20:00 UTC residents in Falmouth Harbor observed a sudden retreat of water out of the harbour, followed by a sudden influx minutes later. The waters continued to surge in and out of the harbour for the next hour.

The entire event was tsunami-like, but there were no reports of any offshore earthquake capable of generating a tsunami. USGS Woods Hole was the first group to publicly suggest that a meteotsunami had hit the East Coast. Shortly thereafter, the NOAA West Coast/Alaska Tsunami Warning Center (WC/ATWC) announced in the Tsunami Bulletin Board that the event was a meteotsunami.

The most detailed description of the event was provided by Brian Coen who observed the event at Barnegat Inlet: "Around 3:30 PM < 19:30 UTC > on Thursday June 13, 2013, Brian Coen was spear fishing near the mouth of Barnegat Inlet... At approximately 3:30, the outgoing tide was amplified by strong currents which carried divers over the submerged breakwater (normally 3–4 feet deep). This strong outrush continued for 1–2 min and eventually the rocks in the submerged breakwater were exposed. At this point, Brian noticed a large wave coming in, approximately 6 feet peak-to-trough and spanning across the inlet. The upper 2 feet of the wave was breaking. This wave occurred in conjunction with a reversal of the current. This surge carried the divers back over the submerged reef and into the inlet from where they were picked up. On the south jetty three people were swept off the rocks which were 5 to 6 feet above sea level at the time. At least two were injured requiring medical treatment. There was no more strong activity after about 5 min."

The US media paid much attention to this event. As was said (WCBS-TV broadcasting): "This was the first tsunami observed on the Atlantic coast of the United States in 80 years" (in fact, in 84 years: since the 1929 Grand Banks tsunami). NOAA initially examined the June 2013 event to determine if the source was seismic, but no evidence was found of an earthquake that might trigger a tsunami. An underwater landslide was ruled out after the NOAA R/V *Okeanos Explorer* examined the Hudson River Canyon and determined that there was no sediment displacement. Post-event analyses (cf. Knight et al. 2013) confirmed that an earlier weather system moving offshore was the source of the tsunami and the event may be considered as a validated meteotsunami.

Bailey et al. (2014) and Wertman et al. (2014) thoroughly examined this event. A huge area of the Mid-Atlantic coast was found to be affected: tsunami-like waves were recorded by $\sim$ 40 tide gauges, from Cutler Farris Wharf (Maine) in the north to Oyster Landing (South Carolina) in the south (Wertman et al. 2014); this meteotsunami was also measured by DART 44402[6] located in the open Atlantic about 240 km off the New Jersey mainland coast and it was also recorded by the Bermuda tide gauge. The strongest oscillations were on the coasts of New Jersey and southern Massachusetts, i.e. specifically in the regions most impacted by this event. Unlike previously documented East Coast meteotsunami events, in particular the 2012 derecho-induced meteotsunami (Šepić and Rabinovich 2014), the recently installed USA Transportable Array (TA) of meteorological stations recorded atmospheric surface pressure anomalies over the land with high precision and time resolution (1 s). These measurements provided detailed information about the magnitude, dimension, duration and speed of the atmospheric disturbance during the 2013 event (Wertman et al. 2014).

The air pressure data show that between 11:30 UTC and 15:00 UTC on 13 June the MCS (derecho) extended from south of New York City through southeast Pennsylvania, southern New Jersey, Delaware and eastern Maryland to the Washington, DC, metropolitan area, and was accompanied by severe thunderstorms, squall lines and abrupt jumps of atmospheric pressure. These pressure jumps were recorded at a great many stations; the most rapid increases in air pressure were recorded at Ship John Shoal, NJ (3.0 hPa/6 min), Wilmington, DE (2.8 hPa/6 min), Atlantic City Airport (4.7 hPa/18 min), and Philadelphia Airport, PA (3.8 hPa/12 min) (Bailey et al. 2014).

It is known that strong and abrupt pressure jumps

[6] DART = Deep-ocean Assessment and Reporting of Tsunamis, is an effective network of deep-ocean stations, developed by NOAA for monitoring tsunami waves and early tsunami warning.

moving with a resonant speed, constitute the most common reason for meteotsunamis (cf. Monserrat et al. 2006; Thomson et al. 2009). The problem, however, was that the thunderstorm/squall line passages and associated strong pressure jumps occurred at 13:20–15:00 UTC (Bailey et al. 2014), while the meteotsunami affected the Mid-Atlantic coast approximately 4.5–6 h later (Wertman et al. 2014). The numerical model of Wang and Weinstein (2013) explained this time difference. The model showed that the initial meteotsunami disturbance was produced by the pressure anomaly moving eastward over the coastal shelf of New Jersey–New England; this disturbance, as a forced wave, propagated resonantly (amplifying) across the shelf and then reflected from the shelf-break and moved back to the coast as a free wave additionally intensifying in the coastal area due to the depth decrease. The mechanism of such meteotsunami reflection and amplification in this region was examined by Pasquet and Vilibić (2013). The coastal radar systems observed the generated meteotsunami 23 km offshore, 47 min before it arrived at the coast of New Jersey (Lipa et al. 2014). This sequence of events produced the 1.5–2.0-m wave that destructively impacted the New Jersey and Massachusetts coasts. According to Dusek et al. (2019), a total of 16 tide gauges recorded the event with a maximum wave height of 0.58 m at Providence, RI, indicating that gauges, which are deployed only in specific sites, can give underestimated representation of the actual event.

32. *Praia do Cassino, Rio Grande do Sul, Brazil,* 9 *February* 2014; *Type:* GWSW; QI = 3.
On 9 February 2014, the coast of Praia do Cassino, Brazil was stricken by powerful ($\sim$ 1 m) tsunami-like waves that severely affected local beach infrastructure and parked cars. Data analysis of sea level records at Rio Grande, located nearby, showed considerable oscillations with periods of 1–3 h. The analysis of meteorological data showed that this was a meteotsunami and it was associated with a warm barotropic trough from the centre of Brazil and a very sharp pressure gradient (around 6–7 hPa/100 km). Historical reports demonstrate that this was not an exceptional event, as similar events had regularly occurred at this site in the past (a 3-m wave was reported at this site in March 1977; Melo Filho and Rogerio Candella, 2016–2019, Pers. Comm.). This means that this site can be considered as a "hot spot" of the Brazilian coast, however, in contrast to other sites with regular meteotsunamis, Praia do Cassino is situated on the open coast; thus the meteotsunamis at this site are likely due to the sharp resonant conditions over the adjacent shelf.

33. *Panama City, Gulf coast of northern Florida, USA,* 28 *March* 2014; *Type:* GWSW; QI = 3.
On 28 March 2014, unusual non-tidal sea-level oscillations were reported along the northeastern Gulf of Mexico, from New Canal (Louisiana) to Naples (Florida). The sea level reached 1.3 m above its predicted level at Panama City Beach (FL) soon after 17:20 UTC on 28 March. The extreme free-surface elevation was associated with a frontal passage, marked by a squall-line and an abrupt weather change (Olabarrieta et al. 2017). Atmospheric pressure dropped by 5.0 hPa in 2.3 h, followed by a fast increase of 4.0 hPa in less than 24 min as the squall passed. This pressure disturbance propagated eastward at a speed of $\sim$ 19–20 m/s, which is roughly equivalent to the phase speed of long ocean waves at depths of 40 m. The air pressure disturbance initially appeared near Dauphin Island, and its shape remained unchanged between 15:00 and 18:00 UTC on 28 March. At Dauphin Island the meteotsunami amplitude was only a few centimetres, but it increased as propagated toward Panama City, FL. Maximum elevations were recorded at two neighbouring stations, Panama City (Station ID 8729108) and Panama City Beach (ID 8729210) (Fig. 6). The meteotsunami signal at these two stations looked very similar and had the character of a solitary wave, typical for "beach meteotsunamis" (Fig. 3); however, at Panama City the maximum height was $\sim$ 45 cm and the duration $\sim$ 70 min (Fig. 3a), while at Panama City Beach it was 1.35 m and about 25 min (Fig. 3b), indicating strong nonlinear local amplification effects. Witnesses reported that around 12:25 PM (17:25 UTC), right on the

edge of the rain shower, the waves started to crest over the shoreline "barrier" above and moved up the beach picking up the deck chairs, jet skis, rental booths and piling them up inland. The beach infrastructure on the coast of Panama City was strongly damaged.

Air pressure and sea level records along the northeastern coast of the Gulf of Mexico show that the atmospheric disturbance and the meteotsunami travelled in-phase, demonstrating that the meteotsunami propagated as a forced wave, "locked" to the disturbance. However, east of Panama City, the disturbance attenuated quickly and east of Apalachicola the meteotsunami, with strongly diminishing amplitude, propagated as a free wave (Olabarrieta et al. 2017).

34. *Mediterranean/Black Sea chain of events,* 22–27 *June* 2014*; Types:* GWHO and GWSW; QI = 5.

Normally meteotsunamis are local phenomena that affect one specific bay/harbour or a few bays located nearby (except those caused by destructive hurricanes and typhoons, affecting the entire coast along the hurricane/typhoon track). The chain of strong tsunami-like events of 22–27 June 2014 was unique: hazardous waves were observed in many locations in the Mediterranean and Black seas. The first site to record this event was Ciutadella Harbour, Menorca Island (Balearic Islands, Spain), where 10.5 min sea level oscillations with heights up to 1 m occurred during the night hours of 22/23 June. Two days later, on 25 June, tsunami-like waves with heights up to 3 m hit several bays in the Adriatic Sea (Croatia), and in the evening hours of the same day, Mazara del Vallo harbour on the southwestern coast of Sicily (Italy). On the following day, 26 June, tsunami-like waves with maximum heights of ~ 1 m were again observed in the Adriatic Sea and also on the coasts of Greece and Turkey. Finally, on 27 June, a sudden 1.5 m solitary wave swept beaches in Odessa (Ukraine) on the northwestern coast of the Black Sea. Substantial oscillations were also observed at a number of other sites along the Black Sea coast.

Šepić et al. (2015) examined this chain of events and concluded that it was caused by an anomalous atmospheric system ("tumultuous atmosphere") propagating eastward over the Mediterranean/Black Sea region and continuously producing small-scale intense tsunamigenic atmospheric disturbances. The system was characterized by: (1) warm air advected from the southwest at heights of ~ 850 hPa (~ 1500 m), (2) unstable moist air at heights of 700–400 hPa (~ 3000–7000 m), and (3) a strong southwesterly to westerly jet-stream embedded in unstable layers with wind speeds up to 40 m/s. The onset of pronounced sea level oscillations at each particular location coincided with the arrival time of the synoptic system to that location. However, destructive meteotsunamis occurred only in specific regions, where the *Froude number, Fr* (the ratio of the atmospheric gravity wave speed, *U*, to the phase speed of long ocean waves, *c*) was approximately equal to 1.0 (i.e. when the Proudman resonance could occur). Local harbour and shelf resonances strongly amplified the effect.

Particular regions with hazardous observed oscillations associated with this event are considered separately.

34.1. *Adriatic Sea, Croatian coast,* 25–26 *June* 2014*; Type:* GWHO; QI = 5.

Six harbours and bays of the Adriatic coast of Croatia were affected by powerful meteotsunamis on 25–26 June 2014; the strongest events occurred in Vela Luka (Korčula Island) and in Rijeka Dubrovačka Bay (mainland). Vela Luka is the location of the famous 6-m flood of 21 June 1978, the world highest meteotsunami ever observed; the meteotsunami of 15 June 2014 had a maximum wave height of ~ 3 m. The seiche oscillations in Rijeka Dubrovačka Bay had almost the same height 2.5 m; extremely strong currents of approximately 10 knots (5 m/s) were recorded in this bay. Intense oscillations, with wave heights from 0.8 to 1.5 m, were reported at four other sites: Vrboska and Stari Grad (both on Hvar Island), Viganj and Ston (both on the Pelješac Peninsula)

(Šepić et al. 2016). As follows from the catalogue of Adriatic meteotsunamis (Orlić 2015), in the past, all six sites had been affected by similar destructive tsunami-like waves (locally known as *šćiga*). The intensification of high-frequency sea level activity was also observed on the western coast of the Adriatic Sea (68 cm at Ortona, Italy). The analysis of various meteorological data revealed that a series of individual air pressure disturbances characterized by limited spatial extent (~ 50 km), very fast pressure changes (up to 2.4 hPa/5 min),) and high temporal variability propagated over the Adriatic on 25–26 June 2014, being probably responsible for the generated meteotsunamis (Šepić et al. 2016; http://jadran.izor.hr/~sepic/meteotsunami_catalogue/).

34.2. *Mazara del Vallo, Western Sicily, Italy,* 25–26 *June* 2014*; Type:* GWSW; QI = 5.
A major tsunami-like event, locally known as *marrobbio,*[7] impacted the southwestern coast of Sicily on 25–26 June 2014. The synoptic patterns and vertical structure of the atmosphere over Sicily at the time of the event indicated that ducted atmospheric waves were responsible for the generation of tsunamigenic air pressure disturbances that produced extreme sea level responses at certain sites along the Sicilian coast. The strongest oscillations were observed at Mazara del Vallo, where a 1-m *meteotsunami bore*, propagating upstream in the Mazaro River, was generated. The combined effect of the external resonance (the Proudman resonance on the western Sicilian shelf) and internal resonant conditions (bathymetric and topographic characteristics of specific sites, including the V-shaped estuary river entrance) were found to be the key factors that caused the meteotsunami (*marrobbio*) on the coast of Sicily and the meteobore at Mazara del Vallo (Šepić et al. 2018b).

34.3. *Odessa, Black Sea, Ukraine,* 27 *June* 2014*; Type:* GWSW; QI = 5.
On 27 June 2014 a 1–2 m high tsunami-like wave struck beaches at Odessa, the third largest Ukrainian city, and the neighbouring port-town Illichevsk (Northwestern Black Sea). The event became known as the "*Odessa tsunami*". Several people were swept into the sea; six people, including four children, were injured (Nikonov and Fleyfel 2015). Beach structures were tossed about and damaged. Similar waves had never been observed on this coast before (in contrast to the Baleares, Croatia and Sicily, where such waves were common); thus rumours and numerous maladaptive hypotheses appeared, including an underwater explosion, a huge ship passing by, a whirlwind effect, extreme interacting currents, abrupt temperature changes and even 'the great cross of planets'! However, as was demonstrated by Šepić et al. (2015), the Odessa tsunami was not an isolated event, but the last link in the chain of destructive events of 22–27 June 2014 in the Mediterranean and Black Sea regions. Moreover, prominent seiche oscillations were observed on 27 June 2014 in several other ports of the Black Sea. Tsunamigenic synoptic conditions were found over a large part of the Black Sea, stretching from Romania in the west to the Crimean Peninsula in the east. Intense air pressure disturbances, and convective thunderstorm clouds were associated with these conditions; right at the time of the event, a 1.5 hPa air pressure jump was recorded at Odessa, and a few hours earlier in Romania (Šepić et al. 2018a). The numerical experiments confirmed the meteorological origin of the tsunami-like waves on the coast of Odessa and indicated that intensified long-ocean waves in this region were generated via the Proudman resonance

[7] It appears that the first person who described the phenomenon of *marrobbio* in Mazara del Vallo was Admiral Smyth in his "*Memoir descriptive of the resources, inhabitants and hydrography of Sicily*" (London 1824) (Honda et al. 1908).

mechanism over the northwestern Black Sea shelf.

35. *Sault Saint Marie, Lake Superior, USA,* 4 *September* 2014*; Type:* BWHO; QI = 3.
An extreme seiche hit the eastern shoreline of Lake Superior near Sault Saint Marie on 4 September 2014. The wind was blowing out of the southeast for a long time period, pushing water along the shoreline out into Lake Superior. Then a line of strong thunderstorms moved through, and the accompanying west wind blew the water back into shore.
36. *The coast of Chile and Peru, SE Pacific,* 8 *August* 2015*; Type:* BWHO; QI = 3.
An intense storm impacted central Chile on 8 August 2015. The event was characterized by strong winds and storm waves, low atmospheric pressure (as low as 981 hPa) and intense sea-level oscillations that caused six casualties, massive beach erosion, destruction of port facilities and severe infrastructure damage along ~ 500 km of the Chilean coastline (Carvajal et al. 2017). Pronounced coastal seiches associated with this storm were recorded at a great number of tide gauges covering roughly 3000 km of the southwest coast of South America between Pichidangui and Bucalemu in central Chile and Callao in central Peru. The atmospheric origin of these intense tsunami-like waves was confirmed by analysis of high-resolution records of air pressure and wind. The strongest oscillations were observed at Bucalemu (~ 1.3 m) and Arica (~ 1 m); at more than seven other stations sea level oscillations of > 0.7 m were recorded.
37. *Naples, Southwestern Florida, USA,* 17 *January* 2016*; Type:* GWSW; QI = 3.
In the early morning of 17 January 2016, a pronounced meteotsunami impacted Naples on the southwestern coast of Florida. The event had the character of a solitary wave and was clearly measured by the Naples tide gauge; the record showed a sharp dagger-shaped sea level peak, which looked very similar to that recorded at Panama City on 28 March 2014. The meteotsunami was found to be caused by a 2.9 hPa air pressure jump moving eastward with a speed of ~ 24 m/s across the West Florida Shelf (WFS). A 21-cm ocean wave propagating 250 km across the shelf had tripled its amplitude when it reached the Naples Coast Pier (Alfonso-Sosa 2016). Combined with a surge and tide, it reached 2.37 m above mean lower low water (MLLW). This case was a perfect example of how a moderate size meteotsunami (< 1 m), synchronized with high tides and mean high water (MHW) levels, can become a natural hazard for coastal communities.
38. *Tofino, British Columbia, Canada,* 14 *October* 2016*; Type:* BWSW; QI = 3.
Typhoon "Songda", generated on 2 October 2016 southwest of Hawaii, passed along the coast of Japan, and then turned eastward, crossed the Pacific Ocean, arriving on the coast of British Columbia (BC) on 12–16 October 2016. Upon approaching the coast of BC, the typhoon interacted with a northward moving atmospheric cyclone, causing an extremely low atmospheric depression, which, in turn, induced significant sea level oscillations, including strong storm surges, seiches and infragravity waves (Fig. 8). On the west side of Vancouver Island, the record of the tide gauge at Tofino indicated a very sharp "dagger-like" 40-cm increase in sea level with a peak value at 07:01 UTC on 14 October 2016, which was identified as a meteotsunami. Similar sharp sea level peaks (but not as sharp and strong as at Tofino) were also observed at Bamfield, Port Alberni and Port Renfrew. The rise of sea level corresponded to the fall of air pressure and vice versa; moreover, abrupt sea level peaks at some stations (Tofino, Bamfield and Port Alberni) were found to be related to particular shapes of the air pressure variations at these sites. The minimum atmospheric pressure at Tofino was 971.4 hPa, but the sea level response was 1.5–2.5 times larger than the inverted barometer effect, demonstrating the topographic amplification of the sea level in the respective areas.
39. *Balneário Rincão, Santa Catarina, Brazil,* 16 *October* 2016*; Type:* GWSW; QI = 1.
A meteotsunami struck Balneário Rincão, Santa Catarina, Southern Brazil on 16 October 2016. A big wave hit cars and people who were on the

beach of Barra do Torneiro. No one was injured, but several cars were submerged, beach infrastructure was damaged, and some coastal restaurants were flooded.

40. *Dayyer, Persian Gulf, Iran,* 19 *March* 2017*; Type:* GWSW; QI = 5.
On 19 March 2017, destructive tsunami-like waves impacted the northeastern shore of the Persian Gulf (PG). The maximum runup heights of about 3 m were observed at Dayyer in southern Iran, where inundation reached a distance of $\sim$ 1 km and resulted in the deaths of five people and severe damage. The event was totally unexpected, since the PG had always been considered safe from extreme oceanic waves. No earthquakes or landslides were reported at the time of the event, while atmospheric processes on 18–22 March were very active (Salaree et al. 2018). The event was found to be very local; maximum trough-to-crest wave heights were 1.97 m at Dayyer and 2.34 m at Asaluyeh (both in Iran), near-field cities located $\sim$ 80 km from each other; the dominant wave periods were 15–20 min. At all other sites, located along the entire coast of the Persian Gulf and in the Gulf of Oman, the observed wave heights were < 35 cm. Therefore, the event of 19 March 2017 was identified as a meteorological tsunami (Salaree et al. 2018; Heidarzadeh et al. 2019). The atmospheric conditions over the PG were highly favourable for generation of meteotsunamis, and very similar to those that caused a chain of strong meteotsunamis in the Mediterranean and Black Sea regions on 22–27 June 2014 (Šepić et al. 2015). The estimated speed and direction of the atmospheric disturbances observed during the Dayyer event were $\sim$ 26 m/s and 77°, respectively. The Froude number, *Fr*, estimated as the ratio of the speed of the air disturbance and the long wave speed, on 19 March 2017 specifically in the area of Dayyer/Asaluyeh was close to resonance, $Fr \sim 0.9$–1.1, which is most favourable for meteotsunami generation (Heidarzadeh et al. 2019).

41. *Durban, RSA,* 12 *March* 2017*; Type:* GWSW; QI = 1.
A series of strong tsunami-like waves impacted the Durban beachfront, Republic of South Africa (RSA) on 12 March 2017. The waves rolled over the promenade, flooded restaurants and many beach stores, and then inundated the car park lot. It appears that this event was not exceptional in South Africa: in 1969 a strong meteotsunami occurred in Dwarskersbos, on the Atlantic coast of RSA (Fig. 1a) (Okal et al. 2014); also strong seiches, significantly affecting port activities were frequently recorded on the coast of RSA in the past (Wilson 1972; Rabinovich 2009).

42. *The Netherlands,* 29 *May* 2017*; Type:* BWSW; QI = 4.
In the early morning of 29 May 2017, a number of totally unexpected strong tsunami-like waves were observed on the coast of the Netherlands. Numerous videos showed a tsunami-like bore propagating along the Dutch coast in the NE direction. The observed height of these waves was 1.5–2.0 m. The waves impacted the beaches and damaged coastal infrastructure. Serious injuries of people were avoided only because of the timing–the wave hit beaches at about 5:45 AM (3:45 UTC) (Assink et al. 2018). These high waves, identified as a meteotsunami (https://www.youtube.com/watch?v=CjQk_xt_WU0; https://www.youtube.com/watch?v=zGy3FXLEJ2Q)), were found to be caused by an active line of thunderstorms (squall lines). At some places air pressure disturbances of more than 5 hPa were measured. Around 5:00 AM (3:00 UTC) the squall line was located at the Zeeland coast, with the most active part remaining offshore. Afterwards, the thunderstorm moved parallel to the coast towards South, then North Holland at a speed of approximately 20 m/s. From some point of view, this event was unique: typically, meteotsunamis are local events or a chain of local events; the event of 29 May 2017 affected almost the entire coast of the Netherlands, with evident propagation of the major wave (tsunami bore) along the sea coast. *Deltares Company* simulated this meteotsunami (https://www.youtube.com/watch?v=crE3ZMmUCQc). The simulation, based on a schematic and simplified wind and air pressure gradient, showed the generation and propagation of the waves along

the coast. The factor responsible for this meteotsunami was the resonant interaction of propagating squall lines and long ocean waves on the shelf of the Netherlands that has typical depths of 30–50 m.

43. *Florida, USA, Hurricane "Irma",* 10–11 *September* 2017*; Type:* BWHO; QI = 3.
On 10–11 September 2017 hurricane Irma destructively affected the coast of Florida, USA. This was the most powerful, Category 5, Atlantic hurricane in recorded history (Berkowitz et al. 2017). Its winds in Florida reached 115 knots, the minimum atmospheric pressure in the center of the hurricane was 930 hPa (Cangialosi et al. 2018). The total estimated damage in the United States alone was approximately $50 billion (Cangialosi et al. 2018). The hurricane induced significant seiches along the coast of Florida. Dusek et al. (2019) examined meteotsunamis associated with this event along the Atlantic coast of Florida and found a maximum wave height of 0.97 m at Port Canaveral. However, strong seiches were also recorded on the Gulf coast of Florida. What is especially important, hurricane Irma produced extreme storm surges: The National Hurricane Center posted a storm surge warning from Florida up to South Carolina, estimating that water levels would rise more than 4.5 m in some areas. The low-lying areas on the western and southwestern coasts of Florida (in particular, Fort Myers, Naples and Marco Island) were particularly affected by the surges (Berkowitz et al. 2017). Specific sites, e.g. Naples, Trident Pier and Port Canaveral, had "double jeopardy", being impacted by a storm surge combined with a meteotsunami.

44. *Southern and eastern Kyushu Island and Sikoku Island, Japan,* 8–9 *January* 2018*; Type:* BWHO; QI = 3.
Powerful tsunami-like oscillations (*abiki*) were observed on the southern coast of Kyushu Island. The maximum wave height of 1.6 m was recorded at Makurazaki at approximately midnight, where such oscillations are typical in the winter season and in early April. What was unusual is that large amplified oscillations were also recorded on 9 January on the Pacific coasts of Kyushu and Shikoku islands, which is very rare. The maximum recorded heights were at Cape Muroto (0.8 m) and at Tosa Shimizu (1.0 m) (Tanaka 2019). A series of pressure disturbances of 0.5–1.0 hPa were observed passing over this region. A significant jump of atmospheric pressure of 1.5–2.0 hPa/10 min was measured at the Muroto meteorological observatory. The air pressure disturbances propagated with high speed (30–40 m/s) in the southwest direction (i.e. from Sikoku to Kyushu), creating the Proudman resonance effect on the shelf of Tosa Bay. Specifically, this effect appears to have been responsible for the significant amplification of sea level oscillations in this region.

45. *Lake Michigan, USA,* 13 *April* 2018*; Type:* GWHO; QI = 3.
On 13 April 2018 a powerful meteotsunami (of $\sim$ 2 m) occurred in Lake Michigan. The meteotsunami waves were generated by high-amplitude (up to 10 hPa) atmospheric gravity waves propagating across the lake with a speed matching the long-wave speed of ocean waves, which were, thus, significantly amplified by the Proudman resonance. The waves strongly impacted the eastern coast of the lake. The main affected area was near Ludington, in the central part of the eastern Lake Michigan coast: the harbour walls were overtopped, and shoreline houses and boat docks were severely damaged (Anderson and Mann 2019). The recorded water level in Ludington rose by about 0.5 m, however this measurement was taken in a harbour that was sheltered and the actual water level rise could have been much higher along the open Michigan coast.

46. *Alcudia, Mallorca Island, Spain,* 16 *July* 2018*; Type:* GWHO; QI = 2.
On 16 July 2018 tsunami-like waves (*rissaga*) hit the coasts of Mallorca and Menorca, Balearic Islands. The strongest oscillations were observed at Alcudia, a city located on the northeastern end of Mallorca Island, on the opposite side of the strait from Ciutadella, Menorca Island (Fig. 4b). A meteotsunami wave was seen smashing into the boardwalk at Puerto de Alcudia, flooding

restaurants and covering beachside roads. A 53-year-old German tourist was swept away by the wave and killed while walking with his family in Cala Egos, near the southeastern coast of Mallorca Island (https://www.euronews.com/2018/07/18/tourist-dies-after-meteotsunami-hits-spanish-island). Videos made during the event show boats detached from their ropes and dragged around ports by strong currents (https://www.youtube.com/watch?v=R7cxRp5gYAo; https://www.youtube.com/watch?v=WBoUrLC5m_c).

According to local media, waves at Alcudia were as high as 1.5 m; strong oscillations flooding beaches, seafront promenades, bars and terraces were also reported at several other cities on Mallorca Island, including Porto Colom and Puerto de Andraitx (Fig. 4b). Beach chairs, boats and parasols were swept away, as well as a historic sailing boat used by the Mallorca Island Council.

The Spanish meteorological agency, Agencia Estatal de Meteorología, on Sunday, 15 July 2018, issued a "yellow warning" (*alerta amarilla*) for Ciutadella, Menorca Island, which was increased to an "orange alert" (*alerta naranja*) on Monday 16 July, signifying a very high risk of meteotsunamis (https://watchers.news/2018/07/17/meteotsunami-floods-islands-of-mallorca-and-menorca-spain/). A wall of water hit Cituadella on 16 July, causing widespread flooding. On the west coast of Menorca *rissaga* waves of > 1 m crashed into touristic beaches, damaging boats and beach infrastructure.

47. *Southwestern Florida, USA,* 20 *December* 2018*; Type:* GWSW; QI = 2.

A totally unexpected meteotsunami hit Southwest Florida (the Gulf coast) on 20 December 2018 and sent the surf rushing up beach, seafronts and walkways. The meteotsunami had the character of a solitary wave. The coastal infrastructure was severely damaged. The most affected sites were Naples, Captiva and Sanibel islands, where sea level was almost 1.5 m higher than normal. The meteotsunami was recorded by numerous videos, the most spectacular one made on Captiva Island: https://www.wtsp.com/video/weather/weather-explainer/mini-tsunami-washes-ashore-on-captiva-island/67-8382112.

The meteotsunami appears to be related to the passing of a cold front; at 1:30 PM (18:30 UTC) on 20 December the temperature dropped by nearly 5 °C in Southwest Florida, while the barometric pressure momentarily rose by 1.6 hPa. The sea level recorded at Naples increased very quickly by ~ 60 cm, before rapidly going down. The meteotsunami affected almost the same area as that on 17 January 2016.

48. *Lake Erie, USA,* 14 *April* 2019*; Type:* GWSW; QI = 2.

A strong meteotsunami occurred in Lake Erie on 14 April 2019 at 07:17 PM (23:17 UTC). According to eyewitness reports, there were 3 consequent waves with estimated heights of about 3.5 m. The entire event lasted about 10 min. Concrete blocks weighing 2 tons were moved for several meters, boat houses and some homes along the beach were flooded. The Geneva-on-the-Lake, Ohio marina sustained considerable damage to the docks, as the water level rose about 1.2 m and receded, all within about 10 min (Kirk Lombardy 2019, Pers. Comm.). The meteotsunami was also responsible for serious damage and depositing of debris at the Madison Township Park beach. According to Lombardy, the radar loops of the event showed that the meteotsunami had the character of an undular bore.

49. *Puerto Soller, Mallorca Island, Spain,* 21 *June* 2019*; Type:* GWHO; QI = 1.

On 21 June 2019, a meteotsunami (*rissaga*) was observed at Puerto Soller, a small city located on the northwestern coast of Mallorca Island (Fig. 4b). The maximum wave height, according to local fishermen was 1.74 m, the highest *rissaga* at this site since 1999 (https://www.majorcadailybulletin.com/news/local/2019/06/22/55837/meteotsunami-puerto-soller.html).

Similar but weaker oscillations were also reported at a number of other sites on Mallorca and Menorca islands. At Ciutadella, Menorca, the maximum wave height was 92 cm. The phenomenon did not cause any noticeable damage or delays to ferries.

6. *Discussion and Conclusions*

The catastrophic tsunamis of 2 September 1992 off Nicaragua and the 12 December 1992 tsunami off Flores Island, as well as the destructive meteotsunami of 4 July 1992 at Daytona Beach, marked the beginning of *"modern tsunami science era"* (Okal 2019; Kânoğlu et al. 2019). Modern technology, which provides remarkable volumes of high-quality instrumental data, combined with comprehensive post-event field surveys, real-time precise numerical models, and broad international cooperation have facilitated major progress in tsunami research.

In recent years, meteorological tsunamis have begun to attract ever greater attention (cf. Pattiaratchi and Wijeratne 2015; Vilibić et al. 2016). Several destructive, even catastrophic, events occurred during 2013–2019. It is now obvious that this phenomenon is not a local exotic feature of a few particular locations but can occur throughout the world oceans. What is important is that even moderate meteotsunamis (< 50 cm) can produce strong currents and be dangerous both for port operations and human safety. Coastal regions of high touristic activity, including Spain, Croatia, Italy, Australia, and the East Coast and the Gulf of Mexico in the United States, are under major threat from meteorological tsunamis.

The current study appears to be the first attempt to catalogue meteotsunami events worldwide. Only the strongest, most destructive and/or best-known events have been included in the list, with the corresponding locations shown in Fig. 1. Some of them have been the subject of intensive study and research; for these events, it was possible to present detailed and comprehensive descriptions. For others, only scarce information was found. Altogether, 51 events were selected, including the three individual events of 25–27 June 2014 in the Adriatic (#34.1), Sicily (#34.2) and Odessa (#34.3), produced by the same eastward propagating anomalous atmospheric system. A global examination of the available long-term 1-min sea level data series (cf. Pasquet et al. 2013; Vilibić and Šepić 2017; Dusek et al. 2019) could considerably help to complete this list and bring additional important information about powerful meteotsunami events. This would also make it possible to include moderate events in the list.

Cataloguing of meteorological tsunami events aids in our understanding of the origin of these events and provides crucial information for meteotsunami science. At the same time, it identifies "hot spots" along the coast that should help mitigate the hazardous effects of such events. The key question, however, is: *How can we identify meteorological tsunamis?* In the case of "classical" tsunamis, identification is relatively straightforward; any longwave oscillations generated by an earthquake, landslide or volcanic eruption can be recognized as a "tsunami". A similar approach for meteotsunamis makes no sense because almost all longwave sea level oscillations, except for tides and seismic tsunamis, have an atmospheric origin, It would not be useful to consider all of these to be "meteorological tsunamis". That is why, when describing meteotsunamis, it is common to use such terms as "*anomalous*", "*extreme*" and "*destructive*". But such definitions are subjective and cannot be applied to consistent and reliable automatic detection of meteotsunamis in sea level records.

Rabinovich and Monserrat (1996) used the criterion of 40 cm for sea level trough-to-crest wave heights to select *rissaga* events in Ciutadella and identified ten events for the period 1989–1990, including three with wave heights above 1 m. However, their study indicated that such threshold criterion cannot be considered as universal for all regions. Bechle et al. (2016), identified possible meteotsunamis in the Great Lakes using a threshold of 30 cm for wave-heights associated with strong surface wind and/or barometric pressure perturbations observed at a nearby weather station. In their effort to detect meteotsunami events on the coast of Australia, Pattiaratchi and Wijeratne (2014, 2019) used two criteria: (1) a 40-cm threshold (to select the most powerful events), and (2) a "4-sigma" criterion to detect extreme events, which are four times higher than the background noise level. They examined 5-year records of sea level time series (December 2009–December 2014) from 13 sites along the coast of Australia, as well as a few instruments located offshore, and identified 214 events, or roughly 25 events per year. Pasquet et al. (2013) also used a threshold of 40 cm to detect meteotsunami events on the East Coast of the United States for the period of 2006–2011. The original series were de-tided and residual series were high-pass filtered with a 6-h Kaiser–Bessel filter (cf. Thomson and Emery 2014). Pasquet et al. (2013) selected only those events that had been recorded by at least three tide gauges, leading to a total of nine events. For these events, they estimated the principal statistical parameters, including the duration of an event, defined as the time during which the wave height exceeded a threshold of 10 cm, similar to the criterion introduced by Rabinovich and Monserrat (1996). The strongest oscillations, with maximum wave heights of $\sim$ 1 m, were observed at Atlanta City (NJ) and Boston (MA).

Dusek et al. (2019) used a combined approach to detect meteotsunamis on the same coast as Pasquet et al. (2013) for the period 1996–2017. Their approach included the joint application of a "20-cm threshold", wavelet analysis of the 0.5–5 cph frequency band with a "6-sigma" criterion and air pressure perturbation analysis ($>$ 0.9 hPa per 6 min). Dusek et al. (2019) identified a total of 548 meteotsunamis, or about 25 per year, of which 30 events had observed wave heights were $>$ 0.6 m, including several $>$ 1 m. They also indicated the "hottest spots" along the East Coast of the United States: Atlantic City (NJ), Duck (NC), Myrtle Beach (NC), Providence (RI) and Port Canaveral (FL).A problem with the present state of meteotsunami research is that different teams of investigators (Pasquet et al. 2013; Pattiaratchi and Wijeratne 2014, 2019; Vilibić and Šepić 2017; Romero et al. 2019; Dusek et al. 2019) have used a variety of methods and high-pass windows to prepare the corresponding series and different criteria to identify meteotsunami events. However, these differences are not important for strong events, for which all approaches are acceptable. Improving our ability to reliably detect weaker events could help complete the present cataloguing list of meteorological tsunamis.

Remarkable progress has been achieved in meteotsunami research over the 27 years since the 1992 Daytona event. This progress can be summarized as follows:

- A crucial factor advancing meteotsunami science is the major enhancement in measurement capabilitys. High-resolution sea level and air pressure monitoring systems installed along the coasts of the world oceans now enable researchers to collect vast amounts of precise data and to provide evidence that non-seismic tsunami-like events are, in fact, meteorological tsunamis generated by atmospheric processes.
- As meteorological tsunamis have been reported and investigated throughout the entire World Ocean, it has become evident that the phenomenon—once considered confined to a few specific regions—is actually quite common throughout the world.
- Similar to regular systematic post-tsunami field surveys (Arcos et al. 2019; Okal 2019), destructive meteotsunamis are now the subject of runup and inundation examination along the affected coasts (cf. Salaree et al. 2018). This is essential for the long-term forecast of future events and for the estimation of meteotsunami risk for specific regions (Kânoğlu et al. 2019).
- Effective numerical models of meteotological tsunamis have been recently developed (cf. Vilibić et al. 2008; Orlić et al. 2010; Šepić et al. 2018a); these models can reliably reproduce actual events. Moreover, the models can be successfully used for real time meteotsunami warning (Denamiel et al. 2019).
- Although meteorological tsunamis still remain one of the most mysterious of natural phenomena, the general physics of such events in most cases has become more clear. Specifically, our better understanding of this phenomenon allows scientists to reconstruct past events and to predict incoming meteotsunamis.
- In some countries strongly affected by meteorological tsunamis, such as Spain (Marcos et al. 2009; Renault et al. 2011) and Croatia (Vilibić et al. 2016; Denamiel et al. 2019), preliminary meteotsunami warning systems have already been developed. Similar systems are under development in some other countries, including the United States, Portugal and Japan.

It is now safe to state that *meteorological tsunamis* have beome an important component of the "tsunami family". They are considered, along with seismically, landslide and volcanically generated tsunamis, as one of the main tsunami types (Tsunami glossary 2016).

Figure 1 shows specific regions of the world oceans, where meteotsunamis are most common and dangerous:

1. The Mediterranean, especially the coasts of the Balearic Islands and the Adriatic Sea (Fig. 1b, c);
2. The Yellow Sea and the northwestern part of the East China Sea (Fig. 1c);
3. The East Coast of the United States (Fig. 1d);
4. The northeastern coast of the Gulf of Mexico (Fig. 1d);
5. The Great Lakes (Fig. 1d).

It is evident that for these "hot" regions meteotsunamis are under major threat and should attract major public attention. Two directions of applied research are most important for these regions: (a) *Long-term meteotsunami forecasting*, i.e., the estimation of meteotsunami risk and possible maximum wave heights along the coast, and identification of "hot spots", where hazardous meteotsunamis are most probable; and (2) *Short-term meteotsunami forecasting*, i.e., development of operative real-time meteotsunami warning and prediction. However, recent destructive and unexpected events in various regions of the world oceans, in particular, in the Persian Gulf, The Netherlands, South Africa, Brazil, and Chile (Fig. 1a) indicate the overworld significance of this phenomenon and the necessity of its thorough study.

Acknowledgements

The original idea for this overview of meteorological tsunamis was suggested to the author by Emile Okal (Northwestern University, Evanston, IL, USA). In fact, the present paper can be considered as a continuation of his overview of seismically generated tsunamis for the period 1992–2018 (Okal 2019). Many of the contributions to meteotsunami research included in this overview were performed by the author in close cooperation with Sebastian Monserrat (University of the Balearic Islands, Palma de Mallorca, Spain), Ivica Vilibić and Jadranka Šepić (both at the Institute of Oceanography and Fisheries, Split, Croatia), and Richard Thomson and Isaac Fine (at the Institute of Ocean Sciences, Sidney, BC, Canada); their support, numerous useful discussions and the provision of data used in this overview are greatly appreciated. The author gratefully acknowledges Richard Thomson, Fred Stephenson (the Institute of Ocean Sciences, Sidney, BC, Canada), Ivica Vilibić and Emile Okal for valuable comments and suggestions. This work was partially supported by the Russian State Assignment of IO RAS #0149-2019-0005 and #0128-2019-0010, and by the FCT-funded project FAST (PTDC/CTA-MET/32004/2017).

References

Alasset, P.-J., Hébert, H., Maouche, S., Calbini, V., & Meghraoui, M. (2006). The tsunami induced by the 2003 Zemmouri earthquake (Mw = 6.9, Algeria): Modelling and results. *Geophysical Journal International, 166*(1), 213–226. https://doi.org/10.1111/j.1365-246x.2006.02912.x.

Alfonso-Sosa, E. (2016). *Naples Meteotsunami on January 17, 2016*. Ocean Physics Education, 16-Feb-16, Technical Report; https://doi.org/10.13140/rg.2.2.26518.75840.

Anderson, E. J., Bechle, A. J., Wu, C. H., Schwab, D. J., Mann, G. E., & Lombardy, K. A. (2015). Reconstruction of a meteotsunami in Lake Erie on 27 May 2012: Roles of atmospheric conditions on hydrodynamic response in enclosed basins. *Journal of Geophysical Research—Oceans, 120,* 8020–8038. https://doi.org/10.1002/2015JC010883.

Anderson, E. J., & Mann, G. E. (2019). Atmospheric and hydrodynamic simulation of a gravity wave induced meteotsunami in Ludington, Michigan. In *The First World Conference on Meteotsunamis*, Split, Croatia, 8–11 May 2019, Book of Abstracts, p. 36.

Arcos, N. P., Dunbar, P. K., Stroker, K. J., & Kong, L. S. L. (2019). The impact of post-tsunami surveys on the NCEI/WDS global historical tsunami database. *Pure and Applied Geophysics, 176*(7), 2809–2829. https://doi.org/10.1007/s00024-019-02191-7.

Asano, T., Yamashiro, T., & Nishimura, N. (2012). Field observations of meteotsunami locally called ''abiki'' in Urauchi Bay, Kami-Koshiki Island, Japan. *Natural Hazards, 64,* 1685–1706. https://doi.org/10.1007/s11069-012-0330-2.

Assink, J., Evers, L., Smink, M., & Apituley, A. (2018). *High-resolution Observations of a Meteo-tsunami*. Geophysical Research Abstracts, Vol. 20, EGU2018-11848, 2018 EGU General Assembly, Vienna.

Bailey, K., DiVeglio, C., & Welty, A. (2014). *An Examination of the June 2013 East Coast Meteotsunami Captured by NOAA Observing Systems* (p. 42). Maryland: NOAA Technical Report NOS CO-OPS 079, U.S. Department of Commerce, NOAA National Ocean Service, Silver Spring.

Bechle, A. J., Kristovich, D. A. R., & Wu, C. H. (2015). Meteotsunami occurrences and causes in Lake Michigan. *Journal of*

Geophysical Research—Oceans, 120, 8422–8438. https://doi.org/10.1002/2015JC011317.

Bechle, A. J., Wu, C. H., Kristovich, D. A. R., Anderson, E. J., Schwab, D. J., & Rabinovich, A. B. (2016). Meteotsunamis in the Laurentian Great Lakes. *Scientific Reports, 6*(37832), 1–8. https://doi.org/10.1038/srep37832.

Belušić, D., Grisogono, B., & Klaić, Z. B. (2007). Atmospheric origin of the devastating coupled air–sea event in the east Adriatic. *Journal of Geophysical Research, 112,* D17111. https://doi.org/10.1029/2006jd008204.

Belušić, D., & Strelec Mahović, N. (2009). Detecting and following atmospheric disturbances with a potential to generate meteotsunamis in the Adriatic. *Physics and Chemistry of the Earth, 34,* 918–927.

Berkowitz, B., Muyskens, J., Meko, T., Emamdjomeh, A., Lu, D., Steckelberg, A., et al. (2017). What Irma's wind and water did to Florida. *The Washington Post.* Retrived September 12, 2017 from https://www.washingtonpost.com/graphics/2017/national/hurricane-irma-florida/.

Bouhadad, Y., Nour, A., Slimani, A., Laouaimi, N., & Delhai, D. (2004). The Boumerdes (Algeria) earthquake of May 21, 2003 (Mw = 6.8): Ground deformation and intensity. *Journal of Seismology, 8*(4), 497–506. https://doi.org/10.1007/s10950-004-4838-0.

Candela, J., Mazzola, S., Sammari, C., Limeburner, R., Lozano, C. J., Patti, B., & Bonnano, A. (1999). The ''Mad Sea'' phenomenon in the Strait of Sicily. *Journal of Physical Oceanography, 29,* 2210–2231.

Cangialosi, J. P., Latto, A. S., & Berg, R. (2018). *Tropical Cyclone Report: Hurricane Irma,* 30 August–12 September 2017, National Hurricane Center, Miami, FL, NOAA/NWS, 111 pp. https://www.nhc.noaa.gov/data/tcr/AL112017_Irma.pdf.

Carvajal, M., Contreras-López, M., Winckler, P., & Sepúlveda, I. (2017). Meteotsunamis occurring along the southwest coast of South America during an intense storm. *Pure and Applied Geophysics, 174*(8), 1313–1323.

Chapman D. C. & Giese, G. S. S. (2001). In J. H. Steele, K. K. Turekian & S. A. Thorpe (Eds.), *Encyclopedia of Ocean Sciences* (Vol. 5, pp. 2724–2731). London: Academic Press.

Cho, K.-H., Choi, J.-Y., Park, K.-S., Hyun, S.-K., Oh, Y., & Park, J. Y. (2013). A synoptic study on tsunami-like sea level oscillations along the west coast of Korea using an unstructured-grid ocean model. *Journal of Coastal Research, 65,* 678–683. https://doi.org/10.2112/si65-115.1.

Choi, B.-J., Hwang, C., & Lee, S.-H. (2014). Meteotsunami-tide interactions and high-frequency sea level oscillations in the eastern Yellow Sea. *Journal of Geophysical Research—Oceans, 119,* 6725–6742. https://doi.org/10.1002/2013JC009788.

Churchill, D. D., Houston, S. H., & Bond, N. A. (1995). The Daytona Beach wave of 3–4 July 1992: A shallow water gravity wave forced by a propagating squall line. *Bulletin of the American Meteorological Society, 76*(1), 21–32.

Colucci, P., & Michelato, A. (1976). An approach to the study of the ''Marrubbio'' phenomenon. *Bollettino di Geofisica Teorica ed Applicata, 19,* 3–10.

De Jong, M. P. C., & Battjes, J. A. (2004). Low-frequency sea waves generated by atmospheric convection cells. *Journal of Geophysical Research, 109,* C01011. https://doi.org/10.1029/2003jc001931.

De Jong, M. P. C., Holthuijsen, L. H., & Battjes, J. A. (2003). Generation of seiches by cold fronts over the southern North Sea. *Journal of Geophysical Research, 108*(C4), 3117. https://doi.org/10.1029/2002jc001422.

Defant, A. (1961). *Physical Oceanography* (Vol. 2, p. 729). Oxford: Pergamon Press.

Denamiel, C., Šepić, J., Ivanković, D., & Vilibić, I. (2019). The Adriatic Sea and coast modelling suite: Evaluation of the meteotsunami forecast component. *Ocean Modelling, 135,* 71–93. https://doi.org/10.1016/j.ocemod.2019.02.003.

Drago, A. F. (1999). Sea level variability and the 'Milghuba' seiche oscillations in the northern coast of Malta, Central Mediterranean. *Physics and Chemistry of the Earth, 34*(17–18), 948–970. https://doi.org/10.1016/j.pce.2009.10.002.

Drago, A. F. (2009). *A Study on the Sea Level Variations and the 'Milghuba' Phenomenon in the Coastal Waters of the Maltese Islands.* Ph.D. thesis, University of Southampton.

Dusek, G., DiVeglio, C., Licate, L., Heilman, L., Kirk, K., Paternostro, C., & Miller, A. (2019). A meteotsunami climatology along the U.S. East Coast. *Bulletin of the American Meteorological Society, 100*(7), 1329–1345. https://doi.org/10.1175/bams-d-18-0206.1.

Eom, H., Byun, K.-Y., You, S.H., Kim, M.-S., & Woo, S.-B. (2019). Developing an early warning system for meteotsunami in South Korea (KMA). *The First World Conference on Meteotsunamis,* Split, Croatia, 8-11 May 2019, Book of Abstracts, p. 80.

Ewing, M., Press, F., & Donn, W. J. (1954). An explanation of the Lake Michigan wave of 26 June 1954. *Science, 120,* 684–686.

Frère, A., Daubord, C., Gailler, A., & Hébert, H. (2014). Sea level surges of June 2011 in the NE Atlantic Ocean: Observations and possible interpretation. *Natural Hazards, 74*(1), 179–196. https://doi.org/10.1007/s11069-014-1103-x.

Fukuzawa, K., & Hibiya, T. (2019). The amplification mechanism of the meteo-tsunami originating off the western coast of Kyushu Island in Japan in the winter of 2010. In *The First World Conference on Meteotsunamis,* Split, Croatia, 8–11 May 2019, Book of Abstracts, p. 55.

Gomis, D., Monserrat, S., & Tintoré, J. (1993). Pressure-forced seiches of large amplitude in inlets of the Balearic Islands. *Journal of Geophysical Research, 98,* 14437–14445.

Gusiakov, V. (2019). Seismically generated tsunamis, meteotsunamis and rogue waves: Problems of identification, parameterization and cataloguing. In *The First World Conference on Meteotsunamis,* Split, Croatia, 8–11 May 2019, Book of Abstracts, p. 70.

Gusiakov, V. K., Dunbar, P. K., & Arcos, N. (2019). Twenty-five years (1992–2016) of global tsunamis: Statistical and analytical overview. *Pure and Applied Geophysics, 176*(7), 2795–2807. https://doi.org/10.1007/s00024-019-02113-7.

Heidarzadeh, M., Šepić, J., Rabinovich, A., Allahyar, M., Soltanpour, A., & Tavakoli, F. (2019). Meteorological tsunami of 19 March 2017 in the Persian Gulf: Observations and analyses. *Pure and Applied Geophysics.* https://doi.org/10.1007/s00024-019-02263-8.

Hibiya, T., & Kajiura, K. (1982). Origin of 'Abiki' phenomenon (kind of seiches) in Nagasaki Bay. *Journal of Oceanographical Society of Japan, 38,* 172–182.

Honda, K., Terada, T., Yoshida, Y., & Isitani, D. (1908). Secondary undulations of oceanic tides. *Journal of the College of Science, Imperial University of Tokyo, 24,* 1–110.

Jansà, A., Monserrat, S., & Gomis, D. (2007). The rissaga of 15 June 2006 in Ciutadella (Menorca), a meteorological tsunami. *Advances in Geosciences, 12,* 1–4.

Kânoğlu, U., Tanioka, Y., Okal, E. A., Baptista, M. A., & Rabinovich, A. B. (2019). Introduction to "Twenty five years of modern tsunami science following the 1992 Nicaragua and Flores Island tsunamis, Volume I". *Pure and Applied Geophysics, 176*(7), 2757–2769. https://doi.org/10.1007/s00024-019-02266-5.

Kharif, Ch., Pelinovsky, E., & Slunyaev, A. (2009). *Rogue Waves in the Ocean* (p. 216). Dordrecht: Springer.

Kim, M.-S., Kim, H., Kim, Y.-K., Yoo, S.-H., Eom, H.-M., & Woo, S.-B. (2017). Determination of accurate arrival time of meteotsunami event in Yellow Sea. *Journal of Coastal Research, 79,* 149–153.

Kim, M.-S., Kim, H., Kim, Y.-K., Gu, B.-H., Lee, H.-J., Woo, S.-B. (2016). Double resonance effect at Daeheuksando Port caused by air pressure disturbances in Yellow Sea on 31 March 2007. *Journal of Coastal Research, 75*(spl), 1142–1146. https://doi.org/10.2112/SI75-229.1.

Knight, W.R., Whitmore, P., Kim, Y., Wang, D., Becker, N. C., Weinstein, S., et al. (2013). The US East Coast meteotsunami of June 13, 2013. In *AGU 2013 Fall Meeting Abstracts* (Vol. 1, p. 1740).

Kulikov, E. A., & Medvedev, I. P. (2013). Variability of the Baltic sea level and floods in the Gulf of Finland. *Oceanology, 53*(2), 145–151.

Lander, J. F. (1996). *Tsunamis Affecting Alaska, 1737–1996* (p. 195). Boulder: USDC/NOAA.

Lander, J. F., & Lockridge, P. A. (1989). *United States Tsunamis (Including United States Possessions), 1690–1988* (p. 265). Boulder: National Geophysical Data Center, USDC/NOAA.

Lin, L.-C., & Liang, M.-C. (2017). Meteotsunamis produced by high frequency atmospheric pressure forcing. *Terrestrial Atmospheric and Oceanic Sciences, 28*(6), 1033–1040. https://doi.org/10.3319/tao.2017.03.20.01.

Lindzen, R. S. (1974). Wave-CISK in the tropics. *Journal of the Atmospheric Sciences, 31,* 156–179.

Lipa, B., Parikh, H., Barrick, D., Roarty, H., & Glenn, S. (2014). High-frequency radar observations of the June 2013 US East Coast meteotsunami. *Natural Hazards, 74,* 109–122. https://doi.org/10.1007/s11069-013-0992-4.

Liu, P. L.-F., Monserrat, S., Marcos, M., & Rabinovich, A. B. (2003). Coupling between two inlets: Observation and modeling. *Journal of Geophysical Research, 108*(C3), 3069. https://doi.org/10.1029/2002jc001478.

Marcos, M., Monserrat, S., Medina, R., Orfila, A., & Olabarrieta, M. (2009). External forcing of meteorological tsunamis at the coast of the Balearic Islands. *Physics and Chemistry of the Earth, 34,* 938–947. https://doi.org/10.1016/j.pce.2009.10.00.

Miles, J., & Munk, W. (1961). Harbor paradox. *Journal of the Waterways and Harbors Division, American Society of Civil Engineers, 87*(3), 111–130.

Monserrat, S., Ibberson, A., & Thorpe, A. J. (1991). Atmospheric gravity waves and the "rissaga" phenomenon. *Quarterly Journal of the Royal Meteorological Society, 117,* 553–570.

Monserrat, S., Rabinovich, A. B., & Casas, B. (1998). On the reconstruction of the transfer function for atmospherically generated seiches. *Geophysical Research Letters, 25,* 2197–2200.

Monserrat, S., & Thorpe, A. J. (1996). Use of ducting theory in an observed case of gravity waves. *Journal of the Atmospheric Sciences, 53*(12), 1724–1736.

Monserrat, S., Vilibić, I., & Rabinovich, A. B. (2006). Meteotsunamis: Atmospherically induced destructive ocean waves in the tsunami frequency band. *Natural Hazards and Earth System Sciences, 6,* 1035–1051.

Munk, W. H. (1962), Long waves. In *The Sea* (pp. 647–663). New York: Wiley.

Nikonov, A. A., & Fleyfel, L. D. (2015). Tsunami in Odessa: Natural or man-made phenomenon? *Priroda, 4,* 36–43. **(in Russian)**.

Nomitsu, T. (1935). A theory of tsunamis and seiches produced by wind and barometric gradient. *Memoirs of the College of Science; Kyoto Imperial University Series A, 18,* 201–214.

O'Loughlin, K. F., & Lander, J. F. (2003). *Caribbean Tsunamis: A 500-Year History from 1498–1998* (p. 263). Dordrecht: Springer.

Okal, E. A. (2019). Twenty-five years of progress in the science of "geological" tsunamis following the 1992 Nicaragua and Flores events. *Pure and Applied Geophysics, 176*(7), 2771–2793. https://doi.org/10.1007/s00024-019-02244-x.

Okal, E. A., Visser, J. N., & de Beer, C. H. (2014). The Dwarskersbos, South Africa local tsunami of August 27, 1969: Field survey and simulation as a meteorological event. *Natural Hazards, 74*(1), 251–268. https://doi.org/10.1007/s11069-014-1205-5.

Olabarrieta, M., Valle-Levinson, A., Martinez, C. J., Pattiaratchi, C., & Shi, L. (2017). Meteotsunamis in the northeastern Gulf of Mexico and their possible link to El Niño Southern Oscillation. *Natural Hazards, 88,* 1325–1346. https://doi.org/10.1007/s11069-017-2922-3.

Orlić, M. (1980). About a possible occurrence of the Proudman resonance in the Adriatic. *Thalassia Jugoslavica, 16,* 79–88.

Orlić, M. (2015). The first attempt at cataloguing tsunami-like waves of meteorological origin in Croatian coastal waters. *Acta Adriatica, 56,* 83–95.

Orlić, M., Belušić, D., Janeković, I., & Pasarić, M. (2010). Fresh evidence relating the great Adriatic surge of 21 June 1978 to mesoscale atmospheric forcing. *Journal of Geophysical Research, 115,* C06011. https://doi.org/10.1029/2009JC005777.

Pasquet, S., & Vilibić, I. (2013). Shelf edge reflection of atmospherically generated long ocean waves along the central U.S. East Coast. *Continental Shelf Research, 66,* 1–8. https://doi.org/10.1016/j.csr.2013.06.007.

Pasquet, S., Vilibić, I., & Šepić, J. (2013). A survey of strong high-frequency sea level oscillations along the US East Coast between 2006 and 2011. *Natural Hazards and Earth System Sciences, 13,* 473–482. https://doi.org/10.5194/nhess-13-473-2013.

Pattiaratchi, C. B., & Wijeratne, E. M. S. (2014). Observations of meteorological tsunamis along the southwest Australian coast. *Natural Hazards, 74*(1), 281–303. https://doi.org/10.1007/s11069-014-1263-8.

Pattiaratchi, C. B., & Wijeratne, E. M. S. (2015). Are meteotsunamis an underrated hazard? *Philosophical Transactions of the Royal Society, A, 373*(20140377), 1–23.

Pattiaratchi, C.B., & Wijeratne, S. (2019). Observations of meteotsunamis in Australia. In *The First World Conference on Meteotsunamis*, Split, Croatia, 8–11 May 2019, Book of Abstracts, p. 26.

Paxton, L. D. (2016). *Development of a Forecast Process for Meteotsunami Events in the Gulf of Mexico.* Master Thesis, School of Geosciences, College of Arts and Sciences, University of South Florida, 76 p.

Paxton, C. H., & Sobien, D. A. (1998). Resonant interaction between an atmospheric gravity wave and shallow water wave along Florida's west coast. *Bulletin of the American Meteorological Society, 79*(12), 2727–2732.

Pellikka, H., Rauhala, J., Kahma, K. K., Stipa, T., Boman, H., & Kangas, A. (2014). Recent observations of meteotsunamis on the Finnish coast. *Natural Hazards, 74*(1), 197–215. https://doi.org/10.1007/s11069-014-1150-3.

Platzman, G. W. (1965). The prediction of surges in the southern basin of Lake Michigan. Part I. The dynamical basis for prediction. *Monthly Weather Review, 93*(5), 275–281.

Proudman, J. (1929). The effects on the sea of changes in atmospheric pressure. *Geophysical Supplement, Monthly Notices of the Royal Astronomical Society, 2,* 197–209.

Pugh, D., & Woodworth, P. (2014). *Sea-level Science: Understanding Tides, Surges, Tsunamis and Mean Sea-level Changes* (p. 395). Cambridge: Cambridge Univ. Press.

Rabinovich, A. B. (1997). Spectral analysis of tsunami waves: Separation of source and topography effects. *Journal of Geophysical Research, 102*(C6), 12663–12676.

Rabinovich, A. B. (2009). Seiches and harbor oscillations. In Y. C. Kim (Ed.), *Handbook of Coastal and Ocean Engineering* (Chap. 9, pp. 193–236). Singapore: World Scientific Publ.

Rabinovich, A. B., & Monserrat, S. (1996). Meteorological tsunamis near the Balearic and Kuril Islands: Descriptive and statistical analysis. *Natural Hazards, 13,* 55–90.

Rabinovich, A. B., & Monserrat, S. (1998). Generation of meteorological tsunamis (large amplitude seiches) near the Balearic and Kuril Islands. *Natural Hazards, 18,* 27–55.

Rabinovich, A. B., Monserrat, S., & Fine, I. V. (1999). Numerical modeling of extreme seiche oscillations in the region of the Balearic Islands. *Oceanology, 39*(1), 16–24.

Rabinovich, A. B., & Šepić, J. (2016). Meteorological tsunamis: What is this? *Priroda., 1,* 12–26. **[in Russian]**.

Ranguelov, B., Tinti, S., Pagnoni, G., Tonini, R., Zaniboni, F., & Armigliato, A. (2008). The nonseismic tsunami observed in the Bulgaria Sea on 7 May 2007—Was it due to a submarine landslide? *Geophysical Research Letters, 35,* L18613.

Renault, L., Vizoso, G., Jansá, A., Wilkin, J., & Tintoré, J. (2011). Toward the predictability of meteotsunamis in the Balearic Sea using regional nested atmosphere and ocean models. *Geophysical Research Letters, 38,* L10601. https://doi.org/10.1029/2011gl047361.

Romero, R., Vich, M.D.M., & Ramis, C. (2019). A pragmatic approach for the numerical prediction for meteotsunamis in Ciutadella Harbour (Balearic Islands). In *The First World Conference on Meteotsunamis*, Split, Croatia, 8–11 May 2019, Book of Abstracts, p. 76.

Salaree, A., Mansouri, R., & Okal, E. A. (2018). The intriguing tsunami of 19 March 2017 at Bandar Dayyer, Iran: Field survey and simulations. *Natural Hazards, 90*(3), 1277–1307. https://doi.org/10.1007/s11069-017-3119-5.

Sallenger, A. H., List, J. H., Gelfenbaum, G., Stumpf, R. P., & Hansen, M. (1995). Large wave at Daytona Beach, Florida, explained as a squall-line surge. *Journal of Coastal Research, 11,* 1383–1388.

Šepić, J., Međugorac, I., Janeković, I., Dunić, N., & Vilibić, I. (2016). Multi-meteotsunami event in the Adriatic Sea generated by atmospheric disturbances of 25–26 June 2014. *Pure and Applied Geophysics, 173,* 4117–4138. https://doi.org/10.1007/978-3-319-55480-8_23.

Šepić, J., & Rabinovich, A. B. (2014). Meteotsunami in the Great Lakes and on the Atlantic coast of the United States generated by the "derecho" of June 29–30, 2012. *Natural Hazards, 74,* 75–107. https://doi.org/10.1007/s11069-014-1310-5.

Šepić, J., Rabinovich, A. B., & Sytov, V. N. (2018a). Odessa tsunami of 27 June 2014: Observations and numerical modelling. *Pure and Applied Geophysics, 175*(4), 1545–1572. https://doi.org/10.1007/s00024-017-1729-1.

Šepić, J., Vilibić, I., & Belušić, D. (2009). Source of the 2007 Ist meteotsunami (Adriatic Sea). *Journal of Geophysical Research, 114,* C03016. https://doi.org/10.1029/2008JC005092.

Šepić, J., Vilibić, I., Rabinovich, A. B., & Monserrat, S. (2015). Widespread tsunami-like waves of 23-27 June in the Mediterranean and Black Seas generated by high-altitude atmospheric forcing. *Scientific Reports, 5*(11682), 1–5. https://doi.org/10.1038/srep11682.

Šepić, J., Vilibić, I., Rabinovich, A., & Tinti, S. (2018b). Meteotsunami ("Marrobbio") of 25–26 June 2014 on the Southwestern Coast of Sicily, Italy. *Pure and Applied Geophysics, 175*(4), 1573–1593. https://doi.org/10.1007/s00024-017-1827-8.

Shiga, T., Ichikawa, M., Kusumoto, K., & Suzuki, H. (2007). A statistical study on seiche around Kyushu and Satsunan Islands. *Papers in Meteorology and Geophysics, 74,* S139–S162.

Soloviev, S.L., & Go, Ch. N. (1974). *Catalogue of Tsunamis on the Western Shore of the Pacific Ocean*, Nauka, Moscow, pp. 309. [in Russian; English Translation: Canadian Transl. Fish. Aquatic Sci., No. 5077, Ottawa, 1984, 439 pp].

Soloviev, S.L., & Go, Ch. N. (1975). *Catalogue of Tsunamis on the Eastern Shore of the Pacific Ocean*, Nauka, Moscow, pp. 204. [in Russian; English Translation: Canadian Transl. Fish. Aquatic Sci., No. 5078, Ottawa, 1984, 285 pp.].

Tanaka, K. (2010). Atmospheric pressure-wave bands around a cold front resulted in a meteotsunami in the East China Sea in February 2009. *Natural Hazards and Earth System Sciences, 10,* 2599–2610. https://doi.org/10.5194/nhess-10-2599-2010.

Tanaka, K. (2012). On meteotsunamis around Tsushima Strait generated by the Baiu front. *Natural Hazards, 63*(2), 805–822. https://doi.org/10.1007/s11069-014-1103-x.

Tanaka, K. (2019). Meteotsunamis propagating along continental slope in the northwestern Pacific Ocean: A case study of the January 2018 event in Shikoku area. In *The First World Conference on Meteotsunamis*, Split, Croatia, 8–11 May 2019, Book of Abstracts, p. 35.

Tanaka, K., Gohara, S., Koga, T., Yamaguchi, R., & Yamada, F. (2014). Abiki oscillations in Sakitsu Bay, west Kyushu, Japan. *Natural Hazards, 74*(1), 233–250. https://doi.org/10.1007/s11069-013-0959-5.

Tappin, D. R., Sibley, A., Horsburgh, K., Daubord, C., Cox, D., & Long, D. (2013). The English Channel 'tsunami' of 27 June 2011—A probable meteorological source. *Weather, 68*(6), 144–152.

Thomson, R. E., & Emery, W. J. (2014). *Data Analysis Methods in Physical Oceanography* (3rd ed., p. 716). New York: Elsevier.

Thomson, R. E., Rabinovich, A. B., Fine, I. V., Sinnott, D. C., McCarthy, A., Sutherland, N. A. S., & Neil, L. K. (2009). Meteorological tsunamis on the coasts of British Columbia and Washington. *Physics and Chemistry of the Earth, 34,* 971–988. https://doi.org/10.1016/j.pce.2009.10.003.

Thomson, R. E., Rabinovich, A. B., & Krassovski, M. V. (2007). Double jeopardy: Concurrent arrival of the 2004 Sumatra tsunami and storm-generated waves on the Atlantic coast of the United States and Canada. *Geophysical Research Letters, 34,* L15607. https://doi.org/10.1029/2007GL030685.

Thomson, R.E., Rabinovich, A.B., & Šepić, J. (2019). The extreme typhoon "Songda" event of 14 October 2016 on the coasts of British Columbia and Washington State. In *The First World Conference on Meteotsunamis*, Split, Croatia, 8–11 May 2019, Book of Abstracts, p. 62.

Tintoré, J., Gomis, D., Alonso, S., & Wang, D. P. (1988). A theoretical study of large sea level oscillations in the Western Mediterranean. *Journal of Geophysical Research, 93,* 10797–10803.

Tsunami Glossary (2006). Intergovernmental Oceanographic Commission of UNESCO, International Tsunami Information Centre, UNESCO 2006, IOC/INF-1221, Hawaii, pp. 36.

Tsunami Glossary (2016). Intergovernmental Oceanographic Commission. Third Edition, Paris, UNESCO, IOC Technical Series, 85, pp. 44.

Vich, M., & Monserrat, S. (2009). The source spectrum for the Algerian tsunami of 21 May 2003 estimated from coastal tide gauge data. *Geophysical Research Letters, 36,* L20610. https://doi.org/10.1029/2009GL039970.

Vilibić, I. (2008). Numerical simulations of the Proudman resonance. *Continental Shelf Research, 28,* 574–581.

Vilibić, I., Domijan, N., Orlić, M., Leder, N., & Pasarić, M. (2004). Resonant coupling of a traveling air pressure disturbance with the east Adriatic coastal waters. *Journal of Geophysical Research—Oceans, 109,* C10001. https://doi.org/10.1029/2004JC002279.

Vilibić, I., Horvath, K., Strelec Mahović, N., Monserrat, S., Marcos, M., Amores, Á., & Fine, I. (2014a). Atmospheric processes responsible for generation of the 2008 Boothbay meteotsunami. *Natural Hazards, 74*(1), 25–53. https://doi.org/10.1007/s11069-013-0811-y.

Vilibić, I., Monserrat, S., & Rabinovich, A. B. (2014b). Meteorological tsunamis on the US East Coast and in other regions of the World Ocean. *Natural Hazards, 74*(1), 1–9. https://doi.org/10.1007/s11069-014-1350-x.

Vilibić, I., Monserrat, S., Rabinovich, A. B., & Mihanović, H. (2008). Numerical modelling of the destructive meteotsunami of 15 June 2006 on the coast of the Balearic Islands. *Pure and Applied Geophysics, 165,* 2169–2195.

Vilibić, I., & Šepić, J. (2009). Destructive meteotsunamis along the eastern Adriatic coast: Overview. *Physics and Chemistry of the Earth, 34,* 904–917. https://doi.org/10.1016/j.pce.2009.08.004.

Vilibić, I., & Šepić, J. (2017). Global mapping of nonseismic sea level oscillations at tsunami timescales. *Scientific Reports, 7*(40818), 1–6. https://doi.org/10.1038/srep40818.

Vilibić, I., Šepić, J., Rabinovich, A. B., & Monserrat, S. (2016). Modern approaches in meteotsunami research and early warning. *Frontiers in Marine Science, 3*(57), 1–7. https://doi.org/10.3389/fmars.2016.00057.

Vilibić, I., Šepić, J., Ranguelov, B., Strelec Mahović, N., & Tinti, S. (2010). Possible atmospheric origin of the 7 May 2007 western Black Sea shelf tsunami event. *Journal of Geophysical Research, 115,* C07006. https://doi.org/10.1029/2009jc005904.

Wang, X., Li, K., Yu, Z., & Wu, J. (1987). Statistical characteristics of seiches in Longkou Harbour. *Journal of Physical Oceanography, 17,* 1963–1966.

Wang, D., Becker N.C., Weinstein, S., Whitmore, P., Knight, W., Kim, Y., et al. (2013). June 13, 2013 US East Coast meteotsunami: Comparing a numerical model with observations. In *AGU 2013 Fall Meeting Abstracts* (Vol. 1, p. 02), Abstract: NH52A-02.

Wertman, C. A., Yablonsky, R. M., Shen, Y., Merrill, J., Kincaid, C. R., & Pockalny, R. A. (2014). Mesoscale convective system surface pressure anomalies responsible for meteotsunamis along the U.S. East Coast on June 13th, 2013. *Scientific Reports, 4*(7143), 1–9. https://doi.org/10.1038/srep07143.

Whitmore, P., & Knight, B. (2014). Meteotsunami forecasting: sensitivities demonstrated by the 2008 Boothbay, Maine, event. *Natural Hazards, 74*(1), 11–23. https://doi.org/10.1007/s11069-014-1056-0.

Wilson, B. (1972). Seiches. *Advances in Hydrosciences, 8,* 1–94.

Yoo, J., Lee, D.-Y., Ha, T.-M., Cho, Y.-S., & Woo, S.-B. (2010). Characteristics of abnormal large waves measured from coastal videos. *Natural Hazards and Earth System Sciences, 10,* 947–956. https://doi.org/10.5194/nhess-10-947-2010.

(Received July 23, 2019, revised September 17, 2019, accepted October 17, 2019, Published online November 7, 2019)

Pure Appl. Geophys. 177 (2020), 1231–1259

https://doi.org/10.1007/s00024-019-02263-8

Pure and Applied Geophysics

Meteorological Tsunami of 19 March 2017 in the Persian Gulf: Observations and Analyses

Mohammad Heidarzadeh,[1] Jadranka Šepić,[2] Alexander Rabinovich,[3,4] Mohammadreza Allahyar,[5] Ali Soltanpour,[6] and Farokh Tavakoli[7]

Abstract—On 19 March 2017, destructive tsunami-like waves impacted the northeast shore of the Persian Gulf (PG). The maximum surveyed runup of about 3 m was observed at Dayyer in southern Iran, where damaging waves inundated the land for a distance of ~ 1 km and resulted in the deaths of five people. Because the PG has always been considered safe from extreme oceanic waves, the event was totally unexpected. In this study, we examined sea level data from 12 stations across the PG and a variety of meteorological information, including satellite imagery, high-altitude isohypse maps and high-resolution air pressure records from 47 instruments along the PG. Our results show that the event was very local, with recorded maximum trough-to-crest wave heights of 197 cm at Dayyer and 234 cm at Asaluyeh, near-field cities in Iran located ~ 80 km apart. The dominant wave periods were in the range of 15–20 min. At all distant tide gauges, the observed wave heights were < 35 cm. No earthquakes or landslides were evident at the time of the event. On the other hand, atmospheric processes during 18–22 March were very active and 10 distinctive tsunamigenic air pressure disturbances were observed propagating over the PG, suggesting that the event of 19 March 2017 was a "*meteorological tsunami*". Atmospheric conditions over the PG were highly favourable for the generation of meteotsunamis and very similar to those that caused a chain of strong meteotsunamis in the Mediterranean and Black Sea regions during 23–27 June 2014. Based on the 500 hPa wind, we evaluated that the disturbances had propagation speeds of 21–38 m/s, with the disturbance at Dayyer having a speed of ~ 26 m/s toward 77° True. The Froude number, *Fr* (estimated as the ratio of the air disturbance speed to the long wave speed), on 19 March 2017 in the Dayyer/Asaluyeh region was close to resonance, *Fr* ~ 0.9 to 1.1, which is highly favourable for meteotsunami generation. Our findings indicate that the Dayyer/Asaluyeh area is a "hot spot" that is highly vulnerable to extreme, weather-induced tsunami-like waves.

Key words: Persian Gulf, meteotsunami, extreme waves, atmospheric waves, tide gauge data, wavelet analysis, remote sensing.

[1] Department of Civil and Environmental Engineering, Brunel University London, Uxbridge UB8 3PH, UK. E-mail: mohammad.heidarzadeh@brunel.ac.uk
[2] Institute of Oceanography and Fisheries, Split, Croatia.
[3] P.P. Shirshov Institute of Oceanology, Russian Academy of Sciences, Moscow, Russia.
[4] Institute of Ocean Sciences, Fisheries and Oceans Canada, Sidney, BC, Canada.
[5] Port and Maritime Organization, Tehran, Iran.
[6] Hydrography and Tidal Analysis Department, National Cartographic Center, Tehran, Iran.
[7] Geodetic Department, National Cartographic Center, Tehran, Iran.

1. Introduction

The Persian Gulf (PG) is an extensive inland sea, located between the Arabian Peninsula and Iran, that connects to the Indian Ocean (Gulf of Oman) through the relatively narrow Strait of Hormuz. This is a shallow basin: mean and maximum water depths in the PG are 50 m and 90 m, respectively (Fig. 1). Unlike the neighbouring Gulf of Oman, which is at risk to hazardous cyclones (cf. Fritz et al. 2010) and damaging tsunamis (e.g. the 1945 Makran tsunami killed 300 people; Heidarzadeh et al. 2008), the Persian Gulf has always been considered as relatively safe. Major storm surges or detectable tsunamis have never been observed in this basin (El Sabh and Murty 1989; Lin and Emanuel 2016). Moreover, there are no documented substantial crustal faults in the Persian Gulf region capable of producing major tsunamigenic earthquakes ($M > 7.5$), while tsunami waves from the Indian Ocean either have not penetrated into the PG or have been strongly attenuated once inside the basin (Rabinovich and Thomson 2007; Ambraseys 2008; Heidarzadeh et al. 2008). As a consequence, the highly destructive tsunami-like waves that impacted the northern shore of the PG on 19 March 2017 were totally unexpected.

Waves associated with the March 2017 event reached a maximum runup height of approximately 3 m at the area of Dayyer in southern Iran (Fig. 1)

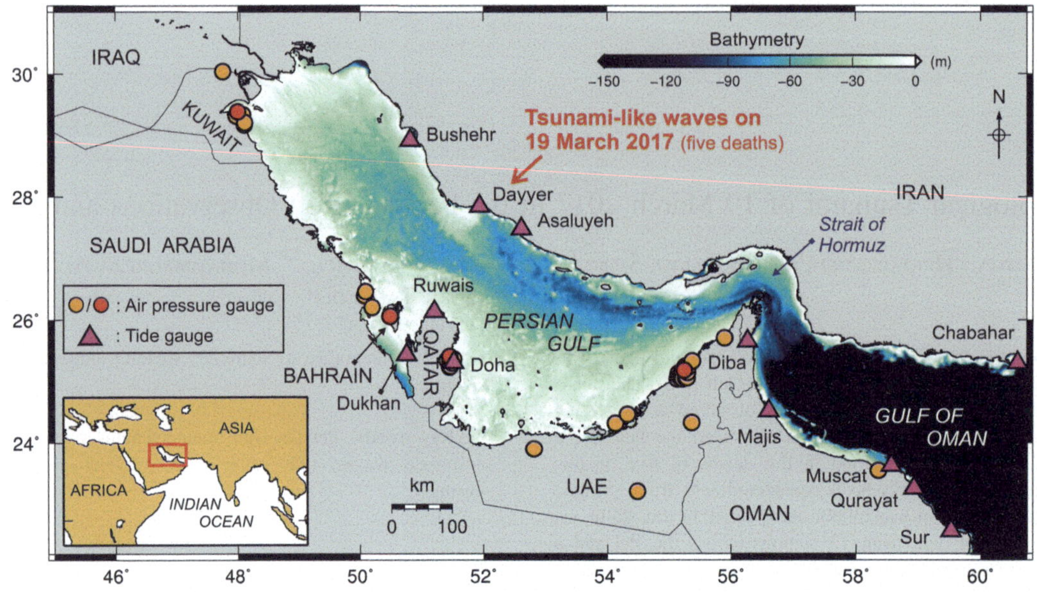

Figure 1
Map showing Dayyer and other regional tide gauges and air pressure stations located within and outside the Persian Gulf. Dayyer was the site of deadly tsunami-like waves on 19 March 2017. Air pressure gauges shown in red are those used for wavelet analysis

and flooded inland for ~ 1 km, resulting in severe damage and five deaths. Salaree et al. (2018) provided a field survey of the affected coastline (~ 40 km) and collected witness reports that enabled them to reconstruct a preliminary picture of the entire event, including estimates of runup and inundation along the coast. The main focus of their study was to identify the physical mechanism responsible for these extreme waves. First, they checked the seismic activity at the time of the event and, because there were no noticeable earthquakes in the area at the time, ruled out a seismic origin for the observed tsunami-like waves. Salaree et al. (2018) then considered the possibility of a submarine landslide in the vicinity of Dayyer. A numerical simulation of possible landslide-generated waves demonstrated that a submarine slide was a highly unlikely cause of the event. Having eliminated all other generation mechanisms, Salaree et al. (2018) concluded that the 2017 wave event at Dayyer had an atmospheric origin, i.e. that the event was a *meteorological tsunami.*

Meteorological tsunamis ("meteotsunamis") are tsunami-like waves that have the same temporal and spatial scales as ordinary tsunami waves but are generated by atmospheric processes rather than seismic sources or landslides (Rabinovich and Monserrat 1996; Monserrat et al. 2006). Meteotsunamis can affect coastal regions in a destructive way, similar to seismic tsunamis, causing loss of life and severe damage to local infrastructure. Meteotsunamis reported throughout the World Ocean, were either in the form of "*harbour meteotsunamis*" (multiple harbour seiche oscillations) in harbours/bays with high Q-factors[1] (Monserrat et al. 2006; Rabinovich et al. 2009), or in the form of "*beach meteotsunamis*" (solitary waves observed near straight beaches). Destructive harbour meteotsunamis have occurred in Ciutadella Harbour in Spain, Vela Luka and Stari Grad bays in Croatia, Nagasaki Bay in Japan, Mazara del Vallo Harbour in Sicily, Italy, and in certain bays in the Great Lakes in North America (Ewing et al. 1954; Hibiya and Kajiura 1982; Šepić et al. 2015, 2018a). A spectacular example of a beach meteotsunami is the 1992 "Daytona Beach event" in Florida, USA (Churchill et al. 1995; Sallenger et al. 1995). Other examples are the 2014 events in Praya

[1] Quality (Q) factor is a measure of energy damping in the system: the higher Q is, the slower the energy decays, and the stronger wave amplification is within the harbour (Rabinovich 2009).

de Cassino (Brasil), Panama City (Florida, USA), and Odessa (Ukraine) (Vilibić et al. 2016; Šepić et al. 2015, 2018b).

The crucial factor responsible for the generation of strong meteotsunamis is the need for the speed of an atmospheric disturbance (U) to match the phase speed (c) of the long ocean waves (the so-called "*Proudman resonance*" condition; Proudman 1929):

$$U \sim c, \tag{1}$$

where $c = \sqrt{gh}$, g is the gravity acceleration and h is the water depth. The governing parameter determining the sea level response to the atmospheric forcing is the *Froude number*, defined as:

$$Fr = \sim U/c. \tag{2}$$

Resonance occurs when $Fr \sim 1.0$ (Vilibić 2008; Šepić et al. 2015). Most often, meteotsunamis are observed on the "windward coasts" (*onshore* disturbance direction), however they can also be induced on "leeward coasts" (*offshore* disturbance direction)—the situation caused by reflections from the opposite coast (Ewing et al. 1954), or from the shelf break (Pasquet and Vilibić 2013; Šepić et al. 2018b). The *Proudman resonance* normally plays the key role in formation of prominent meteotsunamis (Vilibić 2008), but in specific situations of *alongshore* propagating disturbances, significant meteotsunamis can be generated due to the *Greenspan resonance* (Munk et al. 1956; Greenspan 1956). Salaree et al. (2018) assumed that favourable conditions for the Proudman resonance existed in the PG on 19 March 2018, and that was the main reason for the extreme tsunami-like waves affecting Dayyer. Based on the results of numerical experiments, Salaree et al. (2018) found that the atmospheric source model provides much better agreement with the results of the field survey than any landslide model. However, their model could not adequately explain the distribution of wave amplitudes along the coast. The lack of atmospheric and sea level data in the area of the event did not allow Salaree et al. (2018) to investigate this problem in detail and to relate the observed destructive surge waves to the specific atmospheric disturbance with their particular propagation parameters.

Our study of the Dayyer event is based on sea level records from 12 stations, including the Dayyer tide gauge (Fig. 1) and comprises statistical, wavelet, and spectral analyses of the corresponding data. In addition, we also examined 47 high-resolution air pressure records from several countries located in this region (Kuwait, Saudi Arabia, Bahrain, Qatar, UAE and Oman) (Fig. 1), and various satellite data. These numerous data enabled us to understand the general atmospheric situation in the PG region and to estimate the propagation characteristics of atmospheric waves at the time of the Dayyer event.

2. *Description of the Event*

On 19 March 2017 at around 8:00 AM local time (4:30 UTC), large waves impacted the northern shore of the PG. According to media reports and field surveys by Salaree et al. (2018), the maximum runup heights near Dayyer (Figs. 1, 2) reached 3 m; the inundated distance was up to 1 km. The waves affected about 40 km of the coastline, causing five deaths and injuring 20 people. Local authorities reported damage to $\sim$ 300 fishing and recreational boats and 85 coastal properties. The event was largely referred to as a *tsunami* in the Persian media, since its wave heights, periods, inundation patterns and the destructive power resembled typical tsunami waves. Numerous videos recorded bore-like waves attacking the coast; a few images from two of these videos are shown in Fig. 2.

The Dayyer event of 19 March 2017 did not have the character of a "harbour meteotsunami", and affected a long segment ($\sim$ 40 km) of a fairly straight beach. In fact, this was a classical "beach meteotsunami", similar to the 1992 Daytona Beach and 2014 Odessa events. As discussed by Salaree et al. (2018), the 2017 Dayyer event took place during a calm and cloudy day, preceded by a significant thunderstorm occurring a few hours before the tsunami. This is typical for meteotsunamis in that they commonly are not associated with extreme weather conditions, but rather with relatively calm weather. This was also the situation for the 1992 Daytona Beach, 2014 Odessa and 2014 Praya do Cassino meteotsunamis, as well as for a number of events on the coast of British Columbia (Thomson et al. 2009) and in the Mediterranean Sea (Monserrat et al. 2006).

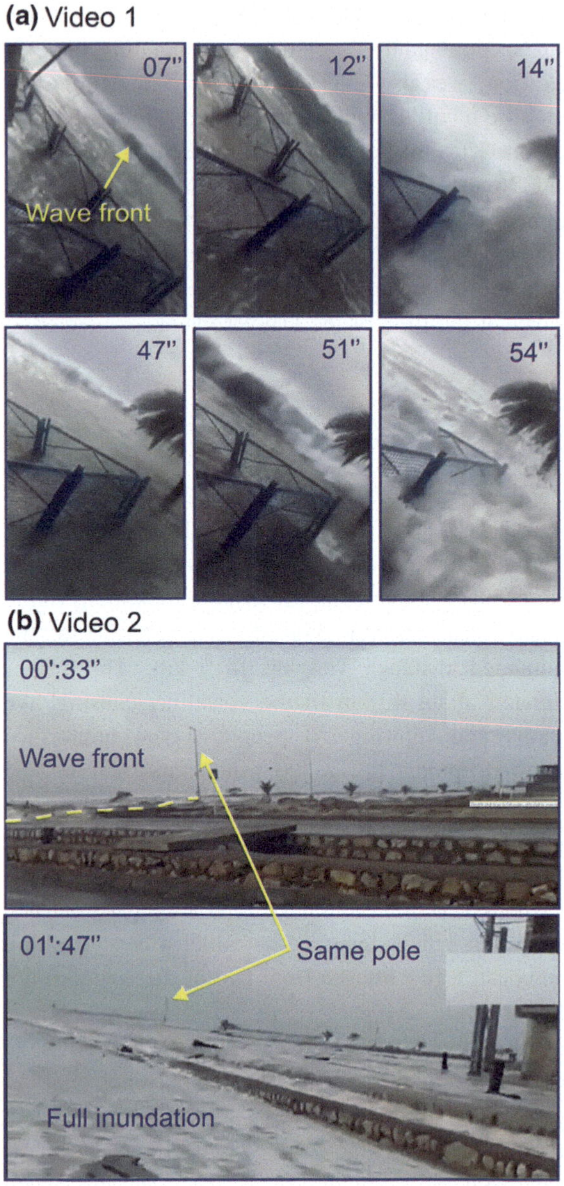

Figure 2
Snapshots of videos showing the inundation of the 19 March 2017 Dayyer tsunami-like waves. The times shown on the corners of the panels are relative to the start of each video. The sources for videos **a**, **b** are: http://jamejamonline.ir/nama and http://www.eghtesadonline.com/, respectively

The 2017 Dayyer event is the first known example of recorded destructive waves in the Persian Gulf, which caused severe damage and killed people. The PG is the main hub for worldwide oil supply as giant oil producers (Saudi Arabia, Iran, UAE, Iraq, Qatar, Kuwait and Bahrain, Fig. 1) export most of their oil through this gulf and the Strait of Hormuz. In addition, the PG coasts host some large cities, such as Dubai, Abu Dhabi and Doha (Fig. 1). Therefore, thorough insight into this event is vitally important.

The satellite data analysis revealed a sand storm that was formed in North Africa (in the area of Libya) and propagated over the PG (Fig. 3) around the time of the Dayyer event (18–19 March 2017). It is likely that this sand storm and the air pressure disturbances, that appear to have been responsible for the Dayyer extreme waves, are associated with the same atmospheric system. The entire situation looks comparable to the one observed over the Mediterranean-Black Sea region on 23–27 June 2014, when an atmospheric synoptic pattern ("tumultuous atmosphere"), propagating eastward, supported the generation of numerous intense small-scale atmospheric disturbances, which subsequently resonantly triggered meteotsunamis in coastal areas with $Fr \sim 1.0$ (Šepić et al. 2015). Similar atmospheric disturbances are seen in most of the air pressure records in the PG region (Sect. 5), while local surges and strong seiche oscillations are detected in individual tide gauge records across the region (Qatar, Oman, UAE and Iran), with the extreme waves observed specifically at the area of Dayyer (Sect. 4).

3. Data and Methodology

3.1. Sea Level Data and Analysis

Sea level data from 12 tide gauges were used in this study (Fig. 1; Table 1). The Dayyer station is located at the centre of the region affected by the devastating waves. The gauge, which is operated by the National Cartographic Center (NCC) of Iran, is based on ultrasonic transducers; sea levels are recorded at 5-min intervals. The instrument was originally deployed for tidal measurements; thus the 5-min sampling interval was quite adequate. However, such a sampling rate is not optimum to measure short-period events, like the destructive waves that occurred on 19 March 2017. Two more Iranian tide gauge records, Asaluyeh and Bushehr, that also had a sampling interval of 5 min, were provided by the Port and Maritime Organization (PSO) of Iran (https://

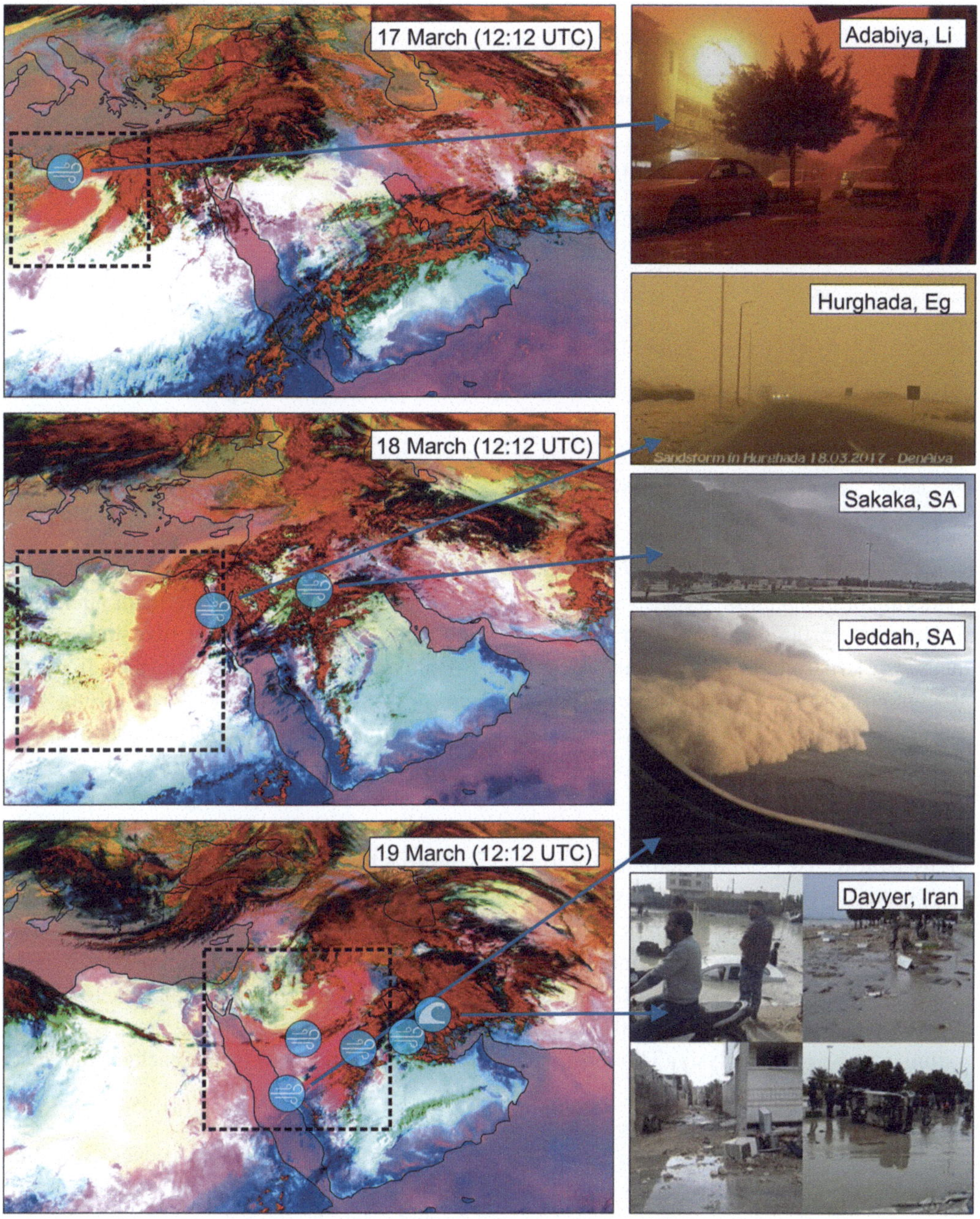

Figure 3
Sequence of satellite images taken over the Middle East during 18–19 March 2017, revealing the development and propagation of a sand storm from Libya, over Egypt and Saudi Arabia, towards the Persian Gulf. The sand storm is represented with purplish colours (within the dashed boxes). Locations of reported sand storm and tsunami events are marked with appropriate symbols, at times closest to event times. Photos taken during these events are shown in the right column

www.pmo.ir/en/home). The three Qatar records at Ruwais, Doha and Dukhan, all with sampling intervals of 10 min, were supplied by the Hydrographic Unit of the Qatar Ministry of Municipality and Environment. The data from six tide gauges (from Oman and Chabahr, Iran) with sampling intervals of 1 min were provided by the Intergovernmental

Table 1

Information of the tide gauge and wave gauge records analysed in this study

Station name (country)	Longitude (°E)	Latitude (°N)	Type[b]	Sampling interval (min)	Max wave height[a] (cm)	Visible period (min)
Dayyer (Iran)	51.94	27.84	WG	5	197	20
Asaluyeh (Iran)	52.62	27.47	TG	5	234	15
Bushehr (Iran)	50.82	28.92	TG	5	26	30–40
Ruwais (Qatar)	51.20	26.13	TG	10	32	40
Doha (Qatar)	51.52	25.30	TG	10	17	20–30
Dukhan (Qatar)	50.75	25.43	TG	10	30	20–30
Diba (Oman)	56.27	25.65	TG	1	33	N/A
Majis (Oman)	56.61	24.52	TG	1	26	10–20
Muscat (Oman)	58.57	23.63	TG	1	25	10–15
Quaryat (Oman)	58.93	23.26	TG	1	24	15
Sur (Oman)	59.52	22.57	TG	1	5	15–20
Chabahr (Iran)	60.60	25.30	TG	1	15	N/A

[a]Trough-to-crest height; [b]WG and TG indicate the wave gauge and tide gauge, respectively, "N/A" means "Not Applicable"

Oceanographic Commission (http://www.ioc-sealevelmonitoring.org/map.php).

Sea level data were de-tided based on the least squares method of harmonic tidal analysis and consequent removing of tides from the original records. Our spectral analysis procedure was similar to that described by Thomson and Emery (2014). To improve the spectral estimates, we used a Kaiser–Bessel (KB) window with half-window overlaps prior the Fourier transform. To examine temporal variations of the observed event in the frequency domain, we used a multiple-filter method, which is based on narrow-band filters with a Gaussian window that isolates a specific centre frequency (cf. Rabinovich and Thomson 2007; Rabinovich et al. 2013). The method is similar to wavelet analysis (Thomson and Emery 2014).

To examine the spectral properties of the tsunami-like oscillations of 19 March 2017, and to compare these properties with those of the background oscillations at the same sites, we separated the records into two parts. The time period preceding the event (approximately 3.5 days) was identified as "normal" and selected for analysis of background signals. For the "tsunami" (event) period following the extreme wave arrivals, we took the time period of 1280 min ($\sim$ 21.3 h). The length of the KB window was chosen to be 640 min, yielding $v = 30$ degrees of freedom for background spectra and $v = 6$ for the tsunami spectra. The spectral resolution for all spectra was $\Delta f \approx 0.094$ cph and the Nyquist frequency $f_N \approx 0.17$ cph. The main attention was paid to analysing the records at two stations, Dayyer and Asaluyeh, that are located in the affected area and, according to Fig. 4, measured strong tsunami-like oscillations. The results of spectral analysis of these two records are shown in Fig. 6a, b, respectively.

Rabinovich (1997) suggested a method to separate the influences of the topography and the source on the observed tsunami spectrum. The method is based on the assumption that background long waves and tsunami waves approaching the coast, are transformed by the topography in a similar way, i.e. that the topographic transfer function for both types of waves is the same. In that case, if we estimate the ratio of the tsunami to the background spectrum

$$R(\omega) = S_{tsu}(\omega)/S_{bg}(\omega), \quad (3)$$

where ω is the angular frequency, we eliminate the influence of topography and obtain a function determined solely by the external forcing. The ratios show not only the spectral structure of the source, but also the frequency diapason of the forcing and the relative amplification of the tsunami signal at various frequencies relative to the background oscillations. The similarity of the computed ratios for different stations normally gives the evidence of the identity of the source (that is why these ratios are also known as the "*source functions*"). This method was successfully used for a great number of tsunami events (cf.

Rabinovich 1997; Rabinovich et al. 2013; Zaytsev et al. 2017). Monserrat et al. (1998) and Marcos et al. (2009) demonstrated that this approach is also effective for meteorological tsunamis.

3.2. *Atmospheric Data and Analysis*

To analyse the general synoptic situation over the Middle East during 17-22 March 2017, we used the ERA-Interim reanalysis data with 0.75° × 0.75° spatial resolution and 6-h time resolution (00, 06, 12 and 18 h UTC). To investigate atmospheric processes in the affected region in more detail, we downloaded surface atmospheric records from the amateur meteorological network *Wunderground* (https://www.wunderground.com/wundermap). Although these records did not have the quality control provided by state meteorological organizations, they can be effectively used for scientific research and studies of atmospheric disturbances responsible for meteorological tsunamis (cf. Šepić et al. 2015). Altogether, we found data of satisfactory quality from 47 meteorological stations (Fig. 1). Various variables are measured at these stations, including air temperature, dew point, humidity, wind velocity and gust, precipitation, atmospheric pressure, etc. Since the main drivers of meteotsunamis are air pressure waves and disturbances (cf. Monserrat and Thorpe 1992; Rabinovich and Monserrat 1996; Monserrat et al. 2006), we have focused our attention on atmospheric pressure measurements, which were recorded with samplings of 5–25 min, depending on the station. Prior to the analysis, all pressure series were interpolated to a 5-min time step. Unfortunately, the available *Wunderground* stations were located on the western and southern coasts of the PG—in Kuwait (6 stations), Saudi Arabia (3), Bahrain (1), Qatar (8), United Arab Emirates (26) and Oman (2)—and none on the eastern or northern PG coasts (e.g. in Iran) (Fig. 1). This means that, sadly, we did not have surface meteorological observations located close to the area affected by the tsunami-like waves of 19 March 2017 and had to use the data from relatively remote stations.

Assuming that the atmospheric pressure disturbance propagates as a plain wave with a constant speed and direction, we can estimate these parameters from atmospheric pressure records by the least squares method (LSM) (cf. Orlić 1980; Thomson et al. 2009). The velocity parameters of the propagating air pressure disturbances were estimated in two variants: (1) based on all stations which measured the corresponding disturbance; and (2) based only on stations located more than 5 km apart from each other. Further on, it was reliably shown that tsunamigenic air pressure disturbances propagate with the speed and direction of a higher altitude/level guiding wind (cf. Monserrat and Thorpe 1992; Šepić et al. 2009; Thomson et al. 2009). In fact, these disturbances are often a surface manifestation of high-altitude processes spanning up to the level of ~ 500-hPa (approximately 5500 m above the ground) (Monserrat and Thorpe 1996). Since measurements at these levels are sparse, we extracted 500-hPa wind speed and direction from the ERA-Interim reanalysis data set, available at 6-h intervals with a spatial resolution of 0.75° × 0.75° and used these data to approximate the velocity parameters of propagating air pressure disturbances.

3.3. *Satellite Data and Analysis*

We have also analysed satellite images of the Persian Gulf and adjacent regions taken with the MVIRI instrument of the EUMETSAT Meteosat7 satellite (https://www.eumetsat.int/website/home/Satellites/CurrentSatellites/Meteosat/index.html). Satellite images were processed to show a composite of the following RGB channels: 05-06; 04-09; 03-01. This composite is regularly used to show convective activity during daylight hours, with convective clouds depicted with yellowish colours (Kerkmann 2005).

4. *Tsunami Waveforms Analysis and Physical Properties of the Observed Sea Waves*

4.1. *Parameters of the Observed Sea Level Records*

The Dayyer wave gauge, which is the closest station to the area impacted by the tsunami, recorded a maximum trough-to-crest wave amplitude of 197 cm at 08:10 AM local time (04:40 UTC) on 19 March 2017 (Fig. 4), consistent with the timing of the event according to the media reports. The actual wave height at Dayyer appears to be higher than 197 cm, but was attenuated due to the 5-min sampling interval of the recorded data, which is too long for short-period processes. At Asaluyeh, located approximately 80 km to the SE from Dayyer, the maximum recorded tide gauge wave height was 234 cm, which also was probably attenuated by the 5-min sampling. The ratio between tide gauge recorded wave heights (H_{tg}) and field survey estimated tsunami runups (H_{ru}) is one of the most important and difficult questions in the tsunami observations. It appears that this ratio depends on a number of factors and, first of all, on particular topographic properties of the region. Nevertheless, if we use an empirical coefficient $H_{ru}/H_{tg} \approx 1.4$, proposed by Soloviev (1978), we find very good agreement between runup estimates from the field survey of Salaree et al. (2018), $\sim$ 3.0 m, and tide gauge measurements at Dayyer and Asaluyeh.

Other regional tide gauges, both inside and outside the PG, at the time of the Dayyer event registered wave heights in the range of 5–32 cm (Figs. 4, 5). Three Qatar tide gauges, located on the southern coast of the PG opposite Dayyer, recorded maximum wave heights of 17–32 cm. Based on the data from these 12 tide gauges (Figs. 1, 4, 5), distributed over a wide area with dimensions of 1000 km (length) × 700 km (width), and also on field survey results by Salaree et al. (2018), we may conclude that the Dayyer tsunami was mainly a very local event. The timing of the surges/waves at different locations (Fig. 4, 5) reveals an approximately west–east propagation track of the storm consistent with the direction of the sand storm propagation (Fig. 3). Based on our visual inspection of these 12 sea level records, the visible period of the observed extreme waves is in the range of 10–40 min, with a dominant wave period of 15–20 min. The latter period appears at most stations.

4.2. *Spectral and Wavelet Analyses*

The considerable difference in energy between the background and event spectra is apparent. The "tsunami" spectrum at both stations (Dayyer and Asaluyeh) occupies a relatively broad frequency band between approximately 0.25 and 5 cph (periods from 4 h to 12 min). A very important feature became evident from the spectral analysis of the Dayyer record. It is well known that the frequencies of tsunami and background spectral peaks at each station are almost the same (cf. Honda et al. 1908). The matching of the spectral frequencies is related to the dominant role of the local/regional topography over the individual characteristics of the tsunami source (Rabinovich 1997). That is why, tsunami spectra from different events at each site are very similar. Recently, this was (again) clearly demonstrated for a great number of Mexican stations that recorded several 2010–2017 tsunamis (Zaytsev et al. 2017). The same is true for meteorological tsunamis (Monserrat et al. 1998; Marcos et al. 2009). The very steady periods of meteotsunamis in various bays and harbours of the world oceans, where such events occur regularly, are well known. These periods are normally associated with the fundamental (Helmholtz) mode of the corresponding basin (Rabinovich 2009): for example, $\sim$ 10.5 min in Ciutadella Harbour, Menorca Island, Spain (Rabinovich and Monserrat 1996), $\sim$ 20 min in Vela Luka Bay, Korčula Island, Croatia (Šepić et al. 2016) and $\sim$ 35 min in Nagasaki Bay, Japan (Hibiya and Kajiura 1982). However, the background spectrum at Dayyer (Fig. 6a) does not have any evident spectral peaks, except a broad peak with a central period of $\sim$ 13.5 min. In contrast, the tsunami (event) spectrum has a number of peaks, in particular at 1.8 h, 55 min and 18 min, while there is no peak at 13.5 min (Fig. 6a). This difference between tsunami and background spectra is important evidence that the observed properties of the tsunami spectrum, including the spectral peaks, are related to the external forcing (i.e. to atmospheric processes) rather than to the resonant features of the local topography, and that the entire event is not associated with "harbour resonance", as at the sites listed above.

The spectra at Asaluyeh (Fig. 6b) have a quite different character than the spectra at Dayyer; the tsunami spectrum has spectral peaks at periods 1.4 h and 49, 22 and 14.5 min, and the same peaks, but much weaker, are also seen in the background spectrum. Thus, significant atmospherically induced ocean waves, incoming from the open part of the PG, generate "harbour oscillations" (cf. Rabinovich 2009) at this station or, more precisely, amplify

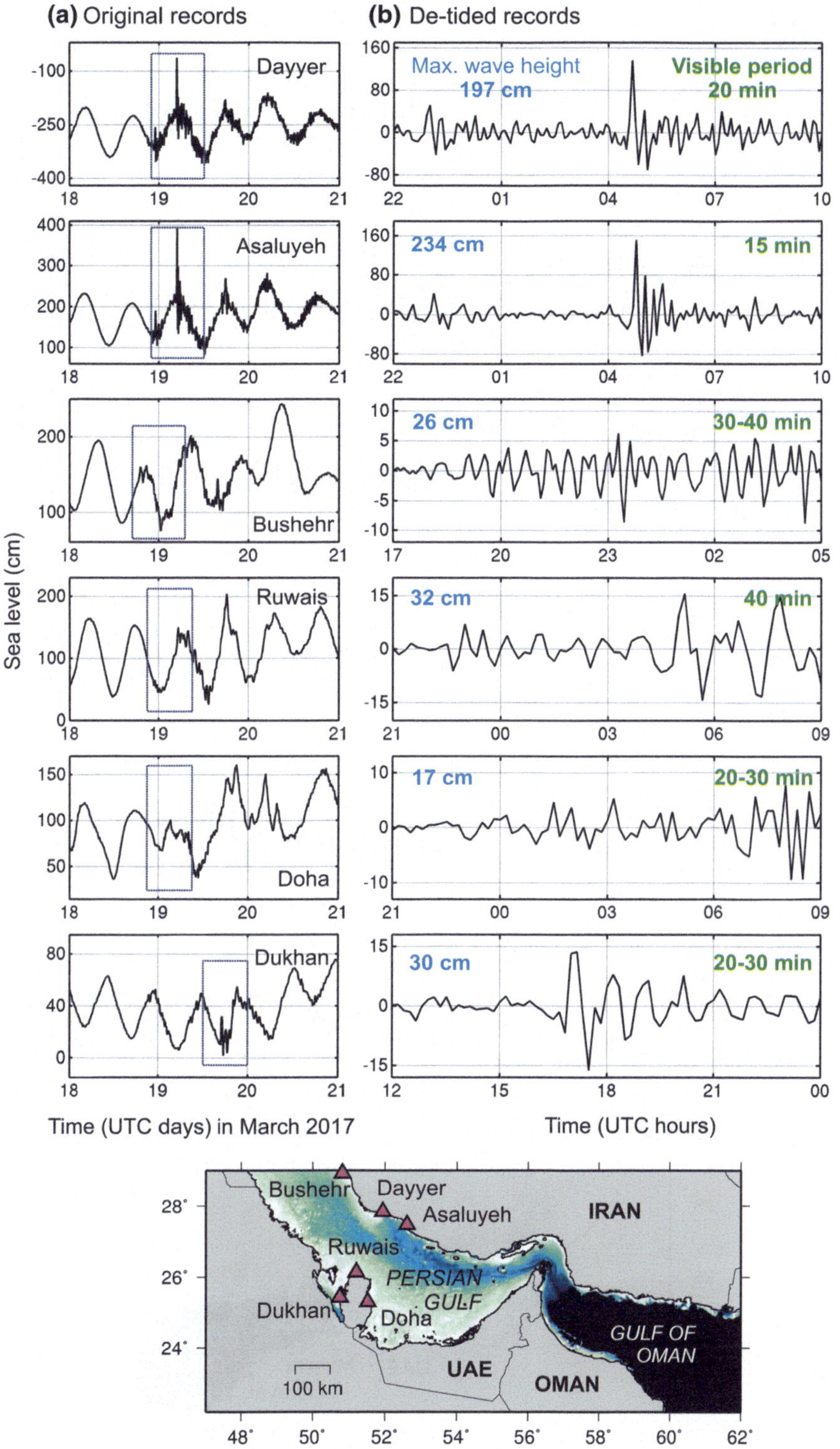

Figure 4

Recorded sea level waveforms for the period 18–21 March 2017 for the tide gauges located inside the Persian Gulf. **a** Original tide gauge records. **b** De-tided records. The blue numbers are the maximum trough-to-crest wave heights. Visible wave periods are denoted by green numbers

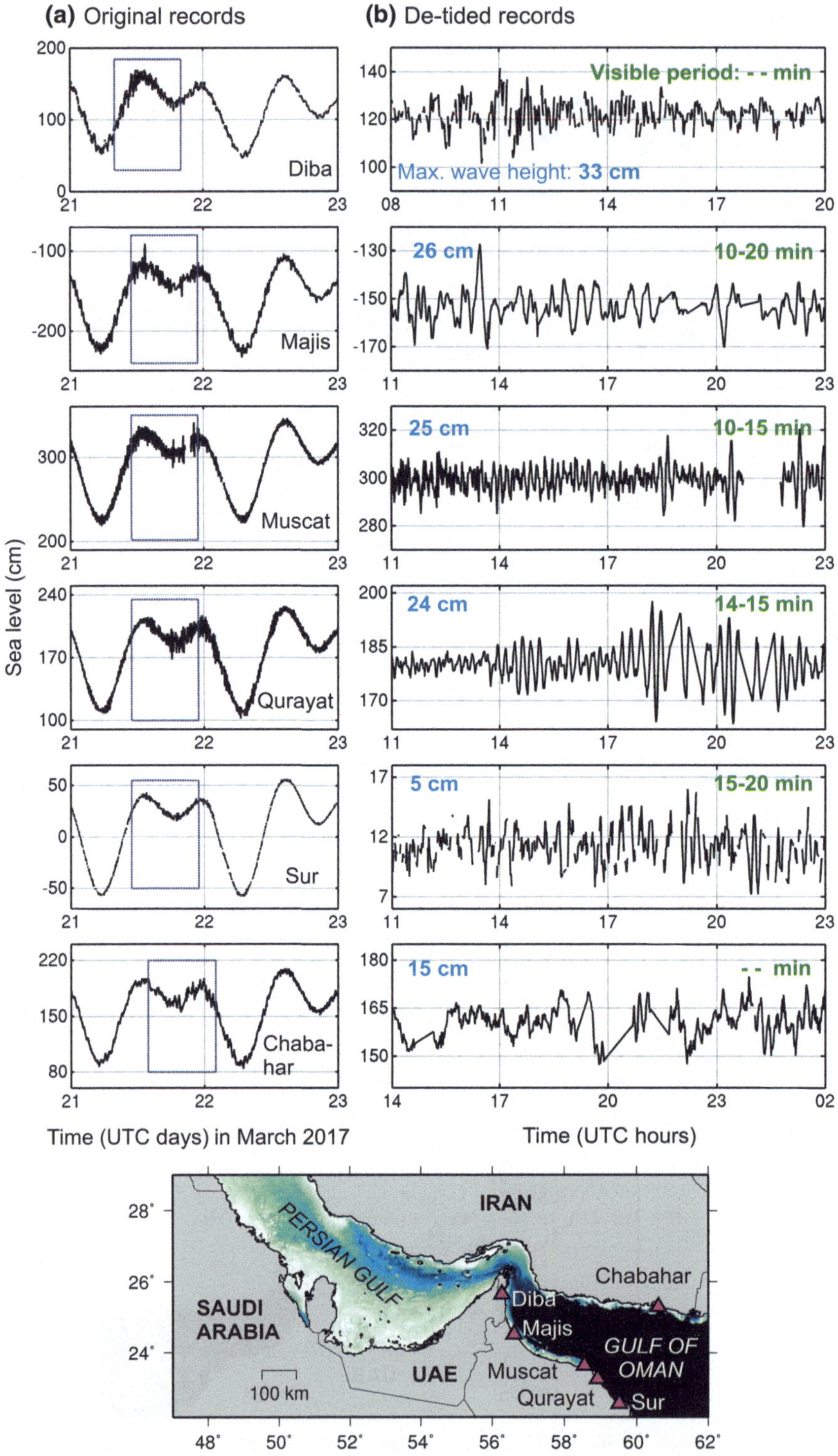

Figure 5
Recorded sea level waveforms for the period 21–23 March 2017 for the tide gauges located outside the Persian Gulf. **a** Original tide gauge records. **b** De-tided records. The blue numbers are the maximum trough-to-crest wave heights

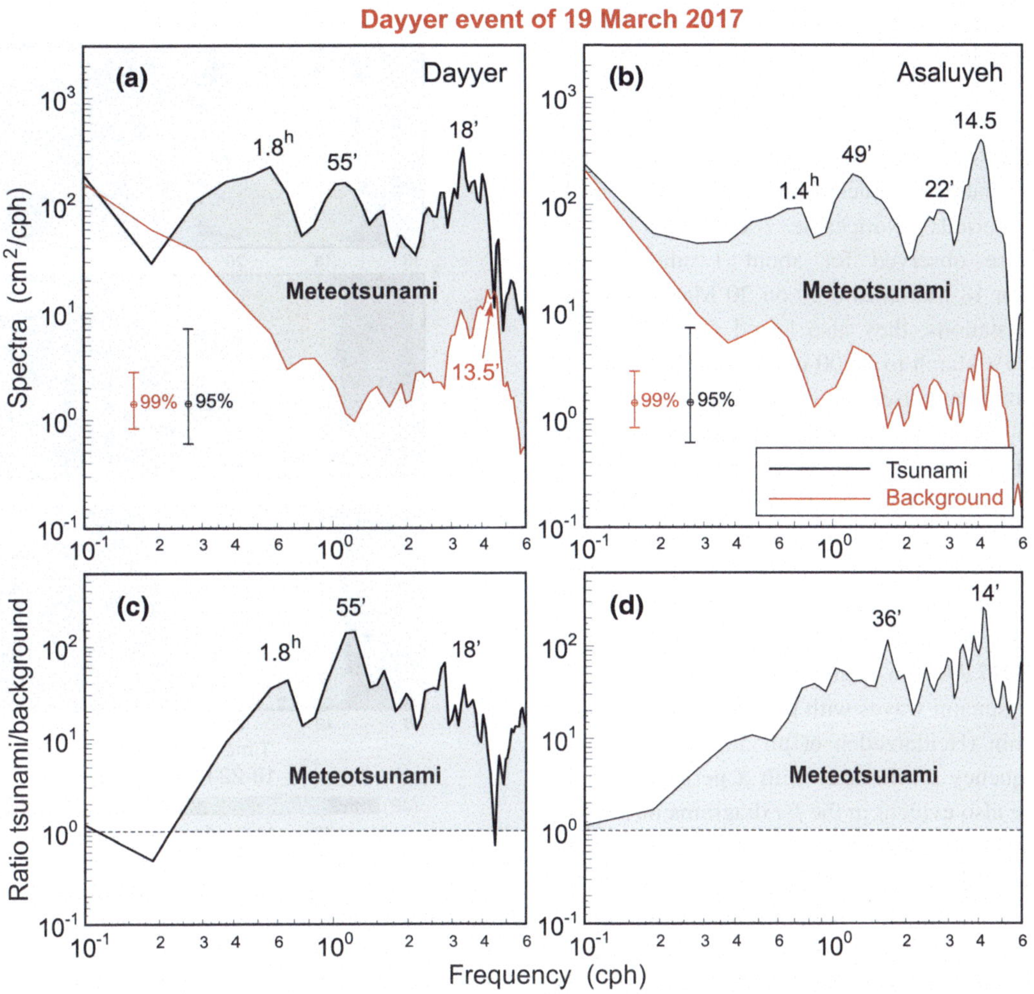

Figure 6
Spectra of the meteotsunami event of 19 March 2017 and the background sea level oscillations at **a** Dayyer and **b** Asaluyeh and spectral ratios (tsunami/background) at **c** Dayyer and **d** Asaluyeh

harbour (seiche) oscillations which already existed at this site. The strongest tsunami oscillations were at periods of 49 and 14.5 min.

Figure 6c, d show the computed tsunami/background ratios, $R(\omega)$, for Dayyer and Asaluyeh, respectively. The general shape and absolute values of these two functions look alike but specific features are different. At Dayyer two ratio peaks, 1.8 h and 55 min, are the same as the peaks of the corresponding tsunami spectrum, but the peak with a period of 18 min is relatively much weaker (Fig. 6a, c). These results support the assumption that the observed spectral peaks at this station are caused by the particular properties of the external (atmospheric) forcing. The situation at Asaluyeh is dissimilar: three marked peaks in the tsunami spectrum at this station, 1.4 h, 49, and 22 min (Fig. 6b) are absent in the spectral ratio, while there is a prominent peak with a period of 36 min. It is evident that the latter peak is associated with the spectral properties of the forcing. There is only one peak that has a similar period in both the tsunami spectrum and in the spectral ratio: 14.5 and 14 min, respectively (Fig. 6b, d). The closeness of these two peaks indicates the resonant interaction between the forcing oscillations with this period (14 min) and the natural (eigen) period (14.5 min) of the Asaluyeh harbour.

The results of the wavelet (f–t) analysis of three Iranian sea level records, Bushehr, Dayyer and Asaluyeh (Fig. 7), give us some additional information about the properties of the measured oscillations. At Bushehr (the upper plot) these oscillations had relatively chaotic character and occupied a wide range of periods. Noticeable oscillations at this station were observed for about 1.5 days (from ~ 18:00 on 18 March to 6:00 on 20 March). At the two other stations, they also lasted for ~ 1.5 days (00:00 on 19 March to 12:00 on 20 March), however the extreme event itself was very short, only about 30–40 min (compare with Fig. 4b). According to the results of f–t analysis, the periods of extreme oscillations were about 14–18 min, in good agreement with the visible dominant periods at the respective records (Fig. 4b). These periods are comparable with those of moderate earthquake-generated tsunamis (e.g. Heidarzadeh et al. 2015). For instance, the July 2017 M_w 6.6 earthquake in Turkey–Greece generated tsunami waves with peak energy at periods of 7–13 min (Heidarzadeh et al. 2017). Additional lower frequency oscillations with a period of about 50 min are also evident in the f–t diagrams of Dayyer and Asaluyeh, in agreement with what was seen in the spectra at the corresponding stations (Fig. 6a, b).

In general, the f–t diagrams for Dayyer and Asaluyeh are very similar, both in time and frequency (Fig. 7), demonstrating high correlation of the extreme events at these two sites. At the same time, these diagrams, together with the records themselves (Fig. 4) and the results of their spectral analysis (Fig. 6a, b), indicate that the physical properties of generated meteotsunamis at these two sites were slightly different. The first arriving waves at Dayyer and Asaluyeh look almost identical, but then we see the difference (see two upper plots in Fig. 4b). The meteotsunami at Dayyer had the character of a *solitary wave* ("solitone"), typical for "beach meteotsunamis". Only one *impulse-type* oscillation is seen at this site, similar to the wave that was recorded at Panama City, Florida on 28 March 2014 (Olabarrieta et al. 2017) and the wave which impacted Odessa beaches, Ukraine on 27 June 2014 (Šepić et al. 2018b). At Asaluyeh, the initial soliton wave came into the harbour and generated seiches within the harbour. One of the dominant periods of

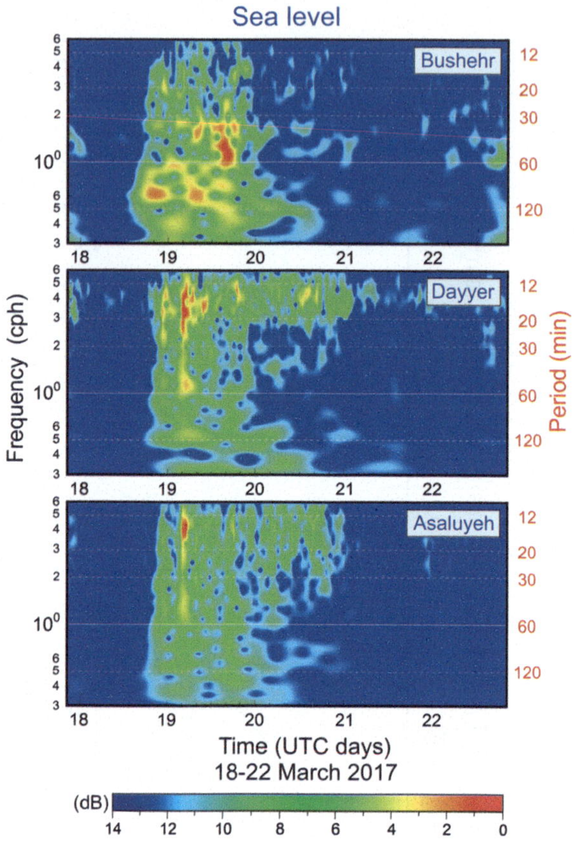

Figure 7
Frequency–time plots (f–t diagrams), which are similar to wavelets, of sea level oscillations on 18–23 March 2017 at Bushehr, Dayyer and Asaluyeh stations located along the Iranian coast of the Persian Gulf. The colormap shows high wave energy with red and low energy with dark blue colours

the external forcing (14 min) almost coincided with one of the resonant periods of this particular site (14.5 min), i.e. with one of the harbour modes. Thus, at this site the event had some features of a "harbour meteotsunami", but without pumping additional energy from the gulf, the generated harbour oscillations decayed very quickly (the second plot in Fig. 4b and the respective f–t diagram). The situation where at one site the event had the character of a "beach meteotsunami" and at another site the character of a "harbour meteotsunami" is not unique: during the chain of meteotsunami events of 23–27 June 2014 in the Mediterranean and Black seas, at Ciutadella Harbour (Šepić et al. 2015) and in bays and harbours of the Croatian coast (Šepić et al. 2016) "harbour meteotsunamis" occurred, while the wave at Odessa (Šepić et al. 2018b) was a typical "beach meteotsunami".

As the sampling interval for all three sea level records was 5 min, the spectral and wavelet analyses only include periods longer than 10 min. Relatively high energy at a period of about 10 min (Fig. 6a, b), corresponding to the Nyquist frequency, indicates the presence of shorter-period oscillations that could probably substantially amplify the harmfulness of the event. Unfortunately, we do not have the data that could be used to examine this effect.

5. *Results of Atmospheric Analysis*

5.1. *Synoptic Setting*

The results presented in Sects. 2–4, as well as findings of Salaree et al. (2018), indicate that the catastrophic event of 19 March 2017 in Dayyer (Iran) was caused by atmospheric processes. However, it remained unclear what was the exact atmospheric phenomenon or pressure disturbance that produced so strong an effect and the physical mechanism responsible for generation of the meteotsunami at this particular site. That is why, we paid specific attention to the analysis of the general synoptic situation over the Persian Gulf and to the examination of all available atmospheric pressure records.

A well-developed cyclone, characterized by a minimum air pressure of ~ 1002 hPa, was centred over eastern Libya during the midday hours of 17 March 2017 (Fig. 8, upper left). Strong surface winds, associated with the cyclone, generated dust storms in Libya (Fig. 3). Estimated surface wind speeds on the back side of the cyclone were up to 16 m/s (not shown). Within next two days, the cyclone advanced eastward, centering over Saudi Arabia on 18 March, and over Iraq one day later. The cyclone's propagation to Egypt and the Arabian Peninsula initiated strong surface winds and disruptive sand storms (Fig. 3). Following the midday hours of 19 March, the propagation direction of the cyclone changed to southeastward, and at 12:00 UTC on 20 March it reached the southeastern part of the Arabian Peninsula (Fig. 8). The propagation of the cyclone was accompanied by a low-level inflow of warm and dry air on its frontal side and cold and humid air on the rear side, resulting in strong temperature and humidity gradients across the area and in convective storms. During the initial stage, the cyclone spread vertically up to the level of 500-hPa (seen as cut-off low on 17 and 18 March) and then its mid-troposphere component deteriorated to an upper-level trough (Fig. 8). A strong jet stream with westerly to southwesterly winds was formed at the southern end of the 500-hPa trough, roughly following the positioning of the low-level cyclone. Wind speeds over the PG reached more than 40 m/s within the jet core, and this lasted throughout the entire period of 19–21 March 2017, i.e. precisely when the tsunami event occurred. A very similar situation was observed over the Mediterranean region on 23–27 June 2014 when the location and timing of individual meteotsunami events were closely correlated with the position and evolution of the jet stream (Šepić et al. 2015).

5.2. *Air Pressure Disturbances Causing Sea Level Oscillations*

Although we did not have high-resolution surface pressure data from the Iranian coast, the air pressure records from other countries located along the PG were found to be very useful. They show that during the period of 17–21 March 2017, which includes the tsunami event, a number of substantial, potentially tsunamigenic, pressure disturbances propagated over the PG. Concurrent with the cyclone's arrival around noon on 18 March 2017 (Fig. 8), the air pressure started to drop rapidly at most stations in the region (Fig. 9); the minimum low pressure occurred approximately 24–48 h later (on 19–20 March), first in Kuwait, Saudi Arabia, Bahrain and Qatar, and later in the United Arab Emirates and Oman (Fig. 9). According to Fig. 9, prominent pressure fluctuations were observed over the entire region. The atmospheric processes during the period 17–21 March 2017 were very active in the PG region and favourable for thunderstorms and strong wind waves. Some destructive effects of such storms across the PG region are shown in Fig. 3.

We used high-pass filtering of the air pressure time series with a 3-h Kaiser–Bessel window (cf. Thomson and Emery 2014) to isolate particular pressure disturbances (Fig. 10). It appears that the cyclone's propagation over the PG was accompanied

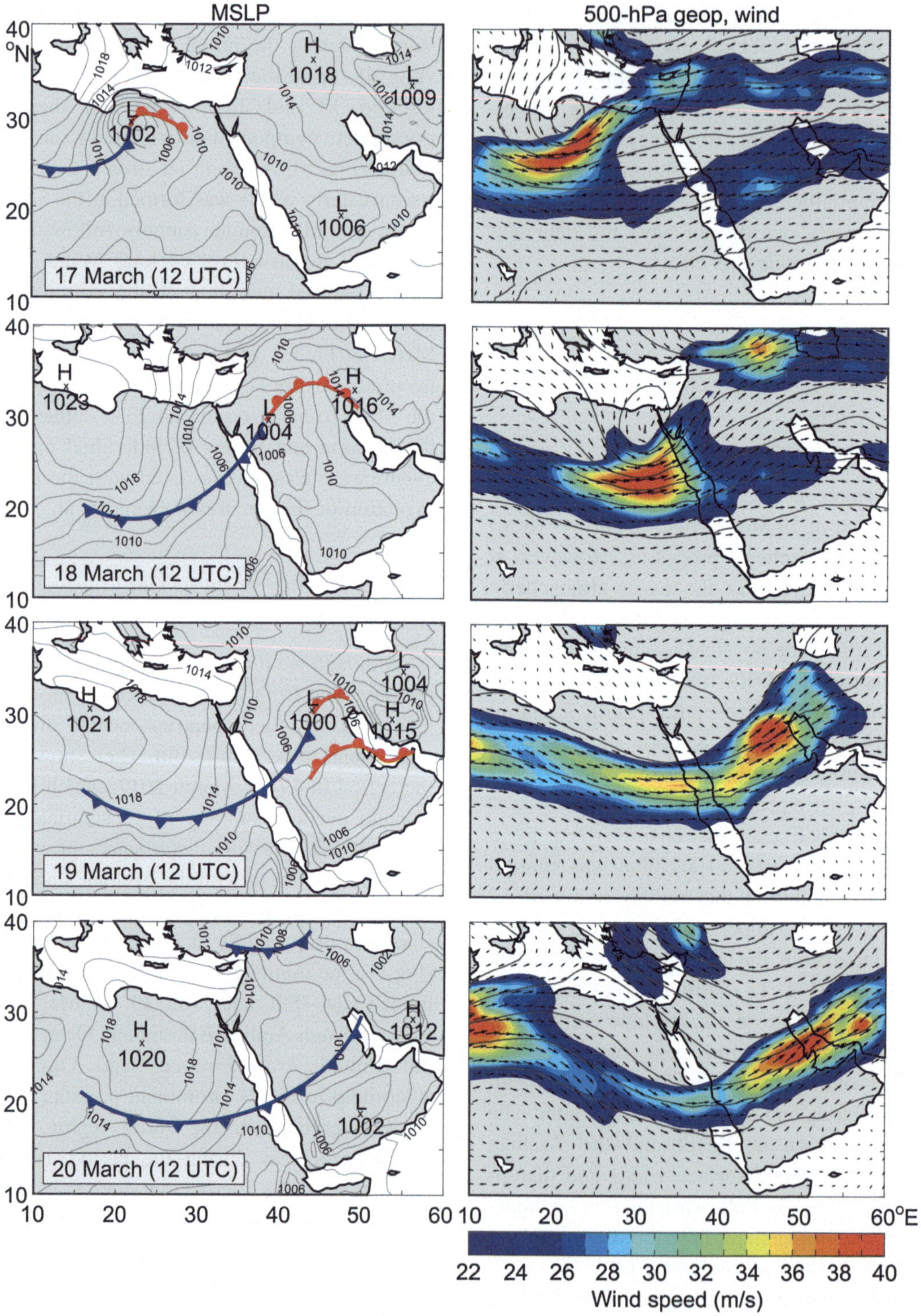

Figure 8
Temporal and spatial evolution of the synoptic situation across the northeastern Africa and Middle East regions for the period of 17–20 March 2017 for: (left column) sea surface pressure and temperature fronts; and (right column) 500-hPa geopotential height and wind fields

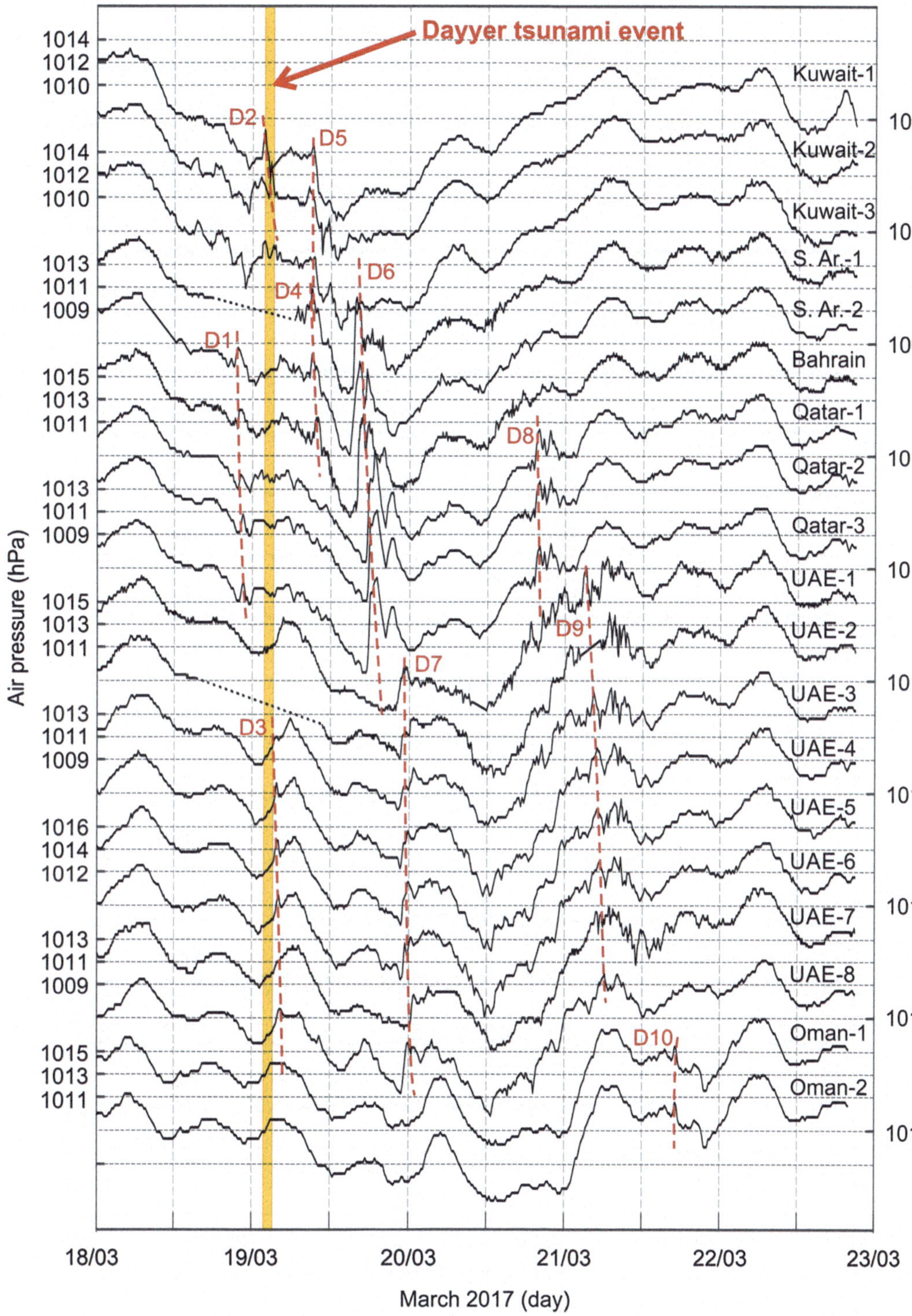

Figure 9

Air pressure time series measured at *Wunderground* meteorological stations. Shaded yellow band indicates the time of the Dayyer tsunami; the red dashed lines denote distinct air pressure disturbances (D1–D10)

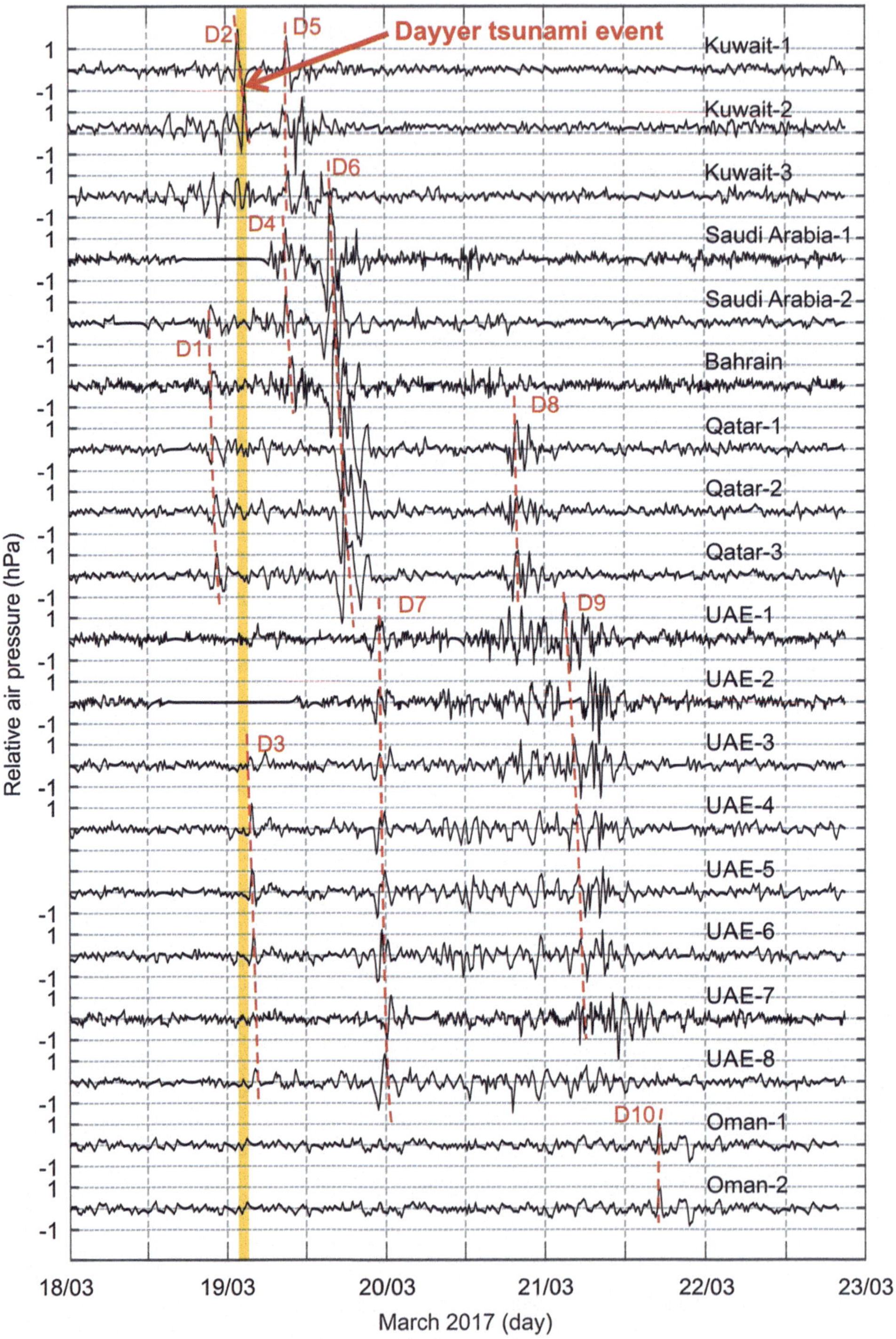

Figure 10
Relative air pressure time series measured at *Wunderground* meteorological stations. The records were high-pass filtered with a 3-h Kaiser–Bessel window. Shaded yellow band indicates the time of the Dayyer tsunami; the red dashed lines denote distinct air pressure disturbances (D1–D10)

Table 2

Main characteristics of atmospheric disturbances recorded during 18–21 March 2017 over the PG

Disturbance	Date (March 2017)	Time (UTC)	Countries	Period (min)	Height (hPa)	RC^{c}_{max} (hPa/5 min)	Type
D1	18	21:00–24:00	Bahrain, Qatar	110	1.6	0.4	Single
D2	19	01:00–04:00	Kuwait	68	3.1	1.3	Single
D3	19	03:00–05:00	UAE[a]	60	1.7	− 0.5	Single
D4	19	06:00–11:00	SA[b], Bahrain	60	2.6	0.5	Multiple
D5	19	08:00–11:00	Kuwait	95	2.6	1.4	Single/double
D6	19	15:00–21:00	SA, Qatar, Bahrain	90	5.7	3.0	Single/double
D7	19–20	22:00–01:00	UAE	65	2.6	0.7	Single/double
D8	20	18:00–23:00	Qatar	57	2.6	1.3	Multiple
D9	20–21	17:00–10:00	UAE	66	3.2	2.7	Multiple
D10	21	16:00–18:00	Oman	77	1.4	0.4	Single

[a]The United Arab Emirates

[b]Saudi Arabia

[c]Maximum 5-min rate of air pressure change

by intense high-frequency air pressure oscillations. These disturbances had higher magnitudes and shorter train durations (one-two oscillations), specifically at the time of the cyclone's passage over the region and lower magnitudes/longer durations (multiple oscillations at each station) when the cyclone was departing from the region. We identified ten distinctive disturbances, i.e. periods with intensified high-frequency activity in air pressure records; they are marked as D1–D10 in Figs. 9 and 10, while their main characteristics are given in Table 2. For the multiple-disturbance events, we marked only the most pronounced disturbance.

The two strongest consequent disturbances were recorded in Saudi Arabia, Bahrain and Qatar in the afternoon hours of 19 March. The first had a maximum peak-to-trough height of 5.7 hPa, a very abrupt air pressure change of 3.0 hPa/5 min, and a duration of $\sim$ 90 min. The second disturbance had a similar duration, but with a lower maximum measured magnitude of 3.0 hPa and a slower rate of 1.1 hPa/5 min. The first disturbance had a substantial tsunamigenic potential, whereas the second disturbance was also potentially tsunamigenic, but slightly weaker. Other important disturbances are: (1) D2 and (2) D5 which were recorded over Kuwait during the morning hours of 19 March, with an air pressure rate of change of up to 1.4 hPa/5 min; (3) a series of disturbances indicated as D8 that were recorded over Qatar during the night time hours of 20 March, which had a rate of up to 1.3 hPa/5 min; and (4) another prolonged series of disturbances measured over the United Arab Emirates (D9) during 20–21 March, with rates as high as 2.7 hPa/5 min.

It is interesting that at the time of the Dayyer tsunami (indicated by the yellow band in Figs. 9, 10), distinct atmospheric pressure disturbances, D2 and D3, were recorded only at the Kuwait and UAE stations, respectively. However, these stations are located more than 400 km away from Dayyer, and it is highly unlikely that these particular disturbances were related to the observed tsunami at Dayyer. It is more probable that the atmospheric disturbances responsible for the Dayyer event were simply not captured by the *Wunderground* measurement network. This is not surprising, since atmospheric disturbances responsible for meteotsunamis have typically small spatial dimensions (20–60 km) and short life cycle (a few hours) (cf. Thomson et al. 2009; Ličer et al. 2017) and thus were not properly captured by available measurement networks, which had no stations in the area of Dayyer. Nevertheless, we may conclude that our analysis of the atmospheric pressure records demonstrates that atmospheric conditions over the PG were highly favourable for generation of tsunamigenic disturbances during 18–21 March 2017, with the most intense disturbances occurring over the middle part of the PG precisely on 19 March.

The indicated properties of the observed atmospheric disturbances in the area of the PG are clearly seen in the *f–t* (wavelet) plots for the atmospheric

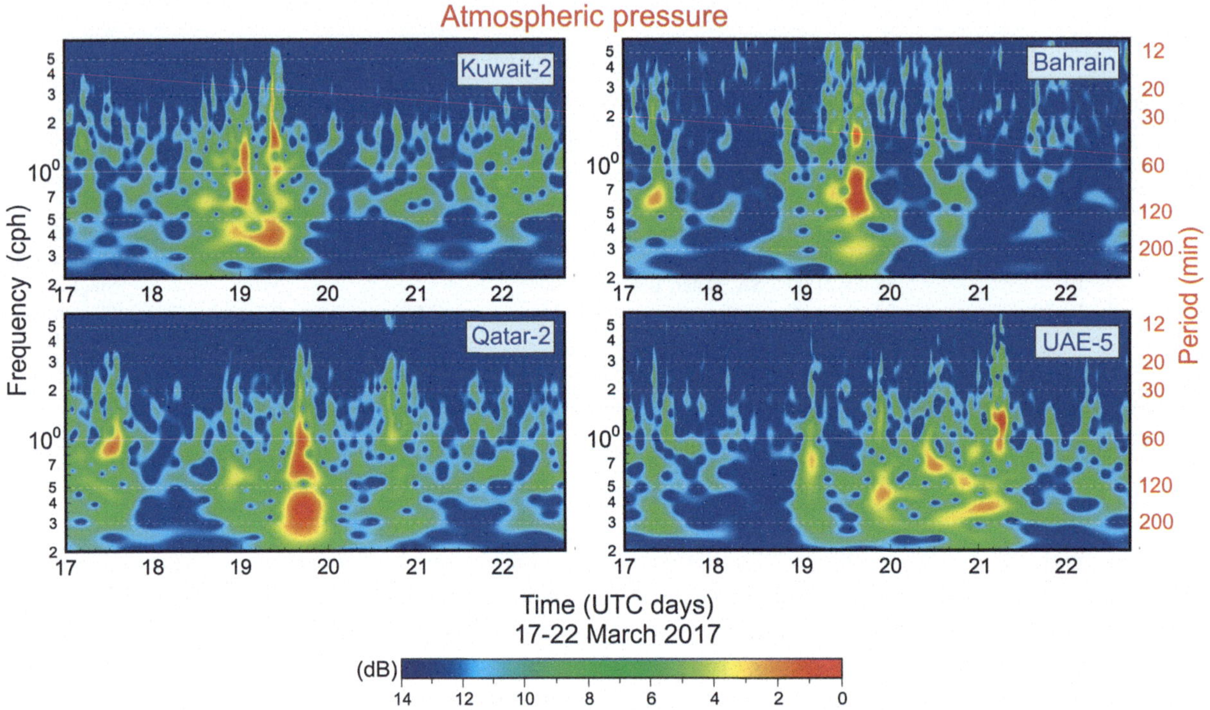

Figure 11
Frequency–time plots (*f–t* diagrams) for the atmospheric pressure records at four stations: Kuwait-2, Bahrain, Qatar-2 and UAE-5 (the corresponding records are shown in Fig. 10)

pressure fluctuations presented in Fig. 11, in particular:

- The entire period of 19–21 March 2017 is characterized by prominent activity of high-frequency atmospheric processes;
- The main frequency band of the corresponding disturbances is 0.3–3.0 cph (i.e. periods of 20–200 min);
- A specific feature of the observed disturbances is high spatial variability (for example, poor correlation between these disturbances at UAE-5 and at other stations, located nearby) and quite short life cycle (only a few hours).

In general, we may say, that although we cannot identify the exact atmospheric disturbance responsible for the Dayyer event, we see that there were several strong disturbances during the period which includes this event, and that the entire atmospheric situation over the Persian Gulf was very similar to the one observed in the Mediterranean/Black Sea region on 23–27 June 2014 ("tumultuous atmosphere"), which triggered a number of devastating meteotsunamis in various parts of the region (Šepić et al. 2015).

5.3. *Propagation Velocity of Atmospheric Pressure Disturbances*

To examine whether the Proudman resonance took place in the central region of the PG (in the vicinity of Dayyer), i.e. if the speed of atmospheric pressure disturbances (U) was equal or close to the speed of long ocean waves (c), we need to determine the propagation speed and direction of the air pressure disturbances in this region. For selected disturbances (D2, D5 and D6), i.e. for those where we could pin-point the arrival time of the same peak/trough at three or more stations, we have estimated propagation speed and direction using the LSM approach to this problem described by Thomson et al. (2009). The results of analysis are presented in Table 3.

Unfortunately, our analysis demonstrated that the velocity parameters estimated from two sets of stations are inconsistent and, therefore, not reliable

Table 3

Velocity parameters of selected air pressure disturbances in the PG region, estimated: (1) for all stations which recorded this disturbance, and (2) for selected stations only

Disturbance	All stations					Selected stations				
	No.	U (m/s)[a]	ε_u (m/s)[a]	γ (°)[a]	ε_γ (°)[a]	No.	U (m/s)	ε_u (m/s)	γ (°)	ε_γ (°)
D2	5	8.1	2.6	150.5	43.7	3	3.9	N/A	90.6	N/A
D5	6	17.7	51.4	70.4	17.2	3	0.7	N/A	71.1	N/A
D6	12	18.2	2.2	138.9	41.9	7	18.0	7.1	145.6	78.8

[a]For each disturbance and estimate, the following parameters are indicated: Number of stations used in the calculation (No.), the computed disturbance speed (U) and the speed error (ε_u), the computed direction (γ—true North) and the direction error (ε_γ); "N/A" indicates that number of stations was insufficient to estimate speed and direction errors. For calculation of disturbance direction, we applied the oceanographic convention, i.e. the direction to a specific azimuth

Table 4

The 500 hPa wind speed (u) and the speed error (ε_u), the direction (γ), and the direction error (ε_γ), estimated from the ERA-Interim dataset at times and points closest to the times and areas of appearance of the PG disturbances and the Dayyer tsunami

Disturbance	Time (date, UTC h)	u (m/s)	ε_u (m/s)	γ (°)[a]	ε_γ (°)
D1	18 March, 21:00–24:00	23.4	1.5	259.4	2.9
D2	19 March, 01:00–04:00	29.4	3.1	240.2	5.4
D3	19 March, 03:00–05:00	21.4	2.2	269.6	4.1
DX (Dayyer)	19 March, 05:00	26.1	2.2	256.6	3.8
D4	19 March, 06:00–11:00	26.8	2.0	250.6	10.2
D5	19 March, 08:00–11:00	35.3	4.6	238.2	4.0
D6	19 March, 15:00–21:00	33.0	1.3	239.0	3.8
D7	19–20 March, 22:00–01:00	27.0	1.7	242.0	7.5
D8	20 March, 18:00–23:00	38.2	3.1	243.7	8.8
D9	20–21 March, 17:00–10:00	31.5	4.4	229.9	12.2
D10	21 March, 16:00–18:00	26.6	3.6	246.0	12.3

[a]For wind direction calculation, we applied the meteorological convention

enough. The main problem is the low time resolutions of the recorded air pressure series: the sampling interval of most stations was too long (5 min for 6 stations, 15 min for 40 stations, and 25 min for one station). Such intervals do not allow us to determine the exact timing of the main peaks of the studied disturbances, which had typical periods from 60 to 110 min (Fig. 10; Table 2). Furthermore, most stations at which any individual disturbances have been recorded, are within a short distance (< 5 km) from each other (Fig. 1; Table 2), i.e. the expected propagation time from one station to another was often within the sampling interval. The only exception is disturbance D6: this event was recorded by a large number of stations (Tables 2, 3; Fig. 10) located along an ~ 200 km path, thus making its speed and direction estimates more reliable and coherent (Table 3). These estimates show that disturbance D6 propagated southeastward, i.e. approximately along the PG axis, with a phase speed of ~ 18 m/s, which is typical for tsunamigenic atmospheric disturbances (Monserrat and Thorpe 1996; Monserrat et al. 2006). However, our estimates cannot give us the exact information about the particular disturbance that caused the Dayyer disaster due to the absence of *Wunderground* pressure instruments in the vicinity of the affected coast.

Additionally for each disturbance (D1–D10), as well as for a probable disturbance responsible for the Dayyer tsunami (DX), we estimated the propagation velocity parameters from the ERA-Interim 500-hPa wind speed using two spatial points, closest to the disturbance area, and two temporal points, closest to the disturbance times. The results of our analysis are given in Table 4. We assume that the observed PG disturbances, as well as the DX disturbance, propagated with speeds and directions comparable with the listed wind parameters (Table 4). Generally, wind speeds were higher than 21 m/s throughout the entire studied period, with the lowest values (21.4 m/s) during the morning of 19 March (D3), and the highest values of 33–35.3 m/s between 08:00 and 21:00 on 19 March (D5–D6), and 38.2 m/s during the D8 event at 18:00–23:00 on 20 March. Throughout the period under study, the modelled 500-hPa winds were from the southwest/west. During the Dayyer event, the 500 hPa wind had an estimated speed of 26.1 m/s and direction (coming from) of 256.6° (Table 4), i.e. it was directed precisely towards the Dayyer coast.

5.4. Remote Sensing

Analysis of satellite images can give us a much-needed insight into the character of recorded atmospheric pressure disturbances. Figure 12 shows these images, which were processed to depict convective activity at the approximate time of the Dayyer tsunami (05:12 UTC), and at the times of the most intensive atmospheric pressure disturbances over the PG (D1–D10), identified in Fig. 10. It appears that most of the recorded disturbances were correlated with either convective clouds (D2, D4, D5 and D6) or with atmospheric gravity waves (D7 and D8); and only a few disturbances (D1, D3 and D9) were not clearly linked to the atmospheric cloud cover. We may assume that the latter events were not directly related to the convective activity; however this does not imply the absence of atmospheric waves, but rather that the gravity waves were either not visible because of the upper level clouds (D9), or because there was not enough moisture in the atmosphere to form the clouds (D1 and D3).

Convective activity over the PG was particularly strong between the evening hours of 18 March 2017 and the late afternoon hours of 19 March 2017, covering areas of up to ~ 1000 km in diameter. In general, the area of high convective activity, moved from the northwest to the southeast, changing its major axis from NW–SE (up to the morning hours of 19 March) to SW–NE (during the afternoon hours of 19 March). Convective clouds can be seen over Dayyer at 21:57 UTC on 18 March 2017 (Fig. 12), i.e. close to the time of the onset of the recorded extreme sea level oscillations at the Dayyer and Asaluyeh tide gauges (Fig. 4). Furthermore, specifically at the time of the Dayyer tsunami, a series of convective clouds, organized in a wave-like pattern, was present over the area impacted by the tsunami waves (Fig. 12, 05:12 UTC). This agrees with eyewitness reports that pointed out that a strong thunderstorm preceded the Dayyer event. Intensive convective activity is a common feature that is known to precede or be concurrent with strong meteotsunamis (cf. Jansà et al. 2007; Wertman et al. 2014; Bechle et al. 2015). This is mainly related to the fact that convective storms, due to strong air updrafts and downdrafts, are frequently associated with abrupt air pressure changes, i.e. with tsunamigenic pressure jumps (Holton 2004). Another reason is the wave-CISK mechanism, responsible for generation, amplification and propagation of atmospheric gravity waves (Powers and Reed 1993). In particular, the second mechanism was found to occur during several previous meteotsunami events (Belušić et al. 2007; Tanaka 2010; Horvath et al. 2018).

6. Link between Air Pressure Disturbances and Sea Level Oscillations

In Fig. 13, we show the maximum heights of high-frequency ($T < 3$ h) atmospheric disturbances and sea level oscillations in the PG region for the simultaneous time intervals. They are overlaid by the 500-hPa wind contours, which can be used to illustrate the presence of a tsunamigenic synoptic pattern: several historical examples demonstrate that meteotsunamis are commonly associated with strong winds at the 500-hPa level (cf. Jansà et al. 2007; Thomson et al. 2009; Šepić et al. 2015). The reasons are two-fold: (1) atmospheric gravity waves can be developed by atmospheric instabilities, occurring at locations of strong spatial gradients of the upper-level winds, in particular, at the jet stream exit regions (Wang and Zhang 2007); and (2) providing a stable atmospheric layer below, a 500-hPa wind can serve as a steering wind for ducted atmospheric gravity waves, allowing their prolonged propagation and, consequently, the meteotsunami generation (Monserrat and Thorpe

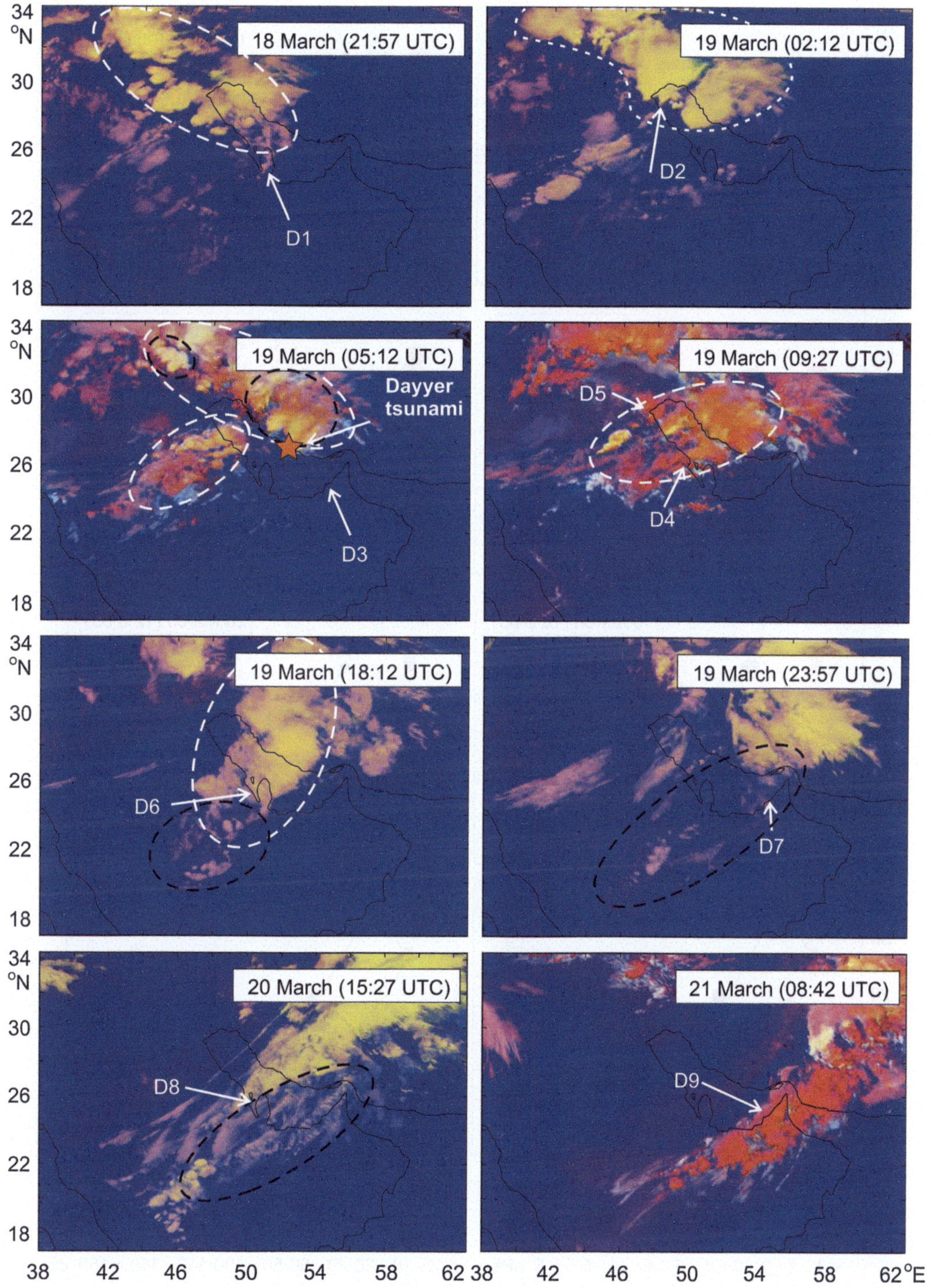

Figure 12
Sequence of satellite images illustrating propagation of clouds over the Middle East during 18–21 March 2017. The areas showing high concentration of convective clouds are encircled by white dashed lines, and the areas with high activity of atmospheric gravity waves are encircled by black dashed lines. Approximate locations of atmospheric disturbances marked in Figs. 9 and 10 (D1–D9) are indicated by white arrows at the respective times

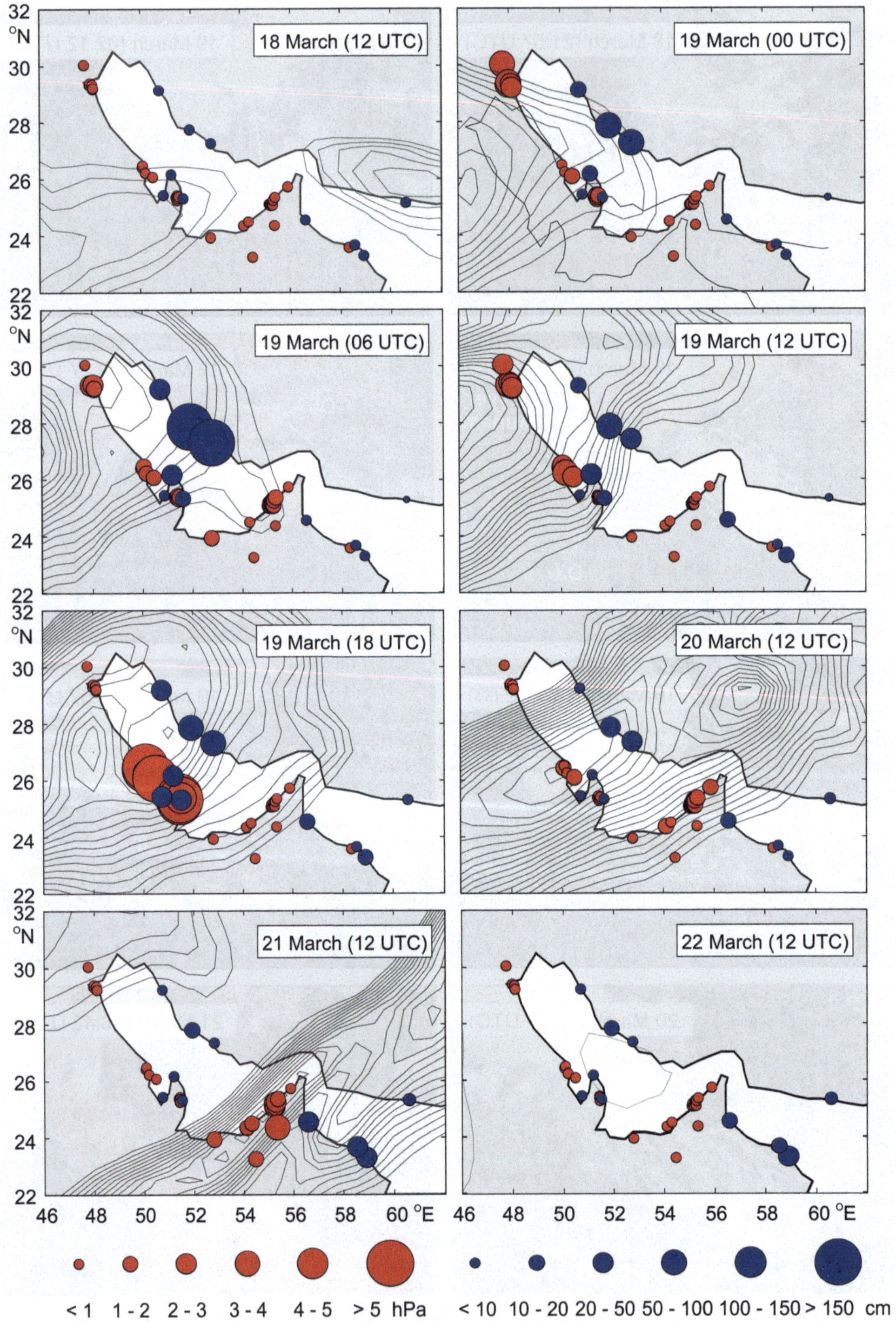

Figure 13
Temporal and spatial evolution of trough-to-crest heights of high frequency (periods of < 3 h) atmospheric pressure (red circles) and sea level (blue circles) oscillations in the PG region. The size of the circles is proportional to the oscillation height. The contours shown are at 1 m/s intervals

1996). Figure 13 shows that: (a) the intensification of atmospheric pressure activity is concurrent with the intensification of high-frequency sea level oscillations; and that (b) both are concurrent with the arrival of a favourable tsunamigenic synoptic pattern to the PG region. In particular, a day before the Dayyer tsunami (at 12:00 UTC on 18 March), no tsunamigenic atmospheric system was noticeable over the PG, and there was no high-frequency activity of either air pressure or sea level. Then, 12 h later (at 00:00 UTC of 19 March), a favourable synoptic system moved into the PG region, and was accompanied by high-frequency air pressure and sea level oscillations (Figs. 4, 7, 10, 11), in particular, in Kuwait (air pressure) and Iran (sea level stations), which were, at that time, located precisely at the leading edge of the jet stream. At the following examined time (06:00 UTC on 19 March), the region was still under the influence of a tsunamigenic synoptic pattern, and during this interval (00:00–06:00), the deadly tsunami waves impacted Dayyer. During the next 12 h, high-frequency activity of both air pressure and sea level continued over the central PG region, reaching peak strength at some specific locations/times. In particular, high atmospheric activity was observed over Saudi Arabia, Bahrain and Qatar at 18 UTC on 19 March 2017. Later, on 20–21 March, the tsunamigenic atmospheric system kept slowly propagating eastward. This translation was accompanied by intensification of atmospheric and sea level oscillations at eastern stations: at 12 UTC on 21 March, the strongest air pressure disturbances were observed over the United Arab Emirates and the largest sea level oscillations in the Gulf of Oman. One day later, the tsunamigenic atmospheric system has traversed further to the east, out of the area of our interest, and high frequency sea level and atmospheric oscillations stopped being recorded in the PG region.

The co-occurrence of a tsunamigenic synoptic pattern and high-frequency atmospheric/sea level oscillations implies that air pressure disturbances were the main generating force of extreme sea level oscillations in the PG region. However, despite the atmospheric activity during the afternoon hours of 19 March, which was strong all over the PG region, especially over Saudi Arabia, Bahrain and Qatar, destructive tsunami waves occurred only at a limited stretch of the coast. Whereas recorded sea level oscillations at the central Iranian coast were up to 234 cm (while eye-witness reports and land marks indicate runup heights of > 3 m, see Salaree et al. 2018), the sea level oscillations at non-Iranian stations had a maximum height of only 33 cm (Table 1). For effective generation of a meteotsunami, the atmospheric disturbance must propagate with a near-resonant speed: $U \sim c$, $Fr \sim 1.0$. In that case, the ocean waves begin to actively accumulate energy from the atmosphere and amplify rapidly. Such resonantly amplified waves can then devastatingly impact the endangered coast. Due to the low-resolution sampling interval of the available air pressure measurements, and their total unavailability over Iran, our best attempt at guessing the speed of air pressure disturbances was to approximate it with a 500-hPa wind speed. In Fig. 14, we show the Froude number, defined as the ratio of the ERA-Interim wind and the phase speed of long ocean waves, at two ERA-Interim time periods (00:00 and 06:00 UTC on 19 March 2017) surrounding the tsunami event at $\sim$ 05:00 UTC. It appears that before the tsunami, there was a wide area in front of Dayyer with Froude number, $Fr \sim 0.9$ to 1.1. Combined with the wind directed towards the Iranian coast, these are ideal conditions for meteotsunami generation, if the air pressure disturbance is strong enough. It is also indicative that the Froude number for the southern coast of the PG was far from unity, and in this region the governing upper-level winds were directed along the coast. Therefore, atmospheric pressure disturbances propagating over the southern coast of the PG (Kuwait, Saudi Arabia, Bahrain, Qatar, UAE, and Oman), could intensify high-frequency sea level oscillations, but could not produce really hazardous waves. The only area where such waves could be formed was the Dayyer region, i.e. exactly the region where the tsunami occurred. The available tide gauge data along the eastern and southern coasts of the PG (Table 1; Fig. 4) support this conclusion: the high frequency sea level activity was recorded at the corresponding locations exactly at the arrival times of the air pressure disturbances, but only in the area of Dayyer were the oscillations of extreme height.

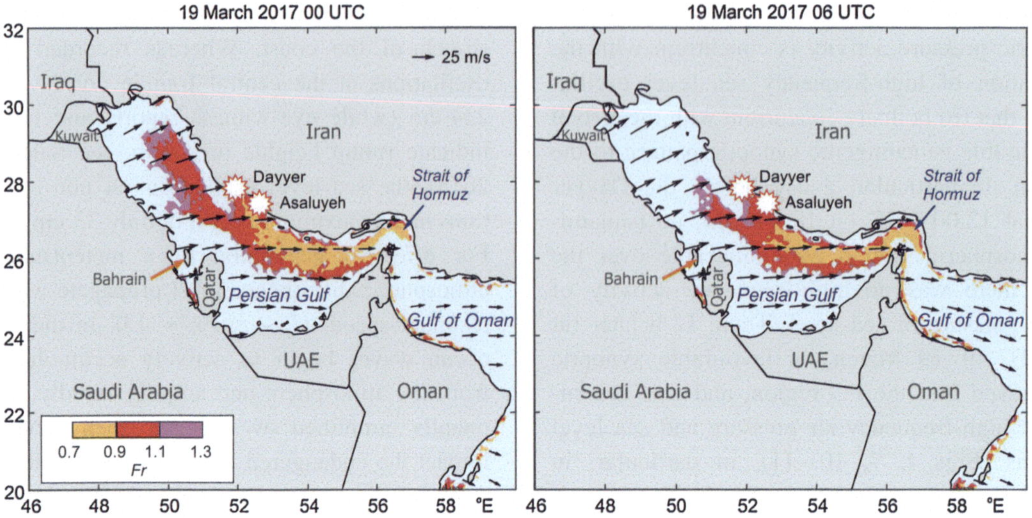

Figure 14
The region of the Persian Gulf with meteotsunami-favourable areas of $0.7 < Fr < 1.3$, where Fr is the Froude number, defined as the ratio of the wind speed at a height of 500 hPa and the phase speed of long ocean waves. **a** At 00:00 UTC and **b** at 06:00 UTC on 19 March 2017. Multibeam white stars indicate the locations of the highest measured meteotsunami waves; 500-hPa wind vectors are superimposed over the sea

7. Discussions

Our understanding of meteotsunamis—destructive atmospherically generated tsunami-like waves and associated hazards—has considerably advanced during the last 15 years (Monserrat et al. 2006; Rabinovich et al. 2009; Bailey et al. 2014; Šepić et al. 2015, 2018a, b; Horvath et al. 2018). The phenomenon has been reported and investigated throughout the entire World Ocean (Pattiaratchi and Wijeratne 2015; Vilibić et al. 2016). There are several "hot spots" around the world, where strong meteotsunamis occur regularly, such as Ciutadella Harbour in Spain, Vela Luka and Stari Grad harbours in Croatia, Nagasaki and Makurazaki bays in Japan. The hazardous meteotsunami events at these places are known, expected and theoretically can be forecast. In fact, preliminary meteotsunami warning systems have already been created in certain affected countries, in particular, in Spain (Marcos et al. 2009; Renault et al. 2011) and Croatia (Vilibić et al. 2016). Such systems are under development in some other countries, e.g. in USA, Portugal and Japan.

The difficult thing about the Dayyer event of 19 March 2017 is that it occurred in a place where such tsunami-like waves had never been reported before. Therefore, a question arises: *Is this place also a "hot spot", where similar events took place in the past, but remained unnoticed, or was the event totally accidental?* The question is quite important and urgent. Our findings indicate that the main reasons for this catastrophic event were a specific combination of *external factors*, i.e. particular resonant parameters of atmospheric disturbances, propagating across the Persian Gulf, and *internal factors*, i.e. the local topography/depth distribution, favourable for the local resonance and strong amplification of incoming waves. *Can such a situation repeat once again, or is the probability of such a coincidence negligible?*

The best way to address this question is to collect and examine high-resolution sea levels at Dayyer and Asaluyeh. Have any extreme events been observed at these sites before? How strong were sea level oscillations recorded at these stations in the past? Unfortunately, the authors could not find such information. However, the absence of historical reports about tsunami-like destructive events in the PG, obviously indicates that the Dayyer meteotsunami of 19 March 2017 should be considered as a rare phenomenon for this region. At the same time, if such a destructive event occurred at this coast once, we may assume that sooner or later it will occur there again.

As it follows from the results of the field survey of Salaree et al. (2018) and from our analysis of the tide gauge data, the Dayyer tsunami was a local event; it affected only a relatively short segment of the Iranian coast, roughly between Dayyer and Asaluyeh. The reason is explained in our Fig. 14; specifically, for the PG shelf adjacent to this coast the atmospheric disturbance speed, U, was close to the speed of long ocean waves, c: $U \sim c$, $Fr \sim 1.0$. There were no other regions in the Persian Gulf where these conditions were satisfied and there were no other coastal areas where strong sea level oscillations were reported on 18–20 March 2017. The area of $Fr \sim 1$ was of $\sim$ 100 km width. If we assume that the atmospheric disturbance responsible for the Dayyer meteotsunamis had a speed of $\sim$ 26 m/s, and properties similar to that of the strongest disturbance (D6) recorded over the PG during 18–21 March (Table 2), i.e. height of $\sim$ 6 hPa and rate of air pressure change up to 3.0 hPa/5 min, and if we assume that its ocean wave period was $\sim$ 14 min (Fig. 6), we can estimate its wave length to be $\sim$ 22 km. Then, following the numerical exercise by Vilibić (2008), we can estimate the height of long ocean waves generated by such an atmospheric disturbance at the end of the 100-km path over which $Fr \sim 1$, to be $\sim$ 70 to 80 cm. Over this path an ocean wave propagates as a forced wave locked to the atmospheric disturbance. Once the ocean wave leaves the area of $U \sim c$, $Fr \sim 1$ and enters the shallower area of $U > c$, $Fr > 1$, it starts propagating as a free wave. However, due to the Greens law, its amplification continues all the way to the coast. The Green law states that (Lamb 1932):

$$H_2 = \sqrt[4]{\frac{d_1}{d_2}} H_1 \tag{4}$$

where H_1 and H_2 are wave heights of ocean waves at locations with corresponding water depths d_1 and d_2. Assuming that $d_1 = 70$ m, $d_2 = 1$ m, and $H_1 = 80$ cm, it follows that $H_2 = \sim 230$ cm, in good agreement with the maximum height of wave recorded in the region of Dayyer (Table 1). Additional amplification can be expected at the coast due to final shoaling and local bathymetry and topography characteristics. The question, however, remains: *How common is this situation?*

As was emphasized by Monserrat et al. (2006), a "classic" meteotsunami is a resonant phenomenon, mainly associated with the *Proudman resonance*, when $U \sim c$ (cf. Vilibić 2008). Monserrat and Thorpe (1992, 1996) indicated that tsunamigenic atmospheric disturbances have a typical speed $U \sim$ 20 to 30 m/s. This means that the resonant generation of meteotsunamis should occur in regions where long ocean waves have the same speed: $c \sim$ 20 to 30 m/s. Figure 15 shows the distribution of $c = \sqrt{gh}$ in the Persian Gulf. It is seen that the "resonant region" is the northeastern part of the gulf adjacent to the Dayyer/Asaluyeh coast, which was destructively affected by the 2017 Dayyer tsunami. This means that it is not an accident that the 2017 event occurred in this particular region. Actually, the Dayyer/Asaluyeh coast may be considered as a "hot zone", which is potentially under risk of similar events that can impact this coast in the future.

Local meteotsunamis are normally produced by intense, but relatively small-scale and short-lived, tsunamigenic disturbances (Thomson et al. 2009). A number of such disturbances was recorded on 18–22 March 2017 in the PG region (Figs. 9, 10, 11). Unfortunately, the lack of high-resolution atmospheric measurements in the affected region did not allow us to identify the exact disturbance responsible

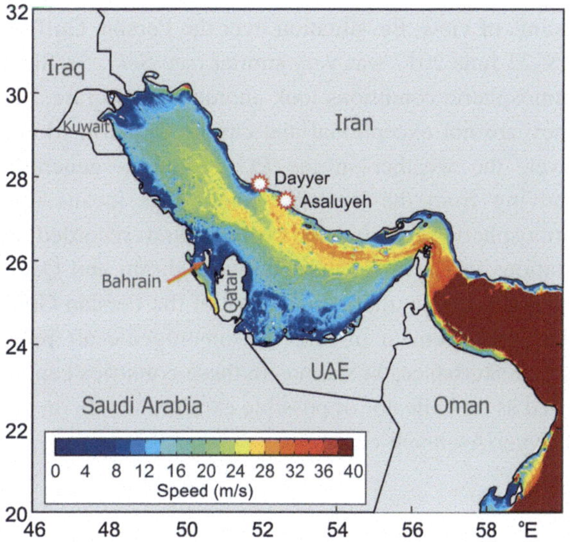

Figure 15
The distribution of long wave speed (colour map) according to Eq. (1) in the Persian Gulf region

for the 2017 Dayyer tsunami and to estimate its characteristics. However, certain information about the forcing can be obtained based on analysis of tide gauge data. Stations Dayyer and Asaluyeh are located on the northeastern coast of the Persian Gulf, only 80 km from each other (Fig. 1). Nevertheless, the spectral ratios ("source functions") are noticeably different for the two stations (Fig. 6c, d). This means that either the extreme waves at these two stations were generated by two independent small-scale atmospheric disturbances (the results of Thomson et al. 2009 and Šepić et al. 2015 showed that typical sizes of such disturbances are a few tens of kilometres) or that the propagating disturbance had substantially different parameters in the cross-travel direction (the atmospheric systems during the event propagated in the northeastern direction, i.e. normal to the affected Iranian coast of the PG—see Sect. 5).

Šepić et al. (2015) indicated the generation of numerous intense, small-scale tsunamigenic atmospheric disturbances over the Mediterranean/Black Sea region on 23–27 June 2014 was associated with anomalous atmospheric conditions, including (1) inflow of warm and dry air from Africa; (2) a strong high-altitude jet stream; and (3) the presence of unstable atmospheric layers characterized by a small Richardson number,[2] $Ri < 0.25$. Values of the Richardson number smaller than 0.25 imply the presence of unstable atmospheric layers. From many points of view, the situation over the Persian Gulf on 18–22 June 2017 was very similar (see Sect. 5). Such atmospheric conditions look anomalous and rare, but they are not exceptional and can be forecast. Moreover, the weather in the PG region is generally moving from the western sector. This means that atmospheric disturbances will be first recorded at stations in Kuwait, Saudi Arabia, Bahrain and Qatar and only then at the Iranian coast of the Persian Gulf. The measurement of intense tsunamigenic air pressure disturbances at stations in these countries can be used as an indicator of possible extreme waves on the Dayyer/Asaluyeh coast.

A long list of damaging meteotsunami events observed recently in various regions of the world (Brazil, South Africa, Australia, Netherlands, USA, Chile, Japan, Republic of Korea, Bulgaria, UK, France, Portugal and Ukraine) demonstrates the substantial threat of this phenomenon for coastal regions. A devastating event at Dayyer, i.e. in a new region for meteotsunamis, requires a worldwide knowledge of this natural hazard. Despite multitudinous information on this phenomenon accumulated recently, meteotsunamis still present a challenge for researchers and emergency managers. Unlike tectonic tsunamis, which are preceded by major submarine earthquakes, meteotsunamis are hardly predictable and can be considered as *silent* and *unexpected* events. We showed that the Froude number (Eq. 2) is a useful index for identification of "hot spots" and forecasting possible meteotsunamis. We used this approach for the Dayyer meteotsunami of 19 March 2017 and results indicate the high perspective of this method. Based on this successful experience, we can recommend a similar approach for some other potentially risky regions of the world oceans. High-resolution monitoring of air pressure disturbances, satellite imaging of atmospheric processes, continuous inspection of the high-altitude jet-stream regime, combined with estimation of the current Froude number for "hot-spot" regions, are probably the main directions for the meteotsunami warning and mitigation strategies in the future.

8. Conclusions

A destructive event occurred on 19 March 2017 on the northeastern coast of the Persian Gulf: the extreme tsunami-like waves with runup heights of ~ 3 m strongly impacted the area of Dayyer (Iran) and caused five deaths, several tens of people injured, and severe damage. The thorough examination of this event provided by Salaree et al. (2018), excluded the possibility that these waves were produced by a seismic source or by an underwater landslide; consequently, the authors assumed the atmospheric origin of the event, that can be termed as a *meteotsunami*. To support this assumption, the authors of the present study used the data from 12

[2] The Richardson number is a dimensionless parameter which expresses the ratio of the atmospheric buoyancy term to the flow shear term (vertical shear of horizontal wind speed).

tide gauges, located along the coasts of the Persian Gulf and the Gulf of Oman, and various meteorological information, including satellite images, high-altitude isohypse maps and 47 high-resolution air pressure recorders. The main findings of our study are the following:

- The Dayyer event of 19 March 2017 was a *local event*; maximum trough-to-crest wave heights of 197 and 234 cm were recorded at two near-field cities, Dayyer and Asaluyeh (both in Iran, 80 km from each other), respectively. At all other sites these heights were < 35 cm.
- The observed wave periods were in the range of 10–40 min; the dominant wave periods were 15–20 min.
- No earthquakes or visible landslides were reported at the time of the event which indicates that the event was due to atmospheric processes and thus may be considered as a "*meteorological tsunami*". This conclusion is supported by various atmospheric and satellite data.
- Analysis of 47 air pressure records resulted in identification of 10 distinctive, potentially tsunamigenic, pressure disturbances in the PG region in the period 18–22 March 2017. The atmospheric conditions over the PG were found to be highly favourable for generation of meteotsunamis and very similar to those that caused a chain of strong meteotsunamis in the Mediterranean and Black Sea regions in 23–27 June 2014.
- Based on the 500-hPa wind speed, we estimated the propagation speed of the atmospheric disturbances to be in the range of 21–38 m/s; the wind speed and direction during the Dayyer event were $\sim$ 26 m/s and 257°, respectively.
- The Froude number, *Fr*, estimated as the ratio of the speed of the air disturbance and the long wave speed, on 19 March 2017 was close to resonance, $Fr \sim 0.9$ to 1.1, specifically in the area of Dayyer/Asaluyeh, which is the most favourable for meteotsunami generation.

In general, our findings indicate that the north-eastern coast of the Persian Gulf should be considered as a "hot spot", highly vulnerable to extreme atmospherically induced tsunami-like waves. The atmospheric conditions of 18–22 March 2017 are anomalous, but they are not exceptional and can occur again in the future. However, taking into account the general eastward propagation direction of the atmospheric processes, we may assume that these tsunamigenic weather conditions can be forecast based on the information from stations in Kuwait, Saudi Arabia, Bahrain and Qatar, located to the west of the risky region.

Acknowledgements

Tide gauge data used in this study were provided by various agencies in the Persian Gulf region and international organisations: the Port and Maritime Organization (PSO) of Iran (https://www.pmo.ir/en/home), the Hydrographic Unit of the Qatar Ministry of Municipality and Environment (http://www.mme.gov.qa), the National Cartographic Center (NCC) of Iran (http://www.ncc.org.ir), and the Intergovernmental Oceanographic Commission, UNESCO (http://www.ioc-sealevelmonitoring.org/). We are grateful to Mr Vladan Janković (Qatar Ministry of Municipality and Environment) for providing us the Qatar sea level data. Air pressure data are from the amateur meteorological network *Wunderground* (https://www.wunderground.com/wundermap). The satellite data were downloaded from the EUMETSAT Meteosat7 satellite (https://www.eumetsat.int/website/home/Satellites/CurrentSatellites/Meteosat/index.html). The authors gratefully acknowledge Richard Thomson and Fred Stephenson (both from the Institute of Ocean Sciences, Sidney, BC, Canada) for valuable comments and suggestions. MH is funded by the Brunel University London through the Brunel Research Initiative and Enterprise Fund 2017/18 (BUL BRIEF). The work for AR was partially supported by the Russian State Assignment of IO RAS # 0149-2019-0005 and by the FCT-funded project FAST (PTDC/CTA-MET/32004/2017). The work of JS has been supported by the Croatian Science Foundation under the project MESSI (UKF Grant no. 25/15). This article benefited from constructive review comments from Emile A. Okal (Northwestern University, USA) and Rachid Omira (Instituto Português do Mar e da Atmosfera, Portugal) for which we are sincerely grateful.

Compliance with Ethical Standards

Conflict of interest The authors declare that they have no competing interests regarding the work presented in this article.

References

Ambraseys, N. N. (2008). Descriptive catalogues of historical earthquakes in the Eastern Mediterranean and the Middle East. In J. Fréchet, M. Meghraoui, & M. Stucchi (Eds.), *Historical seismology. Modern approaches in solid earth sciences* (Vol. 2, pp. 25–39). Dordrecht: Springer.

Bailey, K., DiVeglio, C., & Welty, A. (2014). An examination of the June 2013 East Coast meteotsunami captured by NOAA observing systems. *NOAA Techical Report NOS CO-OPS 079*, p. 56.

Bechle, A. J., Kristovich, D. A. R., & Chin, H. W. (2015). Meteotsunami occurrences and causes in Lake Michigan. *Journal of Geophysical Research Oceans, 120,* 8422–8438. https://doi.org/10.1002/2015jc011317.

Belušić, D., Grisogono, B., & Bencetić Klaić, Z. (2007). Atmospheric origin of the devastating coupled air-sea event in the east Adriatic. *Journal of Geophysical Research Atmospheres, 112,* D17111. https://doi.org/10.1029/2006jd008204.

Churchill, D. D., Houston, S. H., & Bond, N. A. (1995). The Daytona Beach wave of 3–4 July 1992: A shallow water gravity wave forced by a propagating squall line. *Bulletin of the American Meteorological Society, 76,* 21–32.

El Sabh, M. I., & Murty, T. S. (1989). Storm surges in the Arabian Gulf. *Natural Hazards, 1,* 371–385.

Ewing, M., Press, F., & Donn, W. L. (1954). An explanation of the Lake Michigan wave of 26 June 1954. *Science, 120,* 684–686.

Fritz, H. M., Blount, C. D., Albusaidi, F. B., & Al-Harthy, A. H. M. (2010). Cyclone Gonu storm surge in Oman. *Estuarine Coastal and Shelf Science, 86*(1), 102–106.

Greenspan, H. P. (1956). The generation of edge waves by moving pressure distributions. *Journal of Fluid Mechanics, 1,* 574–592.

Heidarzadeh, M., Necmioglu, O., Ishibe, T., & Yalçiner, A. C. (2017). Bodrum-Kos (Turkey–Greece) Mw 6.6 earthquake and tsunami of 20 July 2017: A test for the Mediterranean tsunami warning system. *Geoscience Letters, 4,* 31. https://doi.org/10.1186/s40562-017-0097-0.

Heidarzadeh, M., Pirooz, M. D., Zaker, N. H., Yalçiner, A. C., Mokhtari, M., & Esmaeily, A. (2008). Historical tsunami in the Makran Subduction Zone off the southern coasts of Iran and Pakistan and results of numerical modeling. *Ocean Engineering, 35*(8), 774–786.

Heidarzadeh, M., Satake, K., Murotani, S., Gusman, A. R., & Watada, S. (2015). Deep-water characteristics of the trans-Pacific tsunami from the 1 April 2014 M_w 8.2 Iquique, Chile Earthquake. *Pure and Applied Geophysics, 172*(3), 719–730.

Hibiya, T., & Kajiura, K. (1982). Origin of the 'Abiki' phenomenon (a kind of seiche) in Nagasaki Bay. *Journal of the Oceanographical Society of Japan, 38,* 172–182.

Holton, J. R. (2004). *An introduction to dynamic meteorology* (p. 528). New York: Elsevier Academic Press.

Honda, K., Terada, T., Yoshida, Y., & Isitani, D. (1908), An investigation on the secondary undulations of oceanic tides. *Journal of the College of Science*, Imperial University of Tokyo, p. 108.

Horvath, K., Šepić, J., & Prtenjak, M. T. (2018). Atmospheric forcing conducive for the Adriatic 25 June 2014 Meteotsunami event. *Pure and Applied Geophysics, 175*(11), 3817–3837.

Jansà, A., Monserrat, S., & Gomis, D. (2007). The rissaga of 15 June 2006 in Ciutadella (Menorca), a meteorological tsunami. *Advances in Geosciences, 12,* 1–4.

Kerkmann, J. (2005). *Applications of METEOSAT second generation (MSG)—RGB images: Part 04, RGB composites with channels 01-11 and their interpretation, EUMETSAT.* http://oiswww.eumetsat.org/WEBOPS/msg_interpretation/msg_channels.php.

Lamb, H. (1932). *Hydrodynamics* (6th ed., p. 768). Cambridge: Cambridge University Press.

Ličer, M., Mourre, B., Troupin, C., Krietemeyer, A., Jansà, A., & Tintoré, J. (2017). Numerical study of Balearic meteotsunami generation and propagation under synthetic gravity wave forcing. *Ocean Modelling, 111,* 38–45.

Lin, N., & Emanuel, K. (2016). Grey swan tropical cyclones. *Nature Climate Change, 6*(1), 106–111.

Marcos, M., Monserrat, S., Medina, R., Orfila, A., & Olabarrieta, M. (2009). External forcing of meteorological tsunamis at the coasts of the Balearic Islands. *Physics and Chemistry of the Earth, 34,* 938–947.

Monserrat, S., Rabinovich, A. B., & Casas, B. (1998). On the reconstruction of the transfer function for atmospherically generated seiches. *Geophysical Research Letters, 25,* 2197–2200.

Monserrat, S., & Thorpe, A. J. (1992). Gravity-wave observations using an array of microbarographs in the Balearic Islands. *Quarterly Journal of the Royal Meteorological Society, 118,* 259–282.

Monserrat, S., & Thorpe, A. J. (1996). Use of ducting theory in an observed case of gravity waves. *Journal of the Amtospheric Sciences, 53*(12), 1724–1736.

Monserrat, S., Vilibić, I., & Rabinovich, A. B. (2006). Meteotsunamis: Atmospherically induced destructive ocean waves in the tsunami frequency band. *Natural Hazards and Earth System Sciences, 6,* 1035–1051.

Munk, W. H., Snodgrass, F. E., & Carrier, G. F. (1956). Edge waves on the continental shelf. *Science, 123*(3187), 127–132.

Olabarrieta, M., Valle-Levinson, A., Martinez, C. J., Pattiaratchi, C., & Shi, L. (2017). Meteotsunamis in the northeastern Gulf of Mexico and their possible link to El Niño Southern Oscillation. *Natural Hazards, 88,* 1325–1346. https://doi.org/10.1007/s11069-017-2922-3.

Orlić, M. (1980). About a possible occurrence of the Proudman resonance in the Adriatic. *Thalassia Jugoslavica, 16*(1), 79–88.

Pasquet, S., & Vilibić, I. (2013). Shelf edge reflection of atmospherically generated long ocean waves along the central US, East Coast. *Continental Shelf Research, 66,* 1–8.

Pattiaratchi, C. B., & Wijeratne, E. M. S. (2015). Are meteotsunamis an underrated hazard? *Philosophical Transactions of the Royal Society A, 373,* 20140377.

Powers, J. G., & Reed, R. J. (1993). Numerical simulation of the large-amplitude mesoscale gravity-wave event of 15 December 1987 in the central United States. *Monthly Weather Review, 121,* 2285–2308.

Proudman, J. (1929). The effects on the sea of changes in atmospheric pressure. *Geophysical Supplements to the Monthly Notices of the Royal Astronomical Society, 2,* 197–209.

Rabinovich, A. B. (1997). Spectral analysis of tsunami waves: Separation of source and topography effects. *Journal of Geophysical Research, 102*(C6), 12663–12676.

Rabinovich, A. B. (2009). Seiches and harbor oscillations. In Y. C. Kim (Ed.), *Handbook of coastal and ocean engineering* (Vol. 9, pp. 193–236). Singapore: World Scientific Publ.

Rabinovich, A. B., & Monserrat, S. (1996). Meteorological tsunamis near the Balearic and Kuril Islands: Descriptive and statistical analysis. *Natural Hazards, 13,* 55–90.

Rabinovich, A. B., & Thomson, R. E. (2007). The 26 December 2004 Sumatra tsunami: Analysis of tide gauge data from the World Ocean. Part 1: Indian Ocean and South Africa. *Pure and Applied Geophysics, 164*(2/3), 261–308.

Rabinovich, A. B., Thomson, R. E., & Fine, I. V. (2013). The 2010 Chilean tsunami off the west coast of Canada and the northwest coast of the United States. *Pure and Applied Geophysics, 170*(9–10), 1529–1565. https://doi.org/10.1007/s00024-012-0541-1.

Rabinovich, A. B., Vilibić, I., & Tinti, S. (2009). Meteorological tsunamis: Atmospherically induced destructive ocean waves in the tsunami frequency band. *Physics and Chemistry of the Earth, 17*(34), 891–893.

Renault, L., Vizoso, G., Jansá, A., Wilkin, J., & Tintoré, J. (2011). Toward the predictability of meteotsunamis in the Balearic Sea using regional nested atmosphere and ocean models. *Geophysical Research Letters, 38,* L10601. https://doi.org/10.1029/2011gl047361.

Salaree, A., Mansouri, R., & Okal, E. A. (2018). The intriguing tsunami of 19 March 2017 at Bandar Dayyer, Iran: Field survey and simulations. *Natural Hazards, 90*(3), 1277–1307.

Sallenger, A. H., List, J. H., Gelfenbaum, G., Stumpf, R. P., & Hansen, M. (1995). Large wave at Daytona Beach, Florida, explained as a squall-line surge. *Journal of Coastal Research, 11,* 1383–1388.

Šepić, J., Međugorac, I., Janeković, I., Dunić, N., & Vilibić, I. (2016). Multi-meteotsunami event in the Adriatic Sea generated by atmospheric disturbances of 25–26 June 2014. *Pure and Applied Geophysics, 173,* 4117–4138.

Šepić, J., Rabinovich, A. B., & Sytov, V. N. (2018a). Odessa tsunami of 27 June 2014: Observations and numerical modelling. *Pure and Applied Geophysics, 175*(4), 1545–1572.

Šepić, J., Vilibić, I., & Belušić, D. (2009). Source of the 2007 Ist meteotsunami (Adriatic Sea). *Journal of Geophysical Research, 114,* C03016. https://doi.org/10.1029/2008jc005092.

Šepić, J., Vilibić, I., Rabinovich, A. B., & Monserrat, S. (2015). Widespread tsunami-like waves of 23–27 June in the Mediterranean and Black Seas generated by high-altitude atmospheric forcing. *Scientific Reports, 5*(11682), 1–5. https://doi.org/10.1038/srep11682.

Šepić, J., Vilibić, I., Rabinovich, A., & Tinti, S. (2018b). Meteotsunami (" Marrobbio") of 25–26 June 2014 on the Southwestern Coast of Sicily, Italy. *Pure and Applied Geophysics, 175*(4), 1573–1593.

Soloviev, S. L. (1978). Tsunamis. *The assessment and mitigation of earthquake risk (Natural Hazard, 1)* (pp. 118–139). Paris: UNESCO.

Tanaka, K. (2010). Atmospheric pressure-wave bands around a cold front resulted in a meteotsunami in the East China Sea in February 2009. *Natural Hazards and Earth System Sciences, 10,* 2599–2610.

Thomson, R. E., & Emery, W. J. (2014). *Data analysis methods in physical oceanography* (3rd ed., p. 728p). New York: Elsevier Science.

Thomson, R. E., Rabinovich, A. B., Fine, I. V., et al. (2009). Meteorological tsunamis on the coasts of the British Columbia and Washington. *Physics and Chemistry of the Earth, 34,* 971–988.

Vilibić, I. (2008). Numerical simulations of the Proudman resonance. *Continental Shelf Research, 28,* 574–581.

Vilibić, I., Šepić, J., Rabinovich, A. B., & Monserrat, S. (2016). Modern approaches in meteotsunami research and early warning. *Frontiers in Marine Science, 3*(57), 1–7. https://doi.org/10.3389/fmars.2016.00057.

Wang, S., & Zhang, F. (2007). Sensitivity of the mesoscale gravity waves to the baroclinity of jet-front systems. *Monthly Weather Review, 135,* 670–688.

Wertman, C. A., Yablonsky, R. M., Shen, Y., Merrill, J., Kincaid, C. R., & Pockalny, R. A. (2014). Mesoscale convective system surface pressure anomalies responsible for meteotsunamis along the US East Coast on June 13th, 2013. *Scientific Reports, 4,* 7143. https://doi.org/10.1038/srep07143.

Zaytsev, O., Rabinovich, A. B., & Thomson, R. E. (2017). The 2011 Tohoku tsunami on the coast of Mexico: A case study. *Pure and Applied Geophysics, 174*(8), 2961–2986. https://doi.org/10.1007/s00024-017-1593-z.

(Received March 14, 2019, revised June 20, 2019, accepted June 22, 2019, Published online July 3, 2019)

Pure Appl. Geophys. 177 (2020), 1261–1266

https://doi.org/10.1007/s00024-020-02437-9

Pure and Applied Geophysics

Global Occurrence of Large Tsunamis and Tsunami-like Waves Within the Last 120 years (1900–2019)

Viacheslav K. Gusiakov[1]

Abstract—The run-up catalogs of two global tsunami databases maintained by the NCEI/WDC NOAA and NTL/ICMMG SD RAS are examined to compile the list of annual maximum runups observed or measured in the oceanic, marine and inland basins during the last 120 years (from 1900 to 2019). All the retrieved annual maximum runups were divided into four groups according to four main types of tsunami sources (seismogenic, landslide-generated, volcanic, and meteorological). Their distribution over the type of sources shows that of the 120 maximum runups only 78 (65%) resulted from seismogenic sources, while the remaining 42 runups were divided between landslide-generated (19%), volcanic (8%), and meteorological (7.5%) sources. The analysis of geographical distribution of source locations demonstrates that tsunamis are not exclusively a marine hazard—over 15% of all maximum runups were observed in coastal and inland water basins (narrow bays, fiords, lakes, and rivers). Temporal distribution of the collected runups shows that annual occurrence of large tsunamis was more or less stable throughout the twentieth century and only demonstrates some increase during the last 27 years (since 1992) when the practice of post-event surveys of all damaging tsunamis was implemented. This paper also outlines the existing problems with data compilation, cataloguing, and distribution, and discusses incompleteness of runup and wave-form data for a considerable number of non-damaging tsunamis, even those resulting from the strong (magnitude higher than 7.5) submarine earthquakes.

Keywords: *Tsunami*, tsunami sources, tsunami occurrence, historical catalogs, historical databases, data processing, tsunami hazard, early warning.

1. Introduction

The significance of catalogs and databases, especially those that cover not only the instrumental period but historical and pre-historical periods as well, becomes increasingly important both for tsunami hazard assessment and for early warning. In the problem of hazard assessment, historical data are needed for correct evaluation of recurrence rates of tsunamigenic earthquakes and for determination of a maximum possible event in a particular tsunamigenic zone. In the operational forecast they are needed for determining tsunamigenic zones threatening coastlines and for rational selection of magnitude thresholds for the issuing of tsunami warnings. Therefore, the completeness and quality of historical catalogs, the correctness of their parameterization, and the transfer of this information into the databases are of critical importance for many areas of tsunami research.

A comprehensive analysis of the problems associated with tsunami data compilation, cataloguing and parameterization was given in Gusiakov (2009). Currently, two global historical tsunami databases exist and are separately maintained by the NOAA National Centers for Environmental Information (NCEI) and the Novosibirsk Tsunami Laboratory (NTL) of the Institute of Computational Mathematics and Mathematical Geophysics of the Siberian Division of the Russian Academy of Sciences (ICMMG SD RAS). In this paper, these databases are referenced as NCEI database (NCEI/WDS 2020) and NTL database (NTL/ICMMG 2020), respectively. For the instrumental period (20th and 21th centuries), the two databases are similar in content in terms of the number of events and runup observations. The difference is mainly connected with event classification (e.g. type of sources, validity index). Another difference is that most of the events in the NTL database are provided with their intensity on the Soloviev-Imamura scale, enabling events to be compared by their intensity value, which is a good proxy for their energy scale (Gusiakov 2011). In this study, the

[1] Institute of Computational Mathematics and Mathematical Geophysics, Novosibirsk, Russian Federation. E-mail: gvk@sscc.ru

NCEI database is used to determine general statistics on tsunami occurrences (temporal and spatial distribution of tsunamigenic events) and the statistics of run-up observations. The NTL database is used to retrieve the annual maximum runups during this period and determine statistics related to tsunami classification by the types of sources. In cases where minor differences between the NCEI and NTL databases are not important, they are both referenced as the Global Tsunami Database (GTDB).

Recently, several papers (Okal 2019; Gusiakov et al. 2019; Arcos et al. 2019) reviewing global occurrence of tsunamis in the last 25 years that mark the "high-instrumental" period of tsunami observations have been published. In this paper, a similar analysis is made for the last 120 years (1900–2019). The starting year of this period (1900) is pretty close to the median date (1902) that divides (by number of events) the global tsunami dataset into two equal parts. This 120-year period also roughly coincides with the "instrumental" period in seismological and tsunami measurements. The results of this analysis will provide users with a better understanding of the completeness of the global tsunami dataset and the level of uncertainty of the available data.

An important parameter of both databases is the validity index that varies from − 1 to 4 in the NCEI database and from 0 to 4 in the NTL database. The validity index relates not only the degree of confidence that a particular event was a tsunami or tsunami-like phenomenon, but also to its relationship to the indicated source location and indicated date. In this analysis we include tsunamis and tsunami-like events that resulted from of all types of tsunamigenic sources and that have positive validity indices, thus excluding only those events that are currently considered as erroneous entries.

2. Data Selection and Description

The Global Tsunami Database contains a total of 1270 tsunamigenic events that have been observed in oceanic, marine, and inland basins during the period from 1900 to 2019; this gives an average global tsunami occurrence of 10.6 events per year. The "size" (energy scale) of these events is very different and varies from very minor and local, hardly visible only on a single tide-gauge record, up to the catastrophic trans-oceanic mega-tsunamis affecting an entire oceanic basin that have thousand or more runup measurements. The vast majority of the registered tsunamis were weak events that are seen only on tide-gauge records. However, of the total 1,270 tsunamis there were 401 damaging or potentially damaging (with $H_{max} > 1$ m) events and among them there were 140 destructive tsunamis resulted in human fatalities.

Of these 1270 events, 412 do not have measured or estimated run-up or wave height measurements. The absence of quantitative data on a tsunami manifestation does not necessarily mean that they were all weak and had negligibly small runups or wave heights. Some of these events could have had a damaging impact (e.g. May 26, 1946 tsunami in Peru which resulted from a destructive M_w 8.2 earthquake) but historical catalogs (e.g. Soloviev and Go 1975) do not supply quantitative data on the tsunami effect. The remaining 858 events have at least one measurement or witness report on the runup height. These events were analyzed in order to obtain a set of annual maximum runups resulting from all types of tsunamigenic sources. Selection of annual maximum runups was done on the basis of a special computational procedure built-in in the PDM/TSU (Parametric Data Manager for Tsunami) graphic shell developed in the NTL/ICMMG for the database maintenance.

Results, presented in Fig. 1 show the temporal distribution of annual maximum tsunami runups observed or measured at oceanic and marine coasts, and in inland waters, during the last 120 years (1900–2019). For further analysis, all tsunamigenic sources were divided into four main groups: seismogenic, landslide-generated, volcanic, and meteorological. It is important to note that in Fig. 1 non-seismic events are indicated only when their runup heights are larger than those of seismogenic events for the same year.

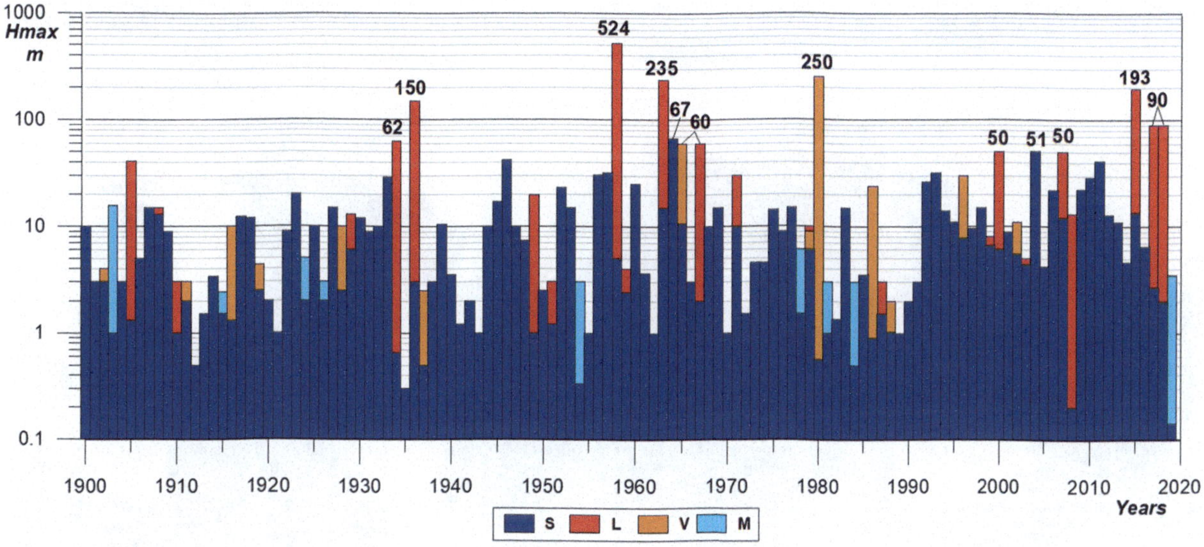

Figure 1
Maximum tsunami runup heights per years observed or measured globally during the last 120 years (from 1900 to 2019). Colors indicate the type of sources: *S* seismogenic, *L* landslide-generated, *V* volcanic, *M* meteorological. Non-seismic events are indicated only when their runup heights are larger than those of seismogenic events. For runups higher than 50 m their exact values are indicated

It should be further noted that due to the complex nature of tsunami generation mechanisms, the event classification by type of source can present a certain problem and in many cases requires expert judgment, so its result can be subjective. For example, the input of slide mechanism into the total tsunami energy can vary from very limited [as in the case of the 1964 Alaska tsunami where coastal and shallow-water landslides produced several of the highest (more than 25 m), but localized runups observed before the main tectonic tsunami arrived at the coast (Lander 1996)] to almost 100% [as in the case of the 1958 Lituya Bay tsunami where the large subaerial landslide was triggered by a strong (M_w 7.8) earthquake which itself did not contribute to the generation of the huge water wave in the bay that climbed up to a height of 524 m on a steep slope of its northern coast (Miller 1960)]. A similar situation exists for volcanic tsunamis—their generation mechanism almost always involves a slope failure or volcano flank collapse caused by the preceding volcanic activity.

The average annual maximum runup during this 120-year period is 25.4 m. In consideration of this value, one should take into account that this number is largely affected by a few years with the highest run-ups from several large landslide-generated tsunamis, e.g. July 10, 1958 Alaska 524 m, May 18, 1980 Spirit Lake, Washington 250 m, October 9, 1963 Vajont Dam, Italy 235 m. The average annual maximum runup for seismogenic tsunamis is just 9.2 m, for landslide tsunamis—65.4 m, for volcanic tsunamis—30.4 m, and for meteorological tsunamis—4.0 m.

Geographical distribution of the sources of these 120 tsunamigenic events with annual maximum runup heights is shown in Fig. 2. Historically, the Pacific remains the main tsunamigenic region of the World Ocean. Within its basin, 91 events with annual maximum runups occurred. The remaining events are divided between the Indian Ocean (6 events), the Atlantic Ocean (13 events), and the Mediterranean (10 events) regions.

Based on the data presented in Fig. 1 we can estimate that on the global level an average return period of the tsunamis and tsunami-like events with $H_{max} > 100$ m is 24 years, $H_{max} > 50$ m is 10 years, $H_{max} > 20$ m is 4 years, $H_{max} > 10$ m is 2.2 years, $H_{max} > 5$ m is 1.6 years, $H_{max} > 2$ m is 1.2 year, $H_{max} > 1$ m is 1.1 years.

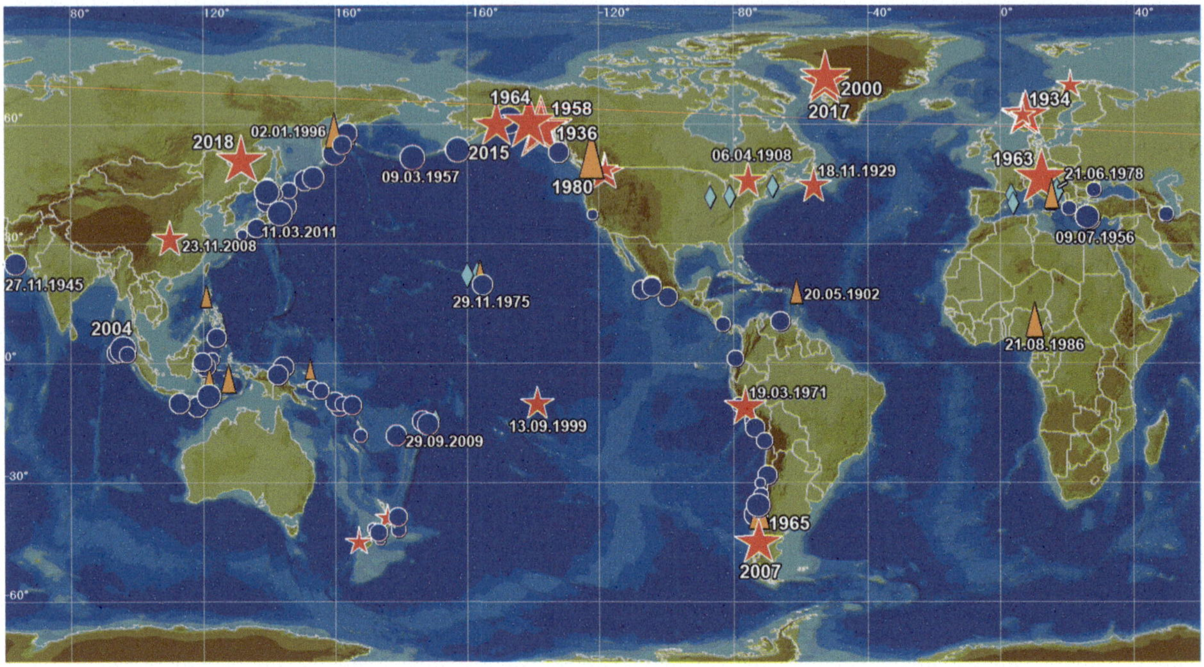

Figure 2

Map showing the locations of tsunamigenic events presented in Fig. 1. Symbols indicate the type of tsunami source: (circles) seismogenic, (stars) landslide-generated, (triangles) volcanic, and (diamonds) meteorological. The symbol size is proportional to the runup height of the corresponding tsunami. The sources of the events with runup values $H_{max} > 50$ m are marked by the year of the occurrence; some other selected events are marked by their full date (dd.mm.yyyy)

3. Discussion and Conclusions

The first and foremost conclusion that follows from the diagram in Fig. 1 is that among the 120 annual maximum runups, only 76 (63.3%) resulted from seismotectonic tsunamis generated by submarine earthquakes. The rest of the maximum runups are divided between 23 landslide-generated events (19.2%), 12 volcanic tsunamis (10%), and 9 meteorological events (7,5%). Taking into account that for the vast majority of volcanic events the actual generation mechanism of destructive water waves is a volcanic slope failure, the fraction of landslide-generated tsunamis in the oceanic and marine basins and tsunami-like waves in coastal and inland waters increases up to 29%.

Despite the fact that these numbers do not reflect the actual breakdown of tsunami occurrence by the type of source (in Fig. 1 we count only events that produced an annual maximum runup and ignore all other events with smaller runups), they are quite interesting because they clearly demonstrate the importance of non-seismic events in the assessment of overall tsunami hazard. In total they constitute 36.7% (that is more than one third) of annual maximum tsunami run-ups. However, with a few exceptions, the contribution of these events to tsunami hazard maps constructed according to the PTHA methodology, is not taken into account, and early warning of such events presents a problem for the existing Tsunami Warning Systems.

All seven trans-oceanic mega-tsunamis that occurred within the last 120-years (1946 Aleutians, 1952 Kamchatka, 1957 Aleutians, 1960 Chile, 1964 Alaska, 2004 Indonesia and 2011 Japan) gave annual maxima in runup heights in their respective years. The average of their maximum run-ups is 40.2 m. Based on this and on a wide extent and a variety of types of coast affected by mega-tsunamis, we can consider this value as the upper limit of maximum runup possible for seismogenic tsunamis. Any substantial excess in a reported height of a seismogenic tsunami above this level at some stretched coastline (ignoring possible localized runups into small steep-sided valleys) must be considered as an indication to the involvement of the submarine or coastal slides in the tsunami generation (as was a case for the 1964 Alaska tsunami).

Another important conclusion that follows from the source location map presented in Fig. 2 is that tsunamis are not exclusively a marine hazard. Over 15% of all maximum runups were observed in coastal and inland water basins (narrow bays, fiords, lakes, water reservoirs and rivers). All ten extreme (higher than 60 m) runups shown in diagram in the Fig. 1 fall within this category.

Temporal distribution of the collected runups demonstrates that the annual occurrence of large tsunamis was more or less stable during the whole twentieth century with some increase during the last 27 years (since 1992), when the practice of post-tsunami surveys of all damaging tsunamis was implemented. The tsunami occurrence rate was on average 106 events per decade with only one notable exception for the decade of 1981–1990 when 77 tsunamigenic events globally occurred. An even more prominent decrease was observed in the number of strong tsunamis: just three events with runup heights exceeding 3 m. In fact, within this decade only one really damaging and fatal seismogenic event occurred—the Japan Sea tsunami of May 26, 1983 with maximum run-up height of 14.9 m (Shuto 1983). Another fatal event was a rather unique and rare volcanic tsunami with 24-m runup resulting from a "limnic eruption" (also referred to as a "lake overturn") that occurred on August 21, 1986 in Lake Nyos in Cameroon (Gusiakov 2014).

An essential increase in the annual maxima in the run-up heights of seismogenic tsunamis for the last three decades is clearly visible in Fig. 1. An average H_{max} for 1992–2019 is 14.2 m, while for the preceding years (1900–1991) an average H_{max} is only 7.2 m. This is obviously a result of the practice of Post-tsunami Surveys for all damaging tsunamis that was implemented in 1992 (Arcos et al. 2019), as well as a considerable improvement in the tsunami instrumentation that occurred in the same years (Rabinovich and Eblé 2015). Of the total 24,620 runups available in the Global Tsunami Database for the last 120 years, this period (1992–2019) contains 15,307 (62%).

The final comment concerns the year of 2019, the last year in the time period considered. This year looks like the most quiet year for more than a century, at least, as far as "geological" tsunamis are concerned (under this term, following Okal (2019), we understand the tsunamis triggered by earthquakes, landslides and volcanoes). Only six tsunamigenic events are available in the database for this year, all of them being very weak. The maximum reported runup height was only 0.3 m, and it was measured at the coast of Stromboli Island, Italy after a paroxysmal explosion on August 28, 2019. The largest instrumentally recorded height of a seismogenic tsunami was just 14 cm recorded in New Zealand after the M_w 7.2 submarine earthquake in the Kermadek region on June 15, 2019. However, the year of 2019 is marked by at least two considerable meteotsunamis that occurred on April 14, 2019 in Lake Erie (with an estimated height of about 3.5 m) and on June 21, 2019 at the northwestern coast of Mallorca Island, Spain (maximum wave height was 1.74 m) (Rabinovich 2020).

Acknowledgements

Figures in this paper were drawn using built-in options of the PDM/TSU graphic shell (PDM/TSU 2020) developed in the NTL/ICMMG SD RAS as supporting software for the tsunami database. The author wishes to thank Emile Okal and Fred Stephenson for careful reading of the manuscript and making numerous suggestions for its improvement as well as Tamara Kalashnikova and Katia Lyskovskaya for assistance in data processing and figure drawing. The results presented in this paper were supported by the RAS Project no. 0315-2019-0005.

References

Arcos, N., Dunbar, P., Stroker, K., & Kong, L. (2019). The impact of post-tsunami surveys on the NCEI/WDS Global Historical Tsunami Database. *Pure and Applied Geophysics, 176*, 2809–2829. https://doi.org/10.1007/s00024-019-02191-7.

Gusiakov, V. K. (2009). Tsunami history: Recorded, In: Eds. E.N. Bernard and A.R. Robinson, *The Sea*, (Vol.15, pp. 23–53), Cambridge, MA, USA: Harvard University Press.

Gusiakov, V. K. (2011). Relationship of tsunami intensity to source earthquake magnitude as retrieved from historical data. *Pure and Applied Geophysics, 168*(11), 2033–2041. https://doi.org/10.1007/s00024-011-0286-2.

Gusiakov, V. K. (2014) Tsunami impact on the African coast: Historical cases and hazard assessment. In: Eds. A. Ismail-Zadeh, J.C. Fucugauchi, A. Kijko, K. Takeuchi, and I. Zaliapin, *Extreme Natural Hazards, Disaster Risks and Societal Implications*, Cambridge University Press, pp. 225- 233. 10.1017/CBO9781139523905.021.

Gusiakov, V. K., Dunbar, P., & Arcos, N. (2019). Twenty ive years (1992–2016) of global tsunamis: statistical and analytical overview. *Pure and Applied Geophysics, 176*, 2795–2807. https://doi.org/10.1007/s00024-019-02113-7.

Lander, J. F. (1996). *Tsunamis affecting Alaska, 1737–1996* (p. 195). Colorado, National Geophysical Data Center: Boulder.

Miller, D.J. (1960) Giant waves in Lituya Bay, Alaska. *Geological Survey Professional Paper* 354-C, U.S. Government Printing Office, Washington, p.50–85.

NCEI/WDS. (2020). Global Historical Tsunami Database, 2100 BC to Present, 10.7289/V5PN93H7, Accessed 30.11.2018.

NTL/ICMMG SD RAS. (2020). Novosibirsk Tsunami Laboratory of the Institute of Computational Mathematics and Mathematical Geophysics of Siberian Division of Russian Academy of Sciences) Global Tsunami Database, 2100 BC to Present, Available at: https://tsun.sscc.ru/nh/tsunami.php, Accessed 15.12.2018.

Okal, E. (2019). Twenty-five years of progress in the science of "geological" tsunamis following the 1992 Nicaragua and Flores events. *Pure and Applied Geophysics, 176*, 2771–2793. https://doi.org/10.1007/s00024-019-02244-x.

PDM/TSU (Parametric Data Manager for Tsunami data management) graphic shell. (2020). NTL/ICMMG SD RAS, https://tsun.sscc.ru/PDM.htm, Accessed 15.01.2020.

Rabinovich A. (2020) Twenty-seven years of progress in the science of meteorological tsunamis following the 1992 Daytona Beach event, *Pure and Applied Geophysics, 177*, (this issue) https://doi.org/10.1007/s00024-019-02349-3

Rabinovich, A. B., & Eblé, M. C. (2015). Deep-ocean measurements of tsunami waves. *Pure and Applied Geophysics, 172*, 3281–3312. https://doi.org/10.1007/s00024-015-1058-1.

Shuto, N. (1983). The Nihonkai Chubu earthquake tsunami. *Tsunami Newsletter, XVI*(2), 31–40.

Soloviev, S.L., & Go, Ch.N. (1975). *A Catalog of Tsunamis on the Eastern Shore of the Pacific Ocean,* USSR Academy of Sciences, Nauka Publishing House, Moscow, 204 p. Translated from Russian to English by Canadian Institute for Science and Technical Information, No. 5078, National Research Council, Ottawa, Canada, 293 pp, 1984.

(Received January 22, 2020, revised January 29, 2020, accepted January 30, 2020, Published online February 10, 2020)

Pure Appl. Geophys. 177 (2020), 1267–1293

https://doi.org/10.1007/s00024-019-02363-5

Pure and Applied Geophysics

The Large Earthquake (~ M7) and Its Associated Tsunami of 8 November 1905 in Mt. Athos, Northern Greece

I. Triantafyllou,[1] F. Zaniboni,[2] A. Armigliato,[2] S. Tinti,[2] and G. A. Papadopoulos[3]

Abstract—Mt. Athos, Northern Greece, is a monastic community with life since ~ 1000 AD, a UNESCO worldwide cultural heritage site that was repeatedly hit by earthquakes. On 8 November 1905, a large earthquake, with estimated magnitude ranging from 6.8 to 8.3, caused destruction in Mt. Athos, but the event remains little known since only few documentary sources have been utilized so far. We collected a set of sources including contemporary press reports, letters, manuscripts and books. We reconstructed the earthquake impact field and organized a database containing (1) assignment of macroseismic intensity in 44 observation points at epicentral distances up to ~ 660 km, and (2) identification of sites where important co-seismic ground failures were reported, e.g., landslides and rock falls. From magnitude-intensity relationships, a macroseismic magnitude of ~ 7 equivalent to M_s was estimated, which is close to recent instrumental estimates of ~ M_w 7.2. The examined sources revealed for the first time that a local but powerful tsunami with an ~ 3-m run-up was generated by earthquake-triggered landslides in southern Mt. Athos. Eleven persons were reportedly killed. Both the landslide and the tsunami were numerically simulated by a scenario of massive landslide entering the sea. The synthetic marigrams obtained from the simulation are consistent with the historical tsunami descriptions. Another little-known tsunami was reported in Mt. Athos in association to the large 1585 earthquake. We concluded that a comprehensive hazard assessment in Mt. Athos should take into account not only strong earthquakes but also associated phenomena like landslides and tsunamis.

Keywords: Historical earthquakes, tsunamis, landslides, numerical simulation, Mt. Athos, Northern Greece.

[1] Faculty of Geology and Geoenvironment, National and Kapodistrian University of Athens, Athens, Greece. E-mail: ioannatriantafyllou@yahoo.gr

[2] Section of Geophysics, University of Bologna, Bologna, Italy.

[3] International Society for the Prevention and Mitigation of Natural Hazards, Athens, Greece.

1. Introduction

The seismicity in the Aegean Sea and surrounding areas (Fig. 1), eastern Mediterranean basin, is one of the highest in the entire western Eurasia. From historical documents, archaeological and geological signatures as well as from instrumental records, it is evidenced that in the Mediterranean, including the Aegean Sea, large submarine or coastal earthquakes are frequently associated with tsunami waves which propagate at local, regional or even basin-wide distances (e.g., Guidoboni et al. 1994; Ambraseys 2009; Papadopoulos et al. 2014). In the North Aegean Sea, the ~ 50-km-long peninsula of Mt. Athos (Fig. 1) is one of the poorly studied areas in terms of seismic and tsunami hazard assessment. The geomorphology of Mt. Athos Peninsula is dominated by mountainous relief of maximum altitude of 2033 m at its southern part. Mt. Athos is a worldwide renowned monastic community of Orthodox Christianity which attracts thousands of visitors every year and has been declared by UNESCO as a world cultural heritage site. The life of this community goes back to ca. 1000 AD. Currently, there are 20 main monastery building complexes and many other smaller monastic units with a totality of about 2000 monks (e.g., della Dora 2012).

On 8 November 1905, a large earthquake shook the Balkan peninsula (Fig. 1) causing disastrous effects at the monastic communities of Mt. Athos and in nearby towns and villages (Sieberg 1932; Kárnik 1969; Papazachos and Papazachou 1997). However, although this event is listed in various regional and global earthquake catalogues (Table 1), the information available so far about the earthquake impact remains limited because only a few documentary

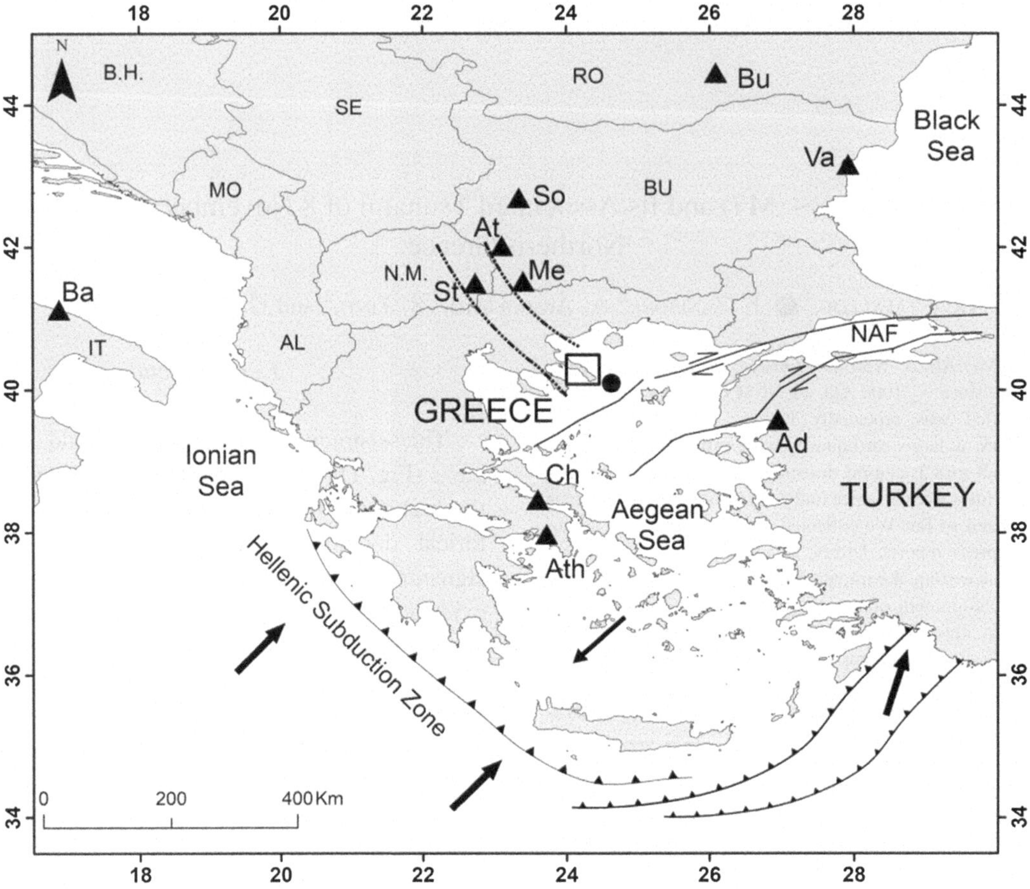

Figure 1
Seismotectonic setting of the study area and macroseismic intensities felt at sites in large epicentral distances (solid triangle) by the earthquake (solid circle) that occurred in Mt. Athos peninsula (square) on 8 November 1905. Details and documentation for the intensities assigned are given in Table 3. All intensities plotted here are of II–III degree in 12-grade modified Mercalli (MM) scale. For sites of higher intensity, see Figs. 2 and 3. Symbol key: arrow = plate motion, thin lines = branches of the North Anatolian Fault (NAF) entering in North Aegean Sea, dashed lines = Serbo-Macedonian composite terrane. Site symbols: *Ad* Adramition, *Ath* Athens, *At* Ano Tzoumayia, *Ba* Bari, *Ch* Chalkis, *Me* Melenikon, *So* Sofia, *St* Stromnitsa, *Va* Varna. Country symbols: *AL* Albania, *B.H.* Bosnia and Herzegovina, *BU* Bulgaria, *N.M.* North Macedonia, *IT* Italy, *MO* Montenegro, *RO* Romania, *SE* Serbia

sources were found and published. Besides, the earthquake size is still controversial since very divergent magnitudes have been estimated (Table 1). As a consequence, despite its large magnitude and the extensive damage caused in a world cultural heritage site such as Mt. Athos, the large 1905 earthquake event remains little known. On the other hand, because of its large magnitude, this earthquake is a key event for better understanding the level of hazard not only in Mt. Athos but also in the North Aegean Sea region.

From a variety of documentary sources, presented for the first time in this paper, we assessed the 1905 earthquake impact and assigned macroseismic intensities felt in 44 observation sites situated not only in Mt. Athos but also in a large part of the Balkan peninsula and other places in eastern Mediterranean. From the reconstruction of the macroseismic field, we were able to independently estimate the earthquake magnitude and compare it with other magnitude estimates. In addition, we revealed a detailed picture of the remarkable ground failures of various types caused by the 1905 earthquake, such as landslides and rock falls. Importantly, from the analyzed documentary sources, we also found for the first time that a local but destructive tsunami wave accompanied the

Table 1

Focal parameters of the Mt. Athos 8 November 1905 earthquake listed in various earthquake catalogues

y	m	d	φ_N^0	λ_E^0	Time (GMT)	h (km)	I_o (MM)	M_S	References
1905	11	8	240 km NE from Athens		23.41.54 (l.t.)				Eginitis (1910)
1905	11	8	40.00	24.00	22:06:12	*n*		7.8	Richter (1958); Duda (1965)
1905	11	8	40.25	24.25	21:41	*n*	X	~ 7	Galanopoulos (1960)
1905	11	8	40.25	24.25	21:41	*n*	X	~ 7	Kárnik (1969)
1905	11	8	40.00	24.00	22.06	*n*	IX–X	7.5	Kárnik (1969)
1905	11	8	40.30	24.40	22:06:00	17		7.4	Shebalin et al. (1974); Makropoulos et al. (2012)
1905	11	8	40.25	24.25			IX–X	7.5	Galanopoulos (1981)
1905	11	8	40.30	24.30	22:06:30	*n*	X	7.5	Comninakis and Papazachos (1982)
1905	11	8	40.00	24.00	22:06	*n*		6.8	Abe and Nogushi (1983)
1905	11	8	40.30	24.40	22:30:30	*n*	X	7.5	Papazachos and Papazachou (1989)
1905	11	8	40.26	24.33	22:30:30	22.4*	10.21*	8.3*	Papazachos and Papaioannou (1997)
1905	11	8	40.00	24.50	22:07			7.24	Ambraseys (2001)
1905	11	8	40.00	24.50	22:30:30	*n*	X	7.5	Papazachos and Papazachou (1997, 2003)
1905	11	8	40.092	24.627	22:06:11	15		7.2 (M_w)	ISC-GEM (2018)

Key: *y* year, *m* month, *d* day, φ_N^0 and λ_E^0 are geographic latitude north and longitude east, respectively; *l.t.* local time, *h* focal depth, *n* shallow earthquake, I_o epicentral intensity in modified Mercalli (MM) scale, M_s surface-wave magnitude, M_w moment magnitude. Parameters marked by an asterisk have been derived by modeling of macroseismic data. For the earthquake origin time, see explanations in the text

1905 earthquake. Based on documentary sources, we developed a cascade scenario incorporating three sequential events, i.e., earthquake, landslide and tsunami, enabling us to numerically simulate the tsunami and compare successfully the synthetic wave run-ups with the reported ones. Finally, we discussed the seismic and tsunami hazards in Mt. Athos in light of our results and the earthquake history of the area.

2. *Seismotectonic Setting and Past Seismicity*

2.1. *Seismotectonics*

The geodynamics of the broad Aegean Sea area is characterized by two main active processes (Fig. 1): (1) the segmented subduction from about SW to NE of the Nubian (African) lithosphere beneath the South Aegean Sea along the Hellenic Arc (e.g., Bocchini et al. 2018 and references therein), and (2) the westward escape of the Anatolian microplate which splits at its northern margin along the major right-lateral North Anatolian Fault (NAF) (e.g., Reilinger et al. 2010 and references therein). The seismotectonics of Mt. Athos are controlled by two main geotectonic units, the first occupying the North Aegean Sea (NAS), and the second the central part of continental North Greece (Fig. 1). The NAS area is tectonically dominated by branches of the NAF extending from east to west into the tensional domain of the Aegean area (Fig. 1) (e.g., Le Pichon et al. 1987; Taymaz et al. 1991; Papadopoulos et al. 2002; Kreemer et al. 2004; Karabulut et al. 2006). However, all evidence indicates the NAF branches terminate in the western NAS area and do not continue to continental Greece (Koukouvelas and Aydin 2002; Ferentinos et al. 2018). The ~ 300-km-long and ~ 1000-m-deep North Aegean Trough (NAT) is the main active morphotectonic structure associated with the westward extension of the NAF into the NAS (Pavlides et al. 1990; Roussos and Lyssimachou 1991; Koukouvelas and Aydin 2002) (Fig. 1). From geodetic data, it results that the NAT is a fast-moving (~ 20 mm/year) strike-slip fault structure (Reilinger et al. 2010; Floyd 2010; Müller et al. 2013; Mouslopoulou et al. 2014) associated with large-magnitude earthquakes. All authors agree that no deep seismicity is expected in the NAS area and continental North Greece since it is situated about 600–700 km away from the Hellenic subduction zone (Fig. 1), where the seismically active part of the subducting slab goes down to no more than ~ 200 km. However, weak intermediate-depth seismicity recorded in the NAS area may indicate a paleo-subduction process dying off in the present geological period (e.g., Papazachos and Papadopoulos 1979).

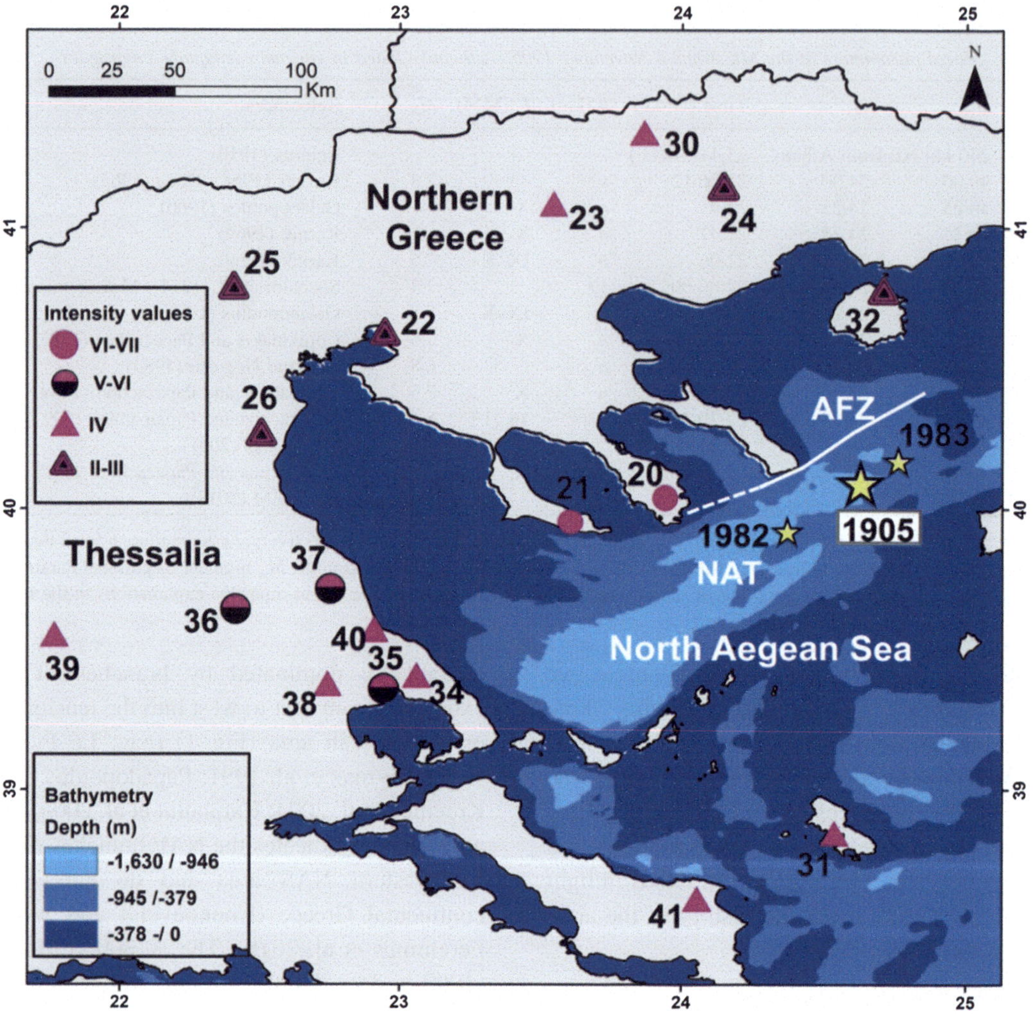

Figure 2
Macroseismic intensities in 12-grade MM scale felt at sites in mid-epicentral distances due to the Mt. Athos earthquake (large star) of 8 November 1905. For intensity documentation and numbering of sites, see Table 3. Symbol key: *AFZ* Athos Fault Zone (white line, redrafted from Roussos and Lyssimachou 1991), *NAT* North Aegean Trough. Small stars show the epicenters of the 18 Jan. 1982 (M_w 6.56) and 6 Aug. 1983 (M_w 6.65) earthquakes mentioned in the Discussion section; earthquake focal parameters are from the ISC-GEM (2018) catalogue

Along the southern edge of Mt. Athos peninsula, but to the north of the NAT, the presence of an active, dextral strike-slip fault zone subparallel to the NAT has been found from detailed interpretation of marine seismic lines (Roussos and Lyssimachou 1991) (Fig. 2). The so-called Athos Fault Zone (AFZ) is over 100 km long and has both transtensional and transpressive features which are observable on the seismic lines (Koukouvelas and Aydin 2002). Roussos and Lyssimachou (1991) suggested that the dextral strike-slip displacement along the AFZ is on the order of 5–7 km. On the other hand, the seismicity in continental Northern Greece is also high, particularly along the NNW–SSE-trending structure known as the Serbo-Macedonian massif or composite terrane as proposed by Koukouvelas (2018). This terrane is an arcuate geological zone approximately 600 km long and 60–100 km wide (Kockel et al. 1977; Psilovikos 1984) (Fig. 1). Tectonically, the Serbo-Macedonian terrane is currently affected by an ~ N–S tensional stress field and is crossed by ~ E–W-trending normal faults (e.g., Pavlides and Kilias 1987). Geographically, Mt. Athos peninsula occupies the southern part of the terrane, but it is not clear if Mt. Athos peninsula is tectonically controlled by E–W normal faulting given that not only normal but also strike-slip faults have been mapped in the area (Kockel et al. 1977; Georgiadis et al. 2007).

2.2. *Earthquake and Tsunami History*

In the historical period, various strong earthquakes were reported in the NAS as well as along the Serbo-Macedonian terrane. The seismic record, however, is incomplete, and the epicentral and magnitude determinations suffer from large uncertainties (e.g., Guidoboni et al. 1994; Ambraseys 2009). During the instrumental era of seismology, the NAS area and Serbo-Macedonian zone were ruptured by several strong or large earthquakes (Voidomatis et al. 1990; Papadopoulos et al. 2002; Evangelidis 2014; ISC-GEM 2018), one of them being the 1905 earthquake under study.

The monastic communities of Mt. Athos have an established tradition of nearly 1000 years in keeping archives of various types including marginal notes and memories regarding earthquakes felt in Athos. Such archives were collected and/or utilized in several papers and books aiming at understanding the seismicity of the area (Critikos 1932, 1933; Sieberg 1932; Maravelakis 1938; Galanopoulos 1960; Papazachos and Papazachou 2003; Guidoboni and Comastri 2005; Ambraseys 2009). In Mt. Athos peninsula, at least 15 strong earthquakes were felt since the thirteenth century. However, our knowledge about the locations of earthquake sources is limited since the archives only rarely provide relevant information. The rate of tsunami generation in the NAS area is very low as compared to that in the South Aegean, the Hellenic Arc and other active zones in the eastern Mediterranean (Papadopoulos and Chalkis 1984; Ambraseys 2009; Papadopoulos et al. 2014). Only very few local tsunami events have been historically reported in the NAS. A possible explanation is that the strike-slip faulting dominating the seismotectonics of the NAS area does not favor tsunami generation.

3. *Previous Studies About the 1905 Event*

From a few publications containing information about the 1905 earthquake and its impact, we learn that it caused great damage to several monasteries, e.g., the Iviron Monastery, eastern coast of Mt. Athos (Fig. 3) (Eginitis 1910; Critikos 1932, 1933; Sieberg 1932; Galanopoulos 1955; Kadas 1996; Papazachos and Papazachou 2003). In addition, due to rockfalls, monastic cells were ruined and some persons were killed. Damage was also caused in towns and villages of the area, while the earthquake was felt in various places across the Balkan peninsula as far as Athens (Greece), Sofia and the west coast of the Black Sea (Bulgaria), Bucharest (Romania), Adramition (NW Turkey) and Bari (eastern Italy) (Figs. 1, 2 and 3). Sieberg (1932), who dated the event at 9 November 1905, has been the only author who tried to draw isoseismals, although based only on very few intensity observation points. Voidomatis et al. (1990) determined rupture zones of strong earthquakes occurring within the Serbo-Macedonian composite terrane, including an arbitrary rupture zone of the 1905 event based on very few data.

The 1905 earthquake has been listed in several descriptive and/or parametric, global or regional earthquake catalogues (Table 1). However, there is confusion regarding the earthquake origin time. In most catalogues, the earthquake origin time is listed either as 21:41 GMT or as 22:06 GMT. Kárnik (1969) listed two large earthquake events occurring on 8 November 1905, one at 21:41 GMT and another at 22:06 GMT. An explanation may come out from the records of the Agamennone-type seismograph installed at the Athens station in 1898. This seismograph, which was an instrument of intermediate response period introduced by G. Agamennone, Professor of Seismology in Rome, recorded the first seismic wave arrival of the 1905 earthquake at 23:41:54 local time which should correspond to 21:41:54 GMT. This may explain why several authors adopted origin time at 21:41 (Table 1).

To explain why other authors adopted 22:06 GMT, we performed a careful examination of the Agamennone records for various Greek earthquakes listed in the Annals of the National Observatory of Athens (NOA) for the period 1900–1910. We found

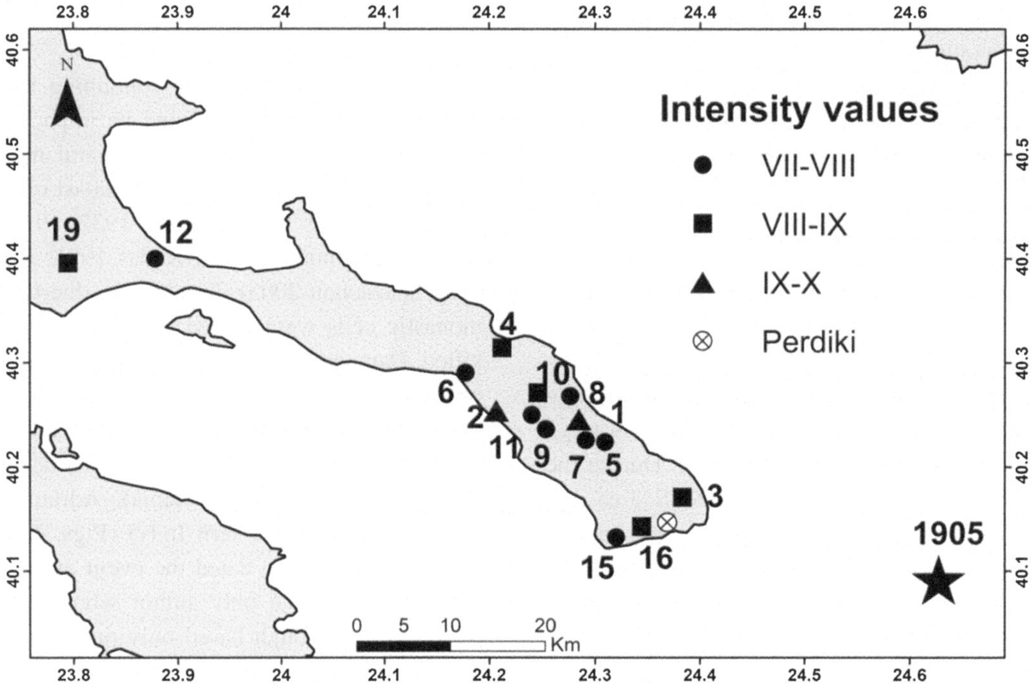

Figure 3
Macroseismic intensities in the 12-grade MM scale felt at sites of Mt. Athos peninsula due to the 8 Nov. 1905 earthquake. For intensity documentation and numbering of sites, see Table 3; star = earthquake epicenter. Earthquake parameters are from the ISC-GEM (2018) catalogue

that the time at which the Agamennone instrument was set up, according to the exact astronomical time at Athens, is equal to GMT + (1 h 34 min 53 s) (Eginitis 1910) instead of GMT + (2 h) which corresponds to the time zone of Greece. The instrumental sources collected by the International Seismological Centre (ISC) and used to analyze the 1905 event (see ISC-GEM 2018 catalogue) do not include records corresponding to an event dated 8 Nov. 1905, 21:41 GMT. However, European stations offer plenty of measurements for an event occurring at 22:06 GMT (D. Di Giacomo, ISC, personal communication, 12 December 2018). We concluded that the earthquake origin time, t_0, was 8 November 1905, 22:06 GMT. However, the local origin time should be shifted at t_0+ 2 h, that is at 9 November 1905, 00:06, which justifies the dating by Sieberg (1932). On the other hand, the astronomical origin time of the earthquake was 8 November 1905, 23:41:54 as determined by Eginitis (1910; see Table 1).

Various estimates of the 1905 earthquake epicenter are consistent with only small deviations from the most recent one calculated in the ISC-GEM (2018) catalogue (Table 1). Various catalogues, however, provided widely divergent magnitude estimates ranging between 6.8 and 8.3 depending on the different magnitude scales and estimation methods used. Aiming to contribute in resolving the ambiguity regarding the earthquake magnitude, we calculated a macroseismic magnitude as explained in detail in a following section.

4. *Documentary Sources*

To further study the 1905 earthquake and tsunami, we collected and evaluated a wealth of new information from a variety of documentary sources, most of which remained unknown to the seismological tradition so far. The sources collected include contemporary press reports, letters, manuscripts and marginal notes, as well as various articles and books containing descriptions about the earthquake, its impact and the associated phenomena. From the

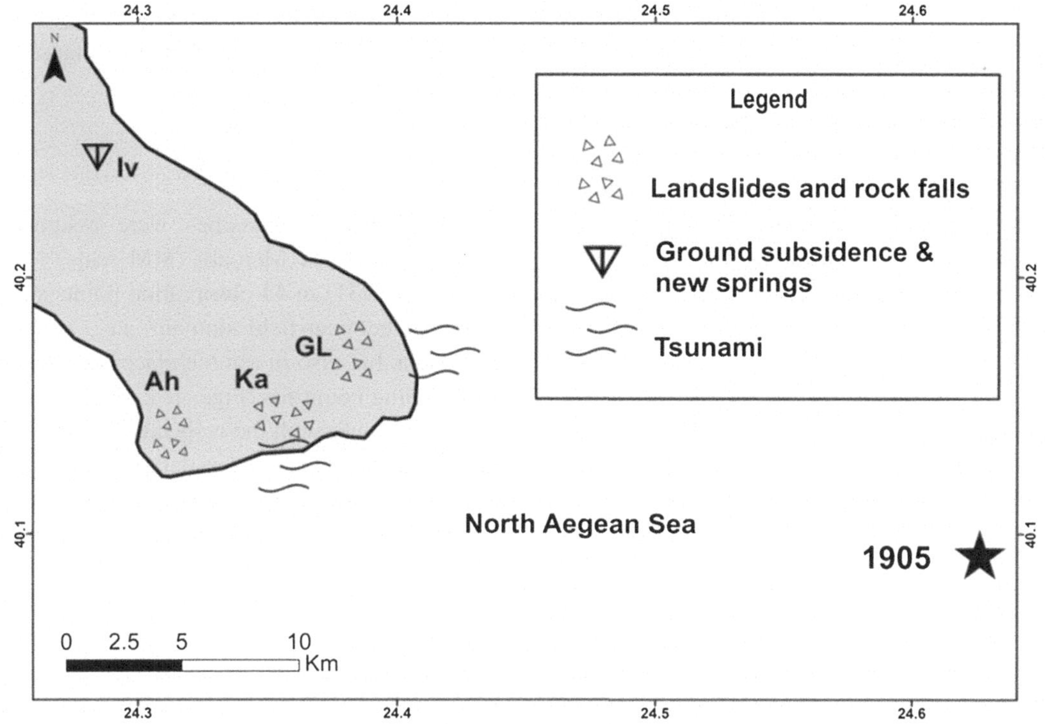

Figure 4
Main co-seismic ground failures reported in association to the 1905 Athos earthquake (star). Key for localities: *Iv* Iviron Monastery, *GL* Great Lavra Monastery, *Ka* Kafsokalivion Monastery, *Ah* St. Anna hermitage. The tsunamis reported offshore Ka and GL were in association to the 1905 and 1585 earthquakes, respectively

various sources collected, a database was organized to document the earthquake impact in both the built and natural environment (Table 3). Seismic intensities assigned in 44 observation sites are also included in Table 3 and mapped in Figs. 1, 2 and 3, while ground failures and other co-seismic phenomena reported are illustrated in a separate map (Fig. 4). In the next subsections, we present briefly the various kinds of documentary sources collected and summarize the most important findings. The full list of documentary sources and reproduction of relevant texts can be found in the Appendix. Dates according to Old Style (Julian) and New Style (Gregorian) calendars are noted by [O.S.] and [N.S.], respectively. The terms *arsanas* and *skiti*, which are in use in some of the documents examined, mean *small port* and *small monastery unit*, respectively. It is noticeable that the building infrastructure existing in Mt. Athos in 1905, when the large earthquake struck the area, was nearly the same as the infrastructure today, i.e., consisting of 20 main monastery building complexes and many other smaller monastic units (e.g., skitis), cells, churches and chapels.

We collected and examined a set of contemporary press reports (CPRs) from newspapers published in various cities mainly in Athens, the capital city of Greece, as well as in the Thessalia Province, Central Greece (Fig. 2). An unpublished manuscript called "Book of Earthquakes" found at NOA and written in the aftermath of the 1905 earthquake was also examined. Two manuscripts as well as a marginal note archived in Mt. Athos monasteries provided valuable information. The first manuscript (ATHMS-1), dated 1 [O.S.]/14 [N.S.] November 1905, was written by an eyewitness and has been published in a religious book by Rodostolos (2004) (BOOK-1). The author of BOOK-1 also performed an investigation of personal accounts among survivors of the 1905 catastrophic event. The second manuscript (ATHMS-2) was written in Mt. Athos during 1915, but still remains unpublished. The marginal note (MARN-1) found in a religious book was published by the historian Kadas (1996). Our collection of sources containing information about the 1905 earthquake also includes two relevant letters published by the eyewitness Bishop Chrysostomos Lavriotis (1905), the religious book by Kaloutsis (2000) (BOOK-2), an article (ART-1) published by the eminent architect Mylonas (1995) as well as his commentary inserted in the book of the Ukrainian monk Barksy (2009) (BOOK-3) who visited Mt. Athos during the eighteenth century.

5. *Earthquake Impact and Associated Phenomena*

According to the CPRs, the earthquake activity started around midnight of 26 October [O.S.] or 8 November [N.S.] 1905 and continued during the night as well as the next day. The manuscript ATHMS-1 says that aftershocks were being felt for at least 6 days. In the marginal note published by Kadas (1996), it is stated that the aftershock activity lasted for at least 6 months, which is anticipated for a shallow and large-magnitude earthquake event.

5.1. *Earthquake Impact in the Built Environment*

From the CPRs as well as by the ATHMS-1 manuscript, it results that the damage area was extended along the Athos peninsula (Fig. 3) where the Iviron Monastery was rendered nearly useless. Important damage was also noted in the monasteries of Megisti (Great) Lavra, Vatopedion, Karakallou, Philotheou, Stavronikita, Koutloumousiou as well as at the Russian monastery (see also ATHMS-2). In Karyes town, many cells collapsed. Less damage was caused in Ierissos town, where one church and 150 houses collapsed. In Gomati, St. Paraskevi and Sykia villages of Chalkidiki peninsula, many houses collapsed too (Fig. 2). In the article ART-1, Mylonas (1995) published a photograph dated either 1869 or 1870, as well as an aquarelle dated 1900, both illustrating the St. Michael Chapel that collapsed with the 1905 earthquake and was replaced by a new church in 1911. In his commentary inserted in BOOK-3 (Barksy 2009), P. Mylonas noted that the chapels of St. Michael and St. Paraskevi both situated in Great Lavra Monastery, were in good form according to Barsky's testimony during his visit to Mt. Athos in 1745. However, both collapsed because of the 1905 earthquake.

5.2. *Macroseismic Intensities*

Macroseismic intensities were assigned in the 12-grade modified Mercalli (MM scale) (Wood and Neumann 1931) in 44 observation points situated not only in the near-field domain, i.e., in Mt. Athos peninsula, but also in remote places of Greece and of surrounding countries (Figs. 1, 2 and 3). An intensity inventory along with the respective impact documentation can be found in the database of Table 3. For each observation point, we preferred to determine an intensity range consisting of two consecutive intensity values since most of the descriptions available are not precise enough to allow assignment of single intensity values. Damage caused by co-seismic ground failures, such as landslides and rock falls, were not taken into account for seismic intensity assignment for the reason that this kind of damage is not directly due to the Earth shaking.

From Figs. 1, 2 and 3, it becomes clear that the meizoseismal area with intensity up to IX–X degree was observed at the southern side of Athos peninsula. Away from Athos peninsula, little damage was reported only in localities of the Thessalia Province, Central Greece (Fig. 2), while the earthquake was just felt in many remote places in Greece, in other sites of the Balkan peninsula as well as in NW Turkey and SE Italy (Fig. 1). Far from Athos peninsula, the seismic intensities did not exceed VI–VII. The perceptibility radius was approximately 660 km, which is the maximum epicentral distance at which the earthquake was reportedly felt, i.e., in the city of Bari, Italy (Fig. 1).

5.3. *Co-Seismic Ground Failures and Tsunami*

A variety of co-seismic ground failures and other earthquake-associated phenomena were observed. The localities where the co-seismic phenomena were reported are illustrated in Fig. 4. According to various documentary sources, including CPRs, the manuscripts ATHMS-1 and ATHMS-2, the document LETT-1 and the BOOK-1 and BOOK-2, because of the seismic shaking, the mountain partly detached from a place lying uphill at a distance of at least 300 m above sea level and descended into the sea. At the same time, big stones rolled downhill. The massive landslide covered half a mile distance along the rocky coast near the Kafsokalivion Monastery (site no. 16, Fig. 3), particularly at the arsanas by the locality Perdiki (Fig. 3), at the southern side of Mt. Athos (Fig. 4). At the same locality, seven cells collapsed because of the co-seismic rock falls, while the skiti of Kafsokalivion filled up with stones that fell from above. The landslide crashed upon 13 persons on board fishing boats which submerged and were taken away by the sea that overtopped the arsanas with a run-up of 3 m. Very likely, only 2 out of 13 persons survived. Five out of 13 persons were standing in the arsanas when the event happened. One of the survivors, monk Isidoros by name, described his personal experience to the author of BOOK-1. From this account, we learn that Isidoros suddenly found himself in the sea, and while floating with the help of a wooden beam, a strong sea wave threw him violently up to the mountain cliff near Perdiki. Eventually, monk Isidoros survived since the water ebbed and was unable to drift him back to sea. He was also able to save another monk who was found jammed in the coastal rocks.

According to document LETT-1, eyewitnesses reported that with the Earth shaking, they felt a dull din and saw a flame coming out from the seashore, while the landslide was accompanied by smoke. One possible interpretation is that the flame was a case of co-seismic earthquake light, but no further evidence is available. One may not rule out, however, that flames were caused by the high friction exercised in rocks during the landslide process. The same document describes that sponge fishermen examined the sea bottom offshore of the arsanas and reported that up to a depth of ten yards, they observed oversized stones that either rolled down or were thrown from above. They claimed also that at the same depth, rocks were thrown from the sea bottom upwards, apparently due to the strong earthquake shaking. In the area of the Iviron Monastery, the appearance of new water springs as well as a local ground subsidence was reported by ATHMS-1 and LETT-1 documents, respectively (Table 3).

5.4. Human Victims

Based on all the evidence, humans were not victims due to building collapse, but only due to the massive landslides, the rock falls and the local tsunami. Four documents provided relevant information with the total number of victims estimated as equal to 9, 10 (in two documents) and 11. We have concluded that the total number of victims was very likely 11, and that 6 of them were killed by the landslide, while the other 5 were killed by the tsunami.

5.5. The Earthquake and Tsunami of 1585

Investigation of other earthquakes that caused tsunamis in the area of Mt. Athos indicated that the author of LETT-1 added the next comment: "*…Perhaps the uncountable rocky debris in Perdiki were originated either by an earlier earthquake or by the 1585 one….*" This is the case of the 18 [O.S.]/28 [N.S.] June 1585 earthquake which is mentioned in various documentary sources reviewed by Ambraseys (2009). The 1585 earthquake caused extensive damage in the eastern side of the peninsula, such as in the Monastery of Great Lavra. Ambraseys (2009) quoted an English translation of a Greek document which indicates that a local tsunami was generated. We found the original Greek text of this account which is contained in Codex Ω 147 of the Great Lavra Monastery and published in a religious journal by Bishop Gennadios (1917). A more faithful English translation of the relevant passage reads as follows: "[after the earthquake]…*the sea grew rough and then drew back up to the mouth of the port….*" Very likely, this happened at the arsanas of the Great Lavra Monastery (Fig. 4), which is not coincident with the arsanas near Perdiki where the 1905 tsunami was observed. However, no other relevant information is available to us about the 1585 tsunami.

6. Magnitude Determination

The magnitude determination of earthquakes occurring in the very early era of instrumental seismology is quite challenging. For the 1905 Athos event, based on a variety of techniques, several magnitude determinations range from 6.8 to 8.3 (Table 1). From trace amplitudes of surface-wave maxima recorded by undamped Milne seismographs contained in the original worksheets for Seismicity of the Earth (Gutenberg and Richter 1954), Richter (1958) was able to calculate surface-wave magnitude, M_s, for global earthquakes. For the 1905 earthquake, he found $M_s = 7.8$, which later was also adopted by Duda (1965). Abe and Nogushi (1983) used additional independent data sets, performed calibration tests for the effective gains of Milne seismographs and found that for the period 1904–1906, Richter's (1958) magnitudes are systematically overestimated by an average excess of 0.5 from the real M_s. For the 1905 earthquake, the magnitude recalculated by Abe and Nogushi (1983) is as low as M_s= 6.8. Galanopoulos (1960) adopted magnitude of $\sim$ 7, but it remains unclear how this value was derived and which magnitude scale was used. Kárník (1969) also endorsed magnitude $\sim$ 7 for an event occurring on 8 November 1905, 21:41 GMT, but he provided a magnitude 7.5 for another event occurring in the same area at 22:06 GMT of the same date. As already explained earlier, we have good reason to believe that Kárník (1969) is very likely referring to a single event. The magnitude 7.5 listed by Kárník (1969) was later considered as M_s and uncritically inserted in several Greek earthquake catalogues (e.g., Galanopoulos 1981; Comninakis and Papazachos 1982; Papazachos and Papazachou 1989, 1997, 2003). Shebalin et al. (1974) compiled an earthquake catalogue for the Balkan region and by combining estimations from previous sources assigned $M_s = 7.4$. This magnitude was adopted by Makropoulos et al. (2012) who also included a second magnitude estimation of $M_w = 7.1$, the source of which is likely the ISC catalogue. From M_s/logM_o empirical relationships, Ambraseys (2001) calculated a seismic moment log M_o= 26.93 and M_s= 7.24. As regards the ISC-GEM (2018) catalogue, an earthquake magnitude of $M_s = 7.2 \pm 0.1$ was initially calculated on the basis of five station records: POT, OSA, JEN, LEI, GTT (D. Di Giacomo, ISC, personal communication, 17 Sept. 2018). Then, from empirical M_s/M_w relationships found by Di Giacomo et al. (2015), M_w proxy equal to 7.24 ± 0.1 was calculated.

For the magnitude determination from macroseismic observations, we used empirical magnitude-intensity relationships. A variety of such relationships have been investigated based on data sets from instrumental earthquakes (e.g., Gutenberg and Richter 1954; Galanopoulos 1961; Kárnik 1969; Street and Turcotte 1977; Nuttli et al. 1979; Tinti et al. 1986; Sibol et al. 1987; Ambraseys 1988; Bakun and Wentworth 1997; Gasperini and Ferrari 2000; Gutdeutsch et al. 2002; Bakun and Scotti 2006). The general form of magnitude-intensity relationships can be expressed as

$$M = f(\text{Int}) \tag{1}$$

where M is magnitude and f(Int) is a function of the macroseismic intensity field. The element Int could be either the maximum (epicentral) intensity, I_o or the area, A, within the contour of intensity of i degree or a function of both I_o and A.

For Greece, several papers have been published on this issue (e.g., Galanopoulos 1961; Kárnik 1969; Drakopoulos 1978; Ambraseys 1988; Papazachos and Papazachou 2003; just to mention a few). Papazachos and Papaioannou (1997) elaborated 356 macroseismic maps of the Balkan area and produced empirical relationships for intensity attenuation and macroseismic magnitude determination. For the 1905 event, Papazachos and Papaioannou (1997, their Table 1) found macroseismic magnitude M_m= 8.3 from the macroseismic field modelling. However, Trifunak (1998) argued that Papazachos and Papaioannou (1997) introduced possible biases in their analysis resulting from the inhomogeneous database, conversion of local intensity scales and differences in regional geology (for a reply, see Papazachos and Papaioannou 1998).

A serious problem is the consideration of the epicenter location, particularly in the physiographic province of Greece which is characterized by a mainland with a highly complex coastal configuration surrounded by hundreds of islands (Fig. 1). For example, the establishment of empirical magnitude-intensity formulas is strongly influenced by the geographical positions of the epicenters either on land or offshore. The issue regarding historical epicenters at sea or on land was discussed by Guidoboni and Ebel (2009, p. 322). To overcome such difficulties,

Papadopoulos (2011) advanced further an earlier effort by Fokaefs et al. (2004), used magnitude and intensity data for shallow Greek earthquakes for the time period of 1911 up to 2005 and, through linear regression testing, introduced empirical formulas of the type

$$M = a + bI_o \quad (2)$$

for earthquakes occurring on land, in coastal areas and in the open sea, where a and b are coefficients determined by the data. Papadopoulos (2011) found that the best fit between M and A, after validation with data of the period 2006–2011, is provided by simple linear formulas of the form

$$M = a + b\log A_i \quad (3)$$

A set of 18 such formulas were produced for intensity degree i ranging from IV to VIII in the modified Mercalli (MM) intensity scale for earthquakes occurring on land, in coastal areas and offshore. Coastal area means a zone of less than 20 km on either side of the shoreline. Magnitudes of tens of historical earthquakes occurring along the Hellenic Arc were calculated with the application of formulas (2) and (3) (Papadopoulos 2011).

A suitable subset of eight formulas, 2 of type (2) and 6 of type (3) (Table 2) were applied for the magnitude determination of the Athos 1905 earthquake. Although all the earthquake catalogues are consistent in placement of the 1905 epicenter offshore, we tested relationships for both the offshore and coastal domains, thus allowing for errors involved in the epicentral determinations. I_{max} was taken as high as IX or X (Table 3). As regards formulas of type (3), they were applied only when at least five observation points of the respective intensity value were available. A formula with area A_V has not been used since less than five intensity points of V degree were available.

All the macroseismic magnitudes calculated (Table 2) are equivalent to M_s. The average of the eight values obtained in Table 2 is equal to M_m= 6.71 ± 0.32. If, however, we consider only formulas with correlation coefficient $r^2 > 0.5$, then we get M_m= 6.83 ± 0.31. On the other hand, one may observe that for formulas of the coastal domain magnitudes obtained are lower than the ones obtained for the offshore domain. This is an additional indication that the epicenter was very likely placed offshore. Then, by considering only the two magnitudes calculated from formulas with $r^2 > 0.5$ for the offshore domain, we get average M_m = 6.99 ± 0.51, which is our preferred macroseismic magnitude for the 1905 Athos earthquake.

7. Tsunami Simulation

7.1. Tsunami Source

The historical documents analyzed earlier leave no doubt that a cascade effect occurred with the 1905 earthquake, which triggered mass land collapses turning to tsunami generation along the Mt. Athos southern coast. Due to the lack of instrumental records, the landslide and the tsunami events have

Table 2

Macroseismic magnitude, M_m, calculated from empirical relationships established from Greek macroseismic data (Papadopoulos 2011)

No.	Empirical relationships	r^2	I_{max}	log (A_i)	M_m, mean (M_m)
1	C: M_m = (4.125 ± 0.256) + (0.250 ± 0.03)·I_{max}	0.412	IX/X		6.38/6.63, 6.50
2	O: M_m = (4.553 ± 0.692) + (0.233 ± 0.08)·I_{max}	0.345	IX/X		6.65/6.88, 6.77
3	C: M_m = (2.917 ± 0.363) + (0.665 ± 0.033)·log (A_{IV})	0.491		4.99	6.24
4	O: M_m = (0.653 ± 0.921) + (1.197 ± 0.196)·log (A_{IV})	0.773		4.99	6.63
5	C: M_m = (3.799 ± 0.243) + (0.631 ± 0.06)·log (A_{VI})	0.571		4.74	6.79
6	O: M_m = (2.607 ± 0.919) + (1.001 ± 0.221)·log (A_{VI})	0.804		4.74	7.35
7	C: M_m = (4.265 ± 0.236) + (0.604 ± 0.07)·log (A_{VII})	0.552		3.78	6.55
8	C: M_m = (4.453 ± 0.703) + (0.638 ± 0.228)·log (A_{VIII})	0.663		3.70	6.81

Key: *C* coastal, *O* offshore, I_{max} maximum intensity in MM scale; A_i area (in km^2) within the contour of intensity of i degree, with i taking values IV and VI, VII, VIII (for explanation, see the text); r^2 correlation coefficient

Table 3

Impact of the 8 November 1905 earthquake based on various documentary sources

No.	Observation site	Lat. N^0	Long. E^0	Impact on humans and built environment	Ground failures	MM	Δ (km)	Source
1	Iviron Mn.	40.24554	24.284648	Rendered nearly uninhabited		IX–X	31	6
				Many buildings and monastic cells partly destroyed				12
				Nearly all the rooms rendered uninhabitable, the Cathedral fissured in many points, three chapels collapsed. The bell tower, the oven, the synodiko and the fiali all suffered great fissures				5
				Rendered nearly uninhabitable. Many parts of the mn. collapsed or fissured including the church of Assumption of Mary	In the mn. courtyard, the ground subsided			8
				Intensity of X degree in Rossi-Forel scale				14, 15
					Water springs appeared			8
				Three chapels collapsed and all the rooms rendered useless	The ground around the mn subsided in many places			7
					In St. Peter hermitage, the area became susceptible to subsidence			11
2	Russian Mn.	40.2419	24.196793	Great damage		IX–X	38	3
3	Great Lavra Mn.	40.170869	24.383114	Important damage		VIII–IX	20	6
				Part of the wall and some of the rooms collapsed				5
				Damage				8
				Many buildings and monastic cells partly destroyed				12
				St. Michael Chapel destroyed entirely				5
					Big stones were thrown on cells in Palamas locality			11
4	Vatopedion Mn.	40.314242	24.212042	Important damage		VIII–IX	41	6
				Great damage				4
				Damage				5
5	Karakallou Mn.	40.223885	24.309577	Important damage		VII–VIII	28	6
				Damage				5
				One cell collapsed				8
				Damage in Provata cells				11
6	Kastamonitou Mn.	40.288241	24.173947	Many buildings and cells partly destroyed		VII–VIII	42	12

Table 3 *continued*

No.	Observation site	Lat. N^0	Long. E^0	Impact on humans and built environment	Ground failures	MM	Δ (km)	Source
7	Filotheou Mn.	40.225715	24.291246	Important damage		VII–VIII	30	6
				Damage				5
8	Stavronikita Mn.	40.268052	24.276671	Important damage		VII–VIII	33	6
				Damage				5
9	Koutloumousiou Mn.	40.253232	24.247799	Damage		VII–VIII	35	5
				Important damage				6
10	Russian skiti, St. Andreas	40.26245	24.244964	Important damage		VIII–IX	35	6
11	Karyes town	40.25728	24.245261	Many cells collapsed		VII—VIII	35	6
				Great damage in the church				4
				Small fissures, chimneys fell				11
12	Ierissos town	40.399708	23.878219	150 houses and the church collapsed		VII–VIII	70	6
13	Kafsokalyvion Mn.	40.130491	24.340208	Great damage from the big stones detached	Big stones detached			4
					Stones detached from the mountain and crashed 7 cells, others damaged seriously by the mountain that landslided into the sea. The sea rose and overtopped the local port (arsanas). Fishing boats crushed. Only 3 out of 13 fishermen survived			5
					Many cells collapsed by the falling stones			12
	Kafsokalyvion skiti			Great damage by the falling stones	Filled up by stones that fell from above			11
					Falling stones			8
	St. Anna hermitage	40.140565	24.296628	Many cells collapsed by the falling stones	Falling stones			12
				Great damage from the big stones detached	Big stones detached			4
				The same happened as in Kafsokalyvia	Stones fell from the mountain			11
	St. Basil hermitage	40.126931	24.308232	Many cells collapsed by the falling stones	Falling stones			12
				A great rock rolled down, crushed the cell of a monk and killed him	A great rock rolled down			10
	Romanian St. John hermitage	40.132899	24.35398	Many cells collapsed by the falling stones	Falling stones			12
	St. Pantes (All Saints) cells	40.276518	24.165855	Ten monks crushed alive and killed by the stones. The same happened to another monk in Katounakia.	Falling stones			5
					Din was felt, flame was seen coming out from the seashore			11

Table 3 *continued*

No.	Observation site	Lat. N^0	Long. E^0	Impact on humans and built environment	Ground failures	MM	Δ (km)	Source
14	Perdiki cells	40.13049	24.340209		At sea depth up to 10 yards, very big stones fallen from uphill were observed. Many big stones were thrown from the sea bottom upwards			11
					11 people were on board in 6 fishing boats. All of them but one were submerged along with the boats which were taken away by the sea which rose up to 3 m. The mountain uphill was thrown down covering a length of half a mile			8
					7 monks and 2 seculars, totally 9 persons, were crushed and sunk because part of the mountain detaching and being thrown into the sea			12
	Perdiki coast				10–11 people in 3 fishing boats were killed			4
15	Prodromos cells	40.132721	24.320271	Buildings damaged, the church greatly fissured		VII–VIII	26	11
16	Romanian skiti and church	40.132892	24.353975	Very important damage		VIII–IX	23	9
17	Kerasia, Panagia, Koryfi cells	40.134536	24.316965	No damage was observed				11
18	Mylopotamos cells	40.238693	24.302259	No damage				11
19	Gomati village	40.395425	23.795012	Many houses collapsed		VIII–IX	76	6
20	Sikia village	40.041488	23.942826	Damage		VI–VII	57	6
21	St. Paraskevi village	39.952993	23.608222	Many houses damaged, school collapsed		VI–VII	86	6
22	Thessaloniki	40.630828	22.947301	Felt		II–III	155	6
23	Serres	41.087457	23.547368	Felt		II–III	144	6
24	Drama	41.150919	24.145474	Felt		II–III	123	6
25	Giannitsa	40.793082	22.411469	Felt		II–III	204	6
26	Katerini	40.271331	22.508738	Felt		II–III	182	6
27	Stromnitsa	41.436634	22.637318	Felt		II–III	224	6
28	Ano Tzoumagia	42.020632	23.099298	Felt		II–III	249	6
29	Melenikon	41.523597	23.395084	Felt		II–III	192	6
30	Nevrokopi	41.340901	23.868212	Felt		II–III	149	6
31	Skyros Isl.	38.845331	24.536968	Intensity degree IV (RF)		IV	134	13
32	Thassos Isl.	40.778085	24.706111	Felt		II–III	75	6
33	Alli-Meria	39.370475	22.982738	Intensity degree IV (RF)		IV	161	12
34	Anacassia	39.382018	22.971949	Intensity degree IV (RF)		IV	161	12
35	Volos city	39.359289	22.94415	Church bells rang due to Earth shaking, damage in houses		V–VI	165	1
				Strong?				12

Table 3 *continued*

No.	Observation site	Lat. N^0	Long. E^0	Impact on humans and built environment	Ground failures	MM	Δ (km)	Source
36	Larissa	39.642305	22.417668	Less intense compared to Volos		V–VI	194	2
37	Ayia town	39.715719	22.754729	Strongly felt, but no damage noted		V–VI	163	2
				Intensity degree IV (RF)				12
38	Velestino town	39.379106	22.745491	Intensity degree IV (RF)		IV	178	12
39	Trikala city	39.554273	21.767675	Intensity degree IV (RF)		IV	251	12
40	Keramidion village	39.572722	22.910786	Intensity degree IV (RF)		IV	156	12
41	Androniani	38.614065	24.05654708	Intensity degree IV (RF)		IV	171	12
42	Athens city	37.973381	23.718203	Felt		II–III	253	13, 14
43	Chalkis city	38.463758	23.596543	Intensity degree (IV RF)		III–IV	202	12
44	Sofia city (Bulgaria)	42.695793	23.332544	Felt		II–III	310	13, 14
45	Black Sea, Varna, Bulgaria	43.161081	27.919755	Felt		II–III	438	13, 14
46	Bucharest city (Romania)	44.408381	26.103446	Felt		II–III	497	13, 14
47	Bari city	41.131028	16.870051	Felt		II–III	660	13, 14
48	Adramition town	39.576583	26.944354	Felt		II–III	213	13, 14

Mn./mn. monastery, *MM* intensity in MM scale, *RF* Rossi-Forel scale, *Δ* epicentral distance; *Lat and Long* are geographical latitude and longitude of the observation site

been documented only in historical accounts. Consequently, for the tsunami numerical simulation, one has to rely only on the historical reports and on elements of the present morphology of the area. The tsunami occurrence has been reported from only one observation point, which is the arsanas, the small port situated close to the site Perdiki at the southern tip of the peninsula. According to the eyewitness accounts, the local tsunami was generated from the co-seismic landslide. This provides evidence that very likely no tectonic tsunami was produced in association with co-seismic fault displacements. This view is also supported by the fact that no tsunami was reported from other coastal places, meaning that a local tsunami was generated, a typical feature of landslide-generated ones. For these reasons, we have considered a landslide scenario for the tsunami generation.

7.2. Numerical Techniques

The numerical scheme adopted consists of two basic steps: simulation of the falling mass and simulation of the ensuing tsunami. The former simulation is performed by means of the code UBO-BLOCK1, developed by the University of Bologna tsunami research team. It assumes a Lagrangian approach, dividing the mass into deformable portions and allowing one to compute the time-changing thickness of the mass during the motion. The code needs as input the undisturbed sliding surface, the initial top surface of the sliding body, the path followed by the blocks' center of mass (CoM hereafter), and the lateral boundaries accounting for the slide's lateral spreading. A full description of the model can be found in Tinti et al. (1997).

The tsunamigenic impulse, i.e., the vertical water displacement induced by the landslide at the sea surface, is evaluated by an intermediate code, UBO-TSUIMP. In practice, the sea bottom vertical change due to slide motion is transferred to the sea surface by a filtering function that depends on the local water depth and cuts the higher frequencies. Further, a space-time interpolator maps the sliding dynamics on the nodes of the tsunami's computational grid.

The propagation of the tsunami, which is the second step of the numerical scheme, is simulated by solving the shallow-water equations, through a finite-difference technique that implements the leap-frog scheme with the staggered-grids approach. The impulse given by the slide enters as a perturbation term in the continuity equation. On the grid boundaries, a full transparency condition is implemented in the open sea, while the interaction with the coastal boundary can be tuned. If the user aims at assessing land inundation, as is the case for the simulations run in this work, a moving boundary condition is adopted. Instead, if the no-inundation option is selected, a total reflection condition is imposed. All these tasks are implemented in the code UBO-TSUFD, described in details by Tinti and Tonini (2013). The above numerical procedure has been tested and validated in numerous cases of landslide-induced tsunamis. For some examples, one can refer to Tinti et al. (2011), Zaniboni et al. (2014a, b, 2016, 2019) and further references therein.

7.3. Computational Grid

The numerical grid to treat the case under study has been built by using bathymetric and topographic datasets. Mapping data (scale 1:50,000) produced by the Geographical Survey of the Greek Army have been used to cover the Mt. Athos peninsula on land. As for the bathymetry, due to the lack of more detailed data, we used the EMODnet database (http://www.emodnet.eu/) which provides bathymetry of grid resolution of ~ 115 m. In addition, the shoreline has been extracted by the public database GSHHG (https://www.ngdc.noaa.gov/mgg/shorelines/). The grid obtained by merging the different datasets is shown in Fig. 5. It covers a 15 × 10-km area with 20 × 20-m regular cells, for a total of more than 370,000 nodes. The grid resolution is higher than that of the original topo-bathymetric datasets, the refinement being the result of a linear interpolation process.

7.4. Landslide Scenario Definition and Simulation

The landslide that, according to the historical accounts, caused the local tsunami happened in the area of the arsanas shown in Fig. 5. On the basis of the present morphology as well as of satellite photos, two possible source areas have been hypothesized for the tsunamigenic landslides. The first, marked by the red line in Fig. 5, is placed about 1 km east of the arsanas. In Google Earth images, one may observe an evident non-vegetated scar, more than a half kilometer wide. Scar 1 extends from the seashore up to about 580 m above the sea level. It is of interest to recall that, according to the documentary sources, the mountain fell down, covering an area of half a mile, while the massive landslide occurred in the mountain cliff up to at least 300 m above the sea level. Therefore, one may argue that scar 1 is the result of multiple and progressive sliding episodes, a major one very likely occurring with the 1905 earthquake. The second area is a smaller scar with exposed rock bed located ~ 1.7 km west of the arsanas, marked by a blue line in Fig. 5, and affecting the mountain flank up to ~ 200-m altitude.

The first landslide scenario (S1) regards the motion of a mass that we obtained by "filling" the scar 1 area from the sea level up to 300 m a.s.l. as depicted by the purple boundary in Fig. 6, upper panel. The consideration of an altitude of 300 m is consistent with the eyewitness's accounts. The landslide initial thickness attains a maximum of 30 m and is portrayed in Fig. 6 in a yellow–red scale. The reconstructed volume is ~ 2.7 million m^3, covering an area of 0.28 km^2 for a mean thickness of almost 10 m. The landslide deposit at the end of the simulation may be seen also in Fig. 6, upper panel, as the light green-dark green area. The slide thickness reduces considerably to a maximum of 12 m. After ~ 20 s from the initiation of the process, the slide attains a stable high-velocity plateau of over 32 m/s for about 40 s and then decelerates progressively. The blue lines in Fig. 6, lower panel, represent the velocities of the block CoMs vs. time. Since individual blocks show similar velocity time histories, one can conclude that the mass maintains a general coherence during the descent. The motion stops after 150 s, when the mass settles at a sea depth of more than 700 m, with a runout exceeding 4 km.

The second landslide scenario (S2) regards the area of scar 2 (Fig. 5). The landslide consists of a body of 160×10^3 m^3 in volume, with maximum thickness of about 20 m. The slide dynamics is quite similar to case S1. The mass stops between 600 m and 700 m sea depth, reaching the maximum velocity

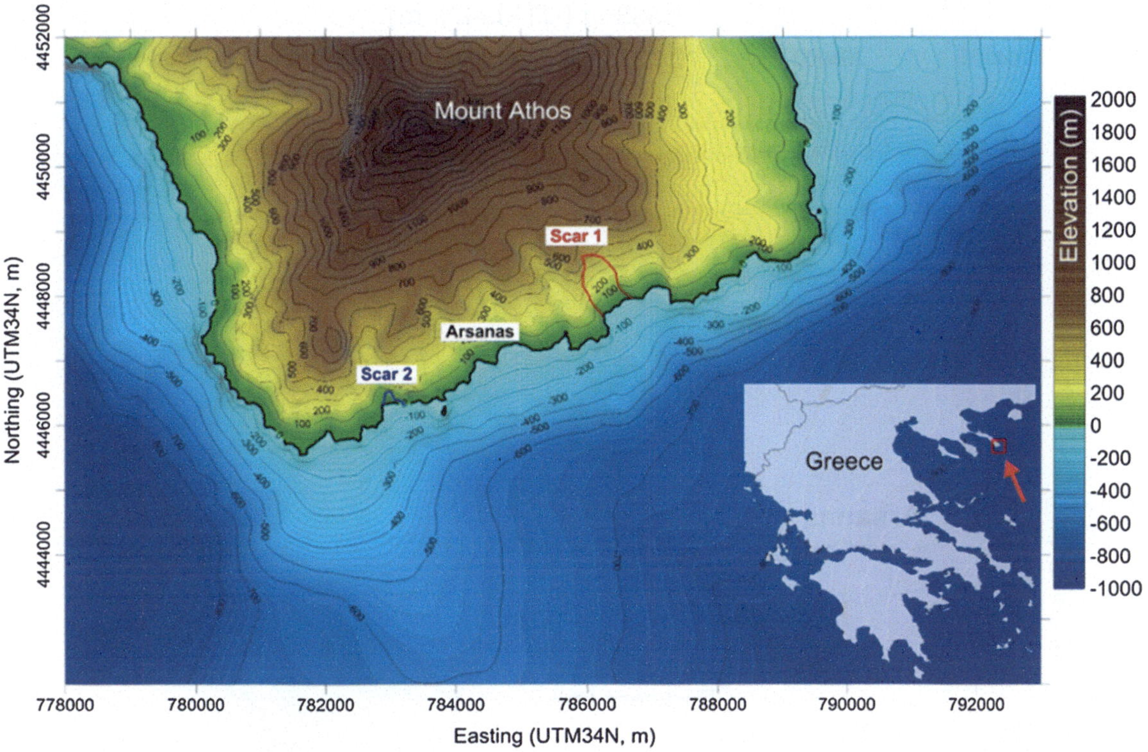

Figure 5
Southern end of the Mt. Athos peninsula and surrounding sea, where the tsunami was reported. The two possible landslide source areas are marked in red (scar 1) and blue (scar 2)

of 25 m/s. However, the mass of S2 is of minor importance as compared to the mass of S1.

7.5. Tsunami Simulation

The tsunami propagation was simulated over the computational grid depicted in Fig. 5. A wider domain has not been considered since tsunami waves generated by moderate-volume landslides usually rapidly vanish with distance (e.g., Papadopoulos and Kortekaas 2003). Figure 7 shows the maximum water elevation obtained at all times from the landslide case S1 for each point of the simulation mesh. So, it is considered that the highest effects of the tsunami are limited to the southern coast of Mt. Athos peninsula, while the eastern coast is only partially affected, at its southern end, by appreciable waves.

The general pattern of the maximum wave height distribution in Fig. 7 shows that most of the tsunami energy directs southeastward. Here, the maximum waves exceed 5 m. In the small bay denoted as the Arsanas, the only observation point where historical descriptions are reported, the water reaches everywhere more than 2 m, and, in some points, 3 m, which means that the hypothesized scar 1 source generates waves compatible with the observations. Moving westward, the tsunami effects reduce considerably, remaining, however, above 1 m in height.

Figure 8 shows the synthetic marigrams computed on four nodes along the coast, west of scar 1, related to scenario S1. Node #1 is the closest to the landslide and registers a first positive oscillation of amplitude ∼ 4 meters, followed by a comparable trough. After that, the oscillations rapidly drop down and within 5 min, the sea level perturbation seems to vanish. Node #2 is located inside a small inlet; here, the wave is much lower, reaching 1 m at maximum, but the oscillations persist longer, at least for about 7 min, probably because of reflection effects. The small bay of the Arsanas (node #3) is reached by the tsunami in slightly more than 1 min, and is affected

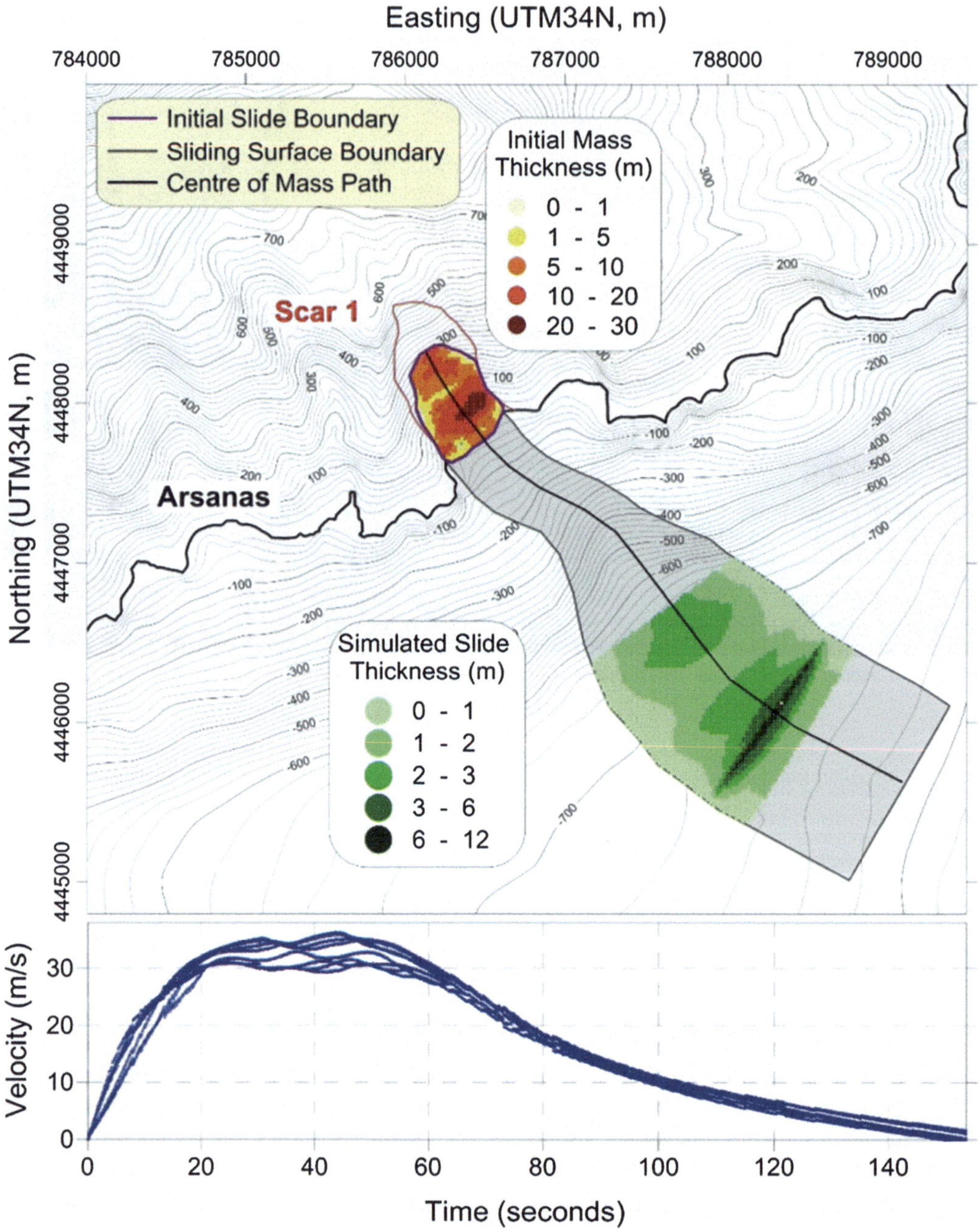

Figure 6

Upper panel: initial sliding mass assumed for case S1 (area within the purple boundary inside the scar area delimited by the red line, and slide thickness given in the yellow–red scale). Predefined center of mass (CoM) trajectory and lateral boundaries are drawn as black and grey curves, respectively. The final deposit thickness is represented by the different scales of green, from light to dark. Lower panel: block CoM velocity vs. time

by a series of four relevant waves, the third of which reaches almost 2 m. The small discrepancy with the 3-m waves reported by eyewitnesses can be easily explained by the fact that the virtual tide gauge is located in the water, at about 2-m water depth, and not on the coast where the waves were observed. Besides, we should also consider the uncertainty involved in the eyewitness's accounts. It is also

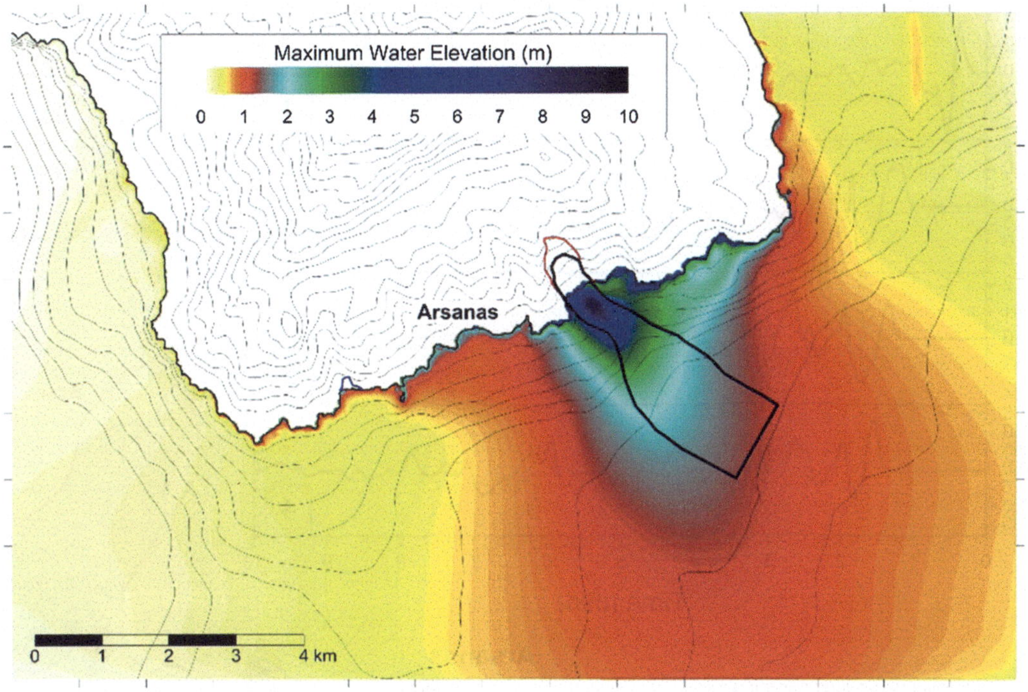

Figure 7
Maximum water elevation for scenario S1. The red line marks the scar 1 boundary. The black boundary evidences the sliding surface

interesting to estimate the wave period. The crest-trough-crest sequence is approximately 40 s long. Finally, node #4 is placed west of the Arsanas; here, the first wave is the highest, exceeding 1 m, followed by a series of crests and troughs for almost 7 min.

As concerns scenario S2, the generated tsunami effects are limited to a small coastal stretch, where waves were at most about 1 m high. In the Arsanas, tsunami amplitudes are only a few tens of cm, and can be considered negligible if compared to the available historical documentation.

8. Discussion

The 8 November 1905 earthquake is an early instrumental event and, therefore, its study becomes quite challenging. However, because of its large magnitude and the cascade effects caused, this earthquake is a key event for better understanding not only the seismicity and seismotectonics of the central part of the North Aegean Trough (NAT), but also the level of earthquake and associated hazards in the Mt. Athos and adjacent areas.

There is a general consensus that the earthquake epicenter was placed in the central segment of the NAT to the south of Mt. Athos peninsula. In this study, we have adopted the most recent epicentral determination by ISC-GEM (2018) (Table 1). In the instrumental period, this area was ruptured by two large earthquakes occurring on 18 January 1982 and 6 August 1983 with magnitudes $M_w = 6.56$ and $M_w = 6.65$, respectively, with their epicenters located to the south of the one adopted for the 1905 earthquake (Papadopoulos et al. 2002; ISC-GEM 2018) (Fig. 2). Fault plane solutions imply that both events were associated with dextral strike-slip motion (Rocca et al. 1985). We may suggest that the 1905 earthquake was also associated with such faulting type which is the dominating tectonic style along the entire NAT. However, one may not exclude the case that the earthquake occurred along the transtensional structure of the Athos Fault Zone (AFZ) trending nearly ENE–WSW and recognized on seismic lines along the southern edge of the Mt. Athos peninsula (Roussos and Lyssimachou 1991) (Fig. 2). Koukouvelas and Aydin (2002) and later Caputo et al. (2013) denoted the AFZ as active; its length is more than

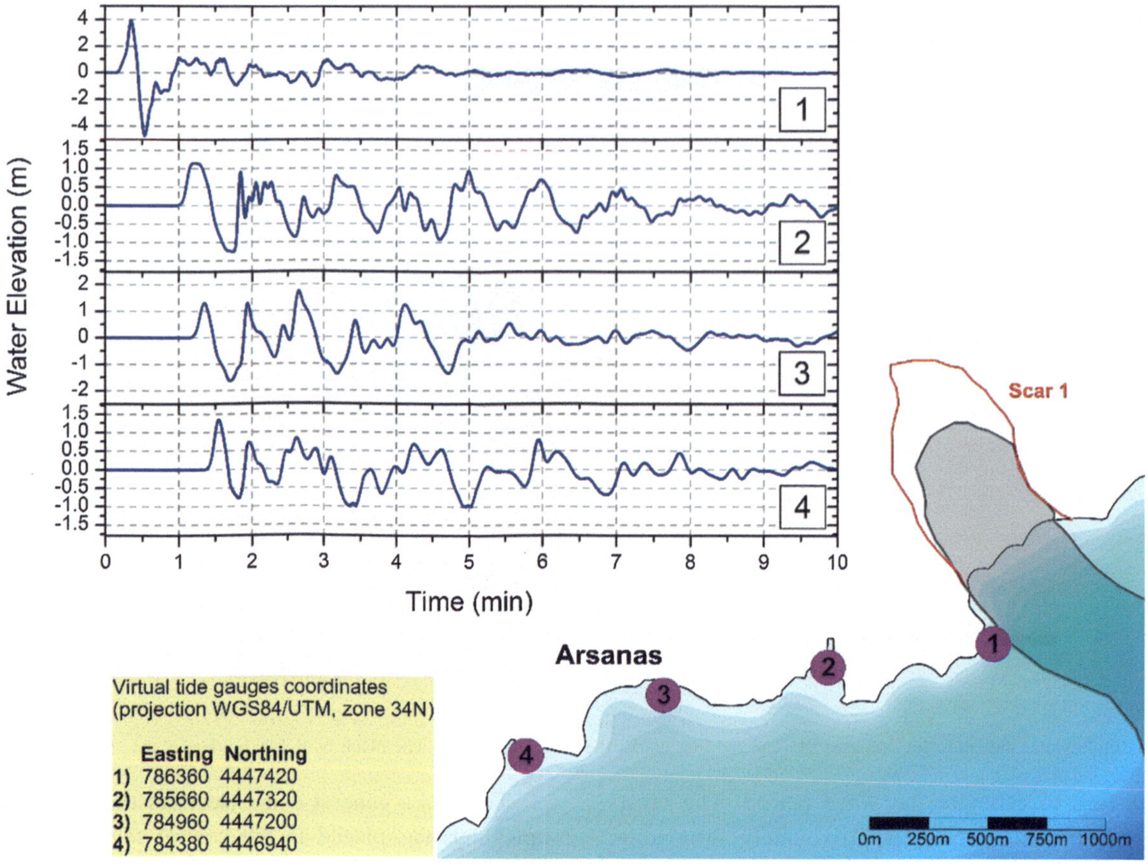

Figure 8
Virtual tide gauges at the nodes 1, 2, 3 and 4 to the west of scar 1, and synthetic tide gauge records obtained for the tsunami generated by landslide scenario S1. Grey area represents the sliding surface

70 km and affects the sea bottom either by positive or negative flower structures. In the AFZ, the predominant fault motion is dextral strike-slip too. An alternative is that the 1905 causative fault was placed offshore the eastern side of Mt. Athos where a possibly active fault zone has been recognized from seismic lines (Papanikolaou et al. 2006). The normal faulting detected there belongs very likely to the nearly N–S tensional field characterizing the Serbo-Macedonian terrane. This solution, however, requires a drastic shift of the epicenter location by at least 30–40 km to the north.

The earthquake size determination is still debated. Various techniques were used, and the magnitude estimates range widely from 6.8 to 8.3. We calculated macroseismic magnitude as high as ~ 7 in scale equivalent to M_s. This magnitude is close to the one of ~ M_w= 7.24 determined by reliable techniques on the basis of instrumental records (Ambraseys 2001; ISC-GEM 2018).

The new documentary sources that we collected and analyzed leave no doubt that the earthquake caused lethal associated phenomena, i.e., massive landslides and rock falls. A flame that was reportedly seen to come out from the sea and occurring with the earthquake may imply that either an earthquake light was observed or a fiery phenomenon was generated due to the high mechanical friction during the landslide process. The landslide episode at the southernmost side of Mt. Athos triggered a local but powerful tsunami. Although the NAS is of low potential for tsunami generation (e.g., Papadopoulos et al. 2014), landslides should not be neglected as a potential source of local tsunamis (Janin et al. 2018). However, this is the first time that such remarkable cascade effects are revealed in Mt. Athos, posing an important threat for the monastic community and the thousands of people visiting the area every year. Relevant evidence comes also from the documentation regarding the 28 June 1585 large earthquake and its associated local tsunami in Mt. Athos regardless that the tsunami generation mechanism is not known yet.

As regards the tsunami wave observed in association with the 1905 large earthquake, despite the several uncertainties involved, which include the historical reconstruction, the lack of detailed data, the arbitrariness of the slide shape and the intrinsic approximations and simplifications of the numerical models, the scenario of landslide S1 involving the scar east of the arsanas, situated at the southern tip of Mt. Athos, appears as a very good candidate source for the wave generation. The simulated waves are compatible with the existing historical descriptions, meaning that a relatively thin layer of material, mobilized by seismic shaking, was able to set up sea waves affecting a small area around the source. Moreover, our landslide simulations suggest that the mass deposits might be found at a sea depth of about 700 m, providing interesting hints for future bathymetric surveys in search of evidence for the 1905 event.

9. Conclusions

The large earthquake occurring on 8 November 1905 offshore of the south of Mt. Athos, North Aegean Sea, Greece, was an event that shook a large part of the Balkan peninsula. From the collection of a variety of documentary sources, several of them being unknown so far to the seismological community, we were able to reconstruct the earthquake's macroseismic field and organize a database of the earthquake's impact. The earthquake caused extensive damage in several monasteries, mainly along the southern side of Mt. Athos peninsula as well as in towns and villages in nearby areas. The maximum intensity in 12-grade MM scale was estimated as high as IX–X degree. Magnitude estimates published by previous authors are quite divergent, but the most reliable ones based on instrumental records indicated $M_s \sim M_w \sim 7.24$. Our preferred macroseismic magnitude determination is equal to $M_s \sim 7.0$.

From the new documentary sources collected, we concluded that important cascade effects were caused by the 1905 earthquake. Namely, the earthquake produced landslides and rock falls that moved massively from an altitude of at least 300 m downhill to the sea covering a distance of nearly 1 km along the coast. This turned to the generation of a local but powerful tsunami of $\sim$ 3 m in height in the arsanas (small port) of the locality Perdiki at the Kafsokalivion Monastery, southern side of Mt. Athos. From the documentary sources, we concluded that 11 people were killed, likely 6 by the landslide and 5 by the tsunami. One of the documents found reported that a local tsunami was also observed at the eastern side of Mt. Athos in association with the large earthquake of 28 June 1585. However, no evidence is available about the generation mechanism of that tsunami. Other co-seismic effects caused by the 1905 event included local subsidence, the appearance of new water springs and a fiery phenomenon.

The numerical simulation of the 1905 landslide and ensuing tsunami was based on two landslide scenarios by following the historical accounts and the current geomorphology of the area. Two alternatives were considered, the first involving a scar situated east of the arsanas with a landslide volume of $\sim$ 2.7 million m^3 and the second being of only 160×10^3 m^3 to the west of the arsanas. The simulated tsunami waves with the first scenario (S1) provided results consistent with the existing historical descriptions. Therefore, it is a very good candidate source for the tsunami generation. On the contrary, the second scenario (S2) underestimated drastically the tsunami height historically reported.

It is possible that the 1905 earthquake was associated with the dextral strike-slip fault system of the NE–SW-trending North Aegean Trough. An alternative, however, is that the event was generated along the nearly ENE–WSW-trending dextral strike-slip Athos Fault Zone, which is situated closer, and in parallel, to the southern edge of the Mt. Athos peninsula. Regardless the seismotectonic interpretation, the 1905 earthquake remains a key event for better understanding the hazards associated with earthquakes and related phenomena, such as landslides, rock falls and tsunamis, in an important worldwide cultural heritage site like Mt. Athos.

Acknowledgements

We are grateful to Father Patapios who kindly provided to us a photograph of the manuscript of Codex (p. 185) of the Kafsokalivion Monastery where he leads a monastic life.

Appendix

To better understand the large Mt. Athos 1905 earthquake and its impact in the built and natural environment, we collected and examined various types of documentary sources including contemporary press reports, with publication time ranging from 27 October [O.S.]/9 November [N.S.] to 7 November [O.S.]/20 November [N.S.] 1905, manuscripts and letters as well as various scientific publications and books; [O.S.] and [N.S.] stand for Old Style (Julian) and New Style (Gregorian) calendars, respectively.

In the next lines, we have inserted the list of sources collected as well as relevant texts translated in English by the last author of this paper. Numbering of sources is the same inserted in Table 3. Short explanatory notes in brackets have been inserted by us.

Contemporary Press Reports (All Dates in O.S.)

(a) Central Greece newspapers

1. ΕΦΗΜΕΡΙΣ ΤΗΣ ΘΕΣΣΑΛΙΑΣ (EFIMERIS THS THESSALIAS), 27 Oct. 1905
2. ΕΦΗΜΕΡΙΣ ΤΗΣ ΘΕΣΣΑΛΙΑΣ (EFIMERIS THS THESSALIAS), 28 Oct. 1905
3. Η ΜΙΚΡΑ (I MIKRA), 30 Oct. 1905

(b) Athens newspapers

4. ΝΕΟΝ ΑΣΤΥ (NEON ASTY), 3 Nov. and 6 Nov. 1905
5. ΧΡΟΝΟΣ (CHRONOS), 6 Nov. 1905
6. ΕΜΠΡΟΣ (EMPROS), 7 Nov. 1905

The information contained in the above press reports about the earthquake impact is included in Table 3. A short review of some Thessalia Province newspaper reports was published by Papaioannou (2018). We have also collected additional contemporary press reports from newspapers published in Athens, in Thessalia Province and in Constantinople (Istanbul). However, they are not mentioned here since they just repeat information contained in the reports listed above.

Manuscripts

(7) National Observatory of Athens Manuscript (NOA-MS)

After the establishment of the Institute of Geodynamics at NOA as the seismological service of the country in 1892, its scientific staff started the systematic record of earthquakes occurring in Greece from 1892 up to 1915 inclusive on the basis of macroseismic observations. The earthquakes were listed in chronological order in an unpublished manuscript called *"Book of Earthquakes."* In *"Book of Earthquakes, Vol. 2,"* covering the time period from 1899 to 1915, there is a short section about the 1905 Athos earthquake (NOA-MS) which reads as follows:

> "In the seismograph [of Agamennone type] at Athens, maximum amplitude of 100 mm was recorded at 23 h 47 min 50 s but the record continued until 23 h 55 min. The astronomical pendulum Fenon stopped at 23 h 53 min 38 s…According to press reports, that earthquake was destructive in Mt. Athos. Since 1585, no such a destructive earthquake happened. In Iviron monastery three chapels collapsed, and all the rooms rendered useless. The ground around the monastery subsided in many places. In general, all the monasteries suffered destruction. Rocks thrown from the mountain destroyed 8 monastic cells and crushed 11 monks.

(8) Manuscript-1 archived in Mt. Athos (ATHMS-1)

This manuscript, dated 1 [O.S.]/14 [N.S.] November 1905, was written on 1 Nov. 1905 [O.S.] by an eyewitness monk named Avimelech Mikragiannanitis. It was published later in the religious book by Rodostolos (2004) (BOOK-1). However, the manuscript remained unknown to the seismological community. The points of interest to this study read as follows:

> In the night of 26th to 27th October [O.S.], around midnight, the monks of Athos awoke by terrible but distant, submarine roars and the frightful shaking of the entire peninsula...The entire cone [mountain] of Athos reshaped due to the so many falls of marble rocks...Great disaster happened in the Iviron monastery which rendered nearly uninhabited since many parts of it collapsed including the church of the Assumption of the Virgin Mary built in the 11th century. The ground of the monastery was shaken and new water springs were created. From the rest of 19 monasteries the Great Lavra one was damaged...as well as the cells in the skiti of Kafsokalivia. In the nearby arsanas of Perdiki five seculars and six monks were situated onboard six fishing boats. All of them but one monk submerged taken away by the sea which rose up to 3 m. For it happened that the uphill part of the mountain near St. Peter thrown down covering a distance of half a mile [along the beach]...One monk survived since he rest upon a wooden beam. He was also able to save a secular who jammed in the rocks...6 days after the terrible earthquake we still feel several shocks.

(9) Manuscript-2 archived in Mt. Athos (ATHMS-2)

This document is an unpublished manuscript of the Codex (p.185) of the Kafsokalivion Monastery written during 1915. A photograph of the manuscript was kindly provided to us by Father Patapios who currently gets monastic life at that monastery (Fig. 9). In this document the 1905 earthquake and its effects are described as follows:

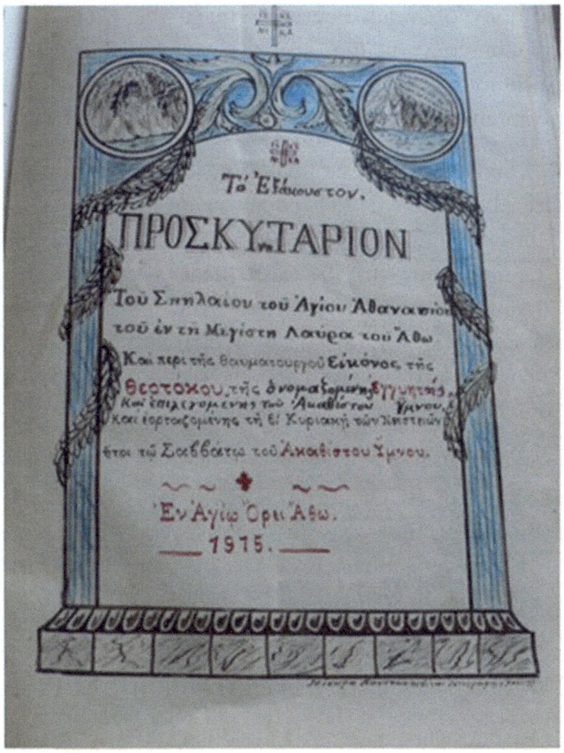

Figure 9
Photograph of the front page of the manuscript of the Codex (p.185) stored at Kafsokalivion Monastery

> During 1905 in the entire Mt. Athos a catastrophic earthquake happened which destroyed partly many buildings and monastic cells at the monasteries of Iviron, Great Lavra and Kastamonitou. The rest monasteries suffered little damage. In particular, damage was caused at the Kafsokalivion monastery where many cells collapsed by the falling stones. The same happened at cells of St. Anna, St. Basil as well as of the Romanian St. John and at other hermitages. The most horrible event is that seven monks and two seculars, totally nine persons, were crushed and sunk. This happened while those persons were fishing close to Perdiki [near Kafsokalivion Monastery]...where part of the mountain detached and thrown into the sea... On 6 o'clock evening [midnight] a cave, where four fathers were praying, was shaken violently, and big stones fell in an extremely steep ground...carrying away the cells situated outside the cave up to the sea, which is lying at a distance of 300 m downhills.

(10) The book by Rodostolos (2004) (BOOK-1).

This book not only reproduces the manuscript ATHMS-1, but also contains testimonies by survivors of the 1905 event collected by the book's author. Here, we reproduce a passage regarding the testimony of monk Isidoros who at the event time was standing in the arsanas near Perdiki. This testimony very likely refers to the survival of the two monks mentioned in the document ATHMS-1.

> ...Suddenly Isidoros found himself floating on the sea being rest on a wooden beam...a tidal wave thrown him violently up to the mountain and jammed him in the cliff near Perdiki site... Eventually monk Isidoros survived since the water ebbed away was uncapable to drift him back to the sea...Isidoros narrated also the surviving of one of his colleagues...

Another testimony of interest is the one provided to the author of BOOK-1 by the monk Amvrosios who at the event time was in Great Lavra Monastery: "*...A colleague, monk Varnavas by name, was killed in his Skiti in St Basil by a rolling big stone that crushed his cell...*".

(11) Two contemporary letters (LETT-1 & LETT-2)

Bishop Chrysostomos Lavriotis (1905), being an eyewitness of the 1905 earthquake, sent a comprehensive letter to the Ecumenical Patriarch Joachim III in Constantinople (Istanbul) narrating the destructive effects of the earthquake in Mt. Athos. LETT-1 dated 3 [O.S.]/16 [N.S.] November 1905 was published in a contemporary religious journal and later reproduced in BOOK-1. Of relevance is also a postscript (LETT-2) dated 4 [O.S.]/17 [N.S.] November 1905 and inserted at the end of LETT-1. The main points of LETT-1, which are of interest to our study, have as follows:

> During that ill-omened night, five fishing boats with monks and seculars were situated in arsanas of St. Pantes [All Saints]. Suddenly they felt a dull din and saw a flame coming out from the seashore. At the same time the entire part of the mountain lying above Kafsokalivia uphill to Kria Nera rolled down to the sea accompanied by smoke. The sea rose and overtopped arsanas, the boats crushed or shattered. Only 3 out of 13 persons survived, 5 of them were standing in arsanas...No damage happened in the skiti of Kafsokalivia; however, it filled up with stones that fell from above...The same happened in St. Anna... In Provatas several cells suffered either small or great damage...In Karyes town small wall fissures happened and chimneys were fallen. In Kerasia almost nothing happened, while Panagia and Koryfi remained intact. The area of St. Peter rendered susceptible to subsidence...Amazingly oversized stones were thrown from mountain summits near Palamas...

Document LETT-1 concluded with a postscript (LETT-2) referring to the large 1585 earthquake.

Scientific Publications (See Main Text and Table 3)

(12) Eginitis (1910); (13) Critikos (1932); (14) Critikos (1933); (15) Mylonas (1995); (16) Kadas (1996); (17) Kaloutsis (2000) (BOOK-2); (18) Barksy (2009) (BOOK-3).

References

Abe, K., & Nogushi, S. (1983). Revision of magnitudes of large shallow earthquakes, 1897–1912. *Physics of the Earth and Planetary Interiors, 33*, 1–11.

Ambraseys, N. N. (1988). Engineering seismology. *Earthquake Engineering and Structural Dynamics, 17*, 1–105.

Ambraseys, N. N. (2001). Reassessment of earthquakes, 1900–1999, in the Eastern Mediterranean and the Middle East. *Geophysical Journal International, 145*, 471–485.

Ambraseys, N. N. (2009). *Earthquakes in the Mediterranean and Middle East, a multidisciplinary study of seismicity up to 1900* (p. 947). Cambridge: Cambridge University Press.

Bakun, W. H., & Scotti, O. (2006). Regional intensity attenuation models for France and the estimation of magnitude and location of historical earthquakes. *Geophysical Journal International*. https://doi.org/10.1111/j.1365-246x.2005.02808x.

Bakun, W. H., & Wentworth, C. M. (1997). Estimating earthquake location and magnitude from seismic intensity data. *Bulletin of the Seismological Society of America, 87*(6), 1502–1521.

Barksy, V. G. (2009). *The travels of Vasily Grigorovich-Barsky in Mt Athos, 1725–1726 and 1744–1745* (p. 715). Athens: Benaki Museum. **(in Greek)**.

Bocchini, G. M., Brüstle, A., Becker, D., Meier, T., van Keken, P. E., Ruscic, M., et al. (2018). Tearing, segmentation, and backstepping of subduction in the Aegean: New insights from seismicity. *Tectonophysics, 734–735*, 96–118. https://doi.org/10.1016/j.tecto.2018.04.002.

Caputo, R., Chatzipetros, A., Pavlides, S., & Sboras, S. (2013). The Greek Database of Seismogenic Sources (GreDaSS): state-of-the-art for northern Greece. *Annals of Geophysics*. https://doi.org/10.4401/ag-5168.

Comninakis, P. E., & Papazachos, B. C. (1982). A catalogue of earthquakes in Greece and the surrounding area for the period 1901–1980. Publ University of Thessaloniki. *Geophysics Laboratory, 5,* 146.

Critikos, M. N. (1932). Sur la Seismicité de Macédoine. *Annales de l'Observatoire National d'Athènes, 12,* 149–159.

Critikos, M. N. (1933). Sur la Seismicité de Macédoine. *Gerl Beiträge zur Geophysik, 40,* 371–379.

della Dora, V. (2012). Setting and blurring boundaries: pilgrims, tourists, and landscape in Mount Athos and Meteora. *Annals of Tourism Research, 39,* 951–974. https://doi.org/10.1016/j.annals.2011.11.013.

Di Giacomo, D., Bondar, I., Storchak, D. A., Engdahl, E. R., Bormann, P., & Harris, J. (2015). ISC-GEM: global instrumental earthquake catalogue (1900–2009), III. Re-computed M_S and m_b, proxy M_W, final magnitude composition and completeness assessment. *Physics of the Earth and Planetary Interiors, 239,* 33–47. https://doi.org/10.1016/j.pepi.2014.06.005.

Drakopoulos, J. (1978). Attenuation of intensities with distance for shallow earthquakes in the area of Greece. *Bolletino Geofisica Teorica et Applicata, 20,* 114–130.

Duda, S. J. (1965). Secular seismic energy release in the circum-Pacific belt. *Tectonophysics, 2,* 409–452. https://doi.org/10.1016/0040-1951(65)90035-1.

Eginitis, D. (1910). Catalogue des tremblements de Terre observes en Gréce pendant les années 1904–1908. *Annales de l'Observatoire National d'Athènes, 10,* 560–585.

Evangelidis, C. (2014). Imaging supershear rupture for the 2014 M_w 6.9 Northern Aegean earthquake by back projection of strong motion waveforms. *Geophysical Research Letters, 42,* 307–315. https://doi.org/10.1002/2014gl062513.

Ferentinos, G., Georgiou, N., Christodoulou, D., Geraga, M., & Papatheodorou, G. (2018). Propagation and termination of a strike slip fault in an extensional domain: The westward growth of the North Anatolian Fault into the Aegean Sea. *Tectonophysics*. https://doi.org/10.1016/j.tecto.2018.08.003.

Floyd, M. A., et al. (2010). A new velocity field for Greece: Implications for the kinematics and dynamics of the Aegean. *Journal of Geophysical Research*. https://doi.org/10.1029/2009jb007040.

Fokaefs, A., Roussopoulou, C. & Papadopoulos, G.A. (2004). Magnitudes of historical Greek earthquakes estimated by magnitude/intensity relationships. In Proceedings of 5th International Symposium on Eastern Mediterranean Geology, Thessaloniki (pp.568-571), 14–20 Apr. 2004, 2.

Galanopoulos, A. G. (1955). The seismic geography of Greece. *Annales Geologique des Pays Hellenique, 6,* 83–121.

Galanopoulos, A. G. (1960). *A catalogue of shocks with $I_0 \geq$ VI or $M \geq 5$ for the years 1801–1958. Seismol* (p. 119). Athens: Laboratory in University of Athens.

Galanopoulos, A. G. (1961). On magnitude determination by using macrosesmic data. *Annali di Geofisica, 14,* 225–253.

Gasperini, P., & Ferrari, G. (2000). Deriving numerical estimates from descriptive information: The computation of earthquake parameters. *Annali di Geofisica, 43*(4), 729–746.

Gennadios, Bishop. (1917). Marginal notes in Mt Athos Codecis. *Grigorios Palamas Religious Journal, 1,* 457–472. **(in Greek)**.

Georgiadis, G. A., Tranos, M. D., & Mountrakis, D. M. (2007). Late- and post-alpine tectonic evolution of the southern part of the Athos Peninsula, Northern Greece. *Bulletin of the Geological Society of Greece, 40,* 309–321.

Guidoboni, E., & Comastri, A. (2005). *Catalogue of Earthquakes and tsunamis in the Mediterranean Area from the 11th to the 15th Century* (p. 1037). Bologna: INGV-SGA.

Guidoboni, E., Comastri, A., & Traina, G. (1994). *Catalogue of ancient earthquakes in the Mediterranean Area up to the 10th century* (p. 504). Bologna: ING-SGA.

Guidoboni, E., & Ebel, J. E. (2009). *Earthquakes and tsunamis in the past—A guide to techniques in historical seismology* (p. 590). Cambridge: Cambridge University Press.

Gutdeutsch, R., Kaiser, D., & Jentzsch, G. (2002). Estimation of earthquake magnitudes from empirical intensities and other focal parameters in Central and Southern Europe. *Geophysical Journal International, 151,* 824–834. https://doi.org/10.1046/j.1365-246X.2002.01804.x.

Gutenberg, B., & Richter, C. F. (1954). *Seismicity of the earth and associated phenomena* (2nd ed., p. 310). Princeton: Princeton University Press.

ISC-GEM. (2018). The ISC-GEM global instrumental earthquake catalogue, version 4.0. (http://www.isc.ac.uk/iscgem/download.php). Accessed 26 Jan 2017

Janin, A., Rodriguez, M., Sakellariou, D., Lykousis, V., & Gorini, Ch. (2018). Tsunamigenic potential of a Holocene submarine landslide along the North Anatolian Fault (North Aegean Sea, off Thasos Island): insights from numerical modeling. *Natural Hazards and Earth System Sciences*. https://doi.org/10.5194/nhess-2017-405.

Kadas, S. N. (1996). *Marginal notes in codecis of Dionysiou monastery, Mt Athos* (p. 417). Mt Athos: Dionysiou monastery. **(in Greek)**.

Kaloutsis, S.D. (2000). The Virgin Mary of Mirtidiotissa (4th ed.). Kythira: Kytherian Publications, p. 126 (in Greek).

Karabulut, H., Roumelioti, Z., Benetatos, C., Multu, A. K., Özalaybey, S., Aktar, M., et al. (2006). A source study of the 6 July 2003 (Mw 5.7) earthquake sequence in the Gulf of Saros (Northern Aegean Sea): Seismological evidence for the western continuation of the Ganos fault. *Tectonophysics, 412*(3–4), 195–216. https://doi.org/10.1016/j.tecto.2005.09.009.

Kárnik, V. (1969). *Seismicity of the European Area* (1st ed., p. 364). Dordrecht: D. Reidel Publ. Company.

Kockel, F., Mollat, H., & Walther, H. W. (1977). *Erlauterungen zür Geologischen Karte der Chalkidhiki und angrenzender Gebiete 1:100.000 (Nord-Griechenland)* (p. 119). Hannover: Bundesanstalt fur Geowisseschaften und Rohstoffe.

Koukouvelas, I. K. (2018). *The Geology of Greece* (p. 343). Athens: Liberal Books. **(in Greek)**.

Koukouvelas, I. K., & Aydin, A. (2002). Fault structure and related basins of the North Aegean Sea and its surroundings. *Tectonics, 21*(5), 1046. https://doi.org/10.1029/2001TC901037.

Kreemer, C., Chamot-Rooke, N., & Le Pichon, X. (2004). Constraints on the evolution and vertical coherency of deformation in the Northern Aegean from a comparison of geodetic, geologic and seismologic data. *Earth and Planetary Science Letters, 225*(3–4), 329–346. https://doi.org/10.1016/j.epsl.2004.06.018.

Lavriotis, Ch. (1905). A letter to Ecumenical (Universalis) Patriarch Joakim III. *Religious Truth, 29,* 522–524. **(in Greek)**.

Le Pichon, X., Lybéris, N., & Alvarez, F. (1987). Discussion on the subsidence of the North Aegean Trough: An alternative view. *Journal of Geological Society, 144*(2), 349–351. https://doi.org/10.1144/gsjgs.144.2.0349.

Makropoulos, K., Kaviris, G., & Kouskouna, V. (2012). An updated and extended earthquake catalogue for Greece and adjacent areas since 1900. *Natural Hazards and Earth Systems Sciences, 12,* 1425–1430.

Maravelakis, M. I. (1938). *Contribution a l'étude des séismes du Mont Athos par les "mémoires". Sci. Yearbook Fac. Phys* (4th ed., pp. 133–147). Thessaloniki: Aristotelian University of Thessaloniki.

Mouslopoulou, V., Saltogianni, V., Gianniou, M., & Stiros, S. (2014). Geodetic evidence for tectonic activity on the Strymon fault system, northeast Greece. *Tectonophysics, 633*(1), 246–255.

Müller, M. D., Geiger, A., Kahle, H.-G., Veis, G., Billiris, H., Paradissis, D., et al. (2013). Velocity and deformation fields in the North Aegean domain, Greece, and implications for fault kinematics, derived from GPS data 1993–2009. *Tectonophysics, 597–598,* 34–49. https://doi.org/10.1016/j.tecto.2012.08.003.

Mylonas, P. M. (1995). Nineteenth century representational paintings and photographs of Athos monasteries and their interpretation. *Archaeology, 54,* 6–19. **(in Greek with Engl. Abstr.)**.

Nuttli, O., Bollinger, G. A., & Griffiths, D. W. (1979). On the relation between modified Mercalli intensity and body-wave magnitude. *Bulletin of the Seismological Society of America, 69,* 893–909.

Papadopoulos, G. A. (2011). *A seismic history of crete: earthquakes and tsunamis, 2000 BC–AD 2010* (p. 415). Athens: Ocelotos Publications.

Papadopoulos, G. A., & Chalkis, B. (1984). Tsunamis observed in Greece and the surrounding area from antiquity up to the present times. *Marine Geology, 56,* 309–317.

Papadopoulos, G. A., Ganas, A., & Plessa, A. (2002). The Skyros earthquake (M_w 6.5) of 26 July 2001 and precursory seismicity patterns in the North Aegean Sea. *Bulletin of the Seismological Society of America, 92,* 1141–1145.

Papadopoulos, G. A., Gràcia, E., Urgeles, R., et al. (2014). Historical and pre-historical tsunamis in the Mediterranean and its connected seas: Geological signatures, generation mechanisms and coastal impacts. *Marine Geology, 5,* 44. https://doi.org/10.1016/j.margeo.2014.04.014.

Papadopoulos, G. A., & Kortekaas, S. (2003). Characteristics of landslide generated tsunamis from observational data. In J. Locat & J. Mienert (Eds.), *Submarine mass movements and their consequences* (pp. 367–374). The Netherlands: Kluwer.

Papaioannou, I. (2018). The seismic activity in Thessalia from 1900 to 1950. *Thessaliko Imerologio, 73,* 289–314. **(in Greek)**.

Papanikolaou, D., Alexandri, M., & Nomikou, P. (2006). Active faulting in the north Aegean basin. In Y. Dilek & S. Pavlides (Eds.), *Postcollisional tectonics and magmatism in the Mediterranean region and Asia* (pp. 189–209). Washington: Geological Society of America. https://doi.org/10.1130/2006.2409.

Papazachos, B.C. & Papadopoulos, G.A. (1979). Deep tectonics and associated ore deposits in the Aegean area. In Proceedings of 6th Colloquium on the Geology of the Aegean Region, Athens (pp.1071–1080), 1977, 3.

Papazachos, C., & Papaioannou, Ch. (1997). The macroseismic field of the Balkan area. *Journal of Seismology, 1,* 181–201.

Papazachos, C., & Papaioannou, Ch. (1998). Further information on the macroseismic field of the Balkan area. *Journal of Seismology, 2,* 363–375.

Papazachos, B. C., & Papazachou, C. (1989). *The Earthquakes of Greece* (p. 347). Thessaloniki: Ziti Publications. **(in Greek)**.

Papazachos, B. C., & Papazachou, C. (1997). *The earthquakes of Greece* (p. 304). Thessaloniki: Ziti Publications.

Papazachos, B. C., & Papazachou, C. (2003). *The Earthquakes of Greece* (p. 273). Thessaloniki: Ziti Publications. **(in Greek)**.

Pavlides, S., & Kilias, A. A. (1987). Neotectonic and active faults along the Serbomacedonian zone (SE Chalkidiki, northern Greece). *Annales Tectonicae, 1*(2), 97–104.

Pavlides, S., Mountrakis, D., Kilias, A. A., & Tranos, M. (1990). The role of strike-slip movements in the extensional area of Northern Aegean (Greece): A case of transtensional tectonics. *Annales Tectonicae, 4,* 196–211.

Psilovikos, A. (1984). Geomorphological and structural modification of the Serbomacedonian massif during the neotectonic stage. *Tectonophysics, 110,* 27–45.

Reilinger, R., McClusky, S., Paradissis, D., Ergintav, S., & Vernant, P. (2010). Geodetic constraints on the tectonic evolution of the Aegean region and strain accumulation along the Hellenic subduction zone. *Tectonophysics, 488*(1–4), 22–30. https://doi.org/10.1016/j.tecto.2009.05.027.

Richter, Ch. (1958). *Elementary seismology* (p. 768). San Francisco: Freeman.

Rocca, A Ch., Karakaisis, G. F., Karacostas, B. G., Kiratzi, A. A., Scordilis, E. M., & Papazachos, B. C. (1985). Further evidence on the strike-slip faulting of the northern Aegean trough based on properties of the August–November 1983 seismic sequence. *Bolletino Geofisica Teorica et Applicata, 27,* 101–109.

Rodostolos, B. C. (2004). *Ode to unwithered: to Athos* (p. 671). Mt Athos: Lavriotikon Holly Cell of All Saints. **(in Greek)**.

Roussos, N., & Lyssimachou, T. (1991). Structure of the Central North Aegean Trough: an active strike-slip deformation zone. *Basin Research, 3,* 39–48.

Shebalin, N. V., Kárník, V., & Hadzievski, D. (Eds.). (1974). *Catalogue of earthquakes, part II, prior to 1901* (p. 69). Skopje: UNDP/UNESCO.

Sibol, M. S., Bollinger, G. A., & Birch, J. B. (1987). Estimation of magnitudes in Central and Eastern North America using intensity and felt area. *Bulletin of the Seismological Society of America, 77*(5), 1635–1654.

Sieberg, A. (1932). Erdbebengeographie (p. 1202). Berlin: Verlag von Gebrüder Borntraeger

Street, R. L., & Turcotte, F. T. (1977). A study of northeastern North American spectral moments, magnitudes, and intensities. *Bulletin of the Seismological Society of America, 67*(3), 599–614.

Taymaz, T., Jackson, J., & McKenzie, D. (1991). Active tectonics of the north and central Aegean Sea. *Geophysical Journal International, 106*(2), 433–490. https://doi.org/10.1111/j.1365-246X.1991.tb03906.x.

Tinti, S., Bortolucci, E., & Vannini, C. (1997). A block-based theoretical model suited to gravitational sliding. *Natural Hazards, 16,* 1–28.

Tinti, S., Chiocci, F. L., Zaniboni, F., Pagnoni, G., & de Alteriis, G. (2011). Numerical simulation of the tsunami generated by a past catastrophic landslide on the volcanic island of Ischia, Italy. *Marine Geophysical Research, 32,* 287–297. https://doi.org/10.1007/s11001-010-9109-6.

Tinti, S., & Tonini, R. (2013). The UBO-TSUFD tsunami inundation model: validation and application to a tsunami case study focused on the city of Catania, Italy. *Natural Hazards and Earth System Sciences, 13,* 1795–1816. https://doi.org/10.5194/nhess-13-1795-2013.

Tinti, S., Vittori, T., & Mulargia, F. (1986). Regional intensity-magnitude relationships for the Italian territory. *Tectonophysics, 127,* 129–154.

Trifunak, M. D. (1998). Comment on "The macroseismic field of the Balkan area" by C. Papazachos and C. Papaioannou. *Journal of Seismology, 2,* 359–362.

Voidomatis, Ph, Pavlides, S., & Papadopoulos, G. A. (1990). Active deformation and seismic potential in the Serbomacedonian zone, northern Greece. *Tectonophysics, 179,* 1–9.

Wood, H. O., & Neumann, F. (1931). Modified Mercalli Intensity Scale of 1931. *Bulletin of the Seismological Society of America, 21,* 277–283.

Zaniboni, F., Armigliato, A., Pagnoni, G., & Tinti, S. (2014a). Continental margins as a source of tsunami hazard: the 1977 Gioia Tauro (Italy) landslide-tsunami investigated through numerical modelling. *Marine Geology, 357,* 210–217. https://doi.org/10.1016/j.margeo.2014.08.011.

Zaniboni, F., Armigliato, A., & Tinti, S. (2016). A numerical investigation of the 1783 landslide-induced catastrophic tsunami in Scilla, Italy. *Natural Hazards, 84,* S455–S470. https://doi.org/10.1007/s11069-016-2461-3.

Zaniboni, F., Pagnoni, G., Armigliato, A., Elsen, K., & Tinti, S. (2014b). Investigations on the possible source of the 2002 landslide tsunami in Rhodes, Greece, through numerical techniques. In G. Lollino, A. Manconi, J. Locat, Y. Huang, & M. C. Ardilas (Eds.), *Engineering geology for society and territory* (p. 4). Berlin: Springer. https://doi.org/10.1007/978-3-319-08660-6_17.

Zaniboni, F., Pagnoni, G., Gallotti, G., Paparo, M. A., Armigliato, A., & Tinti, S. (2019). Assessment of the 1783 Scilla landslide–tsunami's effects on the Calabrian and Sicilian coasts through numerical modelling. *Natural Hazards and Earth Systems Sciences, 19,* 1585–1600. https://doi.org/10.5194/nhess-19-1585-2019.

(Received July 4, 2019, revised October 9, 2019, accepted November 6, 2019, Published online November 18, 2019)

Pure Appl. Geophys. 177 (2020), 1295–1313

https://doi.org/10.1007/s00024-019-02410-1

Pure and Applied Geophysics

The Chios, Greece Earthquake of 23 July 1949: Seismological Reassessment and Tsunami Investigations

Nikolaos S. Melis,[1] Emile A. Okal,[2] Costas E. Synolakis,[3,4] Ioannis S. Kalogeras,[1] and Utku Kânoğlu[5]

Abstract—We present a modern seismological reassessment of the Chios earthquake of 23 July 1949, one of the largest in the Central Aegean Sea. We relocate the event to the basin separating Chios and Lesvos, and confirm a normal faulting mechanism generally comparable to that of the recent Lesvos earthquake located at the Northern end of that basin. The seismic moment obtained from mantle surface waves, $M_0 = 7 \times 10^{26}$ dyn cm, makes it second only to the 1956 Amorgos earthquake. We compile all available macroseismic data, and infer a preference for a rupture along the NNW-dipping plane. A field survey carried out in 2015 collected memories of the 1949 earthquake and of its small tsunami from surviving witnesses, both on Chios Island and nearby Oinousses, and on the Turkish mainland. While our results cannot help discriminate between the two possible fault planes of the 1949 earthquake, an important result is that both models provide an acceptable fit to the reported amplitudes, without the need to invoke ancillary sources such as underwater landslides, in contrast to the case of other historical tsunamis in the Aegean Sea, such as the 1956 Amorgos and 1948 Rhodos events.

Keywords: Aegean Sea, historical earthquakes, tsunami surveys.

1. Background and Previous Investigations

This paper provides a modern seismological investigation of the earthquake of 23 July 1949 on the Greek island of Chios, in the Aegean Sea. This event was given a "Pasadena" magnitude $M_{PAS} = 6\frac{3}{4}$, reported by various agencies, but not included in Gutenberg and Richter's (1954) generally authoritative catalog. This magnitude makes it the second largest instrumentally recorded historical earthquake in the Central Aegean Sea after the 1956 Amorgos event (Okal et al. 2009), a region broadly defined as limited to the South by the Cretan–Rhodos subduction arc and to the north by the western extension of the Northern Anatolian Fault system. The earthquake resulted in considerable damage, with several hundred houses destroyed in the northeastern villages of Chios Island and on the nearby Karaburun Peninsula in Turkey. As such, it ranks among the more destructive of the past hundred years in the Eastern Aegean Sea. Miraculously, only ten casualties were reported, as compiled from a detailed review of news reports performed during the present study.

The earthquake was accompanied by a small tsunami in northern Chios and the nearby island of Oinousses, with additional wave activity reported in Çeşme and on the Karaburun Peninsula (Fig. 1).

The 1949 earthquake was the subject of a study by Galanopoulos (1954), and more recently of a review by Altınok et al. (2005). Galanopoulos (1954) described the effects of the event on both the Greek and Turkish sides. In Greece, macroseismic observations consisted of filled questionnaires and telegrams, used to report back information to the National Observatory of Athens (NOA) from dedicated locally appointed personnel, especially on Chios Island Mr. Kreatsas, a physicist, who was at the time Director of the Chios Gymnasium. This information, and reports from Turkey, were then evaluated by Galanopoulos (1954) in terms of Mercalli-Sieberg Intensities (MSI) (Musson et al. 2010), and compiled into a map of the devastation on both sides of the

[1] Geodynamics Institute, National Observatory of Athens, P.O. Box 20048, 11810 Athens, Greece.

[2] Department of Earth and Planetary Sciences, Northwestern University, Evanston, IL 60208, USA. E-mail: emile@earth.northwestern.edu

[3] Department of Civil Engineering, University of Southern California, Los Angeles, CA 90089, USA.

[4] Environmental Engineering School, Technical University of Crete, 73100 Chania, Greece.

[5] Department of Aerospace Engineering, Middle East Technical University, 06800 Çankaya, Ankara, Turkey.

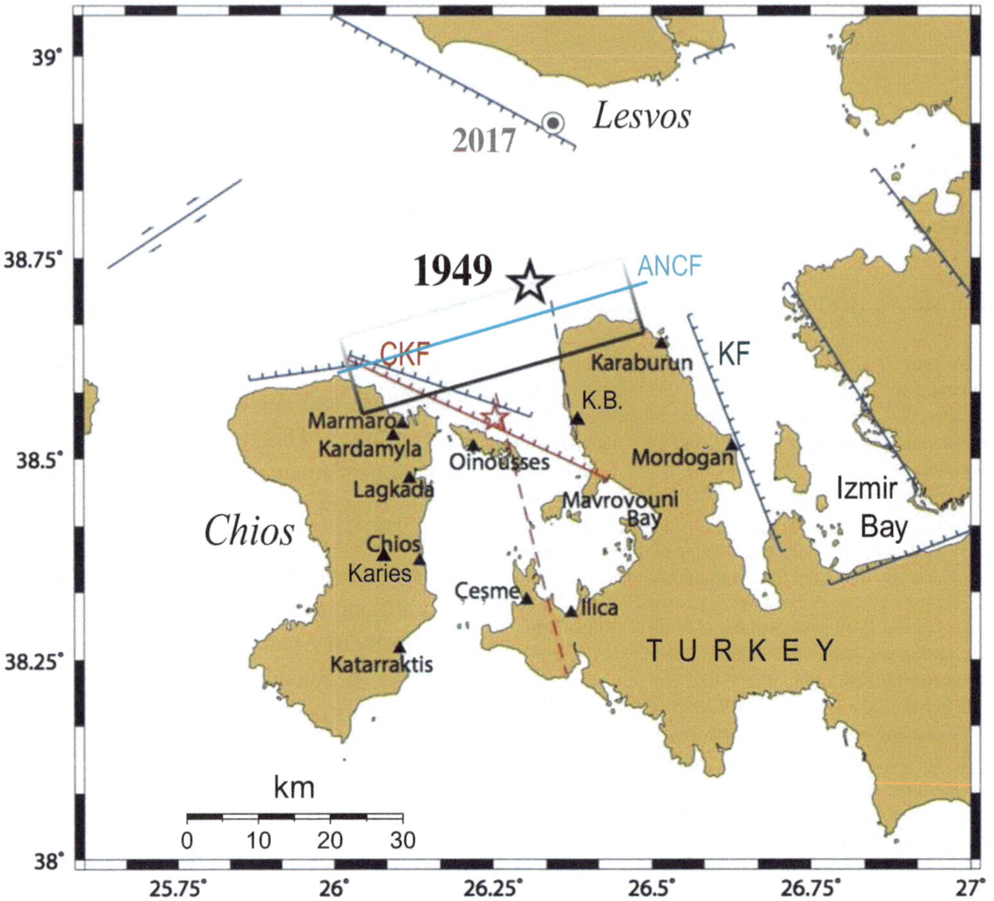

Figure 1
General location map of the 1949 Chios earthquake (large star). The rectangle shows the projection of our preferred rupture area. Also shown are the principal faults recognized or proposed in the area, notably the Chios–Karaburun fault (CKF), proposed by Galanopoulos (1954), the Karaburun Fault (KF) defining the Western boundary of the Izmir Bay, the unnamed offshore fault hosting the 2017 Lesvos earthquake, and in red dashes, another fault proposed by Galanopoulos (1954) as a possible site of the 1949 earthquake; the red star is his epicenter. The dark blue dashed line is the fault proposed by Altınok et al. (2005) along the northwestern coast of Karaburun Peninsula. Localities described in the text are shown as small triangles. *K.B.* Küçükbahçe Bucağı, site of an VIII$\frac{1}{2}$ MSI report on the Turkish side. The light blue line is the Ayasmaton-Northern Chios Fault (ANCF) identified in the GreDaSS database (Caputo et al. 2012). Adapted from Galanopoulos (1954), Altınok et al. (2005), Chatzipetros et al. (2013; Fig. 5a) and Kiratzi (2018; Fig. 2)

border, offered as Figure 2 of Galanopoulos (1954). Unfortunately, little information could be gleaned from local instrumental data, since at the time, there was only one seismic station (Athens, ATH) in operation in Greece; although it was well-equipped under the world-wide instrumental standards of the day, with one Mainka instrument (two horizontal components; $T_p = 5.8$ s), one horizontal two-component ($T_p = 9$ s) and one vertical ($T_p = 4$ s) Wiechert seismographs, their records were all saturated by the mainshock. Hence, Galanopoulos (1954) relied on the manual drafting of isoseismals, together with some knowledge of the local geomorphology and structural geology, to define the main fault responsible for the event and its relation to the epicenter. He favored a "Chios–Karaburun" WNW–ESE normal fault dipping towards NNE, with its hypothetical surface trace lying North of the coast of Chios and Oinousses, to the Western coast of the Karaburun Peninsula into the Mavrovuni (Gerence) Bay (Fig. 1).

Altınok et al. (2005) described the damage for both the 1881 and the 1949 events that took place South and North of the Chios-Çeşme and Chios–Karaburun areas, respectively, but mostly focused their reports on the Turkish side. Regarding the 1949 event, they used mainly a review of Galanopoulos (1954) and incorporated material from Pinar (1950) and Papazachos and Papazachou (1997), as well as from Turkish newspapers, which reported earthquake effects and damages nearly exclusively on the Turkish side, and in particular episodes of water jets gushing on the port of Çeşme. Altınok et al. (2005) adopted Galanopoulos' (1954) proposed Chios–Karaburun fault, which they complemented with an almost NW-SE striking hypothetical fault mapped in the sea and consistent with the geomorphology of the western coast of the Karaburun Peninsula. They do not add standard information on flooding (inundation, run-up) on the Turkish side.

In this context, the present paper presents a new, comprehensive, re-investigation of the 1949 Chios earthquake, including the building and modeling of a macroseismic database, a quantification of the seismic moment of the source and of its radiated energy, based on waveform analysis, and a tsunami simulation to model the results of a survey which involved the interview of elderly witnesses on both the Greek and Turkish sides. The latter sought to provide an independent validation of quantitative reports of inundation and flooding. As discussed in Sects. 4 and 5, its results were moderate, but can help to put general constraints on the seismic source of the 1949 event.

2. *Relocation and Background Seismicity*

We relocated the 1949 Chios earthquake based on arrival times listed in the International Seismological Summary (ISS), using Wysession et al.'s (1991) interactive, iterative method, which defines a confidence ellipse based on a Monte Carlo algorithm injecting Gaussian noise into the dataset; for an event in 1949, we give the noise a standard deviation $\sigma_G = 3$ s. Our relocation converges to 38.71° N, 26.32° E, approximately 25 km NNE of the Island of Oinousses (Fig. 2a). This solution is obtained at a constrained depth of 10 km. Attempts to keep the depth floating could not produce a stable positive hypocentral depth. This is confirmed by a weak, but systematic increase in r.m.s. residuals σ with hypocentral depth for constrained-depth relocations (Fig. 2b); at the same time, increasing constrained depth results in a northerly moveout of the epicenter (small dots on Fig. 2a), into an area close to Lesvos where only two small, poorly located events ($m_b \approx 3$) are claimed to have occurred deeper than 70 km in the past 50 years. We conclude that the 1949 earthquake was most probably shallow ($h \leq 30$ km), in agreement with trends in modern seismicity, as defined by NEIC solutions (1970–2015) reported within 30 km of our preferred epicenter and with at least one magnitude $M \geq 4.5$ (such events can be considered reliably located), for which the maximum depth is 31 km. Our results also agree with Galanopoulos' (1954) suggestion of a depth of 12–17 km for the 1949 earthquake, using the rapid decay of felt intensities with distance, as interpreted in the framework of Gutenberg and Richter's (1942) empirical relations.

Figure 2a also compares our solution with the epicenters originally given by the ISS, the U.S. Geological Survey and the Jesuit Seismological Association (JSA), those proposed by Erkman (1949), Labrouste and Pinar (1953), Galanopoulos (1954), Ambraseys (1998), or obtained by modern computerized relocations such as McKenzie's (1972), Ambraseys' (2001), or the ISC-GEM relocated epicenter (Storchak et al. 2013). Note that the earthquake is not part of Gutenberg and Richter's (1954) otherwise authoritative catalog (even though a "Pasadena" magnitude appears in the ISC bulletin), nor was it relocated as part of Engdahl and Villaseñor's (2002) Centennial catalog.

We note that modern relocations [such as by the ISC, McKenzie (1972), Ambraseys (2001) or this study] move the epicenter about 30 km North of the group of more ancient estimates, such those of the ISS, Erkman (1949), Labrouste and Pinar (1953), Galanopoulos (1954) or even Ambraseys (1988). In this context, it seems difficult to associate the earthquake with the proposed fault running WNW–ESE from North of Chios and Oinousses to the Southern tip of the Karaburun Peninsula (Galanopoulos 1954,

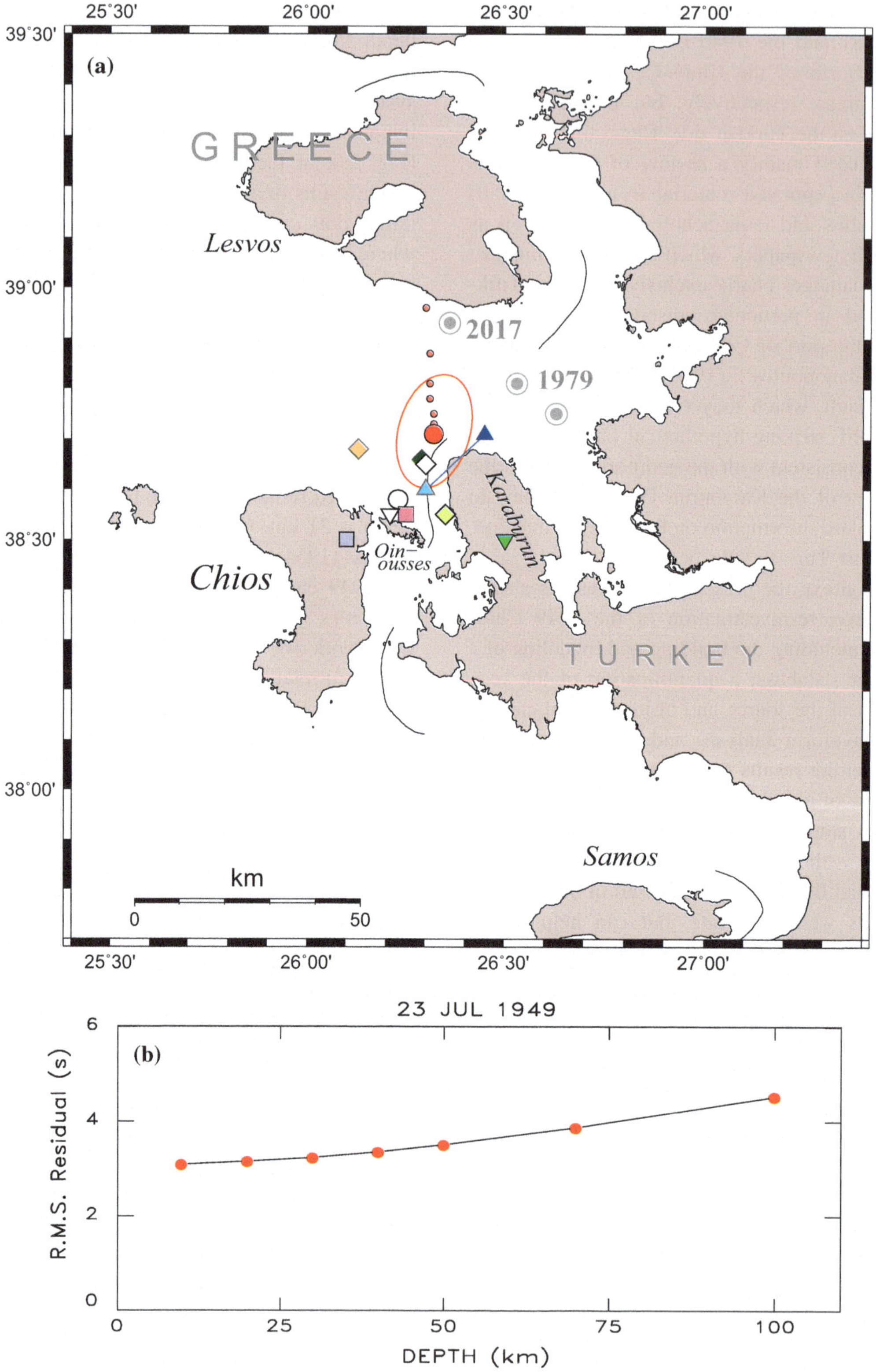
25°30'
26°00'
26°30'
27°00'
39°30'
39°00'
38°30'
38°00'
(a)
GREECE
Lesvos
2017
1979
Karaburun
Chios
Oin–
ousses
TURKEY
Samos
km
0
50
23 JUL 1949
(b)
R.M.S. Residual (s)
DEPTH (km)
0
2
4
6
0
25
50
75
100

◀Figure 2
a Relocation of the 1949 Chios earthquake. Our relocated epicenter is shown as the solid red dot, with Monte carlo ellipse; small dots indicate the moveout of the epicenter as depth is increased from 10 to 100 km. Other epicenters shown are: original ISS (light blue triangle); relocated ISC (dark blue triangle); USGS (green inverted triangle); JSA (light purple square); Galanopoulos (1954) (magenta square); Papazachos and Papazachou (1997) (light brown diamond); Labrouste and Pinar (1953) (yellow diamond); Erkman (1949) (open circle); Ambraseys (1988) (white open inverted triangle); Ambraseys (2001) (white open diamond); and McKenzie (1972) (dark green diamond). For reference, the 1979 and 2017 Lesvos earthquakes are shown as the grey bull's eye symbols. **b** R.M.S. residual as a function of constrained depth for relocations of the 1949 Chios earthquake

Figure 2). Rather, the emerging picture would associate the source with the faults named Ayasmaton and Northern Chios in the Greek Database of Seismogenic Sources (Caputo et al. 2012), and shown as the greenish blue line (ANCF) on Fig. 1.

Table 1 also lists the result of relocations of eight "associated" earthquakes originally listed as occurring in the Aegean Sea during the years 1949–1950. Of those, only Events A03 and A06 can be considered genuine aftershocks of the Chios earthquake. This meager number prevents an interpretation of aftershocks in terms of the extent of the fault zone of the mainshock. Event A05 is poorly located (with only 6 stations), but its Monte Carlo solutions run with unconstrained depths are peaked around 126 km, approximately 120 km South of the mainshock; similarly, Event A04, originally undetermined by the ISS, relocates to the Peloponese-Crete subduction system, with a most probable depth of 70 km. We have verified that modern seismicity exists at the proposed 3-D locations of Events A04 and A05. Two more events (F01 and A08) relocate on Lesvos, in the vicinity of the 2017 epicenter, and the remaining two earthquakes (A07 and A09) relocate in mainland Turkey. Figure 3 summarizes these results; no magnitude estimates are available, except $M_s = 5.1$, recomputed by the ISC for Event A09, and largely irrelevant to the present study.

3. *Focal Mechanism and Quantification of the Source*

3.1. *Focal Mechanism*

McKenzie (1972) listed a mostly normal focal mechanism ($\phi = 141°$, $\delta = 65°$, $\lambda = 321°$), based on Wickens and Hodgson's (1967) catalog of focal solutions (Fig. 4). We verified this mechanism through an independent examination of historical seismograms, which allowed us to obtain 14 first motion polarities. These are plotted as large symbols

Table 1

Relocations carried out in the present study

Code	Date	Relocation								Notes
(*)	D M (J) Y	Origin time	Epicenter		Depth		Nr. of stations		σ	
			(°N)	(°E)	(km)	(**)	Read	Kept	(s)	
F01	21 MAY (141) 1949	17:41:18.5	39.05	26.07	67	F	16	13	4.96	Lesvos
M02	**23 JUL (204) 1949**	**15:03:32.5**	**38.71**	**26.32**	**10**	**C**	**132**	**129**	**3.10**	**Mainshock**
A03	30 JUL (211) 1949	17:47:12.6	38.73	26.35	60	F	35	32	4.13	Genuine aftershock
A04	01 AUG (213) 1949	15:27:36.8	35.32	22.31	67	F	21	11	3.98	Southern Greece
A04	01 AUG (213) 1949	15:27:36.8	35.32	22.31	67	F	21	11	3.98	Southern Greece
A05	01 AUG (213) 1949	22:42:26.9	37.77	26.30	126	C	7	6	2.89	Aegean Sea; Deep
A06	23 NOV (327) 1949	16:51:01.4	38.60	26.23	33	F	54	51	3.32	Genuine aftershock
A07	03 MAY (123) 1950	07:13:43.1	38.66	27.16	10	C	28	25	4.02	Gediz Valley, Turkey
A08	08 JUL (189) 1950	07:07:24.2	39.10	26.31	10	C	13	11	3.72	Lesvos
A09	28 NOV (332) 1950	17:53:20.5	39.69	28.17	33	F	32	30	3.77	Balıkesir, Turkey
				For reference						
L10	12 JUN (163) 2017	12:28:04.9	38.81	26.2	12					2017 Lesvos Event

(*) *F* Foreshock, ***M* Mainshock**, *A* Aftershock, *L* Lesvos event (2017)

(**) *C* Constrained depth, *F* Floated depth

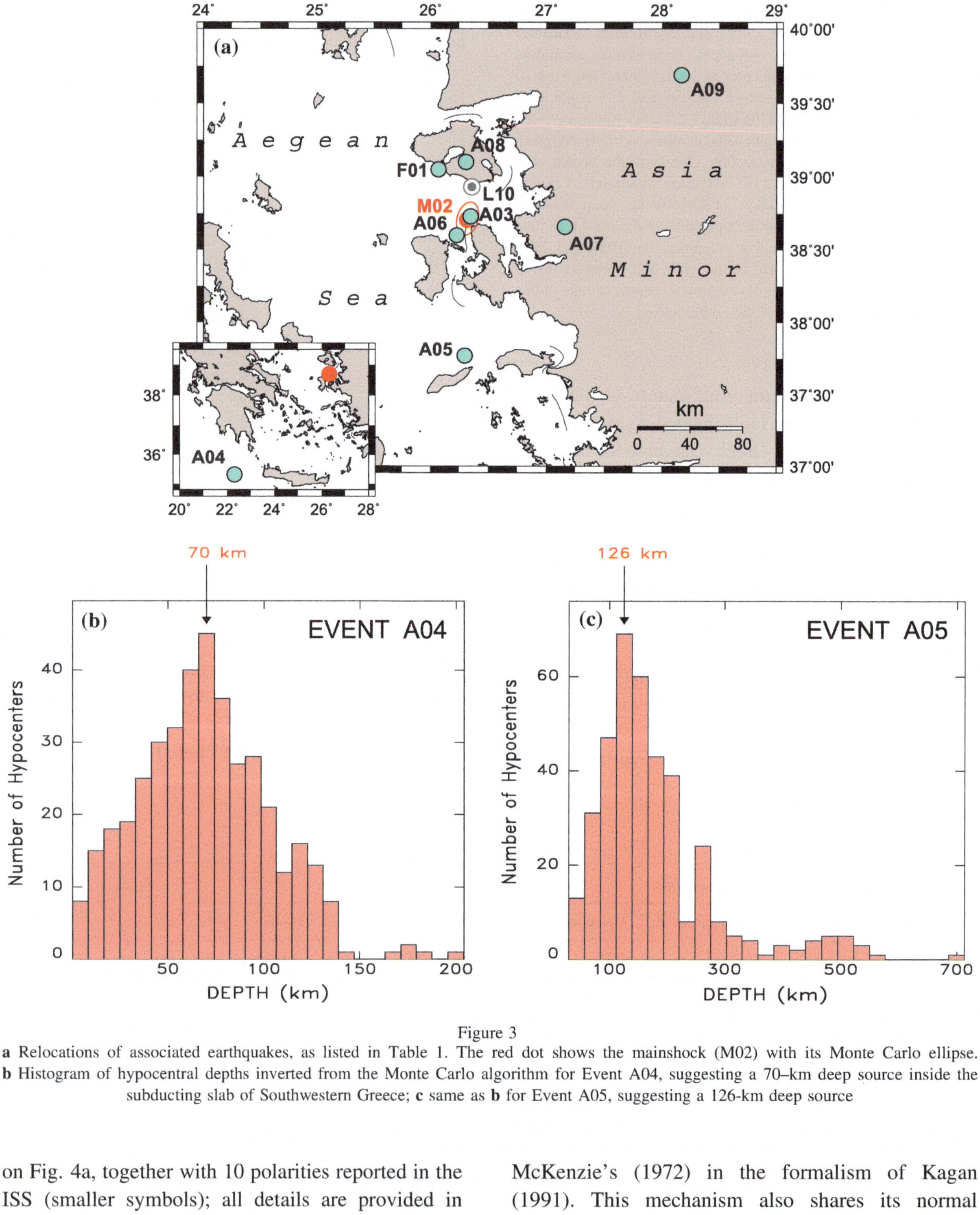

Figure 3

a Relocations of associated earthquakes, as listed in Table 1. The red dot shows the mainshock (M02) with its Monte Carlo ellipse. **b** Histogram of hypocentral depths inverted from the Monte Carlo algorithm for Event A04, suggesting a 70–km deep source inside the subducting slab of Southwestern Greece; **c** same as **b** for Event A05, suggesting a 126-km deep source

on Fig. 4a, together with 10 polarities reported in the ISS (smaller symbols); all details are provided in Table 2. On this basis, we propose the mechanism $\phi = 254°$, $\delta = 57°$, $\lambda = 218°$ (conjugate mechanism $\phi = 141°$, $\delta = 59°$, $\lambda = 320°$), rotated only 6° from McKenzie's (1972) in the formalism of Kagan (1991). This mechanism also shares its normal faulting character with the composite solution ($\phi = 249°$, $\delta = 47°$, $\lambda = 259°$), proposed by Karakostas et al. (2010; Figure 6 and Table 1) for a

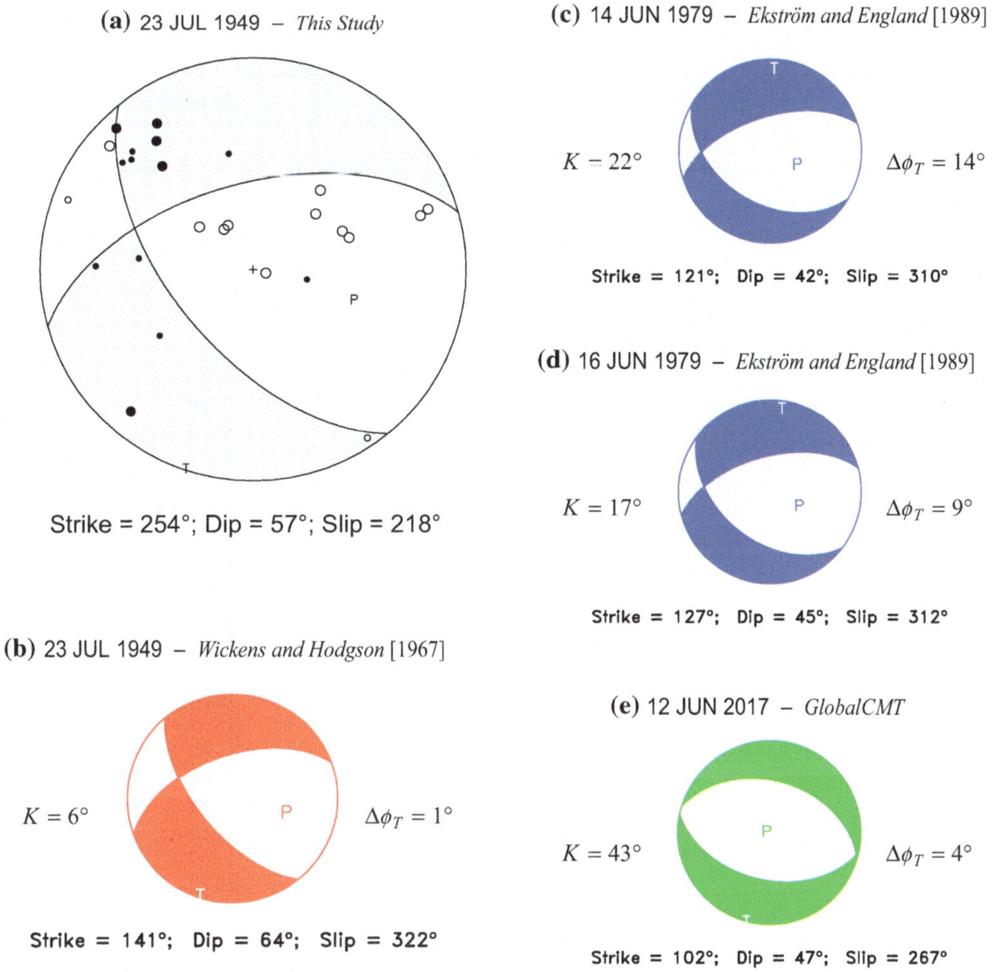

Figure 4

Top left: focal mechanism obtained in this study for the 1949 Chios earthquake. Full symbols are anaseismic arrivals, open ones kataseismic; large symbols read as part of this study, small ones transcribed from the ISS, as listed in Table 2. Bottom left: mechanism proposed by Wickens and Hodgson (1967). Right: published mechanisms for neighboring earthquakes in the vicinity of Lesvos. For the four referenced mechanisms, we give, with respect to our solution, the solid rotation angle K in the formalism of Kagan (1991), and the difference $\Delta\phi_T$ in the azimuths of tensional axes (always sub-horizontal)

cluster of recent low-level seismicity centered about 15 km South of our relocated epicenter.

On Fig. 4, we further compare the 1949 mechanism with those of the nearby 2017 Lesvos earthquake, and of two 1979 earthquakes, studied by Ekström and England (1989), $\sim$30 km to the Southeast of Lesvos. All feature normal faulting, and a mechanism very close to that of the 1949 earthquake, as measured by the angle K of solid rotation separating them in the formalism of Kagan (1991). Even more remarkable is the agreement between tensional axes **T**, all being sub-horizontal and differing at most 14° in azimuth.

This general pattern of consistent tensional axes in the region was pointed out by Taymaz et al. (1991) from a compilation of generally smaller events. It expresses the deformation of the Aegean Basin upon the transformation of strike-slip motion between Asia Minor and Eurasia, notably along the North Anatolian Fault, to the collisional regime between the African plate and the Aegean basin at the Hellenic Arc to the South.

Table 2

First motion data used in the focal mechanism study

Station Name	Code	Polarity (*)	Distance (°)	Azimuth (°)
Read on original seismograms				
Athens, Greece	ATU	k	2.17**	250.9
Istanbul, Turkey	IST	a	3.11**	40.5
Zagreb, Croatia	ZAG	a	10.43	316.2
Trieste, Italy	TRI	k	11.63	311.0
Prague, Czech Republic	PRA	a	14.16	327.2
Tbilisi, Georgia	TIF	k	14.42	72.1
Göttingen, Germany	GTT	a	17.20	323.6
De Bilt, The Netherlands	DBN	a	19.88	319.2
Yekaterinberg, Russia	SVE	k	28.85	39.9
Toshkent, Uzbekistan	TAS	k	32.73	71.5
Almaty, Kazakhstan	AAA	k	37.95	66.4
Irkutsk, Russia	IRK	k	53.65	48.1
Harvard, Massachusetts	HRV	k	70.01	308.9
Tucson, Arizona	TUC	k	98.71	324.4
Pasadena, California	PAS	k	100.22	330.7
Wellington, New Zealand	WEL	k	155.79***	106.2
Additional data reported by the ISS				
Belgrade, Serbia	BEO	k	7.51**	326.3
Rome Monte Porzio, Italy	RMP	k	10.84	290.9
Zürich, Switzerland	ZUR	a	15.55	309.5
Stuttgart, Germany	STU	a	15.90	314.8
Strasbourg, France	STR	a	16.61	312.5
Algiers, Algeria	ALG	a	18.47	271.3
Uppsala, Sweden	UPP	a	21.86	348.2
Cartuja, Spain	CRT	a	23.57	275.7
Tamanghaset, Algeria	TAM	a	23.82	234.2
Jakarta, Indonesia	BAT	a	86.46	100.8

Station names reflect present spelling and political boundaries

(*) a: anaseismic (first motion up); k: kataseismic (first motion down)

(**) Station maps onto upper focal hemisphere

(***) Core phase (PKP)

3.2. Seismic Moment

We quantified the seismic moment of the 1949 Chios earthquake from records of mantle surface waves at the teleseismic stations listed in Table 3, with representative examples of waveforms shown on Fig. 5. The records were digitized and processed through the M_m algorithm (Okal and Talandier 1989), with results shown on Fig. 6 in the form of the variation with frequency of the mantle magnitude, corrected for focal mechanism and depth, related to seismic moment through:

$$M_c = \log_{10} M_0 - 20 \qquad (1)$$

where M_0 is in dyn cm. Figure 6 shows a slight growth of moment with decreasing frequency (with a slope of -0.07 logarithmic units per mHz) and suggests a value of 7×10^{26} dyn cm at the longest sampled periods (170–200 s).

3.3. Macroseismic Modeling

We resolved the indeterminacy between the two fault planes on Fig. 4a by modeling the field of macroseismic data. Galanopoulos (1954, Figure 2) provided an original compilation of isoseismals and described widespread destruction corresponding to Mercalli-Sieberg Intensity VIII in the Northern part

Table 3

Surface wave records used in the moment determination

Station	Code	Distance (°)	Azimuth (°)	Instrument type	Mantle waves used
De Bilt, The Netherlands	DBN	19.9	319.2	Golitsyn	G_1, R_1
San Juan, Puerto Rico	SJG	80.7	285.9	Wenner	G_1, R_1
Tucson, Arizona	TUC	98.7	324.4	Benioff 1-77	R_1
Pasadena, California	PAS	100.2	330.8	Benioff 1-90	G_1

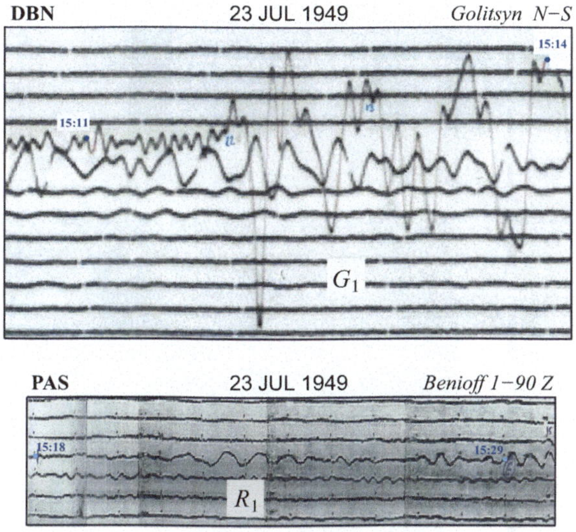

Figure 5
Representative waveforms used in the determination of the seismic moment of the 1949 Chios earthquake. Top: Love wave recorded on the Golitsyn N–S instrument at De Bilt, The Netherlands. Bottom: Rayleigh wave recorded on the Benioff 1–90 vertical instrument at Pasadena

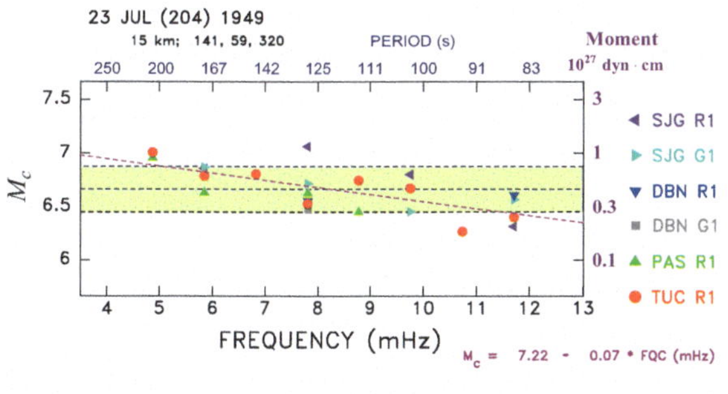

Figure 6
Mantle magnitudes computed for the 1949 Chios earthquake. The horizontal scale is linear in frequency. The horizontal dashed line is the average magnitude corrected for focal mechanism, M_c (Okal and Talandier 1989) and the yellow band its 1–σ confidence interval. Equivalent moments (logarithmic scale) shown at right. The purple dashed line (and equation at *bottom right*) is a linear regression of M_c with frequency

of Chios (reaching IX in Kardamyla), with several villages devastated. He also documented a comparable level of destruction in the Northern part of the interior of the Karaburun Peninsula in Turkey.

In this study, we re-examined all telegrams, completed questionnaires and newspaper clippings and other unpublished material available as part of the NOA Bulletin Library. At each reporting location, a new Modified Mercalli Intensity (MMI) was assigned, and the material converted into a homogeneous dataset of macroseismic intensities, listed in Table 4 (see "Appendix"). The resulting dataset was smoothed using a kriging technique (Schenková et al. 2007; Linkimer 2008), and is presented in Fig. 7a, b. We note that, even though the earthquake was reported felt on Lesvos, 40 km to the North, there were no reports of destruction, indicating an intensity of only VI on the Mercalli-Sieberg or Modified Mercalli scales.

Based on the long-period seismic moment of 7×10^{26} dyn cm, we use Geller's (1976) scaling laws to derive a rupture length $L = 40$ km, a width $W = 20$ km, and a seismic slip $\Delta u = 1.2$ m. For each of the two possible fault planes, we position the top of the fault at a depth of 5 km below the Earth surface, and use the Scenario Shake Map Calculator (Field et al. 2005) to compute a field of maximum ground accelerations, which are then converted to Modified Mercalli Intensities.

Results are shown on Fig. 7c, d. It is clear that the use of the NNW-dipping plane ("Model 1"; $\phi = 254°$, $\delta = 57°$, $\lambda = 218°$; Fig. 7c) adequately reproduces the concentration of maximum intensities in the Northern part of Chios, with similar intensities in Northern Karaburun. By contrast, the SW-dipping plane ("Model 2"; $\phi = 141°$, $\delta = 59°$, $\lambda = 320°$; Fig. 7d) results in weaker intensities, not exceeding VII in Northern Chios, with the field of Intensity VIII

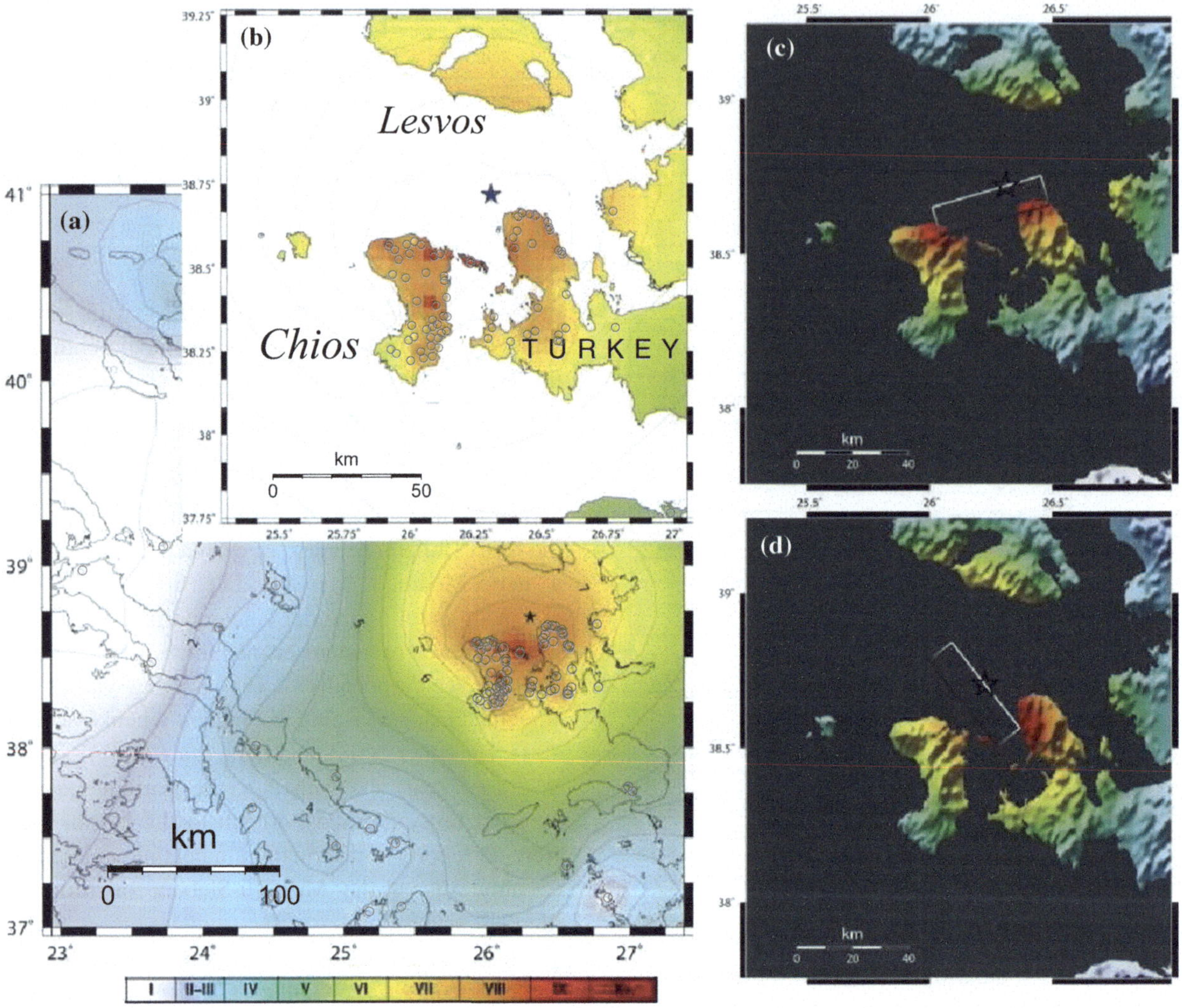

Figure 7

Macroseismic dataset and modeling. **a** Map of the dataset obtained in this study. Individual points shown as circles color-coded according to their MSI value. The background shading is the result of the kriging procedure for the entire Aegean Sea. **b** Close-up of **a** with shading restricted to land masses. Note highest MSI values (VIII$\frac{1}{2}$ to IX) restricted to Northern Chios, one inland location (Karies), Oinousses, and one location on the Western Coast of Karaburun (Küçükbahçe Bucağı), and low values (VI$\frac{1}{2}$ to VII) at Lesvos. **c** Field of MSI values predicted for Model 1 by the Scenario Shake Map Calculator. The rectangle shows the projection of the rupture area and the star denotes the epicenter. **d** Same as **c** for Model 2. Note better agreement of Model 1 with dataset in **b**

restricted to Karaburun, in contradiction with the field data (Fig. 7a). Note that both predict weak intensities, not exceeding VI on the Southern coast of Lesvos. On this basis, we prefer to interpret the NNW-dipping plane as the fault plane.

We note that this plane defines the Southern boundary of the Lesvos Basin, which reaches a depth of 800 m between the islands of Lesvos and Chios. The basin is limited to the North by the unnamed SSW-dipping normal fault along which the 2017 Lesvos earthquake took place (Kiratzi 2018). The combination of the two facing normal faults takes up the local tensional regime in the general "telephone book" model of Taymaz et al. (1991).

Finally, given the poor depth control of our relocations, we have examined the effect on our macroseismic modeling of deepening the seismic source by 20 km, thus keeping it within the range of acceptable solutions suggested by Fig. 2b, but at the same time keeping all other source parameters (seismic moment, focal mechanism, source dimensions). Not surprisingly, we found that the modeled intensities are significantly lower (by about two units) for the deep source, with a maximum of MMI VII, which does not reproduce the intensities compiled in Table 4 and described by Galanopoulos (1954). These results thus confirm the shallow character of the source.

3.4. Source Slowness and Parameter Θ

We further investigated the possible existence of slowness in the source of the 1949 Chios earthquake through the parameter Θ introduced by Newman and Okal (1998) in the framework of Boatwright and Choy's (1986) methodology:

$$\Theta = \log_{10} \frac{E^E}{M_0} \qquad (2)$$

where E^E is the estimated seismic energy radiated into teleseismic body waves, which is obtained by integrating the energy flux at a receiver station. Earthquakes following scaling laws are expected to feature $\Theta = -4.90$; lower values, typically $\Theta < -5.80$, are characteristic of source slowness, as featured by "tsunami earthquakes" (Kanamori 1972; Newman and Okal 1998); by contrast, higher values, typically $\Theta > -4.50$ are typical of "snappy" earthquakes, in particular intraplate events featuring higher stress drops. The method can be successfully applied to historical earthquakes (e.g., Okal and Kirby 2002; Okal and Borrero 2011), as long as adequate short-period teleseismic records are available. In the present case, we obtained short-period records at Harvard (HRV) and Tucson (TUC). At TUC, the epicentral distance of 98.7° warrants a special correction; following Okal and Saloor (2017), we use a nearby reference event, in this case the Lesvos earthquake of 12 June 2017, whose epicenter is only 25 km from the 1949 one and with a normal faulting mechanism rotated only $K = 39°$ (out of a possible 120°) in the formalism of Kagan (1991). We compute a value of $\Theta = -4.78$ for the Lesvos event, globally averaged using stations at standard distances, and then use regional values from a dense network at greater distances in Western North America, to define an empirical distance correction beyond 85° ($Corr. = 0.0923 \times \Delta - 8.04$), which we then apply to the 1949 record at TUC ($Corr. = 1.07$). We use the instrument response documented for 1947 at TUC by Okal and Saloor (2017) to obtain a final value $\Theta = -4.44$, shown as the centered blue square on Fig. 8. On the other hand, the station at HRV is at an appropriate distance (70.0°), but the gain of the instrument suffers from uncertainty (Ishii, pers. comm., 2018); the value of 25,000 reported by Charlier and van Gils (1953) would yield $\Theta = -4.87$, a significantly lower value than at TUC. At any rate, none of the two available records confirms the trend towards slowness suggested by the mantle wave study and which could affect the tsunami generation. The tentative value at HRV (-4.87) fits world-wide averages (-4.80; -4.98) and the theoretical value (-4.90) predicted by scaling laws Boatwright and Choy 1986; Newman and Okal 1998); the one at TUC (-4.44), probably more reliable on account of better constraints on instrumental response, would characterize the 1949

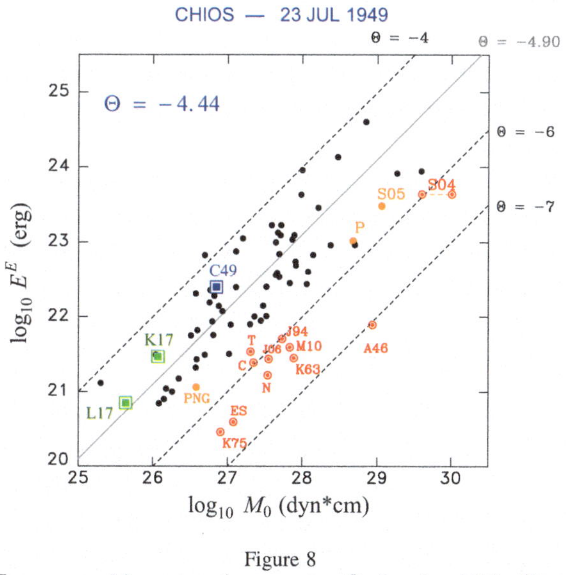

Figure 8
Energy vs. Moment and parameter Θ for the 1949 Chios earthquake (C49; shown in blue). The diagonal lines correspond to constant Θ. The points in the background constitute a selection of significant earthquakes of the past 25 years; "tsunami earthquakes" featuring Θ deficient by as much as 1.5 logarithmic unit, are shown in red. See Okal and Saloor (2017; Figure 13) for a complete list of labels. The 2017 Aegean events at Lesvos (L17) and Kos (K17) are shown in green

earthquake as "snappy", and is in general agreement with global values obtained in this study for the nearby 2017 Lesvos and Kos earthquakes (− 4.78 and − 4.59, respectively; see Fig. 8), which also featured comparable focal mechanisms. The apparent discrepancy between our surface wave results, which would advocate a trend towards slowness, and the energy estimated from *P* waves, which supports a snappy character for the source, can be reconciled by the presence of a high-frequency wavetrain (peaked at 2 Hz) in the TUC record, arriving about 40 s into the time series of the *P* wave, the resulting interpretation being a complex source with a later, more impulsive component.

4. Tsunami Field Survey

A field survey was conducted on 14–16 July 2015, which aimed at recording memories of elderly witnesses of the 1949 earthquake and tsunami. Its methodology followed our experience in previous surveys of historical tsunamis (e.g., Okal et al. 2002, 2009, 2015). We were able to interview a total of about 30 witnesses in the towns of Chios, Lagkada, Marmaro, Katarraktes (Chios); Oinousses; and Çeşme, Ilıca and Mordoğan (Karaburun Peninsula, Turkey); these locations are shown on Fig. 1. In the days following the event, local newspapers provided descriptions of wave activity, mentioned by Galanopoulos (1954); this information was later reproduced in Ambraseys (1988), Papazachos and Papazachou (1997), Altınok et al. (2005) and Solov'ev et al. (2013). Through the systematic interview of elderly witnesses, we sought to confirm, or as the case would be infirm, these reports.

On the island of Chios and in very general terms, our witnesses had vivid memories of the destruction caused by the earthquake, and accurately reported its main effects as having occurred in the northeastern villages, but none of them described flooding in the port of Chios, contrary to the description of a 2-m run-up by the national newspaper *Akropolis* (27 July 1949), later transcribed by Papazachos and Papazachou (1997), and referred to by Altınok et al. (2005), but revealed as incorrect by Papadopoulos (2015).

Similarly, none of our witnesses had any recollection of a significant subsidence of the port of Chios city, contrary to the report in *Vatan* (35 cm) mentioned by Altınok et al. (2005). We note that Model 1, preferred by our macroseismic study, would predict a slight *uplift* of only a few cm at that location, while Chios City would sit on a neutral line experiencing less than 1 cm vertical displacement under Model 2. It is thus probable that the report in *Vatan* relates to a destruction of port infrastructure by the shaking, rather than to a static subsidence.

The only reference to wave activity on the island of Chios was the confirmation of the oscillation of water, with an amplitude of 70 cm, at Marmaro, originally mentioned by Galanopoulos (1954), and reproduced in later publications. It was interpreted as seiching of the local bay, ~ 1.8 km by 0.7 km in dimensions by Altınok et al. (2005).

On the Karaburun peninsula, we were able to interview a number of elderly residents in the villages of Ilıca and Mordoğan, as well as in Çeşme. One of our witnesses (75 years old in 2015, or 9 years at the time of the event), described structural damage to houses in Alaçatı where she lived at the time, and reddish water sprouting to a height of about 6 m on Ilıca beach, a phenomenon reported in the local press at the time of the event, and mentioned by Altınok et al. (2005). However, our witnesses mentioned no significant flooding or standard wave activity on the beach, a conclusion upheld by a careful examination of the local newspaper *Akşam*, which documents sprouting (which we have confirmed with an original photograph), but not flooding. In the town of Mordoğan, two witnesses (aged 8 and 15 at the time of the event) similarly mentioned co-seismic destruction but no wave activity, even though one of them also described rockslides falling into the sea, which apparently did not create observable waves.

Finally, we visited the island of Oinousses on 16 July 2015, where no reports of tsunami activity were mentioned in the local press at the time of the event. We were able to interview eight elderly witnesses, aged 77–82 years (11–16 in 1949). Based on one testimony, we measured a run-up of 0.32 m at the natural level of the quay along the port (a modern quay has since been built 50 cm taller); another witness who was riding a bicycle during the earthquake, along the coastal road to the Merchant Marine School, about 1 km to the West of the port, described a withdrawal of the sea, with an amplitude of ~35 cm. Yet another witness, who was at the time of the earthquake docking his boat at the Eastern end of the island (estimated location 38.514° N, 26.261° E) described having seen water springing to a reported height of 2 m between the two small islets of Gaidouroniso and Pontikonisi at an estimated position of 38.504° N, 26.270° E); however, we regard this height as tentative, given the distance of 1.4 km at which it was observed. We were unfortunately unable to visit those locations.

The dataset of confirmed tsunami amplitudes thus consists of: (1) a run-up of 32 cm at the port of Oinousses (38.514° N, 26.261° E); (2) a seiche of 70 cm amplitude in the Bay of Marmaro (38.546° N; 26.113° E); (3) no documented reports of wave activity (which we interpret as amplitudes probably less than 30 cm) at Chios City and on the Turkish coast (outside of observations of water locally springing from the ground at heights possibly reaching several meters). While this dataset remains meager, we will see that it can be reasonably well simulated by a model derived from our seismological observations.

5. *Tsunami Simulations*

Tsunami simulations were performed using the ComMIT package which was developed by Titov et al. (2011) as a graphical interface to the MOST finite differences algorithm (Titov and Synolakis 1998; Titov et al. 2016). MOST solves the shallow water approximation of the non-linear equations of hydrodynamics, using the method of alternated steps (Godunov 1959). It has been extensively validated through comparisons with laboratory and field data, per standard international protocols; full details can be found in Synolakis (2003) and Synolakis et al. (2008). In the present context, and since ComMIT allows the simulation of the inundation of initially dry land, time series at our virtual gauges are representative of flow depth at the coast line, as previously used in Greece, e.g., by Melis et al. (2016).

We use a bathymetric grid obtained from the European Marine Observation and Data Network (EMODnet), sampled at 30 m. Seismic sources consist of Models 1 and 2, as defined in the macroseismic study in Sect. 3 above, with fault geometries $\phi = 254°$, $\delta = 57°$, $\lambda = 218°$ for the NNW-dipping plane ("Model 1"), and $\phi = 141°$, $\delta = 59°$, $\lambda = 320°$ for the SW-dipping one ("Model 2"). In both instances we use a slip $\Delta u = 1.2$ m on a 40 km × 20 km fault. Initial displacements are calculated using the static deformation algorithm of Mansinha and Smylie (1971).

We then position three virtual gauges at the harbors of Chios Town, Oinousses and Marmaro. Figure 9 plots in red the time series simulated at the three gauges for Model 1, which uses the WSW–ENE striking plane and is preferred on the basis of the macroseismic data, and in blue for Model 2, which uses the NW–SE striking plane. We note that the

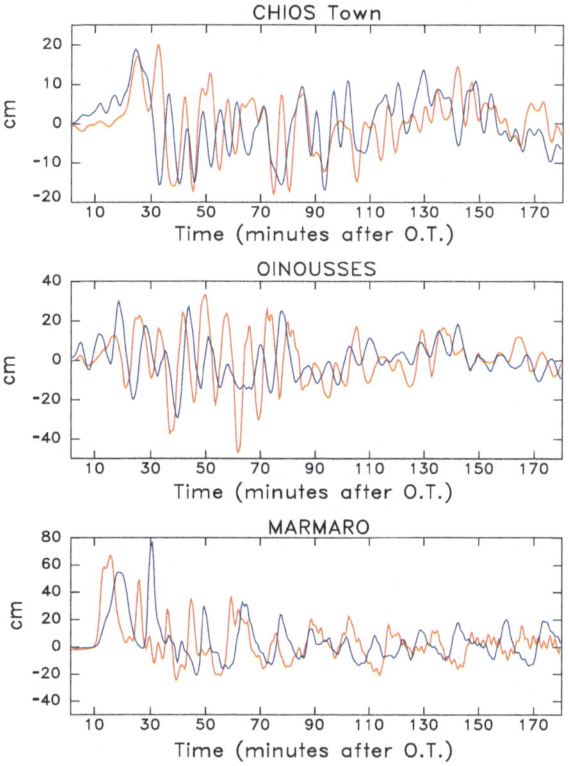

Figure 9
Simulated time series at the virtual gauges in the harbors of Chios Town (top), Oinousses (center), and Marmaro (bottom) for Models 1 (in red) and 2 (in blue). Note that maximum amplitudes for both models are comparable at each location

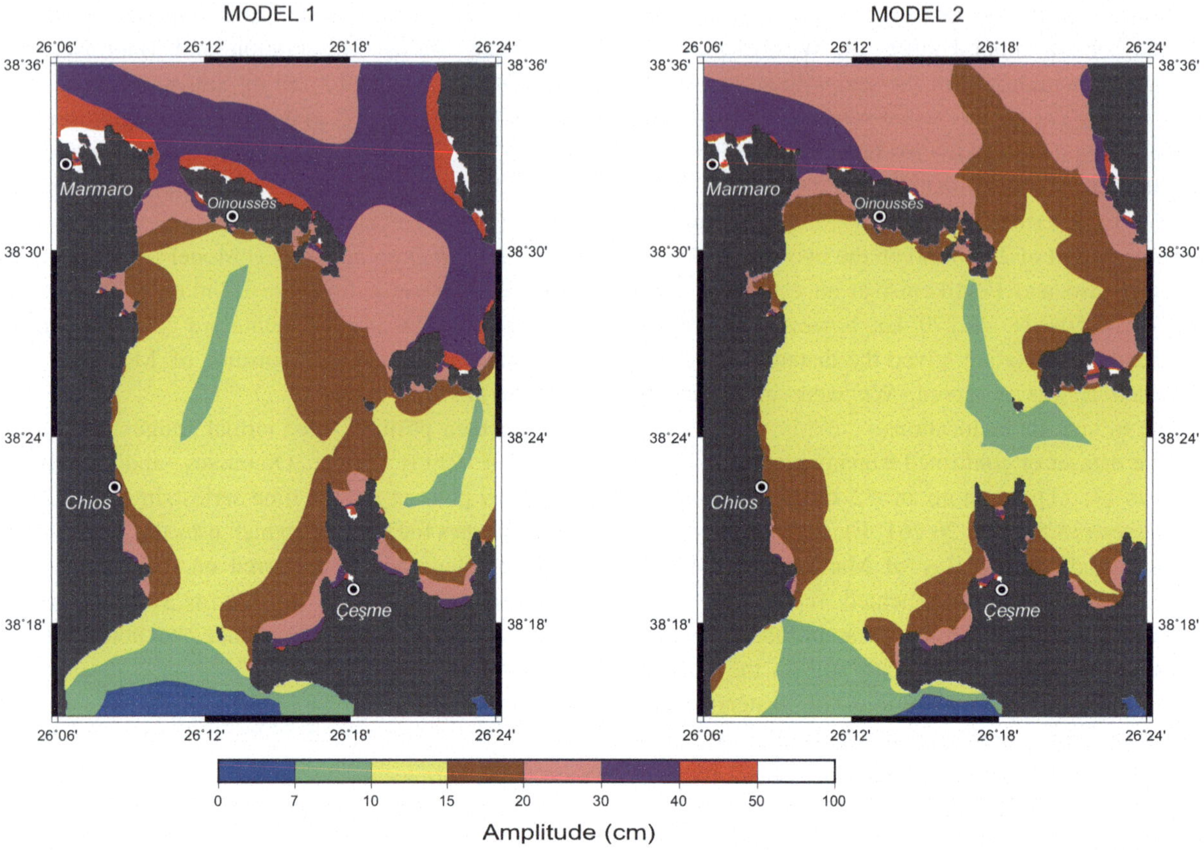

Figure 10
Results of hydrodynamic simulation for Models 1 (left) and 2 (right). In both cases, the field of maximum amplitudes of the tsunami is contoured using the same palet. Note that both models produce comparable amplitude fields, expect for larger η_{max} for Model 1 North of Oinousses and West of the Karaburun Peninsula. The bull's eye symbols identify the locations contributing to the database, occasionally slightly displaced so as not to mask simulation details

resulting amplitudes remain comparable at the three locations, and thus cannot be used to discriminate between the models: both reproduce well the amplitudes resulting from our survey (34 cm at Oinousses and ~70 cm at Marmaro); at Chios Town, the simulated amplitudes (20 and 19 cm, respectively) are generally compatible with the lack of reports by our witnesses, notably given the significant level of destruction by the earthquake.

Figure 10 further shows a map of the maximum simulated wave amplitude η_{max} in our study area. Again, we note that results from the two models are very similar, with differences concentrated on the Northern coast of Oinousses, which is rugged and unpopulated, even to this day, as well as along the Northwestern coast of Karaburun, notably at the 2-km long bay South of Küçükbahçe Bucağı, a location also unpopulated in 1949. In this context, we must conclude that the evidence retrieved during our field survey is insufficient to discriminate between the two fault models. Seventy years after the fact, it is doubtful that additional evidence could be collected in what were, and largely remain, unpopulated areas.

Finally, we compare the 1949 tsunami with that generated by the recent Bodrum-Kos earthquake of 20 July 2017, approximately 240 km to the Southeast. Despite a significantly smaller moment of only 1.2×10^{26} dyn cm, the latter generated a locally stronger tsunami, with run-up reaching 1.9 m near Bodrum, and inundation up to 60 m. We note however that these effects were constrained to a number of specific locations involving small bays, coves and estuaries

(Doğan et al. 2019). By contrast, amplitudes recorded on tidal gauges did not exceed 20 cm, as reported and successfully modeled by Heidarzadeh et al. (2017).

In this context, the larger 2017 tsunami amplitudes are the results of site response effects, and not directly attributable to a different genesis at the source.

6. Discussion and Conclusions

We conducted a complete, modern seismological reassessment of the 1949 Chios earthquake, the second largest event of the past 100 years in the Central Aegean Sea. The event relocates to the basin separating Chios and Lesvos, in the general area of other modern relocations, but about 20 km NNE of the older ones, such as Galanopoulos' (1954); in this context, it is difficult to associate the 1949 event with the proposed Chios–Karaburun Fault. On the basis of individually read first motions, we confirm a normal faulting mechanism, expressing the "telephone book" extension during the transition from the strike-slip motion along the North Anatolian Fault to the East to the collision in the Southern Aegean to the Southwest. Mantle surface waves obtained at a number of teleseismic stations suggest a low-frequency moment of 7×10^{26} dyn cm, making this the second largest instrumentally recorded event in the Aegean. A complete recompilation of all available macroseismic data yields a distribution favoring rupture along the NNW-dipping plane. That structure complements the SSW-dipping fault which ruptured during the 2017 Lesvos earthquake to the North, on the other side of the Lesvos Basin (Kiratzi 2018). While this interpretation is in agreement with the description of the Northern Chios fault by Sboras (2011), as and listed by Karakaisis et al. (2010), the latter was described by Karakostas et al. (2010) as dipping to the South based on recent low-level seismicity.[1]

A field survey was carried out in order to collect memories of the event, and in particular of its tsunami, from surviving elderly witnesses. While most of them remembered the destruction caused by the earthquake, we could gather only a limited amount of quantitative data regarding the tsunami, which cannot help discriminate between the two possible fault planes of the 1949 earthquake. However, an important result is that both models provide an acceptable fit to the reported amplitudes. In this respect, the latter are compatible with the order of magnitudes of the maximum amplitudes η_{max} simulated from the dislocative sources derived from our seismological study, and we do not have to invoke ancillary sources such as underwater landslides. This is in contrast to the case of other historical tsunamis in the Aegean Sea, such as the 1956 Amorgos and 1948 Rhodos events, for which instances of spatially isolated excessive run-up amplitudes required generation by landslides (Okal et al. 2009; Ebeling et al. 2012).

Acknowledgements

We are grateful to Brian Ferris (Lower Hutt), Marjan Herak (Zagreb), and Eliška Zábranová (Prague) for access to historical records. Some figures were drafted using the GMT software (Wessel and Smith 1991). We thank two anonymous reviewers for their comments on the original draft of the paper. We salute the memory of Neşe Kânoğlu who shared with us the surveying in Turkey before her untimely passing in 2017.

Appendix

See Table 4.

[1] We note here an inconsistency in the numbering by Karakostas et al. (2010) of several of the clusters, between their Figs. 6 and 7 and their Table 1. This makes any reference to their work difficult and potentially unreliable.

Table 4

Macroseismic dataset compiled in this study

Number	Location		Latitude (°)	Longitude (°)	Epicentral distance (km)	Modified Mercalli intensity
	Town	Country				
1	Kardamyla	GR	38.538	26.085	28	IX
2	Karies	GR	38.391	26.093	41	IX
3	Oinousses	GR	38.524	26.222	22	IX
4	Egrigoros	GR	38.575	25.917	38	VIII$\frac{1}{2}$
5	Inousses	GR	38.515	26.221	23	VIII$\frac{1}{2}$
6	Küçükbahçe Bucağı	TR	38.561	26.390	18	VIII$\frac{1}{2}$
7	Amades	GR	38.571	26.037	29	VIII
8	Anbarseki Köyü	TR	38.614	26.523	21	VIII
9	Bozköy	TR	38.657	26.468	14	VIII
10	Chios	GR	38.358	26.121	43	VIII
11	İnecik	TR	38.551	26.567	28	VIII
12	Karaburun	TR	38.638	26.513	19	VIII
13	Keramos	GR	38.556	25.936	38	VIII
14	Kösedere	TR	38.554	26.554	27	VIII
15	Kourounia	GR	38.566	25.912	39	VIII
16	Marmaro	GR	38.544	26.108	26	VIII
17	Pispilounta	GR	38.527	25.951	38	VIII
18	Saip	TR	38.622	26.519	20	VIII
19	Spartounta	GR	38.545	25.992	34	VIII
20	Tepeboz	TR	38.661	26.447	12	VIII
21	Anavatos	GR	38.403	26.020	43	VII$\frac{1}{2}$
22	Armolia	GR	38.253	26.036	56	VII$\frac{1}{2}$
23	Birgi	TR	38.303	26.555	50	VII$\frac{1}{2}$
24	Çeşme	TR	38.324	26.303	43	VII$\frac{1}{2}$
25	Chalkio	GR	38.334	26.098	46	VII$\frac{1}{2}$
26	Coste	TR	38.356	26.311	39	VII$\frac{1}{2}$
27	Dapnonas	GR	38.347	26.080	45	VII$\frac{1}{2}$
28	Diefcha	GR	38.473	25.978	40	VII$\frac{1}{2}$
29	Eğlenhoca	TR	38.543	26.568	28	VII$\frac{1}{2}$
30	Exo Didima	GR	38.270	26.074	53	VII$\frac{1}{2}$
31	Flatsia	GR	38.239	26.079	56	VII$\frac{1}{2}$
32	Germiyan	TR	38.314	26.466	46	VII$\frac{1}{2}$
33	Hasseki	TR	38.664	26.417	10	VII$\frac{1}{2}$
34	Kalamoti	GR	38.234	26.046	58	VII$\frac{1}{2}$
35	Kallimasia	GR	38.293	26.101	50	VII$\frac{1}{2}$
36	Kampia	GR	38.572	25.982	33	VII$\frac{1}{2}$
37	Katarraktis	GR	38.265	26.103	53	VII$\frac{1}{2}$
38	Lagkada	GR	38.478	26.123	31	VII$\frac{1}{2}$
39	Ildır	TR	38.383	26.477	39	VII$\frac{1}{2}$
40	Neochori	GR	38.308	26.113	48	VII$\frac{1}{2}$
41	Parlak	TR	38.612	26.398	13	VII$\frac{1}{2}$
42	Pitious	GR	38.488	26.055	34	VII$\frac{1}{2}$
43	Salman	TR	38.592	26.384	14	VII$\frac{1}{2}$
44	Sarpıncık	TR	38.654	26.403	10	VII$\frac{1}{2}$
45	Sikiada	GR	38.464	26.124	32	VII$\frac{1}{2}$
46	Thimiana	GR	38.314	26.132	47	VII$\frac{1}{2}$
47	Tholopotami	GR	38.295	26.073	51	VII$\frac{1}{2}$
48	Varvasi	GR	38.357	26.135	42	VII$\frac{1}{2}$
49	Vrontados	GR	38.414	26.132	37	VII$\frac{1}{2}$
50	Yaylaköy	TR	38.574	26.455	19	VII$\frac{1}{2}$
51	Zifias	GR	38.327	26.080	47	VII$\frac{1}{2}$
52	Agios Georgios	GR	38.319	26.057	49	VII
53	Barbaros	TR	38.323	26.581	49	VII

Table 4 *continued*

Number	Location		Latitude (°)	Longitude (°)	Epicentral distance (km)	Modified Mercalli intensity
	Town	Country				
54	Çiftlik	TR	38.292	26.288	47	VII
55	Lithi	GR	38.331	26.001	50	VII
56	Foça	TR	38.670	26.758	38	VII
57	Reisdere	TR	38.304	26.437	46	VII
58	Uzunkuyu Bucağı	TR	38.285	26.551	51	VII
59	Viki	GR	38.580	26.012	30	VII
60	Volissos	GR	38.483	25.928	42	VII
61	Zeytinler	TR	38.284	26.561	52	VII
62	Alaçatı	TR	38.283	26.374	48	VI$\frac{1}{2}$
63	Balıklıova	TR	38.423	26.584	39	VI$\frac{1}{2}$
64	Elata	GR	38.287	25.989	55	VI$\frac{1}{2}$
65	Pirgi	GR	38.227	25.997	61	VI$\frac{1}{2}$
66	Vessa	GR	38.297	26.010	53	VI$\frac{1}{2}$
67	Mesta	GR	38.260	25.922	61	VI
68	Olimpi	GR	38.248	25.942	61	VI
69	Urla	TR	38.325	26.767	58	VI
70	Myrina	GR	39.881	25.083	168	V$\frac{1}{2}$
71	Patmos	GR	37.330	26.544	155	V
72	Samos	GR	37.772	26.974	119	V
73	Vathy	GR	37.746	27.004	123	V
74	Andros	GR	37.826	24.935	156	IV$\frac{1}{2}$
75	Karystos	GR	37.998	24.362	188	IV$\frac{1}{2}$
76	Ermoupoli	GR	37.440	24.933	186	IV
77	Ierissos	GR	40.415	23.833	285	IV
78	Kalymnos	GR	36.970	26.982	202	IV
79	Naxos	GR	37.103	25.391	196	IV
80	Ioulis	GR	37.657	24.336	209	III
81	Milos	GR	36.682	24.433	280	III
82	Mykonos	GR	37.455	25.350	163	III
83	Paros	GR	37.075	25.164	208	III
84	Seriphos	GR	37.163	24.477	236	III
85	Tinos	GR	37.538	25.172	165	III
86	Kymi	GR	38.652	24.102	193	II$\frac{1}{2}$
87	Skyros	GR	38.882	24.506	159	II$\frac{1}{2}$
88	Chalkida	GR	38.462	23.630	236	I
89	Istiaia	GR	38.965	23.148	277	I
90	Kassandreia	GR	40.062	23.384	294	I
91	Leros	GR	37.153	26.833	179	I
92	Samothraki	GR	40.456	25.577	204	I
93	Skiathos	GR	39.172	23.470	252	I
94	Skopelos	GR	39.097	23.714	230	I
95	Sykia	GR	40.034	23.938	252	I
96	Thasos	GR	40.768	24.691	268	I

References

Altınok, Y., Alpar, B., Özer, N., & Gazioğlu, C. (2005). 1881 and 1949 earthquakes in the Chios-Çeşme Strait (Aegean Sea) and their relation to tsunamis. *Natural Hazards and Earth System Sciences*, *5*, 717–725.

Ambraseys, N. N. (1998). Engineering seismology. *Earthquake Engineering and Structural Dynamics*, *17*, 1–105.

Ambraseys, N. N. (2001). Reassessment of earthquakes, 1900–1999, in the Eastern Mediterranean and the Middle East. *Geophysical Journal International, 145*, 471–485.

Boatwright, J., & Choy, G. L. (1986). Teleseismic estimates of the energy radiated by shallow earthquakes. *Journal of Geophysical Research, 91*, 2095–2112.

Caputo, R., Chatzipetros, A., Pavlides, S., & Sboras, S. (2012). The Greek database of seismogenic sources (GreDaSS): State-of-the-art for northern Greece. *Annals of Geophysics, 55*, 859–894.

Charlier, C., & van Gils, J.-M. (1953). *Catalogue des stations séismologiques mondiales*. Uccle: Observatoire Royal de Belgique.

Chatzipetros, A., Kiratzi, A., Sboras, S., Zouros, N., & Pavlides, S. (2013). Active faulting in the north-eastern Aegean Sea islands. *Tectonophysics, 507*, 106–122.

Doğan, G. G., Annunziato, A., Papadopoulos, G. A., Güler, H. G., Yalçıner, A. C., Çakır, T. E., et al. (2019). The 20th July 2017 Bodrum-Kos tsunami field survey. *Pure and Applied Geophysics, 176*, 2925–2949.

Ebeling, C. W., Okal, E. A., Kalligeris, N., & Synolakis, C. E. (2012). Modern seismological reassessment and tsunami simulation of historical Hellenic Arc earthquakes. *Tectonophysics, 530*, 225–239.

Ekström, G., & England, P. (1989). Seismic strain rates in regions of distributed continental deformation. *Journal of Geophysical Research, 94*, 10231–10257.

Engdahl, E. R., & Villaseñor, A. (2002). Global seismicity: 1900–1999. *International earthquake and engineering seismology Part A* (pp. 665–690). New York: Elsevier.

Erkman, H. K. (1949). *23-7-1949 Karaburun (İzmir) zelzelesı* (19 pp.). Istanbul: Kandilli Observatory.

Field, E. H., Seligson, H. A., Gupta, N., Gupta, V., Jordan, T. H., & Campbell, K. (2005). Loss estimates for a Puente Hills Blind-Thrust earthquake in Los Angeles. *Earthquake Spectra, 21*, 329–338.

Galanopoulos, A. (1954). Die Seismizität der Insel Chios. *Gerlands Beiträge zur Geophysik, 63*, 253–264.

Geller, R. J. (1976). Scaling relations for earthquake source parameters and magnitudes. *Bulletin of the Seismological Society of America, 66*, 1501–1523.

Godunov, S. K. (1959). Finite difference methods for numerical computations of discontinuous solutions of the equations of fluid dynamics. *Matematicheskii Sbornik, 47*, 271–295.

Gutenberg, B., & Richter, C. F. (1942). Earthquake magnitude, intensity, energy and acceleration. *Bulletin of the Seismological Society of America, 32*, 163–191.

Gutenberg, B., & Richter, C. F. (1954). *Seismicity of the earth and associated phenomena* (310 pp.). Princeton: Princeton University Press.

Heidarzadeh, M., Necmioğlu, O., Ishibe, T., & Yalçıner, A. C. (2017). Bodrum–Kos (Turkey–Greece) $M_w = 6.6$ earthquake and tsunami of 20 July 2017: A test for the Mediterranean tsunami warning system. *Geoscience Letters, 4*(31), 11.

Kagan, Y. Y. (1991). 3-D rotation of double-couple earthquake sources. *Geophysical Journal International, 106*, 709–716.

Kanamori, H. (1972). Mechanism of tsunami earthquakes. *Physics of the Earth and Planetary Interiors, 6*, 346–359.

Karakaisis, G. F., Papazachos, C. B., & Scodilis, E. M. (2010). Seismic sources and main seismic faults in the Aegean and surrounding area. *Bulletin of the Geological Society of Greece, 43*, 2026–2042.

Karakostas, V. G., Papadimitriou, E. E., Tranos, M. D., & Papazachos, C. B. (2010). Active seismotectonic structures in the area of Chios Island, North Aegean Sea, revealed from microseismicity and fault plane solutions. *Bulletin of the Geological Society of Greece, 43*, 2064–2074.

Kiratzi, A. A. (2018). The 12 June 2017 $M_w = 6.3$ Lesvos Island (Aegean Sea) earthquake: Slip model and directivity estimated with finite-fault inversion. *Tectonophysics, 724*, 1–10.

Labrouste, H., & Pinar, N. (1953). Etude microséismique des tremblements de terre du 23 juillet 1949 et du 13 août 1951 en Turquie. *Bull Inform UGGI, 2*, 267–269.

Linkimer, L. (2008). Application of the kriging method to draw isoseismal maps of the significant 2002–2003 Costa Rica earthquakes. *Revista Geologica de America Central, 38*, 119–134.

Mansinha, L. A., & Smylie, D. E. (1971). The displacement fields of inclined faults. *Bulletin of the Seismological Society of America, 61*, 1433–1440.

McKenzie, D. P. (1972). Active tectonics of the Mediterranean region. *Geophysical Journal of the Royal Astronomical Society, 30*, 109–185.

Melis, N. S., Barberopoulou, A., Frentzos, E., & Krassanakis, V. (2016). Scenario based tsunami wave height estimation towards hazard evaluation for the Hellenic coastline and examples of extreme inundation zones in South Aegean. *Geophysical Research Abstracts, 18*, 1. (EGU2-16-11285-1, Vienna [abstract]).

Musson, R. M., Grünthal, G., & Stucchi, M. (2010). The comparison of macroseismic intensity scales. *Journal of Seismology, 14*, 413–428.

Newman, A. V., & Okal, E. A. (1998). Teleseismic estimates of radiated seismic energy: The E/M_0 discriminant for tsunami earthquakes. *Journal of Geophysical Research, 103*, 26885–26898.

Okal, E. A., & Borrero, J. C. (2011). The "tsunami earthquake" of 22 June 1932 in Manzanillo, Mexico: Seismological study and tsunami simulations. *Geophysical Journal International, 187*, 1443–1459.

Okal, E. A., Fritz, H. M., Hamzeh, M. A., & Ghasemzadeh, J. (2015). Field survey of the 1945 Makran and 2004 Indian Ocean tsunamis in Baluchistan, Iran. *Pure and Applied Geophysics, 172*, 3343–3356.

Okal, E. A., & Kirby, S. H. (2002). Energy-to-moment ratios for damaging intraslab earthquakes: Preliminary results on a few case studies. *USGS Open File Reports, 02–328*, 127–131.

Okal, E. A., & Saloor, N. (2017). Historical tsunami earthquakes in the Southwest Pacific: An extension to $\Delta > 80°$ of the energy-to-moment parameter Θ. *Geophysical Journal International, 210*, 852–873.

Okal, E. A., Synolakis, C. E., Fryer, G. J., Heinrich, P., Borrero, J. C., Ruscher, C., et al. (2002). A field survey of the 1946 Aleutian tsunami in the far field. *Seismological Research Letters, 73*, 490–503.

Okal, E. A., Synolakis, C. E., Uslu, B., Kalligeris, N., & Voukouvalas, E. (2009). The 1956 earthquake and tsunami in Amorgos, Greece. *Geophysical Journal International, 178*, 1533–1554.

Okal, E. A., & Talandier, J. (1989). M_m: A variable period mantle magnitude. *Journal of Geophysical Research, 94*, 4169–4193.

Papadopoulos, G. A. (2015). *Tsunamis in the European-Mediterranean region: From historical record to risk mitigation* (271 pp.). Amsterdam: Elsevier.

Papazachos, B., & Papazachou, C. (1997). *The earthquakes of Greece* (304 pp.). Thessaloniki: Ziti.

Pinar, N. (1950). Etude géologique et séismologique du tremblement de terre de Karaburun (Izmir) du 23 juillet 1949. *Revue de la Faculté des Sciences de l'Université d'Istamboul, 15*, 363–375.

Sboras, S. (2011). *The Greek Database of Seismogenic Sources: Seismotectonic implications for North Greece*. PhD Thesis, University of Ferrara, 274 pp.

Schenková, Z., Schenk, V., Kalogeras, I., Pichl, R., Kottnauer, P., Papatsimba, C., et al. (2007). Isoseismal maps drawing by the kriging method. *Search Results, 11*, 345–353.

Solov'ev, S. L., Solov'eva, O. N., Go, Ch. N., Kim, Kh. S., & Shchetnikov, N. A. (2013). *Tsunamis in the Mediterranean Sea, 2000 B.C.–2000 A.D.* (308 pp.). Dordrecht: Elsevier.

Storchak, D. A., Di Giacomo, D., Bondár, I., Engdahl, E. R., Harris, J., Lee, W. H. K., et al. (2013). Public release of the ISC-GEM global instrumental earthquake catalogue (1900–2009). *Seismological Research Letters, 84*, 810–815.

Synolakis, C. E. (2003). Tsunami and seiche. In W.-F. Chen & C. Scawthron (Eds.), *Earthquake Engineering Handbook* (pp. 9_1–9_90). Boca Raton: CRC Press.

Synolakis, C. E., Bernard, E. N., Titov, V. V., Kânoğlu, U., & González, F. (2008). Validation and verification of tsunami numerical models. *Pure and Applied Geophysics, 165*, 2197–2228.

Taymaz, T., Jackson, J., & McKenzie, D. (1991). Active tectonics of the north and central Aegean Sea. *Geophysical Journal International, 106*, 433–490.

Titov, V. V., Kânoğlu, U., & Synolakis, C. E. (2016). Development of MOST for real-time tsunami forecasting. *Journal of Waterway, Port, Coastal, and Ocean Engineering, 142*(6), 03116004.

Titov, V. V., Moore, C. W., Greenslade, D. J. M., Pattiaratchi, C., Badal, R., Synolakis, C. E., et al. (2011). A new tool for inundation modeling: Community Modeling Interface for Tsunamis (ComMIT). *Pure and Applied Geophysics, 168*, 2121–2131.

Titov, V. V., & Synolakis, C. E. (1998). Numerical modeling of tidal wave runup. *Journal of Waterway, Port, Coastal, and Ocean Engineering, 124*, 157–171.

Wessel, P., & Smith, W. H. F. (1991). Free software helps map and display data. *Eos Transactions American Geophysical Union, 72*, 441, 445–446.

Wickens, A. J., & Hodgson, J. H. (1967). Computer re-evaluation of earthquake mechanism solutions, 1922–1962. *Publications of the Dominion Observatory, Ottawa, 33*(1), 71.

Wysession, M. E., Okal, E. A., & Miller, K. L. (1991). Intraplate seismicity of the Pacific Basin, 1913–1988. *Pure and Applied Geophysics, 135*, 261–359.

(Received May 28, 2019, revised December 18, 2019, accepted December 27, 2019, Published online January 30, 2020)

Pure Appl. Geophys. 177 (2020), 1315–1333
© 2019 Springer Nature Switzerland AG
https://doi.org/10.1007/s00024-019-02246-9

Pure and Applied Geophysics

Tsunami Potential of Moderate Earthquakes: The July 1, 2009 Earthquake (M_w 6.45) and its Associated Local Tsunami in the Hellenic Arc

Gian Maria Bocchini,[1,3] T. Novikova,[1] G. A. Papadopoulos,[1] A. Agalos,[1] E. Mouzakiotis,[1] V. Karastathis,[1] and N. Voulgaris[2]

Abstract—On July 1, 2009, a M_w 6.45 earthquake ruptured south of Crete Island (Greece) along the Hellenic Arc triggering a local tsunami. Eyewitness reported the tsunami from Myrtos and Arvi (south-eastern Crete) and from the north of Chrisi Islet, located to the southeast of Crete. The earthquake occurred offshore, about 80 km south of Crete, where routine earthquake locations are poor. The hypocentre is relocated using a 2-D velocity model and several local 1-D velocity models. Epicentral locations obtained by using the different velocity models show very minor variations. Instead, relocated hypocentres can be grouped into two sets of solutions: (1) those with a shallower depth (depth < 12 km) obtained with the 2-D velocity model and the 1-D velocity models having a shallower Moho at less than 30 km, and (2) those with a larger depth (depth of 28 and 40 km) obtained with the velocity models having a Moho at about 40 km. Shallower hypocentres are more consistent with the tsunamigenic nature of the earthquake as also supported by tsunami numerical simulations. In fact, shallow sources (depths < 12 km) are capable of generating tsunami waves, while it is not the case for deeper sources (depth > 25 km) either in the upper-plate or along the plate interface. Models accounting for either homogeneous or heterogeneous slip on the causative fault are tested, with the heterogeneous one better reproducing the observations in terms of number of tsunami waves reaching the shoreline and duration of the sea disturbance. The short travel time, about 10 min, of the first tsunami arrival at the southern coast of Crete represents a big challenge for tsunami early warning systems operating in the area.

Key words: Hellenic subduction zone, tsunamigenic earthquakes, tsunami hazard, earthquake relocation, tsunami early warning systems.

Electronic supplementary material The online version of this article (https://doi.org/10.1007/s00024-019-02246-9) contains supplementary material, which is available to authorized users.

[1] Institute of Geodynamics, National Observatory of Athens, Lofos Nymfon 1, 11810 Athens, Greece. E-mail: bocchini@noa.gr; Gian.Bocchini@rub.de

[2] Department of Geology and Geoenvironment, National and Kapodistrian University of Athens, Athens, Greece.

[3] Faculty of Geosciences, Institute of Geology, Mineralogy, and Geophysics, Ruhr University of Bochum, Bochum, Germany.

1. Introduction

The Hellenic subduction zone (HSZ; Fig. 1) is seismically the most active structure in the eastern Mediterranean Sea. The high seismicity is associated with the convergence between the Nubian Plate and the Aegean region occurring at about 35–40 mm/year (McClusky et al. 2000; Reilinger et al. 2006).

Historical and instrumental seismicity catalogues for Greece, covering more than 2000 years (e.g., Guidoboni and Comastri 2005; Papazachos and Papazachou 2003; Ambraseys 2009), report the occurrence of several large earthquakes of magnitude larger than 7. At least two earthquakes with estimated magnitude larger than 8, namely the 365 CE and the 1303 CE ones, ruptured the western and eastern segments of the HSZ, respectively (Guidoboni and Comastri 1997; Shaw et al. 2008; Papadopoulos 2011). These two earthquakes triggered large tsunamis that inundated many coastal localities in the entire basin of the eastern Mediterranean causing loss of life and devastation (Papadopoulos 2011; Papadopoulos et al. 2014). The largest documented earthquake and associated tsunami occurred in the HSZ, namely the 365 CE ($M \sim 8.3$) event, very likely ruptured the overriding plate southwest of Crete and not the main plate interface as would be expected for such large earthquakes in a subduction zone (Shaw et al. 2008). More recently, a destructive tsunami was also generated by the surface wave magnitude (M_s) 7.4 normal faulting earthquake of the 9th of July 1956 that occurred in the back-arc area (Fig. 1; Galanopoulos 1957). The high seismogenic and tsunamigenic hazard associated with the active margin to the south of Crete, as highlighted by

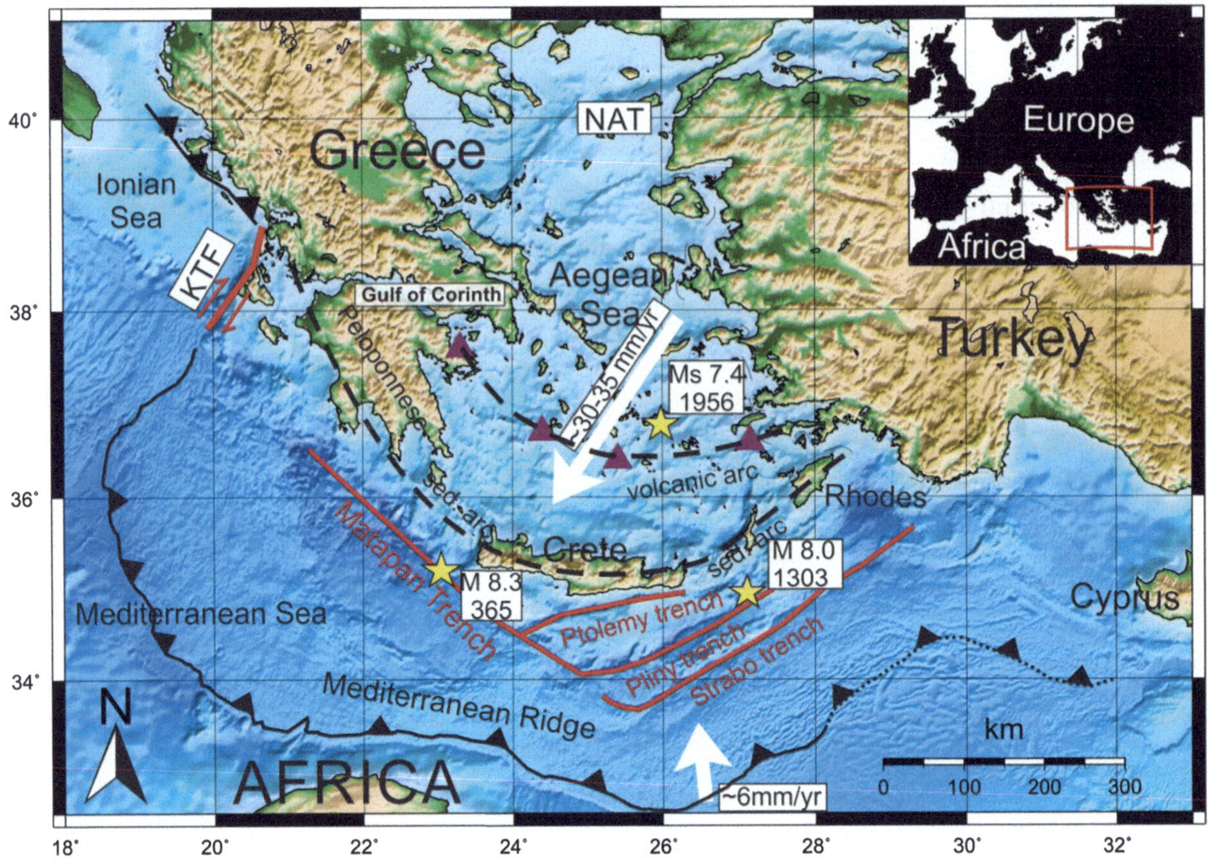

Figure 1

Map view of the Hellenic subduction zone. Line with triangles (in black) indicates the deformation front. Yellow stars indicate the epicentral location of some of the largest tsunamigenic earthquakes that have occurred in the Aegean region, namely the $M \sim 8.3$ 365 CE, the $M \sim 8$ 1303 CE and the more recent M_s 7.4 1956 Amorgos earthquakes (see text for references). White arrows indicate the motion of the Nubian Plate (lower) and of the Aegean region (upper) with respect to the stable Eurasian Plate. The purple triangles in the central Aegean indicate the position of the volcanic arc. *KTF* Kefalonia Transform Fault; *NAT* North Aegean Trough; *sed. arc.* sedimentary arc

historical and instrumental records (Papadopoulos 2011), poses interest in the study of seismogenic and tsunamigenic sources in the area.

On July 1, 2009, UTC 09:30, a moderate magnitude earthquake of moment magnitude (M_w) 6.45 ruptured the southern offshore margin of Crete (Fig. 3) causing a local tsunami of about 0.3 m in height, as reported by eyewitnesses. This earthquake provided a good opportunity to further investigate the seismogenic and tsunamigenic sources in the HSZ and particularly the south of Crete segment. We used various velocity models to relocate the event, while tsunami numerical simulations were performed for different fault scenarios reproducing the different locations obtained. Synthetic and observed tsunami waves were compared aiming to constrain the best fault solution. Moreover, the implications for the seismogenic and tsunamigenic potential in the area are discussed. Yet, although of moderate magnitude and not damaging, the study of such kind of earthquakes is very useful for testing the performance of early warning systems in the Mediterranean Sea.

2. *Geotectonic Setting and Seismic Imaging*

The forearc and the inter-plate seismogenic zone in the area of Crete have been widely investigated through temporary local seismic networks (e.g., Delibasis et al. 1999; Meier et al. 2004; Becker et al.

2010) and active seismic experiments (e.g., Bohnhoff et al. 2001). To the south of Crete, strong micro-seismicity occurs along the subduction plate interface from about 20 to 40-km depth and in the upper crust mainly along morphotectonic features, namely the Ptolemy, Pliny and Strabo trenches (e.g., Delibasis et al. 1999; Meier et al. 2004; Becker et al. 2010).

Eurasia-Africa convergence at the HSZ is accommodated by low-angle thrust faults (dip angles < 20°), dipping towards the volcanic arc, that are located at depths between about 15 and 45 km along the subduction plate interface (e.g., Taymaz et al. 1990; Kiratzi and Louvari 2003; Shaw and Jackson 2010; Fig. 1). Higher-angle reverse faults with dip angles larger than 30°, also dipping towards the volcanic arc, are located at shallower depths, between about 5 and 30 km, and have been suggested to splay off from the main plate interface (Taymaz et al. 1990; Shaw and Jackson 2010). Repeated activity along such a high-angle reverse fault has been proposed to produce pronounced bathymetric features on the ocean floor (Shaw et al. 2008; Mouslopoulou et al. 2015).

The accuracy of earthquake locations is of crucial importance for reliably constraining seismic sources. Routine earthquake locations in the outer forearc, where most of the seismicity occurs (Fig. 2), are usually poor due to the low station density along the sedimentary arc, the poor azimuthal coverage, and the presence of a velocity structure hardly reproduced by the simple 1-D velocity models used. Sachpazi et al. (2016) compared locations obtained by a local network composed by ocean bottom seismometers (OBSs) and land stations with those obtained from the National Observatory of Athens (NOA, http://bbnet.gein.noa.gr/HL/) in the outer forearc, between southern Peloponnese and western Crete. The study revealed errors up to several tens of kilometres in the revised seismicity catalogue of NOA. Karastathis et al. (2015) showed how the adoption of a 2-D velocity model, resembling the upper crust thickening from the plate boundary towards the inner forearc, significantly improved the location accuracy and the spatial distribution of the seismicity in the area of the Kefalonia Transform Fault, when compared to the revised hypocentral solutions from NOA. In fact, a crustal thickness variation of more than 10 km corresponds to additional travel times of up to 1 s, leading to location errors up to several tens of kilometres if not taken into account in the location algorithms (Koulakov and Sobolev 2006).

Results from a wide-angle seismic experiment along three profiles onshore and offshore of Crete showed that the lateral variation of the crustal and sedimentary thickness is stronger along the N–S direction than along the E–W direction (Bohnhoff et al. 2001). A maximum crustal thickness of about 32.5 km is observed to the north of central Crete and decreases to 28–30 km towards west and to 24–26 km towards east (Bohnhoff et al. 2001). North of Crete the crust thins to about 15 km below the Cretan Sea and to the south it reaches about 17 km at the plate boundary, located approximately 100 km south of Crete below the Mediterranean Ridge (Bohnhoff et al. 2001; Fig. 1). The crustal thickness beneath Crete, inferred by wide-angle seismic data, is consistent with results from other independent geophysical studies based on receiver function techniques (e.g., Li et al. 2003; Sodoudi et al. 2006) or combined seismic and gravimetric data (e.g., Makris et al. 2013).

3. The July 1, 2009 Earthquake

The M_w 6.45 July 1, 2009 earthquake struck offshore, about 80 km to the south of Crete (Fig. 3), where hypocentral coordinates are poorly constrained, as discussed in the introduction. The shock was slightly felt in several places of Crete and as far as Cairo in Egypt and Nicosia in Cyprus.

NOA, the official agency providing real-time earthquake solutions in Greece, calculated an epicentre at 34.35N/25.40E and a focal depth of 30 km. Different locations were obtained from the European Mediterranean Seismological Centre (EMSC), which calculated an epicentre at 34.14N/25.42E and a focal depth of 30 km, and by the International Seismological Centre (ISC), which calculated an epicentre at 34.15N/25.54E and a focal depth of 13.7 km. The three aforementioned solutions show a maximum difference of about 30 km in epicentral coordinates and 16 km in focal depth. The discrepancy in hypocentral coordinates can be inferred to: (1) the

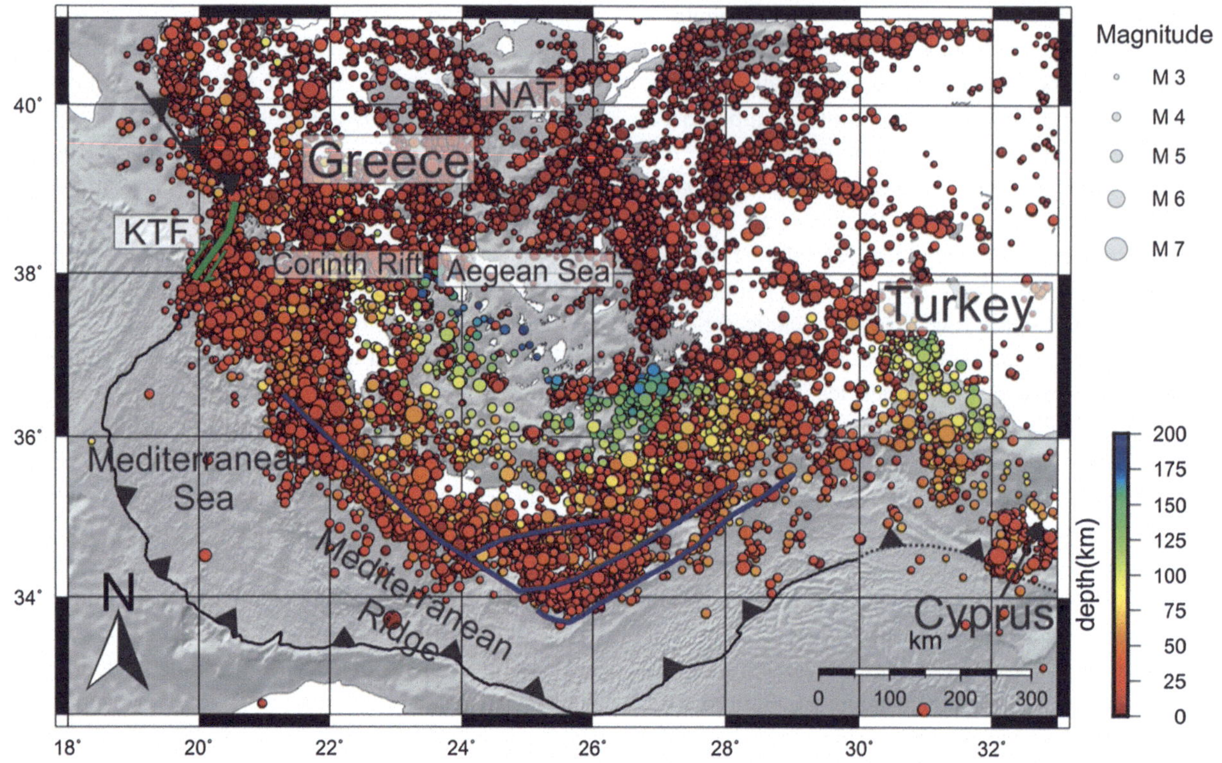

Figure 2
Distribution of instrumental seismicity at the Hellenic subduction zone for the period 1964–2016 (ISC 2017). Hypocentres are colour-coded according to depth and sized according to magnitude. Only earthquakes with horizontal and vertical errors smaller than 10 km are shown. *KTF* Kefalonia Transform Fault, *NAT* North Aegean Trough. Other tectonic elements are described in Fig. 1

different number of seismic stations used; (2) the type of phases used, namely at local, regional or teleseismic distances; and (3) the velocity models used. The NOA solution was obtained using phases recorded at stations from the Hellenic Unified Seismic Network (HUSN) only, while EMSC and ISC also included phases from stations at regional and teleseismic distances that allow the reduction of the azimuthal gap in the solution but do not necessarily lead to a better constrained hypocentre.

We collected and analysed the focal mechanisms published from several agencies to further understand the characteristics of the event (Fig. 3). Although they exhibit significant differences in terms of centroid coordinates, the fault plane solutions are quite similar, showing an almost pure reverse faulting for both nodal planes (Fig. 3). A minor strike-slip component is observed in few focal mechanism solutions (NOA and UPSL; see Fig. 3 for explanations). One of the nodal planes dips to N-NNE (NE for AUTH) with an angle larger than 30° with the only exception being represented by the IPGP solution showing a lower-angle fault of 14° (Fig. 3). The other nodal plane dips S-SSW with an angle larger than 40°–50° (Fig. 3). Zahradnik et al. (2009) analysed the nodal planes of the earthquake combining them with the available hypocentral solutions soon after the earthquake and indicated the N-NNE-dipping nodal plane as the most likely one. The N-NNE-dipping nodal plane is consistent with the seismotectonic setting of the area and was also suggested by Shaw and Jackson (2010). The latter authors suggested that the July 1, 2009 earthquake occurred on a splay fault in the overriding plate whose projection on the sea bottom corresponds to a clear bathymetric scarp.

According to preliminary solutions aiming to define the causative fault plane (Zahradnik et al. 2009), previous literature about the event (Shaw and

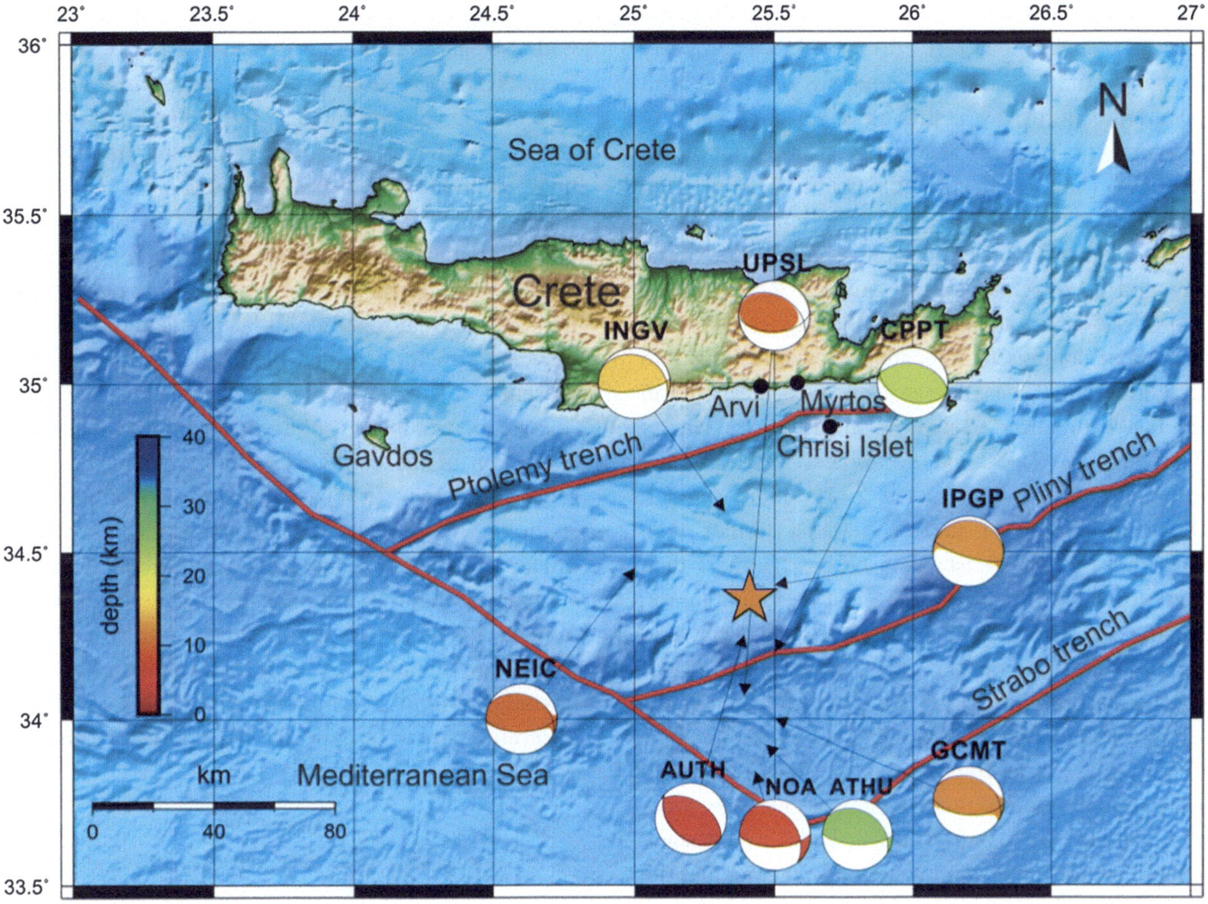

Figure 3

Focal mechanisms for the July 1, 2009 earthquake: *NEIC* National Earthquake Information Center, *INGV* Istituto Nazionale di Geofisica e Vulcanologia, *UPSL* University of Patra Seismic Laboratory, *CPPT* Centre Polynesien de Prevention des Tsunamis, *AUTH* Aristoteles University of Thessaloniki, *ATHU* University of Athens Seismic Laboratory, *NOA* National Observatory of Athens, *IPGP* Institut de Physique du Globe de Paris, *CMT* centroid moment tensors of Harvard. The focal mechanisms are coloured according to focal depth. The arrows starting from the beach balls point toward the epicentral coordinates of the earthquake's centroid obtained from the various agencies. The star indicates the epicentral location of the July 1, 2009 earthquake according to the relocation performed in this study using the 2-D velocity model. The three locations at which the tsunami was observed, namely Myrtos, Arvi, and Chrisi Islet, are indicated in the map

Jackson 2010), and the tectonic setting of the area (e.g., Taymaz et al. 1990; Shaw and Jackson 2010), the NNE-dipping nodal plane is assumed as the nodal plane of the event. Harvard's GCMT solution for M_w and for the focal parameters was preferred to other solutions mainly because it is the result of a world-wide-acknowledged methodology based on the inversion of teleseismic waveforms (Dziewonski et al. 1981; Ekström et al. 2012).

3.1. *Earthquake Relocation*

The July 1, 2009 south of Crete earthquake was relocated by using either: local 1-D velocity models (Delibasis et al. 1999; Meier et al. 2004; Becker et al. 2010; the velocity model used by NOA at the time of the earthquake; Fig. 1S); or a 2-D velocity model created to better resemble the crustal structure of the area (Fig. 2S). The 2-D velocity model was extracted from a refraction profile crossing the island of Crete from NNE–SSW (Bohnhoff et al. 2001). It extends from Antikythira to Karpathos in the x-direction and from latitude of 33.85N to Santorini in the y-direction (Fig. 3S). The original refraction profile was properly adapted to the N–S direction and was extended a bit toward the north, by assuming a constant velocity structure, to include also the seismic station in Santorini (SANT) in the relocation. The original seismic refraction profile was also extended down to 60-km depth assuming a constant dipping angle of the slab and a constant velocity towards larger depths (Fig. 2S). We tried to include the highest possible number of stations to relocate the event taking into account regional crustal heterogeneities, especially in the E–W direction (Bohnhoff et al. 2001; Makris et al. 2013). N–S heterogeneities were in fact accounted by the constructed 2-D velocity model. Thus, we could not include stations farther than Karpathos to the east as well as farther than Antikythira to the west, or in the Peloponnese, because the constructed 2-D velocity model would not be valid. A constant V_p/V_s ratio of 1.78 was adopted according to the existing velocity models for the area (Meier et al. 2004).

To make all the hypocentral solutions relative to the different velocity models comparable, they were all calculated by using phase arrivals at the same seismic stations, the same weighting scheme, and the same software.

The local 1-D velocity models retrieved from the literature and used in the relocation (Fig. 1S) were obtained during microseismicity experiments employing mainly temporary land stations (Delibasis et al. 1999; Meier et al. 2004), and in some cases, also OBSs (Becker et al. 2010). We did not use station corrections provided by the original studies because of the different network configurations. The RMSs and the associated vertical and horizontal errors were calculated to rank the quality of the different solutions and select those to test in the tsunami simulations.

The relocations were performed by using the non-linear location (NLLoc) algorithm (Lomax et al. 2000). The non-linear location methods provide more reliable solutions and hypocentre error estimates than linearized location methods in case of ill-conditioned location problems as in this study (Lomax et al. 2009).

A total of ten stations (Fig. 3S) from the HUSN with 14 phases including P- and S-arrivals available from NOA were used to relocate the earthquake. In case of 1-D velocity models, the inclusion of stations at larger distances was observed to not improve the final solution and, therefore, they were not used. Karastathis et al. (2015) showed how the inclusion of phases at stations at distances larger than the critical distance lowers the quality of the hypocentral solutions when major structural heterogeneities in the velocity structure, as in the Hellenic forearc, are present.

The hypocentral solutions obtained from the relocation of the earthquake (Table 1) can be separated into two groups according to depth: (1) shallower hypocentres with depth shallower than 13 km are obtained by using 1-D velocity models having a Moho at about 30 km, and the 2-D velocity model built in this study; (2) deeper hypocentres of 28 and 40 km are obtained by using 1-D velocity models having a Moho at about 40 km (Table 1). These differences are to be expected because the earthquake

Table 1

Hypocentral solutions for the July 1, 2009 earthquake obtained by using different velocity models

Lat. (°)	Long. (°)	Depth (km)	Obs.	RMS (s)	H. err. (km)	V. err. (km)	Velocity model
34.36	25.41	11.7	14	0.2	2.8	2.9	2-D Bohnhoff et al. (2001)
34.39	25.40	12.1	14	0.27	3.9	3.8	1-D Meier et al. (2004)
34.38	25.38	12.6	14	0.23	3.0	3.6	1-D Delibases et al. (1999)
34.37	25.41	28	14	0.24	4.0	2.9	1-D NOA 2009
34.48	25.39	39.6	14	0.31	4.9	5.5	1-D Becker et al. (2010)

The locations and error estimates reported in the table have been obtained by using NLLoc

H. err. horizontal error, *V. err.* vertical error, *Obs.* number of observations occurred about 80 km away from the closest station, where the head-wave (Pn) represents the first arrival. Thus, Moho depth differences in the velocity model have a first-order control on the hypocentral solution. The epicentral coordinates are observed to be less affected by the use of different velocity models (Table 1).

The 2-D velocity model provides the best results in terms of RMS and vertical and horizontal errors (Table 1). The maximum likelihood hypocentre and the posterior density function obtained by using the 2-D velocity model in NLLoc are reported in Fig. 4.

4. *Modelling of the Local Tsunami*

The July 1, 2009 south of Crete earthquake was followed by a small local tsunami which reportedly inundated the south-eastern coast of Crete and the northern coast of Chrisi Islet. To approach the best source model that produced the tsunami wave, we compared reported tsunami observations with synthetic tsunami features obtained through numerical simulations based on various homogeneous and heterogeneous slip models for the causative seismic fault.

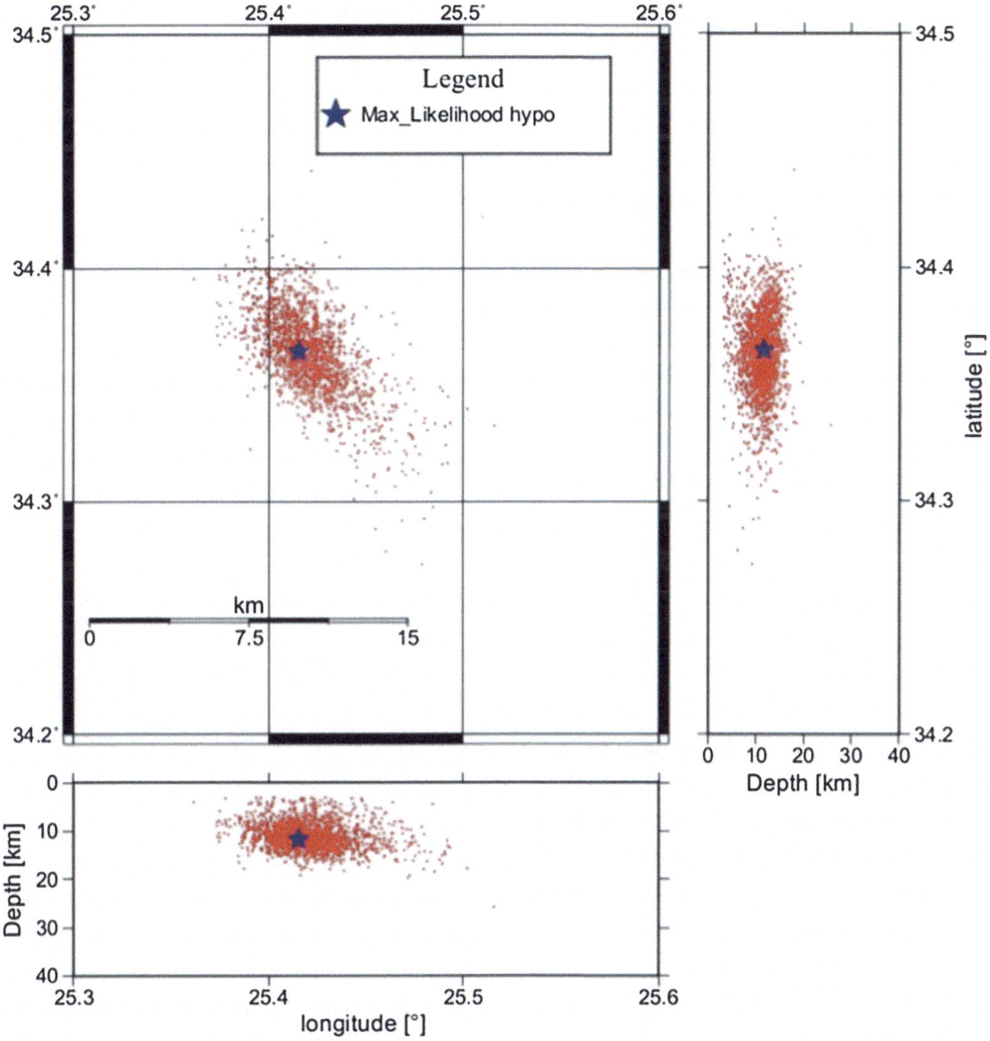

Figure 4
The maximum likelihood hypocentre (blue star) and the posterior density function (red cloud) of the July, 1 2009 south of Crete earthquake. The solution is obtained by using NLLoc, the 2-D velocity model and the seismic stations described in Fig. 3S

4.1. Tsunami Observations

The tsunami was not recorded at tide gauges, but it was observed by many eyewitnesses given its occurrence in summer, at 12.30 pm local time, when many people were on the beach. According to eyewitnesses accounts reported in local press, the tsunami was observed in the coastal localities of Myrtos and Arvi (south-eastern coast of Crete) and to the north of Chrisi Islet (Fig. 3). In Arvi, after a water retreat of about 1 m, four or five wave arrivals with amplitudes of up to 30 cm were reported. A video shoot recorded by an amateur and communicated to us shows the strong water withdrawal in Arvi. The sea disturbance lasted for about 1 h in Arvi and to the north of Chrisi Islet.

4.2. Homogeneous Fault Slip

The dimension of the causative seismic fault to use in the tsunami simulations with a homogeneous slip model was estimated by averaging the results from five empirical relationships all applicable to the analysed tectonic context. In fact, a direct estimation of the fault dimensions from the aftershocks distribution has not been possible due to the poor quality of aftershock determinations. We used empirical and semi-empirical relationships between the fault length (L) and width (W) and M_w or the seismic moment (M_0). We used M_0 obtained from the Harvard GCMT ($M_0 = 5.85 \times 10^{25}$ dyne cm) from which we calculated an M_w 6.45 applying the formula proposed by Hanks and Kanamori (1979):

$$M_w = 2/3 \times \mathrm{Log}_{10} M_0 - 10.73. \quad (1)$$

These relationships included: (1) the well-known empirical relationship for reverse-fault earthquakes of Wells and Coppersmith (1994); (2) the empirical relationship for inter-plate earthquakes of Blaser et al. (2010); (3) the semi-empirical relationship for dip-slip earthquakes of Wang and Tao (2003); (4) the empirical relationship for dip-slip events of Leonard (2010); and (5) the empirical relationship for earthquakes in the Mediterranean Sea region of Konstantinou et al. (2005). The aforementioned empirical and semi-empirical relationships are reported in Table 1S. Concerning the fault length (L), all the relationships give values ranging between 21 and 21.7 km, with the exception of the empirical relationship of Konstantinou et al. (2005) providing a length of 36 km. The obtained average value for the length is 24.2 km. The fault width (W) ranges from 11.1 to 13.5 km, with an average value of 12.3 km. Thus, the dimensions of our causative fault are 24.2 × 12.3 km leading to a rupture area A = 298 km^2. Applying the equation proposed by Hanks and Bakun (2002), relating the rupture area A with the M_w:

$$M_w = \mathrm{Log}A + 3.98(\pm 0.03), \quad (2)$$

we obtain a M_w of 6.45 ± 0.03 supporting the estimated L and W values.

After the determination of the fault dimensions (L and W) we derived the average slip on the fault (D) by using the fundamental seismic moment equation introduced by Aki (1966):

$$M_0 = \mathrm{GDA}. \quad (3)$$

We obtained an average slip (D) of 65.5 cm using a rigidity modulus (G) of 30 GPa.

We placed the fault centroid at the epicentral coordinates estimated by relocating the July 1, 2009 earthquake with the 2-D velocity model and at varying depths according to the relocation results (Table 1). In particular, we used centroid depths of 25 km, 26 km, and 9.7 km to reproduce the scenario of an earthquake occurring: (1) along the inter-plate seismogenic zone, where the depth of the plate interface and its dipping angle beneath the calculated epicentral location were estimated from the microseismicity cross sections reported in Becker et al. (2010) and Bocchini et al. (2018); (2) at a hypocentral depth indicated in the revised NOA catalogue, similar to the hypocentral depth obtained herein using the same velocity model; (3) at a hypocentral depth obtained with the 2-D velocity model (Table 1). Centroid depths for scenarios 2 and 3 were estimated by assuming a unilateral up-dip propagation of the rupture from the hypocentre on a fault plane dipping 32°. The NNE-dipping nodal plane of the Harvard GCMT was used to define the strike 295°, dip 32°, and rake 108° of the causative fault.

We reproduced two additional scenarios involving the use of the heterogeneous and of the homogeneous slip model on the same fault plane determined by the inversion of teleseismic data as described more in detail in Sect. 4.3.

4.3. Heterogeneous Fault Slip

To compute the spatial and temporal evolution of the slip on the causative fault, the kinematic, non-negative, least squares inversion technique by Hartzell and Heaton (1983) and Mendoza and Hartzell (2013) was applied. The finite-fault waveform inversion scheme developed by Hartzell and Heaton (1983) involves a trial-and-error procedure to obtain the best smoothing weight that identifies the simplest solution that still fits the data (Mendoza and Hartzell, 2013). The plot of residual norms $||Kx - b||$, where K is the sub-fault synthetics matrix, b is the matrix of observations, and x is the solution vector containing the slip required to reproduce the observations of the regularized solutions versus their corresponding smoothing norms $||Lx||$, forms an L-curve that quantifies the balance between fitting the data and meeting the smoothing constraints (Mendoza and Hartzell 2013). A rectangular fault plane, discretized into uniform cells (sub-faults) was constructed, and each point of the source response was computed with a code based on the generalized ray theory (Langston and Helmberger 1975). Synthetics were constructed following the discussion in the study of Heaton (1982). The calculated elementary synthetics were convolved with an attenuation operation under the assumption that $t^* = 1$ s, where t^* is the attenuation parameter of teleseismic body waves that represents the total body wave travel time divided by Q along the ray path for P waves (Stein and Wysession, 2003).

We used P-phases from 30 stations at teleseismic distances ranging from 30° to 90° with good azimuthal coverage (Fig. 4S). The waveform data were downloaded from the IRIS Data Management Centre (https://ds.iris.edu/ds/nodes/dmc/). All waveforms were pre-processed to remove the mean offset and instrument response before the inversion. They were also band-pass filtered between 0.03 and 0.09 Hz using a Butterworth filter, re-sampled to 0.2 samples/s and finally integrated in time to obtain displacements.

4.3.1 Fault Parameterization

Several values of source velocity, varying from 2.5 to 3.5 km/s, were examined. Rise time, fault dimensions, position and depth of the hypocentre, and time lag were also changed during the inversions. The source of the elementary synthetics was taken as a trapezoidal shape and the width of the source was chosen to be short enough, i.e., 0.5 s, compared to the total rise time on the fault using siz time windows. We performed the inversion to find the slip distribution assuming the NNE-dipping fault and a simple 1-D velocity model. The fault was discretised by 108 sub-faults, 18 of them along the strike and 6 along the dip. After several inversions, we ended up with a fault length of 27 km starting from the earth surface and going down to 12-km depth with a hypocentre at 9-km depth. Starting from the GCMT faulting parameters, it was found that the strike of the fault which gave us better solutions was 305° with a dip of 32°. During the inversion we left the rake to change from 70° to 130°.

The co-seismic slip distribution of the earthquake, representing the movement of the hanging wall with respect to the foot wall, is illustrated in Fig. 5. The time evolution of slip is presented in each case by six snaps using nearly 2.3-s intervals. The slip distribution is smoothed in 2-D, and a cone-shaped filter was applied.

The fitting between teleseismic waveforms and synthetics is shown in Fig. 5S. As observable from Fig. 5, the rupture started at 9 km depth and was strongly unilateral, showing a strong directivity towards the west with a velocity of 2.6 km/s. The rupture started with small values of slip (< 10 cm) and after about 4 s the slip moved upward and westward with larger values (up to more than 1 m). The M_0 calculated from the inversion $M_0 = 6 \times 10^{25}$ dyne cm is consistent with that determined by Harvard's GCMT solution. The earthquake's total rupture duration was shorter than 14 s. Maximum calculated co-seismic slip was 1.07 metres at about 13 km west from the hypocentre (Fig. 5).

The above described heterogeneous slip model was used to run tsunami numerical simulations (scenario 4, Table 2) in addition to the homogeneous models. We modelled an additional case, namely

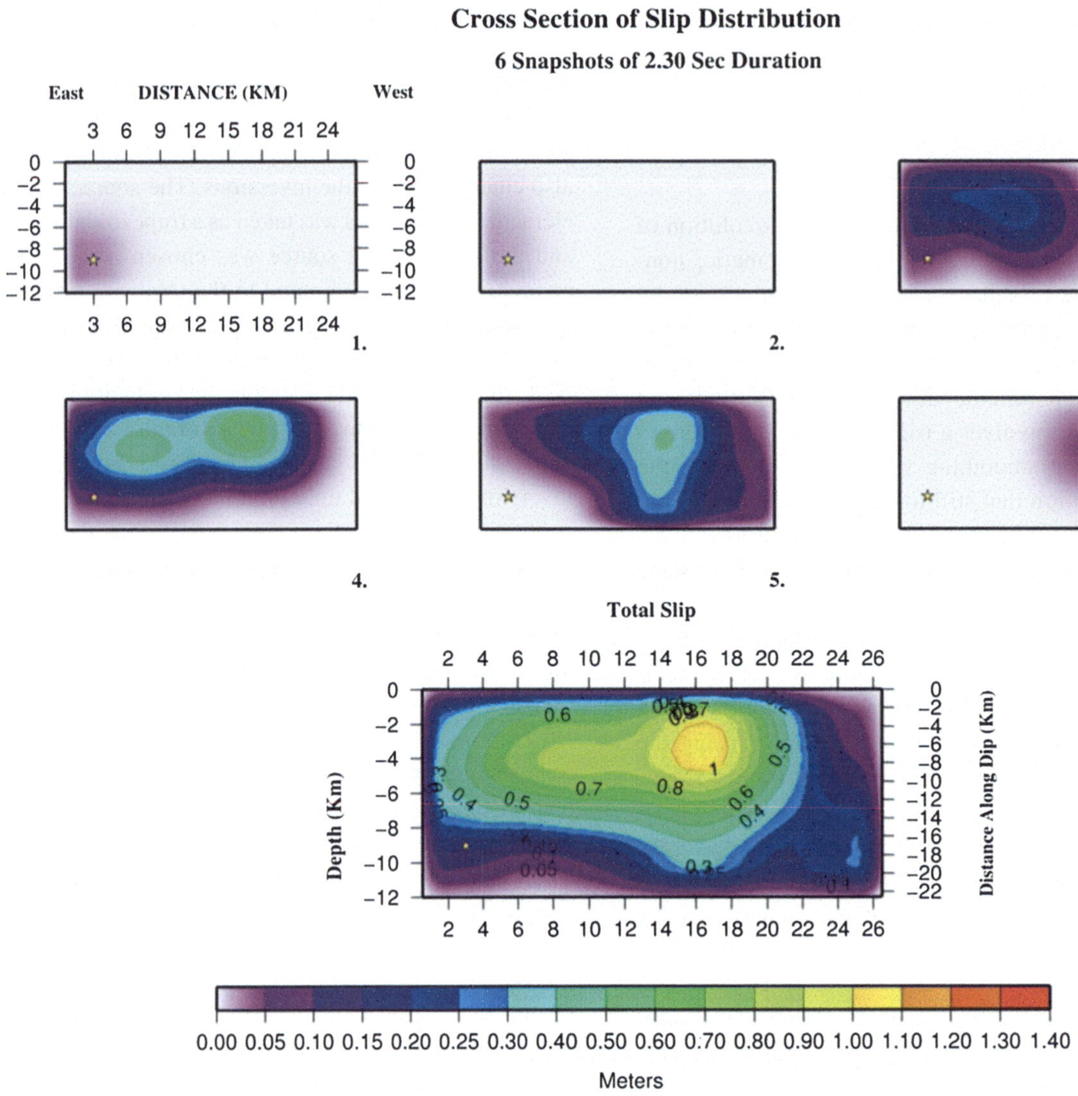

Figure 5
Cross section showing the temporal distribution of slip along the fault plate (1–6). The total slip on the fault is shown at the bottom part of the figure

scenario 5 (Table 2), to better compare the differences between homogeneous and heterogeneous slip models. In the latter case, we used the same fault parameters as for the heterogeneous fault model, but we averaged the slip above the fault plane.

4.4. Modelling of the Surface Deformation

The surface deformation due to a finite rectangular source on top of an elastic half-space was obtained through Okada (1985) modelling.

In particular, we employed a homogeneous (Sect. 4.2) and a heterogeneous fault slip model (Sect. 4.3) and further made a comparative analysis of the wave amplitudes calculated for the two cases. The cosesimic displacement produced by the heterogeneous slip model was obtained by applying the Okada's dislocation model (Okada 1985) to each of the sub-faults and summing up their contribution. A total amount of 108 sub-faults was used, 18 along strike and 6 along dip, with size of 1.5×3.6 km.

Table 2

Input parameters used for the tsunami simulations

	Scenario 1	Scenario 2	Scenario 3	Scenario 4	Scenario 5
Lat. (°)	34.4	34.4	34.4	34.4	34.4
Long. (°)	25.4	25.4	25.4	25.4	25.4
Centroid depth (km)	25[a]	26	7.7	[b]	6
Strike (°)	295	295	295	305	305
Dip (°)	15	32	32	32	32
Rake (°)	108	108	108	108	108
Average slip (m)	0.66	0.66	0.66	Heterogeneous	0.36
Length (km)	24.2	24.2	24.2	27	27
Width (km)	12.3	12.3	12.3	22	22
Shear modulus (GPa)	30	30	30		

[a]Depth of the plate interface beneath the earthquake epicentre

[b]The fault is discretised in sub-faults located at different centroid depths

The main geometrical and focal parameters describing the five scenarios resulting from the five different fault configurations, discussed in Sects. 4.2 and 4.3, are summarised in Table 2. The resulting deformation for all the modelled scenarios, representing the input for the tsunami modelling, is shown in Fig. 6. Maximum calculated vertical deformation was 5, 6, 25, 24, and 20 cm for scenario 1, 2, 3, 4, and 5, respectively.

The vertical component of the deformation obtained for the scenarios reproducing the case of an inter-plate earthquake (scenario 1, Table 2) and of the hypocentral solution obtained from NOA (scenario 2, Table 2), shows almost negligible deformation with a maximum vertical deformation of few cm (Fig. 6a, b), unable to explain the tsunamigenic nature of the earthquake. The aforementioned scenarios consider centroid depths of 25 km for the inter-plate earthquake (scenario 1) and of 26 km for NOA hypocentral solution (scenario 2). On the other hand, a maximum vertical deformation at surface of about 20–25 cm is observed for the scenarios employing a shallower centroid depth (scenarios 3–5, Fig. 6c–e). Thus, the shallower sources obtained by relocating the earthquake using the 2-D velocity model (scenario 3), and using the fault geometry constrained from the heterogeneous fault slip model (scenarios 4–5) produced vertical deformation at the surface compatible with the tsunamigenic nature of the earthquake.

4.5. Tsunami Modelling

Tsunami modelling was performed using the software package GEOWAVE (Watts et al. 2003), which is a combination of TOPICS (Tsunami Open and Progressive Initial Conditions System; Grilli and Watts 1999) and FUNWAVE (Wei and Kirby 1995; Wei et al. 1995). TOPICS uses a variety of curve fitting techniques to estimate the initial conditions, namely surface elevations and water velocities for tsunami propagation. FUNWAVE performs wave propagation simulation based on Boussinesq equations for the description of fully non-linear and dispersive waves, allowing the estimation of accurate run-ups and inundation at the same time. FUNWAVE includes also well-calibrated dissipation models for wave breaking and bottom friction. For the present study, however, we noticed that Boussinesq and non-linear shallow water (NLSW) models show similar results (Fig. 6S).

Time histories of the tsunami waves were calculated at synthetic tide gauges placed close to the locations where the tsunami was observed, namely: (1) Arvi; (2) Myrtos; and (3, 4) to the North of the Chrisi Islet (Fig. 7a). Unfortunately, as already mentioned, no operating tide gauges were in place at the time of the earthquake in the areas hit by the tsunami. Thus, a direct comparison between observed and calculated tsunami wave amplitudes at tide gauges was not possible. Therefore, we calculated time histories of tsunami waves in proximity of the sites where the tsunami was observed (Fig. 7a) with

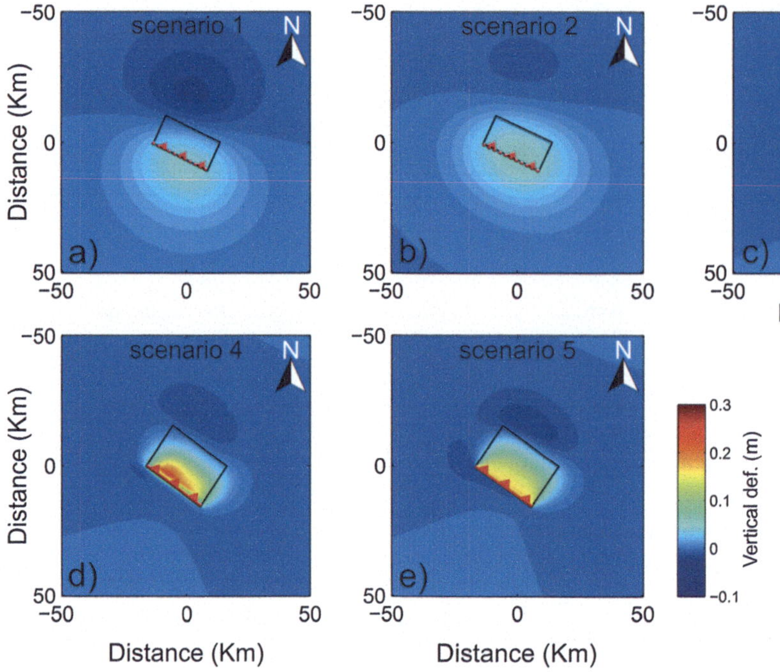

Figure 6
Vertical component of the deformation vector calculated for the faults with characteristics summarised in Table 3. The surface deformation is obtained by using the analytical solution of Okada (1985). The black rectangles within the subplot indicate the surface projection of the fault modelled in each scenario. The red lines indicate location of the up-dip limit of the fault while the barbs the dipping direction of the fault. In case of scenarios 1, 2 and 3 the red lines is dotted because by construction the fault did not reach to the sea bottom as in the case of scenarios 4 and 5. The centre of the plot (0, 0) coincides with the fault centroid

the aim to compare results from tsunami simulations with eyewitnesses' observations. The time step used for the tsunami simulations was of 0.1243 s.

The reference grid to model tsunami propagation in the sea was constructed by using the bathymetry grid from the European Marine Observation and Data Network (EMODnet) project (EMODnet DTM version released in 2018, http://www.emodnet.eu/), which has a resolution of about 115 m. Since the EMODnet bathymetry does not resolve areas above the sea level (a standard positive number is assigned to all the areas lying above the sea level), we used the topography grid from the Shuttle Radar Topography Mission with a resolution of 30 m (SRTM30; https://dwtkns.com/srtm30m/) to better resolve areas above the sea level and to account for possible interactions of the tsunami waves with the shoreline. The SRTM30 topography grid was resized and merged with the EMODnet bathymetry grid by using the Global Mapper tool v. 13. Both grids were freely available online. The final grid employed for tsunami wave modelling is shown in Fig. 7a. The coastline geometry employed in tsunami modelling (Fig. 7a) shows good agreement with the coastline observed in Google Earth for the same region (Fig. 7b).

The mareograms calculated at the synthetic tide gauges (Fig. 7a) are shown in Fig. 8. A synthesis of the results with the absolute maximum amplitudes calculated at each tide gauge for the five scenarios are reported in Table 3.

The implementation of a low-resolution bathymetry grid, i.e., $\sim$ 115 m, prevents a direct comparison between calculated and observed tsunami wave amplitudes at the shore. In fact, although we use the best bathymetry source available to us, possible amplification or attenuation phenomena associated to tsunami waves at the shoreline are not well reproduced because of the low bathymetry resolution (an amplification factor of 2–5 on wave heights could locally occur). Thus, we limit the

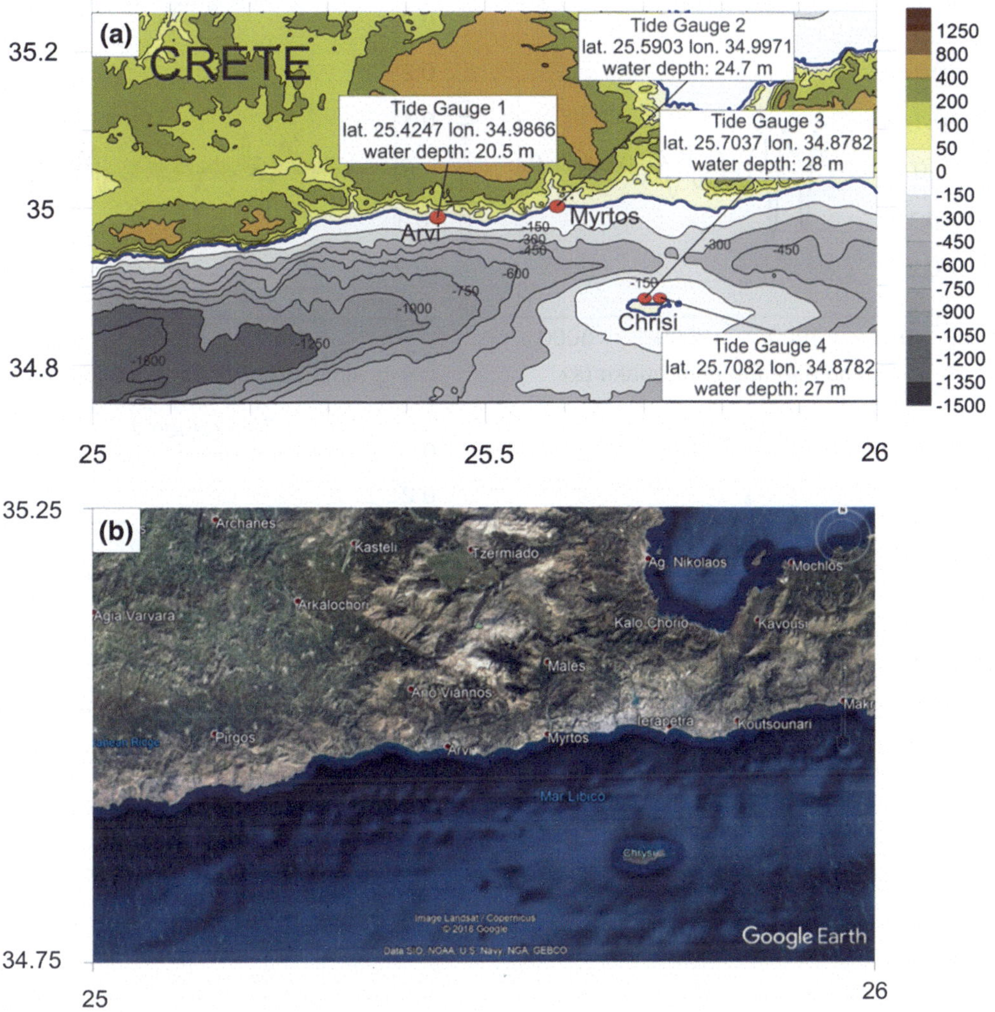

Figure 7

a Plot of the grid file and of the tide gauges used for tsunami modelling. Water depth and coordinates of the synthetic tide gauges are reported. The bathymetry is taken from EMODNet DTM 2018 while the topography from the SRTM (see text for the description of the grid files). **b** Google map screen shot of the region shown in sub-figure a

description of the results to the general trends of calculated mareograms. Tide gauges are located at similar water depths, between 20.5 and 28 m (Fig. 7a), for a relative comparison of the synthetic mareograms. We could not place tide gauges at shallower water depths because results would become more instable.

From the synthetic mareograms in Fig. 8 it is possible to distinguish two sets of solutions: (1) those showing longer-period waves with lower amplitudes (scenarios 1–2); and (2) those showing shorter-period waves with larger amplitudes (scenarios 3–5). The first set of solutions (red and black lines, Fig. 8), showing low-amplitude waves, are obtained by using deeper centroid depths (25 km for scenario 1 and 26 km for scenario 2) and homogeneous slip on the causative fault (see Table 2 for input parameters). In addition, calculated wave amplitudes for scenarios 1 and 2 do not show significant variations at the different tide gauges as instead observed for the other three scenarios (scenarios 3–5, Fig. 8; Table 3). The Okada analytical solution for displacement provides

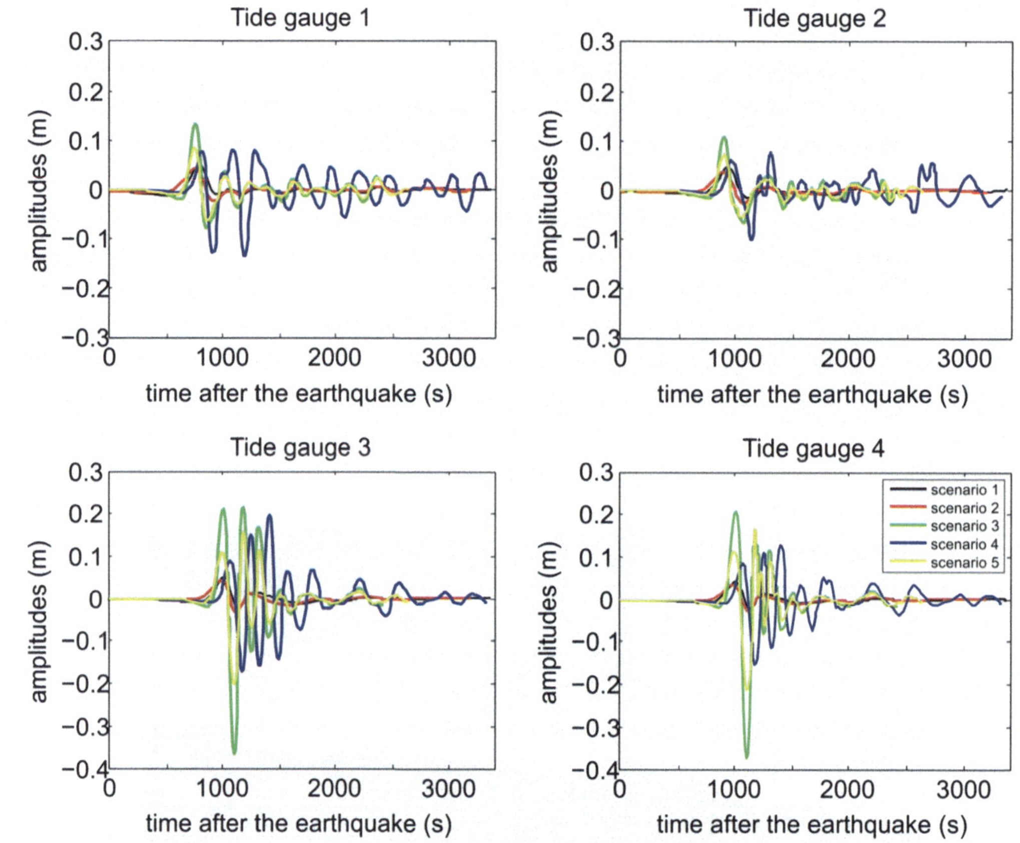

Figure 8
Mareograms calculated for the five different scenarios detailed in Table 2 at the synthetic tide gauges shown in Fig. 7a. Maximum and minimum synthetic tsunami waves are reported in Table 3

negligible vertical deformation associated with the aforementioned scenarios (scenarios 1–2, Fig. 6).

The other set of solutions, showing larger-amplitude waves were obtained by using shallower centroid depths (6 or 7.7 km) and both homogeneous and heterogeneous slip on the causative fault (scenarios 3–5, see Table 2 for input parameters). A clear oscillation in the signal indicates the passage of the tsunami waves starts about 500–600 s after the earthquake origin time (Fig. 8). The largest tsunami wave amplitudes are calculated at synthetic tide gauges located to the north of Chrisi Islet (scenarios 3–5, Fig. 8; Table 3). We can neither confirm nor exclude such result because we do not have official reports of tsunami waves observed at the northern shoreline of Chrisi Islet. We speculate that a possible explanation for the larger wave amplitudes calculated to the north of Chrisi Islet, with respect to those close to the sites of Arvi and Myrtos (Fig. 7a, Table 3), could be due to the fact that the wave was trapped to the north of the Islet, because of the morphology of the shoreline (Fig. 9), and amplified by multiple reflection.

Scenario 3, namely the one calculated for a fault having a centroid depth of 7.7 km and homogeneous slip, generates the largest wave amplitudes (Table 3). Scenarios 4 and 5, namely the scenario reproducing heterogeneous slip on the causative fault and the scenario reproducing homogenous slip on the same fault (Table 2), generate similar zero-to-peak amplitudes at tide gauges located close to Arvi and Myrtos. Larger zero-to-trough amplitudes at the same tide gauges are observed for scenario 4 (Table 3). At tide gauges located to the north of Chrisi Island (tide gauges 3 and 4; Fig. 7a), scenario 5 produces zero-to-trough amplitudes larger than those observed for

Table 3

Zero-to-peak (+) and zero-to-trough (−) tsunami wave amplitudes (m) at the synthetic tide gauges shown in Fig. 7. Values are taken from Fig. 8

Tide gauge #	Synthetic tsunami wave amplitudes (m)									
	Scenario 1		Scenario 2		Scenario 3		Scenario 4		Scenario 5	
	+	−	+	−	+	−	+	−	+	−
Tide gauge 1	0.05	0.01	0.04	0.02	0.134	0.077	0.082	0.134	0.087	0.064
Tide gauge 2	0.05	0.01	0.04	0.02	0.108	0.065	0.076	0.098	0.073	0.054
Tide gauge 3	0.05	0.02	0.04	0.03	0.215	0.365	0.199	0.175	0.16	0.2
Tide gauge 4	0.04	0.02	0.03	0.03	0.21	0.37	0.13	0.153	0.17	0.21

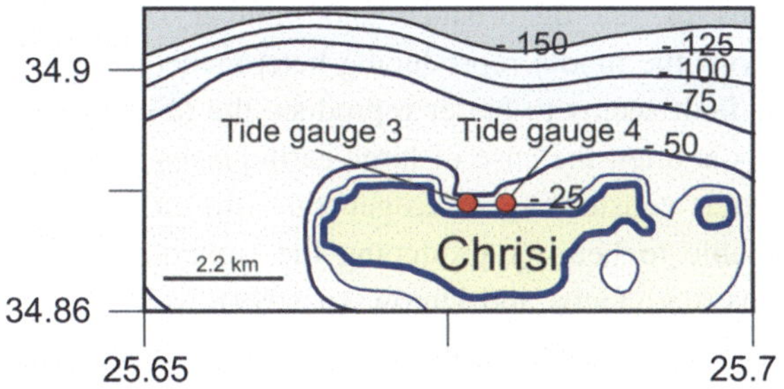

Figure 9
Closer view of Chrisi Islet and surrounding bathymetry. Red circles indicate the location of synthetic tide gauges. The thick blue line defines the shoreline

scenario 4 (Table 3), while zero-to-peak amplitudes are larger for scenario 5 at one tide gauge (tide gauge 4, Table 3) and for scenario 4 at the other tide gauge (tide gauge 3, Table 3). The scenario reproducing the heterogeneous slip on the fault, namely scenario 4, shows synthetic tsunami wave oscillations lasting longer with respect to the other scenarios (Fig. 8). In fact, after about 1200–1500 s, synthetic waves calculated for scenarios 1 and 2 are not visible anymore, while they are still visible but with very small amplitudes for scenarios 3 and 5. According to eyewitnesses, the sea disturbance lasted about 1 h in Arvi and to the north of Chrisi Islet. Thus, the scenario involving the heterogeneous slip model on the fault seems to better reproduce the observed sea disturbance among all considered scenarios. The same is also valid for the number of tsunami waves reaching the south-eastern coast of Crete. Eyewitnesses reported four or five tsunami waves in Arvi. At the synthetic tide gauge close to Arvi, at least three significant waves are observed for scenario 4 (tide gauge 1, Fig. 8). Instead, at the same tide gauge, for scenarios 3 and 5, a first arrival with quite large amplitude is observable, and it is followed by only one other clear wave, however, of much smaller amplitude (tide gauge 1, Fig. 8).

Interestingly enough, for all the modelled scenarios, the tsunami wave oscillations calculated at the several synthetic tide gauges start with negative amplitude (Fig. 8), which implies water retreat. This negative amplitude, although very small, is consistent with the water withdrawal described by eyewitness before the first tsunami wave arrival in Arvi and verified by a video recorded soon after the earthquake occurrence.

5. *Discussion: Implications on Tsunami Hazard Assessment and Early Warning*

Results from the relocation of the July 1, 2009 earthquake and its tsunamigenic nature indicate a shallow depth of the seismic source. The modelling of both the surface deformation according to Okada's approach (Okada 1985) and of the tsunami waves allow exclusion of the deep source (scenario 2) and the inter-plate nature of the event (scenario 1). An upper-plate origin of the event, as already proposed by Shaw and Jackson (2010), is favoured. The upper-plate origin of the event is also supported by: (1) the calculated hypocentral depth of 11.7 km, that is shallower than the depth of the plate interface beneath its epicentral location, as indicated by microseismicity studies (Becker et al. 2010; Bocchini et al. 2018); (2) the event focal mechanism solutions (Fig. 3) exhibiting for the most likely nodal plane, namely the one dipping towards either N or NNE, a reverse fault with a dip angle larger than 30°. The fault dip angle suggested by most of the focal mechanisms is observed to be steeper than the dip angle of the plate interface in the area ($\sim$ 15°) inferred by the distribution of microseismicity (Becker et al. 2010; Bocchini et al. 2018).

The 2-D velocity model provides a better constraint of the hypocentral solution. Smaller vertical and horizontal errors as well as RMS with respect to the 1-D velocity models are observed. This underlies the importance of accounting for crustal thickness variations in the overriding plate when the sources to be located are situated, with respect to the receivers, at distances larger than the critical distance (e.g., Karastathis et al. 2015). However, the final hypocentre obtained relocating the event with some of the 1-D velocity models (i.e., Delibasis et al. 1999; Meier et al. 2004) is comparable with the hypocentre obtained using the 2-D velocity model. This is likely due to the configuration of the stations used in the inversion, which are located in regions with crustal thickness and velocity structure compatible with the aforementioned 1-D velocity models.

Scenario 4 reproducing the heterogeneous slip fault model, as well as scenarios 3 and 5 reproducing shallow centroid depths and homogeneous slip on the fault, are able to generate tsunami waves. Although we observe clear differences in the synthetic mareograms, especially between the scenarios involving homogeneous slip on fault and the one involving heterogeneous slip, we are unable to conclude which of them reproduces better the observations. A finer bathymetry grid, with resolution of about 20 m, would be needed to model tsunami waves at shore and perform a direct comparison with the maximum observed tsunami heights ($\sim$ 30 cm in Arvi). According to the general trends of synthetic mareograms, scenario 4 seems to better reproduce the observations with respect to scenarios 3 and 5 in terms of duration of the sea disturbance and number of observed tsunami waves. In fact, the scenario reproducing the heterogeneous slip on fault shows a longer disturbance of the sea and a larger number of tsunami waves.

Due to various sources of uncertainties, it is hard to draw deeper conclusions between observed and estimated tsunami wave amplitudes. Sources of uncertainty include the absence of tide gauge records, the low resolution of the bathymetry grid, and the characterization of the causative fault. Furthermore, uncertainties involved in eyewitness observations are also to be taken into account.

The results suggest that for moderate-magnitude earthquakes, as in the case of the July 1, 2009 earthquake M_w 6.45, the tsunami modelling performed by using the homogeneous slip model on the fault is comparable with the heterogeneous one in terms of maximum tsunami wave amplitudes if the fault parameters are carefully estimated. However, in terms of sea disturbance and number of tsunami waves, the model reproducing heterogeneous slip on the fault seems to better reproduce the observations. This is often the case of large earthquakes where the models considering heterogeneous slip on faults are capable to better reproducing the tsunami observations (e.g., Geist and Dmowska 1999).

This study highlights the tsunamigenic potential of upper-plate faults as observed also for recent tsunamigenic earthquakes in the Aegean (Heidarzadeh et al., 2017; Yalciner et al., 2017; Papadimitriou et al., 2018). In fact, it shows how even moderate-magnitude ($M_w < 6.5$), shallow earthquakes (depth < 12 km) rupturing relatively steep reverse faults (dip > 30°) can trigger tsunamis. Upper-plate reverse faults in the HSZ have been suggested to coincide with pronounced bathymetric features on the ocean floor (Shaw et al. 2008; Mouslopoulou et al. 2015), indicating their facility to displace the sea bottom and, therefore, their high tsunamigenic potential. Due to the shallower centroid depths and the steeper dip angles, the tsunamigenic potential of upper-plate reverse faults in the outer forearc is expected to be higher than that of inter-plate earthquakes of similar magnitude. Given the steeper dip angle, splay faults above the plate interface tend to produce a sharper vertical deformation which is more prone to generate a tsunami with respect to inter-plate earthquakes of similar magnitude (e.g., Wendt et al. 2009). In fact, possible inter-plate earthquakes with magnitude as large as 6.5 (1972-05-04) or even 7.0 (1952-12-17) have occurred south of Crete and have not triggered tsunamis (Papadopoulos 2011). Also, no tsunami observations were reported for the M_w 6.85 February 14, 2008 earthquake in southwestern Peloponnese, the strongest instrumentally well-recorded inter-plate earthquake in Greece (e.g., Howell et al. 2017). A very tiny tsunami wave of 3 cm (6 cm peak to trough) was recorded at the tide gauge in Kasteli (Crete) in the aftermath of the M_w 6.76 inter-plate earthquake (Howell et al. 2017) occurred west of Crete on October 12, 2013 (Dr. Charalampakis M. pers. comm., NOA). Despite the larger magnitude with respect to the July 1, 2009, the October 12, 2013 earthquake generated a much smaller tsunami because of the larger depth ($\sim$ 45 km) and of the lower dip angle ($\sim$ 12°) of the causative fault.

Tsunami simulations show the very short time ($\sim$ 10 min) needed by the first tsunami waves to reach the southern coast of Crete, posing a big challenge for early warning systems. These small tsunamigenic earthquakes are quite challenging to test and calibrate capabilities of tsunami early warning systems. Currently, the Hellenic National Tsunami Warning Center (http://hl-ntwc.gein.noa.gr/) and other tsunami service providers operating in the North East Atlantic and Mediterranean Tsunami Warning System under the coordination of IOC/UNESCO have recommended 10 min as the target time for the release of the first tsunami message. A posteriori scenarios of early warning can be reproduced to understand drawbacks and strengths of the operational early warning systems.

6. Conclusions

The M_w 6.45 July 1, 2009 tsunamigenic earthquake ruptured the overriding plate to the south of Crete as indicated by the fault dipping angle retrieved by focal mechanism solutions, the relocated hypocentre calculated with a 2-D velocity model, and by the modelling of the small ensuing tsunami.

The use of a 2-D velocity model provides a slightly better constrain on the solution as compared to the 1-D velocity models since it accounts for the strong crustal thickness variations existing in the N–S direction across Crete, from the plate boundary towards the back-arc region.

The comparison of synthetic tsunami waves with eyewitness observations suggests that for earthquakes with magnitudes similar to the July 1, 2009 event, the heterogeneous slip model on the causative fault generates better results than the homogeneous slip model in terms of number of tsunami waves reaching the coast and of duration of the sea disturbance. However, although we observe different maximum amplitudes between scenarios involving heterogeneous and homogeneous slip on the causative fault, we cannot conclude which of them fits better with the observed tsunami heights.

This study highlights the high tsunamigenic potential associated with high-angle reverse faults in the upper-plate above the plate interface, higher than that of inter-plate earthquakes of similar or even slightly larger moment magnitude.

Tsunami simulations show the very short time, about 10 min, needed by the first tsunami waves to hit the southern coast of Crete, posing a big challenge for tsunami early warning systems.

Acknowledgements

Gian Maria Bocchini has been funded by the People Program (Marie Curie Actions) of the European Union's 7th Framework Programme FP7-PEOPLE-2013-ITN under REA grant agreement no. 604713. We are grateful to Dr. Mansour Ioualalen for his comments and suggestions in the construction of the heterogeneous slip model. We thank two anonymous reviewers and the Guest Editor Prof. M. A. Baptista (ISEL, Lisboa) for the useful comments that helped to improve the initial version of the manuscript. Several figures have been generated with the Generic Mapping Tools (Wessel et al. 2013) and the background bathymetry from the GMRT dataset.

References

Aki, K. (1966). Generation and propagation of G waves from the Niigata Earthquake of June 16, 1964. Part 2. Estimation of earthquake movement, released energy, and stress–strain drop from the G wave spectrum. *Bulletin. Earthquake Research Institute, University of Tokyo, 44,* 73–88.

Ambraseys, N. (2009). *Earthquakes in the Mediterranean and Middle East: A multidisciplinary study of seismicity up to 1900.* Cambridge: Cambridge University Press. https://doi.org/10.1017/CBO9781139195430.

Becker, D., Meier, T., Bohnhoff, M., & Harjes, H.-P. (2010). Seismicity at the convergent plate boundary offshore Crete, Greece, observed by an amphibian network. *Journal of Seismology, 14,* 369–392. https://doi.org/10.1007/s10950-009-9170-2.

Blaser, L., Krüger, F., Ohrnberger, M., & Scherbaum, F. (2010). Scaling relations of earthquake source parameter estimates with special focus on subduction environment. *Bulletin of the Seismological Society of America, 100,* 2914–2926.

Bocchini, G. M., Brüstle, A., Becker, D., Meier, T., van Keken, P. E., Ruscic, M., et al. (2018). Tearing, segmentation, and backstepping of subduction in the Aegean: New insights from seismicity. *Tectonophysics, 734–735,* 96–118. https://doi.org/10.1016/j.tecto.2018.04.002.

Bohnhoff, M., Makris, J., Papanikolaou, D., & Stavrakakis, G. (2001). Crustal investigation of the Hellenic subduction zone using wide aperture seismic data. *Tectonophysics, 343,* 239–262. https://doi.org/10.1016/S0040-1951(01)00264-5.

Delibasis, N., Ziazia, M., Voulgaris, N., Papadopoulos, T., Stavrakakis, G., Papanastassiou, D., et al. (1999). Microseismic activity and seismotectonics of Heraklion area (central Crete Island, Greece). *Tectonophysics, 308,* 237–248. https://doi.org/10.1016/S0040-1951(99)00076-1.

Dziewonski, A. M., Chou, T., & Woodhouse, J. H. (1981). Determination of earthquake source parameters from waveform data for studies of global and regional seismicity. *Journal of Geophysical Research: Earth, 86,* 2825–2852.

Ekström, G., Nettles, M., & Dziewoński, A. M. (2012). The global CMT project 2004–2010: Centroid-moment tensors for 13,017 earthquakes. *Physics of the Earth and Planetary Interiors, 200,* 1–9.

Galanopoulos, A. G. (1957). The seismic sea wave of July 9, 1956. *Prakt Academy Athens, 32,* 90–101.

Geist, E. L., & Dmowska, R. (1999). Local tsunamis and distributed slip at the source. *Pure and Applied Geophysics, 154,* 485–512. https://doi.org/10.1007/s000240050241.

Grilli, S. T., & Watts, P. (1999). Modeling of waves generated by a moving submerged body. Applications to underwater landslides. *Engineering Analysis with Boundary Elements, 23,* 645–656.

Guidoboni, E., & Comastri, A. (1997). The large earthquake of 8 August 1303 in Crete: Seismic scenario and tsunami in the Mediterranean area. *Journal of Seismology, 1,* 55–72. https://doi.org/10.1023/A:1009737632542.

Guidoboni, E., & Comastri, A. (2005). *Catalogue of earthquakes and tsunamis in the Mediterranean area from the 11th to the 15th century.* Rome: Istituto nazionale di geofisica e vulcanologia.

Hanks, T. C., & Bakun, W. H. (2002). A bilinear source-scaling model for M-log A observations of continental earthquakes. *Bulletin of the Seismological Society of America, 92,* 1841–1846. https://doi.org/10.1785/0120010148.

Hanks, T. C., & Kanamori, H. (1979). A moment magnitude scale. *Journal of Geophysical Research: Solid Earth, 84,* 2348–2350.

Hartzell, S. H., & Heaton, T. H. (1983). Inversion of strong ground motion and teleseismic waveform data for the fault rupture history of the 1979 Imperial Valley, California, earthquake. *Bulletin of the Seismological Society of America, 73,* 1553–1583.

Heaton, T. H. (1982). The 1971 San Fernando earthquake: A double event? *Bulletin of the Seismological Society of America, 72,* 2037–2062.

Heidarzadeh, M., Necmioglu, O., Ishibe, T., & Yalciner, A. C. (2017). Bodrum-Kos (Turkey–Greece) Mw 6.6 earthquake and tsunami of 20 July 2017: A test for the Mediterranean tsunami warning system. *Geoscience Letters.* https://doi.org/10.1186/s40562-017-0097-0.

Howell, A., Jackson, J., Copley, A., McKenzie, D., & Nissen, E. (2017). Subduction and vertical coastal motions in the eastern Mediterranean. *Geophysical Journal International, 211*(1), 593–620. https://doi.org/10.1093/gji/ggx307.

ISC, 2017. Reviewed International Seismological Centre (ISC) Bulletin. http://www.isc.ac.uk. Accessed 7 Jan 17.

Karastathis, V. K., Mouzakiotis, E., Ganas, A., & Papadopoulos, G. A. (2015). High-precision relocation of seismic sequences above a dipping Moho: The case of the January–February 2014 seismic sequence on Cephalonia island (Greece). *Solid Earth, 6,* 173.

Kiratzi, A., & Louvari, E. (2003). Focal mechanisms of shallow earthquakes in the Aegean Sea and the surrounding lands determined by waveform modelling: A new database. *Journal of Geodynamics, 36,* 251–274. https://doi.org/10.1016/S0264-3707(03)00050-4.

Konstantinou, K. I., Papadopoulos, G. A., Fokaefs, A., & Orphanogiannaki, K. (2005). Empirical relationships between aftershock area dimensions and magnitude for earthquakes in the Mediterranean Sea region. *Tectonophysics, 403,* 95–115.

Koulakov, I., & Sobolev, S. V. (2006). Moho depth and three-dimensional P and S structure of the crust and uppermost mantle in the Eastern Mediterranean and Middle East derived from tomographic inversion of local ISC data. *Geophysical Journal International, 164,* 218–235.

Langston, C. A., & Helmberger, D. V. (1975). A procedure for modelling shallow dislocation sources. *Geophysical Journal International, 42,* 117–130.

Leonard, M. (2010). Earthquake fault scaling: Self-consistent relating of rupture length, width, average displacement, and moment release. *Bulletin of the Seismological Society of America, 100,* 1971–1988.

Li, X., Bock, G., Vafidis, A., Kind, R., Harjes, H. P., Hanka, W., et al. (2003). Receiver function study of the Hellenic subduction zone: Imaging crustal thickness variations and the oceanic Moho of the descending African lithosphere. *Geophysical Journal International, 155,* 733–748. https://doi.org/10.1046/j.1365-246X.2003.02100.x.

Lomax, A., Michelini, A., & Curtis, A. (2009). Earthquake location, direct, global-search methods. In R. A. Meyers (Ed.), *Encyclopedia of complexity and systems science* (pp. 2449–2473). New York: Springer. https://doi.org/10.1007/978-0-387-30440-3_150.

Lomax, A., Virieux, J., Volant, P., & Berge-Thierry, C. (2000). Probabilistic earthquake location in 3D and layered models. In C. H. Thurber & N. Rabinowitz (Eds.), *Advances in seismic event location* (pp. 101–134). Dordrecht: Springer. https://doi.org/10.1007/978-94-015-9536-0_5.

Makris, J., Papoulia, J., & Yegorova, T. (2013). A 3-D density model of Greece constrained by gravity and seismic data. *Geophysical Journal International, 194,* 1–17.

McClusky, S., Balassanian, S., Barka, A., Demir, C., Ergintav, S., Georgiev, I., et al. (2000). Global positioning system constraints on plate kinematics and dynamics in the eastern Mediterranean and Caucasus. *Journal of Geophysical Research: Solid Earth, 105,* 5695–5719. https://doi.org/10.1029/1999JB900351.

Meier, T., Rische, M., Endrun, B., Vafidis, A., & Harjes, H.-P. (2004). Seismicity of the Hellenic subduction zone in the area of western and central Crete observed by temporary local seismic networks. *Tectonophysics, 383,* 149–169. https://doi.org/10.1016/j.tecto.2004.02.004.

Mendoza, C., & Hartzell, S. (2013). Finite-fault source inversion using teleseismic P waves: Simple parameterization and rapid analysis. *Bulletin of the Seismological Society of America, 103,* 834–844.

Mouslopoulou, V., Nicol, A., Begg, J., Oncken, O., & Moreno, M. (2015). Clusters of megaearthquakes on upper plate faults control the Eastern Mediterranean hazard. *Geophysical Research Letters, 42,* 10282–10289. https://doi.org/10.1002/2015GL066371.

Okada, Y. (1985). Surface deformation due to shear and tensile faults in a half-space. *Bulletin of the Seismological Society of America, 75,* 1135–1154.

Papadimitriou, P., Kassaras, I., Kaviris, G., Tselentis, G. A., Voulgaris, N., Lekkas, E., et al. (2018). The 12th June 2017 Mw = 6.3 Lesvos earthquake from detailed seismological observations. *Journal of Geodynamics, 115,* 23–42. https://doi.org/10.1016/j.jog.2018.01.009.

Papadopoulos, G. A. (2011). *A seismic history of Crete: Earthquakes and tsunamis, 2000 B.C.–A.D. 2010.* Athens: Ocelotos Publications.

Papadopoulos, G. A., Gràcia, E., Urgeles, R., Sallares, V., De Martini, P. M., Pantosti, D., et al. (2014). Historical and prehistorical tsunamis in the Mediterranean and its connected seas: Geological signatures, generation mechanisms and coastal impacts. *Marine Geology, 354,* 81–109. https://doi.org/10.1016/J.MARGEO.2014.04.014.

Papazachos, B. C., & Papazachou, C. (2003). *The earthquakes of Greece.* Thessaloniki: Ziti Publications. **(in Greek)**.

Reilinger, R., McClusky, S., Vernant, P., Lawrence, S., Ergintav, S., Cakmak, R., et al. (2006). GPS constraints on continental deformation in the Africa–Arabia–Eurasia continental collision zone and implications for the dynamics of plate interactions. *Journal of Geophysical Research: Solid Earth.* https://doi.org/10.1029/2005JB004051.

Sachpazi, M., Laigle, M., Charalampakis, M., Sakellariou, D., Flueh, E., Sokos, E., et al. (2016). Slab segmentation controls the interplate slip motion in the SW Hellenic subduction: New insight from the 2008 Mw 6.8 Methoni interplate earthquake. *Geophysical Research Letters, 43*(18), 9619–9626. https://doi.org/10.1002/2016GL070447.

Shaw, B., Ambraseys, N. N., England, P. C., Floyd, M. A., Gorman, G. J., Higham, T. F. G., et al. (2008). Eastern Mediterranean tectonics and tsunami hazard inferred from the AD 365 earthquake. *Nature Geoscience, 1,* 268–276. https://doi.org/10.1038/ngeo151.

Shaw, B., & Jackson, J. (2010). Earthquake mechanisms and active tectonics of the Hellenic subduction zone. *Geophysical Journal International, 181,* 966–984. https://doi.org/10.1111/j.1365-246X.2010.04551.x.

Sodoudi, F., Kind, R., Hatzfeld, D., Priestley, K., Hanka, W., Wylegalla, K., et al. (2006). Lithospheric structure of the Aegean obtained from P and S receiver functions. *Journal of Geophysical Research: Solid Earth.* https://doi.org/10.1029/2005JB003932.

Stein, S., & Wysession, M. (2003). *An introduction to seismology, earthquakes, and earth structure.* Oxford: Blackwell Science. https://doi.org/10.1017/S0016756803318837.

Taymaz, T., Jackson, J., & Westaway, R. (1990). Earthquake mechanisms in the Hellenic Trench near Crete. *Geophysical Journal International, 102,* 695–731. https://doi.org/10.1111/j.1365-246X.1990.tb04590.x.

Wang, H., & Tao, X. (2003). Relationships between moment magnitude and fault parameters: Theoretical and semi-empirical relationships. *Earthquake Engineering and Engineering Vibration, 2,* 201–211.

Watts, P., Grilli, S. T., Kirby, J. T., Fryer, G. J., & Tappin, D. R. (2003). Landslide tsunami case studies using a Boussinesq model and a fully nonlinear tsunami generation model. *Natural Hazards and Earth Systems Sciences, 3,* 391–402.

Wei, G., & Kirby, J. T. (1995). Time-dependent numerical code for extended Boussinesq equations. *Journal of Waterway, Port, Coastal, and Ocean Engineering, 121,* 251–261.

Wei, G., Kirby, J. T., Grilli, S. T., & Subramanya, R. (1995). A fully nonlinear Boussinesq model for surface waves. Part 1. Highly nonlinear unsteady waves. *Journal of Fluid Mechanics, 294,* 71–92.

Wells, D. L., & Coppersmith, K. J. (1994). New empirical relationships among magnitude, rupture length, rupture width, rupture area, and surface displacement. *Bulletin of the Seismological Society of America, 84,* 974–1002.

Wendt, J., Oglesby, D. D., & Geist, E. L. (2009). Tsunamis and splay fault dynamics. *Geophysical Research Letters, 36,* 1. https://doi.org/10.1029/2009GL038295.

Wessel, P., Smith, W. H. F., Scharroo, R., Luis, J., & Wobbe, F. (2013). Generic mapping tools: Improved version released. *Eos, Transactions American Geophysical Union, 94*(45), 409–410.

Yalciner, C.A., Annunziato, A., Papadopoulos, G.A., Guney Dogan, G., Gokhan Guler, H., Eray Cakir, T., Ozer Sozdinler, C., Ulutas, E., Arikawa, T., Suzen, L., Kanoglu, U., Guler, I., Probst, P., & Synolakis, C. (2017). The 20th July 2017 (22:31 UTC) Bodrum/Kos Earthquake and Tsunami; Post Tsunami Field Survey Report.

Zahradnik, J., Sokos, E., & Tselentis, A. (2009). Quick identification of the fault plane for Mw 6.3, earthquake south of Crete, Greece July 1, 2009. Report to EMSC issued on July 1st, 2009 at 20:00 UTC.

(Received October 5, 2018, revised May 10, 2019, accepted June 6, 2019, Published online June 19, 2019)

Pure Appl. Geophys. 177 (2020), 1335–1346

https://doi.org/10.1007/s00024-020-02433-z

Pure and Applied Geophysics

A Multi-fault Model Estimation from Tsunami Data: An Application to the 2018 M7.9 Kodiak Earthquake

M. Jakir Hossen,[1] Anne F. Sheehan,[2] and Kenji Satake[3]

Abstract—In this study, we developed a new search algorithm to find a multi-fault model of a complex earthquake using tsunami data, and applied it to the January 23, 2018 M7.9 Kodiak earthquake. Our method includes a Green's function based time reverse imaging (GFTRI) approach to invert for sea surface displacement using tsunami waveforms, followed by inversion of the sea surface displacement for the earthquake slip distribution. The global CMT focal mechanism for this event indicates that faulting occurred on a steeply dipping fault striking either N–S (right lateral) or E–W (left lateral), while subsequent work reveals a more complex pattern of strike-slip faulting. We carried out a number of source inversions using different combinations of faults to find the model based on an extremum for residual errors. Our results suggest that the rupture occurred on at least three faults oriented in approximately N–S and E–W directions. We further explored the fault-geometry parameters by perturbing them within a range suggested by previous work. We found that the sea surface displacement model is best fit by our preferred three fault-model with the set of parameters (strike, dip, rake): (165°, 60°, 154°) and (265°, 60°, 10°) for N–S and E–W directions, respectively.

Keywords: Tsunami source inversion, multi-fault model, time reverse imaging.

1. Introduction

A large tsunami is usually generated by a megathrust earthquake, for example, the 1964 M 9.2 Great Alaska Earthquake, the 2011 M 9.1 Tohoku-Oki earthquake, or the 2004 M 9.1 Sumatra earthquake. Such earthquakes occur in subduction zones at the boundary of the continental plate over the oceanic plate, and displace a huge amount of water due to vertical movement of the sea-floor. However, intra-plate events at outer-rise regions of the incoming plate can also generate destructive tsunamis. These earthquakes are typically strike-slip and are less likely to cause vertical deformation of the sea-floor and displace a large amount of water. A few outer-rise events have caused tsunamis in the past, for example, the 11 April 2012 M 8.2 Sumatra earthquake (Yue et al. 2012).

On January 23, 2018 at 9:31:42 UTC, an earthquake of M_w 7.9 occurred at 56.046° N, 149.073° W, 300 km southeast of Kodiak Island, Alaska. The event was on the incoming Pacific plate in the outer rise region of the Alaska-Aleutian subduction zone. The focal mechanism solutions indicate that faulting occurred on a steeply dipping fault either N–S striking (right lateral) or E–W striking (left lateral). According to the centroid moment tensor (CMT) solution of US Geological Survey (USGS), the seismic moment is $1.01 \times 10^{21} Nm$, Magnitude is $M_w = 7.9$ and the centroid depth is 32.0 km; the nodal planes have strike $\psi_1 = 261°$, dip $\phi_1 = 69°$, rake $r_1 = 18°$ and strike $\psi_2 = 164°$, dip $\phi_2 = 74°$, rake $r_2 = 158°$. This single fault-model for the event provides a reasonable fit to teleseismic body and surface waves but the aftershock seismicity shows a broad and complex spatial distribution. Recent studies suggest that the rupture might have occurred on several fault planes instead of a single fault (e.g, Ruppert et al. 2018; Lay et al. 2018; Zhao et al. 2018). Ruppert et al. (2018) used teleseismic P wave back-projection and GPS modeling to find their proposed model consisting of five ruptured

Electronic supplementary material The online version of this article (https://doi.org/10.1007/s00024-020-02433-z) contains supplementary material, which is available to authorized users.

[1] Cooperative Institute of Research in Environmental Sciences, University of Colorado, Boulder, CO, USA. E-mail: md.hossen@colorado.edu

[2] Department of Geological Sciences and Cooperative Institute of Research in Environmental Sciences, University of Colorado, Boulder, CO, USA.

[3] Earthquake Research Institute, The University of Tokyo, Tokyo, Japan.

subfaults, among which a NE–SW trending left-lateral strike slip segment released the most energy. Lay et al. (2018) suggested a four-fault model, dominated by right lateral slip on a westward-dipping fault, is most compatible with most seismic, GPS and tsunami data. Using onshore GPS data, Zhao et al. (2018) also proposed a four-fault model where the seismic moment is mainly released on two WSW oriented fault planes.

There are several ways to estimate tsunami source models using tsunami data, including least-squares (LSQ) (Satake 1987; Wei et al. 2003; Satake et al. 2013; Baba and Cummins 2005; Gusman et al. 2016; Mulia and Asano 2016; Hossen et al. 2015b; Ho et al. 2017), genetic algorithms (Mulia and Asano 2016), time reverse imaging (TRI) (Hossen et al. 2015c; An et al. 2014) and Green's function based time reverse imaging (GFTRI) methods (Hossen et al. 2015a). The GFTRI method involves convolution of a set of Green's functions (GFs) with the time-reversed observed tsunami waveforms. The method was applied successfully to the 2011 Tohoku-Oki megathrust earthquake tsunami and the 2009 Samoa earthquake tsunami (Hossen et al. 2018).

In this paper, we adopted a two-step inversion (TSI) for the 2018 Kodiak earthquake. In the first step, the GFTRI method is used to determine a sea surface displacement (SSD) model. In the second step, we determine the slip distribution of the earthquake from the sea surface displacement, inverting for slip distribution on the fault planes. The estimation of slip distribution from the SSD model is computationally efficient and does not require the user to recompute waveforms (tsunami and seismic)

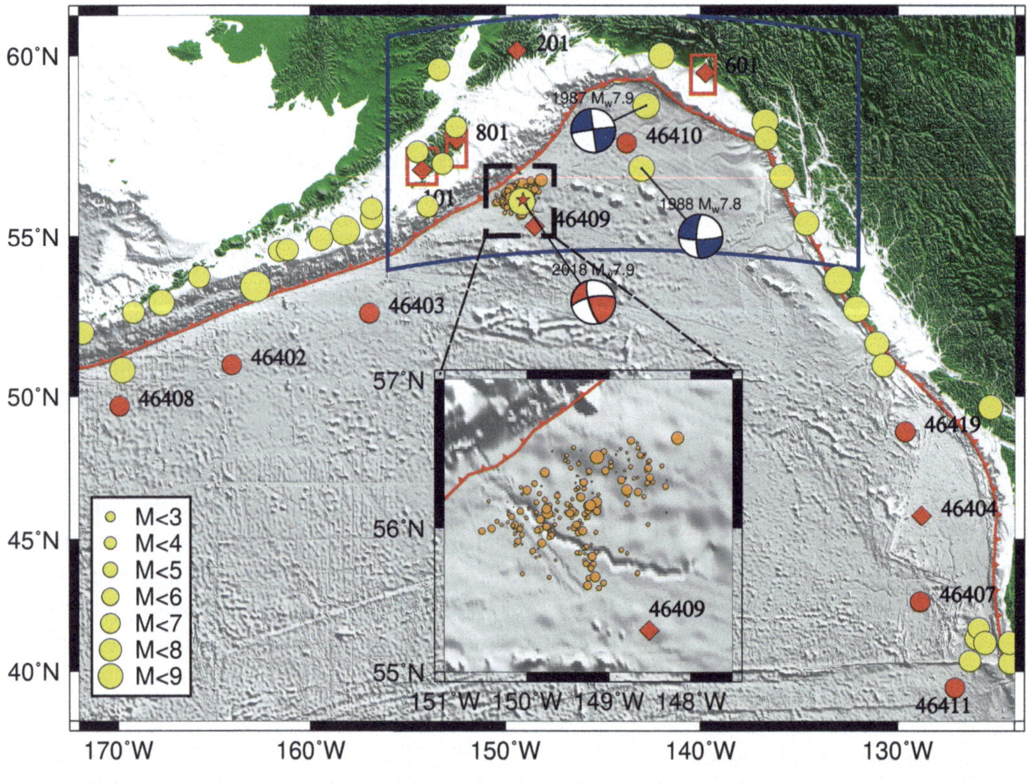

Figure 1
Map of study area, seismicity, and tsunami observing stations. Red star shows hypocenter of the January 23, 2018 M 7.9 offshore Kodiak earthquake. Inset map corresponds to dashed black box and shows aftershock distribution (orange circles) of the offshore Kodiak event. Focal mechanism for the 2018 Kodiak earthquake and two other outer rise earthquakes are also shown. Yellow circles show past earthquakes with $M_w > 5$. Tsunami observation stations are depicted by red circles and diamonds, with red circles used in the tsunami source estimation and all stations used in comparison of model to observations. For the tsunami modeling we used three nested domains with grid-spacing of 60 arc sec (for entire map area), 20 arc sec (blue box) and 20/3 arc sec (red boxes)

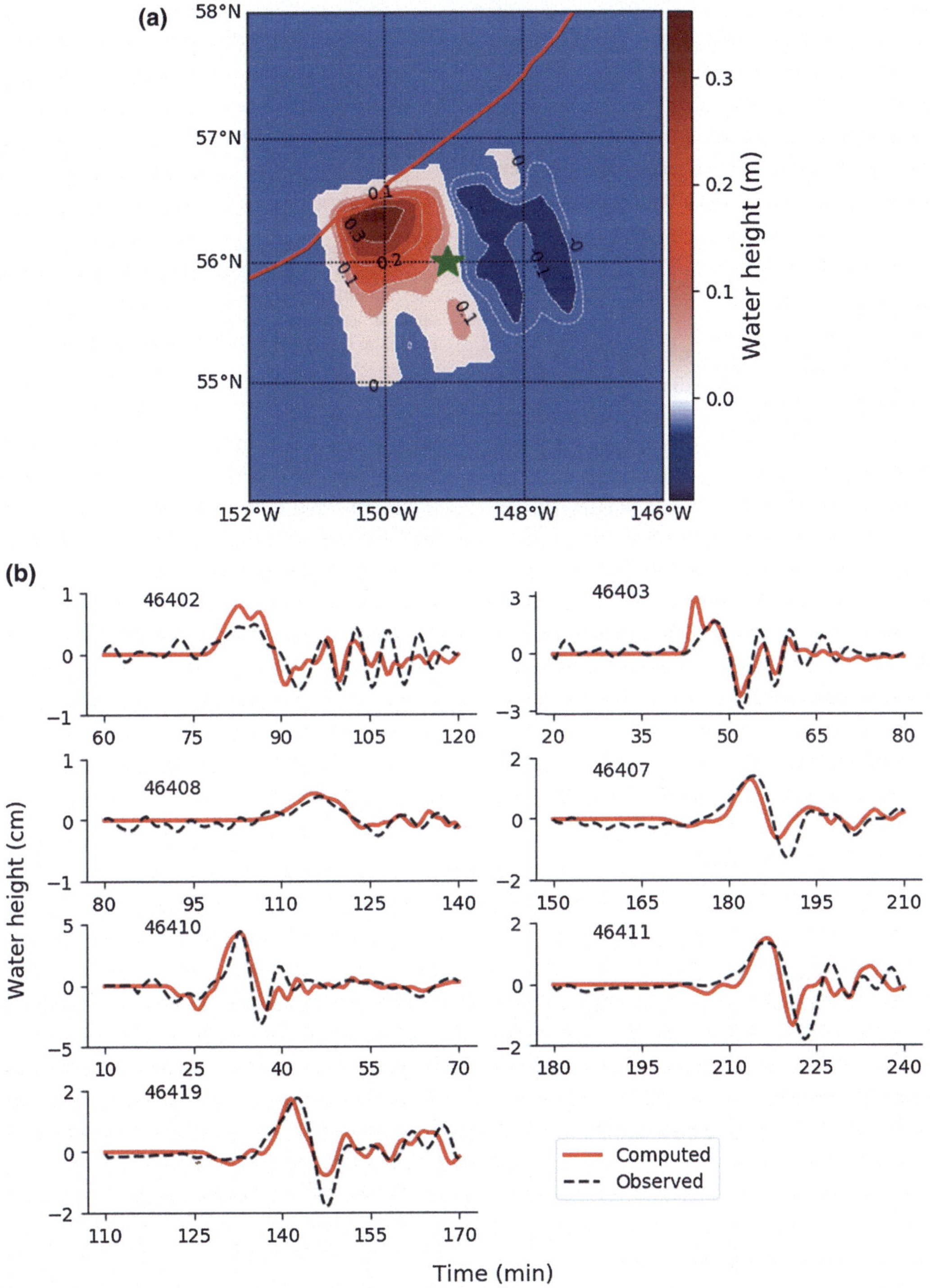

Figure 2

a The initial sea surface displacement model (SSD) for the 2018 offshore Kodiak tsunami obtained from Green's function based time reverse imaging (GFTRI) of tsunami waveform data from DART buoys 46402, 46403, 46407, 46408, 46410, 46411, and 46419. **b** Agreement between observed (dashed black) and computed tsunami waveforms (red) at DART buoys 46402, 46403, 46407, 46408, 46410, 46411, and 46419 as GFs, instead the sea-floor deformation computed by the Okada (1985) model for each sub-fault are used as GFs in the inversion. The TSI approach has been applied to several earthquake events, such as the 2017 Tehuantepec (Mexico) earthquake (Gusman et al. 2018) and the 2009 Samoa earthquake (Hossen et al. 2018) to characterize the earthquake faults.

2. Data

Thirteen DART and tide gauge stations in the NE Pacific with high quality recordings of the 2018 Kodiak earthquake and tsunami were utilized in our analysis (Figure 1, Table T1 in Supplement). DART sea-floor absolute pressure records for the 2018 Kodiak earthquake and tsunami were obtained from the National Oceanic and Atmospheric Administration-National Data Buoy Center (http://www.ndbc.noaa.gov/dart.shtml) (Mungov et al. 2013). Coastal tide gauge records were obtained from the Intergovernmental Oceanographic Commission-United Nations Educational, Scientific and Cultural Organization sea level monitoring (http://www.ioc-sealevelmonitoring.org/index.php). The tsunami waveforms are prepared by resampling, detiding and filtering. The sampling interval of the tide gauge is 60 s, while that of the DART is 15–60 s (event mode vs. continuous). For waveform data processing and inversion purposes, we resampled all of the tsunami time series data to 30 s time interval. For removing tides, we used a low sample rate of 15 min and removed the tidal component by estimating a polynomial fit and removing it from the records. After removing the tidal signal the data were bandpass filtered using a Butterworth filter with cutoff periods from 1.5 h (0.000185185 Hz) to 4 min (0.004166667 Hz).

3. Methodology

In the first step of our inversion, we estimate the SSD model by implementing the Green's function based time reverse imaging (GFTRI) method as described in Hossen et al. (2015a). The GFTRI method uses precomputed Green's functions (GFs) from a unit source to each observation station, convolves the GFs with the time reversed observed waveforms and then scales the reverse wave-field by the GFs to estimate amplitude over each source patch. The source patches are obtained by dividing the source domain; over each source patch a unit source is created by using a box function with cosine tapering. The dimension of each source patch is 25 × 25 km^2, which makes in total 60 source grid points (Supplementary Figure S1) for the 250 × 150 km^2 source area of the 2018 Kodiak Alaska tsunami. The GFTRI has the advantage that it does not require the use of the LSQ method to solve the evolved linear system and additionally does not need regularization constraints (Hossen et al. 2015a). However, the GFTRI method requires a tsunami propagation model to satisfy the reciprocity principle (Korolev 2011). To compute the GFs, we use the JAGURS tsunami propagation model that satisfies the reciprocity principle (see for details, Hossen et al. 2015c; Mulia et al. 2018). JAGURS solves the two dimensional nonlinear shallow water equations with Boussinesq dispersion in a finite difference scheme (Baba et al. 2017). This code is parallelized, hence reducing the computational time significantly.

We set our computational domain to latitude 38° N–61° N and longitude 172° W–124° W in order to cover the epicenter and the locations of the observation stations considered in this work (Fig. 1). We considered three nested grids in our simulation: the coarsest grid resolution (60 arc-sec) bathymetry derived from the General Bathymetry Chart of the Ocean global database (Weatherall et al. 2015) over

Figure 3 ▶
Residual errors obtained using different fault parameterizations in the earthquake slip inversion. **a** Case 1a, residual error for inversion with a single fault (F_1 to F_9) oriented in the N–S direction. **b** Case 1b, residual error for inversion with a single fault (S_1 to S_9) in the E–W direction. For Case 1 (single fault), fault F_4 provides the smallest residual error. **c** Case 2, fault F_4 plus one fault from the N–S direction is used in the inversion; **d** Case 3a, faults (F_4 & S_5) and an additional fault in the E–W direction; **e** Case 3b, faults (F_4 & S_5) and an additional fault in the N–S direction. The smallest error (**d**) is obtained with the set $\{F_4, F_8, S_5\}$. **f** Case 4, faults (F_4, F_8 & S_5) with one additional fault in the N–S direction. In this case (case 4), the improvement in fit (reduction of error) is very slight. In a similar fashion, we considered five and six faults in cases 5 & 6 and find that the error reduction (**g**, **h**) is very slight

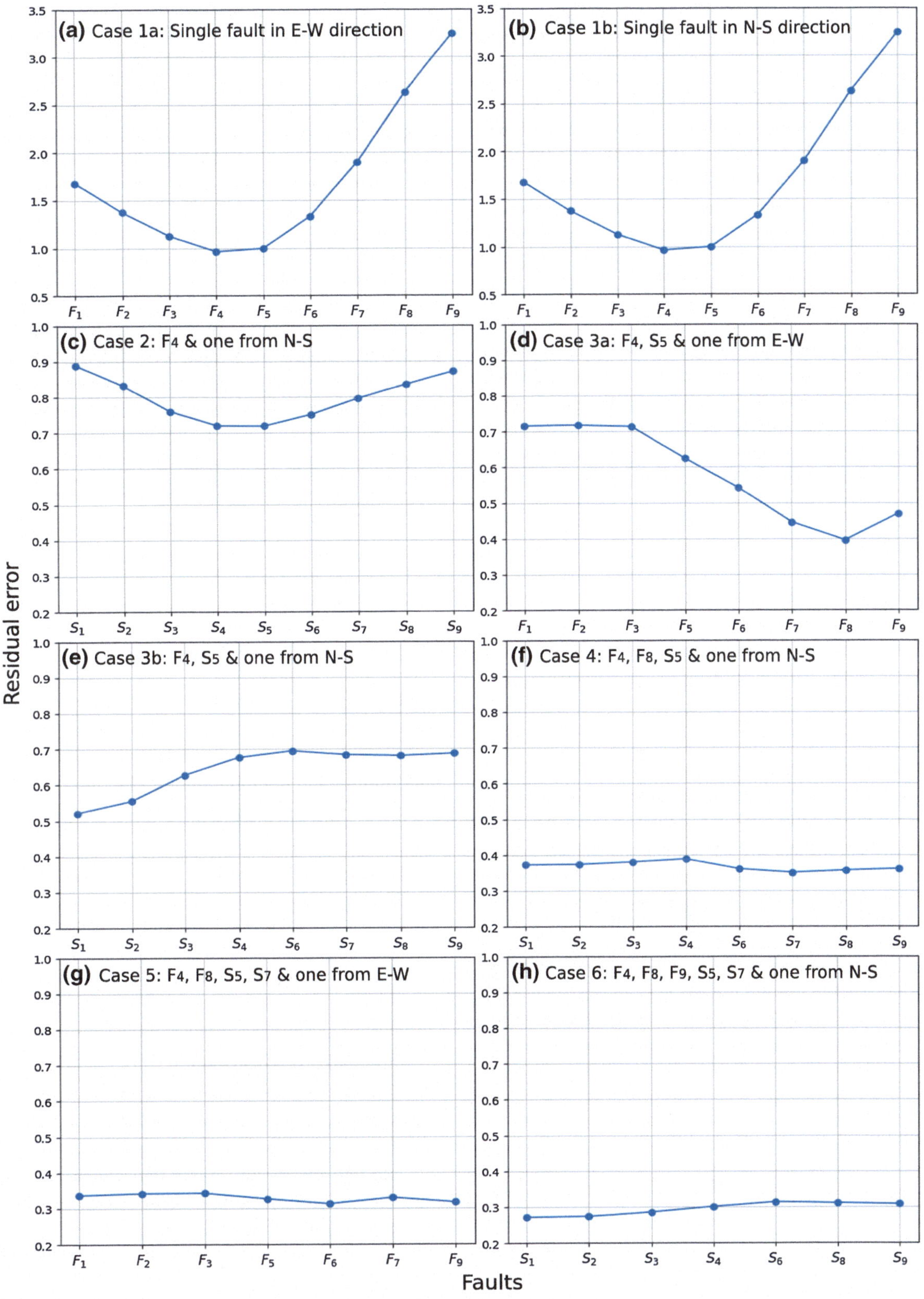
(a) Case 1a: Single fault in E-W direction
(b) Case 1b: Single fault in N-S direction
(c) Case 2: F4 & one from N-S
(d) Case 3a: F4, S5 & one from E-W
(e) Case 3b: F4, S5 & one from N-S
(f) Case 4: F4, F8, S5 & one from N-S
(g) Case 5: F4, F8, S5, S7 & one from E-W
(h) Case 6: F4, F8, F9, S5, S7 & one from N-S
Residual error
Faults

the entire domain, intermediate grid resolution (20 arc-sec) derived from the bathymetry provided by NOAA for the domain (54° N–60.5° N, 156° W–132° W) and the finest grid resolution (20/3 arc sec) derived from the same bathymetry for the domain shown in Fig. 1 (red boxes). The time step for this computation is set as 0.5 s.

In the second step of our inversion, we estimate the earthquake slip distribution by inverting the SSD model. From the earthquake slip distribution, we then compute a co-seismic water surface displacement (WSD) model which can be compared to the input SSD. We determine the slip distribution by solving a linear system $Gm = d$ using a least squares method with regularization constraints (damping and smoothing). The matrix G contains the sea-floor deformation from each sub-fault calculated by Okada (1985) elastic dislocation model and the vector d contains the SSD model as grid points (explained in detail in Sect. 5).

4. Sea Surface Displacement Model

The initial sea surface displacement (SSD) model that we determine using GFTRI is shown in Fig. 2a. In this source analysis, we use seven DART buoys (46402, 46403, 46407, 46408, 46410, 46411, 46419) located in the deep ocean surrounding the earthquake source. These stations are selected because the quality of recordings is excellent, the bathymetry data surrounding these stations is more accurate than at shallow sites and grid resolution near these stations does not have a significant impact on GFs. We consider only 1 h of data after the first arrival of tsunami in order to avoid coastal reflection or local resonance. We do not use DART 46409 because it is too close to the source for reciprocity to hold (Hossen et al. 2015a).

In GFTRI method, the amplitude of each unit source is calculated by

$$a_i = \frac{1}{p}\sum_{j=1}^{p} \frac{1}{|G_{ij}|^2} TR_j(s_i, T - \delta t_i), \qquad (1)$$

where δt_i is the time adjustment (for details, see Hossen et al. (2015a)), p is the number of stations,

Figure 4 ▶
a A schematic diagram of the fault planes considered in the earthquake slip distribution estimation. Fault planes in the N–S direction (purple) have strike 164° and planes in the E–W direction (yellow) have strike 261°. **b**, **c** The fault models for case 1 (one fault in each direction is identified as the ruptured fault); **d** slip distribution for case 2, which has two faults (F_4 and S_5) and **e** slip distribution for case 3, with 3 faults (F_4, F_8 and S_5). Case 3 is our preferred fault model for the 2018 Kodiak earthquake

T is the time length and $TR_j(s_i, \cdot)$ is the reverse wavefield obtained by convolving GFs from the ith source with the jth observation in reverse time order. In this study, we find that time adjustment does not provide a reliable SSD model that fits observed waveforms well. Instead, it is important to include an attenuation factor γ to account for decrease in tsunami height with propagation away from the source. Thus, the equation for calculating the amplitude becomes

$$a_i = \frac{1}{p}\sum_{j=1}^{p} \frac{1}{\gamma |G_{ij}|^2} TR_j(s_i, T) \,. \qquad (2)$$

We find that an attenuation factor of 2 determined by trial and error basis provides a more accurate SSD model than a model with time adjustment. The resulting agreement between observed and simulated waveforms is very good at most of the stations, the period of the waveforms matches very well, and the peak amplitude is close to that observed (Fig. 2b). The resulting displaced surface water model is mostly concentrated between the trench line and the epicenter, with maximum water height of approximately 30 cm (Fig. 2a).

5. Fault Slip Distribution

In the second step of the inversion, we estimate the earthquake slip distribution from the SSD model. The aftershock distribution of the Kodiak event suggests that the rupture likely occurred on both N–S and E–W trending faults, and subsequent studies indicate that the rupture likely took place on a complex distribution of multiple fault planes oriented in the N–S and E–W directions (e.g, Lay et al. 2018; Ruppert et al. 2018; Krabbenhoeft et al. 2018; Zhao et al. 2018). We use the SSD model to find the

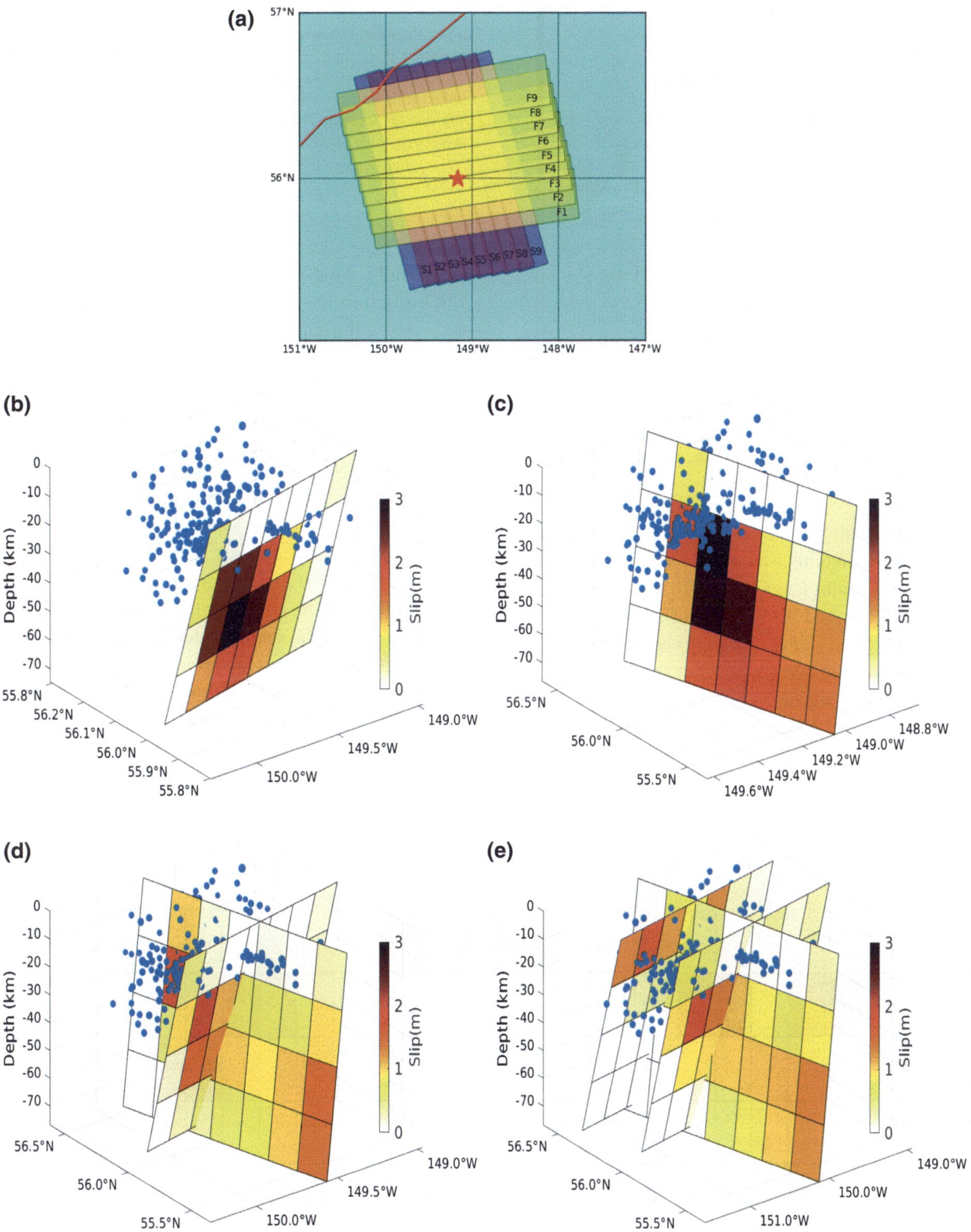
(a)
57°N
56°N
151°W
150°W
149°W
148°W
147°W
F9
F8
F7
F6
F5
F4
F3
F2
F1
S1 S2 S3 S4 S5 S6 S7 S8 S9
(b)
(c)
(d)
(e)
Depth (km)
Slip(m)
0
-10
-20
-30
-40
-50
-60
-70
55.8°N
56.2°N
56.1°N
56.0°N
55.9°N
150.0°W
149.5°W
149.0°W
56.5°N
55.5°N
148.8°W
149.2°W
149.4°W
149.6°W
151.0°W

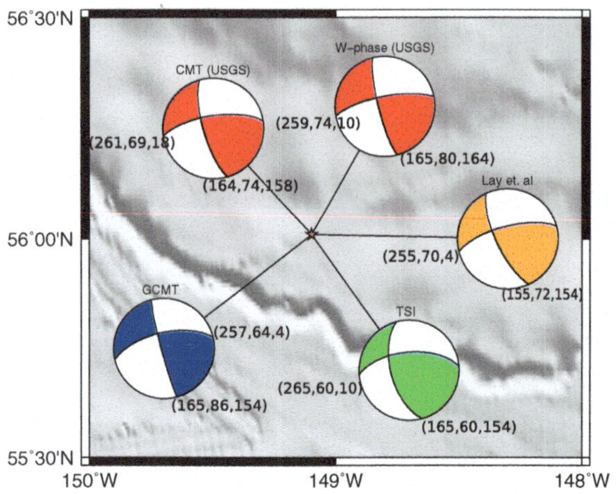

Figure 5
Focal mechanisms determined by USGS (red), GCMT (blue), and Lay et al. (2018) (orange). TSI (green) is our proposed focal mechanism, using the best fitting orientations of N–S and E–W planes

appropriate fault planes associated with the rupture process by searching over a grid of 18 faults with nine parallel fault planes spaced 10 km apart in each direction (N–S and E–W striking). The faults are denoted as $\{F_1, \ldots, F_9\}$ for faults with E–W strike and northward dip and $\{S_1, \ldots, S_9\}$ for faults striking N–S with westward dip (Fig. 4a). Among them, we chose the faults based on the smallest "residual error", defined as the l_2 norm of the difference between the SSD model (determined by inverting tsunami data) and the WSD model (calculated from the earthquake slip distribution) (Fig. 3).

Figure 7 ▶
a Sea-floor deformation calculated from the slip distribution of our preferred three-fault model; **b** (dashed black) observed tsunami waveforms, (red) modeled waveforms from the tsunami SSD model (Fig. 2a), (green) modeled waveforms from the earthquake slip distribution (**a**). The first seven DART buoys are used in the inversion, other DART and tide gauge stations shown are for comparison

Initially, we use fault orientations from the USGS CMT solution (N–S and E–W oriented faults: strike $\psi_1 = 164°$, dip $\phi_1 = 74°$, rake $r_1 = 158°$ and strike $\psi_2 = 261°$, dip $\phi_2 = 69°$, rake $r_2 = 18°$). We approximate each fault using a 7×4 grid of sub-faults of dimension of $20 \times 20\,\text{km}^2$. Using these fault parameters with unit slip, we compute synthetic GFs of sea-floor deformation (different from tsunami GFs of section 4) for each sub-fault using the Okada (1985) model over the same domain as the SSD model. The domain extends from 54° N–58° N and 152° W–146° W (Fig. 2a) with grid spacing of 1 arc min, giving a total number of grid points of 87001 ($= 361 \times 241$). The sea-floor deformation is assumed to be identical to the sea-surface displacement. We store the GFs associated with each sub-fault in each column of the matrix G and the sea-surface displacement as the vector d that yields the system $Gm = d$. The length of each column of G is 87,001 and the number of columns depends on how many fault planes are considered in the source inversion.

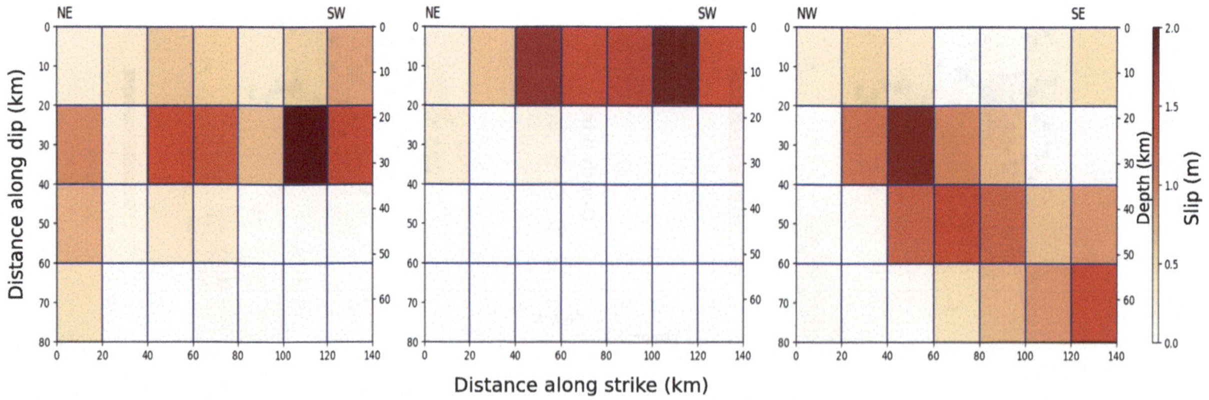

Figure 6
Slip distribution for the preferred model, which has three faults: F_4 (left), F_8 (middle), S_5 (right). Color represents amount of slip in each sub-fault

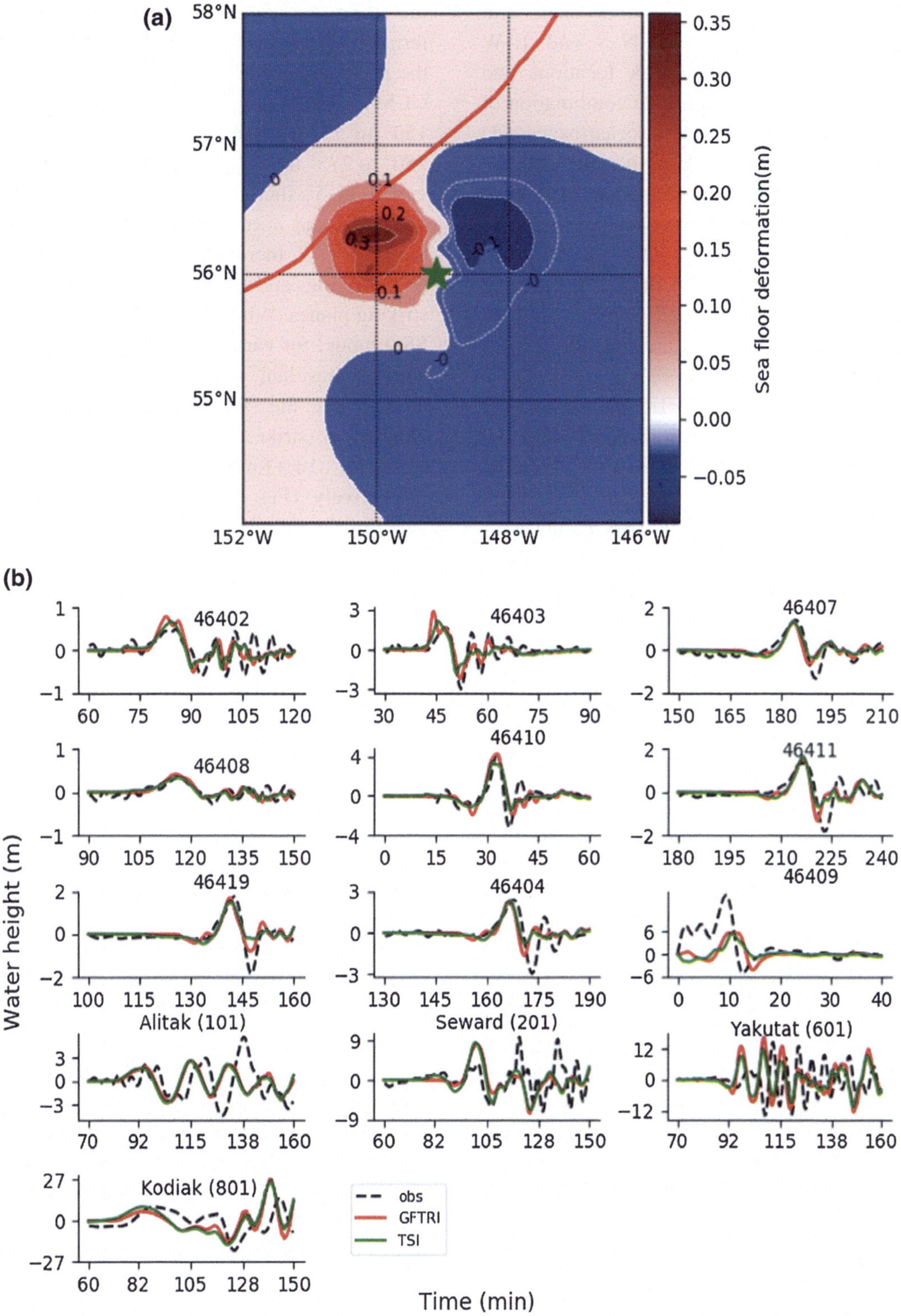

The vector *m* that we solve for represents the slip over the subfaults. The system is solved by using a non-negative least squares method with damping and smoothing constraints- (0.3 and 0.01, respectively) chosen by trial and error in order to obtain a stable solution.

We assume that the rupture is associated with several fault planes oriented in N–S and E–W directions, based on the aftershock locations and previous studies. To find the optimal combination of ruptured faults, we consider several scenarios: case 1) a single fault in either N–S or E–W orientation; case 2) two faults, with a fixed fault from case 1 (the case 1 fault that provides the smallest residual error); case 3) three faults, with two fixed faults determined in case 2; case 4) four faults, with three from case 3; case 5) five faults, with four fixed faults determined in case 4; and finally, case 6) six faults with five fixed faults from case 5.

In case 1, we carried out 18 earthquake slip inversions in total; nine for each direction (Fig. 3a, b). The inversion using fault F_4 in the E–W direction produces the smallest residual error and is thus fixed in the next case 2. In case 2, we fix fault F_4 in case 1 and search over the nine faults in the N–S direction and find that the combination of faults $F_4 \,\&\, S_5$ provides the smallest error (Fig. 3c). These two faults ($F_4 \,\&\, S_5$) are fixed for case 3, where we conducted 16 inversions with one of the remaining 8 faults in both E–W and N–S directions in each inversion. The residual error is presented in Fig. 3d, e; comparing the errors in different fault combinations, we find that the set of $\{F_4, F_8, S_5\}$ produces the smallest residual error. In case 4, we consider these three faults $F_4, F_8 \,\&\, S_5$ with one of the remaining 8 faults in the N–S direction and find that the error varies very slightly. In a similar way, we conducted inversion for the case of 5 and 6 using five and six faults; the resulting error variation is so subtle that it is difficult to distinguish which fault should be considered for improving the result. We concluded that the set of three faults $\{F_4, F_8, S_5\}$ in case 3 is sufficient for recovering the SSD model. The residual error is 0.396 and the slip distribution is displayed in Supplementary Figure S2. In our model the maximum slip occurs on E–W fault F_4 at a depth of 20 km. In this model, the total seismic moment is 9.37×10^{20} Nm, M_w7.9, which is very close to both the USGS and global CMT (GCMT) solution.

Our initial fault plane orientations were based on the USGS CMT solution. However, subsequent studies have suggested a range of possible solutions (Fig. 5 and references in caption). Thus, we perturbed the strike, dip, and rake of the 3 planes of our preferred model (case 3) within a range consistent with the previous studies of Lay et al. (2018), USGS and GCMT. We varied the strike in 5° increments from 150° to 175° for the N–S-striking planes and from 250° to 275° for the E–W striking planes; the dip was varied in 5° increments from 60° to 85° for both westward and northward dipping planes. We varied the rake in 3° increments from 154° to 163° for the N–S-striking and from 10° to 19° for the E–W striking planes. With these variations, we inverted the SSD model for earthquake slip model and found the smallest residual error (misfit) between the SSD model and the WSD model resulted from the parameters (strike, dip, rake): (165°, 60°, 154°) and (265°, 60°, 10°) for the N–S and E–W striking planes, respectively (Fig. 5). We found that the smallest value of the rake in the range was preferred for both planes whereas strike and dip changed by a smaller amount. The slip distribution over the faults with this set of parameters is displayed in Fig. 6 and the WSD model is presented in Fig. 7a.

6. Discussion

After performing our two step inversions (TSI), which first consists of Green's function based time reverse imaging (GFTRI) for the sea surface displacement (SSD) model (Sect. 4), followed by inversion of the SSD model for the fault model (earthquake slip distribution) (Sect. 5), we find a preferred fault model consisting of one N–S oriented fault and two E–W oriented faults (Fig. 6). The coseismic water surface displacement (WSD) model from our preferred earthquake fault model can reproduce the peak of the observed waveform, but not the trough in the leading wave, particularly at stations 46411 and 46419 in Fig. 7b. This misfit indicates that the WSD model cannot recover subsidence of the sea surface completely. On the other hand, our SSD model can explain the observed tsunami waveforms very well at most DART stations including DART station 46404 which is not used in the source inversion (Fig. 7b). DART 46409 is in close proximity to the source and is not well fit. Lay et al. (2018) also modeled lower water height than the observed at station 46409. We believe that the observed wave height at station 46409 may be a combination of the tsunami wave and the seismic wave, such as the Rayleigh wave as it is located very near to the epicenter. In addition, the fit at tide gauges (not used in source inversion) is not as good as the fit at the DART stations. This could be due to in part to the insufficient grid resolution (20/3 arc sec) used in coastal regions for the tsunami simulation.

From the SSD model we determine the multi-fault earthquake slip distribution model by using the TSI method (Sect. 5). In total, 18 faults were considered in both approximately N–S and E–W directions. Our preferred fault model consists of three faults with slip distribution on the E–W oriented faults located in a spatially compact region shallower than 40 km. On the southernmost E–W fault F_4, the major slip is located within the depth range 18 km to 35 km and on F_8, the major slip is shallower than 20 km. In contrast, the slip on the N–S oriented fault is distributed over a wider range of depths, with the major slip at 20–60 km depth.

Our results, obtained from GFTRI of tsunami data, are broadly consistent with the results obtained in other studies which used seismic, GPS, and in some cases tsunami data. The details between the previous studies vary, including the number of faults, the orientation of the segment which released the most energy (NE-SW trending left-lateral strike-slip segments (Ruppert et al. 2018; Zhao et al. 2018), or right-lateral slip on SSE-trending fault Lay et al. (2018). Nevertheless, all studies require faulting on a set of conjugate faults roughly oriented N–S and E–W. (Krabbenhoeft et al. 2018) included analysis of gravity, magnetics, and bathymetry data of the outer-rise of the Alaska subduction zone in their study of the Kodiak event. They find a N–S aligned crustal fabric influenced by plate bending normal faults, N–S magnetic anomalies, and E–W oriented fracture zones.

7. Conclusion

We estimated the tsunami source model (sea surface displacement model or SSD) of the 2018 offshore Kodiak earthquake tsunami by utilizing the Green's function based time reverse imaging (GFTRI) method. The SSD model determined using tsunami data from the seven DART stations provides good waveform fits between observed and simulated waveforms, the period of the tsunami arrival matches well, and the peak amplitudes are compatible with the observed waveforms. We used DART stations located around the source in the Gulf of Alaska and North Pacific Ocean. The performance could possibly be improved by adding far-field stations located in the Pacific Ocean, which would improve the azimuthal coverage though perhaps at the expense of signal to noise ratio.

We inverted the SSD model for the earthquake slip distribution over a range of fault combinations chosen in a systematic manner. Our preferred fault model consists of three faults. We perturbed the fault parameters using bounds from previously published work to further determine optimal fault orientations. Our preferred earthquake slip model produces sea-surface uplift (water surface displacement or WSD) well-matched with the SSD model. Overall, the WSD model from the three fault model is consistent with the SSD model although it is unable to recover the subsidence completely. While seismic and GPS data are of enormous utility and provide important information, we find that with the computationally efficient two-step inversion (TSI) method, which only needs tsunami data as input, we can recover the major aspects of the tsunami source and earthquake source associated with the 2018 offshore Kodiak earthquake.

Acknowledgements

This work utilized the RMACC Summit supercomputer, which is supported by the National Science Foundation (awards ACI-1532235 and ACI-1532236), the University of Colorado Boulder, and Colorado State University. The Summit supercomputer is a joint effort of these two universities. We thank Aaron Sweeney (NOAA) for checking Kodiak tide gauge data, Dmitry Nicolsky (University of Alaska) & Nicolas Arcos (NOAA) for providing bathymetry data covering the Alaska region and Leonardo Ramirez-Guzman (National Autonomous University of Mexico) for providing a MATLAB script for 3D plot. The first author thanks the Cooperative Institute for

Research in Environmental Sciences (CIRES) for supporting this research through a postdoctoral visiting fellowship. This work was supported in part by NSF Award Number ICER 1855090.

References

An, C., Sepúlveda, I., & Liu, P. L. F. (2014). Tsunami source and its validation of the 2014 Iquique, Chile, earthquake. *Geophysical Research Letters*, *41*(11), 3988–3994.

Baba, T., Allgeyer, S., Hossen, M. J., Cummins, P. R., Tsushima, H., Imai, K., et al. (2017). Accurate numerical simulation of the far-field tsunami caused by the 2011 Tohoku earthquake, including the effects of Boussinesq dispersion, seawater density stratification, elastic loading, and gravitational potential change. *Ocean Modelling*, *111*, 46–54. https://doi.org/10.1016/j.ocemod.2017.01.002.

Baba, T., & Cummins, P. R. (2005). Contiguous rupture areas of two Nankai Trough earthquakes revealed by high-resolution tsunami waveform inversion. *Geophysical Research Letters*, *32*(8), L08305. https://doi.org/10.1029/2004GL022320.

Gusman, A. R., Mulia, I. E., & Satake, K. (2018). Optimum sea surface displacement and fault slip distribution of the 2017 Tehuantepec earthquake (Mw 8.2) in Mexico estimated from tsunami waveforms. *Geophysical Research Letters*, *45*(2), 646–653.

Gusman, A. R., Mulia, I. E., Satake, K., Watada, S., Heidarzadeh, M., & Sheehan, A. F. (2016). Estimate of tsunami source using optimized unit sources and including dispersion effects during tsunami propagation: The 2012 Haida Gwaii earthquake. *Geophysical Research Letters*, *43*(18), 9819–9828.

Ho, T. C., Satake, K., & Watada, S. (2017). Improved phase corrections for transoceanic tsunami data in spatial and temporal source estimation: Application to the 2011 Tohoku Earthquake. *Journal of Geophysical Research: Solid Earth*, *122*(12), 10–155.

Hossen, M. J., Cummins, P. R., Dettmer, J., & Baba, T. (2015a). Time reverse imaging for far-field tsunami forecasting: 2011 Tohoku earthquake case study. *Geophysical Research Letters*, *42*(22), 9906–9915.

Hossen, M. J., Cummins, P. R., Dettmer, J., & Baba, T. (2015b). Tsunami waveform inversion for sea surface displacement following the 2011 Tohoku earthquake: Importance of dispersion and source kinematics. *Journal of Geophysical Research: Solid Earth*, *120*(9), 6452–6473.

Hossen, M. J., Cummins, P. R., Roberts, S. G., & Allgeyer, S. (2015c). Time reversal imaging of the tsunami source. *Pure and Applied Geophysics*, *172*(3–4), 969–984.

Hossen, M. J., Gusman, A., Satake, K., & Cummins, P. R. (2018). An adjoint sensitivity method applied to time reverse imaging of tsunami source for the 2009 Samoa earthquake. *Geophysical Research Letters*, *45*(2), 627–636.

Korolev, Y. P. (2011). An approximate method of short-term tsunami forecast and the hindcasting of some recent events. *Natural Hazards and Earth System Sciences*, *11*(11), 3081–3091. https://doi.org/10.5194/nhess-11-3081-2011. https://www.nat-hazards-earth-syst-sci.net/11/3081/2011/.

Krabbenhoeft, A., von Huene, R., Miller, J. J., Lange, D., & Vera, F. (2018). Strike-slip 23 January 2018 MW 7.9 Gulf of Alaska rare intraplate earthquake: Complex rupture of a fracture zone system. *Scientific Reports*, *8*(1), 13706.

Lay, T., Ye, L., Bai, Y., Cheung, K. F., & Kanamori, H. (2018). The 2018 MW 7.9 Gulf of Alaska earthquake: Multiple fault rupture in the Pacific plate. *Geophysical Research Letters*, *45*(18), 9542–9551.

Mulia, I. E., & Asano, T. (2016). Initial tsunami source estimation by inversion with an intelligent selection of model parameters and time delays. *Journal of Geophysical Research: Oceans*, *121*(1), 441–456.

Mulia, I. E., Gusman, A. R., Jakir, Hossen M., & Satake, K. (2018). Adaptive tsunami source inversion using optimizations and the reciprocity principle. *Journal of Geophysical Research: Solid Earth*, *123*(12), 10–749.

Mungov, G., Eblé, M., & Bouchard, R. (2013). DART® tsunameter retrospective and real-time data: A reflection on 10 years of processing in support of tsunami research and operations. *Pure and Applied Geophysics*, *170*(9–10), 1369–1384.

Okada, Y. (1985). Surface deformation due to shear and tensile faults in a half-space. *Bulletin of the Seismological Society of America*, *75*(4), 1135–1154.

Ruppert, N., Rollins, C., Zhang, A., Meng, L., Holtkamp, S., West, M., et al. (2018). Complex faulting and triggered rupture during the 2018 MW 7.9 offshore Kodiak, Alaska, earthquake. *Geophysical Research Letters*, *45*(15), 7533–7541.

Satake, K. (1987). Inversion of tsunami waveforms for the estimation of a fault heterogeneity: Method and numerical experiments. *Journal of Physics of the Earth*, *35*(3), 241–254.

Satake, K., Fujii, Y., Harada, T., & Namegaya, Y. (2013). Time and space distribution of coseismic slip of the 2011 Tohoku earthquake as inferred from tsunami waveform data. *Bulletin of the Seismological Society of America*, *103*(2B), 1473–1492.

Weatherall, P., Marks, K. M., Jakobsson, M., Schmitt, T., Tani, S., Arndt, J. E., et al. (2015). A new digital bathymetric model of the world's oceans. *Earth and Space Science*, *2*(8), 331–345.

Wei, Y., Cheung, K. F., Curtis, G. D., & McCreery, C. S. (2003). Inverse algorithm for tsunami forecasts. *Journal of Waterway, Port, Coastal, and Ocean Engineering*, *129*(2), 60–69.

Yue, H., Lay, T., & Koper, K. D. (2012). En échelon and orthogonal fault ruptures of the 11 April 2012 great intraplate earthquakes. *Nature*, *490*(7419), 245.

Zhao, B., Qi, Y., Wang, D., Yu, J., Li, Q., & Zhang, C. (2018). Coseismic slip model of the 2018 M w 7.9 Gulf of Alaska earthquake and its seismic hazard implications. *Seismological Research Letters*, *90*(2A), 642–648.

(Received August 14, 2019, revised January 24, 2020, accepted January 28, 2020, Published online February 5, 2020)

Pure Appl. Geophys. 177 (2020), 1347–1378

https://doi.org/10.1007/s00024-020-02427-x

Pure and Applied Geophysics

The 2018 Alaska-Kodiak Tsunami off the West Coast of North America: A Rare Mid-plate Tsunamigenic Event

Kejia Wang,[1,2,3] Richard E. Thomson,[1] Alexander B. Rabinovich,[1,4] Isaac V. Fine,[1] and Tania L. Insua[2,3,5]

Abstract—The major (M_w 7.9) earthquake that struck the Gulf of Alaska near Kodiak Island on 23 January 2018 was a rare, mid-plate strike-slip event that triggered a minor trans-Pacific tsunami. An analysis of the simultaneous measurements of tsunami waveforms at 21 open-ocean sites (including three independent arrays of stations) and 27 coastal tide gauges in the Gulf of Alaska and along the coast of North America has enabled us to examine properties of the 2018 tsunami, its transformation over the continental slope and shelf, and its amplification as the waves approached the coast. Results show that the tsunami wave variance decreased monotonically along the west coast from northern British Columbia to southern Oregon. Based on the variance structure, the mean amplification factor for Tofino on the west coast of Vancouver Island (a "beacon" site with a long time series), was $A_{RMS}^{Tof} = 5.3$, in good agreement with corresponding estimates for four major past events; 4.5 (2009 Samoa), 4.3 (2010 Chile), 6.3 (2011 Tohoku) and 5.2 (2012 Haida Gwaii). This variance-derived amplification for Tofino was greater than the amplification factor based on the amplitude ratio ($A^{Tof} = 3.2$). Spectral analysis of the records showed that the tsunami had a relatively large high-frequency content (i.e., was "blueish"), with nearly 90% of the total energy in the open ocean at frequencies > 1.7 cph (periods < 35 min) and with an "integral frequency scale" of 4 cph (period 15 min). Wavelet analysis revealed strong dispersion of the propagating tsunami waves, in agreement with theoretical estimates. The abrupt jump in water depth of about 4 cm detected at DART 46409, located mid-plate about 85 km from the epicenter of the 2018 Kodiak earthquake, appears to have been due to an earthquake-induced seafloor subsidence.

[1] Fisheries and Oceans Canada, Institute of Ocean Sciences, 9860 West Saanich Road, Sidney, BC V8L 4B2, Canada. E-mail: kejiaw7@gmail.com; Richard.Thomson@dfo-mpo.gc.ca

[2] Ocean Networks Canada, University of Victoria, Queenswood Campus, 2474 Arbutus Road, Victoria, BC V8N 1V8, Canada.

[3] School of Earth and Ocean Sciences, University of Victoria, Bob Wright Center A405, Victoria, BC V8P 3E6, Canada.

[4] Russian Academy of Sciences; P.P. Shirshov Institute of Oceanology, 36 Nakhimovsky Prosp, Moscow 117997, Russia.

[5] Office of Research Services, University of Tasmania, 301 Sandy Bay, Sandy Bay, TAS 7001, Australia.

Keywords: 2018 Kodiak earthquake and tsunami, bottom pressure recorders (BPRs), coastal tide gauges, DART, Mendocino Fracture Zone, tsunami hazard regions, mid-plate earthquakes.

1. *Introduction*

At 09:31:42 UTC on 23 January 2018, the Gulf of Alaska was shaken by a major (M_w 7.9) earthquake. The epicenter of the earthquake (56.004° N, 149.166° W; 14.1 km depth) was located roughly 270 km southeast of Kodiak Island and 90 km from the Aleutian Trench (Lay et al. 2018) (Fig. 1). Historically, numerous earthquakes have occurred in the nearby Alaska-Aleutian subduction zone, including the Great Alaska Earthquake of 27 March 1964 (M_w 9.2), the second largest earthquake ever instrumentally recorded. The Alaska-Aleutian megathrust subduction zone is considered one of the most seismically active zones on earth (Lander 1996; Freymueller et al. 2013; Suleimani et al. 2013), with the potential to generate highly destructive earthquakes and tsunamis.

Due to the widespread destruction caused by the 1964 Alaska tsunami—a catastrophic trans-Pacific event that resulted in 122 deaths—considerable attention was paid to the 2018 event. Shaking from the earthquake was felt in the middle of the night by many people in nearby coastal communities. Based on the substantial magnitude and source location, the U.S. National Tsunami Warning Center (NTWC), Palmer, Alaska (AK), of the National Oceanic and Atmospheric Administration (NOAA) issued a "Tsunami Warning" approximately 4 min after the main shock for the coasts of the Aleutian Islands, Alaska and British Columbia, and a "Tsunami

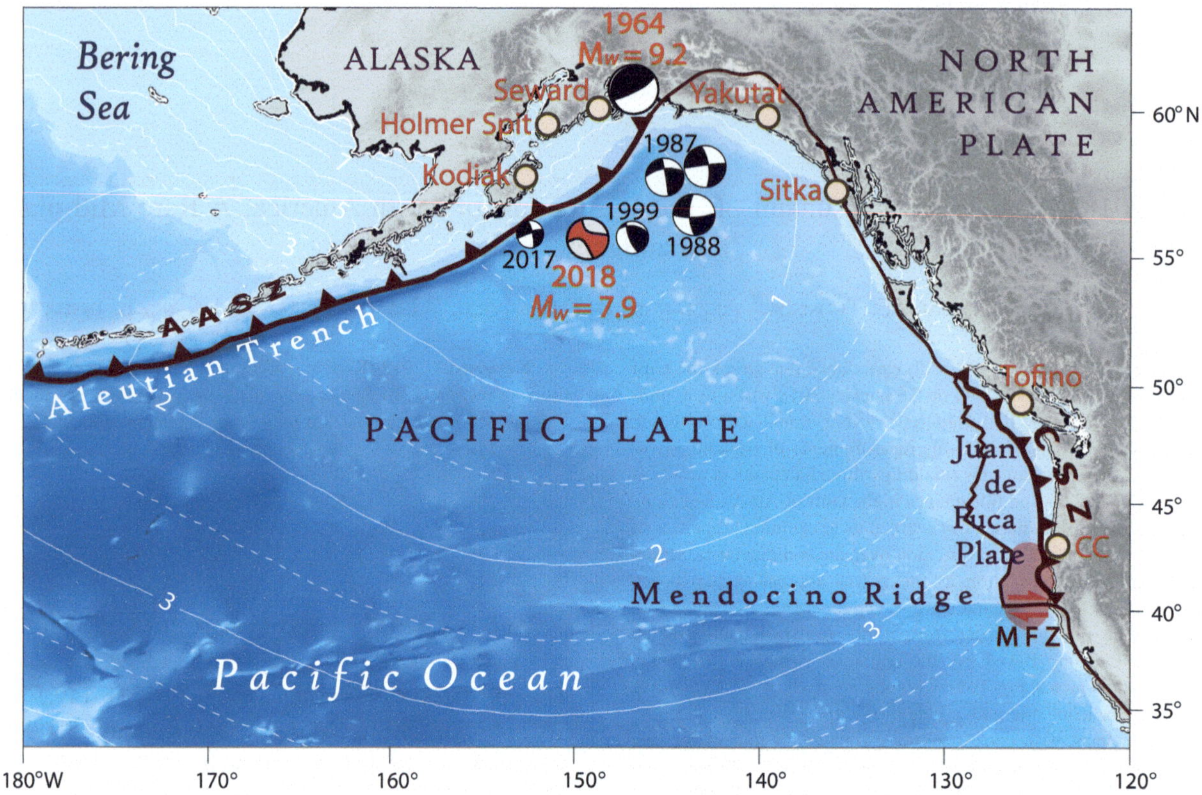

Figure 1

Epicenters of the 2018 (red) and other selected events are denoted by beachballs (USGS earthquake catalog; Lay et al. 2018; Ruppert et al. 2018). The size of each beachball is proportional to the magnitude of the corresponding earthquake. The dark brown contour indicates the boundary of the Pacific Plate and the main subduction zones, with the Alaska Aleutian Subduction Zone and Cascadia Subduction Zone labeled as AASZ and CSZ, respectively. The Mendocino Fracture Zone is marked with red two-way arrows with abbreviation MFZ. Shaded red area marks possible source region for reflected waves. The white curves are the isochrones relative to the epicenter: the solid curves indicate hours and the dashed curves 30-min intervals. *CC* Crescent City

Watch" for the coasts of Washington, Oregon and California. Overnight evacuation took place in many Alaska towns, such as Kodiak, Homer Spit and parts of Seward, and as far south as Tofino, British Columbia (Fig. 1). People awoke to alarming sirens, rushed to high ground and sheltered in local school evacuation centers until NTWC cancelled the Tsunami Warning at 12:12 UTC after it became clear that no significant tsunami was being observed along the coasts of Alaska and British Columbia.

In contrast to powerful megathrust tsunamigenic earthquakes, like the 1964 Great Alaska Earthquake (Satake and Atwater 2007; Freymueller et al. 2013; Gusiakov 2014), the 2018 Kodiak event was of relatively small scale and caused by strike-slip faults originating at a shallow depth of $\sim$ 14 km within the Pacific Plate (cf. Lay et al. 2018; Ruppert et al. 2018). Seismically generated tsunamis originate from the vertical motions of the seafloor associated with thrust-fault earthquakes. Although strike-slip faults can sometimes trigger hazardous underwater landslides and associated tsunamis, such faults cause mostly horizontal displacements of the seafloor and are, therefore, not conducive to seismically generated tsunamis, even during large earthquakes. This appears to be the reason that the 2018 M_w 7.9 Kodiak earthquake did not generate a substantial tsunami.

This tsunami was generated by one of the strongest strike-slip earthquakes on record and spread throughout the Pacific Ocean, where it was recorded by many coastal tide gauges. The tsunami was recorded along the entire coasts of British Columbia

(BC), Washington State (WA), Oregon (OR) and California (CA). What is especially relevant to this study is that the tsunami was also clearly measured by many open-ocean bottom pressure recorders (BPRs) located off the west coast of North America (Fig. 2). In addition to DARTs,[1] situated along the main seismically active zones of the Pacific Ocean (the so-called "Rim of Fire"), the 2018 Kodiak tsunami was also recorded by two arrays of bottom cable stations: (1) the NEPTUNE-Canada array (part of Ocean Networks Canada, ONC), installed in the deep ocean and on the continental margin off southwestern Vancouver Island; and (2) the Cabled Axial Seamount array (part of the US Ocean Observatories Initiative, OOI, USA), deployed in the open ocean, west of northern Oregon. The instruments on these two arrays are located at a variety of depths and offshore distances and, in combination with DARTs and coastal tide gauges from the adjacent coasts, provided us with a unique opportunity to detail the wave characteristics of the 2018 tsunami waves, as well as the evolution and transformation of the waves as they approached the coast. Comparison of the open-ocean and coastal parameters of the tsunami waves has enabled us to examine topographic amplification of the arriving waves and to contribute to future applications of tsunami forecasting models.

In the outer rise region, seaward of the Alaska-Aleutian subduction zone, where the 2018 Kodiak earthquake occurred, major earthquakes are highly uncommon and are generally attributed to plate-bending stresses (cf. Dmowska et al. 1988; Craig et al. 2014; Lay et al. 2018). Since 1900, only a few relatively strong intraplate earthquakes have originated in the northeastern Gulf of Alaska (Fig. 1). These include a sequence of strike-slip events from 17 November 1987 (M_w 7.1), 30 November 1987 (M_w 7.8), and 6 March 1988 (M_w 7.7) along a north–south trend in the seafloor seaward of the Yakutat terrane (Lay et al. 2018). The only historically significant earthquake that took place in the vicinity of the 2018 epicenter was the reverse-faulting M_w 6.0 earthquake of 16 January 1999. There was also a smaller strike-slip M_w 4.9 event in 2017 (Fig. 1) that had a similar focal mechanism to the 2018 M_w 7.9 earthquake (Ruppert et al. 2018). As emphasized by Ruppert et al. (2018) and Lay et al. (2018), the 2018 Kodiak earthquake was an extraordinary event, with a source consisting of multiple rupture faults and strike-slip segments. Because of the complicated earthquake source, it is difficult to reconstruct a reliable tsunami model for this event based on seismic data. This makes tsunami data for this event from open-ocean and coastal instruments in the Gulf of Alaska and along the west North American coast highly valuable. Analysis of the data can assist in the development of effective tsunami source models and in the provision of important information on the characteristics of the source itself.

Organization of the paper is as follows. Section 2 covers the observations and data processing methods used in the study. Basic tsunami statistics and properties are presented in Sect. 3. Wavelet and spectral analysis of the tsunami waves are provided in Sects. 4 and 5, respectively. Tsunami wave propagation is addressed in Sect. 6. Sections 7 and 8 contain discussion and summaries of the results.

2. *Observations and Data Analysis Methods*

The 2018 Kodiak tsunami was measured by a number of high-quality deep-ocean BPRs and Canadian Hydrographic Service (CHS) and NOAA digital tide gauges in the northeast Pacific (Fig. 2), making this one of the most well-recorded tsunamis in the region. Although the present focus is on the coasts of British Columbia and the adjacent states of Washington and Oregon, wave measurements from other instruments in the northeast Pacific are also used to support our analyses.

2.1. *Open-Ocean Instruments*

Three types of BPRs recorded the 2018 Alaska Kodiak tsunami in the northeast Pacific.

[1] DART = Deep-ocean Assessment and Reporting of Tsunamis, is a network of deep-ocean stations, established by NOAA for monitoring tsunami waves and early tsunami warning (cf. Mofjeld 2009; Mungov et al. 2013; Rabinovich and Eblé 2015).

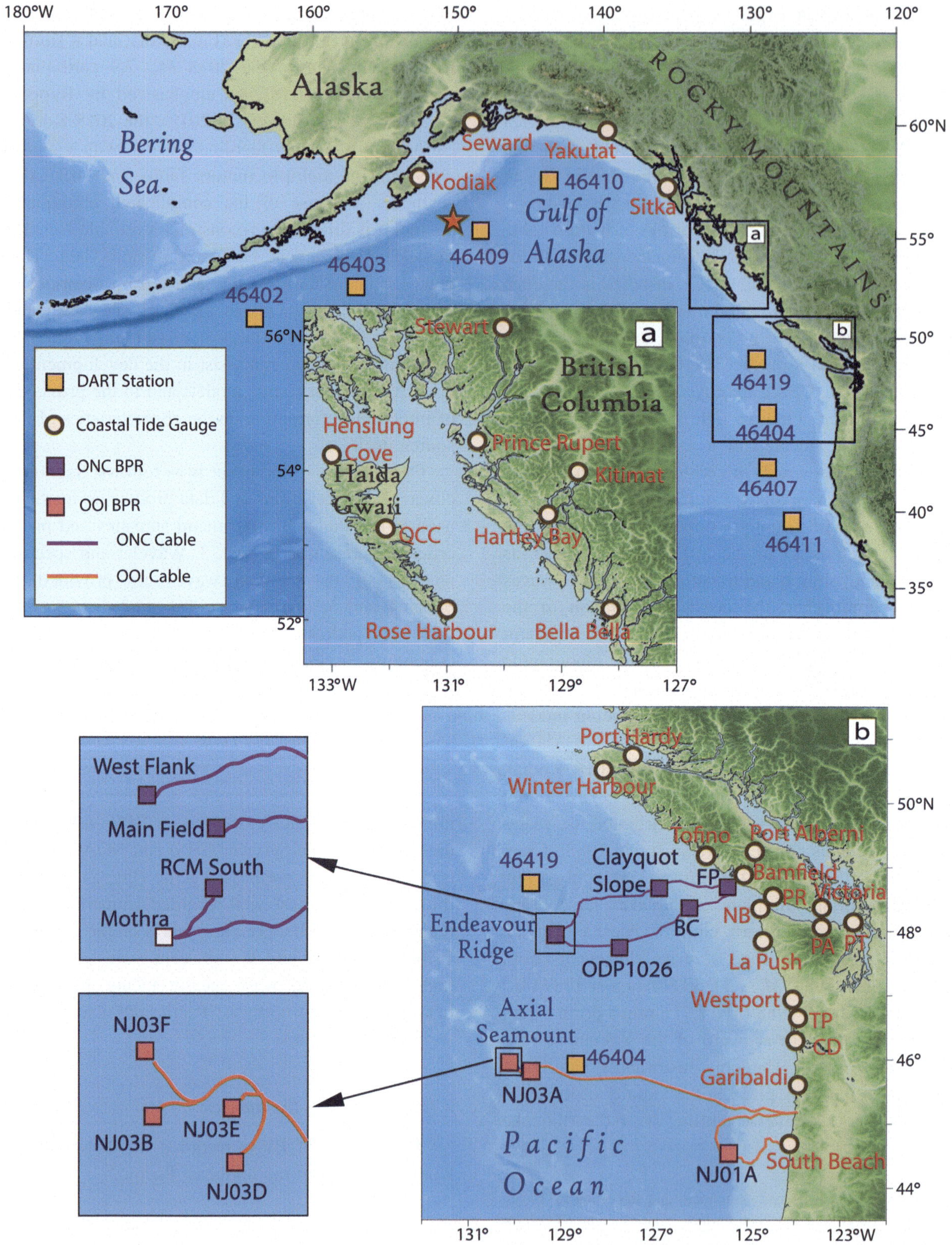
180°W
170°
160°
150°
140°
130°
120°
60°N
55°
50°
45°
40°
35°
Alaska
Bering
Sea
ROCKY MOUNTAINS
Seward
Yakutat
Kodiak
46410
Gulf of
Alaska
Sitka
46409
46403
46402
46419
46404
46407
46411
a
b
DART Station
Coastal Tide Gauge
ONC BPR
OOI BPR
ONC Cable
OOI Cable
56°N
54°
52°
133°W
131°
129°
127°
Stewart
British
Columbia
Henslung
Cove
Prince Rupert
Haida
Gwaii
Kitimat
QCC
Hartley Bay
Rose Harbour
Bella Bella
West Flank
Main Field
RCM South
Mothra
NJ03F
NJ03B
NJ03E
NJ03D
Port Hardy
Winter Harbour
50°N
Tofino
Port Alberni
46419
Clayquot
Slope
FP
Bamfield
PR
Victoria
NB
Endeavour
Ridge
BC
48°
PA
PT
ODP1026
La Push
Axial
Seamount
Westport
TP
CD
46404
46°
NJ03A
Garibaldi
Pacific
Ocean
NJ01A
South Beach
44°
131°
129°
127°
125°
123°W

◀Figure 2
Locations of instruments used in the present study. The epicenter of the 2018 event is marked by a red star. The ONC Mothra BPR failed to record valid data but is mentioned in this study is marked by a white square. **a**, **b** Enlarged versions of the selected areas. Tide gauge abbreviations: *QCC* Queen Charlotte City, *PR* Port Renfrew, *PT* Port Townsend, *PA* Port Angeles, *CD* Cape Disappointment and *TP* Toke Point; BPRs: *BC* Barkley Canyon and *FP* Folger Passage; *ODP* Ocean Drilling Program

2.1.1 DARTs

Open-ocean DART stations employ advanced modern technology for monitoring tsunamis and are an effective component of the Tsunami Warning System (Mofjeld 2009; Rabinovich and Eblé 2015). The DART System transmits real-time data in two modes: the standard operating mode and the event mode. In the standard operating mode, four samples per hour are transmitted from an autonomous bottom pressure transducer through an acoustic channel to a surface buoy at the same station and then through satellite to the Tsunami Warning Centers and the National Data Buoy Center (NDBC). In the event mode, real-time data are transmitted for a few minutes at a 15-s sampling rate and then at a 1-min rate (Mungov et al. 2013; Rabinovich and Eblé 2015). The event mode can be triggered either manually by an operator or automatically by a passing wave that reaches a certain threshold (Mofjeld 2009). Event mode duration typically varies from a couple of hours to 1 day. In addition to the real-time data, the higher resolution 15 s data that are stored internally in the BPRs are recovered during regular maintenance. All DART data used in this study are of the highest resolution available.

All four DART Stations in the Gulf of Alaska (46402, 46403, 46409 and 46410) (Figs. 2, 3) were switched to event mode upon the occurrence of the 23 January 2018 event and 1-min data recorded at these stations were available from 08:00 UTC on the same day. DART 46409 (55.294° N 148.509° W) was located within tens of kilometers of the earthquake epicenter. Among the four DARTs in our main study region (west coast of BC, WA and OR) (Fig. 2), DART 46419 (48.796° N; 129.619° W), positioned off the west coast of Vancouver Island, recorded 5 h of 1-min real-time event data and DART 46411 (39.333° N; 127.067° W), located ~ 150 nautical miles northwest of Mendocino Bay, recorded 6 h of 1-min real-time event data; DARTs 46404 and 46407 failed to provide real-time data of the tsunami waves. Higher resolution 15-s data were later recovered directly from four BPRs: DARTs 46403 and 46409 in the Gulf of Alaska (Fig. 3), and DART 46404 and 46407 in the northeast Pacific. These data were downloaded from NOAA's National Centers for Environmental Information (NCEI) https://www.ngdc.noaa.gov/hazard/dart/ and used in our analyses.

2.1.2 ONC BPRs

Ocean Networks Canada (ONC) is a Canadian universities initiative centered at the University of Victoria. It operates cabled ocean observatories in the northeast Pacific Ocean, Salish Sea and Arctic Ocean. The ONC tsunami network began in September 2009 when the organization deployed a cabled seafloor observatory of seven high precision bottom pressure recorders (BPRs) that transmit bottom pressure values at 1 Hz to the Data Management and Archiving System (DMAS) at the University of Victoria. The roughly 800-km long fiber-optic seafloor cable loop extends westward from Port Alberni over the Cascadia Subduction Zone and covers a large area of the Juan de Fuca Plate (Fig. 2b). This is a real-time reporting network that has proven to be effective for permanently monitoring tsunami waves and for providing early observations of the tsunami for the Tsunami Warning System. On 29 September 2009, 3 days after its installation, the network successfully recorded the 2009 Samoa tsunami (Thomson et al. 2011). Several other tsunamis were also detected by the network, including the 2010 Chilean tsunami (Rabinovich et al. 2013), the 2011 Tohoku tsunami (Fine et al. 2013), and the 2012 Haida Gwaii tsunami (Fine et al. 2015). The ONC BPRs are located at various depths and offshore distances off the southwest coast of Vancouver Island. The combined data from these instruments have allowed us to examine the transformation and amplification of open-ocean tsunami waves as they travel across the flat abyssal plain of Cascadia Basin, propagate over the continental shelf and approach the coast of Vancouver

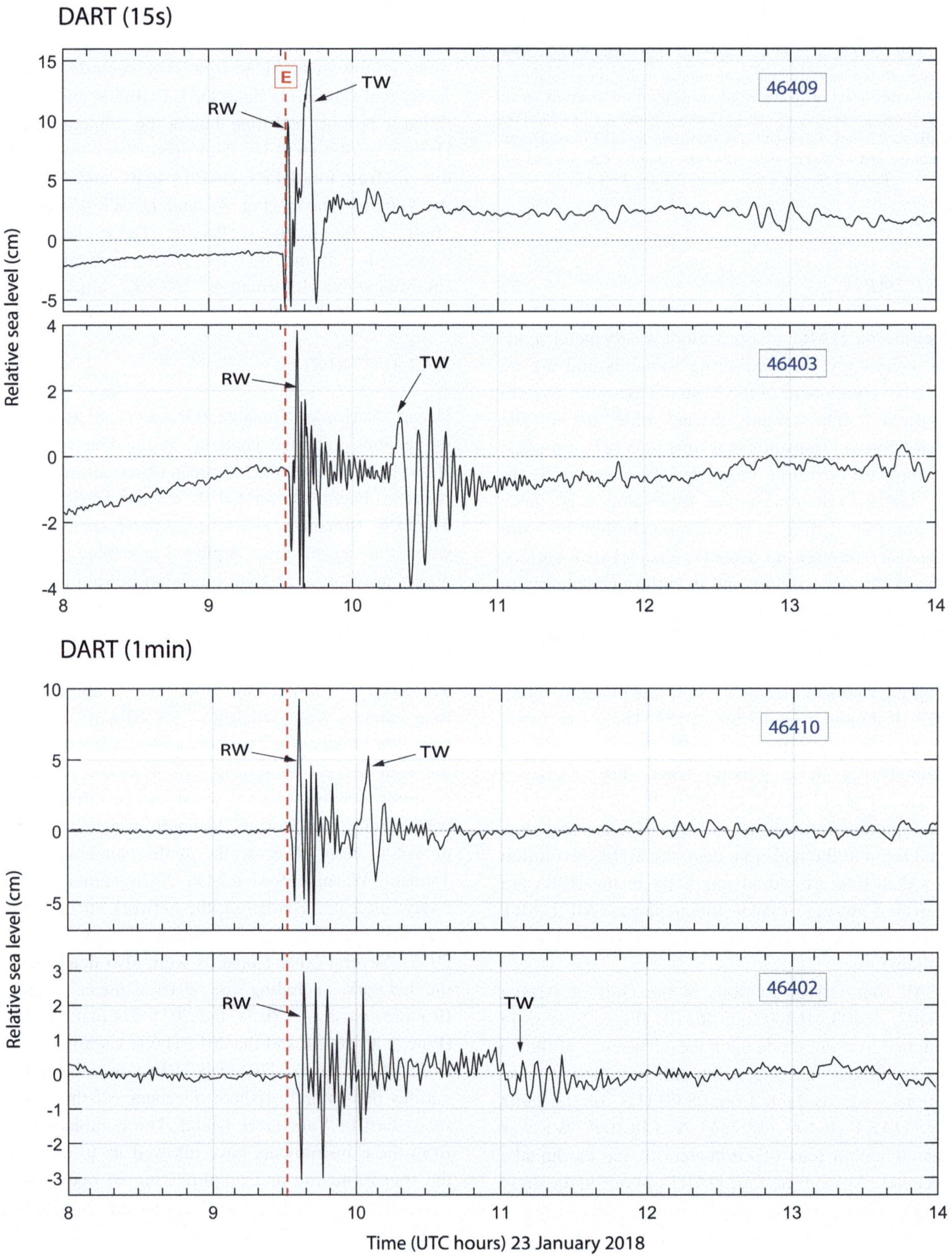
DART (15s)
Relative sea level (cm)
E
RW
TW
46409
RW
TW
46403
DART (1min)
Relative sea level (cm)
RW
TW
46410
RW
TW
46402
Time (UTC hours) 23 January 2018

◀Figure 3
Tsunami waves recorded by DARTs in the Gulf of Alaska. The dashed red vertical line indicates the time of the earthquake. Mean sea level is marked by "0". Labels "RW" and "TW" denote Rayleigh waves and tsunami waves, respectively. The 15-s data from DARTs 46409 and 46403 downloaded from the NCEI website were low-pass filtered with a 2-min Kaiser–Bessel window to remove the Rayleigh waves

Island (cf. Thomson et al. 2011; Rabinovich et al. 2013).

The 2018 Alaska tsunami was recorded by seven ONC BPRs moored at depths of 95–2660 m. Four of the BPRs are located in the vicinity of the Endeavour Ridge (ER) hydrothermal venting sites along the western segment of the cable loop (Fig. 2b). The three BPRs located in the axial valley of the ridge are positioned a few kilometers from each other at vent sites Main Field (ER-Main), West Flank (ER-West) and RCM South (ER-South) at depths of 2195, 2360 and 2228 m, respectively. All three BPRs yielded nearly identical recordings of the leading train of waves. The fourth BPR, at the Mothra vent site (2275 m depth), did not function during the event. The most distinct tsunami waves recorded by the ONC array were from Ocean Drilling Program (ODP) site 1026, located at a depth of ~ 2600 m in a flat area of the Cascadia Basin off the British Columbia-Washington coast (Fig. 2b). Waves were also recorded by three shallower stations over the Vancouver Island slope (Fig. 2a); specifically, Clayquot Slope Bullseye (CSB), Barkley Canyon Upper Slope (BCUS), and Folger Passage (FP) at the entrance to Barkley Sound. Instrument depths were 1258, 392 and 95 m, respectively. The five deep BPRs, ODP1026, ER-Main, ER-West, ER-South and CSB, produced high-quality open-ocean records of the tsunami. Although wave measurements by the two shallow BPRs (BCUS and FP) were "contaminated" by wave interaction with the complex bathymetry and nearshore processes, they provided useful information on the arrival and reflection of waves over the continental shelf.

We downloaded 4-day records of the 1-s sampled bottom pressure time series from the ONC BPRs spanning the day of the event. The time series were first de-tided using harmonic analysis and then low-pass filtered with a 31-s Kaiser–Bessel window (cf. Thomson and Emery 2014) to create 15 s data for spectral and wavelet analyses. To suppress long-period wave motions and isolate the tsunami displacements for plotting and estimation of tsunami parameters, a 2-h high-pass Kaiser–Bessel window was then applied (Fig. 4).

2.1.3 *OOI BPRs*

In addition to data from the ONC BPRs, bottom pressure data for the 2018 tsunami were available from a network of geophysical ocean bottom observatories from the area of Axial Seamount (ASM), located approximately 480 km west of northern Oregon. BPR measurements at the seamount started in 1986 (Eblé and González 1991; González et al. 1991) and a great number of tsunamis have been recorded by these instruments, including the 2004 Sumatra tsunami (Rabinovich et al. 2011). The 900-km long ASM cable array system is operated through the Ocean Observatories Initiative (OOI). It provides two-way communication from a shore station (in this case, Pacific City in Oregon) to scientific sensors on the seafloor (Fig. 2b). The system has two sub-arrays: a shorter cable that curves southward over the continental shelf and slope, and a longer cable that stretches westward to Axial Seamount. Near the summit of Axial Seamount, there are four BPR sites (NJ03B, NJ03D, NJ03E and NJ03F; Fig. 2b) that continually record at a rate of 20 Hz. Installed in 2012, these BPRs are primarily for monitoring underwater volcanic activity (https://www.pmel.noaa.gov/eoi/rsn/index.html). BPR NJ03A, located at the seamount base and NJ01A near the continental slope, both record at 15-s sampling rates. The lead wave of the 2018 Kodiak tsunami was recorded by BPRs near the seamount summit almost simultaneously with the wave arrival first recorded at site ODP1026 and within minutes at Endeavour Ridge.

Data from the BPRs near Axial Seamount summit (NJ03B, NJ03D, NJ03E, and NJ03F) are available from https://ooinet.oceanobservatories.org/. The median values of every 20 points was first taken to convert the data to 1-s records. The 1-s records were

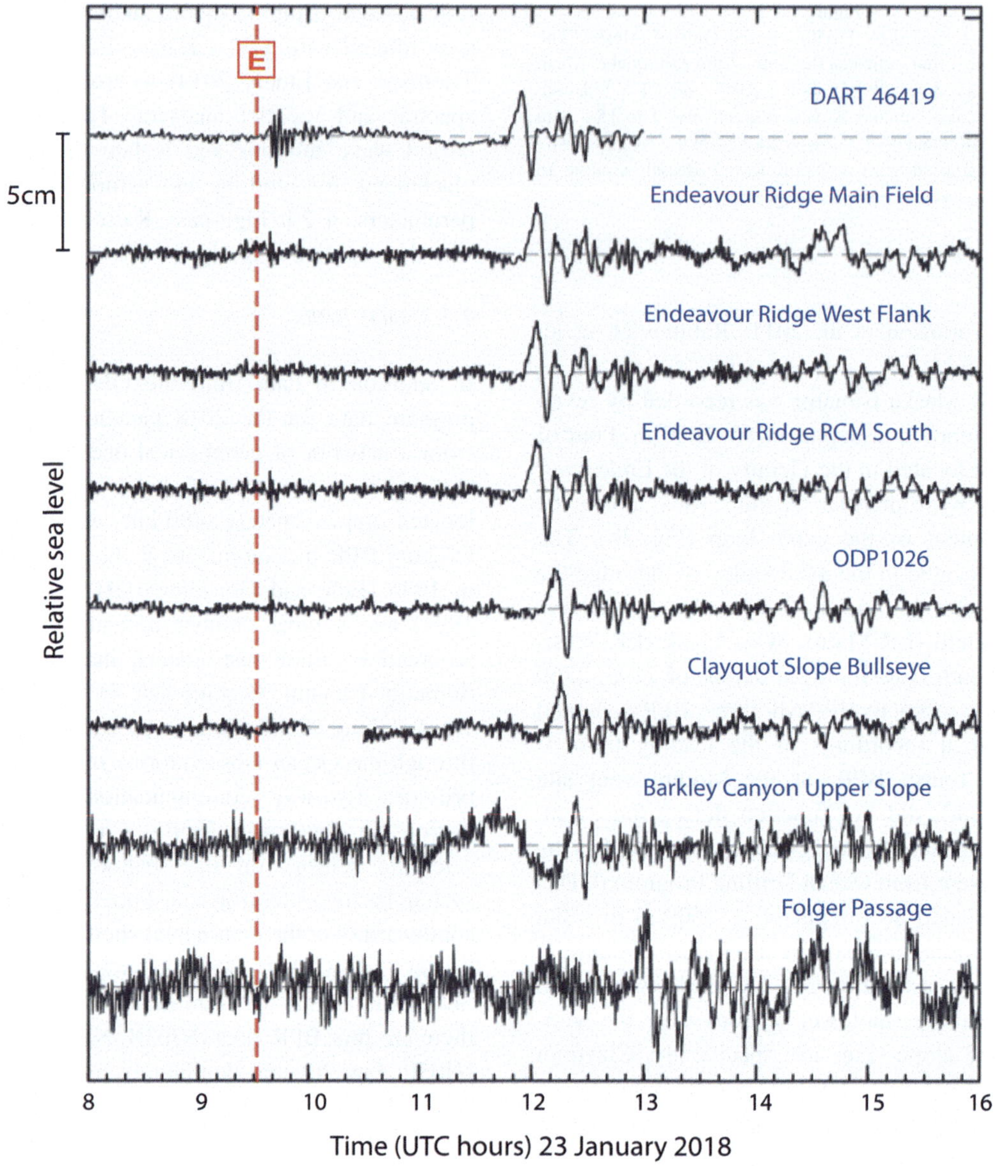

Figure 4
Tsunami waves recorded by DART 46419 and ONC BPRs off the west coast of Vancouver Island during the 23 January 2018 Kodiak tsunami. The dashed red line indicates the time of the earthquake

then low-pass filtered using a 31-s Kaiser–Bessel window to generate 15-s time series that matches the 15-s sampling rates of the BPRs at off-seamount site NJ03A and continental margin site NJ01A (see Fig. 2 for locations). Nine-day duration 15-s records, starting at mid-night on 21 January 2018, were selected for our study. As with the ONC time series, the 15 s data were first de-tided using harmonic analysis and then high-pass filtered with a 2-h Kaiser–Bessel window to further isolate the tsunami signal (Fig. 5).

2.2. *Coastal Tide Gauges*

Despite the relatively small amplitude of the 2018 tsunami and the presence, at the time of the event, of considerable background noise caused by local

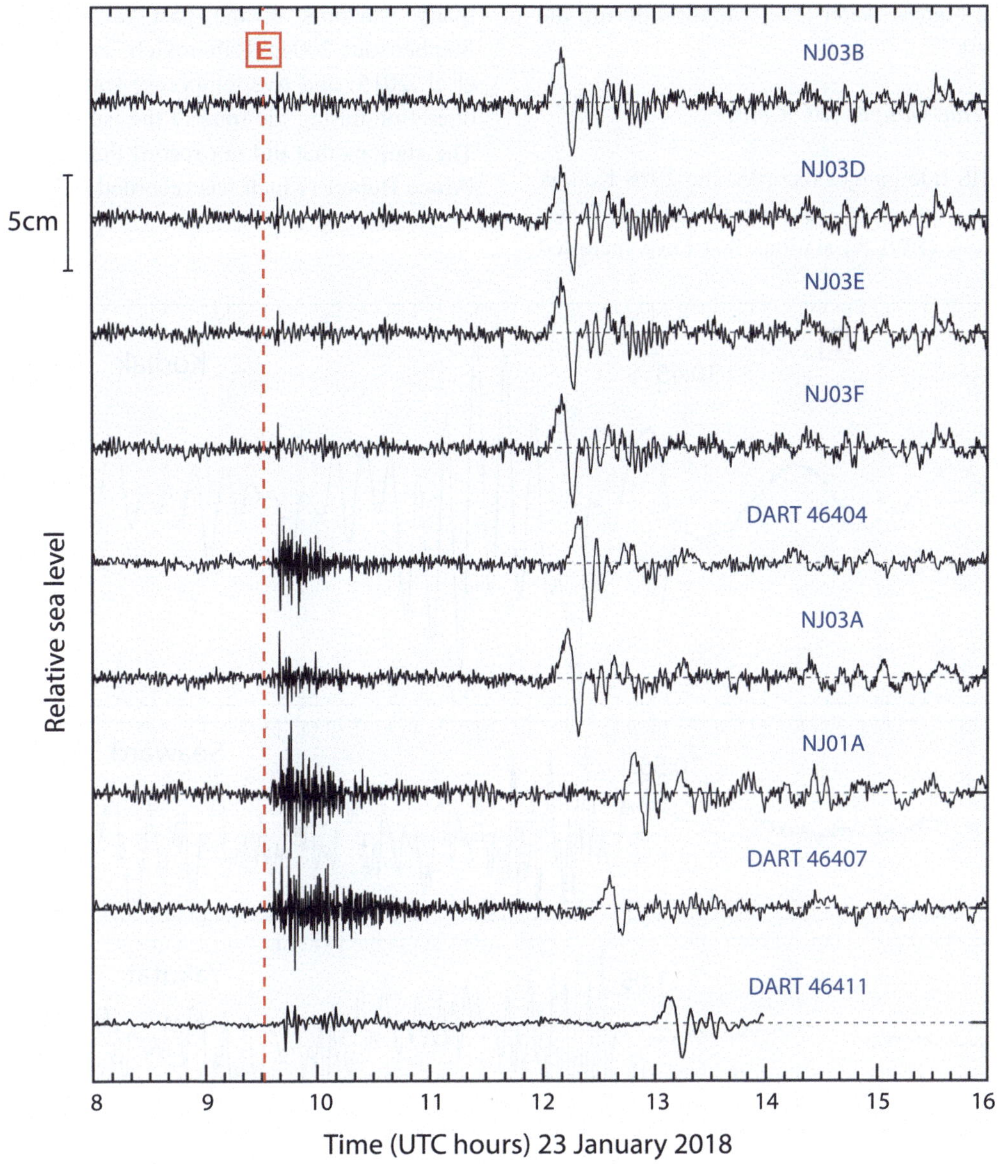

Figure 5

Tsunami waves recorded by DARTs 46407, 46411 and OOI BPRs off the west coast of Washington and Oregon during the 23 January 2018 Kodiak tsunami. The dashed red line indicates the time of the earthquake

atmospherically induced seiches and infragravity waves, the tsunami was clearly recorded by a large number of tide gauges along the Gulf of Alaska and Northwest Pacific coast. All tide gauge measurements were at 1-min sampling rate. These high-resolution data enabled us to conduct a thorough examination of the weak tsunami waves that arrived at the coast.

2.2.1 NOAA Tide Gauges (Alaska)

The tsunami waves were recorded by numerous NOAA tides gauges around the Gulf of Alaska. Four stations (Fig. 2) were selected for the purpose of this study: Kodiak, Seward, Yakutat and Sitka. The Kodiak gauge was the closest to the source and recorded the highest water level variation during the event (Fig. 6).

2.2.2 CHS Tide Gauges (BC Coast)

Fourteen CHS tide gauges recorded the 2018 Kodiak event. These consist of the nine Permanent Water Level Network (PWLN) stations that have recorded many tsunamis in the past (cf. Rabinovich and Stephenson 2004; Rabinovich et al. 2013; Fine et al. 2015) and five temporary stations that were in operation during the time of the tsunami (Fig. 2a, b). The stations that did not record the current event are Prince Rupert (which has recorded only major large-

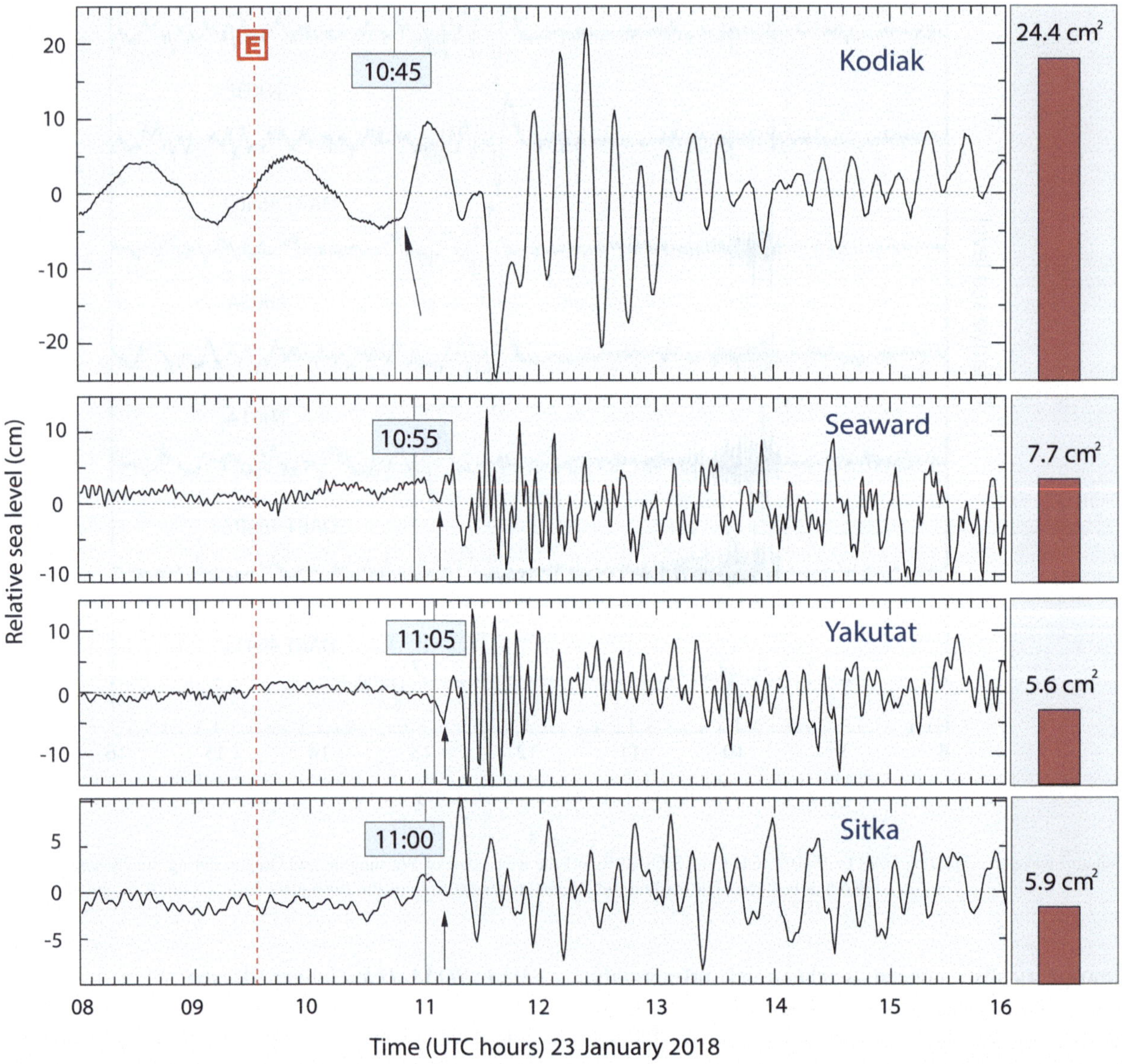

Figure 6
Tsunami waves recorded by four NOAA tide gauges in the northern part of the Gulf of Alaska. The dashed red line denoted by "E" indicates the time of the earthquake. The solid vertical lines and labels indicate the theoretical arrival time (in UTC). Arrows point to the observed arrival time derived from the de-tided sea level records. The vertical bars in red denote the total variance (in cm^2) computed for the first 12 h after the first wave arrival

scale tsunamis in the past; cf. Rabinovich et al. 2019) and those located well inside the Strait of Georgia.

2.2.3 NOAA Tide Gauges (Washington and Oregon)

All NOAA coastal data used in this study were downloaded from the NCEI website, https://www.ngdc.noaa.gov/hazard/tide/. The NCEI data archive provides 1-min de-tided data that are prepared especially for tsunami research. For the present study, we used records from seven gauged stations located along the coast of Washington (WA) and two located along the coast of Oregon (OR) (Fig. 2b). These data, together with the CHS coastal data, enabled us to examine the characteristics of tsunami waves propagating along the west coast of North America.

3. Tsunami Properties

The tsunami waves measured by most of the open-ocean instruments had amplitudes of less than 3 cm and similar wave forms. In contrast, the wave amplitudes and forms observed at the 27 coastal sites used in the study varied markedly from location to location. Stations more directly exposed to the ocean tended to have higher wave heights and more rapid energy decay compared to those located in inlets and bays, which typically had lower wave heights and longer tsunami ringing times.

3.1. Open-Ocean Measurements

The four DART stations in the northern Gulf of Alaska recorded 1-min real-time event data over a 6-h duration starting at 08:00 UTC on 23 January 2018. The Rayleigh waves associated with the earthquake are evident in the de-tided sea level displacements at all four sites (Fig. 3). DART 46409, located ~ 85 km from the earthquake epicenter, recorded the highest tsunami waves, with trough-to-crest heights of over 20 cm. The waves measured by DARTs 46410, 46403 and 46402 had approximate trough-to-crest maximum heights of 9, 5.5 and 1.5 cm, respectively (Fig. 3). Multiple groups of secondary waves were also observed: a second train of waves with heights greater than 1 cm was observed at DARTs 46409 and 46410 beginning around 11:30 and 12:00 UTC, respectively. At around 13:00 and 13:40 UTC, another group of secondary waves with amplitudes ranging from 1 to 2 cm arrived at DARTs 46409 and 46403.

Waves from the 2018 Kodiak tsunami were recorded off Vancouver Island by seven NEPTUNE stations and by DART 46419 (Fig. 4; Table 1). These eight stations (Fig. 2b) present a unique open-ocean array of instruments that has enabled us to examine the transformation of tsunami waves as they cross the continental shelf to the coast. For all eight stations, the measured tsunami waves were markedly higher than the background noise level, simplifying investigation of the wave characteristics. One of the most important properties of the recorded waves was their consistency: the primary wave train looks similar at all stations, with the lead wave being the highest (amplitude, 1.9–3.6 cm) and preceded by a small wave trough. The period of the lead wave at different stations was very steady at 14–16 min (Table 1), while the trailing waves had much smaller heights (~ 1 cm) and much shorter periods of 4–5 min. The frontal wave arrived at DART 46419 off central Vancouver Island at 11:47 UTC, 2 h and 15 min after the main earthquake shock. Nine minutes later, the waves arrived at the Endeavour Ridge stations, followed by ODP-1026 and Clayquot Slope Bullseye (12:06 UTC) and then Barkley Canyon (12:17 UTC). As it propagated shoreward, the tsunami arrived at the shallowest ONC station, Folger Passage, at 12:55 UTC, 68 min later than at DART 46419 and 59 min later than at the Endeavour Ridge stations (Table 1). At about the same time (12:50–13:00 UTC), the tsunami reached the southwestern coast of Vancouver Island (Table 2). This means that, for Alaskan tsunamis, DART 46419 and the most remote ONC stations provide roughly 1-h advance arrival times for coastal sites like Tofino and Bamfield, which is important for early tsunami warning.

The maximum amplitudes and wave heights associated with the lead tsunami wave at DART 46419 and the six deeper ONC stations were very similar: 1.9–2.4 cm and 3.4–4.4 cm, respectively. However, at Folger Passage (FP), the shallowest (95 m) ONC site and the site closest to the Vancouver coast and the area of Bamfield, the maximum

Table 1

Parameters of the primary train of tsunami waves from the M_w = 7.9 Kodiak (Alaska) tsunami of 09:32 UTC, 23 January 2018 as recorded by open ocean instruments off the coasts of British Columbia, Washington and Oregon

Station	First wave			Max wave			Visible period (min)
	Arrival time (UTC)	Travel time (h:mm)	Amplitude (cm) sign	Max amplitude (cm)	Time (UTC) of max amplitude	Max wave height (cm)	
DART 46419	11:47	2:15	+ 2.1	2.1	11:54	3.8	15, 6
ER-M	11:56	2:24	+ 2.4	2.4	12:02	4.3	15, 4
ER-W	11:56	2:24	+ 2.4	2.4	12:02	4.3	15, 4
ER-S	11:56	2:24	+ 2.3	2.3	12:08	4.1	15, 4
ODP 1026	12:06	2:34	+ 1.9	1.9	12:12	3.8	16, 5
CSB	12:06	2:34	+ 2.3	2.3	12:15	3.4	15, 4
BC-UpSlope	12:17	2:45	+ 2.4	2.4	12:23	4.4	14, 5
FP	12:55	3:23	+ 3.6	3.6	13:02	6.5	16, 8, 40
NJ03B	12:03	2:31	+ 3.2	3.2	12:11	5.8	16, 4
NJ03D	12:03	2:31	+ 3.2	3.2	12:17	5.9	17, 4
NJ03E	12:03	2:31	+ 3.1	3.1	12:10	5.9	16, 5
NJ03F	12:04	2:32	+ 3.0	3.0	12:11	6.0	16, 5
DART 46404	12:12	2:30	+ 2.4	3.1	12:26	5.5	15, 5
NJ03A	12:05	2:33	+ 2.7	2.7	12:14	5.7	18, 4
NJ01A	12:43	3:11	+ 3.1	3.1	12:50	4.5	17, 4
DART 46407	12:29	2:57	+ 1.8	1.8	12:36	3.1	14, 8, 6
DART 46411	13:01	3:29	+ 1.3	1.3	13:07	3.2	16, 6

Wave arrival times are in hours (h) and minutes (min). See Fig. 2 insets for site locations

Table 2

Parameters of the Kodiak (Alaska) tsunami of 23 January 2018 recorded by CHS tide gauges on the British Columbia coast (Main shock: M_w = 7.9 at 09:32 UTC)

Station	First wave			Max wave			Visible period (min)
	Arrival time (UTC)	Travel time (h:mm)	Amplitude (cm) sign	Max amplitude (cm)	Time (UTC) of max amplitude	Max wave height (cm)	
Henslung Cove (Langara I)	11:21	1:49	+ 12.5	12.5	11:29	22.5	25, 13
Stewart	13:12	3:40	+ 2.6	2.6	13:37	5.0	50
Queen Charlotte City	13:33	4:01	+ 1.4	3.6	19:22	5.1	15
Hartley Bay	14:22	4 50	+ 0.9	1.9	18:46	4.0	40, 4
Kitimat	14:37	5:05	+ 1.5	2.4	17:49	5.0	25, 8
Rose Harbour	11:39	2:07	+ 7.0	9.4	12:26	16.3	25, 90
Bella Bella	12:58	3:26	+ 2.7	3.4	17:39	7.0	55
Winter Harbour	12:13	2:41	+ 6.3	6.3	12:22	12.1	50, 35
Port Hardy	12:56	3:24	+ 2.7	3.0	14:08	5.0	17
Tofino	12:51	3:19	+ 6.1	6.1	12:57	10.0	30, 25
Bamfield	13:03	3:31	+ 2.6	3.6	18:35	7.2	12
Port Renfrew	13:13	3:41	+ 5.5	5.7	15:31	10.5	40
Port Alberni	13:36	3:43	+ 3.0	8.3	20:55	15.1	105
Victoria	13:49	4:17	+ 4.5	6.1	07:03[a]	11.5	48, 20

[a]24 January 2018

amplitude (3.6 cm) and wave heights (6.6 cm) were approximately 1.5 times greater than at the open-ocean sites (Table 1). The primary wave train, which consisted of the first (maximum) wave followed by several succeeding smaller waves, had a duration of about 1 h.

An important feature of all ONC records was the second train of waves that arrived about 2 h after the first train (Fig. 4). Based on this observation, it appears that, while the first train came directly from the source area, the second train was produced by reflection of the main tsunami train from the nearby coast of Vancouver Island. At the Main Field at Endeavour Ridge, the secondary waves that arrived around 14:30 UTC had a well-defined "double-crest" shape, while at site ODP1026 the shape of the highest waves in the secondary wave train closely resembled that of the first wave train. At Folger Passage, the second wave train had much higher wave heights (~ 4.5 cm) and much longer periods (~ 40 to 45 min) than at the offshore sites, likely from the influence of Barkley Sound.

The tsunami waves recorded by the OOI instruments near the summit of Axial Seamount (Fig. 5; Table 1) were generally higher than those recorded by the ONC gauges (Fig. 4), which was unexpected considering that the seamount instruments were farther from the source area. Specifically, the leading waves reached trough-to-crest heights of up to 6 cm at Axial Seamount, compared with 3.4–4.4 cm at the ONC stations. At Station NJ01A, the site closest to the Oregon coast (Fig. 2b), the recorded maximum amplitudes (3.1 cm) and wave heights (4.5 cm) had a similar range to those recorded at the ONC sites, Barkley Canyon (2.4 cm, 4.4 cm) and Folger Passage (3.6 cm, 6.5 cm), over the continental slope and shelf (Fig. 5). The general character of the tsunami waves recorded by the two arrays, ONC and OOI, was very similar (Figs. 4, 5). In particular, the lead wave was a crestal wave and had the highest amplitude of the first wave train. The lead wave had a period of 16–17 min (Table 1) and was followed by several lower waves with periods of 5–10 min and wave heights of 2–3 cm. This first wave train had a duration of ~ 1 h and was followed 1.5–2 h later by the second wave train (Fig. 5). In our main study region, the earliest arrival of the first wave was registered at DART 46419 at 11:47 UTC, then near Endeavour Ridge area at 11:56 UTC. Within 9 min (at 12:03–12:05 UTC), the first wave next arrived in the vicinity of Axial Seamount, which was 7 min before its arrival at DART46404 ~ 100 km east to NJ03A. DART stations 46407 and 46411, about 370 km apart located on the opposite sides of Mendocino Ridge, registered the first waves at 12:29 and 13:01 UTC, respectively. One hour later, the event mode recording was turned off at DART46411, ~ 120 km south to Mendocino Ridge, the southernmost instrumented site in this study.

3.2. *Coastal Measurements*

Relatively high tsunami waves were recorded at the four stations located in the Gulf of Alaska close to the source area, Kodiak, Seward, Yakutat and Sitka (Figs. 2, 6). Pronounced sea level oscillations lasted for a few hours after the first wave arrival. The tide gauge at Kodiak recorded the largest wave variations and the wave energy received at this site was significantly higher than at other sites in the gulf. The maximum trough-to-crest wave height at Kodiak was ~ 43 cm and the maximum zero-to-crest amplitude was 22 cm. The first wave arrived at this station at 10:48 UTC, i.e., 1 h and 16 min after the main shock. The fifth wave, rather than the lead wave, was the highest (Fig. 6). At other Alaska stations the maximum wave heights (amplitudes) were significantly smaller: 22.5 (12) cm (Seward), 30.5 (14) cm (Yakutat) and 16 (10) cm (Sitka). The first wave arrived at these three stations almost simultaneously, at 11:09–11:10 UTC (97–98 min after the main shock).

Our primary focus in this study involves analysis of the 2018 Kodiak tsunami waves recorded on the coast of British Columbia. The tsunami signal was identified in 14 permanent and temporary CHS stations located on the coasts of Haida Gwaii, Vancouver Island and the mainland coast (Fig. 2). Of all the stations examined, only Prince Rupert did not detect the waves. As reported in previous studies (cf. Rabinovich and Stephenson 2004; Rabinovich et al. 2019), Dixon Entrance to the west of Prince Rupert acts as an effective low-pass filter that strongly suppresses high-frequency waves

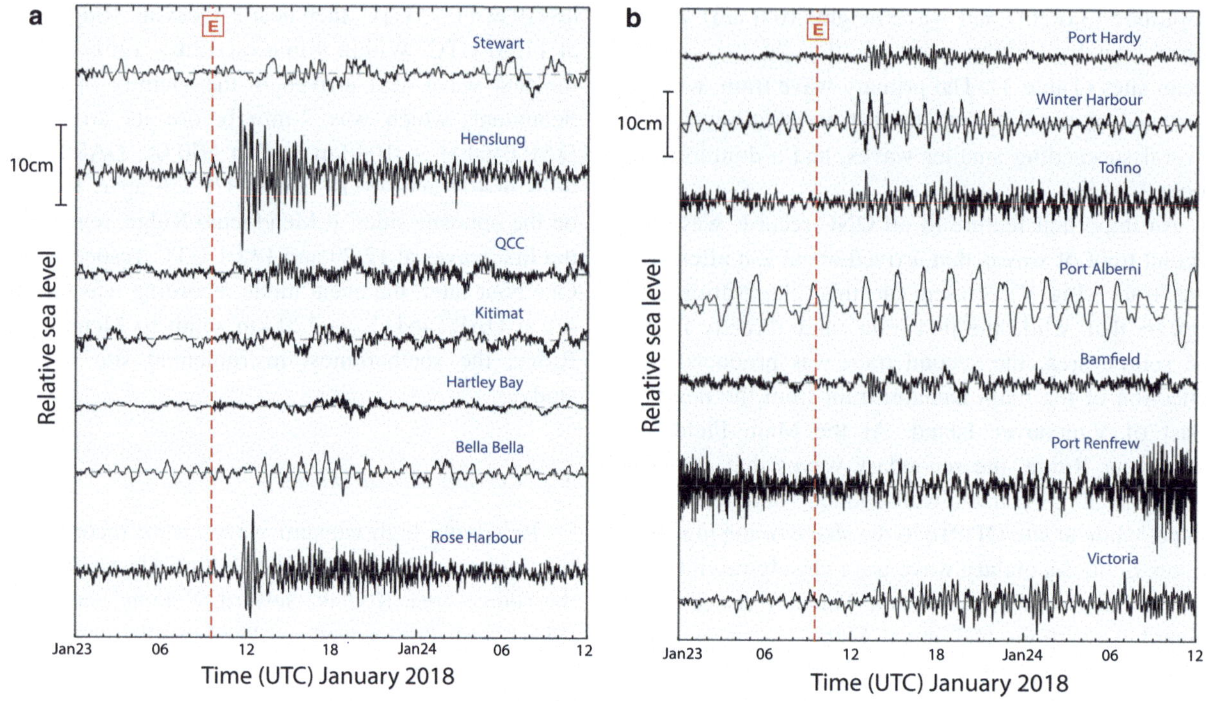

Figure 7
Tsunami waves recorded by CHS tide gauges on the **a** northern and **b** southern British Columbia coasts. Red dashed line indicates the time of the earthquake

(characteristic of the 2018 Kodiak tsunami) incoming to this region.

The statistical parameters of the 2018 tsunami waves recorded at the 14 BC stations are presented in Table 2. In general, tsunami wave heights fell markedly with distance from the source region. The maximum wave heights recorded by the tide gauges at Henslung Cove and Rose Harbour (22.5 and 16.3 cm, respectively) located at the opposite ends of Haida Gwaii in northern British Columbia, were in the same range as those recorded at Seward, Yakutat and Sitka (22.5, 30.5 and 16 cm, respectively). The distances of the above sites to the tsunami source region are also comparable: the travel times to Henslung Cove and Rose Harbour were 1 h 49 min and 2 h 7 min respectively, while the travel times to the Alaskan Stations were between 1 h 30 min and 1 h 40 min. Tsunami waves were also recorded at Bella Bella, Port Hardy (Fig. 7a), Winter Harbour, Tofino, Bamfield and Victoria (Fig. 7b). At other BC stations, the tsunami waves were marginally visible in the wave records. As a consequence, the statistical parameters presented in Table 2 are approximate values only. Queen Charlotte City, Bella Bella (Figs. 2b, 7a) and Port Hardy (Figs. 2a, 7b) recorded weak tsunami waves, but relatively long resonant "ringing" that continued into early January 24. The relatively weak response of Port Alberni (Fig. 2a) to the 2018 event stands in stark contrast to its response to the 1964 Alaska tsunami, during which Port Alberni experienced strong flooding for several hours due to its highly tuned resonance response of the inlet in the tsunami frequency band (Fine et al. 2015). Unexpectedly, it was the tide gauge in Victoria's Inner Harbour, sheltered well inside the ~ 100 km long Juan de Fuca Strait, that recorded one of the strongest tsunami signals (11.5 cm maximum wave height) during the 2018 event (Table 2). At Port Alberni, the first wave arrived at 13:45 UTC, 4 min earlier than it arrived at Victoria (13:49 UTC; Table 2).

We also examined the records of nine NOAA tide gauges located on the coasts of Washington State, WA (7 stations) and northern Oregon, OR (2

Table 3

Parameters of the Kodiak (Alaska) tsunami of 23 January 2018 recorded by NOAA tide gauges on the Washington and Oregon coasts[b] (Main shock: M_w = 7.9 at 09:32 UTC)

Station	First wave			Max wave			Visible period (min)
	Arrival time (UTC)	Travel time (h:mm)	Amplitude (cm) sign	Max amplitude (cm)	Time (UTC) of max amplitude	Max wave height (cm)	
Port Angeles (WA)	13:45	4:15	+ 6.4	6.4	13:55	11.1	23
Neah Bay (WA)	12:58	3:26	+ 5.4	7.2	18:33	12.8	30
La Push (WA)	12:57	3:25	+ 8.5	9.7	00:03[a]	15.5	13
Westport (WA)	13:22	3:50	+ 5.3	5.3	13:25	12.3	13
Toke Point (WA)	13:35	4:03	+ 1.8	2.0	20:20	3.8	13, 30
Cape Disappointment (WA)	13:20	3:48	+ 3.1	5.1	15:09	10.8	15
Garibaldi (OR)	13:19	3:47	+ 2.0	3.3	17:58	6.3	17
South Beach (OR)	13:29	3:57	+ 3.9	6.0	13:57	13.1	23

[a]24 January 2018

[b]Port Townsend is not included because the signal-to-noise ratio was too low

stations). With the exceptions of Port Angeles and Port Townsend, we selected only stations located on the outer coast of these two states. Port Angeles and Port Townsend are located on the southern coast of Juan de Fuca Strait and were used because of their importance to Victoria on the opposite of the channel (Fig. 2). The statistical parameters for the 2018 tsunami waves estimated for eight of the stations (excluding Port Townsend, which had a too low signal-to-noise level) are presented in Table 3; the records are shown in Fig. 8. The tsunami waves first arrived at La Push and Neah Bay at 12:57–58 UTC (3 h 25–26 min after the main shock) and then 20–40 min later at other stations along the outer coast. The maximum tsunami wave heights at these stations were approximately the same as at the BC stations but the tsunami responses varied widely with location (Fig. 8). Garibaldi, Cape Disappointment, La Push and Neah Bay all recorded highest water level changes several hours after the first wave arrival, which was evidently related to the effects of local bathymetry. The lowest detectable tsunami waves were in Port Townsend and Toke Point, both of which are located in protected bays. Despite the low wave heights, the two sites were characterized by a day of marked sea level ringing associated with local standing wave responses (Fig. 8).

The amplitudes of the tsunami waves measured by most of the open-ocean instruments were about 3 cm and showed considerable uniformity among sites. In contrast, wave amplitudes observed at the 22 tide gauge sites that we examined along the coast varied markedly from location to location. Stations more directly exposed to the ocean tended to have relatively high wave heights, whereas those in some inlets had lower wave heights but longer, standing-wave ringing durations. The mean maximum wave amplitude measured at the coast of Vancouver Island was about 6 cm, whereas the maximum open-ocean wave amplitudes measured by the nearby ONC BPRs were about 2 cm. This gives a maximum amplitude ratio of ~ 3, which indicates that there was very little topographic amplification. The observed ratio decreases even further to ~ 2 using the OOI BPR and coast data from Washington and Oregon, which is essentially equivalent to no amplitude amplification. (Note that, for an ideal uniform depth ocean, for which there is no coastward shoaling, wave amplitudes near the coast are doubled because of reflection alone. The theoretical ratio between amplitudes at coastal and open ocean stations would then be 2). The above results contrast with known similar parameters of recent larger events, such as the distant 2011 Tohoku and 2010 Chilean tsunamis. For the Tohoku case, the maximum wave amplitudes recorded by

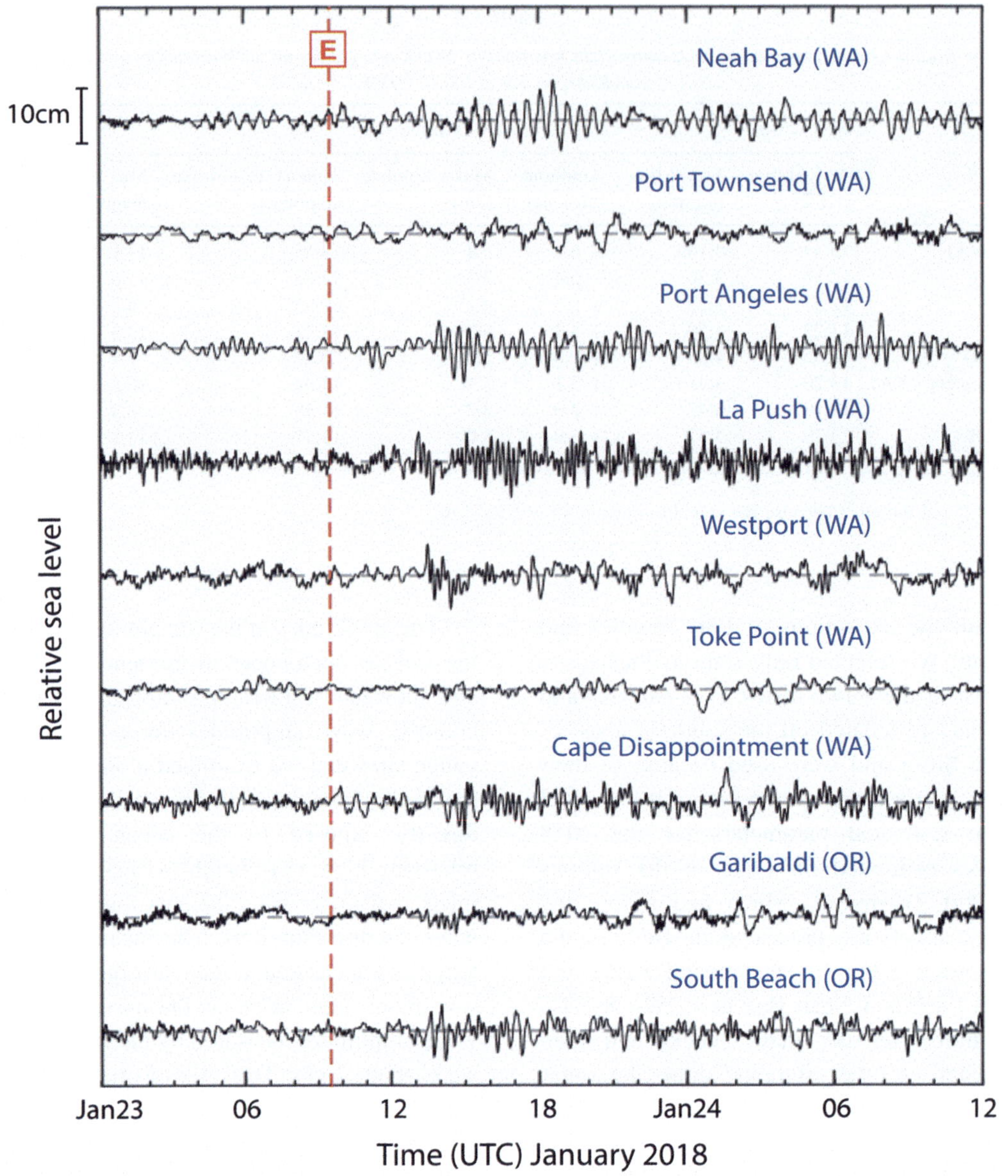

Figure 8
Tsunami waves recorded by NOAA tide gauges on the Washington (WA) and Oregon (OR) coasts. Red dashed line indicates the time of the earthquake

ODP1026 BPR and at the west coast of Vancouver Island were 10 and 40–70 cm, respectively (see tsunami database), corresponding to amplitude amplifications of 4–7 times. For the 2010 Chilean tsunami, maximum wave heights (approximately double the amplitudes) at ODP1026 and at coastal locations were 8 and 37–98 cm, respectively (cf. Rabinovich et al. 2013), corresponding to amplitude amplification factors of 4.5–12. In comparison, the same factor for the 2012 Haida Gwaii tsunami (see Fine et al. 2015 for data) was only around 2, even lower than for the 2018 Kodiak tsunami. These findings are discussed in greater detail in Sects. 5 and 7.

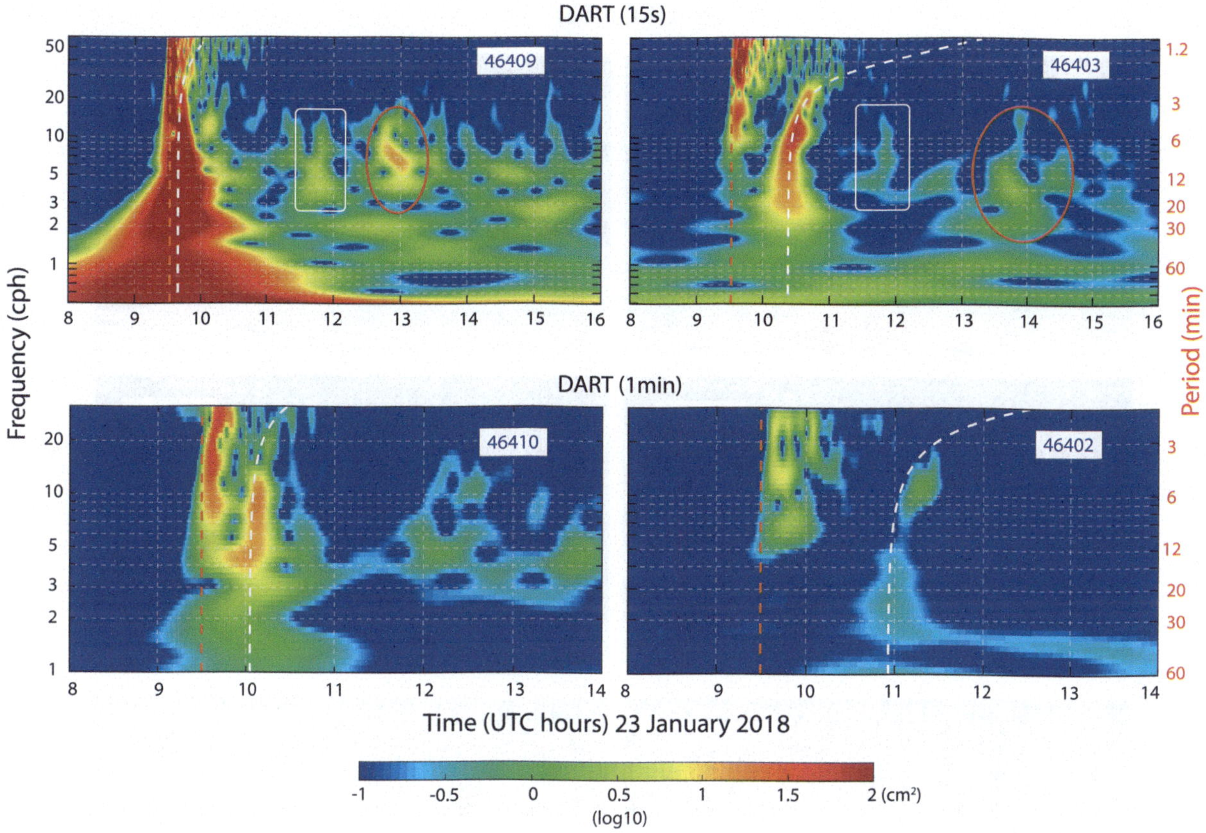

Figure 9
Frequency–time (*f*–*t*) diagrams for four DART stations in the Gulf of Alaska. The earthquake time is marked with a vertical orange dashed line. The dashed white lines in each panel denotes the time of tsunami wave arrival as a function of frequency. The white rectangles mark possible reflected waves from the coast near Kodiak Island; orange ellipses denote possible reflected waves from the Alaska coast near Yakutat and Sitka

4. *Wavelet Analysis*

To examine temporal (*t*) variations of the tsunami waves as a function of frequency (*f*), we have used a Morlet wavelet with a shape parameter equal to 6.0 (cf. Thomson and Emery 2014). The so-called "*f*–*t* diagrams" derived from this analysis reveal how wave energy in different frequency bands vary over time but, more importantly, delineate the degree of frequency-dependent dispersion of the waves during their propagation from the source region. The tsunami arrival times are well-defined and mutually consistent at all sites (Figs. 9, 10), and closely coincide with the computed long-wave arrival times (Fine and Thomson 2013). To estimate the wave dispersion, we first computed the "effective ocean depth", which corresponds to the depth of a flat ocean whose tsunami travel time is equal to observed travel time. We then used the dispersion relation for group velocity (cf. González and Kulikov 1993) for this effective depth to correct the arrival times for the shorter waves. Tsunami waves have significantly different patterns than the background noise; therefore, comparison of wavelet features before and after constructed arrival curves enables us to effectively distinguish the tsunami waves from the background oscillations. The strong dispersion observed in the tsunami waves appears to be related to the relatively short extension of the initial earthquake source area and comparatively short tsunami waves that were generated. Due to wave dispersion, the high-

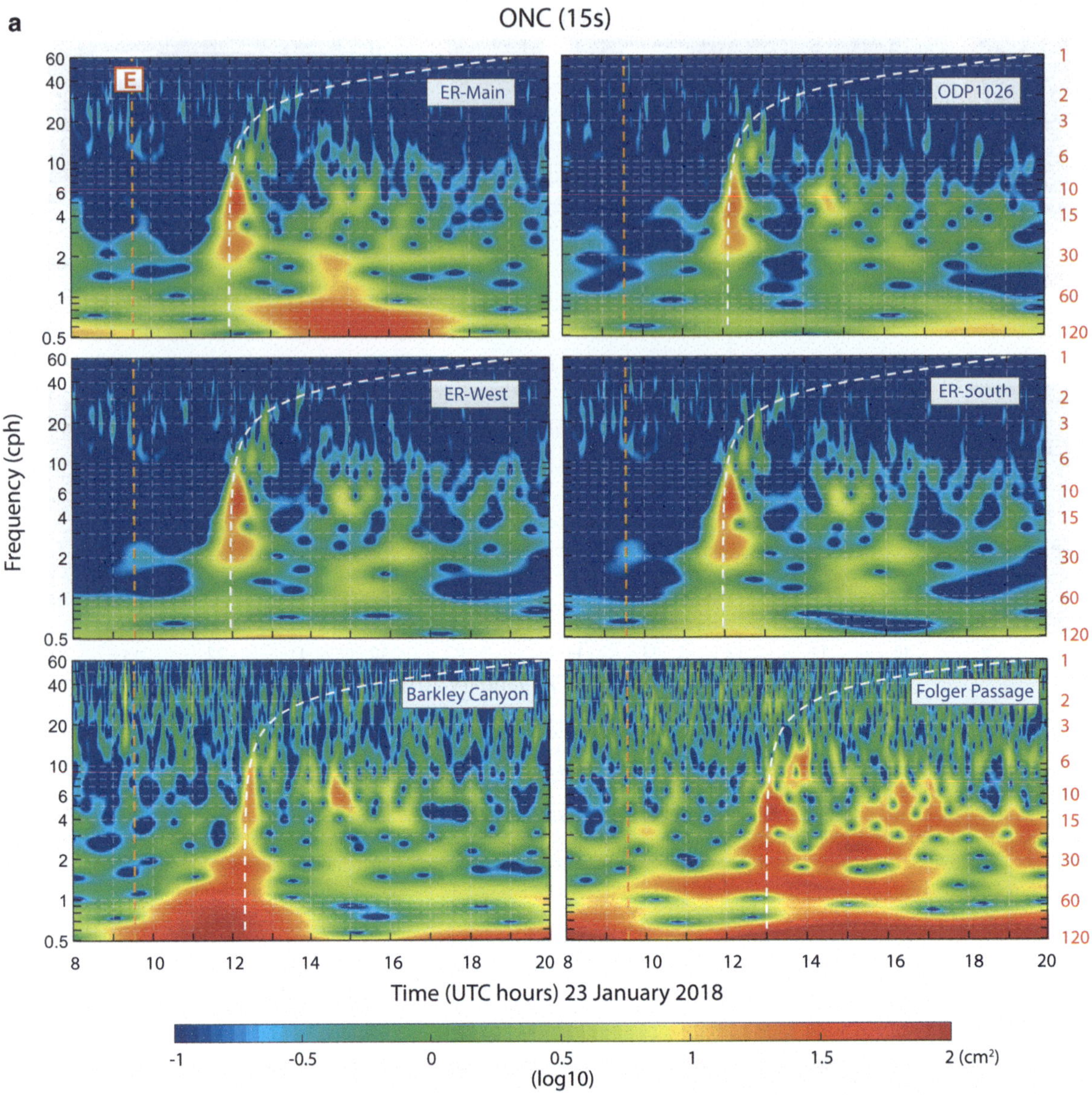

Figure 10

Frequency–time (*f*–*t*) diagrams derived from the 15-s tsunami waveforms recorded by **a** ONC and **b** OOI and DART instruments. The earthquake time is marked with a vertical orange dashed line. The dashed white lines denote the time of tsunami wave arrivals as a function of time. "ER" denotes Endeavour Ridge, and white ellipses mark possible reflections of the primary waves from the Mendocino Fracture Zone

frequency "tail" of the tsunami clearly propagated much slower than the frontal wave.

As seen in all open ocean *f*–*t* diagrams, the primary (frontal) wave train was the most energetic. The primary tsunami waves observed in the vicinity of the source region largely occupied the frequency band of 2–30 cph (periods of 30–2 min). Although the higher frequency tsunami wave components at DART 46409, located ~ 85 km from the earthquake epicenter, were contaminated by the seismic Rayleigh waves, the *f*–*t* diagram of the data (Fig. 9 upper left panel), shows that the tsunami waves also had a lower frequency component of < 2 cph (periods > 30 min), which, as we show later was

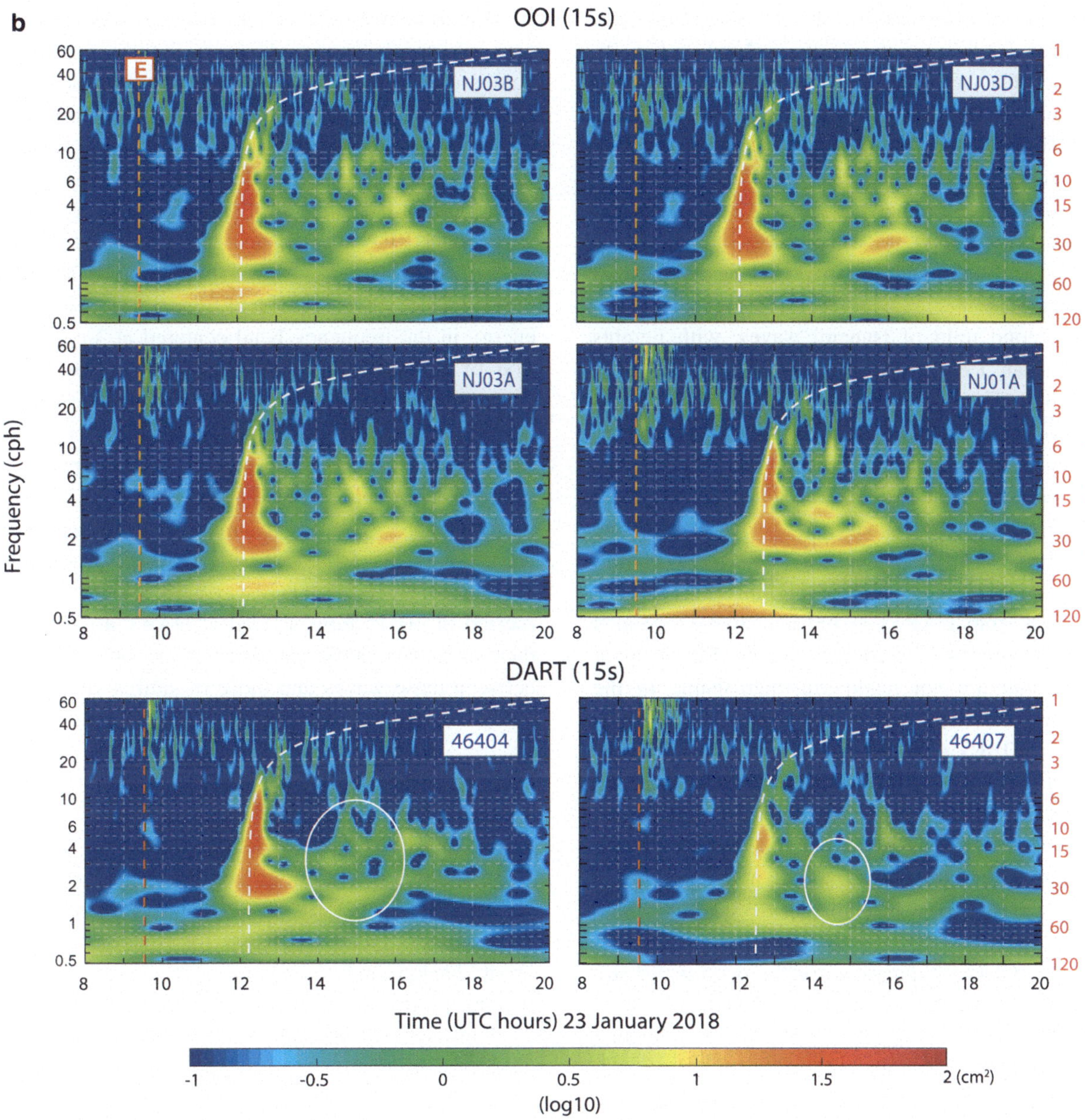

Figure 10
continued

associated with an earthquake-induced jump in the local water depth. The energy of the first waves recorded in the Gulf of Alaska by DARTs 46410 and 46403 (Fig. 9 lower left and upper right panels) and in the weakly detected leading waves recorded by DART 46402 (Fig. 9 lower right panel) were concentrated between 3 and 20 cph (periods between 20 and 3 min).

As shown by the tsunami wave records for the west coast of BC, WA and OR (Figs. 4, 5; Tables 2, 3), the first group of waves was dominated by ~ 15 min motions. Oscillations with periods of 5–10 min were also evident in the open-ocean records. The corresponding *f–t* plots (Fig. 10a, b) show that the most intense tsunami energy was focused in the 2–10 cph frequency band (periods of

30–6 min). Comparison of the *f–t* diagrams with those for DARTs in the Gulf of Alaska (Fig. 9), indicates that the open-ocean waves recorded by the ONC and OOI instruments (Fig. 10a, b) were shifted to a slightly lower frequency band. Moreover, at some sites, especially at those near the continental slope and shelf (namely, Barkley Canyon, Folger Passage, and NJ01A), the wave energy of the primary train in the 3-cph band was markedly weaker than at offshore sites.

A second train of tsunami waves, with heights greater than 1 cm, was observed at DARTs 46409 and 46410 (Fig. 3) beginning around 11:30 and 12:00 UTC, respectively. Their *f–t* diagrams (Fig. 9) confirm this observation and indicate that these waves persisted for about 2 h and occupied a frequency band similar to the primary waves observed at the same sites, namely 3–15 cph (periods of 20–4 min). The *f–t* plots of the higher resolution (15-s) records recovered from DARTs 46409 and 46403 revealed additional groups of secondary waves (Fig. 9 upper panels) that are not readily distinguishable in the tsunami time series (Fig. 3). In particular, the *f–t* plots for DART 46403 show a second train of waves at 11:40 UTC, with frequency and duration that resembled those of the first wave train. An additional group of tsunami waves with frequencies 3–15 cph (periods of 20–4 min) was found to arrive at DARTs 46409 and 46403 approximately 13:00 and 13:40 UTC, respectively.

In the ONC *f–t* diagrams for the deep-water stations (Fig. 10a), the second train of tsunami waves arrived roughly 2.5 h after arrival of the leading waves; at ODP1026, this train arrived a little earlier than at the Endeavour Ridge sites. In contrast, there was no definite demarcation between the primary and secondary wave groups at the shelf stations of Barkley Canyon and Folger Passage, probably due to the multiple reflection from the complex nearshore bathymetry. The presence of long-ringing, high-energy waves in the frequency band 2–10 cph (periods of 30–6 min) in the Folger Passage *f–t* diagram appears to be related to strong tsunami-generated standing waves over the continental shelf, which highlights the potential hazard of persistent, intense tsunami waves in nearby coastal zones.

Due to bathymetric effects, tsunami waves in the area of Axial Seamount were slightly amplified compared to the off-seamount deep-ocean locations of DART 46404 and NJ03A. Specifically, stations on the seamount recorded the arrival of relatively strong, low-frequency (2–3 cph) secondary tsunami waves about 30 min later than similar waves were recorded at station NJ01A located near the continental slope west of South Beach, Oregon (Fig. 10b). At NJ01A, the primary wave train had distinct, albeit low, energy in the frequency band centered at 3 cph (period of 20 min). Two DARTs in the nearby open ocean, 46404 and 46407 (Figs. 2, 9), also recorded detectable secondary waves. The lower frequency tsunami waves of $\sim$ 2 cph (period 30 min) appeared between 14:00 and 15:00 UTC at both DART sites, clearly before waves with similar frequencies were recorded at the OOI stations at Axial Seamount. Starting at $\sim$ 16:00 UTC, waves of about 4–6 cph (period of 10–15 min) arrived at DART 46407, and then at 46404, although there is no obvious link between these waves and those of similar frequency and time period observed at the OOI sites (Fig. 10b).

5. Spectral Analysis

In addition to providing long-duration time series of the 2018 tsunami waves, the ONC and OOI instruments yielded extensive information on pre-event background oscillations. To examine the spectral properties of tsunami oscillations during the 23 January 2018 Kodiak tsunami and to compare these properties with those of the background oscillations at the same sites, we separated records from the ONC and OOI BPR arrays and nearby DARTs into two parts: the 12-h period ending 1.5 h prior to the tsunami arrivals was identified as "normal" and selected for analysis of the background signals, while the 12-h period starting 1.5 h before the tsunami wave arrivals was used to examine the spectra of tsunami waves. Our spectral analysis procedure is based on the 15-s de-tided data and is similar to that described by Thomson and Emery (2014). To improve the spectral estimates, we used a Hamming spectral window of 192-min length with half-window overlaps prior to the Fourier transform and started the

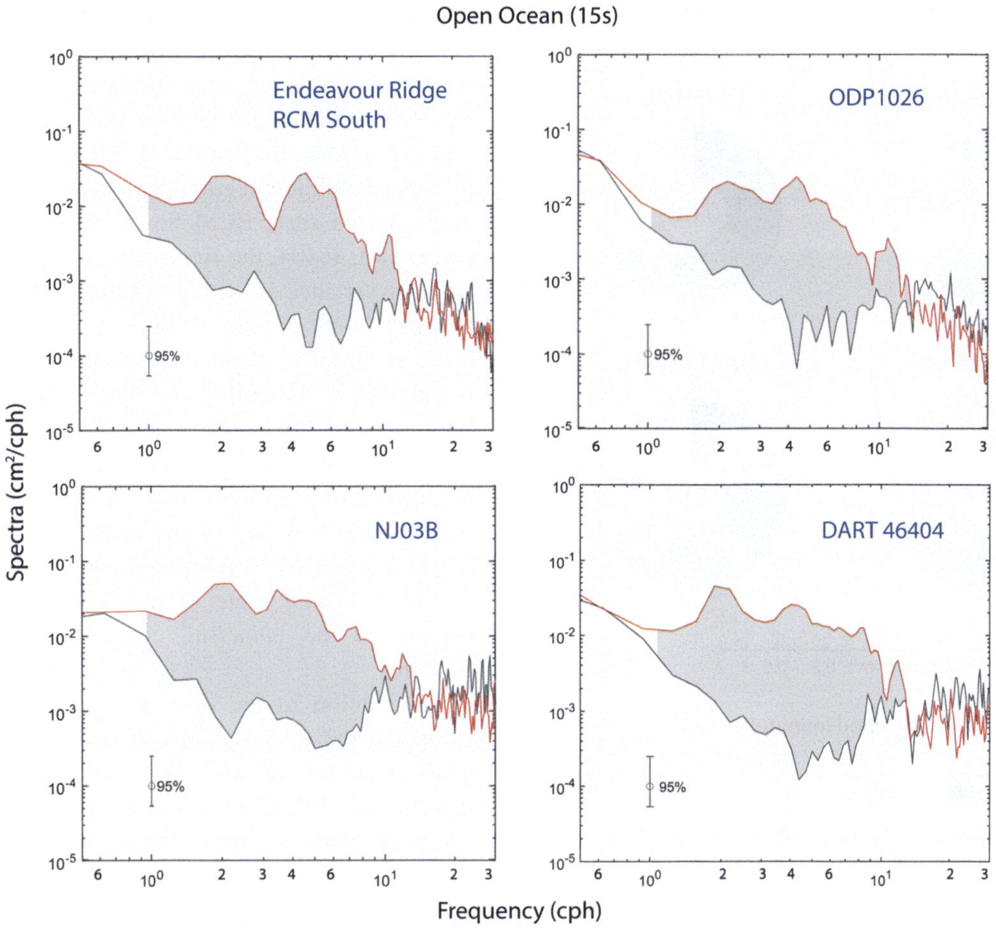

Figure 11
Open-ocean spectra of the 12-h records of background (pre-tsunami) oscillations (black lines) and tsunami waves (red lines). The shaded area indicates tsunami energy in the frequency band of 1–15 cph

tsunami spectral estimation 1.5 h in advance of the tsunami arrival time in order to include all of the energy of the lead waves. The resulting spectral estimates for the open-ocean stations have $\nu \approx 7$ degrees of freedom, the resolution $\Delta f = 0.3125$ cph and the Nyquist frequency $f_N = 120$ cph.

The spectra obtained from the open-ocean tsunami and background BPR records (Fig. 11) show that the tsunami energy was mainly concentrated in the 2–10 cph frequency band. As evident from the tsunami records (Figs. 4, 5; see also Table 1), most of the wave energy was concentrated within the first hours of the tsunami arrival. At almost all stations, the calculated signal-to-noise (tsunami/background) ratio reaches a peak at around 4 cph (period of 15 min) and then falls abruptly to minimum values at about 10 cph (6 min period). The ratios remain low over all higher frequencies, indicating that there was no tsunami energy at shorter periods.

The tsunami and background spectral estimates have enabled us to estimate the "*true*" tsunami spectrum, $E_{tsu}(f)$, at frequency, f, by subtracting the spectrum of the background noise (E_0) from that of the observed tsunami (E_{obs}) (the difference of the red and black curves in each of the plots in Fig. 11), whereby,

$$E_{tsu}(f) = E_{obs}(f) - E_0(f). \qquad (1)$$

As per Zaytsev et al. (2016), integration of the true tsunami spectra over the tsunami frequency

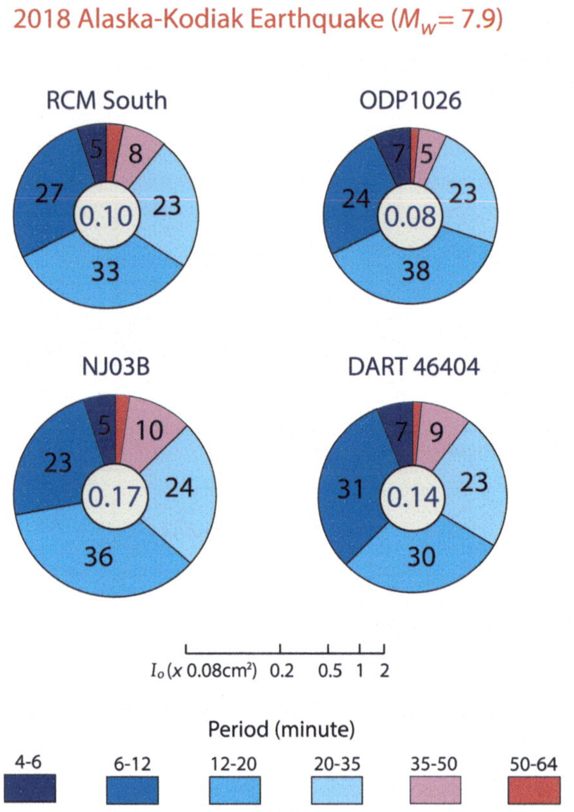

Figure 12
Open-ocean energy and frequency distribution (in percent) of tsunami waves recorded by BPRs at RCM South, ODP1026, NJ03B and DART 46404. The number at the center of each disk indicates the 12-h open-ocean energy, I_O, at the corresponding site. The pie-like segments in the outer part of each disk are proportional (on a log scale) to the open-ocean energy using I_O at site ODP1026 as the unit measure

band, f_1 (1 cph) $\leq f \leq f_2$ (15 cph), yields an estimate of the true integral open-ocean tsunami variance, I_O, corresponding to the shaded areas in Fig. 11, as

$$I_0 = \int_{f_2}^{f_1} E_{tsu}(f)df. \tag{2}$$

The highest open-ocean tsunami variance occurred near Axial Seamount, where $I_O = 0.17$ cm^2, nearly double that at ODP1026 (0.08 cm^2) and at RCM South at Endeavour Ridge (0.10 cm^2) (Fig. 12). We also determined the "*tsunami colour*", a measure of the open-ocean tsunami frequency content (cf. Zaytsev et al. 2016, 2017), by computing the percentage of the variance (band variance) within each of six selected frequency bands. The structures of the frequency distributions for the various open-ocean stations, as denoted by the colored "pies" in Fig. 12, are generally similar, indicating high consistency among the estimates. Despite the marked differences in the integral open-ocean variance at the four selected sites, RCM South, ODP1026, NJ03B, and DART 46404, the frequency content of the total tsunami variance is within a band of wave periods between 6 and 35 min at these sites and nearly the same at all sites: about 83%, 85%, 83% and 84%, respectively. According to the classification of Zaytsev et al. (2016, 2017), the 2018 Kodiak tsunami can be described as "*blueish*", i.e., having a strong dominance of high-frequency (short-period) tsunami waves. The "colour" of the event is linked to the length scale of the respective seismic source region; the small spatial scale of the 2018 source region determined the "*blueish*" colour of the resulting tsunami.

In addition to the open-ocean energy, we also computed the total tsunami variance at the 14 tide gauge sites on the BC, WA and OR coasts that detected the 2018 Kodiak event (Fig. 13). The average total variance from 12 of these tide gauges (excluding Henslung Cove and Rose Harbour, which are located far away from the BPR array) was $Var \approx 2.1$ cm^2. We then used the open–open total variance $I_O \approx 0.08, 0.1, 0.17, 0.14$ cm^2 for ODP1026, RCM South, NJ03B and DART46404, respectively, to estimate the mean RMS (root-mean-square) amplification factor, A_{RMS}, of the tsunami waves that arrived at the coast. Based on the four open-ocean stations,

$$A_{RMS} = \sqrt{Var/I_0} \sim 5.1, 4.6, 3.5, 3.9, \tag{3}$$

which indicates that a 1 cm open-ocean wave at each of the above four sites would be expected to amplify by roughly 3.5–5 times near the coast. The mean amplification factor, $\bar{A}_{RMS}$, for the 2018 Kodiak event, derived from the variance estimation, is higher than the "*traditional*" ratio of maximum amplitudes, which is ~ 3 for Vancouver Island and ~ 2 for WA and OR coast (see Sect. 3.2). The apparent discrepancy between the two kinds of amplification ratios will be further discussed in Sect. 7.3.

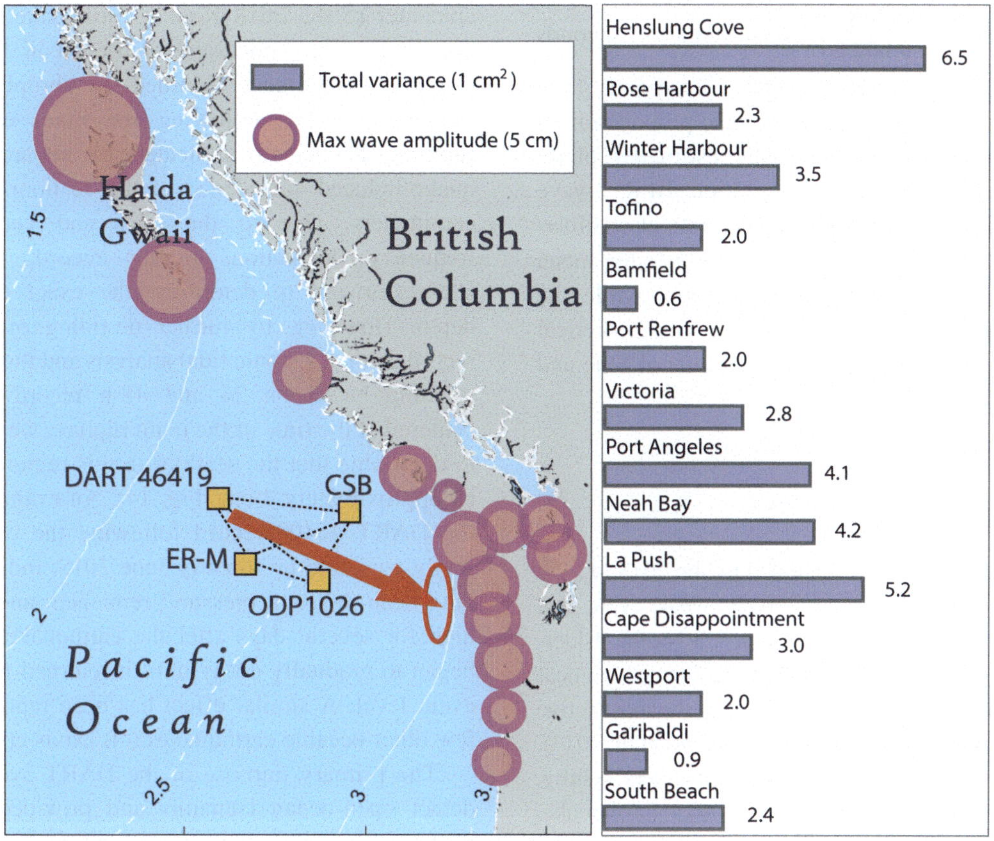

Figure 13

Parameters for the 2018 Kodiak tsunami off the west coast of North America. Disks indicate the maximum wave amplitudes, with sizes measured relative to the maximum wave amplitude recorded at Henslung Cove at the northwest tip of Haida Gwaii (cf. Table 3). The bars on the right denote the total variance (in cm^2) at the specified site, with variance integrated over the 12-h period following arrival of the first wave. The arrow denotes the inverse celerity vector (in s/m); the ellipse shows the 95% confidence interval. The locations of the four open-ocean BPRs used to generate the vector are indicated by the squares; ER-M and CSB are for the Endeavour Ridge Main Field and Clayquot Slope Bullseye sites, respectively. The white curves are the isochrones relative to the epicenter as in Fig. 1; solid curves mark hours and dashed curves mark 30-min intervals

6. *Cross-Correlation and Wave Direction*

Determination of the inverse celerity vector (often referred as "*slowness*" in seismology and other branches of geophysics) for propagating tsunami waves requires high-resolution time series from three or more relatively closely positioned open-ocean BPRs. For this study, we chose four sites for the analysis: DART 46419, ODP1026, Endeavour Ridge Main Field and Clayquot Slope (Fig. 2b). The original data from the three ONC stations had a 1-s sampling rate, while the real-time event data from DART 46419 had a 1-min sampling rate. To match the DART series with the ONC data, we interpolated the 1-min DART data to 1-s using a cubic polynomial. We note that using 1 s data from the three ONC BPRs gives almost the same result, and that adding the fourth DART46419 BPR, only slightly affects the uncertainty in the estimates. Due to the fact that the travel time between DART 46419 site and ONC BPRs is much longer than 1 min (Table 1) and that the primary waves have well-defined waveforms, the error in interpolating the 1 min data to 1 s is minimal. The data from all four series were then de-tided and high-pass filtered with a 30-min Kaiser window to suppress low-frequency oscillations. Cross-correlation analysis performed on the resulting 1-s tsunami records yields an inverse celerity vector

$\mathbf{v} = \begin{pmatrix} v_x \\ v_y \end{pmatrix} = \begin{pmatrix} \mathbf{0.0058} \\ \mathbf{-0.0024} \end{pmatrix}$, with units of seconds per meter (Fig. 13). This vector corresponds to a phase speed, $c \approx 160$ m/s, with wave propagation to the southeast (113° True). This estimated phase speed, derived from the four-station BPR "wave antenna", agrees well with the theoretical shallow-water phase speed, $c = \sqrt{gH} \approx 159$ m/s for mean water depth $H = 2580$ m, in the region, while the direction is nearly perpendicular to the theoretical isochrones computed using the technique of Fine and Thomson (2013) (Figs. 1, 13).

7. Discussion

Although the tsunami generated by the 2018 M_w 7.9 Kodiak earthquake was a relatively weak event and produced no known damage or injury, it is of considerable scientific interest. The earthquake epicenter was near Alaska-Aleutian subduction zone, one of the most active earthquake zones in the world (Freymueller et al. 2013). Many of the previous events with epicenters nearby, including the 1964 M_w 9.2 Great Alaska Earthquake, originated on the North American Plate. In contrast, the 2018 Kodiak Earthquake occurred within the Pacific Plate, where strong earthquakes are rare. Moreover, it was not a thrust-fault event, as are most tsunamigenic earthquakes in this region (Fig. 1), but a shallow-depth (~ 14 km) strike-slip fault earthquake (cf. Lay et al. 2018; Ruppert et al. 2018). Unlike typical strike-slip faults, which do not generate detectable tsunamis, the 2018 Kodiak earthquake generated a tsunami that was sufficiently high to be clearly recorded by both coastal tide gauges and open-ocean BPRs throughout the northeast Pacific. The combination of onshore and offshore measurements has enabled us to provide a comprehensive investigation of the wave properties of the 2018 tsunami, to examine its evolution during propagation over the shelf, and to estimate the amplification of tsunami waves as they approach the coast. Several unexpected results also emerged from our study.

7.1. Earthquake-Induced Seafloor Subsidence

An important unexpected finding was provided by DART 46409, located about 85 km from the epicenter of the 2018 Kodiak earthquake. The 15-s bottom pressure time series retrieved in June 2018 from the instrument recorded an abrupt jump in seafloor pressure following the main earthquake rupture. This jump indicates an apparent earthquake-induced subsidence of the seafloor of a few centimeters. Due to the tides and various low frequency fluctuations in the record, it is not straightforward to determine the exact change in depth. However, by further de-tiding of the 15-s record using harmonic tidal analysis and thorough de-trending based on 24 and 48-h record segments centered at the time of the main rupture, we were able to determine that the seafloor in this region subsided by approximately 4 cm (Fig. 14). An examination of the DART 46409 record following the event, currently available up to early June 2018, indicates that the mean bottom pressure remained anomalously high for several days after the earthquake and then began to gradually decay until it returned to the pre-event level. A similar effect has been reported for a few other oceanic earthquakes (cf. Davis et al. 2001).

The primary purpose of the DART system is to detect open-ocean tsunamis and provide real-time data required for early tsunami warning. The location of each DART station is chosen to avoid "high-risk" earthquake areas, as it is known that any seafloor movement following a tsunamigenic earthquake would seriously complicate the extraction of accurate tsunami displacements and source parameters, which are essential for generating reliable real-time tsunami warnings (Percival et al. 2011). Such movements would also affect real-time tide removal methods (Percival et al. 2014) and post-tsunami analyses of the source distribution using inverse modeling techniques (Satake 1987). In the context of these considerations, it is clear why the vertical seafloor displacement observed at DART 46409 is unexpected and unusual. While this is the first time that DART system recorded earthquake induced seafloor subsidence, a similar phenomenon, consisting of a 10–40 cm seafloor uplift was recorded by two JAMSTEC[2] pressure gauges (PG) during the 2003 Tokachi-oki earthquake (cf. Watanabe et al. 2004).

[2] JAMSTEC = Japan Agency for Marine Earth Science and Technology.

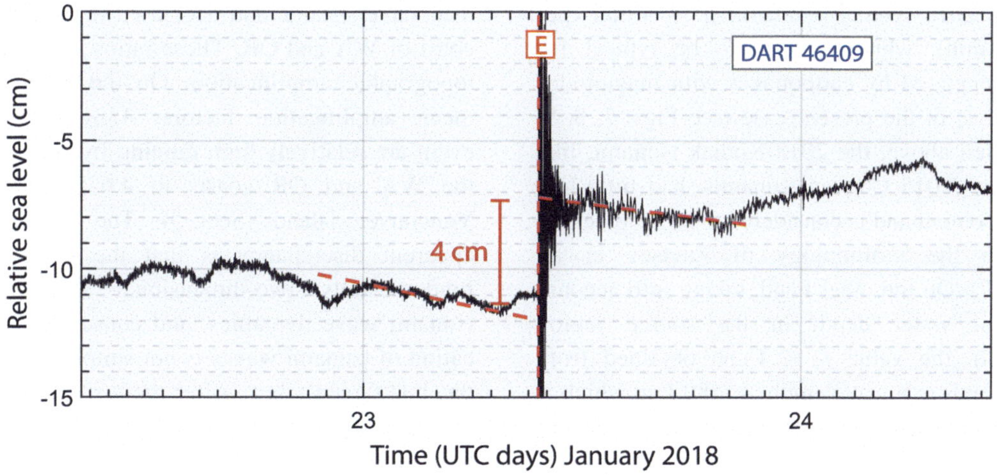

Figure 14
Fifteen second (15-s) de-tided sea level oscillations recorded by DART 46409 for the 50 h centered at the time of the earthquake. The 15 s NOAA record is downloaded from NCEI website; the black curve is obtained by further de-tiding the 15 s NOAA record. Dashed red vertical line marks the time of the earthquake; dashed sloped red lines are the trend for the 12 h before and after the earthquake computed using the least squares method

Our results also present a concrete example demonstrating how some existing data analysis methods and data assimilation based on ideal wave assumptions can fail. On a more positive note, direct information on the vertical seafloor movement during an event can provide important constrains during the formulation of tsunami source models. Recently new methods have been developed to extract wave information from occurrences of permanent seafloor deformation near tsunami source regions (cf. Tsushima et al. 2012; Tanioka 2018).

7.2. Wave Frequency Content

Tsunamis generated by earthquake sources with small spatial extent tend to span a relatively high frequency range, with typical frequencies of 2–30 cph (periods from 30 to 2 min). In this respect, the 2018 Kodiak tsunami, with open-ocean energy confined to the frequency band of 2–15 cph (Figs. 11, 12), was a typical event. Approximately 90% of the total energy of the 2018 tsunami was at frequencies > 1.7 cph (periods < 35 min) (Fig. 12) and about 65% at frequencies > 3.0 cph (periods < 20 min). Thus, as indicated in Sect. 6, the event can be characterized as "*blueish*" (Fig. 12). Other examples of blueish tsunamis, similar to the 2018 Kodiak event, were caused by the 2014 M_w 8.2 Iquique and 2015 M_w 8.3 Illapel Chilean earthquakes (Zaytsev et al. 2016) and the 2012 M_w 7.8 earthquake off Haida Gwaii (Fine et al. 2015). This contrasts with the "*reddish*" Great trans-oceanic tsunamis generated by the 2010 $M_w = 8.8$ Chilean (Maule) and the 2011 $M_w = 9.0$ Tohoku (East Japan) earthquakes (Zaytsev et al. 2016, 2017). The prevalence of low frequency ("reddish") waves in both of these large tsunamis was due to the large spatial extent of their source regions. It is noteworthy that the first waves of the 2018 event, which were the highest waves of the event in the open ocean (Fig. 10), had much lower frequencies of 2–4 cph (periods from 30 to 15 min) than the secondary waves that followed.

The spectra of tsunami waves are directly related to the earthquake source parameters and affects tsunami impact on the coast. As with any distribution, the shape of a spectrum is characterized by its statistical moments. The zero moment of a spectrum is the signal variance, while the first moment characterizes the *integral frequency scale*, f_0, defined by Thomson and Emery (2014) as:

$$f_0 = \frac{\int_{f_1}^{f_2} fS(f)df}{\int_{f_1}^{f_2} S(f)df} \tag{4}$$

For the 2018 Kodiak tsunami, $f_0 \sim 4$ cph (period $\sim$ 15 min), which appears to be typical for tsunamis generated by earthquakes with magnitudes $M_w \sim 7.9$, as in the present case (see Figs. 7, 8, 9, 10). As noted above, the 2018 Kodiak tsunami, like the 2014 and 2015 Chilean tsunamis, had little low frequency content and is considered to be a "blueish" tsunami in the terminology of Zaytsev et al. (2016, 2017). On the other hand, taking into account the unusual water depth in the source region ($\sim$ 4000 m), the value $f_0 \sim 4$ cph obtained from (4) corresponds to a wavelength $\sim$ 180 km, which is quite large for such a moderate tsunami, yet still in normal range. (According to Fujii and Satake (2018), the earthquake rupture length and width were 80 and 20 km, respectively, while the scale law of Wells and Coppersmith (1994) gives corresponding values of 117 and 42 km). In other words, the Kodiak tsunami should have had more high-frequency content (i.e., been more blueish) than it was. The absence of high-frequency tsunami waves is likely related to a smooth tsunami source, which, in turn, can be the result of the strike-slip earthquake mechanism, as occurred in Kodiak case, or a result of a deep fault, as in the case of the 2014 Chilean tsunami. The shallow-end of an earthquake fault, together with a dip-slip movement, creates a sharp response at the seabed interface, which then generates high-frequency tsunami waves, as occurred during the 2012 Haida Gwaii and 2015 Chilean tsunamis.

7.3. Coastal Wave Amplification

The large magnitude of the 2018 Kodiak earthquake triggered NOAA's tsunami warning and tsunami watch. Over-night evacuation took place in certain impacted coastal communities. To further quantify the relation of open-ocean tsunami waves and their impacts at the coast, we examine and compare two types of amplification factors: (a) the "*traditional*" *amplitude amplification factor* A (see Sect. 3.2), the ratio of the maximum wave amplitude near the coast to the maximum wave amplitude in the open-ocean; and (b) the *mean amplification factor* A_{RMS} (3), the ratio of the RMS value at the coast to the RMS value in the open ocean (see Sect. 5). For the 2018 tsunami, A is estimated to be $\sim$ 3.0 near Vancouver Island, and decreased to $\sim$ 2.0 along the coast of WA and OR. These ratios suggest very low topographic amplification. On the other hand, the mean amplification factors, A_{RMS}, for the 2018 event are relatively high ranging from 3.5–3.9 along the WA and OR coasts to 4.6–5.1 on coast of Vancouver Island (Sect. 5). The reason for this apparent discrepancy is that the coastal zone is fundamentally two-dimensional with respect to tsunami wave dynamics, and consequently, amplification of tsunami waves is not simply determined by the 1-D Green's law. After the arrival of a tsunami, the coastal zone accumulates energy by partly trapping waves incoming from the open ocean. The maximum wave amplitude generated at the coast depends on the amount of energy accumulated locally. The time over which the energy builds up depends on the distance from the coastal site to the deep water; the farther a given inner coastal location is from the deep ocean, the longer the time required to build up the tsunami energy on the shelf. For distant tsunamis, like the 2004 Sumatra (Rabinovich et al. 2011), 2010 Chile (Rabinovich et al. 2013) and 2011 Tohoku (Fine et al. 2013) events, tsunami waves continuously arrived at the shelf zone for over 24 h. As a result of the steady accumulation of tsunami energy in the coastal zone, not only the maximum waves at the coast were high but also the highest waves occurred hours after first wave arrival. In contrast, for more proximal events, like the 2009 Samoa, the 2018 Kodiak and the 2012 Haida Gwaii tsunamis, the first wave train was much more energetic than the following waves, so that coastal energy stopped building after the first train had arrived. This caused A between the coastal waves and the open-ocean waves to remain low for the 2009, 2012 and 2018 events. However, when computing the variance-based *mean amplification factor*, A_{RMS}, the results for distant and nearfield earthquake events become more compatible if we choose to integrate over time periods compatible to, or longer than, the basic time for energy to build up at the coast. Although both open ocean and coastal variance estimates are time-dependent, the open ocean variance decreases much faster than the coastal variance because of the effect of coastal energy accumulation. As a result, the coastal/open-ocean variance ratio increases with time. This increase will be nearly linear over a relatively short time span. Here, we chose a fixed integration time of 12 h when comparing different events.

Tofino, one of the largest communities on the west coast of Vancouver Island, is highly vulnerable to incoming tsunami waves and natural hazard management receives considerable public attention (cf. Rabinovich et al. 2019). Following the tsunami warning, the district of Tofino performed an early morning emergency drill. Although there was no substantial tsunami recorded at Tofino, or any other site along the entire outer coast of Vancouver Island (Table 2; Figs. 7b, 13), the tsunami waves recorded in the open ocean and on the coast were sufficiently high to enable us to estimate the amplification of tsunami waves approaching the coast. The maximum recorded wave amplitude of the 2018 tsunami was 1.9 cm at ODP 1026 (depth 2600 m) and 6.1 cm at Tofino, yielding a maximum tsunami wave amplification factor (denoted as A^{Tof}) for this station of 3.2 (Table 4). In Table 4 are listed A^{Tof}, mean amplification factors A_{RMS}^{Tof} and other tsunami parameters for the 2018 Kodiak and four other major past events based on the open-ocean BPR data at site ODP1026 and tide gauge records at Tofino (cf. Thomson et al. 2011; Rabinovich et al. 2013; Fine et al. 2013, 2015): the 2009 Samoa, 2010 Chile (Maule), 2011 Tohoku and 2012 Haida Gwaii tsunamis. The maximum amplitudes at Tofino for the five tsunamis varied by a factor of ten, ranging from 62.0 cm for the 2011 Tohoku tsunami to 6.1 cm for the 2018 Kodiak tsunami, while for same events at ODP 1026 maximum amplitudes varied from 12.8 to 1.9 cm. The corresponding amplification factor, A^{Tof}, varied by a factor of 2.3 for the different events, ranging from 2.1 for the 2012 Haida Gwaii tsunami to 4.8 for the 2011 Tohoku tsunami. The mean amplification factor, A_{RMS}^{Tof}, determined from the spectral variance (3), is a much more stable parameter than A^{Tof}, as it is mostly determined by the general increase of tsunami waves at coastal area and only weakly dependent on the specific characteristics of incoming waves. According to the results shown in Table 4, A_{RMS}^{Tof} varies from 4.3 to 6.3 for Tofino, i.e., by a factor of $\sim$ 1.5. The main factor likely influencing this parameter is the propagation direction of the tsunami waves (Fig. 15).

The incoming wave directions of the 2011 Tohoku tsunami and the 2018 Alaska tsunami were within 5° of one another as they propagated over the continental slope and shelf toward the coast, despite the fact that their source regions were thousands of kilometers apart. The mean amplification factors, A_{RMS}^{Tof}, for these two events and the 2012 Haida Gwaii tsunami have comparable ranges of 6.3, 5.3 and 5.2, respectively (Table 4). Similarly, A_{RMS}^{Tof} for the two southern events are also comparable: 4.3 for the 2010 Chile and 4.5 for the 2009 Samoa tsunamis. In general, we find that A_{RMS}^{Tof} is larger for tsunamis arriving from the northwest and west compared with those arriving from the south and southwest (Table 4; Fig. 15). The similarity of A_{RMS}^{Tof} for waves arriving from the same direction is expected based on the

Table 4

Maximum wave amplitudes and mean RMS observed at Tofino and ODP 1026 and the corresponding amplification ratios

Event	Tofino tide gauge		ODP1026		Ratio	
	Maximum amplitude (cm)	Mean RMS $\sqrt{Var}$ (cm)	Maximum amplitude (cm)	Mean RMS $\sqrt{I}$ (cm)	Maximum amplitude amplification A^{Tof}	Mean amplification A_{RMS}^{Tof}
2018 Kodiak	6.1	1.5	1.9	0.28	3.2	5.3
2012 Haida Gwaii	11.6	3.6	5.5	0.69	2.1	5.2
2011 Tohoku	62.0	21.9	12.8	3.5	4.8	6.3
2010 Chile	18.3	5.9	4.0	1.37	4.6	4.3
2009 Samoa	7.0	2.2	2.5	0.49	2.8	4.5

The mean variance is computed for 12 h after arrival of the lead tsunami wave at each site. RMS denotes the root-mean-square value derived from the variance

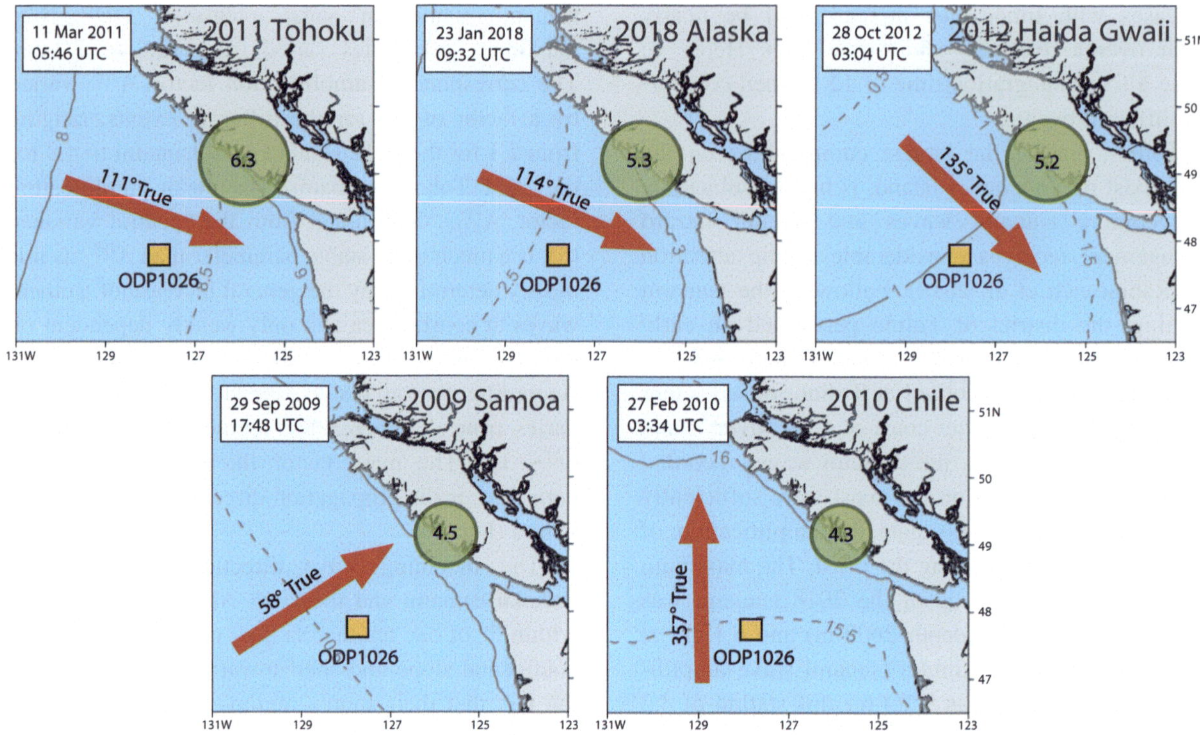

Figure 15

Outgoing wave directions (arrows, in degrees relative to true north) for the five major past events and the corresponding mean amplification factors indicated at the center of each disk, A_{RMS}^{Tof}, computed using the total variance of the 12-h tsunami records at ODP1026 and Tofino (cf. Table 4). The size of each disk is proportional to the magnitude of A_{RMS}^{Tof}. The curves inside each panel are the isochrones of tsunami travel time in hours after the earthquake. The time of the earthquake for each event is provided in the box in the upper left corner

linearity of the wave transformation over shoaling bathymetry. In contrast, the traditional ratio, A^{Tof}, does not demonstrate such a well-defined direction dependency because it is affected by more than one transformation factor.

7.4. *Along-Coast Wave Variance Distribution*

The 1964 Alaska tsunami caused significant damage to Port Alberni, located at the head of long and narrow Alberni Inlet. Destructive flooding in the inlet lasted for several hours. Tsunami wave heights recorded at other sites were also high (cf. Rabinovich et al. 2019), but not as high as in Port Alberni. It is evident that the intense response of Alberni Inlet to the incoming 1964 tsunami waves was the result of local resonance (Fine et al. 2008). In contrast, the 2018 event generated a weak tsunami in Port Alberni with a maximum wave amplitude and height of 3.5 and 5.6 cm, respectively. The tsunami at Bamfield, located at the entrance to Alberni Inlet, was also quite weak, with maximum amplitude and height of 3.6 and 7.2 cm, respectively (Table 2). Thus, while in 1964 the tsunami waves in the Alberni Inlet region were much higher than at other sites along the BC coast, in 2018 they were significantly lower. The different response to the two events is related to the completely different frequency contents of the two events and to the specific bathymetric properties of Alberni Inlet, which is highly tuned to low-frequency waves but not to high-frequency waves (i.e., tuned to "reddish" but not to "blueish" events).

Figure 13 shows the distribution of estimated tsunami variance along the coasts of BC, WA and OR. Our results show that the tsunami impact on the north coast of British Columbia (Haida Gwaii) was comparable to that on the Gulf of Alaska coast (Fig. 6), while the impacts along the rest of the BC coast were similar to those along the Washington and Oregon coast (Fig. 13). The lowest variance among all selected sites of 0.6 cm^2 was observed for Bamfield, while the maximum variance of 6.5 cm^2 was observed for Henslung Cove, Langara Island (northern coast of Haida Gwaii). A high variance of 5.2 cm^2 was also found for La Push, WA, USA.

7.5. Secondary Wave Trains

The first group of waves from the 2018 tsunami was clearly recorded by all operating open-ocean BPRs in the northeast Pacific. Many of these BPRs also recorded much weaker secondary trains of waves, which, unlike the highly consistent primary waves that arrived from the same source region, were observed at different time lags after the first arrival and had no common frequency range. Soon after issuing the tsunami warning, NOAA released a computer animated tsunami propagation video, which provided helpful insight in determining the source regions for the multiple groups of secondary waves. In the Gulf of Alaska, near the source region, several tsunami trains were clearly observed (Fig. 9). Just before 12:00 UTC, secondary waves, with *f–t* distributions similar to those of the first waves, were recorded at DARTs 46403, 46409 (Fig. 9 upper panels, white rectangles) and 46410 (Fig. 9 lower left panel). Based on the arrival times, it is logical to assume that these waves resulted from reflection of the primary waves from the coast near Kodiak Island. Roughly 1.5 h after the earthquake, tsunami waves arrived at Sitka and Yakutat (Fig. 1), causing high-amplitude oscillations at these sites lasting for several hours (Fig. 6). The formation of reflected waves at these two sites is confirmed by the NOAA wave propagation animations, which show that, 2 h after the earthquake, strong reflected waves were radiated from the northeast shelf area of the Gulf of Alaska. The reflected waves from the coastal zone near Yakutat and Sitka were detected by DARTs 46409 and 46403 and clearly shown in the *f–t* plots (Fig. 9 upper panels, orange ellipses): the time difference of $\sim$ 40 min for these reflected waves at these two sites agrees closely with the time difference of the first wave arrivals at these stations.

All BPRs off the west coast of Vancouver Island, Washington and Oregon recorded secondary waves of various strengths. The highest secondary waves were recorded within the expected arrival times for local coastally reflected waves. However, around 14:30 UTC, slightly before the local reflected waves reached the open-ocean ONC and OOI BPRs, detectable waves with frequencies of about 2 cph (period of 30 min) arrived at DARTs 46404 and 46407 (Fig. 10b, bottom panels, white ellipses). Based on the observation that the first waves arrived at Crescent City, California (Fig. 1) at $\sim$ 13:20 UTC, it is apparent that the intense "concentrated" wave train observed at about 14:30 UTC (Fig. 10b, bottom right panel, *f–t* diagram for DART 46407) was the result of reflection of the primary waves from the intersection of the Mendocino Ridge and the continental shelf near the Mendocino Fracture Zone (Fig. 1, shaded red region).

8. Summary and Conclusions

The 2018 Alaska Kodiak tsunami was distinct in that it was generated by a rare, mid-plate strike-slip earthquake of substantial magnitude (M_w 7.9) and complex focal characteristics. Near the earthquake source region, the main rupture instantaneously caused a $\sim$ 4 cm subsidence in the seafloor that was detected by DART 46409. Such occurrences are commonly seen during earthquakes on and near land, but this appears to be one of the first DART-recorded measurements of such an event in the open ocean. The well-defined waves from this event recorded off the west coasts of British Columbia, Washington and Oregon made it possible for us to investigate relations between open-ocean tsunami characteristics and the corresponding coastal response. Due to the complexity of the earthquake faults and limited data currently available, still little is known about the tsunami source region to-date, which makes reconstructing tsunami models a challenge. Multiple groups of secondary waves were also observed at several sites. However, without further numerical modelling, we can only speculate on the sources and nature of these waves based on their frequency components and time frames.

The results and conclusions of this study are summarized as follows:

- The major (M_w 7.9) earthquake that struck the Gulf of Alaska near Kodiak Island on 23 January 2018 occurred not on the North American Plate, where strong tsunamigenic earthquakes commonly take place, but within the Pacific Plate, where such earthquakes are rare. Although it was a mid-plate strike-slip earthquake, which normally do not generate substantial tsunamis, the 2018 Alaska-Kodiak strike-slip earthquake generated detectable tsunami waves that spread throughout the Pacific Ocean and were clearly recorded by a large number of open-ocean and coastal instruments.
- A total of 21 open-ocean BPRs and 27 coastal tide gauges in the northeast Pacific were analyzed. The open-ocean BPRs include three groups of stations: (a) NOAA DARTs (4 offshore of Alaska and 4 offshore of the US West Coast); (b) the ONC NEPTUNE array (7 cable BPRs deployed westward from southern Vancouver Island); and (c) the OOI array (6 cable BPRs located offshore of northern Oregon). We also used tsunami records from all 14 CHS tide gauge stations on the British Columbia coast, 9 NOAA tide gauge stations located on the coasts of Washington and Oregon, and 4 Alaska tide gauges. The simultaneous measurement of tsunami waveforms in offshore and onshore regions enabled us to examine wave properties of the 2018 tsunami and its transformation over the continental slope and shelf. The data also allowed us to estimate the amplification of tsunami waves approaching the coast (i.e., basic information important to tsunami hazard warning and mitigation).
- Spectral analysis showed that nearly 90% of the total energy from the 2018 tsunami in the open ocean occurred at frequencies > 1.7 cph (periods < 35 min) and that about 60% of this energy was at frequencies > 3.0 cph (periods < 20 min). Also, there was little energy at frequencies above 15 cph (periods less than 4 min), which points to a smooth tsunami source. As a result, most of the 2018 Kodiak tsunami variance occupies a narrow frequency band from 2 to 15 cph. The integral frequency scale for this event, based on the first moment of the tsunami spectra, is 4 cph (period 15 min). Following the classification of Zaytsev et al. (2017), this event can be characterized as "blueish", similar to the 2014 and 2015 Chilean and the 2012 Haida Gwaii tsunamis, but contrary to the "reddish" nature of the great tsunamis generated by the 2010 Chilean and 2011 Tohoku earthquakes for which more than 43–46% of the total energy was at frequencies < 1.7 cph (periods from 35 min to ~ 3 h).
- Coastal and open-ocean observations were used to examine the amplification of tsunami waves reaching the coast. The maximum tsunami wave amplification factor for Tofino was $A^{Tof} = 3.2$, while the time-averaged mean amplification factor for this site based on the tsunami wave variance was $A^{Tof}_{RMS} = 5.3$. The same factors estimated for four other major past events—2009 Samoa, 2010 Chile (Maule), 2011 Tohoku and 2012 Haida Gwaii—were 2.8, 4.6, 4.8 and 2.1 (A^{Tof}) and 4.5, 4.3, 6.3 and 5.2 (A^{Tof}_{RMS}), respectively. The mean amplification factor, A_{RMS}, is a much more stable parameter than the corresponding factor, A, estimated from the maximum amplitude ratio, because it is determined mostly by an overall amplification of the tsunami waves in the coastal area and is only weakly dependent on the individual characteristics of the incoming waves. The amplifications A^{Tof}_{RMS} for the three northern tsunamis, Tohoku (6.3), Kodiak (5.3), Haida Gwaii (5.2), are greater than A^{Tof}_{RMS} for distant tsunamis arriving from the south, specifically Samoa (4.3), and Chile 2010 (4.5).
- Strong dispersion of the lead propagating tsunami waves was observed in the wavelet diagrams of all open-ocean gauges. The observed dispersion was found to be remarkably similar to theoretical values for arriving tsunami waves, with dispersion leading to longer periods of the leading tsunami waves and shorter periods of the "tail". The secondary wave trains evident in the wavelet (f–t) diagrams appear to have originated from reflection of the primary waves from the coast of Alaska and the mainland coasts of British Columbia, Washington and Oregon.

- The derived tsunami variance for the west coast decreased monotonically from northern British Columbia to southern Oregon. The variance structure was clearly affected by local topographic effects at some of the tide gauge sites. The lowest variance of 0.6 cm^2 was observed at Bamfield and the maximum variance of 6.5 cm^2 at Henslung Cove, Langara Island. There was also a secondary maximum of 5.2 cm^2 at La Push, Washington.
- The abrupt jump in sea level of about 4 cm detected by DART 46409, located at mid-plate about 85 km from the epicenter of the 2018 Kodiak earthquake, appears to have been related to a vertical earthquake-induced seafloor displacement. Such movements of the seafloor during a tsunamigenic earthquake event can seriously complicate extraction of the tsunami signal and tsunami source parameters that are critical to reliable real-time tsunami warning. On the other hand, measurements of vertical displacements of the seafloor due to earthquakes provide important direct information on the properties of tsunami source region for post-tsunami numerical modelling studies.

Acknowledgements

The authors gratefully acknowledge the Department of Fisheries and Oceans and the Natural Science and Engineering Research Council of Canada for their financial support. We also thank Ocean Networks Canada, Canadian Hydrographic Service and NOAA National Center for Environmental Information for providing open-ocean and coastal tide gauge data. We further acknowledge Professor Kenji Satake (University of Tokyo) and an anonymous reviewer for their helpful and supportive comments. For AR this study work was partially supported by the Russian State Assignment of IORAS #0149-2019-0005.

References

Craig, T. J., Copley, A., & Jackson, J. (2014). A reassessment of outer-rise seismicity and its implications for the mechanics of oceanic lithosphere. *Geophysical Journal International, 197*(1), 63–89. https://doi.org/10.1093/gji/ggu013.

Davis, E. E., Wang, K., Thomson, R. E., Becker, K., & Cassidy, J. F. (2001). An episode of seafloor spreading and associated plate deformation inferred from crustal fluid pressure transients. *Journal of Geophysical Research Solid Earth, 106,* 21953–21963.

Dmowska, R., Rice, J. R., Lovison, L. C., & Josell, D. (1988). Stress transfer and seismic phenomena in coupled subduction zones during the earthquake cycle. *Journal of Geophysical Research, 93*(B7), 7869–7884. https://doi.org/10.1029/JB093iB07p07869.

Eblé, M. C., & González, F. I. (1991). Deep-ocean bottom pressure measurements in the Northeast Pacific. *Journal of Atmospheric and Oceanic Technology, 8*(2), 221–233.

Fine, I. V., Cherniawsky, J. Y., Rabinovich, A. B., & Stephenson, F. E. (2008). Numerical modeling and observations of tsunami waves in Alberni Inlet and Barkley Sound, British Columbia. *Pure and Applied Geophysics, 165*(11/12), 2019–2044.

Fine, I. V., Cherniawsky, J. Y., Thomson, R. E., Rabinovich, A. B., & Krassovski, M. V. (2015). Observations and numerical modeling of the 2012 Haida Gwaii tsunami off the coast of British Columbia. *Pure and Applied Geophysics, 172*(3/4), 699–718. https://doi.org/10.1007/s00024-014-1012-7.

Fine, I. V., Kulikov, E. A., & Cherniawsky, J. Y. (2013). Japan's 2011 Tsunami: Characteristics of wave propagation from observations and numerical modelling. *Pure and Applied Geophysics, 170,* 1295–1307. https://doi.org/10.1007/s00024-012-0555-8.

Fine, I. V., & Thomson, R. E. (2013). A wavefront orientation method for precise numerical determination of tsunami travel time. *Natural Hazards and Earth System Sciences, 13,* 2863–2870. https://doi.org/10.5194/nhess-13-22863-2013.

Freymueller, J. T., Haeussleer, P. J., Wesson, R. L., & Ekström, G. (Eds.). (2013). *Active tectonics and seismic potential of Alaska. Geophysical monograph series, 179.* Washington, DC: American Geophysical Union.

Fujii, Y., & Satake, K. (2018). Alaska Tsunami on Jan. 23, 2018. https://iisee.kenken.go.jp/staff/fujii/Alaska2018/tsunami.html.

González, F. I., & Kulikov, Y. A. (1993). Tsunami dispersion observed in the deep ocean. In S. Tinti (Ed.), *Tsunamis in the world* (pp. 7–16). Dordrecht: Kluwer.

González, F. I., Mader, C. L., Eblé, M. C., & Bernard, E. N. (1991). The 1987–88 Alaskan Bight tsunamis: Deep ocean data and model comparisons. *Natural Hazards, 4*(1–2), 119–139.

Gusiakov, V. K. (2014). Strongest tsunamis in the World Ocean and the problem of marine coastal security. *Izvestiya, Atmospheric and Oceanic Physics, 50*(5), 435–444.

Lander, J. F. (1996). *Tsunamis affecting Alaska 1737–1996.* Boulder: USDC/NOAA.

Lay, T., Ye, L., Bai, Y., Cheung, K. F., & Kanamori, H. (2018). The 2018 M_w 7.9 Gulf of Alaska earthquake: Multiple fault rupture in the Pacific Plate. *Geophysical Research Letters, 45,* 9542–9551. https://doi.org/10.1029/2018gl079813.

Mofjeld, H. O. (2009). Tsunami measurements. In: E. N. Bernard & A. R. Robinson (Eds.), *The sea. Tsunamis*. (Vol. 15, pp. 210–235). Cambridge, USA: Harvard University Press.

Mungov, G., Eblé, M., & Bouchard, R. (2013). DART® tsunameter retrospective and real-time data: A reflection on 10 years of processing in support of tsunami research and operations. *Pure and Applied Geophysics, 170*(9–10), 1369–1384. https://doi.org/10.1007/s00024-012-0477-5.

Percival, D. B., Denbo, D. W., Eblé, M. C., et al. (2011). Extracting of tsunami source coefficients via inversion of DART buoy data. *Natural Hazards, 58,* 567–590.

Percival, D. B., Denbo, D. W., Eblé, M. C., et al. (2014). Detiding DART® buoy data for real-time extraction of source coefficients for operational tsunami forecasting. *Pure and Applied Geophysics, 172*(6), 1653–1678.

Rabinovich, A. B., Candella, R., & Thomson, R. E. (2011). Energy decay of the 2004 Sumatra tsunami in the World Ocean. *Pure and Applied Geophysics, 168*(11), 1919–1950. https://doi.org/10.1007/s00024-01-0279-1.

Rabinovich, A. B., & Eblé, M. C. (2015). Deep-ocean measurements of tsunami waves. *Pure and Applied Geophysics, 172,* 3281–3312. https://doi.org/10.1007/s00024-015-1058-1.

Rabinovich, A. B., & Stephenson, F. E. (2004). Longwave measurements for the coast of British Columbia and improvements to the tsunami warning capability. *Natural Hazards, 32*(3), 313–343.

Rabinovich, A. B., Thomson, R. E., & Fine, I. V. (2013). The 2010 Chilean tsunami off the west coast of Canada and the northwest coast of the United States. *Pure and Applied Geophysics, 170,* 1529–1565. https://doi.org/10.1007/s00024-012-0541-1.

Rabinovich, A. B., Thomson, R. E., Krassovski, M. V., Stephenson, F. E., & Sinnott, D. C. (2019). Five great tsunamis of the 20th century as recorded on the coast of British Columbia. *Pure and Applied Geophysics, 176*(7), 2887–2924. https://doi.org/10.1007/s00024-019-02133-3.

Rabinovich, A. B., Thomson, R. E., Lupton, L. M., & Mundschutz, S. (2018). Historical Tsunamis at the Seal Cove and Victoria Canadian Coast Guard Stations, British Columbia. *Canadian Technical Report of Hydrography and Ocean Sciences, 325,* ix + 44.

Ruppert, N. A., Rollins, C., Zhang, A., et al. (2018). Complex faulting and triggered rupture during the 2018 M_w 7.9 offshore Kodiak, Alaska, earthquake. *Geophysical Research Letters*. https://doi.org/10.1029/2018GL078931.

Satake, K. (1987). Inversion of tsunami waveforms for the estimation of a fault heterogeneity: Method and numerical experiments. *Journal of Physics of the Earth, 35,* 241–254.

Satake, K., & Atwater, B. (2007). Long-term perspectives on giant earthquakes and tsunamis at subduction zones. *Annual Review of Earth and Planetary Sciences, 35,* 349–374.

Suleimani, E. N., Nicolsky, D. J., & Koehler, R. D. (2013). *Tsunami inundation maps of Sitka, Alaska*. Report of Investigations 2013-3. Fairbanks, AK: State of Alaska, Department of Natural Resources, Division of Geological and Geophysical Surveys, p. 76.

Tanioka, Y. (2018). Tsunami simulation method assimilating ocean bottom pressure data near a tsunami source region. *Pure and Applied Geophysics, 175*(2), 721–729. https://doi.org/10.1007/s00024-017-1697-5.

Thomson, R. E., & Emery, W. J. (2014). *Data analysis methods in physical oceanography* (3rd ed., p. 716). Amsterdam: Elsevier Science.

Thomson, R. E., Fine, I. V., Rabinovich, A. B., et al. (2011). Observations of the 2009 Samoa tsunami by the NEPTUNE-Canada cabled observatory: Test data for an operational regional tsunami model. *Geophysical Research Letters, 38,* L11701. https://doi.org/10.1029/2011GL046728.

Tsushima, H., Hino, R., Tanioka, Y., Imamura, F., & Fujimoto, H. (2012). Tsunami waveform inversion incorporating permanent seafloor deformation and its application to tsunami forecasting. *Journal of Geophysical Research, 117,* B03311. https://doi.org/10.1029/2011JB008877.

Watanabe, O., Matsumoto, H., Sugioka, H., et al. (2004). Offshore monitoring system records recent earthquake off Japan's northernmost island. *ESO Earth and Space Science News, 85*(2), 14–15. https://doi.org/10.1029/2004EO020003.

Wells, D. L., & Coppersmith, K. J. (1994). New empirical relationships among magnitude, rupture length, rupture width, rupture area, and surface displacement. *Bulletin of the Seismological Society of America, 84*(4), 974–1002.

Zaytsev, O., Rabinovich, A. B., & Thomson, R. E. (2016). Comparative analysis of coastal and open-ocean records of the great Chilean tsunamis of 2010, 2014 and 2015 off the coast of Mexico. *Pure and Applied Geophysics, 173*(12), 4139–4178. https://doi.org/10.1007/s00024-016-1407-8.

Zaytsev, O., Rabinovich, A. B., & Thomson, R. E. (2017). The 2011 Tohoku tsunami on the coast of Mexico: A case study. *Pure and Applied Geophysics, 174*(8), 2961–2986. https://doi.org/10.1007/s00024-017-1593-z.

(Received October 21, 2019, revised January 13, 2020, accepted January 16, 2020, Published online February 3, 2020)

Pure Appl. Geophys. 177 (2020), 1379–1401

https://doi.org/10.1007/s00024-019-02305-1

Pure and Applied Geophysics

The Tsunami Scenario Database of the Indonesia Tsunami Early Warning System (InaTEWS): Evolution of the Coverage and the Involved Modeling Approaches

Sven Harig,[1] Antonia Immerz,[1] Weniza,[2] Jonathan Griffin,[3,4] Bernd Weber,[5] Andrey Babeyko,[6] Natalja Rakowsky,[1] Dwi Hartanto,[2] Arif Nurokhim,[2] Tri Handayani,[2] and Rikki Weber[3]

Abstract—This study reports on recent developments of the Indonesia Tsunami Early Warning System (InaTEWS), specifically the tsunami modeling components used in the system. It is a dual system: firstly, InaTEWS operates a high-resolution scenario database pre-computed with the finite element model TsunAWI; running in parallel, the system also contains a supra real-time modeling component based on the GPU-parallelized linear long-wave model easyWave, capable of dealing with events outside the database coverage. The evolution of the tsunami scenario database over time is covered in the first sections also touching on the involved capacity building efforts. Starting with a coverage of just the Sunda Arc region, the database now includes scenarios for 15 fault zones. The study is augmented by an investigation of warning products used for early warning; the estimated wave height (EWH) and the estimated time of arrival (ETA). These parameters are determined by easyWave and TsunAWI with model specific approaches. Since the numerical setup of the two models is very different, the extent of variations in warning products is investigated for a number of scenarios, where both pure database scenarios and applications to real events are considered. Finally, the performance of the system in past tsunami events is reviewed to point out major system updates.

Key words: Tsunami, Indonesia, early warning system, numerical modeling, scenario database, warning products.

[1] Alfred Wegener Institute, Helmholtz Centre for Polar and Marine Research, Bremerhaven, Germany. E-mail: sven.harig@awi.de
[2] Badan Meteorologi Klimatologi dan Geofisika, BMKG, Jakarta, Indonesia.
[3] Community Safety Branch, Geoscience Australia, Canberra, ACT, Australia.
[4] Department of Geology, University of Otago, Dunedin, New Zealand.
[5] Gempa GmbH, Potsdam, Germany.
[6] GFZ German Research Centre for Geosciences, Potsdam, Germany.

1. Introduction

Indonesia is very prone to tsunamis, and the December 2004 Indian Ocean tsunami which caused more than 230,000 fatalities demonstrated the necessity of a tsunami early warning system for this region. Started in 2005 as German-Indonesian Tsunami Early Warning System (GITEWS) Project (Münch et al. 2011), the Indonesia Tsunami Early Warning System (InaTEWS) was inaugurated by the President of Indonesia on November 11, 2008, and is fully operational under the Indonesian Agency for Meteorology, Climatology and Geophysics (Badan Meteorologi, Klimatologi dan Geofisika, BMKG). InaTEWS provides national warnings for Indonesia and is also one of the three regional warning providers for the Indian Ocean, along with Australia and India.

Presently, the system contains a multitude of sensors, including 165 broadband seismometers and 137 tide gauges. For warning dissemination purposes 261 Warning Receiver Systems were installed. These are multimode systems containing hardware and applications to receive earthquake information and tsunami warnings in text and map formats from servers at BMKG. Tsunami prone regions of Indonesia were equipped with 52 tsunami sirens (see Fig. 1). The background bathymetry and topography maps used in this and the following map visualizations, where not stated otherwise, are based on GEBCO data (The GEBCO_08 Grid, version 20090202, http://www.gebco.net).

Since a large proportion of the Indonesian coastline faces the Sunda Arc where there is the potential for large earthquakes with epicenters only

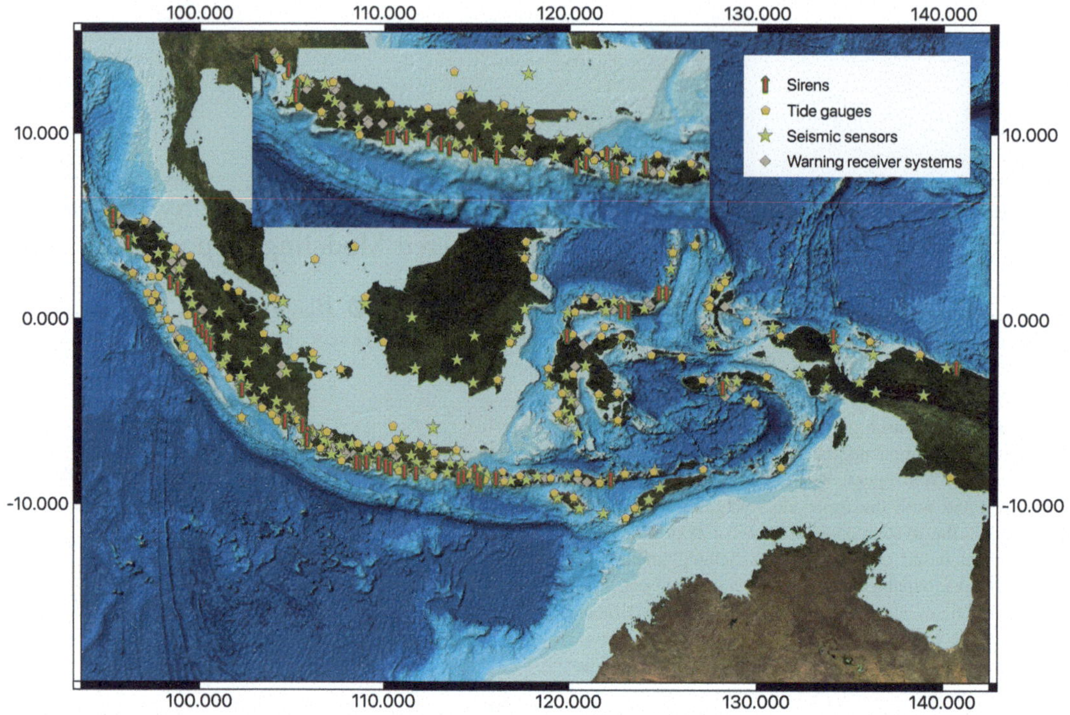

Figure 1
Overview of sensors, sirens and warning receiver systems accessed by InaTEWS

100–300 km offshore, the available time for early warning is very short. Due to national regulations, a warning must be given within 5 min after a potentially tsunamigenic earthquake. Therefore, the warning products for fast response are based on a pre-computed tsunami database covering the main fault zones of the country (see Fig. 2). In the case of an event happening in a location not covered by the database, a real-time model is used to determine the warning products. In the case where the event is covered by the scenario database, real-time models are additionally run once the focal mechanism of the earthquake has been determined accurately.

During the 10-year period of operation InaTEWS has provided bulletins with early warning information and recommendations in numerous cases of tsunamigenic earthquakes both in the national and regional context. One of the most recent updates of the system, InaTEWS 4.0, was launched in September, 2018. The tsunami scenario database available in the InaTEWS has undergone considerable extensions since the first implementation within the GITEWS project (German Indonesian Tsunami Early Warning System, see Lauterjung et al. 2010; Spahn et al. 2010). Initially there was a strong focus on the Sunda Arc region between Bali and the northern tip of Sumatra, whereas now the whole of the Sunda-Banda Arc as well as 14 additional fault zones are included. As a result, the number of available scenarios in the database increased from 4500 in 2010 to 16,529 at the end of 2017, and the most recent update in 2018 resulted in more than 18,000 scenarios in the tsunami database covering all regions of Indonesia. Due to the extended database fast access to warning products is available for the majority of the submarine fault systems in the country.

In the following we present details of the system extension over time as well as the possible impact of the system evolution with respect to the performance of the InaTEWS in a number of past tsunami events. The two model approaches as well as the determination of warning products are investigated.

The paper is organized as follows. Section 2 contains a general description of the InaTEWS

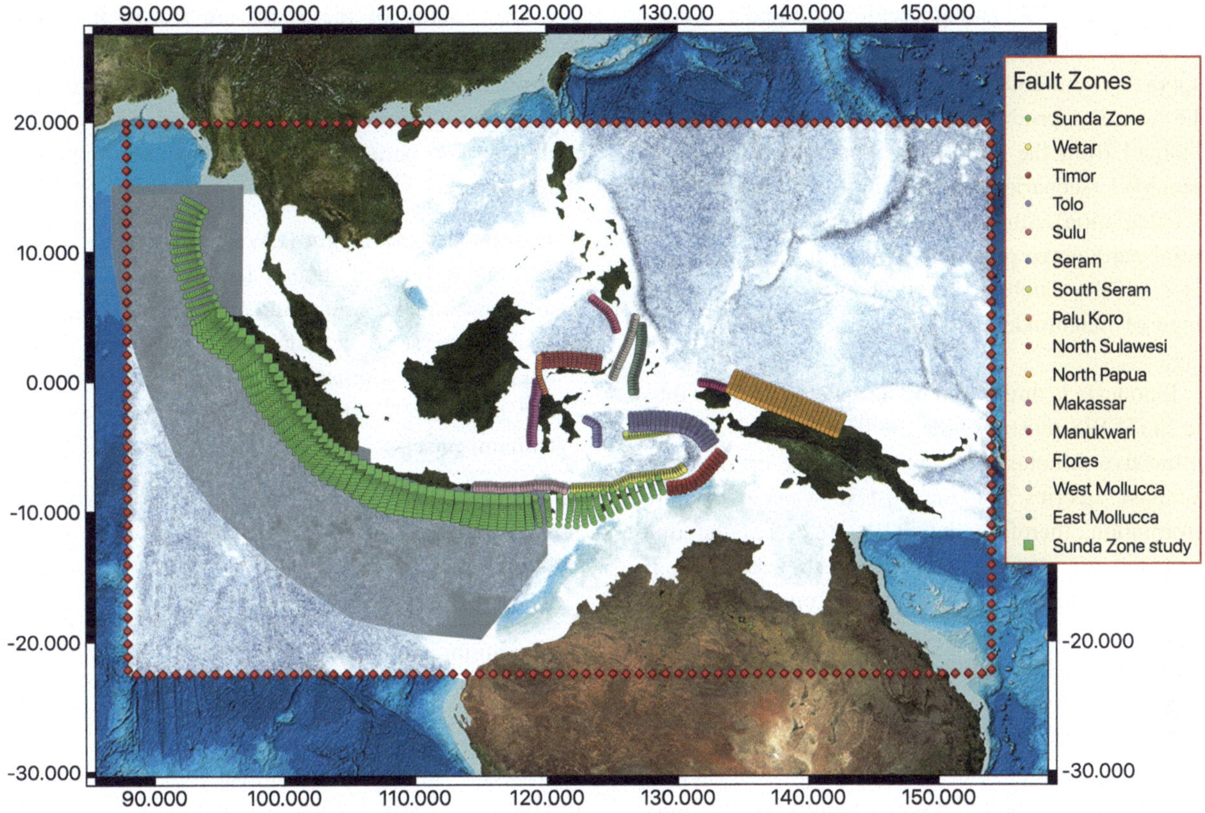

Figure 2
Mesh density for TsunAWI simulations showing the initial operational mesh (2.4 M nodes, dark dots) and most recent mesh (11 M nodes, white dots) together with the scenario earthquake epicenter positions of all fault zones currently covered (big colored dots). The green squares in the Sunda Arc depict the scenarios used in the comparison of Sect. 4.4. Table 2 contains the actual numbers of scenarios calculated in the fault zones. The bounding box of database simulations is shown as red diamonds

system providing an overview and some technical details. Warning products are introduced, along with the decision support software and the modeling approaches used. Section 3 deals with recent system developments, especially the extension of the tsunami scenario database. The evolution over time as well as the current status are presented together with the projects in which the extensions were realized. The presentation includes also the knowledge transfer aspects of the process. Topics of the final Sect. 4 are the performance of the system and a closer look at the model components. Recent events and warnings as well as system estimates of expected wave heights together with available data are presented. This comparison is done for both model components, easyWave and TsunAWI, thus providing insights into the consistency of the two models used within the system.

2. *InaTEWS System Overview*

This section introduces the modeling components of InaTEWS together with more general aspects like the standard operating procedures (SOPs) employed in warning situations, with a strong focus on the rules relevant for the choice of warning products and model results.

2.1. *General Task and SOPs Related to Tsunami Modeling*

The task of the InaTEWS-BMKG as the national tsunami warning center in the case of an earthquake event is to provide information promptly that is as precise and accurate as possible and easily understood by the community. During the 24/7 operation at BMKG seismic activity is continuously monitored.

When a potentially tsunamigenic earthquake occurs, the relevant parameters are forwarded to the decision support system and warning products are generated. The warning products are based on model results obtained from the earthquake parameters passed to TsunAWI (scenario database) or easyWave (on-the-fly calculation). Since there is large uncertainty in the initial earthquake parameters when the event is first detected, the database is queried for scenarios in a range of magnitudes and epicentral locations at this stage.

Indonesia is threatened by near field tsunamis thus the InaTEWS must provide the first tsunami warning within five minutes after the earthquake origin time. Database queries are the fastest way to acquire information, therefore the SOPs state that the first estimates are to be taken from the database if scenarios are available close to the epicenter. Once the moment tensor of the event has been determined real time modeling with easyWave is carried out and updates of the warning are disseminated.

To account for the temporal evolution of the tsunami during an event, a stepwise and area specific approach to disseminating information was recently implemented, augmenting the SOPs in that respect. Originally, a warning was disseminated after 5 min for all potentially threatened Indonesian coastlines with updates in case of new sensor data. This approach is now replaced by a procedure distinguishing between the near field, where a warning has to be issued as fast as possible, and the far field. The near field domain is characterized by an estimated tsunami travel time of 60 min or less based on tsunami simulations. In this area the warning is disseminated as soon as possible (i.e. usually within 5 min) to local authorities. In areas of predicted tsunami arrival time between one and 3 h more information is gathered for an improved warning at a later stage, based on a better knowledge of the actual event. This approach also affects the dissemination of information to the public in order to better inform communities and avoid unnecessary updates of earlier warnings based on limited data.

2.2. *Warning Products*

InaTEWS warning information is provided in warning zones covering all of the Indonesian coastline. Some of the zones are shown in Fig. 4. Data products used by the warning system are inferred from the results of the numerical models, with warning zone values determined by statistical methods. The parameters used and disseminated by the system are:

- Estimated wave height (EWH) given in (m);
- Estimated time of arrival (ETA) given in [min] after earthquake event; and
- Wave time series at tide gauge locations. (These data are not automatically evaluated, rather they serve for visual comparisons and manual picking of tsunami onsets)

Additional products used for information purposes to be displayed on operator screens and in bulletins are:

- Tsunami isochrones: lines of equal arrival time;
- Maximum wave height: global overview of the energy propagation; and
- Snapshots of the sea surface elevation.

Except for minor technical differences these products are provided in identical data formats by the two model components for seamless processing within the warning system.

In each warning zone a warning level is defined. The warning level is based on the estimated wave height. Thresholds for ‘Advisory’, ‘Warning’ and ‘Major Warning’ are shown in Table 1. The corresponding colors are displayed in the warning system. The appropriate warning products are disseminated to the provinces, districts and city governments. For an ‘Advisory’ level the recommendation is that communities immediately move away from the beach and river banks. For ‘Warning’ and ‘Major Warning’ levels additional evacuations and full evacuation are recommended respectively.

Some of the data products are shown in Fig. 3 for a scenario in the North Sulawesi fault zone. Arrival times and isochrones are based on a travel time approximation calculated in the triangular mesh and based on the exact initial conditions of each event. This method provides consistency with the remaining warning products.

Table 1

Warning levels and definitions in InaTEWS

Tsunami category	Warning level	Wave height (wh) range (m)
<none>	<none>	$0.0 \leq \text{WH} < 0.1$
Minor tsunami	Advisory (yellow)	$0.1 \leq \text{WH} < 0.5$
Tsunami	Warning (orange)	$0.5 \leq \text{WH} < 3.0$
Major tsunami	Major Warning (red)	$3.0 \leq \text{WH}$

2.3. Tsunami Situation and Decision Support

The practical determination of warning products and the use thereof for decision support purposes is handled by the decision support software operated in the warning center. Initially the consolidation of different data and observations in the system as well as decision support for the staff was accomplished by the DSS (decision support system, see Strunz et al. 2011, Steinmetz et al. 2010, Raape et al. 2010) developed by the German Aerospace Agency (DLR). Over the course of time these tasks moved to the TOAST (Tsunami Observation And Simulation Terminal) system described in the next section, which was developed by the company gempa GmbH (see www.gempa.de).

SeisComP3, an automatic and interactive seismic analysis system, provides the earthquake hypocenter, magnitude and focal mechanism information to the TOAST decision support system. TOAST supports different tsunami simulation algorithms, allowing the use of both "on-the-fly" tsunami simulation with the

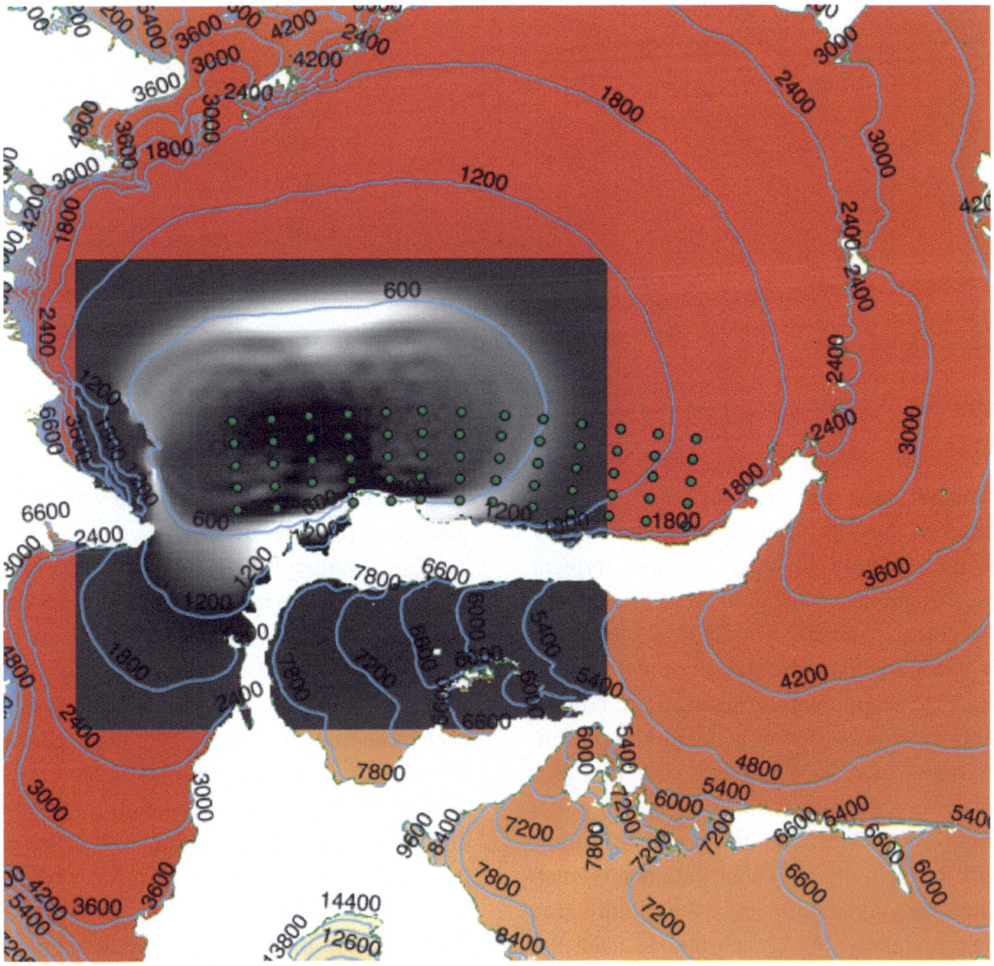

Figure 3
Example of some of the data products derived from the model output. The figure contains arrival times as gridded background field as well as isochrones and a snapshot of sea surface height after 10 min in monochrome

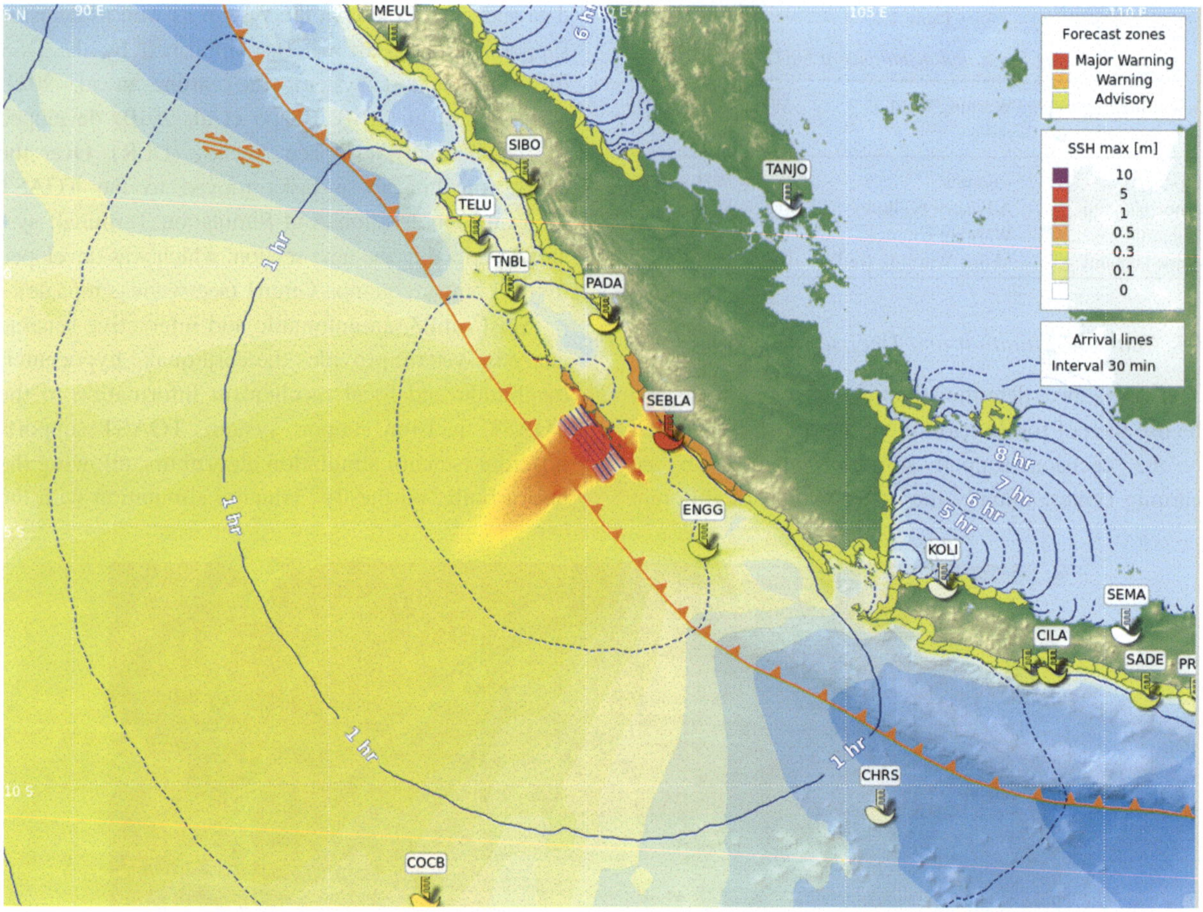

Figure 4
TOAST snapshot displaying source area, SSHMax and isochrones of a database scenario close to the M_w 7.8 Mentawai earthquake on October 25, 2010

easyWave model, and the pre-calculated tsunami database generated by the TsunAWI model. While the first is used mainly for regional and atypical earthquakes, the detailed TsunAWI scenario database is used for local tsunami simulations. Tsunami simulations can be triggered either automatically or manually. In case of automatic triggering pre-configured fault information is used to generate the rupture area or to define the look-up area for the pre-calculated database. In case of manual triggering, the rupture and look-up area can be generated semi-automatically by defining the main parameters such as strike, dip, rake and magnitude, or the rupture area can be drawn manually. With each update from the seismic analysis new simulations are automatically triggered in case of relevant changes in the earthquake parameters. Maps of maximum values of sea surface height (SSHMax) and arrival times are generated (Fig. 4).

The simulations are evaluated against the actual earthquake parameters in terms of hypocenter, magnitude, derived focal mechanism from the seismic system, pre-defined rupture mechanism and tsunami time onsets. An overall match score is generated for each scenario. For the selection an ellipse in strike direction is constructed around the epicenter. The size depends on the magnitude and all scenarios inside the ellipse are considered. The match score depends on database values and observations of hypocenter, magnitude, focal mechanism and tsunami onset. The operator then decides which scenario matches the event best and this is published in the warning product. Tsunami onsets have to be picked manually

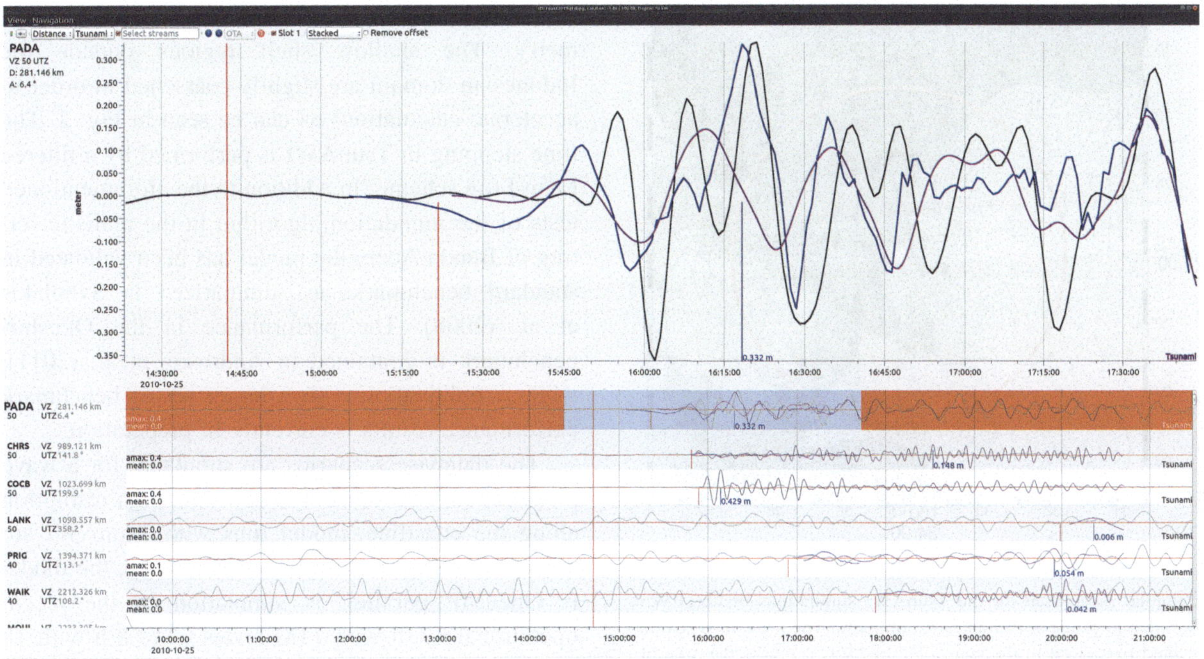

Figure 5

TOAST screenshot with mareograms of the tsunami generated by the 25 October 2010 M_w 7.8 Mentawai earthquake. The tide gauge record (black line) is shown together with the TsunAWI result (blue line) and easyWave solution (magenta line)

on the acquired data from tide gauge or buoy data (Fig. 5).

2.4. *Modeling Approach for the Database: TsunAWI*

As discussed, the database scenarios form the basis of the first warning messages for local events due to the short tsunami travel times. These bulletins contain warning products like estimated time of arrival (ETA) and estimated wave height (EWH) along all the potentially affected parts of the Indonesian coastline. The numerical model used for these calculations is the finite element model TsunAWI (see Rakowsky et al. 2013). The model solves the nonlinear shallow water equations with bottom friction and viscosity terms included. The spatial discretization is realized by triangular meshes, providing high flexibility for realistic representation of coastlines and bathymetric features. Additionally, the model contains a run-up scheme in all of the domain providing realistic wave reflections from the entire coast. An example is shown in Fig. 6 displaying both the ocean area as well as the land nodes covered by the mesh. The bathymetry values interpolated to the mesh nodes are obtained from the General Bathymetric Chart of the Oceans (GEBCO_08 Grid, version 20090202, http://www.gebco.net) dataset augmented in the DMInnovation project (Geoscience Australia led, Australia Aid funded) by the commercial 90 m resolution data set provided by TCarta and several local data sets from ship cruises or local bathymetric charts. On land the data obtained in the Shuttle Radar Topography mission (SRTM, https://www2.jpl.nasa.gov/srtm/) in different resolutions were employed, partly processed by DLR and provided in the GITEWS project (see also Strobl et al. 2007).

The mesh resolution ranges from 20 km in the deep ocean to 300 m along the coast. Initially, three priority areas were resolved at higher resolution (50 m), namely Kuta (Bali), Cilacap (Java) and Padang (Sumatra). These areas were chosen as representatives of specific locations: Kuta represents an area of high importance for tourism, Cilacap is located in a region prone to slow or tsunamigenic earthquakes, and Padang is a location endangered by

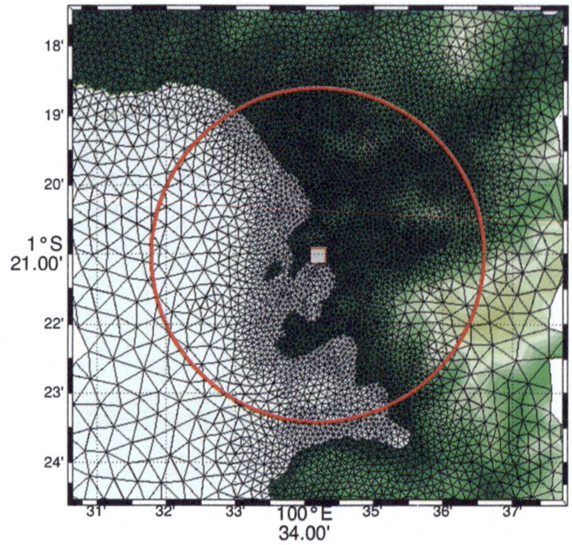

Figure 6
Triangular mesh used for the TsunAWI simulations in InaTEWS near the tide gauge at Painan, West Sumatra. The mesh also covers the land part close to the coast. The red circle marks the area of high resolution (50 m mesh size in a radius of about 4 km) as it is applied in all tide gauge locations

strong earthquakes and particularly low topography making this location especially vulnerable to tsunami flooding. Additionally, the Banda Aceh region has been represented with high resolution in all meshes since this area is used as a validation region for the run-up scheme. The inundation in this area has been investigated in a number of field surveys after the 2004 event (Kongko et al. 2006; Jaffe et al. 2006), and one of the tests for new meshes is the comparison of flow depth values obtained by the model to the measured ones (see also Harig et al. 2008). In the most recent operational mesh, pre-defined tide gauge locations are represented with a higher resolution. By doing so the geographic vicinity of tide gauges is represented realistically and comparisons to data are simplified since computational nodes are close to the actual measurement location. Tide gauge comparisons are exemplified in Fig. 5. Generally, the mesh resolution is based on a combined criterion involving both bathymetry and its gradient:

$$\Delta x \simeq min\left\{c_1\sqrt{gH}, c_2\frac{H}{\nabla H}\right\}$$

where H denotes total water depth, g denotes the acceleration due to gravity and c_1, c_2 are constants depending on the desired time step and local bathymetry. The shallow shelf regions outside the Indonesian domain are slightly coarsened in order to accelerate calculations as can be seen in Fig. 2. The time stepping in TsunAWI is performed by a filtered Leap-Frog scheme. In addition to the aforementioned tests of the inundation algorithm in the realistic setting of Banda Aceh, the model has been validated in standard benchmarks as summarized in Synolakis et al. (2008). The performance in the Okushiri benchmark is contained in Androsov et al. (2011), while a publication with a larger set of benchmark performance results is currently in preparation.

The database scenarios are simulated for a wave propagation time of 12 h. Due to the high resolution along the coastline, model runs with TsunAWI are time consuming. For improved efficiency, the model is OpenMP parallel. A simulation in the recent operational mesh with 11M nodes takes 8 h with 18 threads on an Intel Xeon 18-core CPU E5-2697v4.

2.5. Generating the Warning Products

The TsunAWI model output is provided at mesh resolution, consuming a large amount of disk space, and therefore needs to be post-processed before use in the warning system. Products in several formats are generated which allow for fast retrieval and display in a warning situation. There are gridded products for energy propagation and snapshots of the sea surface height, text files for points of interest (POIs) and sea surface height (SSH) time series in tide gauge locations, as well as json and geojson files for metadata and isochrones. These files are processed by the TOAST system to provide warning information at the warning zone level, for example ETA and EWH information at the POIs. Fig 7 highlights the distinction between forecast points (FPs) used for projections to the coast by easyWave and POIs located directly at the coast which are used for direct calculation of warning products with TsunAWI. The actual warning levels in the warning zones are determined by statistical aggregations of the corresponding values in POIs or forecast points. The estimated arrival time is assigned using the shortest arrival time calculated for coastal points within a warning zone, whereas for EWH alternative metrics

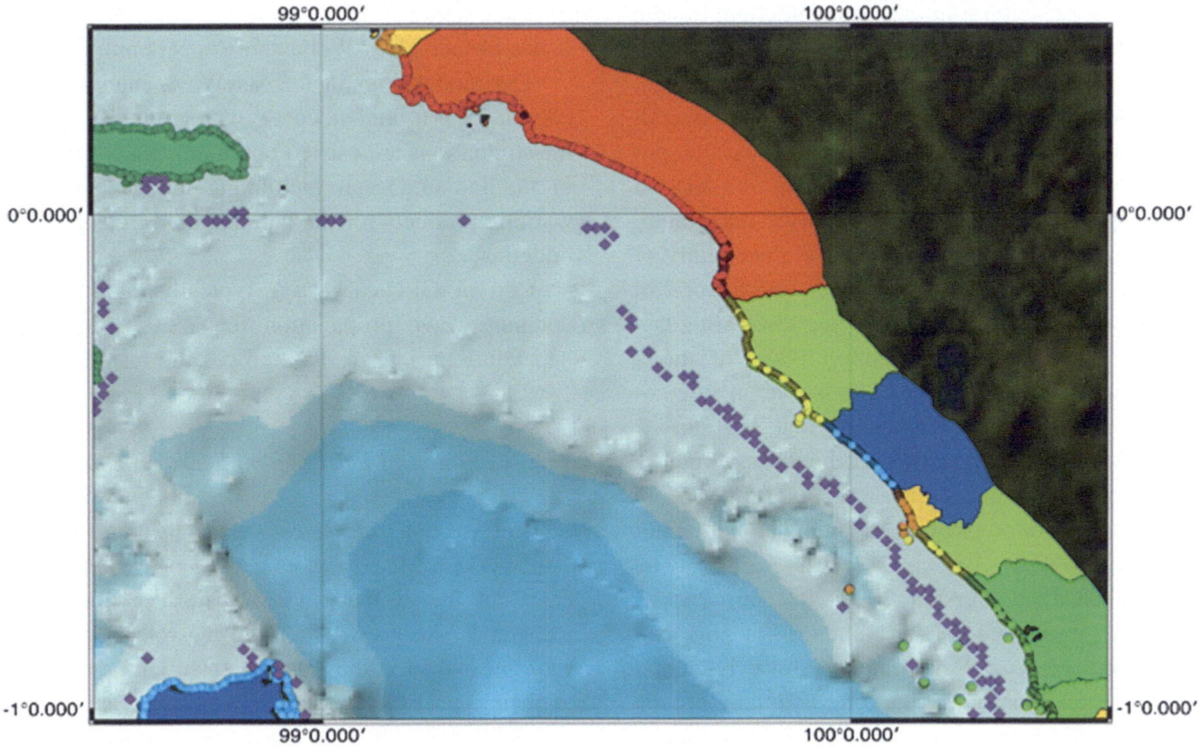

Figure 7
Warning zones (WZs), forecast points (FPs shown in magenta) and points of interest (POIs displayed in WZ color they are assigned to). The model determines warning products like ETA and EWH at these locations. While some of these points are computational nodes of the model, for those that are not, interpolation is performed to calculate values of the relevant quantities at the point

such as the median, or other percentiles, are used to ensure the warning products are not influenced by local numerical instabilities.

2.6. *TsunAWI Database Interface (TsunDaBI)*

Fast assessment of the situation is crucial in an earthquake event. This requires seamless and instantaneous access to the tsunami database. The TsunAWI Database Interface (TsunDaBI) is a web application that was programmed specifically to provide an interface to the set of data products generated from a database of the tsunami simulations described above. TsunDaBI also provides a programmatically accessible interface to the available set of database scenarios from which requesting components can select the best matching tsunami simulations for a given earthquake event. The system also provides a web page via which users can manually select events from the database. Functions for selecting simulations within an ellipse or polygon are available. Integrated functions to administer the set of tsunami simulations within the database facilitate the update of the system.

2.7. *Modeling Approach for Near Real-Time Calculations: easyWave*

As an alternative to the pre-computed scenarios, TOAST offers the possibility to use on-the-fly tsunami simulations. This approach is useful whenever actual earthquake parameters do not match any of the pre-computed scenarios, e.g., by location or by fault mechanism. A GPU-version of the easyWave propagation code is used for such simulations. easyWave follows the classical Leap-Frog staggered-grid finite difference algorithm by Goto et al. (1997) for simulation of linear long-wave propagation in spherical coordinates. The code was designed for rapid (within a few seconds) assessment of the tsunami hazard impact corresponding to fast source models based on preliminary (and continuously updated) SeisComP3 solutions. It is important to note that during the first minutes after an earthquake, source uncertainties (magnitude, focal mechanism, point-source representation) dominate the overall uncertainty of the forecast. Highly accurate computationally extensive propagation and inundation modeling will not balance the large uncertainty within the source itself. For example, uncertainty of only 0.2 magnitude units corresponds to a factor 2 in seismic moment which after applying the scaling law by e.g. Wells and Coppersmith (1994) for thrust faults, leads to about 25% variation in the co-seismic slip and, hence, of the initial sea surface uplift. With easyWave a balance between computational speed and overall accuracy is achieved. In TOAST, tsunami propagation is computed by default on the GEBCO_08 bathymetric grid (The GEBCO_08 Grid, version 20090202, http://www.gebco.net) resampled to 2 arc minutes but grids can be customized. easyWave employs fully reflecting boundary conditions along the coast and therefore the code does not compute inundation and runup in an explicit way. Instead, it employs an indirect metric to assess the coastal tsunami impact. In particular, the code implements the approach adopted by the Japanese Tsunami Early Warning System and described by Kamigaichi (2009). In this approach, the Japan Meteorological Agency suggests to record the wave height at some offshore position, typically at 100-150 m depth, and then project this value to the coastline using the Green's Law (compare Fig. 7). Such projections better mimic the wave shoaling process than direct computation on a coarse grid.

Due to the reduced approximation of the tsunami physics (linear long-wave), easyWave simulates neither nonlinear coastal wave transformation, nor inundation which makes benchmarking in line with Synolakis et al. (2008) impossible. Instead, comparison of easyWave against other propagation codes can be found in Greenslade et al. (2014) and Burbidge et al. (2015).

To initiate easyWave, TOAST dynamically creates an earthquake rupture model based on the observed magnitude and available data. Depth and orientation of the fault plane (strike and dip angles) can be either retrieved from the CMT-solution, or from pre-defined fault lines, whereas fault dimension is determined from Wells and Coppersmith (1994).

The model domain for easyWave calculations is chosen from a predefined set of regions. By default, the simulations are carried out in an area covering all of the Indian Ocean facilitating the generation of warning products for the national as well as the TSP bulletins.

Current GPU-easyWave typically takes 90 s to compute wave propagation for a 20-h simulation depending on intermediate processing results.

3. *Evolution of the Tsunami Scenario Database*

This section addresses the simulation database available at BMKG and its evolution over time. The extent of the database has grown since first implementations (Sects. 3.1, 3.2) and the role that BMKG staff have taken in the database extension has changed as their capacity increased from users of the DSS to active involvement in the modeling process and management of the database into the future (Sect. 3.4).

3.1. *Initial Stages During GITEWS and PROTECTS*

The initial model domain for TsunAWI covered only the part of the Indian Ocean facing the Sunda Arc and stretching about 1000 km into the ocean as shown by the darker part in Fig. 2. In the early stages of the GITEWS project, this approach was motivated by the recent major events in that region and by the limitations with respect to computational resources rendering a coverage of the whole Indonesian domain impossible within the project duration.

The source models for the database were provided by GFZ Potsdam, and the software RuptureGenerator was employed to generate initial conditions for the TsunAWI simulations (for more details see Babeyko et al. 2010).

During the sustainability phase PROTECTS (PROject for Training, Education and Consulting for Tsunami Early Warning Systems, 2011–2014), an extension of the database and model domain was implemented, for the first-time providing warning products for the complete Indonesian domain.

3.2. Recent Database Extension

The restriction of tsunami sources to the Sunda Arc posed a major limitation given the large number of earthquakes in the eastern part of Indonesia. Although no events are known to have occurred here with the same catastrophic death toll as the 2004 Indian Ocean Tsunami in Banda Aceh, historical records show twice as many fatality-causing tsunamis have occurred in eastern Indonesia compared with the Sunda Arc in the past (Latief et al. 2000). The region is tectonically complex, with numerous active faults capable of generating locally and/or regionally significant tsunamis. The role of other source zones was investigated in a national probabilistic tsunami hazard study for Indonesia (Horspool et al. 2014), the results of which emphasized the importance of including sources in eastern Indonesia in the database. This limitation was overcome by the Australian Government funded collaboration between DMInnovation (Geoscience Australia led Australian Aid project), AWI and BMKG between May 2015 and March 2017. In several workshops, the source zones used by Horspool et al. (2014) were incorporated into the InaTEWS database. Additionally, the magnitude coverage was extended such that now the scenarios in the database cover magnitudes from 7.0 to 9.0 in steps of 0.2. In the Sunda Arc the seismic scenarios follow the GFZ plate discretization model with subfaults ranging by length (along trench) 20–55 km and by width (across trench) 5–30 km depending on the megathrust curvature. The down-dip limit after an update in 2010 is 100 km. In all other trenches, the 2010 maximum seismogenic depths are taken from Horspool et al. (2014; see their Table 1), with depths of 50 km for subduction megathrusts and 18–30 km for non-subduction faults. The population of the database is chosen such that it contains a scenario centered in every second subfault in both directions, which, depending on the local curvature of the megathrust, corresponds in average to 75 km along and 21 km across the trench.

Table 2 summarizes the updated database of scenarios for each considered fault source. All simulations except for those in the Sunda Arc were calculated jointly in the recent project, with a large number of the model runs carried out by BMKG staff on HPC systems provided by DMInnovation. Efficiencies were gained where in small trenches neighbouring scenarios of large magnitudes may result in identical initial conditions and thus not all simulations needed to be carried out. The identical runs are represented as referencing links in the database. Therefore, Table 2 shows both the calculated scenarios and the total number of scenarios

Table 2

Current status of tsunami scenarios in the InaTEWS database. More information on the source zones and source generation for the scenarios in all faults except for the Sunda Arc is contained in Horspool et al. (2014). The source modeling for the Sunda Arc is described in Babeyko et al. (2010)

Fault zone	Source type	Sense	Calculated scenarios	Total number (calc. and redundant)
East Mollucca	Crustal	Thrust	247	1350
Flores	Crustal	Thrust	173	1314
Makassar	Crustal	Thrust	105	800
Manokwari	Crustal	Thrust	38	240
North Papua	Subduction	Thrust	674	1896
North Sulawesi	Subduction	Thrust	164	535
Palu Koro	Crustal	Strike-slip	28	180
Seram	Subduction	Thrust	384	1392
South Seram	Crustal	Normal	86	576
Sulu	Crustal	Thrust	67	420
Timor	Crustal	Thrust	174	720
Tolo	Crustal	Thrust	49	336
West Mollucca	Crustal	Thrust	126	768
Wetar	Crustal	Thrust	226	1602
Sunda Zone	Subduction	Thrust	6117	6117
Sum			8658	18,246

available including the referencing links. The geographic location of all scenario epicenters is shown in Fig. 2.

3.3. Limitations of a Scenario Database

Even with a significantly increased number of simulations any scenario database is limited to a finite set of earthquake rupture scenarios. For well-established subduction zones, such as the Sunda Arc, future megathrust earthquakes are expected to be constrained within reasonable bounds of our current knowledge of the interface location, geometry and style of faulting. Even so, other earthquakes not within the scenario database, such as outer-rise earthquakes (e.g. Gusman et al. 2009), may occur. In Eastern Indonesia, our understanding of the major fault systems is still evolving and not well described by current rigid tectonic plate models (Hall 2018). As new geophysical surveys are undertaken, previously unknown faults are identified (e.g. Pownall et al. 2016) with implications for our understanding of the main tsunami sources of the region (Griffin and Davies 2018). Furthermore, known structures may behave in unexpected ways, as most recently demonstrated by the 28 September 2018 M_w 7.5 Palu earthquake and tsunami. This earthquake ruptured the predominantly strike-slip Palu-Koro Fault, yet generated a more significant local tsunami than had been previously expected from strike-slip faulting, with earthquake-triggered landslides likely playing a role in tsunami genesis. It should be again emphasized that the scenario database as currently implemented does not consider volcano or landslide generated tsunami. In consequence it is essential that besides the database the early warning system is equipped with the on-the-fly model component and at any time simulations with refined fault mechanisms can be carried out if necessary.

3.4. Capacity Building—From Trainings to Workshops

Over the course of the years after the inauguration of InaTEWS BMKG staff were trained in a variety of different topics by experts from all over the world, qualifying staff to autonomously run and finally extend the tsunami early warning system. In the following we will focus on the knowledge transfer regarding tsunami modeling and maintaining the tsunami simulation database.

During the PROTECTS project a total of 12 training workshops were held on 8 exchange trips. One series of training workshops focused on installing, maintaining and troubleshooting the scenario database interface, and in another set of workshops BMKG staff were taught the basics of how to run the TsunAWI model and analyze model output.

In the subsequent DMInnovation project mentioned in the previous section the format of knowledge transfer evolved considerably. As described above the scenario database was extended to eastern Indonesia in a joint effort with the Indonesian staff (see Fig. 8). One of the main goals of the project was to enable BMKG to independently extend the database in the future. A total of eight interactive workshops in Jakarta were designed such that participants could continue working on parts of the workflow after being empowered to do so, therefore extending the database over the course of the project. The topics of the eight workshops built on each other and followed the logical organization of the workflow from source generation to the import of data products into TsunDaBI, starting however with a thorough comparison of available decision support and modeling approaches. A major part of the workshops was calculating, understanding and evaluating tsunami simulation data. A short description of the workshops is given in the following.

A methodology was developed with BMKG staff to compare the usability of the decision support approaches DSS (Decision Support System by DLR) and TOAST. Questionnaires were designed in an iterative process to evaluate advantages of both systems. Alongside modeling with TsunAWI participants also learned how to run easyWave. A comparison of the two modeling approaches was performed by evaluating the model results in tide gauge mareograms as well as EWH and ETA in according points of interest along the coast. A tool was developed in the workshop to compare corresponding warning products in TOAST. Source modeling was conducted with tools provided by DMInnovation. Modeling of TsunAWI scenarios was

Figure 8
Workshop members, BMKG staff and representatives of AWI and Geoscience Australia in the headquarters of BMKG, Jakarta 2016

performed on high performance clusters after an introduction to underlying concepts and important commands had been given. A set of post-processing tools was then applied to generate warning products for TOAST. Finally, the import of data products into TOAST was conducted via the scenario database interface TsunDaBI. In addition, sensitivity studies regarding magnitude and epicenter density were conducted.

Besides calculating numerous simulations and therefore contributing substantially to the extension of the scenario database, towards the end of the project BMKG staff also took over the role of teaching their colleagues and jointly discussing topics in Bahasa Indonesia. This represents a remarkable milestone in knowledge transfer since learning new concepts is considerably easier in the first language. Furthermore, this ensures that learnt and applied concepts will be passed on and pursued after the end of the project.

Valuable insights were gained from the experience with the format of the workshops. The interactive design promoted discussions strengthening and deepening the understanding of tsunami modeling in the context of an early warning system on all sides. Regular exchange and repetitive workshops moreover foster a profound understanding of complex topics. Also, the trainers benefited from the regular feedback allowing for the adaptation of workshop structure and contents to improve knowledge transfer. Different technical and cultural backgrounds pose both challenges and opportunities. Challenges lie especially in agreeing on a common vocabulary. Having participants from various technical backgrounds with different paradigms can lead to some initial misunderstandings, which however were reduced considerably during the course of the project (see also Immerz et al. 2018). In return the understanding of the differing paradigms is the source of getting a better grasp of the interlinkage of the different components of the tsunami early warning system. Inarguably, knowledge transfer is crucial for sustainable deployment of the system and tsunami science benefits from the feedback of its main stakeholders.

Most recently in a workshop in August 2018, the possibility of including the inundation estimates into the warning system and the set-up of specific tsunami simulations for high resolution simulations in selected areas was investigated. This approach will be developed further in the future.

4. System Performance and Application

The system has been in operation since 2008 and warnings were issued for a number of tsunamigenic earthquake events based on both database simulations and on-the-fly calculations. The analysis of the system performance for real tsunamigenic events is important to determine weaknesses and infer areas of possible system improvements. In addition, since there are two model components present in the system an investigation of warning product consistency is crucial. The following section focuses on the warnings issued during the operational period and documents the lessons learnt from these events, which in a number of cases informed ongoing updates to the system. This general presentation is augmented by a section dealing more specifically with details of the two model components and their predictions, before the view is widened for a more global view in the concluding subsection.

4.1. Earthquake and Tsunami Events

Table 3 lists tsunami events in the Indonesian region since 2004. For each event information is given regarding the coverage by the database at the time of the event as well as the present state. It also shows if a warning was issued by InaTEWS and remarks on the events.

During the operational period of InaTEWS tsunami warnings were disseminated for most events, however the estimated wave height did not always match the observed values. A large part of this can be attributed to large uncertainties in the earthquake

Table 3

List of tsunami events in the Indonesian region since December 2004

Event date	Event	M_w	Covered by DB? (at that time/now)	Warning issued?	Remarks
2007-09-12	Bengkulu	8.4	NA/yes	Yes	The first event that a warning was issued for. The system components were already installed at BMKG although InaTEWS was not yet officially launched
2009-08-11	Andaman	7.7	No/yes	No	The event epicenter location was too far to the North for NTWC warning
2009-09-02	Java	7.3	Yes/yes	No	Intraslab earthquake, the Sunda arc sources in the database are thrust type
2009-09-30	Padang	7.6	No/yes	No	Intraslab earthquake, in 2009 the database didn't stretch that far inland
2010-04-06	Sinabang	7.8	Yes/yes	Yes	Not covered by the database. Warning based on easyWave calculations
2010-10-25	Mentawai	7.8	Yes/yes	Yes	Slow earthquake. Underestimation of EWH on Mentawai islands
2012-04-11	North Sumatra	8.6	No/no	Yes	Strike slip event. EWH overestimation due to thrust assumption in easyWave calc
2014-11-15	Sangihe	7.1	No/no	Yes	Warning based on easyWave calculations
2016-03-02	SW Sumatra	7.8	No/no	Yes	Strike slip event. Overestimation due to the thrust assumption in easyWave calc
2017-12-15	Java	7.3	Yes/yes	Yes	
2018-08-05	Lombok	7.0	No/yes	Yes	First time use of updated database.
2018-09-28	Palu	7.5	Yes/yes	Yes	Strike slip event. Seismic sources contained in database but landslides are not covered
2018-12-22	Anak Krakatoa	Volcanic	No/no	No	There are earthquake sources in the Krakatau region, however volcanic sources are not covered
2019-04-12	Sulawesi	6.9	No/no	Yes	Warning based on easyWave calculations

source parameters that allow selection of larger magnitude scenarios in the epicentral region. Other important reasons for mismatches are the approximate earthquake sources used for the pre-calculations and deficiencies in bathymetry data available in the model domain. Field surveys were conducted following several of the events producing detailed observations of the actual flow depth and runup caused by the tsunamis. The aforementioned data deficiencies are even more important for the inundation estimates calculated by TsunAWI, with the sensitivity of inundation estimates to elevation data uncertainties shown to be large by Griffin et al. (2015). Tide gauge records and field surveys are an excellent source of data for system evaluation and the observed mismatches in some of the events resulted in system improvements.

4.2. Earthquake and Tsunami Events—Details of the TOAST Predictions and System Updates

In the following the brief remarks from Table 3 are extended to give some more details on the actual system performance and lessons learnt from some selected events are summarized in the following. Up to September 2018 the decision support and warning products were provided by the Decision Support System (DSS) installed during the GITEWS project (Steinmetz et al. 2010). Since September 2018 TOAST has been operational.

M_w 7.6 Padang 2009 This event, accompanied by a strong ground shaking in the city of Padang, was an intraslab earthquake occurring at 90 km depth, rather than a typical tsunamigenic shallow megathrust earthquake. Strong shaking triggered panic and fear of a possible tsunami among the city population. BMKG issued a 'no tsunami potential' bulletin as the epicenter was located below the 'tsunamigenic depth threshold' defined in SOPs. That was the right decision, only minor sea level increase ($<$ 30 cm) was later registered at the tide gauge in Padang.

System update: In the 2010 update of the TsunAWI scenario database (see also next case) down-dip depth limit of virtual megathrust events was extended from 60 to 100 km.

M_w 7.8 Mentawai 2010—Slow 'Tsunami' Earthquake In this case, a tsunami warning was issued, however, modeled wave heights on the Mentawai Islands were strongly underestimated. Hill et al. (2012) and Satake et al. (2013) published tsunami height values of more than 10 m in some locations of Pagai and surrounding islands. A reason for the underestimation was two-fold: (1) the fast body-wave magnitude Mw(mB) (Hanka et al. 2010) underestimated the true geodetic magnitude (7.1 vs. 7.8) which is typical for slow 'tsunami' earthquakes characterized by deficit of high-frequency seismic energy radiation (Kanamori 1972) and (2) tsunami source modeling assumed constant rigidity when doing recalculation of the magnitude value into the coseismic slip. The combined effect of the two factors led to a significant underestimation of the modeled wave heights along the Pagai islands. However, it is important to note, that despite the wave height underestimation, tsunami warning for the Pagai islands was issued by BMKG in time. System update: the RuptureGenerator (a tool which generates initial conditions for TsunAWI) was updated in order to account for 'tsunami' earthquakes. In particular, a variable rigidity profile (after Geist and Bilek 2001) was adopted to simulate higher slip for a given seismic moment near the trench. Additionally, empirical mB magnitude correction was introduced to account for the possible fast magnitude underestimation in case of slow 'tsunami' events. (Note, currently InaTEWS operates with moment tensor). Also, the mesh for the recalculation of the database was refined in the Mentawai region. The improvements are shown in the next section in Table 4 and in Figs. 4 and 5 (Mentawai) and Fig. 9 (Pangandaran).

M_w 8.6 North Sumatra 2012 and M_w 7.8 SW Sumatra—Strike Slip Events These two events were both large strike-slip earthquakes occurring within oceanic crust well offshore from the subduction zone. The epicenters of these events are not covered by the database. The initial warnings issued for these events overestimated the wave heights as the on-the-fly calculation initiated after the event used a thrust fault mechanism as worst case assumption. This initial condition together with the reported magnitude

Table 4

Warning products obtained by easyWave and TsunAWI for tsunami events. Tide gauge observations are included if available. Estimated wave heights are compared in the warning zone with the largest values for both models. For a visual overview three events are shown in Figs. 9, 10, 11

Event	Tide gauge observation data and model estimate from mareograms			Details of model estimates in warning zone with largest EWH values (REGION and place or location are given)	
	Data (m)	easyWave (m)	TsunAWI (m)	easyWave Region (WZ index) EWH in WZ	TsunAWI Region (WZ index) Median of EWH in WZ Max. EWH in POIs of WZ
Pangandaran, 17 July 2006 M_w 7.7	Tide gauge Cocos Island 0.05	0.15	0.29	JABAR (Garut) 2.05 m	JABAR (Tasikmalaya) 4.29 m (median) 8.03 m (max) (scen. with M_w 7.8)
Bengkulu, 12 Sept. 2007 M_w 8.4	Tide gauge Padang 1.2	1.4	0.77	BENGKULU (South) 7.10 m	BENKULU (North) 5.27 m (median) 8.05 m (max)
Padang, 30 Sept. 2009 M_w 7.6	Tide gauge Padang 0.2	0.48	0.2	SUMBAR (Padang-Pariaman) 0.47 m	SUMBAR (Padang-Pariaman) 0.39 m (median) 0.42 m (max)
Sinabang, 6 April 2010 M_w 7.8	Tide gauge Sibolga 0.15	0.55	0.04	NAD (South) 1.40 m	NAD (North) 0.80 m (median) 1.55 m (max)
Mentawai, 25 Oct. 2010 M_w 7.8	Tide gauge Padang 0.3	0.62	0.33	SUMBAR (Pagai) 1.55 m	SUMBAR (Pagai) 1.90 m (median) 7.87 m (max)
North. Sumatra, 11 April 2012 M_w 8.6	Tide gauge Meulaboh 1.1	0.86	NA	NAD (Aceh West) 0.87 m	These events are not covered by the database
Sangihe, 15 Nov. 2014 M_w 7.1	Tide gauge Jailolo 0.09	0.19 (WZ Malut,Halmahera)	NA	SULUT (Minahasa South part) 0.22 m	
SW Sumatra, 2 March 2016 M_w 7.8	Tide gauge Cocos Island 0.05	0.32	NA	SUMUT (Nias West) 0.65 m	
Lombok, 5. Aug. 2018 M_w 6.9	Tide gauge Carik 0.14	0.03	0.14	BALI (Denpasar Pantai Sanur) 0.05 m	NTB (Lombok-Timur North) 0.19 m (median) 0.53 m (max)

results in large wave heights. More realistic wave heights were simulated later once the focal mechanism became available. The difference becomes apparent in the tide gauge position in Meulaboh for North Sumatra 2012. The tide gauge record shows 1.1 m, simulation with thrust assumption result in 5.69 m whereas the strike slip case gives more realistic 0.86 m.

System update: the fault mechanism for easyWave simulations can be chosen by the officer on duty. With epicenter location far offshore (like in case of the M_w 8.6 Wharton Basin event), there will be enough time to wait for the moment tensor solution. Additionally, for this location, a new strike slip fault zone was defined with adjusted strike value as seen in the TOAST screenshot in Fig. 4. The different warning situations are shown in Fig. 11.

M_w 7.5, Sulawesi, Sep. 2018 A devastating local tsunami occurred in Palu Bay after the M_w 7.5 earthquake on 28 September 2018, which had a predominantly strike slip fault mechanism. Eyewitness accounts show the tsunami arriving within a few minutes of the earthquake. A tsunami warning was issued but local wave heights in Palu Bay were underestimated. It appears likely that the tsunami was triggered by a combination of the seismic source and multiple earthquake-triggered landslides observed

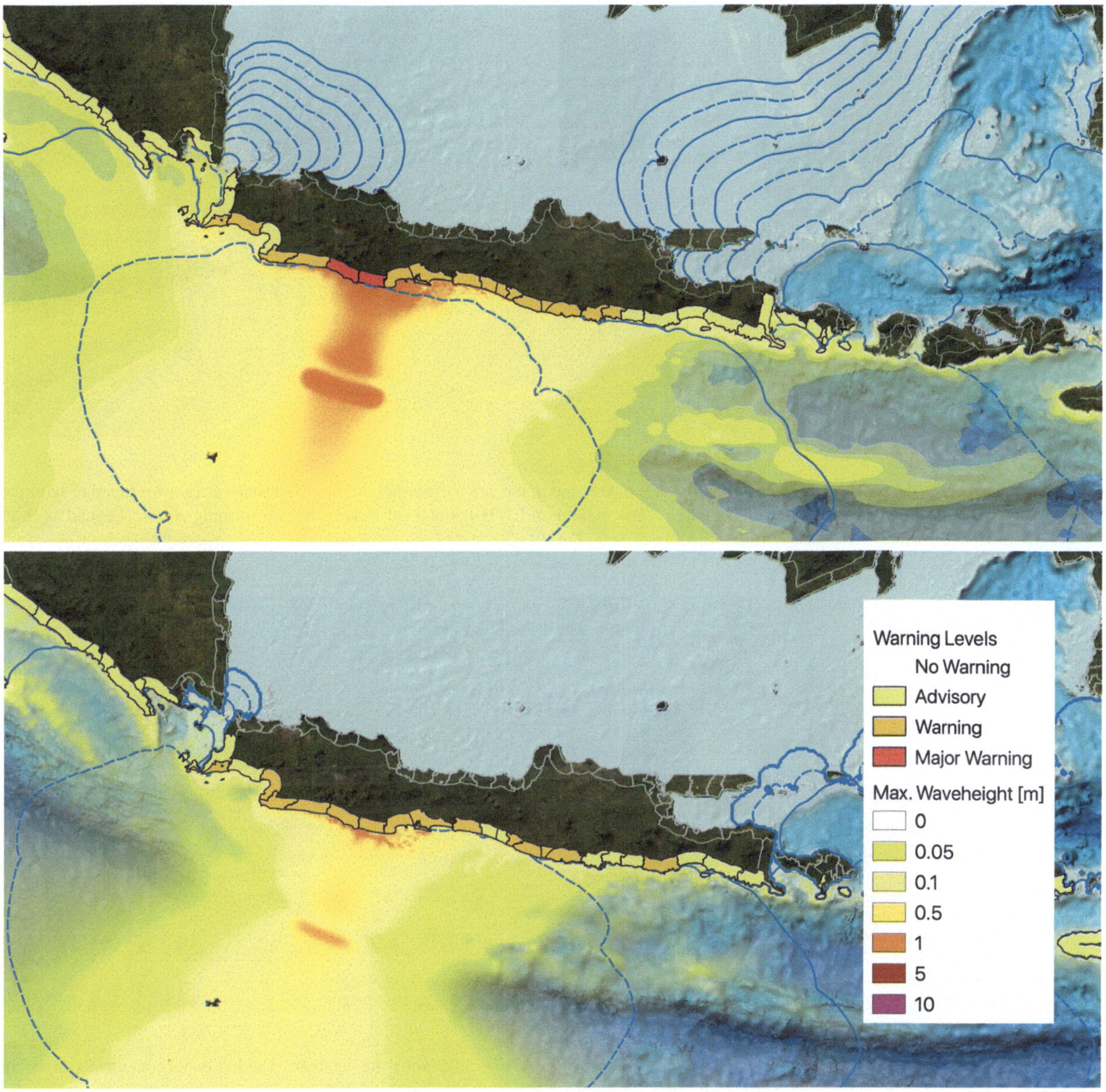

Figure 9
Warning levels for the July 2006 Pangandaran event. Displayed are maximum wave height and warning levels for easyWave (lower panel) and TsunAWI (upper panel). The warning zones with 'Major Warning' for TsunAWI reflect the update in the source modeling which accounts for the low rigidity near the trench. The maximum EWH is obtained in neighboring WZs since the source is slightly shifted (DB sources along the trench are about 75 km apart). Isochrones of ETA are indicated by blue lines (solid lines: full hours, broken lines: half hours)

during the event, however at the time of writing this paper research into the relative contributions of these sources is still ongoing. The pre-calculated wave heights in Palu Bay based on the strike slip source model present in the database correspond well to the published results based on the USGS source model (Heidarzadeh et al. 2018).

Krakatau. Dec. 2018 No warning was issued since no tsunamigenic seismic activity was observed.

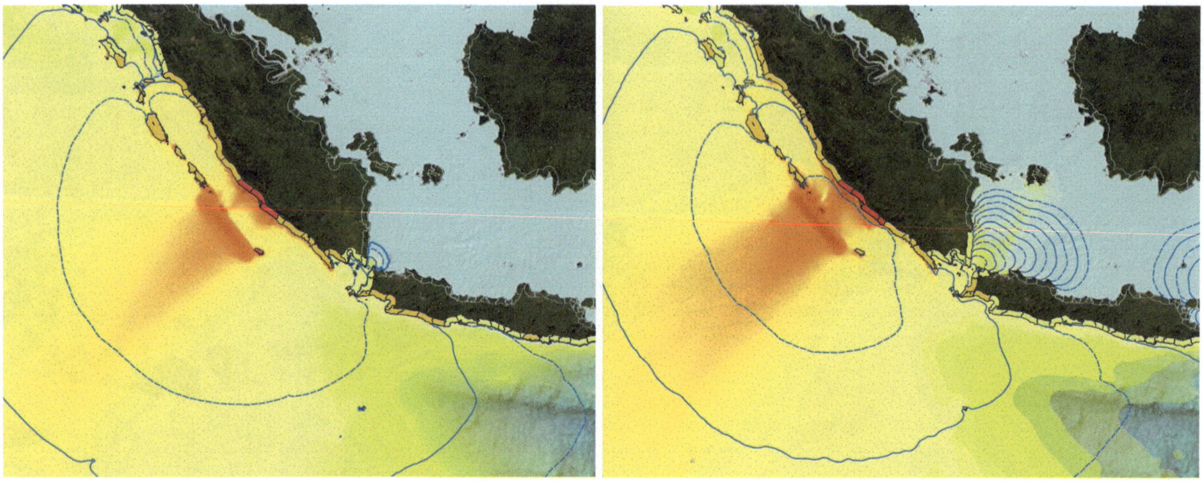

Figure 10
TOAST screenshots for the September 2007 Bengkulu event: The maximum sea surface height and isochrones are shown together with the warning zones for easyWave (left) and TsunAWI (right). The maximum EWH is obtained in neighboring warning zones. (Legend as in in Fig. 9)

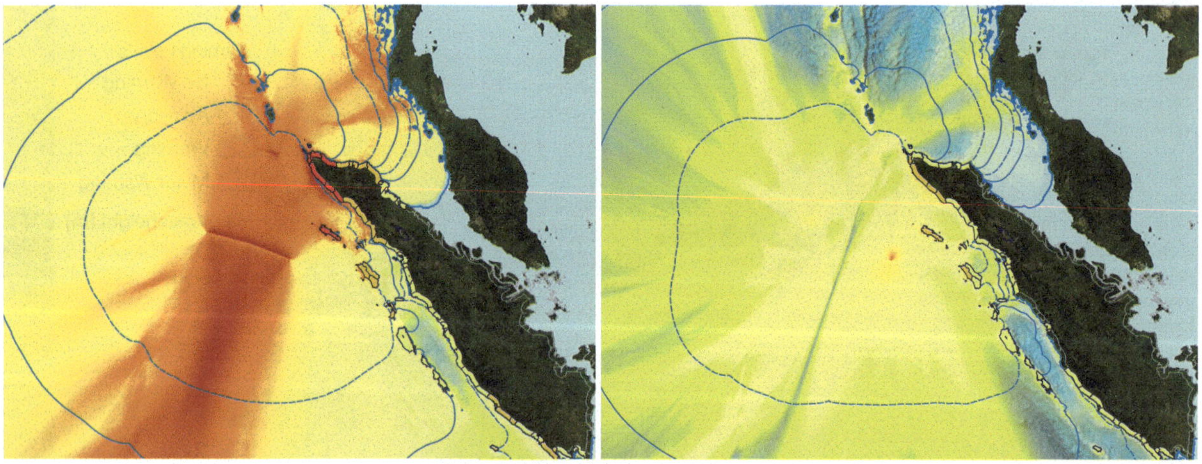

Figure 11
easyWave results for the Northern Sumatra event in April 2012. The area is not covered by the database. The left panel contains the warning situation with adjusted strike angle but worst case thrust assumption, whereas the right panel displays the warning when the strike slip mechanism is assumed. (Legend as in Fig. 9)

Instead the volcanic activity of Anak Krakatau caused a large landslide which generated the tsunami.

4.3. Recent Applications of the System

The aforementioned system updates resulted in a better performance with respect to observations and thus improved preparedness for future events. Table 4 summarizes details of the model predictions for the recent implementation (since August 2018) of the system. Please note that in the below comparison the source models for TsunAWI and easyWave are not identical! In case of the TsunAWI database, the closest available magnitude larger than or equal to the reported magnitude and the closest location to the epicenter is chosen, whereas easyWave is triggered with the exact magnitude and epicenter and uses a fault geometry derived from the initial moment tensor.

More complete views of the warning situation for some of the events in Table 4 are shown in Figs. 9,

10, 11. These illustrations contain the warning data returned by TOAST. Actual TOAST screenshots with the TsunAWI database simulation results for the Mentawai 2010 event are displayed in Figs. 4 and 5.

Of special interest is the warning level issued in the affected regions defined by the EWH values. For realistic runup simulations good knowledge of the actual source as well as high quality topography and bathymetry data are crucial, therefore good agreement cannot be expected for the database predictions or simulations in rather coarse meshes. Nevertheless, the database simulations with adjusted source models show good performance as shown in Table 4.

The situation in the Pangandaran event from 2006 (Fritz et al. 2007, Lavigne et al. 2007) is similar to the one described above for Mentawai 2010. Also, in the Pangandaran case a slow earthquake of magnitude 7.7 resulted in large wave heights at the coast. Fritz et al. (2007) found tsunami height values of around 8 m in several survey locations around Cilacap, on the southern coast of Java. The warning products for Pangandaran 2006 are shown in Fig. 9, the 'Major Warning' in Jabar region issued by TsunAWI is based on the adjusted source model as well as the larger magnitude of 7.8 available in the database. The easyWave results in the same region are lower since (1) the exact magnitude is assumed and (2) on-the-fly easyWave simulations assume constant rigidity value when converting magnitude into co-seismic slip, i.e., do not account for the possible 'tsunami' earthquakes by implementing depth-dependent elasticity. The maximum EWH value obtained by TsunAWI is 8.03 m, the median value in the warning zone is 5.23 m, still in the 'Major Warning' range. For the Mentawai 2010 event, EWH values produced by TsunAWI are also close to the measured values (see Table 4), however, in this case the alert level for the corresponding warning segment is set to 'Warning', not to 'Major Warning'. This is explained by the fact that the median EWH over the whole island (basis for setting an alert level) is significantly lower than the maximum wave height because of the almost absence of tsunami at the opposite side of the Pagai island. An adjusted aggregation rule for the warning zone value is currently investigated.

In the Bengkulu 2007 event (shown in Fig. 10) the predicted wave heights are slightly overestimated for both models. In the warning zones around Bengkulu EWH values higher than 5 m are predicted, Borrero et al. (2009) found runup values of up to 3.5–5 m in the warning segments north and south of Bengkulu.

Finally, Fig. 11 highlights the difference made by the knowledge of the proper fault mechanism in the case of the strike-slip event offshore Northern Sumatra in 2012. The database doesn't stretch that far, therefore only easyWave results are available in this case. If thrust fault is assumed (initial worst-case assumption) with magnitude M_w 8.6 the level 'Major Warning' is issued for many warning zones. With knowledge of the strike-slip nature of the event, the recalculated warning products are much more appropriate.

The differences in estimated times of arrival represented by isochrones in the figures mostly arise from slightly different locations of the source area. This applies especially to events very close to the coast like the September 2009 Padang event.

4.4. *Consistency of Warning Products*

Finally, the investigation of the previous sections is extended to a more global view of the modeling approaches. The aim is to determine whether the estimates of wave height and arrival time obtained by TsunAWI and easyWave for identical sources are consistent.

Since the numerical setup of both models is quite different, deviations between the results are to be expected. However, in the present overview paper, we do not intend to provide a detailed model comparison for specific scenarios. This has to be presented in a separate study. Instead, we aim to investigate the warning products obtained from the two models and their consistency within the warning system. The mismatches of warning levels in a large subset of the database obtained for identical initial conditions are considered. The focus is put on the estimated wave height (EWH) in warning zones.

For a systematic comparison, a subset of the Sunda Arc database scenarios is investigated and compared to the corresponding easyWave results for identical initial conditions. Scenarios used for this study are shown as green squares in the Sunda Arc in

Table 5

Overview of all database scenarios used for the warning level comparison (green squares in Fig. 2). It includes the total number of scenarios for each magnitude, the number of warning zones reached with more than 1 cm, the discrepancies occurring for 'Advisory-Warning' and 'Warning-Major Warning'. The last two columns include the percentage of WZs with an arrival time deviation of more than 10/20 min, in WZs with an earliest arrival time of 90 min or less (denoted DT_{90})

M_w	Nmb. of scenarios	Mean nmb. of reached warning zones ($\overline{NWZ}$)	$\bar{N}_{Ad-W}(M)$ and global mismatch ratio, $R^{glob}_{Ad-W}(M)x100\%$		$\bar{N}_{W-MW}(M)$ and global mismatch ratio, $R^{glob}_{W-MW}(M)x100\%$		Mean nmb. $DT_{90} > 10$ (%)	Mean nmb. $DT_{90} > 20$ (%)
7.0	497	46	1.0	2.2%	0	0	14.1	3.3
7.2	495	61	2.1	3.4%	0	0	14.5	3.6
7.4	486	75	2.6	3.4%	0	0	14.7	4.1
7.6	454	89	4.4	4.9%	1.1	1.2%	13.7	3.9
7.8	412	97	5.2	5.3%	1.7	1.7%	14.2	4.2
8.0	373	101	6.7	6.7%	2.1	2.1%	14.2	4.2
8.2	326	102	9.0	8.7%	2.9	2.8%	14.5	4.1
8.4	271	103	11.2	10.8%	3.9	3.8%	14.4	3.7
8.6	214	103	10.4	10%	5.6	5.4%	14.1	3.4
8.8	142	103	9.8	9.4%	8.5	8.2%	16.4	4.2
9.0	66	104	8.1	7.8%	12.8	12.3%	20.5	7.8

Fig. 2. The specific number of scenarios for each magnitude are given in Table 5 together with the number of warning segments reached by these scenarios (i.e. EWH in WZ for both models is more than 1 cm). In mean magnitude 7.0 scenarios reach 46 warning zones, whereas for the 66 scenarios of magnitude 9.0 a mean value of 104 WZs is obtained. For the determination of warning products, the standard way for TsunAWI (calculation to the POIs) is used, while for easyWave the most corresponding method (calculation to closest coastal point) is employed.

Deviations in warning levels in affected warning zones are investigated and combined in Table 5. We restrict the presentation to mismatches of the types 'Advisory-Warning' (Ad-W: one model predicts EWH $\leq$ 0.5 m the other one calculates 0.5 m $<$ EWH $\leq$ 3.0 m) and the type 'Warning-Major Warning' (W-MW: one model estimates 0.5 m $<$ EWH $\leq$ 3.0 m the other one predicts EWH $>$ 3.0 m). Warning zones are only considered if the prediction of both models is at least 1 cm. For each magnitude global and local mismatch ratios for'Ad-W' are defined as follows:

$$R^{glob}_{Ad-W}(M) = \frac{\bar{N}_{Ad-W}(M)}{\overline{NWZ}(M)} \tag{1}$$

$$R^{WZ}_{Ad-W}(M) = \frac{\bar{N}_{Ad-W}(M, WZ)}{\bar{N}(M, WZ)} \tag{2}$$

In these formulas $\bar{N}_{Ad-W}(M)$ denotes the mean number of Ad-W mismatches over all scenarios of magnitude M and all warning zones whereas $\bar{N}_{Ad-W}(M, WZ)$ is the corresponding number for the single warning zone WZ. $\overline{NWZ}(M)$ denotes the mean number of warning zones reached globally, whereas $\bar{N}(M, WZ)$ is the mean number of scenarios reaching a given warning zone WZ for magnitude M. Ratios for the W-MW case are defined analogously.

The global values $R^{glob}_{Ad-W}(M)$ are summarized in Table 5 for all available values of M. For low magnitudes there are very small inconsistencies. Compared to the number of warning zones reached by a tsunami wave, the 'Advisory-Warning' mismatch ratio is generally less than 10% and less than 12.3% for 'Warning-Major Warning'. There is a systematic overestimation of easyWave as compared to TsunAWI, however the behavior depends on the algorithm used for EWH estimates in the real time model. Several alternative options are available for easyWave, such as the projection methods described in Sect. 2.7, however, such comparisons are not within the scope of this paper.

With respect to arrival times, we investigate those warning zones with the first ETA value below 90 min and deviations of more than 10 and 20 min. In this case, there is no systematic dependence on the magnitude, only for the largest magnitude the mean deviation is slightly larger. The occurring differences can be attributed to the differences in bathymetry used in the two models. For the sake of execution speed easyWave simulations are based on GEBCO values sampled to 2 arc minutes, whereas TsunAWI operates on a higher resolution mesh with additional bathymetry datasets. Therefore, differences are to be expected, nevertheless the deviations are moderate and could be improved by adjusting the bathymetry values as much as possible.

The local ‘Advisory-Warning’ mismatch ratios for three magnitudes are shown in Fig. 12. The figure indicates that significant differences are mostly occurring for large magnitudes and in locations with complicated bathymetric features like the wide shelves in the area of the Sunda Strait and near the Mentawai islands. Since the source model is identical in this comparison, differences are to be attributed to the differences in model numerics and deviations in bathymetry. The areas with large mismatches in the right panel are affected by these two factors. On one hand side the shallow areas are represented differently and with varying resolution, since the triangular mesh resolution is highest in shallow areas, on the other hand side the physical terms for bottom friction, nonlinear advection and viscosity are most significant in the shallow areas, and these terms are missing in the easyWave simulations. These arguments may help to understand the differences, a thorough understanding needs a detailed investigation which is beyond the scope of this paper.

In the warning case these moderate discrepancies are not expected to create large inconsistencies. In

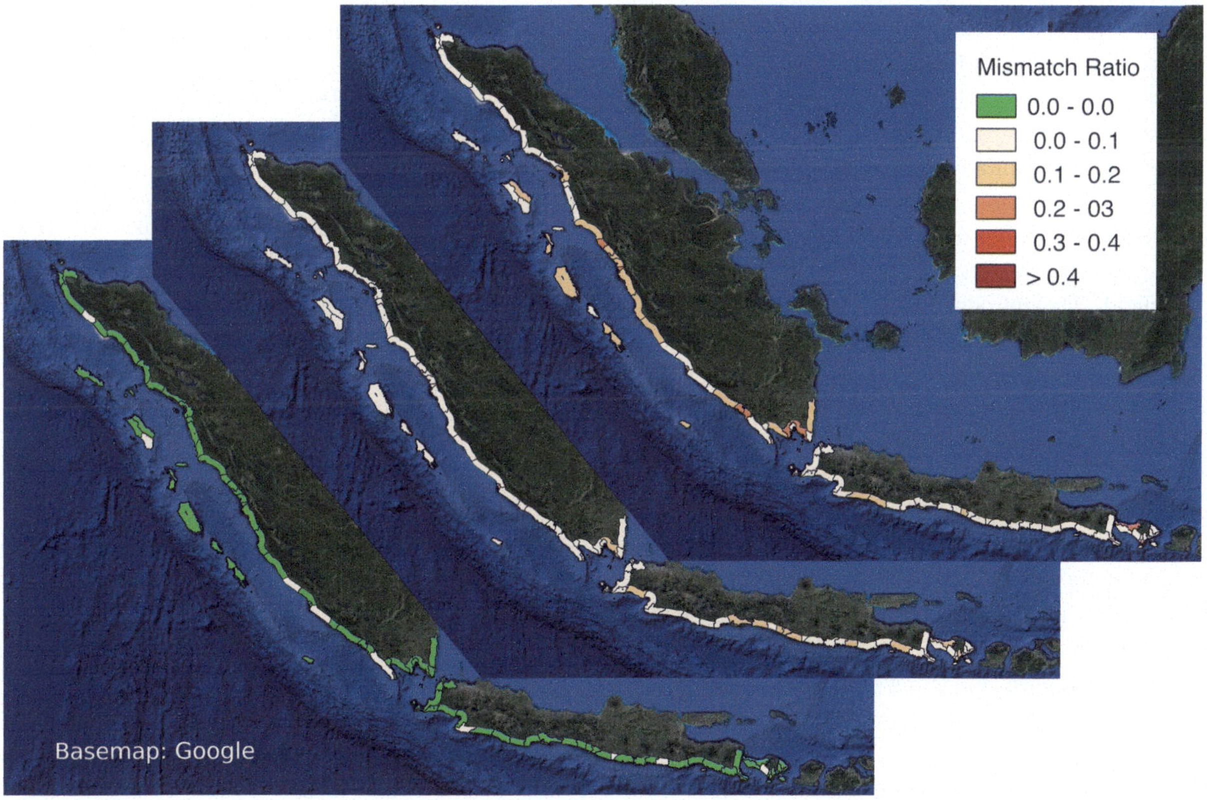

Figure 12
Local ‘Advisory-Warning’ mismatch ratios as defined in (2) for three magnitudes (M_w 7.0, M_w 8.0, M_w 8.8, left to right). Scenarios used for investigation are represented by green squares in Fig. 2. Green areas mark warning zones where both models predict identical warning levels for all scenarios, zones colored in red reflect areas with frequent discrepancies between TsunAWI and easyWave. Basemap Data: Google (c) 2018 Image: Landsat/Copernicus Data SIO, NOAA, U.S. Navy, NGA, GEBCO

locations with database coverage the SOPs define the choice of a TsunAWI scenario in the first moments after an event and at later stages with more accurate source parameters the on-the-fly simulation of easyWave inform the warning. The warning products for database scenarios and on-the-fly simulations are clearly listed and visualized enabling for a fast and detailed analysis in the warning process. In the case of varying predictions, the warning level in individual warning zones may be adjusted by the officer on duty.

5. Summary

We presented the tsunami modeling component of InaTEWS and discussed the evolution of the tsunami scenario database over the course of time with special emphasis on the collaborative approach. The present coverage of 15 fault zones provides simulations for a large fraction of the Indonesian domain prone to tsunamigenic earthquakes, a substantial increase in the coverage and number of communities which now have access to tsunami early warnings informed by the database.

Despite the limitations of such a database being restricted to seismic events with simplified fault mechanisms, the approach offers the fastest way to obtain warning products for coastal regions in the near field. The introduction of the TOAST system means that at any time, real-time simulations are available to improve warning products by using simulations with alternative fault mechanisms.

The two modeling methods were compared against each other, both for real situations with actual events resulting in different source models and in a global approach for identical sources. In both comparisons, the models showed reasonable agreement with respect to the warning products given the large difference in the numerical setups.

This combined modeling approach is a major ingredient for improved warnings to reduce the impact of tsunamis, however, the close connection with monitoring and data comparisons during events and distinctive awareness of the population in tsunami prone regions is essential.

Acknowledgements

The database extension 2015-2017 was funded by the Australian Department of Foreign Affairs and Trade through the DMInnovation project. We are grateful to all involved colleagues at Geoscience Australia, BMKG, gempa GmbH and AWI for the fruitful collaboration. We would like to thank the Head of BMKG and the AWI data centre for their support. Jonathan Griffin and Rikki Weber publish with the permission of the CEO, Geoscience Australia. We would like to thank two reviewers for valuable suggestions leading to considerable improvements of the manuscript.

References

Androsov, A., Behrens, J., Danilov, S. (2011) Tsunami modelling with unstructured grids. Interaction between tides and tsunami waves. In E. Krause et al (Eds.) Computational Science and High Performance Computing IV. NNFM 115, pp. 191–206.

Babeyko, A. Y., Hoechner, A., & Sobolev, S. V. (2010). Source modeling and inversion with near real-time GPS: A GITEWS perspective for Indonesia. *Natural Hazards and Earth System Sciences (NHESS), 10*(7), 1617–1627.

Borrero, J. C., Weiss, R., Okal, E. A., Hidayat, R., Suranto, Arcas, & Titov, dv. (2009). The tsunami of 2007 September 12, Bengkulu province, Sumatra, Indonesia: Post-tsunami field survey and numerical modelling. *Geophysical Journal International*. https://doi.org/10.1111/j.1365-246X.2008.04058.x.

Burbidge, D., Mueller, C., & Power, W. (2015). The effect of uncertainty in earthquake fault parameters on the maximum wave height from a tsunami propagation model. *Natural Hazards and Earth System Science, 3,* 3369–3408.

Fritz, H. M., et al. (2007). Extreme runup from the 17 July 2006 Java tsunami. *Geophysical Research Letters*. https://doi.org/10.1029/2007GL029404.

Geist, E. L., & Bilek, S. L. (2001). Effect of depth-dependent shear modulus on tsunami generation along subduction zones. *Geophysical Research Letters*. https://doi.org/10.1029/2000GL012385.

Goto, C., Ogawa, Y., Shuto, N. & Imamura, F. (1997). IUGG/IOC time project, numerical method of tsunami simulation with the leap-frog scheme. IOC Manuals and Guides 35, Paris.

Greenslade, D. J. M., et al. (2014). An assessment of the diversity in scenario-based tsunami forecasts for the Indian Ocean. *Continental Shelf Research, 79,* 36–45.

Griffin, J. D. & Davies, G. (2018). Earthquake sources of the Australian plate margin: revised models for the 2018 national tsunami and earthquake hazard assessments. Record 2018/31. Geoscience Australia, Canberra. http://dx.doi.org/10.11636/Record.2018.031.

Griffin, J., Latief, H., Kongko, W., Harig, S., Horspool, N., Hanung, R., et al. (2015). An evaluation of onshore digital elevation models for modeling tsunami inundation zones. *Frontiers in Earth Science, 3,* 32. https://doi.org/10.3389/feart.2015.00032.

Gusman, A. R., Tanioka, Y., Matsumoto, H., & Iwasaki, S.-I. (2009). Analysis of the tsunami generated by the great 1977 Sumba earthquake that occurred in Indonesia. *Bulletin of the Seismological Society of America, 99*(4), 2169–2179.

Hall, R. (2018). The subduction initiation stage of the Wilson cycle. *Geological Society, London, Special Publications, 470,* SP470.3.

Hanka, W., Saul, J., Weber, B., Becker, J., Harjadi, P., & Fauzi and GITEWS Seismology Group. (2010). Real-time earthquake monitoring for tsunami warning in the Indian Ocean and beyond. *Natural Hazards & Earth System Sciences, 10,* 2611–2622. https://doi.org/10.5194/nhess-10-2611-2010.

Harig, S., Chaeroni, C., Setiyo Pranowo, W., & Behrens, J. (2008). Tsunami simulations on several scales—Comparison of approaches with unstructured meshes and nested grids. *Ocean Dynamics, 58*(5), 429–440. https://doi.org/10.1007/s10236-008-0162-5

Heidarzadeh, M., Muhari, A., & Wijanarto, A. B. (2018). Insights on the source of the 28 September 2018 Sulawesi tsunami, Indonesia based on spectral analyses and numerical simulations. *Pure and Applied Geophysics*. https://doi.org/10.1007/s00024-018-2065-9.

Hill, E. M., et al. (2012). The 2010 M_w 7.8 Mentawai earthquake: Very shallow source of a rare tsunami earthquake determined from tsunami field survey and near-field GPS data. *Journal of Geophysical Research: Solid Earth, 5,* 45. https://doi.org/10.1029/2012jb009159.

Horspool, N., Pranantyo, I., Griffin, J., Latief, H., Natawidjaja, D. H., Kongko, W., et al. (2014). A probabilistic tsunami hazard assessment for Indonesia. *Natural Hazards and Earth System Sciences.*. https://doi.org/10.5194/nhess-14-3105-2014.

Immerz, A., Harig, S., & Rakowsky, N. (2018). Extending and visualizing the TsunAWI simulation database of the Indonesia tsunami early warning system (InaTEWS). In G. Krause (Ed.), *Building bridges at the science-stakeholder interface* (pp. 101–107). Cham: Springer.

Jaffe, B. E., Borrero, J. C., Prasetya, G. S., Peters, R., McAdoo, B., Gelfenbaum, G., et al. (2006). Northwest Sumatra and offshore islands field survey after the December 2004 Indian Ocean tsunami. *Earthquake Spectra, 22*(S3), S105–S135.

Kamigaichi, O. (2009). Tsunami forecasting and warning. In R. Meyers (Ed.), *Encyclopedia of complexity and systems science* (pp. 9592–9618). New York: Springer. https://doi.org/10.1007/978-0-387-30440-3_568.

Kanamori, H. (1972). Mechanism of tsunami earthquakes. *Physics of the Earth and Planetary Interiors, 6*(5), 346–359.

Kongko, W., Istiyanto, D. C., Irwandi, I. (2006). Tsunami modeling and field observations of December 26 2004 Indian ocean earthquake. Technical report, Coastal Dynamic Research Center, BPPT Indonesia.

Latief, H., Puspito, N., & Imamura, F. (2000). Tsunami catalog and zones in Indonesia. *Journal of Natural Disaster Science, 22,* 25–43.

Lauterjung, J., Münch, U., & Rudloff, A. (2010). The challenge of installing a tsunami early warning system in the vicinity of the Sunda Arc, Indonesia. *Natural Hazards and Earth System Sciences.* https://doi.org/10.5194/nhess-10-641-2010.

Lavigne, F., Gomez, C., Giffo, M., Wassmer, P., Hoebreck, C., et al. (2007). Field observations of the 17 July 2006 Tsunami in Java. *Natural Hazards and Earth System Science, 7*(1), 177–183. **(Copernicus Publications on behalf of the European Geosciences Union)**.

Münch, U., Rudloff, A., & Lauterjung, J. (2011). Postface The GITEWS Project—Results, summary and outlook. *Natural Hazards and Earth Systems Sciences, 11,* 765–769. https://doi.org/10.5194/nhess-11-765-2011.

Pownall, J. M., Hall, R., & Lister, G. S. (2016). Rolling open earth's deepest forearc basin. *Geology, 44*(11), 947–950.

Raape, U., et al. (2010). Decision support for tsunami early warning in Indonesia: The role of OGC standards. In M. Konecny, S. Zlatanova, & T. Bandrova (Eds.), *Geographic information and cartography for risk and crisis management. Lecture notes in geoinformation and cartography.* Berlin: Springer.

Rakowsky, N., Androsov, A., Fuchs, A., Harig, S., Immerz, A., Danilov, S., et al. (2013). *Operational tsunami modelling with TsunAWI—Recent developments and applications.* Nat: Hazards and Earth System Sciences. https://doi.org/10.5194/nhess-13-1629-2013.

Satake, K., Nishimura, Y., Putra, P. S., et al. (2013). *Pure and Applied Geophysics, 170,* 1567. https://doi.org/10.1007/s00024-012-0536-y.

Spahn, H., Hoppe, M., Vidiarina, H. D., & Usdianto, B. (2010). *Experience from three years of local capacity development for tsunami early warning in Indonesia: Challenges, lessons and the way ahead.* Nat: Hazards and Earth System Sciences. https://doi.org/10.5194/nhess-10-1411-2010.

Steinmetz, T., Raape, U., Teßmann, S., Strobl, C., Friedemann, M., Kukofka, T., et al. (2010). *Tsunami early warning and decision support.* Nat: Hazards and Earth System Sciences. https://doi.org/10.5194/nhess-10-1839-2010.

Strobl, C., Kiefl, R., Riedlinger, T., & Strunz, G. (2007). Geodatenmanagement und-dienste am Beispiel des Tsunami-Frühwarnsystems für den Indischen Ozean, Gesellschaft für Informatik in der Land- Forst- und Ernährungswirtschaft e.V. (GIL) **(in German)**.

Strunz, G., Post, J., Zosseder, K., Wegscheider, S., Mück, M., Riedlinger, T., et al. (2011). *Tsunami risk assessment in Indonesia.* Nat: Hazards and Earth System Sciences. https://doi.org/10.5194/nhess-11-67-2011.

Synolakis, C. E., Bernard, E., Titov, V. V., Kanoglu, U., & Gonzalez, F. I. (2008). Validation and verification of tsunami numerical models. *PAGEOPH, 165,* 2197–2228. https://doi.org/10.1007/s00024-004-0427-y.

Wells, D. L., & Coppersmith, K. J. (1994). New empirical relationships among magnitude, rupture length, rupture width, rupture area, and surface displacement. *Bulletin of the Seismological Society of America, 84*(4), 974–1002.

(Received November 10, 2018, revised June 7, 2019, accepted August 14, 2019, Published online September 3, 2019)

Pure Appl. Geophys. 177 (2020), 1403–1423

https://doi.org/10.1007/s00024-019-02252-x

Pure and Applied Geophysics

Automated GNSS and Teleseismic Earthquake Inversion (AutoQuake Inversion) for Tsunami Early Warning: Retrospective and Real-Time Results

KEJIE CHEN,[1] ZHEN LIU,[2] and Y. TONY SONG[2]

Abstract—Rapid finite fault source determination is critical for reliable and robust tsunami early warnings. Near-field Global Navigation Satellite System (GNSS) observations have shown value to constrain the source inversion, but real-time GNSS stations are sparse along most of the active faults. Here we propose an automatic earthquake finite source inversion (AutoQuake Inversion) algorithm jointly using near-field (epicentral distance < 1000 km) GNSS data and mid-range (epicentral distance from 30° to 45°) teleseismic P displacement waveforms. Neither the near-field GNSS nor the mid-range teleseismic data clip or saturate during large earthquakes, while the fast-traveling P-waves are still essential to constrain the source in regions where very few or no GNSS stations are available. Real-time determination of the fault geometry remains to be the main challenge for rapid finite source inversion. We adopt a strategy to use the pre-defined geometry Slab2 for earthquakes within it or to forecast a focal mechanism based on near-by historical events for earthquakes without Slab2 prior. The algorithm has been implemented successfully in the prototype of JPL's GPS-Aided Tsunami Early-Detection system and tested for many real events recently. This article provides the framework of the algorithm, documents the retrospective and real-time results, and discusses remaining challenges for future improvements.

Key words: GNSS data, teleseismic data, Slab2, focal mechanism forecast, fast joint source inversion, tsunami early warning.

1. Introduction

Recent tsunami disasters induced by mega-thrust earthquakes (e.g., the 2004 Sumatra and 2011 Tohoku events) have called for the need for reliable tsunami early warning systems (TEWS), using not only the traditional seismic approach, but also Global Navigation Satellite System (GNSS). Using GNSS to estimate earthquake magnitude for tsunami early warning was first demonstrated by Geoffrey Blewitt and colleagues at the University of Nevada in Reno (Blewitt et al. 2006; Sobolev et al. 2007). Furthermore, Song (2007) demonstrated that coastal GNSS stations can be used to infer seafloor displacements and estimate earthquake-induced oceanic energy and tsunami scales within a few minutes after the quake for early warning. The GNSS approaches have been further demonstrated and improved for more earthquakes (e.g., Ohta et al. 2012; Song et al. 2012; Yang et al. 2014; Chen et al. 2016a; Fu et al. 2017).

To assess the tsunami potential and guide emergency response within minutes after a submarine earthquake, a rapid and accurate estimation of the seismic source is essential (Titov et al. 2005). Currently, most tsunami warning centers, such as the U.S. National Oceanographic and Atmospheric Administration (NOAA)'s Pacific Tsunami Warning Center (NOAA-PTWC), the local German Indonesian Tsunami Early Warning System (GITEWS) and Japan Meteorological Agency (JMA)'s tsunami warning center, use seismic network for earthquake detection, including location and source characterization (Bernard and Titov 2015). As soon as a shallow earthquake (with depth < 100 km) above threshold magnitude is found occurring under sea, tsunami alarms are then triggered and issued to target clients depending on the estimated earthquake magnitude. For example, using real time data from Federation of Digital Seismic Networks (FDSN), the U.S. Geological Survey-National Earthquake

Electronic supplementary material The online version of this article (https://doi.org/10.1007/s00024-019-02252-x) contains supplementary material, which is available to authorized users.

[1] Division of Geological and Planetary Sciences, California Institute of Technology, Pasadena, CA 91125, USA.

[2] Jet Propulsion Laboratory, California Institute of Technology, Pasadena, CA 91109, USA. E-mail: Tony.Song@JPL.NASA.gov

Information Center (USGS-NEIC) is able to determine the origin time, hypocenter location and preliminary estimation of earthquake magnitude worldwide within minutes for all significant events (Hayes et al. 2011). When near-field observations are available, the procedure will take less than tens of seconds (Hoshiba et al. 2011).

While hypocenter can be located with confidence (Husen and Hardebeck 2010), traditional magnitude estimations such as M_L (Richter 1935) and M_S (Boatwright and Choy 1986) using short period seismic waves suffer from saturation for very large earthquakes (Larson et al. 2003). One such example is the underestimate of the magnitude of the 2011 M_w 9.0 Tohoku earthquake for tsunami early warning (JMA 2013). To circumvent this problem, the moment magnitude M_w, defined by Kanamori (1977), which describes total energy required to rupture the entire fault, is considered to be the most appropriate among various magnitudes for tsunami potential evaluation (Blewitt et al. 2006). Rapid determination of M_w from broadband P waveforms (M_{wp}) has been implemented in PTWC officially since 2002 (Hirshorn et al. 2013). In addition, W-phase moment tensor inversion which provides both magnitude and focal mechanism information (Kanamori and Rivera 2008) has also been performed routinely in PTWC (Duputel et al. 2011). Usually, it takes tens of minutes to stabilize a M_w magnitude estimation in such inversion (Kanamori and Rivera 2008; Hayes et al. 2011).

Traditionally, TEWS have been relying on earthquake magnitude to issue tsunami alert or warning levels. While it is true that a larger earthquake magnitude usually indicates a greater tsunami potential, the historical tsunami maximum water heights and earthquake magnitudes show a much more complex correlation between the two (see Fig. 1). For example, with respect to the $M_w \sim 8$ earthquakes, the tsunami maximum water heights vary from less than 3 m to more than 30 m, which poses a challenge for tsunami early warning and suggests that the point source magnitude used in current TEWS needs improvements. Sensitivity analysis conducted by Titov et al. (1999) showed that point source model provides an essential description of far-field tsunami signals. As a result, the NOAA system has been providing regional (distance larger than 1000 km) warnings routinely with success. However, for a local tsunami near Padang of Sumatra Island, Sobolev et al. (2007) demonstrated that different distribution of slip could generate totally different impacts on coastal wave heights in spite of the same magnitude and location. In addition, Weinstein and Lundgren (2008) acknowledged the challenges the warning centers faced when coping with unusual earthquakes (such as 2001 M_w 8.4 Peru, 2004 M_w 9.3 Sumatra, 2006 M_w 7.7 Java), and concluded that improved tsunami wave-height forecasts would be achieved if considering the slip distribution along a fault. Indeed, recent studies of Song et al. (2008, 2017) found that what really matters is not the earthquake magnitude alone. Instead, the size and power of a tsunami depend on the amount of energy transferred to the ocean by the earthquake-induced seafloor displacements. Based on the energy principle, Song (2007) proposed a method for estimating the tsunami energy directly from coastal GNSS stations near the epicenter. More recently, Inazu et al. (2016) developed a near-field TEWS using an automatic CMT estimation, their results showed increasing forecast uncertainties for larger ($M_w > 8$) earthquakes because the tsunami source cannot be treated as a point any more. In a retrospective real-time analysis of the 2011 Tohoku event, Melgar and Bock (2015) highlighted the importance of a reliable finite slip model through a detailed comparison of predicted tsunami's inundation and survey measurements. However, real-time GNSS sites along coastal regions are often not dense enough to provide reliable estimation of the seafloor displacements. So far, many aforementioned GNSS approaches and related publications are based on post-event processed data and implementation. To our knowledge, real-time inversion of earthquakes just using GNSS is only feasible in regions where there is a dense GNSS network (Kawamoto et al. 2016, 2017), while a global GNSS-based earthquake source inversion system has never been reported except post-event demonstration in simulated environment.

To resolve the data sparse problem of real-time GNSS sites along coastal regions, we propose an automated joint inversion based on near-field (< 1000 km) GNSS data and mid-range (epicentral

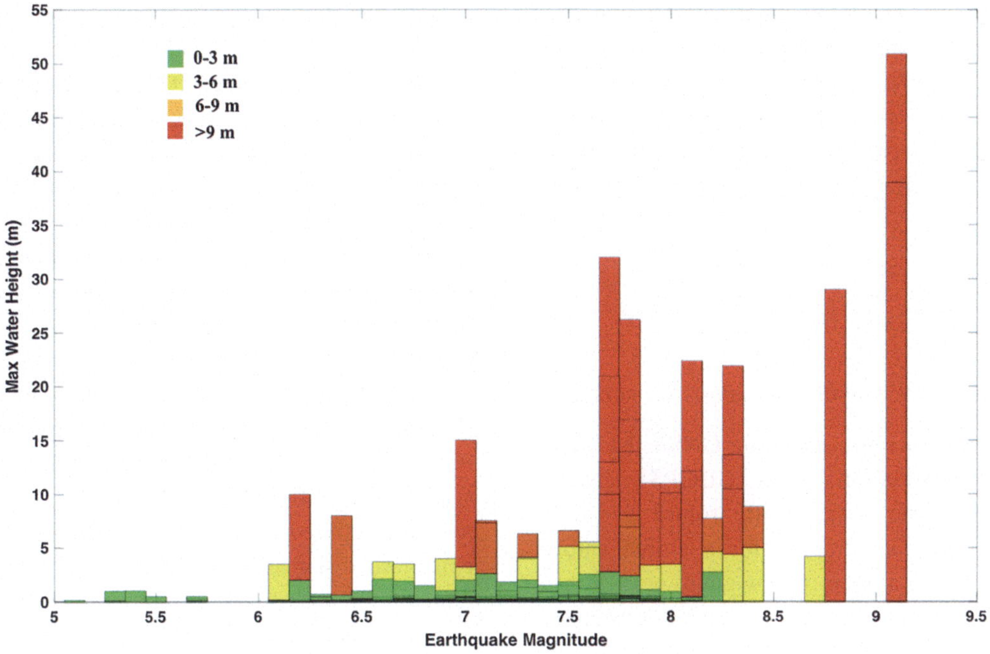

Figure 1
Historical earthquake magnitude and corresponding tsunami maximum water height since 1985. Records are from NOAA tsunami database. Maximum water heights are either runups or buoy records, depending on data availability. Data source: https://www.ngdc.noaa.gov/nndc/struts/form?t=101650&s=166&d=166

distance ranging from 30° to 45°) teleseismic P displacement waveforms. In fact, teleseismic P displacement waveforms, within 6$\sim$8 min away from the epicenter, are still essential to constrain the source in regions where near-field GNSS data are not dense enough. The global deployment of FDSN stations provides publicly available seismic records for all large earthquakes with extensive epicentral distance and azimuthal distributions. In fact, teleseismic P waveforms are among the first datasets proposed to investigate earthquake finite source features in the context of tsunami early warning (e.g., Weinstein and Lundgren 2008; Mendoza and Hartzell 2013). Meanwhile, advances in GNSS have made it possible to measure both static and kinematic ground motions at mm-cm level in real-time without clip (e.g., Bock et al. 2000; Li et al. 2013; Liu et al. 2014; Chen et al. 2016b; Geng et al. 2017). Especially, real-time high-rate ($\geq$ 1 Hz) GNSS has made the boundary between geodetic and seismic sensors blurry. Compared with teleseismic data, GNSS derived kinematic co-seismic signals are more favored for source studies of large earthquakes because of higher spatial–temporal resolution (Yue and Lay 2013).

Besides the real-time data availability issues, other challenges still exist for fast earthquake finite source inversion. One major difficulty is the instant determination of fault geometry. Finite source inversion needs predefined fault geometries, which are traditionally obtained from point-source moment tensor inversions (Kanamori and Rivera 2008). Moment tensor inversions usually rely on long period seismic waveforms, which may have a large latency (20$\sim$30 min). GNSS data have been suggested to estimate the moment tensor or earthquake magnitude by peak ground displacements estimation recently (e.g., Crowell et al. 2012; Melgar et al. 2016), but obtaining sufficient data in real-time is still a challenge globally. Besides, moment tensor solution provides two conjugated planes. Resolving fault plane ambiguity requires additional efforts.

In the following section, we first describe our real-time data source and streaming, and then present approach to deal with fault geometry and finite source inversion strategy. In Sect. 3, we assess the algorithm through both retrospective and real-time analyses of five events in different tectonic settings. We then summarize our work and discuss future improvements to make our system more robust and efficient in Sect. 4. Supplemental materials are provided online.

2. Data and Methods

2.1. Real-Time GNSS and Teleseismic Networks

In this study, we use real-time GNSS data from NASA Global Differential GPS (GDGPS) network, which provides the operationally robust and self-sustainable front-end required for a global natural hazard monitoring system (www.gdgps.net). It is a high-accuracy, high-reliability, 24/7 GNSS augmentation system, developed by JPL to support the real-time positioning, timing, and orbit determination requirements of its customers at NASA, the Air Force, and in industry. By using accurate real-time orbit and clock products, NASA GDGPS employs precise point positioning to estimate site coordinates, which ensures the solutions can be obtained in real time. In order to improve robustness of the system, data filtering and quality control are also applied to the real-time co-seismic displacement waveforms. We are streaming the 1 Hz positioning time series with few centimeters' accuracy for about eight hundred of GNSS continuous tracking stations (see their distribution in Fig. 2) from both regional and global networks, recording ground dynamic displacements right before, during and after earthquakes routinely. In addition, static co-seismic offsets are estimated automatically in near real-time (seconds to minutes latency) from high-rate GNSS displacement waveforms using the methodology by Liu et al. (2014). The near field GNSS stations are subject to coseismic displacement influence when epicentral distance is within $10^{(M/2-0.8)}$ km where M is earthquake magnitude (http://geodesy.unr.edu/). For the megathrust event like the 2011 M_w 9.0 Tohoku earthquake, GNSS stations as far as 3000 km away can experience co-seismic shaking. To avoid long latency associated with far field stations, which can take tens of minutes due to dynamic wave propagation and settlement, in our real-time system we consider only GNSS stations within 1000 km to the epicenter.

As mentioned above, near-field seismograms can saturate during strong shaking caused by large earthquakes. Teleseismic P and S waves are separated from other phases at far-field epicentral distance ($\geq 30°$) and usually used to image earthquake source. Compared with P waves, S waves travel at a relatively slow speed and may delay the system's response. Besides, automated picking of S waves is more challenging. As a result, we adopt teleseismic P waveforms from Global Seismographic Network (GSN) due to its stable real-time access in our system (see station distribution in Fig. 2). Specially, to speed up response time, we require the station epicentral distance to be between 30° and 45°. We remove the instrument response by deconvolution. The sampling rates of the original waveforms are usually larger than 20 Hz. However, this high-frequency information of observations cannot be precisely modeled because of the limitation in 3-D velocity model and inaccuracy of 1-D velocity model. We thus decimate the raw observations to 1 Hz and bandpass filter them with [0.009, 0.4] Hz. We perform the same operation on the synthetic Green's functions. We limit the P wave windows to 120 s enclosing earthquake and expected P wave arrival time. The selected P wave windows could have some contamination from PP wave arrivals for stations at shorter epicentral distances. We neglect the contaminated tail energy after the P wave phase.

Compared with teleseismic stations, the near field GNSS receivers (< 1000 km) have much less latency to get co-seismic signals with just a few seconds data transmission time and provide better constraints on slip distribution. Therefore, in the system we give GNSS data higher priority. If we have more than seven GNSS stations (note that the number seven is quite arbitrary, as it will be discussed later), we will carry out the GNSS inversion. Otherwise, we will include seismic data. The joint usage of GNSS and seismic data raises the issue of relative weighting. This is an ongoing research topic and no thumb of rule exists yet. The trial-and-error method that is frequently used is not suitable for a real-time implementation. Here, we normalize each data type

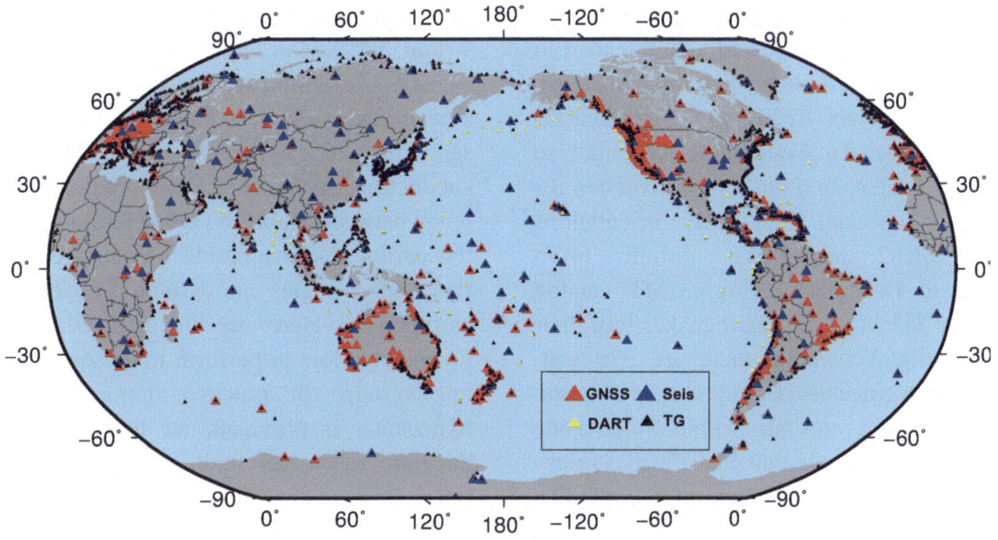

Figure 2
Distribution of real-time GNSS stations processed by JPL (red triangle), global seismograph network (blue triangle), DART buoys (smaller yellow triangle) and tide gauges (smaller black triangle)

by its own norm on a station basis and assign equal weights to them.

2.2. *Fault Geometry Determination*

When modeling an earthquake rupture, inverting heterogeneous slip distribution and fault geometry simultaneously will be nonlinear and intractable. In most cases, we assume a prefixed fault plane in the source region, which is usually discretized into a number of subfaults in the source region. To define the fault plane geometry, seven parameters are needed: the position of reference point (longitude, latitude, and depth), the size (length and width), and the orientation (strike and dip angles). In our system, we set the hypocenter location as the center point (or reference point) of fault plane. With respect to the size, we follow the empirical magnitude-area scaling relationships developed by Wells and Coppersmith (1994). Note that we do not know whether the earthquake is unilateral or bilateral, and the initial seismically estimated magnitude tends to be underestimated. We thus set the source fault dimension large enough to accommodate rupture in either direction. Specially, for $M \geq 7.5$ earthquakes, we set the rupture length as 400 km and width as 200 km. The fault is then subdivided into 20 along-strike and 10-along dip segments.

In addition, we also need to constrain rake angle range to stabilize the inversion results. Strike, dip and rake angles, usually provided as part of focal mechanism, are traditionally derived from centroid moment tensor solution. Currently, focal mechanisms for significant earthquakes are routinely provided by several international organizations (e.g., GCMT, USGS, GFZ) with latencies from tens of minutes to hours. Recently, algorithms based on local and regional GNSS observations are also proposed to speed up the determination of centroid moment tensors (Melgar et al. 2012; Riquelme et al. 2016). However, as mentioned above, real-time availability of GNSS data may severely limit the algorithms' performance for an operational system. Furthermore, focal mechanism solution contains two conjugated nodal planes, and more times are required to resolve nodal plane ambiguity for finite source inversion, which further delays emergency response.

Based on the catalogues of the historic seismicity, bathymetry and gravity data sets, a global subduction zone geometry model called Slab1.0 has been developed (Hayes and Wald 2009; Hayes et al. 2012). Slab1.0 outlines three-dimensional geometry of almost 80% of subduction zones over the world. More recently, Slab1.0 was updated and extended to Slab2 (available at https://github.com/usgs/slab2, also see Fig. 3a), which covers $\sim$ 30% more than Slab1.0 (Hayes et al. 2018). To evaluate its potential to constrain fault geometry in finite source inversion for TEWS, we first checked fault strike orientations inferred from Slab2 and GCMT solutions. From January 1976 to December 2017, GCMT catalog recorded totally 543 $M_w \geq 7.0$ earthquakes (see their distribution in Fig. 3), among them are 370 submarine events with epicenter depth less than 100 km, and 272 of the 370 events are within Slab2 zone coverage. Figure 3 indicates that Slab2 strike orientation agree with GCMT solution quite well. To quantify their consistency, we adopt the criterion proposed by Kagan (1991) (hereafter referred to as Kagan angle). Kagan angle measures differences between the orientations of two seismic moment tensors, it varies from 0° to 120°, and two focal mechanisms are considered to be very similar when the Kagan angle is less than 30° (Kubo et al. 2002). Note that Slab2 is a subduction model, we treat all of these earthquakes as thrusting events with 90° rake angle. The results of comparisons (see Fig. 3b) show that $\sim$ 63% of strike angles in GCMT solutions are within 30° Kagan angle against their Slab2 values, $\sim$ 83% are within 40° and $\sim$ 90% are within 50° Kagan angle, which indicates Slab2 is a good representation of fault geometry for mega-thrust subduction earthquakes. Encouragingly, the consistency between Slab2 and GCMT solution increases as the earthquake magnitude grows (see Fig. 3c), and the Kagan angles are within 40° for all $M_w > 8$ earthquakes. Here, we incorporate Slab2 as fault geometry priors to perform finite source inversion in our system. In practice, immediately after the hypocenter is provided, we begin to search Slab2 database to extract fault geometry. Ideally, for an inter-plate thrust faulting earthquake that takes place in Slab2, its hypocenter depth and corresponding Slab2 depth should be close to each other. Otherwise, even the epicenter falls in the horizontal projection (Fig. 3a) of the Slab2, the earthquake could be another type of faulting, such as intra-plate normal faulting (e.g., the 2018 M_w 8.2 Mexico Tehuantepec earthquake) or back-arc thrust faulting (e.g., the 2018 M_w 7.0 Lombok earthquake discussed later in this paper). To that end, we adopt the depth difference between hypocenter and Slab2 as a criterion to decide

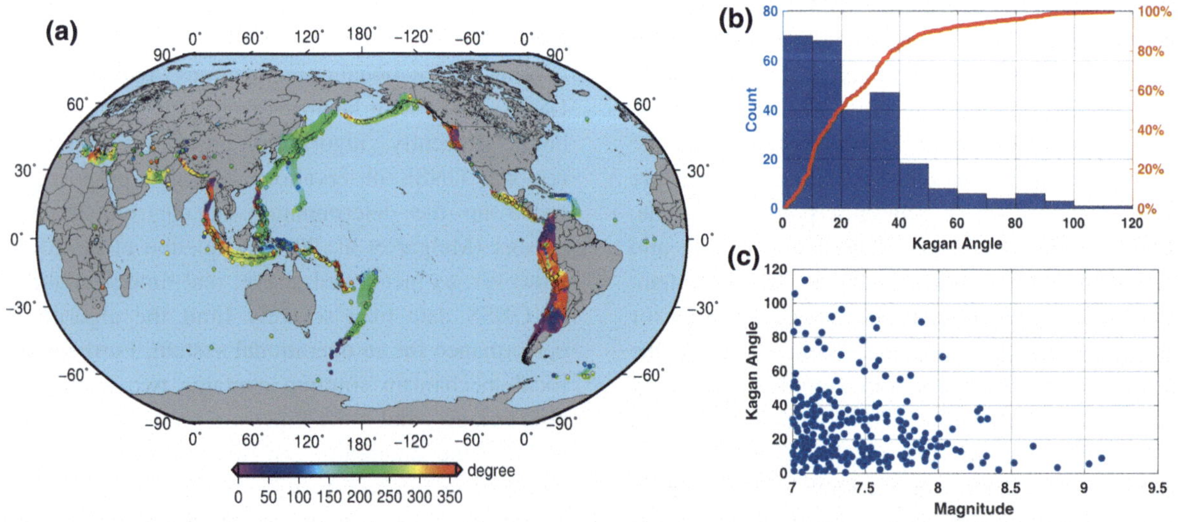

Figure 3
a Strike angles of $M_w > 7$ earthquakes in GCMT from 1976 to 2017 (represented by circles) and that inferred from Slab2 (color). Note that original GCMT solutions provides two strike angles for the two nodal planes, we chose the one closer to Slab2. The strike angle is indicated by the color scale. **b** Histogram and cumulative frequency of similarity between GCMT solutions and Slab2 priors for 272 $M_w \geq 7$ earthquakes in subduction zone. **c** Kagan angle versus earthquake magnitude

whether we can use Slab2-inferrd fault geometry for source inversion. Considering the fact that the hypocenter depth and Slab2 depth can differ up to tens of kilometers within uncertainties, we currently set the threshold of depth difference as 30 km.

Neighboring large earthquakes, if they fall within the same tectonic settings, usually have similar focal mechanisms, suggesting that it is possible to forecast mechanisms of future earthquakes based on nearby historical events (Kagan and Jackson 1994). As a matter of fact, Kagan and Jackson (1994) proposed an interpolation algorithm to predict focal mechanism considering factors such as epicentral distance, magnitude and directivity. Inspired by their ideas, we infer the fault geometries for those earthquakes out of Slab2 by interpolating neighboring $M_w > 5$ historical events (see Fig. S1) as the forecasted focal mechanisms (see Fig. 4a). We validated the predictions by comparing 98 historical earthquakes with $M_w > 7$ against their forecasted values. The comparison statistics show that $\sim$ 75% of them are within 30° Kagan angle, $\sim$ 80% are within 40° Kagan angle (Fig. 4b), and the consistency further improves for larger events (Fig. 4c). These results agree well with Slab2. Of particular note, the forecasted focal mechanism contains two nodal planes, and we run two groups of finite source inversion in parallel to find the favorable one based on the data fits.

2.3. *Rapid Joint Finite Source Inversion*

Finite source inversion can be either static (using permanent offsets) or kinematic (using waveforms). Modelling waveforms can be more time-consuming than static inversion because more record samplings are utilized. Besides, based on our real-time tests, we find that data loss and discontinuity make the kinematic inversion based on high-rate GNSS complicated. As a result, if there are enough (> 7) GNSS offsets with good azimuthal coverage (the epicenter-station azimuth range > 90°) available, we will perform only static inversion to get slip distribution. Note that using station number > 7 and azimuth coverage > 90° as criteria for enough GNSS stations is a bit subjective, but our results have shown that it is able to provide acceptable tsunami forecast for early warning. When GNSS observations are insufficient or absent, we switch to kinematic source inversion based on teleseismic P waveforms and GNSS offsets. To this end, we use the linear multi-time window approach (Ide et al. 1996) with a rupture velocity as 80% of the shear wave speed (Bilek and Lay 1999). However, for "tsunami earthquakes" which generate

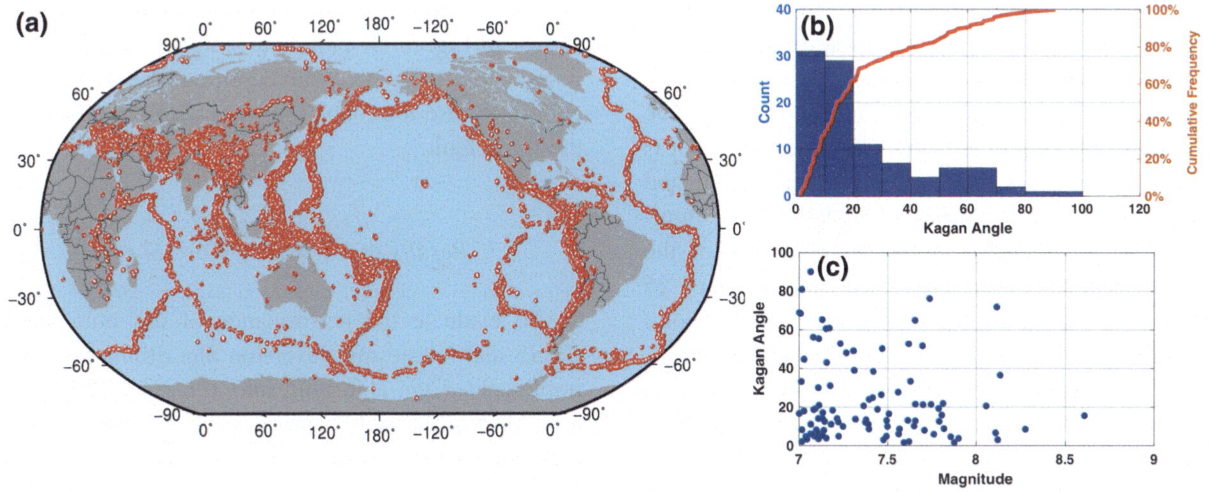

Figure 4

a Focal mechanism forecast based on GCMT catalog from January 1976 to December 2017. Focal mechanisms are shown on 2° × 2° grid. **b** Histogram and cumulative frequency of similarity between GCMT solutions and forecast focal mechanism for 98 $M \geq 7$ earthquakes without Slab2 priors. **c** Kagan angle against magnitude tsunamis much larger than expected based on their magnitudes (Kanamori 1972), the rupture velocity can be as slow as 1.0–1.5 km/s, like the 17 July 2006 M_w 7.7 Java (Ammon et al. 2006) or the 25 October 2010 M_w 7.8 Mentawai (Yue et al. 2014) earthquake. Usually we run iterations with a group of different rupture velocities to identify the reasonable one based on the corresponding data misfits. Unfortunately, teleseismic P waves are less sensitive to the change of rupture velocity, therefore we only run the iteration when GNSS data are included. Furthermore, the choice of window number, rise time length is quite arbitrary. Here, we assume five 4 s long triangle source time functions with 50% overlapping at each sub-fault.

We use frequency wavenumber integration approach (Zhu and Rivera 2002) to calculate Green's functions (GFs) for static GNSS based on the 1-D rigidity structure CRUST 2.0 (http://igppweb.ucsd.edu/~gabi/crust2.html) at the source, and the GFs are computed from 0 to 0.5 Hz for every sub-fault/station pair. The teleseismic Green's functions are generated with a propagator matrix approach (Kikuchi and Kanamori 1982). We adopt CRUST 2.0 for the source side and preliminary reference earth model (PREM) (Dziewonski and Anderson 1981) for the receiver side. Slip distribution inversion is usually an ill-posed problem. To ensure stability, for static source inversion, we employ Laplacian regularization (Hartzell and Heaton 1983) to constrain the total slip at each sub-fault. For multi-window kinematic source inversion, an additional constraint on the amplitude of the neighboring time windows at each sub-fault is added. In this case, we will have both spatial and temporal smoothing factors to be defined. To minimize manual intervention for determination of the optimal regularization values, we run a total of 25 iterations with gradually increasing spatiotemporal smoothing levels and then use the formalism of Akaike's Bayesian Information Criterion (ABIC) (Akaike 1980) to select the most favorable smoothing pair.

2.4. *Tsunami Validation*

We use the open source code GeoClaw (http://clawpack.org) for tsunami simulation. To be specific, we first derive horizontal and vertical sea floor deformation following the planar fault model of Okada (1985), and then obtain total deformation by summing the vertical motion and the vertical component caused by the advection of topography in horizontal directions (Tanioka and Satake 1996). After that, we take the total deformation as an initial condition for a model run and the simulation is conducted by including topography and bathymetry data sets with 15 arc second resolution (Becker et al. 2009). To validate the slip model, we then compare the model-predicted tsunamis with either tsunami run-up measurements or tsunami waves at coastal tide gauges, depending on the data availability. We also run tsunami simulation based on published post-event models by other researchers, To further assess the performance and advantage of our system in the context TEWS, we also run tsunami simulations based on sea floor deformation derived from the currently used standard centroid moment tensor solutions using the so called Rupture Generator developed by Babeyko et al. (2010).

The algorithm has been implemented successfully in the prototype of JPL's GPS-Aided Tsunami Early-Detection (GATED) system. Earthquake hypocenter, origin and initial magnitude are provided by USGS, and NOAA TWCs through operational global seismic network detections (earthquake alert), which trigger our system, and the entire workflow for our system is shown in Fig. 5. Of particular note, considering the initial reported hypocenter depth can be very inaccurate, we use the Slab2 interface depth to construct the finite fault.

3. *Results of Illustrative Case Studies*

To evaluate the performance of our automated GNSS and telesiesmic inversion algorithm, we have implemented the algorithm into JPL's GATED system, and simulated a number of historical large earthquakes ($M \geq 7$) in retrospective mode and processed some moderate earthquakes ($M > 6.5$) in real time. Here we present the results as summarized in Table 1 and evaluate their uncertainties against observations and well-studied published source models as well. The input datasets are either near-

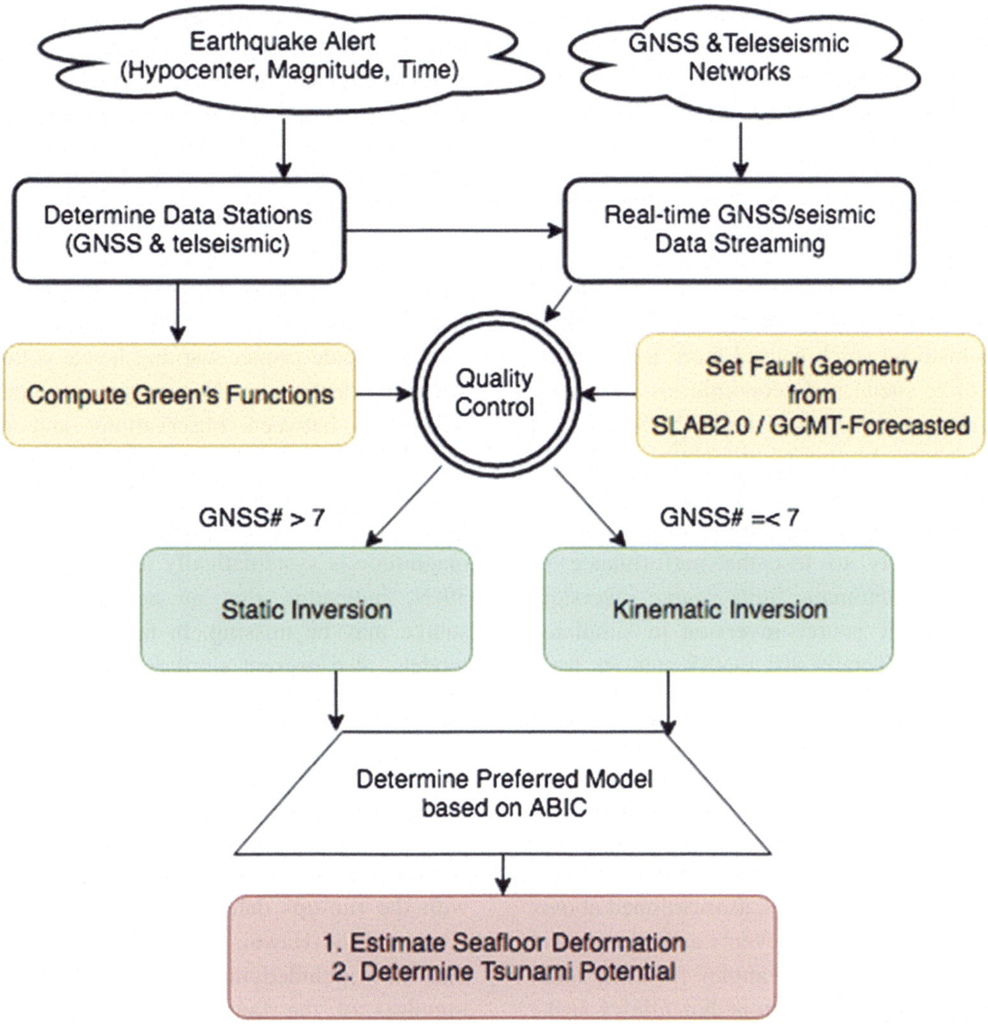

Figure 5
Flowchart of automated GNSS and teleseismic earthquake finite source inversion (AutoQuake Inversion) algorithm implemented at JPL

Table 1

Summary of the AutoQuake inversion in retrospective or real-time mode

Earthquake	Fault geometry	Datasets	Time (min)
2011 M_w 9.0 Tohoku	Slab2	GNSS	~ 7
2015 M_w 8.3 Illapel	Slab2	GNSS + P waveforms	~ 13
2016 M_w 7.8 Solomon Island	Slab2	P waveforms	~ 12
2016 M_w 7.8 Kaikoura	Forecasted focal mechanisms	GNSS	~ 2
2018 M_w 7.0 Lombok	Forecasted focal mechanisms	P waveforms	~ 14

The time is based on a computer with Intel Xeon processor and 2.5 GHz CPU (16 GB RAM) field GNSS offsets or mid-range teleseismic P waveforms or both, accurate or inaccurate fault geometries are inferred from Slab2 or forecasted focal mechanisms, which covers all possible operational scenarios.

3.1. The 11 March 2011 M_w 9.0 Tohoku Event

The 11 March 2011 M_w 9.0 Tohoku earthquake generated one of the most powerful and devastating tsunamis in history, which caused more than 15,000 deaths and huge social and economic losses. Meanwhile, it is also one of the ever best GNSS observed mega-thrust events. As aforementioned, we currently do not have the real-time GNSS data from Japan streamed to our system. Note that this event provides an ideal opportunity to test the performance of GNSS-only-based automatic finite source inversion, we replay automatic source inversion in simulated real-time scenario. Totally 801 GNSS stations from GEONET within 1000 km epicentral distances are adopted.

Fault geometry from Slab2 yields a strike angle as 191°, dip angle as 15° at the hypocenter location, very close to the GCMT solution (strike 203°, dip 10°, rake 88°, M_w 9.1). With respect to the earthquakes happening in Slab2, as mentioned above, we assume they are thrusting events and slip vectors are within rake 45° and 135° angles for each fault, which ensures a thrusting rupture but allows strike components in a non-negative least square inversion.

Automatically-derived slip from the 801 GNSS stations are shown in Fig. 6a. The fit to the GNSS data has a variance reduction (VR) as high as 92%, which indicates that a single plane can represent the overall fault geometries quite well. The largest slip is up to 52 m, and the rupture area covers approximately 250 km along the trench. While major slips concentrate around (38°N, 143°E), slip also extends to the south, which can be interpreted as a second asperity to some extent. The total seismic moment released by the main shock is about 4.1×10^{23} Nm from finite slip inversion, equivalent to M_w 9.0, slightly smaller than the GCMT solution. Besides, the centroid location of the finite slip is about 30 km north to the GCMT solution. Of special importance, the GNSS-based inversion shows a very large and shallow (less than 20 km) slip distribution, which accounts for the huge seafloor deformation and tsunami ensuing from the earthquake.

The simulated maximum tsunami runups from our fast finite slip, Yue and Lay (2013)'s slip model inverted from high-rate GNSS and seismic waveforms, and centroid moment along the coast are shown in Fig. 6b together with the observations. Generally, our finite slip model predicted tsunamis show a good agreement with the records and can be used to decide proper warning levels within minutes after the earthquake. For coastal run-ups, variance reductions between observations and predictions are $\sim$ 80% south of 39°N, which suggests that the major slip distribution around (38°N, 143.25°E) from GNSS offsets is reasonable. However, the runup magnitude is systematically underestimated north of 39°N, indicating that an additional tsunamigenic source may be missing. In fact, many other source models, also present similar mismatches (MacInnes et al. 2013). Tappin et al. (2014) suggests a submarine mass failure (e.g., a submarine landslide) likely contributes to the tsunami as well. However, using the same GNSS data for the seafloor deformation based on Song et al. (2012), Titov et al. (2016) shows that the NOAA tsunami model agrees well with the run-ups data available along the Japanese coast and the Hawaii coast. The major difference is that they included the horizontal force due to the impulses of the faulting continental slope in the tsunami source, as described in Song et al. (2008). Currently, the GeoClaw model does not have the capability to include the horizontal force for tsunami simulation. Modifying the GeoGlaw model is beyond the scope of this work. Nevertheless, the current GeoClaw model serves well for our purpose to quantify our automated finite source inversion algorithm. For example, the GeoGlaw shows quite different tsunami results from our finite source (red in Fig. 6b) and the CMT solution (blue in Fig. 6b), in which the latter fails to characterize the amplitude of the tsunami. In fact, the tsunami amplitude is the most important factor to determine tsunami force and destructiveness for early warning.

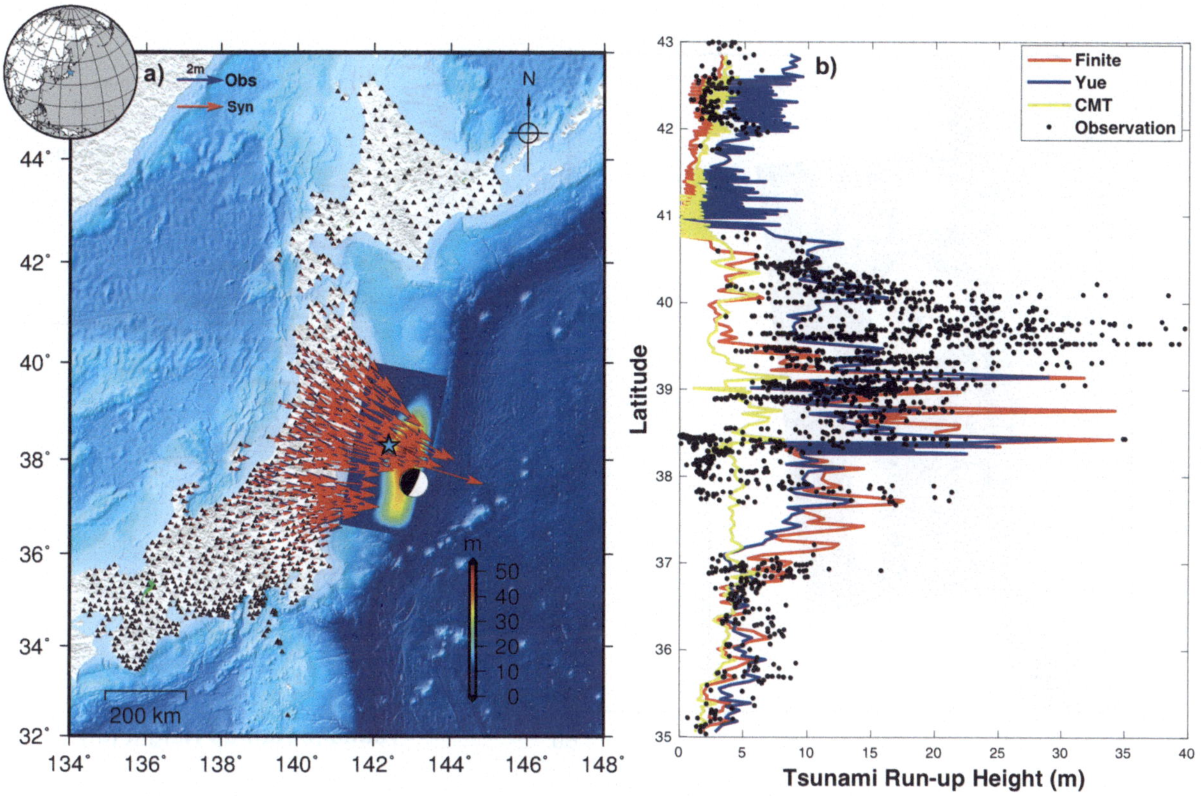

Figure 6

a Co-seismic slip distribution and GNSS data fits for the 2011 M_w 9.0 Japan Tohoku earthquake. The blue star and black beachball denote the preliminary epicenter provided by USGS and GCMT centroid moment, and blue and red vectors are observations and synthetics, respectively. **b** Tsunami run-up height predictions based on our fast co-seismic slip inversion (red), Yue and Lay (2013)'s inversion (blue) using high-rate GNSS and seismic waveforms, and GCMT solution (yellow), run-up observations (black) were provided by Mori et al. (2011)

3.2. The 16 September 2015 M_w 8.3 Illapel Event

As one of the world's most seismically active regions, continuous GNSS network has been deployed in Chile since the early 1990s by several international groups to monitor geodynamics. Following the 2010 M_w 8.8 Maule earthquake, the GNSS network has been improved greatly by increasing spatial density and reducing solution latency, which will also benefit earthquake and tsunami early warnings (Báez et al. 2018). To date, real-time GNSS data at a few stations are streamed to our system. The 16 September 2015 M_w 8.3 Illapel earthquake produced a maximum 11 m local tsunami run-up and all coastal residents (about one million people) were warned to evacuate by Chilean National Emergency Management Office. However, when the evacuation was ordered, it did not contain any information on the tsunami intensity forecast, and it turns out that the tsunami damages were quite localized. Whereas failing to evacuate can cause loss of life, false alarm can undermine the credibility of warning system lead to poor response to future warnings.

For this event, co-seismic offsets at only three GNSS stations are available in our system. As a result, we also included P waves at 11 teleseismic stations (see their distribution in Fig. S2). Strike/dip angles inferred from Slab2 are 7°/15° based on the preliminary hypocenter location, close to the GCMT solution (strike/dip/rake as 7°/19°/109°). Since this is a joint kinematic source inversion, we run iterations using rupture velocities ranging from 1.0 km/s to 3.5 km/s with a step as 0.5 km/s. It turns out that a

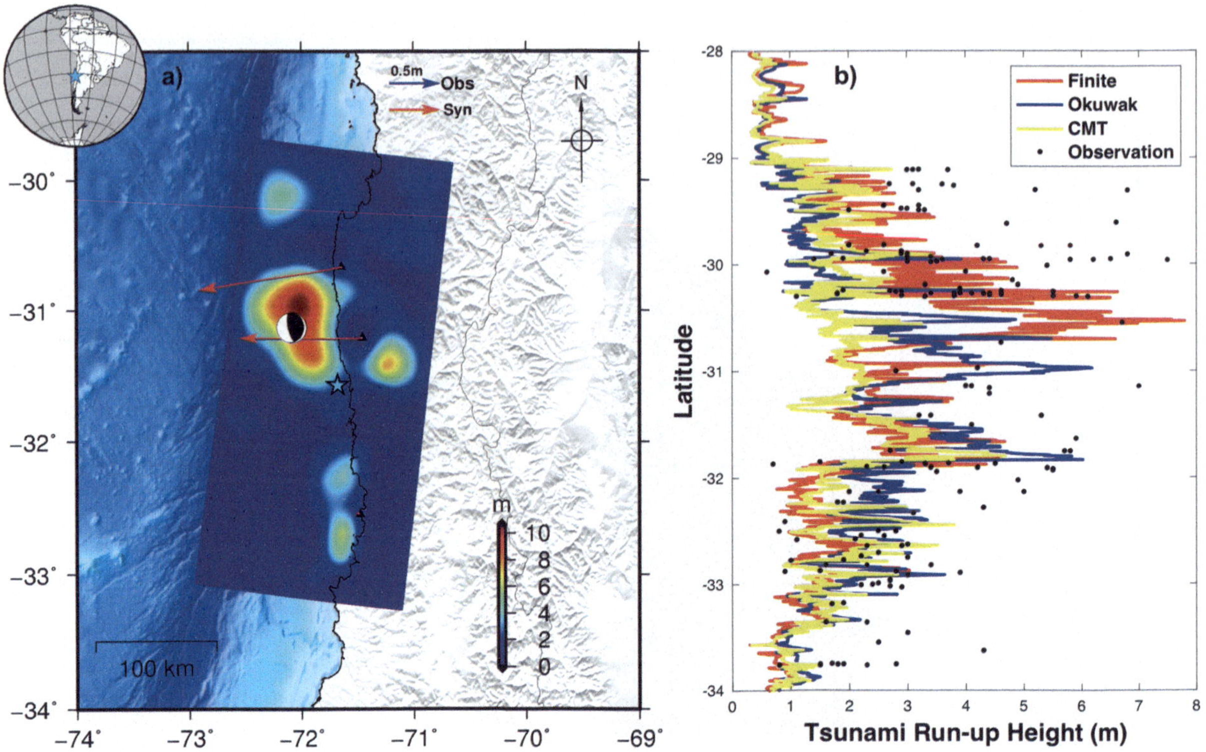

Figure 7

a Slip distribution inverted from GNSS offsets and teleseismic waveforms for the 2015 M_w 8.3 Chile Illapel earthquake. Blue star locates USGS epicenter and beachball denotes GCMT solution, blue are red vectors and observations and fits, respectively. Fits for teleseismic waveforms are shown in Fig. S2. **b** Tsunami run-up predictions from our fast finite source inversion (red), Okuwaki et al. (2016)'s inversion using P waveforms which was constrained by back-projection, and GCMT solution (yellow). Run-up observations (black) were provided by Contreras-López et al. (2016)

rupture velocity of 2.0 km/s gives the minimum misfit of GNSS data and the corresponding slip model (see in Fig. 7a) shows this earthquake ruptured bilaterally along strike, the peak slip reaches about 11 m. The majority asperity occurred about 70 km northwest to the epicenter and consistent with the GCMT centroid location. Besides, there is also a smaller but much shallower rupture asperity. Tsunami runups predicted from the slip model (see in Fig. 7b) indicate a major warning zone from ~ 29.5°S to ~ 32.9°S with the most affected area from ~ 30.2°S to ~ 30.6°S, almost identical with the observations. By contrast, the point source centroid solution fails to forecast the tsunami intensive zone.

3.3. The 8 December 2016 M_w 7.8 Solomon Island Event

Unlike the 2011 M_w 9.0 and 2015 M_w 8.3 Illapel events, we do not have any GNSS data or other nearby observations to constrain finite source inversion for the 8 December 2016 M_w 7.8 Solomon Island earthquake, only teleseismic P waves can be utilized. Worse still, the fault geometry is not very accurate: The Slab2 at hypocenter location indicates strike as 306°, dip as 28°, whereas GCMT solutions shows the strike and dip are 285°, dip 50°, respectively. Besides, the GCMT rake angle is 57°, indicating that thrust components are not overwhelmingly dominant. Tsunami warnings were issued immediately after the earthquake based on its magnitude. However, according to the sea level gauges, the largest observed wave was only around 0.1 m. Here we take it as an

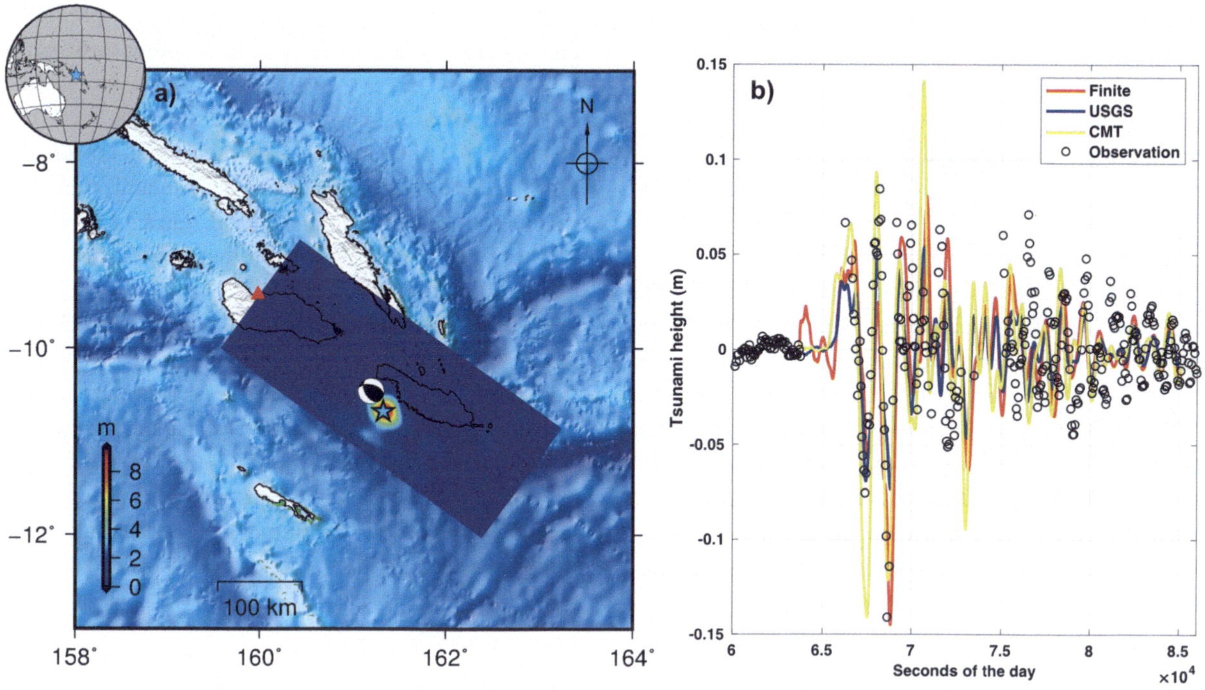

Figure 8

a Slip distribution of the 2016 M_w 7.8 Solomon Island earthquake inverted from teleseismic waveforms based on fault geometry inferred from Slab2, blue star and black beachball show USGS epicenter and GCMT solution, the red triangle is the location of tide gauge SOLO. **b** Predicted tsunamis waves from our fast finite source inversion (red), USGS model (https://earthquake.usgs.gov/earthquakes/eventpage/us20007z80/finite-fault) using P and S waveforms (blue), and GCMT solution (yellow) at coastal tide gauge SOLO. Note that later arriving tsunamis cannot be well modeled due to complex interactions between the coast and shallow bathymetry. Tsunami observations (black) were provided by Global Sea Level Observing System

example to test the system performance under adverse circumstance. The overall P waveform fits are not very good (Fig. S3, VR = 45%), only four stations show pleasant matches between observations and synthetics. The preferred model (Fig. 8a) favors a compact slip around the hypocenter, which is consistent with the P waves. Encouragingly, the predicted tsunami wave (Fig. 8b) at coastal tide gauge re-produce the measurements reliably, the maximum 0.12 m tsunami height indicates a negligible tsunami risk and evacuation to higher land is not required.

We also conducted slip inversion based on the GCMT fault geometry (Fig. S3), while the data fits improve obviously (VR = 82%), the slip distribution remains compact around the epicenter, and corresponding tsunami waves show slight differences compared with result inferred from Slab2 geometry (Fig. 9b). Inaccurate fault geometry increases uncertainties of earthquake source estimation, but still reliable for tsunami early warning.

3.4. The 13 November 2016 M_w 7.8 Kaikoura Event

The earthquakes modelled above are well represented by a single fault. However, earthquake ruptures can be much more complex. For example, as one of the most complicated events ever instrumentally recorded (Hamling et al. 2017), the 2016 M_w 7.8 Kaikoura earthquake in New Zealand ruptured across more than 12 disparate faults with varying focal mechanisms, which raises the question of whether such a complex earthquake can be modeled automatically with confidence to guide tsunami response in real time. To investigate this problem, here we present a retrospective analysis of the Kaikoura earthquake inversion using GNSS offsets. Specially, there is no a priori fault geometry from Slab2, and we inferred the fault geometry based on the forecasted focal mechanisms with strike/dip/rake as 238°/50°/170°, together with an auxiliary plane as 324°/87°/53°. Note that GCMT solution of for the M_w 7.8 Kaikoura event are 226°/33°/141°.

We run inversions based on both nodal planes to identify the preferred fault geometry. Data fits of the fault plane 238°/50°/170° show VR as 52% and is significantly better than fault plane 238°/50°/170° with VR as 31%. The selected slip model and tide gauge predictions are shown in Fig. 9a. The rupture features a maximum slip of ~ 11 m, roughly 110 km northeast of the epicenter, very close to the GCMT centroid location. The derived magnitude is M_w 8.2, which overestimates the magnitude by 0.4 magnitude units compared with the GCMT solution. With respect to the tsunami observations, we find that even just based on a single fault, there is a remarkably good agreement between the predicted and recorded maximum tsunami amplitudes at near-field the tide gauges (see Fig. 9b), even though the far-field tsunamis are a bit amplified. In fact, Bai et al. (2017) showed that two regions of co-seismic seafloor deformation produced the tsunami, of which include a transpressional crustal faulting reaching offshore near Cook Strait, and that is consistent with our main slip patch.

We also synthetized tsunami waves using GCMT solution, and find its fit to amplitude at the nearest tide gauge KAIT is relatively poor, which could be caused by the underestimation of tsunamigenic slip (undersea slip) when expanding the centroid point solution to finite fault. We conclude that for rapid near-field tsunami amplitudes forecasting, even though the single fault model does not account for

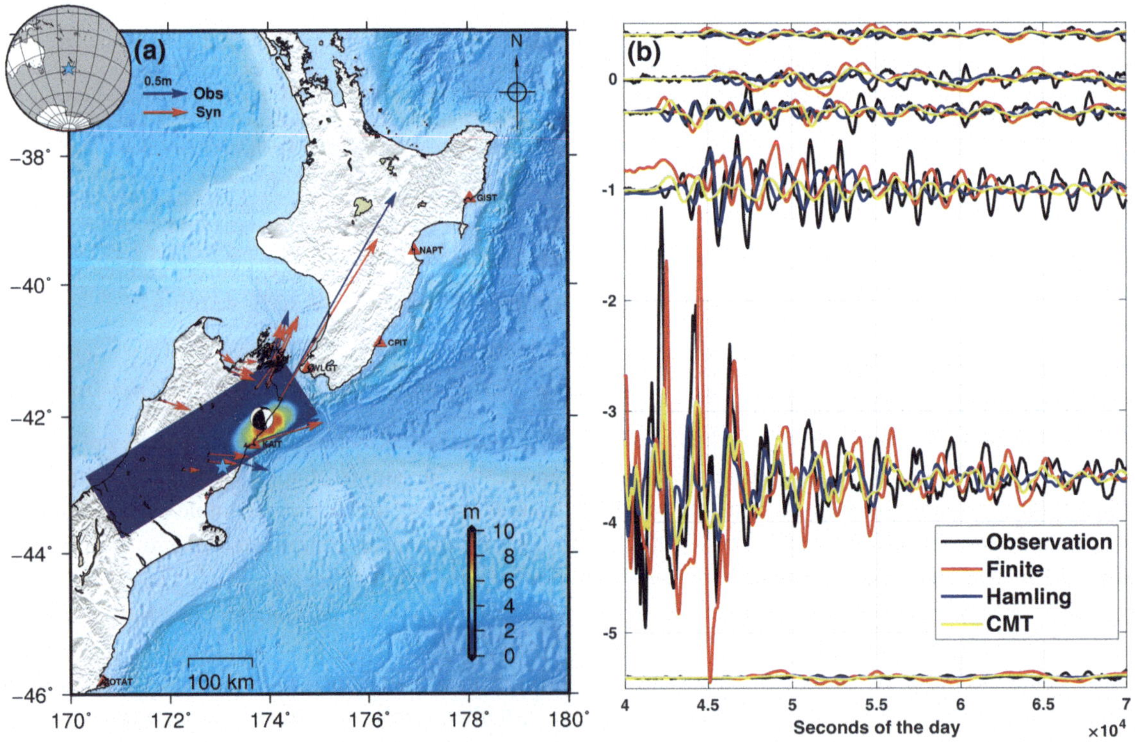

Figure 9

a Slip distribution inverted from a single fault plane for the 2016 New Zealand M_w 7.8 Kaikoura earthquake. USGS epicenter and GCMT solution are denoted by the blue star and black beachball, GNSS observations and fits, are shown by blue and red vectors, respectively. Tide gauges are indicated by red triangles. **b** The tsunami waves at six tide gauges. Black lines show observed wave amplitudes, red and blue are the predicted based our single fault model and multi-fault model provided by Hamling et al. (2017), yellow denotes predictions from the GCMT solution the source complexities, it is still a good-enough average source that provides valuable constraints.

3.5. *The 5 August 2018 M_w 7.0 Lombok Event*

The 5 August 2018 near coast M_w 7.0 Lombok earthquake, which caused more than 560 deaths and triggered a tsunami warning, provided one recent example to test how our system responds in real-time case. Here we present a timeline of the response to this damaging earthquake in the context of tsunami early warning. This earthquake originated at 11:46:35 UTC, and USGS-NEIC released the preliminary epicenter location (10 km default depth) and magnitude (M_L 7.0) about 1 min later, which initialized the automatic finite source inversion on JPL's GATED system. In spite that the epicenter falls in the projection of Slab2, its hypocenter depth as 10 km was too far away from the Slab2 depth 144 km there, which excluded an inter-plate thrusting rupture, and we had to use interpolated focal mechanism: strike/dip/rake either 90/21/96 or 264/69/88. Besides, the epicenter location indicated that there were no near-field GNSS stations available and the inversion had to rely on P waveforms at 10 teleseismic stations within 45° epicentral distance (see Fig. 10a). Once the two fault planes and station distribution were fixed, the computation of Green's functions and formation of normal equations began, which then took ~ 9 min to be accomplished on a desktop computer. At the same time, the system waited for the arrival of P displacement waveforms. Theoretically, it would take ~ 6 min for the P waves travelling to the nearest station PMG (with 30.37° epicentral distance) and ~ 8 min to the furthest station TAU (with 43.96° epicentral distance). P waveforms were automatically collected for inversion at about 11 min after the earthquake origin. The preferable inversion result and corresponding surface deformation were obtained ~ 13 min after the origin.

Moment magnitude of this event is M_w 7.0, data fits favor strike/dip/rake as 264/69/88 and are generally reasonable (see Fig. 10b, VR = 67%) and co-seismic slip distribution (see Fig. 10c) does not suggest a damaging tsunami because most of the ruptures are onshore. For this event, there is no publicly available tsunami data to evaluate the system performance. Alternatively, we conduct a comparison between the synthetic surface deformation and the Sentinel-1 observations (Fig. 10d, e). While the predicted deformations are the same order of magnitude as InSAR observations, it seems that the teleseismic inversion does not produce the slip distribution at the same location as suggested by the Sentinel-1 measurements, which shows the major slip concentrate northwest of the epicenter. This discrepancy is probably a result of the inherent poor spatial resolution of P waves. Nonetheless, the rapid solution excludes the possibility of a devasting tsunami and is very helpful to the local authority for risk management in the first minutes.

4. *Summary and Future Improvements*

4.1. *Summary*

In this paper, we have demonstrated how the two real-time observational networks—GNSS and seismic—can be automated together to have a denser global coverage for earthquake finite-source inversion and for tsunami early warning. The main innovative part of this work is the successful implementation of the proposed inversion algorithm in real-time operation. Though many components of the algorithm have been available for years and published previously as cited in the reference, automating them into an operational system is actually difficult and time consuming. Particularly for earthquakes, which can occur anywhere and anytime, without a global coverage to capture a real event and an automated system to process the event day and night, an algorithm could not be tested sufficient and reliable for confident applications in real time. Our system is currently in operation, and will have more opportunities to gain experience in using GNSS and seismic data. Through a detailed analysis of the five recent tsunamigenic events, we find that our automated finite source inversion jointly using GNSS and teleseismic networks is more reliable and effective than using only one of the two networks for tsunami early warning.

Our main strategy is to take the advantages of near-field (epicentral distance < 1000 km) GNSS

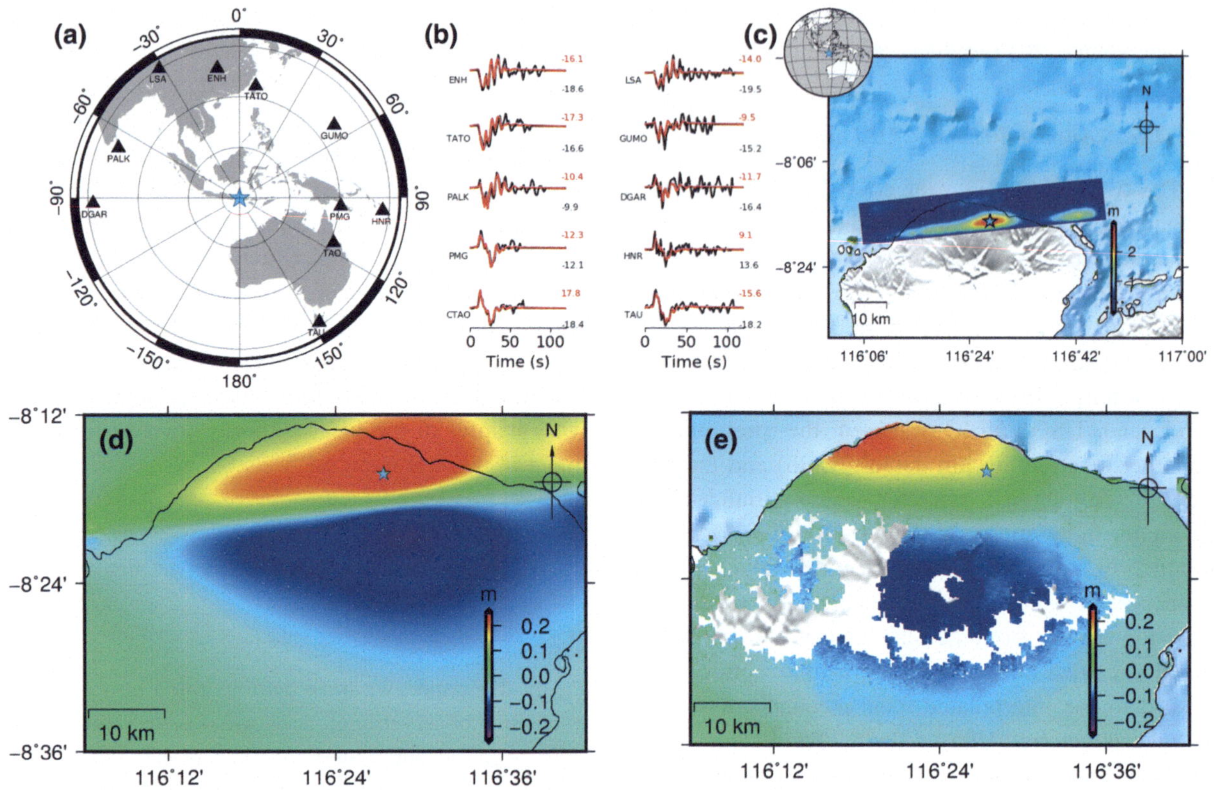

Figure 10
Summary of the 5 August 2018 Lombok event. **a** Distribution of teleseismic stations, blue star locates the epicenter. **b** Observations (in black) and synthetics (in red) of P displacement waveforms. **c** Map view of co-seismic slip distribution from teleseismic observations. **d** Synthetic line of sight deformation based on co-seismic slip distribution. **e** Sentineal-1 line of sight measurements between 30 July and 5 August, 2018

data and mid-range (epicentral distance from 30° to 45°) teleseismic P displacement waveforms for finite source inversion of large earthquakes. Neither the near-field GNSS nor the mid-range teleseismic data clip or saturate during large earthquakes, while the fast-traveling P-waves, within 6 ∼ 8 min away from the epicenter, are still essential to constrain the source in regions where GNSS data are not dense enough. Although many recent studies underscore the advantage of near-field GNSS for tsunami early warning through post-event studies, few applications using real-time GNSS have been reported in operational mode, partly due to its poor coverage and intrinsic accuracy. Jointly augmented with teleseismic data, our system not only takes the benefits of GNSS but also is now operational in real-time. To highlight our strategy, we re-analyzed all of the $M_w > 7.0$ shallow earthquakes (most are tsunamigenic subduction zone earthquakes) from 1995 to the end of 2017 (see Fig. 11), and find that near-filed GNSS data are only available in certain regions; therefore, relying solely on GNSS data would not provide enough tsunami warning information at a global scale (Fig. 11a). By contrast, our system is able to get earthquake source characters within 10 min on a moderate desktop workstation (Fig. 11b: green and purple dots). The computational time will certainly be reduced significantly once the system is matured and ported to parallel high-performance computing systems.

4.2. Future improvements

Clearly, rapid and reliable finite earthquake source inversion is a central part of our ongoing efforts to improve the robustness of tsunami early warning. However, finite source inversion is

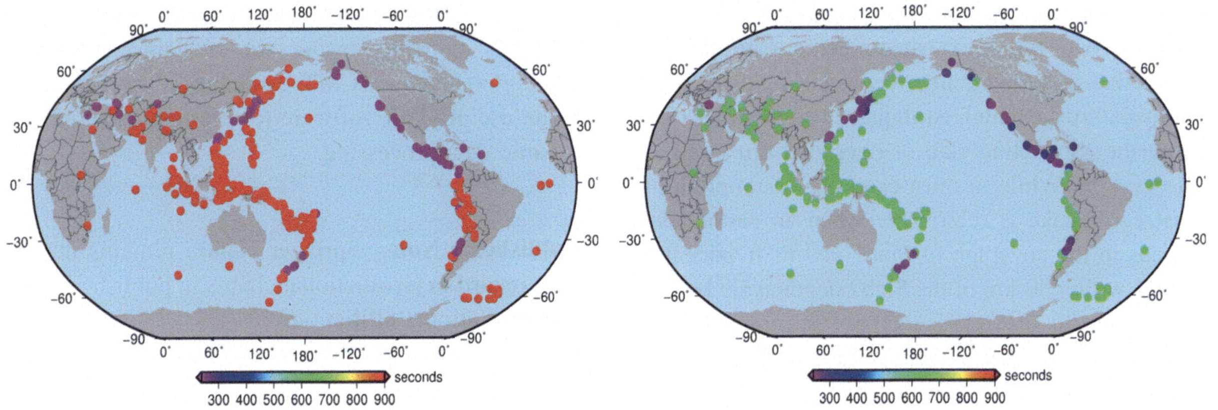

Figure 11
Time needed for earthquake source inversion: (right) using GNSS data only, and (left) using both GNSS and teleseismic data. The synthetic test is for $M_w > 7.0$ earthquakes (color dots on the map) since 1990. The color bar is saturated and > 900 s in right panel means we will never have a solution

inherently ill-posed (Ide 2015). Worse still, due to limited datasets, rapid finite source inversion is subject to even larger uncertainty. To this end, our future work will focus on adopting more real-time datasets and indicating inversion reliability.

Compared with static GNSS offsets, high-rate GNSS displacement waveforms contain timing information of local rupture propagation and can help constrain temporal resolution. What's more, GNSS displacement waveforms have the potential to improve near-trench resolution by reconciling with the depth-dependent megathrust rupture behaviors (Yue and Lay 2011). However, data loss and discontinuity in current high-rate GNSS can be problematic for waveforms in real time. Although the problem will be resolved with the advances in telecommunications, in current situation, the joint GNSS and seismic waveforms inversion is a better strategy for more reliable solutions.

Regional accelerogram waveforms are also suggested to increase the resolution of the slip distribution for large earthquakes and for moderate earthquakes (Chen et al. 2018). Recently, accelerator data have been archived by several large regional networks for public access, such as K-NET and KiK-net (http://www.kyoshin.bosai.go.jp) in Japan, California Strong Motion Instrumentation Program (http://www.consrv.ca.gov/cgs/smip) in United States. However, they are even more sparse than the GNSS data. Co-located with GNSS stations are few. We have gained some experience in using accelerogram waveforms in a moderate earthquake (Chen et al. 2018) and are looking forward to including the strong motion data into our system too.

Jointly using tsunami data for earthquake inversion has also been widely reported in literature. For example, Fujii et al. (2011) found that tsunami data could be used to constrain slip inversions and improve the resolution of near-trench shallow slip. By joint inversion of onshore GNSS data and offshore buoys, Melgar and Bock (2015) demonstrated that the source of 2011 M_w 9.0 Tohoku event can be better imaged which enhances the prediction of tsunami inundation and runup. However, most of the studies using tsunami data are after-event demonstration. Obtaining tsunami data has to wait for the tsunami to pass the observed location. In many cases, it is too late for near-field early warning, which is the focus of tsunami early warning in practice. In addition, the tsunami data inversion is subject to the assumption of the tsunami formation theory, which is a research topic that deserves further study.

A direct evaluation of the uncertainty of the slip model remains challenging since the true source is always unknown for a natural earthquake. While our system does not provide the uncertainty of inverted slip models currently, we acknowledge that there are some epistemic uncertainties which can affect the reliability of the models. For example, there are generally two types of tsunamigenic earthquakes: earthquakes at the plate interface (inter-plate subduction megathrust events), earthquakes at the outer rise, within the subducting slab or overlying crust (intra-plate events). While in most cases the assumption of thrust earthquake is reasonable when an earthquake happens in the coverage of Slab2 and its hypocenter depth is within 30 km of the Slab2 depth as set by our system, due to the relatively poor accuracy of the initial hypocenter depth, there is the possibility that it can be an intra-slab or outer rise normal faulting earthquake (e.g., the 2017 Mexico Tehuantepec M_w 8.2 event), which will lead to poor data fits because of a wrong focal mechanism adopted. Moreover, because teleseismic P waves have low resolution on the absolute location, slip location based on teleseismic P waves may be shifted due to the inaccuracy of an initial epicentral determination as shown in 5 August 2018 M_w 7.0 Lombok event. More research would be needed in this area.

We strongly believe that the research and improvements would be more effective by testing new methods and data in a real-time system, then carefully validating the results after event. With our algorithm implemented in the real-time system, we would have more chances to test it through real events, either large or moderate earthquakes. Our follow-up study will report the real-time performance and uncertainty statistics from the application of our algorithm to more real events.

Acknowledgements

We thank two anonymous reviewers for constructive comments. The research described here was conducted at the Jet Propulsion Laboratory, California Institute of Technology, under contracts with NASA. The Sentineal-1 InSAR measurements for 5 August 2018 M_w 7.0 Lombok event were provided by Advanced Rapid Imaging and Analysis project (ARIA) group at JPL and Caltech. The seismic waveforms were provided by Data Management Center of the Incorporated Research Institutions for Seismology. GNSS data were provided by JPL (www.gdgps.net) and GEONET of Japan. Slab2 is available at https://github.com/usgs/slab2 and Global catenoid moment tensors are available at https://www.globalcmt.org/CMTsearch.html. The work also made use of GMT software for plotting and SAC for seismic data processing.

References

Akaike, H. (1980). Likelihood and the Bayes procedure. *Trabajos de estadistica y de investigacion operativa, 31*(1), 143–166. https://doi.org/10.1007/BF02888350.

Ammon, C. J., Kanamori, H., Lay, T., & Velasco, A. A. (2006). The 17 July 2006 Java tsunami earthquake. *Geophysical Research Letters, 33*(24), 1–5. https://doi.org/10.1029/2006GL028005.

Babeyko, A. Y., Hoechner, A., & Sobolev, S. V. (2010). Source modeling and inversion with near real-time GPS: A GITEWS perspective for Indonesia. *Natural Hazards and Earth Systems Sciences, 10*(7), 1617–1627. https://doi.org/10.5194/nhess-10-1617-2010.

Báez, J. C., et al. (2018). The Chilean GNSS network: Current status and progress toward early warning applications. *Seismological Research Letters, 166*, 29–65. https://doi.org/10.1785/0220180011.

Bai, Y., Lay, T., Cheung, K. F., & Ye, L. (2017). Two regions of seafloor deformation generated the tsunami for the 13 November 2016, Kaikoura, New Zealand earthquake. *Geophysical Research Letters*. https://doi.org/10.1002/2017gl073717.

Becker, J. J., et al. (2009). Global bathymetry and elevation data at 30 arc seconds resolution: SRTM30_PLUS. *Marine Geodesy, 32*(4), 355–371. https://doi.org/10.1080/01490410903297766.

Bernard, E., & Titov, V. (2015). Evolution of tsunami warning systems and products. *Philosophical Transactions of the Royal Society A: Mathematical, Physical and Engineering Sciences., 373*(2053), 20140371. https://doi.org/10.1098/rsta.2014.0371.

Bilek, S. L., & Lay, T. (1999). Rigidity variations with depth along interplate megathrust faults in subduction zones. *Nature, 400*(6743), 443–446. https://doi.org/10.1038/22739.

Blewitt, G., Kreemer, C., Hammond, W. C., Plag, H.-P., Stein, S., & Okal, E. (2006). Rapid determination of earthquake magnitude using GPS for tsunami warning systems. *Geophysical Research Letters, 33*(11), L11309. https://doi.org/10.1029/2006GL026145.

Boatwright, J., & Choy, G. L. (1986). Teleseismic estimates of the energy radiated by shallow earthquakes. *Journal of Geophysical Research, 91*(B2), 2095. https://doi.org/10.1029/JB091iB02p02095.

Bock, Y., Nikolaidis, R. M., de Jonge, P. J., & Bevis, M. (2000). Instantaneous geodetic positioning at medium distances with the Global Positioning System. *Journal of Geophysical Research, 105*(B12), 28223. https://doi.org/10.1029/2000jb900268.

Chen, K., Babeyko, A., Hoechner, A., & Ge, M. (2016a). Comparing source inversion techniques for GPS-based local tsunami forecasting: A case study for the April 2014M 8.1 Iquique, Chile, earthquake. *Geophysical Research Letters, 43*(7), 3186–3192. https://doi.org/10.1002/2016gl068042.

Chen, K., Ge, M., Babeyko, A., Li, X., Diao, F., & Tu, R. (2016b). Retrieving real-time co-seismic displacements using GPS/GLONASS: A preliminary report from the September 2015 Mw 8.3 Illapel earthquake in Chile. *Geophysical Journal International, 206*(2), 941–953. https://doi.org/10.1093/gji/ggw190.

Chen, K., Liu, Z., Liang, C., & Song, Y. T. (2018). Towards the application of seismogeodesy in central Italy: A case study for the 2016 August 24 Mw 6.1 Italy earthquake modelling. *Geophysical Journal International, 213*(3), 1647–1658. https://doi.org/10.1093/gji/ggy089.

Contreras-López, M., et al. (2016). Field survey of the 2015 Chile tsunami with emphasis on coastal wetland and conservation areas. *Pure and Applied Geophysics, 173*(2), 349–367. https://doi.org/10.1007/s00024-015-1235-2.

Crowell, B. W., Bock, Y., & Melgar, D. (2012). Real-time inversion of GPS data for finite fault modeling and rapid hazard assessment. *Geophysical Research Letters*. https://doi.org/10.1029/2012gl051318.

Duputel, Z., Rivera, L., Kanamori, H., Hayes, G. P., Hirshorn, B., & Weinstein, S. (2011). Real-time W phase inversion during the 2011 off the Pacific coast of Tohoku Earthquake. *Earth, Planets and Space, 63*(7), 535–539. https://doi.org/10.5047/eps.2011.05.032.

Dziewonski, A. M., & Anderson, D. L. (1981). Preliminary reference earth model. *Physics of the Earth and Planetary Interiors, 25*(4), 297–356. https://doi.org/10.1016/0031-9201(81)90046-7.

Fujii, Y., Satake, K., Sakai, S., Shinohara, M., & Kanazawa, T. (2011). Tsunami source of the 2011 off the Pacific coast of Tohoku Earthquake. *Earth, Planets and Space, 63*(7), 815–820. https://doi.org/10.5047/eps.2011.06.010.

Fu, Y., Song, Y. T., & Gross, R. S. (2017). Linking oceanic tsunamis and geodetic gravity changes of large earthquakes. *Pure and Applied Geophysics, 174*(8), 3029–3041. https://doi.org/10.1007/s00024-017-1510-5.

Geng, J., Jiang, P., & Liu, J. (2017). Integrating GPS with GLONASS for high-rate seismogeodesy. *Geophysical Research Letters*. https://doi.org/10.1002/2017gl072808.

Hamling, I. J., et al. (2017). Complex multifault rupture during the 2016 M w 7.8 Kaikōura earthquake, New Zealand. *Science*. https://doi.org/10.1126/science.aam7194.

Hartzell, S. H., & Heaton, T. H. (1983). Inversion of strong ground motion and teleseismic waveform data for the fault rupture history of the 1979 Imperial Valley, California, earthquake. *Bulletin of the Seismological Society of America, 73*(6), 1553–1583.

Hayes, G. P., Earle, P. S., Benz, H. M., Wald, D. J., & Briggs, R. W. (2011). 88 hours: The U.S. Geological survey national earthquake information center response to the 11 March 2011 Mw 9.0 Tohoku earthquake. *Seismological Research Letters, 82*(4), 481–493. https://doi.org/10.1785/gssrl.82.4.481.

Hayes, G. P., Moore, G. L., Portner, D. E., Hearne, M., Flamme, H., Furtney, M., et al. (2018). Slab2, a comprehensive subduction zone geometry model. *Science (80-.)*. https://doi.org/10.1126/science.aat4723.

Hayes, G. P., & Wald, D. J. (2009). Developing framework to constrain the geometry of the seismic rupture plane on subduction interfaces a priori—A probabilistic approach. *Geophysical Journal International, 176*(3), 951–964. https://doi.org/10.1111/j.1365-246X.2008.04035.x.

Hayes, G. P., Wald, D. J., & Johnson, R. L. (2012). Slab1.0: A three-dimensional model of global subduction zone geometries. *Journal of Geophysical Research: Solid Earth, 117*(1), 1–15. https://doi.org/10.1029/2011jb008524.

Hirshorn, B., Weinstein, S., & Tsuboi, S. (2013). On the application of Mwp in the near field and the March 11, 2011 Tohoku Earthquake. *Pure and Applied Geophysics, 170*(6–8), 975–991. https://doi.org/10.1007/s00024-012-0495-3.

Hoshiba, M., Iwakiri, K., Hayashimoto, N., Shimoyama, T., Hirano, K., Yamada, Y., et al. (2011). Outline of the 2011 off the Pacific coast of Tohoku Earthquake (M w 90)—Earthquake Early Warning and observed seismic intensity. *Earth, Planets and Space, 63*(7), 547–551. https://doi.org/10.5047/eps.2011.05.031.

Husen, S., & Hardebeck, J. (2010). Earthquake location accuracy. *Community Online Resource Static Seismic Analysis*. https://doi.org/10.5078/corssa-55815573.

Ide, S. (2015). Slip inversion. In G. C. Beroza, & H. Kanamori, (Eds.), *Earthquake Seismology, Treatise on Geophysics* (vol. 4, pp. 215–241). Elsevier: Amsterdam, The Netherlands.

Ide, S., Takeo, M., & Yoshida, Y. (1996). Source process of the 1995 kobe earthquake: Determination of spatio-temporal slip distribution by bayesian modeling. *Bulletin of the Seismological Society of America, 86*(3), 547–566.

Inazu, D., Pulido, N., Fukuyama, E., Saito, T., Senda, J., & Kumagai, H. (2016). Near-field tsunami forecast system based on near real-time seismic moment tensor estimation in the regions of Indonesia, the Philippines, and Chile 4. *Seismology, Earth, Planets and Space*. https://doi.org/10.1186/s40623-016-0445-x.

JMA. (2013). Lessons learned from the tsunami disaster caused by the 2011 Great East Japan Earthquake and improvements in JMA's tsunami warning system. http://www.data.jma.go.jp/svd/eqev/data/en/tsunami/LessonsLearned_Improvements_brochure.pdf.

Kagan, Y. Y. (1991). Correlations of earthquake focal mechanisms. *Geophysical Journal International, 143*(3), 881–897. https://doi.org/10.1046/j.1365-246X.2000.00281.x.

Kagan, Y. Y., & Jackson, D. D. (1994). Long-term probabilistic forecasting of earthquakes. *Journal of Geophysical Research, 99*(B7), 13613–685700. https://doi.org/10.1029/94jb00500.

Kanamori, H. (1972). Mechanism of tsunami earthquakes. *Physics of the Earth and Planetary Interiors, 6*(5), 346–359. https://doi.org/10.1016/0031-9201(72)90058-1.

Kanamori, H. (1977). The energy release in great earthquakes. *Journal of Geophysical Research, 82*(20), 2981–2987. https://doi.org/10.1029/JB082i020p02981.

Kanamori, H., & Rivera, L. (2008). Source inversion of W phase: Speeding up seismic tsunami warning. *Geophysical Journal International, 175*(1), 222–238. https://doi.org/10.1111/j.1365-246X.2008.03887.x.

Kawamoto, S., Hiyama, Y., Ohta, Y., & Nishimura, T. (2016). First result from the GEONET real-time analysis system (REGARD): The case of the 2016 Kumamoto earthquakes. *Earth, Planets and Space*. https://doi.org/10.1186/s40623-016-0564-4.

Kawamoto, S., Ohta, Y., Hiyama, Y., Todoriki, M., Nishimura, T., Furuya, T., et al. (2017). REGARD: A new GNSS-based real-time finite fault modeling system for GEONET. *Journal of Geophysical Research: Solid Earth, 122*(2), 1324–1349. https://doi.org/10.1002/2016JB013485.

Kikuchi, M., & Kanamori, H. (1982). Inversion of complex body waves. *Bulletin of the Seismological Society of America, 72*(2), 491–506.

Kubo, A., Fukuyama, E., Kawai, H., & Nonomura, K. (2002). NIED seismic moment tensor catalogue for regional earthquakes around Japan: Quality test and application. *Tectonophysics, 356*(1–3), 23–48. https://doi.org/10.1016/S0040-1951(02)00375-X.

Larson, K. M., Bodin, P., & Gomberg, J. (2003). Using 1-Hz GPS data to measure deformations caused by the Denali fault earthquake. *Science, 300*(5624), 1421–1424. https://doi.org/10.1126/science.1084531.

Li, X., Ge, M., Guo, B., Wickert, J., & Schuh, H. (2013). Temporal point positioning approach for real-time GNSS seismology using a single receiver. *Geophysical Research Letters, 40*(21), 5677–5682. https://doi.org/10.1002/2013GL057818.

Liu, Z., Owen, S., & Moore, A. (2014). Rapid estimate and modeling of permanent coseismic displacements for large earthquakes using high-rate global positioning system data. *Seismological Research Letters, 85*(2), 284–294. https://doi.org/10.1785/0220130174.

MacInnes, B. T., Gusman, A. R., LeVeque, R. J., & Tanioka, Y. (2013). Comparison of earthquake source models for the 2011 Tohoku event using tsunami simulations and near-field observations. *Bulletin of the Seismological Society of America, 103*(2B), 1256–1274. https://doi.org/10.1785/0120120121.

Melgar, D., & Bock, Y. (2015). Kinematic earthquake source inversion and tsunami runup prediction with regional geophysical data. *Journal of Geophysical Research: Solid Earth, 120*(5), 3324–3349. https://doi.org/10.1002/2014JB011832.

Melgar, D., Bock, Y., & Crowell, B. W. (2012). Real-time centroid moment tensor determination for large earthquakes from local and regional displacement records. *Geophysical Journal International, 188*(2), 703–718. https://doi.org/10.1111/j.1365-246X.2011.05297.x.

Melgar, D., LeVeque, R. J., Dreger, D. S., & Allen, R. M. (2016). Kinematic rupture scenarios and synthetic displacement data: An example application to the Cascadia subduction zone. *Journal of Geophysical Research: Solid Earth, 121*(9), 6658–6674. https://doi.org/10.1002/2016JB013314.

Mendoza, C., & Hartzell, S. (2013). Finite-fault source inversion using teleseismic P waves: Simple parameterization and rapid analysis. *Journal of Geophysical Research: Solid Earth, 103*(2A), 834–844. https://doi.org/10.1785/0120120069.

Mori, N., Takahashi, T., Yasuda, T., & Yanagisawa, H. (2011). Survey of 2011 Tohoku earthquake tsunami inundation and runup. *Geophysical Research Letters*. https://doi.org/10.1029/2011gl049210.

Okada, Y. (1985). Surface deformation due to shear and tensile faults in a half-space. *Bulletin of the Seismological Society of America, 75*(4), 1135–1154.

Ohta, Y., Kobayashi, T., Tsushima, H., Miura, S., Hino, R., Takasu, T., et al. (2012). Quasi real-time fault model estimation for near-field tsunami forecasting based on RTK-GPS analysis: application to the 2011 Tohoku-Oki earthquake (Mw 9.0). *Journal of Geophysical Research, 117*, B02311. https://doi.org/10.1029/2011JB008750.

Okuwaki, R., Yagi, Y., Aránguiz, R., González, J., & González, G. (2016). Rupture Process During the 2015 Illapel, Chile Earthquake: Zigzag-Along-Dip Rupture Episodes. *Pure and Applied Geophysics*. https://doi.org/10.1007/s00024-016-1271-6.

Richter, C. F. (1935). An instrumental earthquake magnitude scale. *Bulletin of the Seismological Society of America, 25,* 1–32.

Riquelme, S., Bravo, F., Melgar, D., Benavente, R., Geng, J., Barrientos, S., et al. (2016). W phase source inversion using high-rate regional GPS data for large earthquakes. *Geophysical Research Letters, 43*(7), 3178–3185. https://doi.org/10.1002/2016GL068302.

Sobolev, S. V., Babeyko, A. Y., Wang, R., Hoechner, A., Galas, R., Rothacher, M., et al. (2007). Tsunami early warning using GPS-Shield arrays. *Journal of Geophysical Research, 112*(B8), B08415. https://doi.org/10.1029/2006JB004640.

Song, Y. T. (2007). Detecting tsunami genesis and scales directly from coastal GPS stations. *Geophysical Research Letters, 34*(19), L19602. https://doi.org/10.1029/2007GL031681.

Song, Y. T., Fu, L. L., Zlotnicki, V., Ji, C., Hjorleifsdottir, V., Shum, C. K., et al. (2008). The role of horizontal impulses of the faulting continental slope in generating the 26 December 2004 tsunami. *Ocean Modelling, 20*(4), 362–379. https://doi.org/10.1016/j.ocemod.2007.10.007.

Song, Y. T., Fukumori, I., Shum, C. K., & Yi, Y. (2012). Merging tsunamis of the 2011 Tohoku-Oki earthquake detected over the open ocean. *Geophysical Research Letters*. https://doi.org/10.1029/2011gl050767.

Song, Y. T., Mohtat, A., & Yim, S. C. (2017). New insights on tsunami genesis and energy source. *Journal of Geophysical Research: Oceans*. https://doi.org/10.1002/2016jc012556.

Tanioka, Y., & Satake, K. (1996). Tsunami generation by horizontal displacement of ocean bottom. *Geophysical Research Letters, 23*(8), 861–864. https://doi.org/10.1029/96GL00736.

Tappin, D. R., Grilli, S. T., Harris, J. C., Geller, R. J., Masterlark, T., Kirby, J. T., et al. (2014). Did a submarine landslide contribute to the 2011 Tohoku tsunami? *Marine Geology, 357,* 344–361. https://doi.org/10.1016/j.margeo.2014.09.043.

Titov, V. V., Gonzalez, F. I., Bernard, E. N., Eble, M. C., Mofjeld, H. O., Newman, J. C., et al. (2005). Real-time tsunami forecasting: Challenges and solutions. *Natural Hazards, 35*(1), 35–41. https://doi.org/10.1007/s11069-004-2403-3.

Titov, V. V, Mofjeld, H. O., & González, F. I. (1999). Offshore forecasting of Hawaiian tsunamis generated in Alaskan-Aleutian subduction zone, *NOAA Tech*.

Titov, V., Song, Y. T., Tang, L., Bernard, E. N., Bar-Sever, Y., & Wei, Y. (2016). Consistent estimates of tsunami energy show promise for improved early warning. *Pure and Applied Geophysics, 173*(12), 3863–3880. https://doi.org/10.1007/s00024-016-1312-1.

Weinstein, S. A., & Lundgren, P. R. (2008). Finite fault modeling in a tsunami warning center context. *Pure and Applied Geophysics, 165*(3–4), 451–474. https://doi.org/10.1007/s00024-008-0316-x.

Wells, D. L., & Coppersmith, K. J. (1994). New empirical relationships among magnitude, rupture length, rupture width, rupture area, and surface displacement. *Bulletin of the Seismological Society of America, 84*(4), 974–1002.

Yang, Y.-M., Meng, X., Komjathy, A., Verkholyadova, O., Langley, R. B., Tsurutani, B. T., et al. (2014). Tohoku-Oki earthquake caused major ionospheric disturbances at 450 km altitude over Alaska. *Radio Science, 49*(12), 1206–1213. https://doi.org/10.1002/2014RS005580.

Yue, H., & Lay, T. (2011). Inversion of high-rate (1 sps) GPS data for rupture process of the 11 March 2011 Tohoku earthquake

(Mw 9.1). *Geophysical Research Letters, 38,* L00G09. https://doi.org/10.1029/2011gl048700.

Yue, H., & Lay, T. (2013). Source Rupture Models for the Mw 9.0 2011 Tohoku Earthquake from Joint Inversions of High-Rate Geodetic and Seismic Data. *Bulletin of the Seismological Society of America, 103*(2B), 1242–1255. https://doi.org/10.1785/0120120119.

Yue, H., Lay, T., Rivera, L., Bai, Y., Yamazaki, Y., Cheung, K. F., et al. (2014). Rupture process of the 2010 M w 7.8 Mentawai tsunami earthquake from joint inversion of near-field hr-GPS and teleseismic body wave recordings constrained by tsunami observations. *Journal of Geophysical Research: Solid Earth, 119*(7), 5574–5593. https://doi.org/10.1002/2014jb011082.

Zhu, L., & Rivera, L. A. (2002). A note on the dynamic and static displacements from a point source in multilayered media. *Geophysical Journal International, 148*(3), 619–627. https://doi.org/10.1046/j.1365-246X.2002.01610.x.

(Received March 23, 2019, revised May 30, 2019, accepted June 7, 2019, Published online June 18, 2019)

Pure Appl. Geophys. 177 (2020), 1425–1436
© 2019 Springer Nature Switzerland AG
https://doi.org/10.1007/s00024-019-02377-z

Pure and Applied Geophysics

Evaluation of Australian Tsunami Warning Thresholds Using Inundation Modelling

Diana J. M. Greenslade,[1] Burak Uslu,[2] Stewart C. R. Allen,[1] Claire L. Kain,[3] Kaya M. Wilson,[4] and Hannah E. Power[4]

Abstract—Tsunami warnings issued by the Joint Australian Tsunami Warning Centre (JATWC) are derived from a database (T2) consisting of more than two thousand pre-computed tsunami scenarios. Following any potentially tsunamigenic earthquake, warnings are issued for individual coastal zones with three different levels of threat: Land Threat, Marine Threat or No Threat. The decision is based on the 95th percentile (P95) of the maximum wave amplitudes (over time) of the relevant T2 scenario within each coastal zone. Threshold values for P95 have previously been derived through analysis of observed impacts for recent events. Given that historical records are available for only a short time period and no observations exist for which a Land Threat would have been issued for Australia, it has been difficult to determine the appropriate threshold for a Land Threat. Several recent tsunami hazard assessment studies have used inundation models nested within T2 scenarios. These modelling results are used to evaluate the threshold values for JATWC tsunami warnings and provide guidance on a possible further warning tier—Major Land Threat. The optimal Land Threat threshold for P95 is found to be 48.5 cm, however, it is not recommended that any changes are made from the existing operational threshold of 55 cm. The optimal threshold for P95 a Major Land Threat is found to be 150.5 cm.

Keywords: Tsunami, tsunami warning, inundation modelling.

1. Introduction

After the Indian Ocean Tsunami of 26 December 2004, the Australian Government committed funding to the development of the Australian Tsunami Warning System (ATWS). The Joint Australian Tsunami Warning Centre (JATWC) was formally established as part of the ATWS in 2008 to provide Australia with an independent capability to detect, monitor, verify and warn the community of possible threats to Australian coastal locations and offshore territories. The JATWC is also one of three Tsunami Service Providers (TSPs) for the Indian Ocean, providing threat information to National Tsunami Warning Centres (NTWCs) of other Indian Ocean countries to help them determine the level of threat and warnings they will issue for their communities.

JATWC tsunami threat assessments are derived from the ‘T2’ scenario database consisting of more than two thousand pre-computed tsunami scenarios (Greenslade et al. 2009; Greenslade et al. 2010; Simanjuntak et al. 2011). Following any potentially tsunamigenic earthquake, the most relevant T2 scenario is inspected and warning requirements are assessed. The T2 scenarios are only computed in water depths greater than 20 m, so they are unable to directly simulate tsunami impacts at the coastline. Therefore, the decision to issue a warning is based on comparing wave amplitudes of the relevant T2 scenario within coastal zones (which extend 60 nautical miles offshore) to pre-determined threshold values. This allows the T2 scenarios to serve as proxies for coastal impacts of tsunamis.

There are currently three different levels of threat: Land, Marine or No threat. Further details of this technique are described in Sect. 2.2. Threshold amplitude values for each of these tiers of warning have been derived through analysis of observed impacts for recent events (Allen and Greenslade 2010; Greenslade and Allen 2019). Given that historical records are available for only a short period, there are no known events for which a Land Threat would have been issued for Australia. Thus it has been difficult to determine the appropriate threshold for a Land Threat and this is currently set using a conservative approach (see Sect. 2.2).

[1] Bureau of Meteorology, Melbourne, Australia. E-mail: diana.greenslade@bom.gov.au
[2] OMC International, Melbourne, Australia.
[3] Department of State Growth, Hobart, Australia.
[4] University of Newcastle, Newcastle, Australia.

In recent years, a number of tsunami hazard assessment studies have been undertaken for the Australian coastline. Some of these studies have used inundation models nested within T2 scenarios (e.g. New South Wavles (NSW) State Emergency Service and Office of Environment and Heritage (2012); Wilson et al. 2018). Under the assumption that the models are appropriately benchmarked (Synolakis et al. 2008) and the inundation modelling predictions are accurate, the model results can be used to evaluate the Marine and Land Threat threshold values for the JATWC tsunami warnings. Initial work towards this is described in Uslu and Greenslade (2013). In the present study, that work is extended with further inundation modelling results, and the possibility of an additional threat level is investigated.

This paper presents the results of using inundation modelling to assess the JATWC warning strategy. Section 2 presents some background, describing the relevant features of the T2 scenario database and the warning strategy. Section 3 gives an overview of the four inundation modelling studies considered in this work. Section 4 describes the method used to assess the warning thresholds, and Sect. 5 presents the results. These results are discussed in Sect. 6 and conclusions are presented in Sect. 7.

2. Background

2.1. T2 Scenario Database

The T2 scenario database was developed to provide forecast guidance for the JATWC and contains pre-computed tsunami propagation results (amplitudes and currents) from tsunamis generated on subduction zones in the Pacific, South Atlantic (i.e. South Sandwich trench) and Indian Oceans. The scenarios are computed with the MOST (Method of Splitting Tsunami) model (Titov and Synolakis 1998; Titov et al. 2016) where the earthquake sources are modelled with the static and instantaneous displacement of the free water surface assumption from Okada (1985). The model grid has a spatial resolution of 4 arc minutes and a reflective boundary at a depth of 20 m. The database currently consists of 2069 scenarios and with appropriate scaling, can provide guidance for any thrust earthquake from Mw = 6.5 to Mw = 9.5 along the major fault lines. Figure 1 shows the source locations of the T2 database scenarios used in this study.

2.2. Warning Thresholds

During a tsunami event, the JATWC issues one of three levels of threat for each Australian Marine Forecast Zone, i.e. coastal zone. The coastal zones used for tsunami warnings are the same as those used for the Bureau of Meteorology's coastal wind warnings. Those relevant to the present study are reproduced in Fig. 2. Maximum amplitudes (over time) from the T2 scenario with the nearest epicentre and magnitude to the earthquake are used. If necessary, the wave amplitudes from that scenario are linearly scaled to the appropriate seismic magnitude (Greenslade et al. 2009). Maximum computed amplitudes at each grid point within each coastal zone are ranked and the 95th percentile (P95) is computed for each zone. If P95 is less than 20 cm, "No Threat" is assigned to the zone. "Marine Threat" is assigned if P95 is between 20 cm and 55 cm, and "Land Threat" is issued for anything larger. A "Marine Threat" means the tsunami may cause potentially dangerous waves and strong ocean currents in the marine environment, but does not require evacuations, whereas a "Land Threat" anticipates possible inundation and a major hazard for low lying coastal areas that will require evacuation. The recommended actions in the event of a Land Threat are that people should move at least 1 km inshore, or at least 10 m above sea level.

The derivation of the warning thresholds (20 cm, 55 cm) is described in Allen and Greenslade (2010). In summary, observed coastal impacts for 9 past events led to retrospective or "ideal" warning schemes being designed for each event. The P95 values from the relevant T2 scenario within the designated coastal zones were examined and thresholds that produced the best match for the ideal schemes were selected, leading to the 20 cm and 55 cm thresholds. This analysis was revisited in Greenslade and Allen (2019) with the addition of 8 further events (and the removal of one) and a more quantitative technique used to derive the threshold

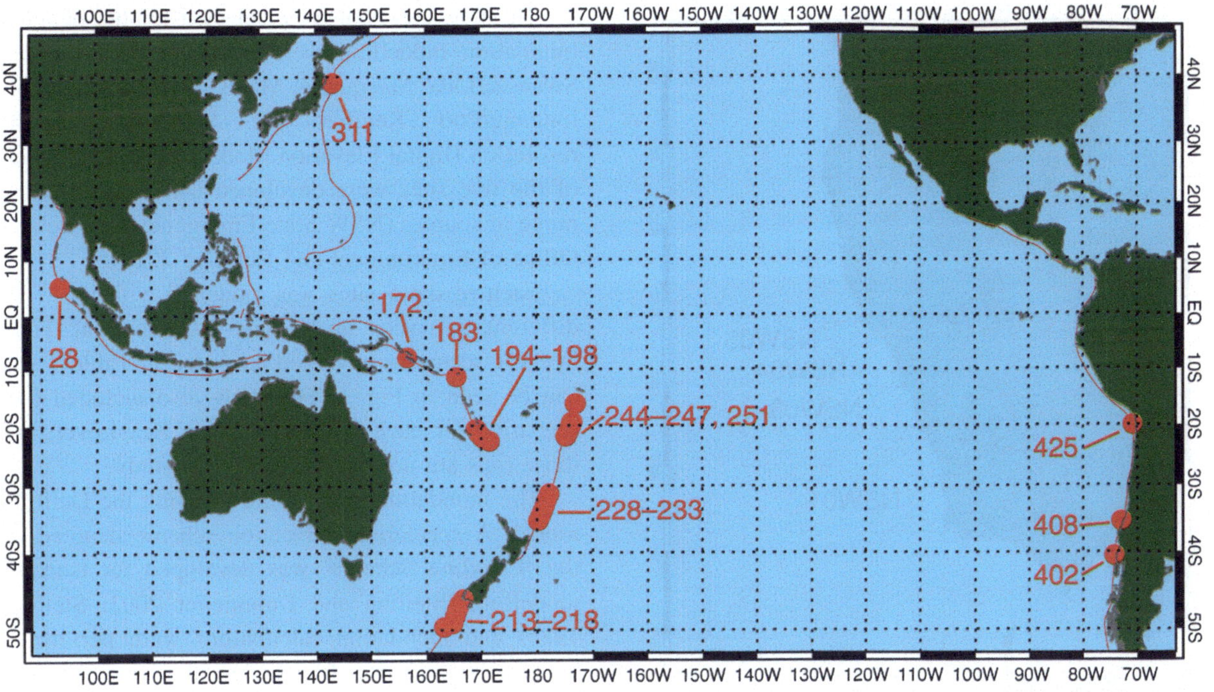

Figure 1
T2 database scenario source locations used in this study (red dots). Scenario numbers are indicated

values. The results of that study suggest that the existing thresholds were appropriate for mainland zones.

Notably, no zones from any of the 16 past events examined were deemed to have experienced significant or widespread inundation of land areas requiring the issuance of a Land Threat. This means that the method provided no specific guidance about where to set the threshold for a Land Threat, so a conservative approach was taken and the Land Threat threshold was set just above the highest Marine Threat value.

One of the issues with the current warning structure is that the Land Threat is 'one-size-fits-all'. The recommended actions for a Land Threat assume that there will be significant and widespread inundation within the coastal zone, requiring extensive evacuation. This will not necessarily be the case. Through ongoing engagement with stakeholders, including emergency services, an additional tier in the threat levels has been proposed, namely a 'Major Land Threat'. The present work allows an assessment of the P95 value required for this new level.

3. Tsunami inundation modelling studies

As noted in the Introduction, a number of tsunami hazard assessment studies have recently been undertaken using inundation models nested within T2 scenarios. Two of these were focussed on locations on the NSW coast (NSW State Emergency Service and Office of Environment and Heritage 2012; Wilson et al. 2018). A further study was also recently undertaken for southeast Tasmania (Kain et al. 2018) and while that study did not use T2 scenarios for the hazard assessment, two scenarios were subsequently run with T2 boundary conditions in order to contribute to the present study. Finally, a pilot inundation study has been undertaken at the Bureau of Meteorology as a precursor to a potential operational forecast system (Allen and Greenslade 2016). All of these studies can contribute to the evaluation of the threshold values used for determining the level of threat. In this section, we provide an overview of these inundation modelling studies.

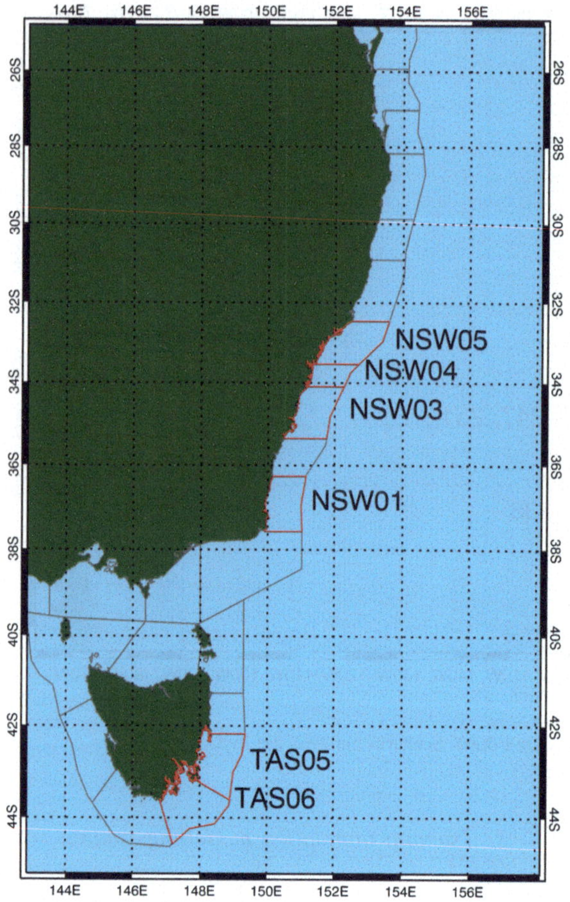

Figure 2
The Australian Marine Forecast Zones used in this study are shown in red polygons

3.1. New South Wales (NSW) Study

The NSW Tsunami Risk Assessment project was funded under the Attorney–General's Department; Natural Disaster Mitigation Program and managed by the NSW Office of Environment and Heritage and the NSW State Emergency Service. It was undertaken in two stages, with the first stage being a broad risk assessment of the entire NSW coast (Somerville et al. 2009) and the second stage comprising detailed inundation modelling of a number of areas that stage one had identified as being the most vulnerable. The environmental services company Cardno was engaged to undertake the inundation modelling and risk assessment (NSW State Emergency Service and Office of Environment and Heritage 2012).

The five sites that were selected for the detailed inundation modelling were, from North to South, Swansea/Lake Macquarie, Manly, Botany Bay, Wollongong/Port Kembla and Merimbula. High-resolution Digital Elevation Models (DEMs) for each of the five sites were developed using data from a range of sources (NSW State Emergency Service and Office of Environment and Heritage 2012).

Each of the 5 sites was modelled with up to 19 different scenarios at Highest Astronomical Tide. Specific scenarios modelled are listed in Table 1 and can be seen in Fig. 1. The study also included five scenarios for each site using Mean Sea Level, but those runs are not considered in this study.

The inundation was computed using the Delft3D model. A specialised advection scheme denoted as the 'Flooding Scheme' was developed for tsunami modelling (Stelling and Duinmeijer 2003; Stelling 1984) and inundation calculations were performed using a wetting and drying algorithm. Forcing was provided at a coarse outer grid via time series extracted from the T2 scenario database and the wave was passed through 3 to 5 progressively higher resolution nested grids until a 10 m resolution grid, where the inundation calculations were performed. Extensive model calibration, verification and examination of various model sensitivities can be found in NSW State Emergency Service and Office of Environment and Heritage (2012).

3.2. Sydney Study

Tsunami inundation modelling was carried out by the University of Newcastle for Sydney Harbour (Port Jackson) as a centre of low-lying development in an estuary with complex geomorphology. Modelled scenarios include tsunamis generated from earthquakes of magnitude 7.5, 8.0, 8.5 and 9.0 Mw from both the Puysegur and New Hebrides trenches as well as representations of the historic Chile event of May, 1960 and the Tohoku event of March, 2011 (see Table 1). Scenarios were modelled with dynamic tides and represent situations where the peak tsunami wave arrives on both a low tide and a high tide, during spring tides. The hydrodynamic model ANUGA (Nielsen et al. 2005) was used for the inundation modelling. Further details of the

Table 1

Summary of inundation modelling studies and T2 scenarios

Zone (number of scenarios)	Site	Study	Scenarios (magnitude)
NSW 5 (17)	Swansea	Cardno	194d (9.0, 9.1, 9.2, 9.3), 196c (8.7), 213d (9.1), 215c (8.7), 216d (9.0, 9.1, 9.2), 217c (8.7), 232d (9.0), 233d (8.9, 9.2), 245d (9.2, 9.3), 402d (9.3)
NSW 4 (28)	Manly	Cardno	194d (9.0, 9.1, 9.2, 9.3), 196c (8.7), 215c (8.6, 8.7), 216d (8.8, 8.9, 9.0, 9.1), 228d (9.3), 231d (9.1, 9.2), 232d (9.0), 245d (9.1, 9.2), 402d (9.3)
	Sydney	Wilson et al.	194d (9.0), 197c (8.5), 198b (8.0), 213b (8.0), 214a (7.5), 216d (9.0), 217c (8.5), **311c (9.1),** 401d (9.5)
	Botany Bay	Cardno	194d (9.0, 9.1, 9.2, 9.3), 196c (8.7), 215c (8.6, 8.7), 216d (8.8, 8.9, 9.0, 9.1), 228d (9.3), 231d (9.2), 232d (9.2), 233d (9.1), 244d (9.2), 245d (9.1), 402d (9.2)
NSW 3 (23)	Port Kembla	Allen and Greenslade	28d (9.1), 172b (8.1), 183b (7.9), 214a (7.4), 216d (9.0), **218a (7.8),** 251b (8.1), **311c (9.0), 408c (8.8),** 425b (8.2)
		Cardno	194d (8.9, 9.0, 9.1, 9.2, 9.3), 196c (8.6), **215c** (8.6, **8.8**), 216d (8.9, 9.0, 9.1), 231d (9.2), 232d (8.9), 233d (9.1, 9.2), 245d (9.3), 247d (9.2), 402d (9.3)
NSW 1 (19)	Merimbula	Cardno	194d (8.9, 9.0, 9.1, 9.2, 9.3), 195c (8.7), 215c (8.6), 216c (8.4, 8.7), 216d (8.8, 8.9, 9.0), 229d (9.2, 9.3), 230d (9.0), 231d (9.2), 246d (9.1, 9.2), 402d (9.3)
TAS 5 (2)	Multiple	Kain et al.	216c (8.7), 194d (9.0)
TAS 6 (2)	Multiple	Kain et al.	216c (8.7), 194d (9.0)

Scenarios in bold refer to scenarios that are scaled beyond operational protocols and thus not used in this study. Number of scenarios refers to those used in the evaluation, i.e. excludes duplicates and 'non-operational' scenarios

modelling setup and the results can be found in Wilson et al. (2018).

3.3. Port Kembla Study

The Bureau of Meteorology recently undertook a pilot tsunami inundation modelling study (Allen and Greenslade 2016). The aim of the work was to assess the feasibility of developing operational tsunami inundation modelling as part of the JATWC warning system. Inundation modelling was undertaken for just one site, Port Kembla on the NSW coast. This site was chosen as it was one of the 5 sites identified as 'at risk' in the NSW study described in Sect. 3.1 and the Digital Elevation Model (DEM) was available from that study. In addition, high quality tide gauge observations from within the harbour were available.

Nine historical events and one hypothetical event were modelled (see Table 1). Only the hypothetical event showed any inundation, but in the present study, all model results are evaluated, because 'No threat' information is still valuable for assessing the threshold values. The Method Of Splitting Tsunami (MOST) model (Titov and Synolakis 1998; Titov et al. 2016) was used for the inundation modelling. The inundation model consisted of three nested grids with the smallest grid having dimensions of approximately 11 km by 13 km and a spatial resolution of 30 m.

3.4. Southeast Tasmania Study

A hazard assessment study has recently been undertaken for south-east Tasmania, focussing on a 'maximum credible' scenario (Kain et al. 2018). That study did not use T2 scenarios for boundary forcing, but two scenarios were subsequently run with T2 boundary conditions in order to contribute to the present study. Kain et al. (2018) describes the development of a high-quality DEM using recently available bathymetric and topographic data. Modelling was performed with the ANUGA model (Nielsen et al. 2005) and the grid used was an unstructured triangular mesh, with resolutions varying from 400 m (triangle side length) in open ocean areas, to 200 m in waters of depth 10 m to 30 m, 50 m for non-urban coastal areas, 20 m for urban areas of interest and two specific locations of interest had grids of 10 m. A variable coefficient of roughness was used according to surface type.

For the present study two T2 scenarios were selected to provide boundary conditions for the inundation modelling (see Table 1). These specific scenarios were selected as their P95 values were close to the existing Land Threat thresholds for the JATWC system, and thus would be able to provide good guidance for the present study.

4. Method

Since the inundation modelling studies described in the previous section use the same underlying data as the JATWC warning system, i.e. the T2 scenario database, the outcomes provide valuable information that can be applied to the warning system. In particular, as noted previously, there has been a lack of guidance on the value at which to set the Land Threat threshold, given that there have been no major inundation events in recent Australian history. The proposed Major Land Threat threshold has not previously been defined. Under the assumption that the inundation modelling results provide accurate predictions, they can be used to assess the validity of the Land and Marine Threat threshold values for the JATWC tsunami warnings, and provide guidance for the Major Land Threat threshold value.

In the present work, maps of maximum runup, flow depths, and inundation extents from each of the model runs described in the previous section were inspected. These maps show detailed land use of each site, indicating what type of infrastructure, if any, might be impacted in the event of a tsunami. For each site and each scenario modelled, a threat level (i.e. No threat, Marine Threat, Land Threat, Major Land Threat) was assigned for each event. In general, a Major Land Threat was assigned if there was significant and widespread inundation with flow depth greater than 2 m; a Land Threat was assigned if there was some inundation, and maximum flow depth was generally less than 2 m; A Marine Threat was assigned if there was no or little inundation, but low level wave activity was evident in marine areas, and No Threat was assigned for negligible impacts.

These threat levels as determined by inspection have a one-to-one correspondence with a specific T2 scenario and coastal zone. Therefore, the P95 value within the relevant coastal zone can be determined, and a comparison can be drawn between the threat level determined through inspection and the P95 values.

The assessment of the threat level was undertaken subjectively, i.e. through inspection. The reason for this is to ensure that the threat level is related to the potential impact, which is dependent on the existence of local population and/or infrastructure. This means that the assessment of the threat level was straightforward for some cases (i.e. Major Land Threat for those cases where there was significant inundation) but in many cases it was borderline or difficult to assess. This is discussed further in Sect. 6.

Some of the model runs described in the previous section are duplicates. In most cases, this was due to two separate studies considering the same scenario for a different site within the same tsunami warning zone. For example, it can be seen from Table 1 that the Wilson et al. study considered scenario 194d at Mw = 9.0 for Sydney, and the Cardno study also considered this scenario for their Manly and Botany Bay sites. All these sites are within NSW zone 4, so they are duplicates. In these cases, we use the worst-case inundation of the duplicate model runs to determine the threat level for that zone and scenario. It was found that apart from the Tasmanian zones, identical threat levels were assigned for the duplicate cases.

A further complication relates to the T2 scenario selection. Since we are developing thresholds for an operational system, we want to base the technique on T2 scenarios that are as close as possible to what is used operationally. The JATWC incorporates an operational protocol that limits the degree to which amplitudes from T2 scenarios can be scaled away from their 'base' scenario. Greenslade and Allen (2019) discuss this in some detail. For the purposes of this paper, it is sufficient to note that 5 of the scenarios from the studies described in Sect. 3 are outside operational protocols, so they are not relevant to the development of operational warnings and thus will not be used here. These are noted in bold in Table 1.

In all, once the duplicates and 'non-operational' scenarios are removed, there are a total of 91 separate inundation model results available.

5. *Results*

Two examples of the inundation maps and assigned threat levels for Manly, Sydney, NSW are shown in Fig. 3. The top panel of this figure shows an example that has been assigned as a Land Threat. It can be seen that there is significant inundation which would require evacuation, but it is constrained to a few small areas. The bottom panel shows an example that has been designated a Major Land Threat. In this case, the inundation is considerably more widespread, and were this event to occur, extensive evacuations would be required. Note that with the present system, in which there is no Major Land Threat tier, the community advice provided in the tsunami warnings would be the same for both situations depicted here.

Figure 4 presents a summary of the results. In this figure, the P95 values for each scenario and each zone are plotted with colours representing the assigned threat level. Dashed horizontal lines show existing JATWC threshold values for the Marine Threat (20 cm) and Land Threat (55 cm).

Some key features are immediately apparent from this figure. Firstly, it can be seen that in general, Major Land Threats (red dots) have higher P95 values than the Land Threats (yellow) which are in turn generally higher than the Marine Threats (blue) and No threats (open black circles). However, there are several anomalies. For example, in NSW zone 1, there is a scenario that has been designated as Major Land Threat, which has a P95 value of 114 cm (the lowest red dot), whereas a scenario that was designated as a Land Threat (higher yellow dot) has a P95 value of 123 cm. In this case, the higher offshore tsunami amplitudes have not led to more severe onshore impacts.

The two scenarios run for the Tasmanian coastline have resulted in assigned Major Land Threats for zone 5, and one Major Land Threat and one Land Threat for zone 6. This is somewhat unexpected because as mentioned earlier, the scenarios were specifically selected because their amplitude values were close to the existing Land Threat thresholds (the lower two dots in Fig. 4 are near 55 cm). One possible reason for the high assigned threat levels relates to the complex geomorphology of the southeast Tasmanian coast compared to NSW. For example, one scenario produces inundation at only two locations across the coastal zone, with flow depths of up to 2 m. In the case of these two locations, modelling animations and nearshore time series data suggest that these isolated inundation levels are due to the interaction of the tsunami waves with the complex geometry of the Tasmanian coast. In particular, the presence of several large offshore islands and numerous peninsulas has resulted in local amplification and funnelling effects that subsequently affected the inundation levels at some locations in the simulation.

At a glance, the existing thresholds appear to be approximately in the right place for NSW zones 1 and 3 and 5, but for NSW zone 4, it can be seen that there are a number of scenarios that have been designed as Marine Threat, but the P95 values are less than 20 cm, so a No threat would be issued. Conversely, there are a number of scenarios in this zone which have been designated Marine Threat, but their P95 values are above 55 cm, so a Land Threat would be issued. Furthermore, three of the four scenarios run for the Tasmanian zones have been designated as requiring a Major Land Threat despite, in two cases, having relatively low P95 values. This will be discussed in more detail in the next section.

A technique for determining the optimal values for the thresholds has been developed by Greenslade and Allen (2019). This is based on considering a contingency table framework, and analysing the number of 'hits', 'misses' and 'false alarms' for a range of threshold values, where a 'hit' is a warning issued at the correct level, a 'miss' is a warning missed, or issued at a lower level than required, and a 'false alarm' is a warning issued at a higher level than required. More specifically, values of the threshold are sought for which the Threat Score is maximised, where:

$$\text{Threat Score} = \frac{\text{hits}}{\text{hits} + \text{misses} + \text{false alarms}} \quad (1)$$

For the present study, we apply this technique to the overall data set and also to each zone individually. The results are presented in Table 2. In most cases, the maximum Threat Score was obtained for a range of threshold values.

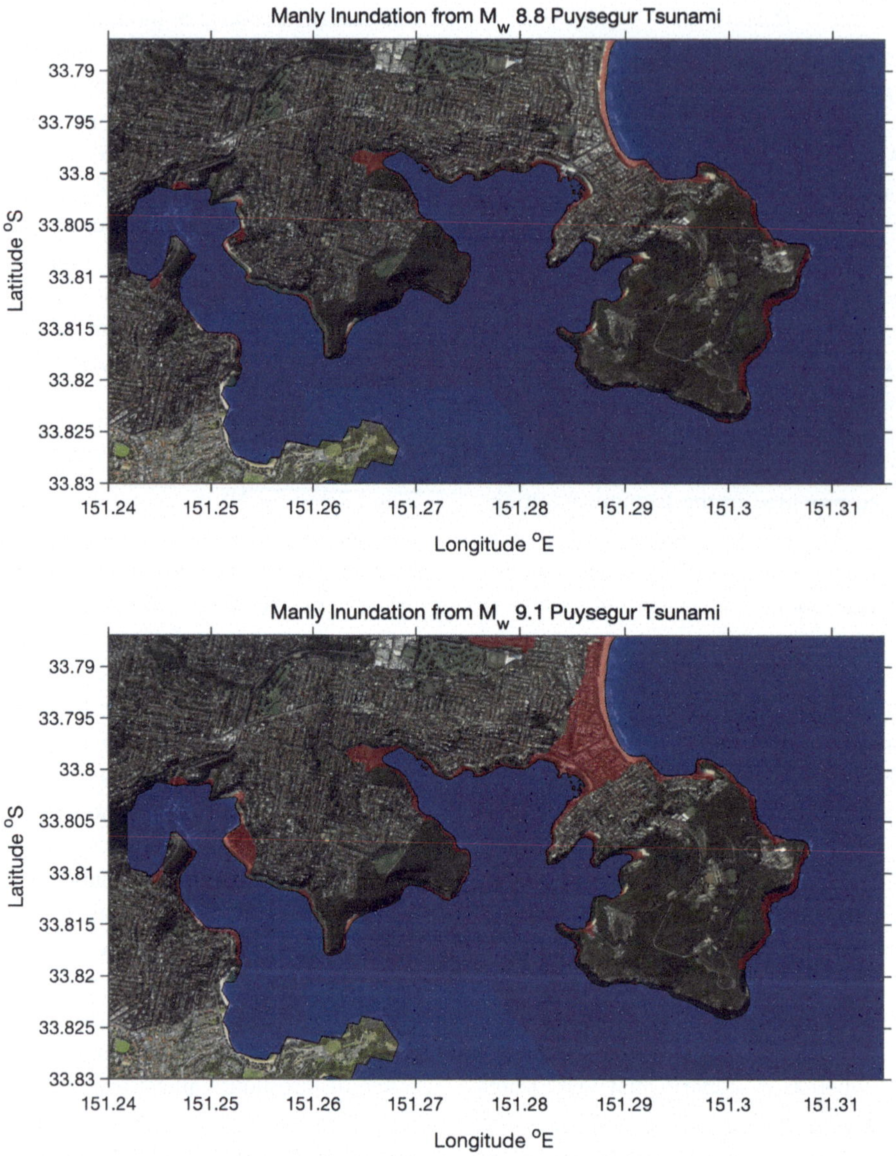

Figure 3
Top: example of an event that has been designated a Land Threat; bottom: example of an event that has been designated a Major Land Threat. The location shown (Manly) is a beachside suburb of Sydney and is in the NSW4 zone

The results for the zones treated as one data set (top row of Table 2) suggest that the optimal threshold for Marine Threat is 7 to 8 cm, for Land Threat is 48.5 cm, and for Major Land Threat is between 150.5 and 162 cm.

6. Discussion

We consider first the Marine Threat. The optimal threshold value found here of 7 to 8 cm is substantially lower than that currently used in the operational system (20 cm) and results in a slightly improved Threat Score of 0.95 compared to 0.93 for the operational threshold of 20 cm. The reason for this low

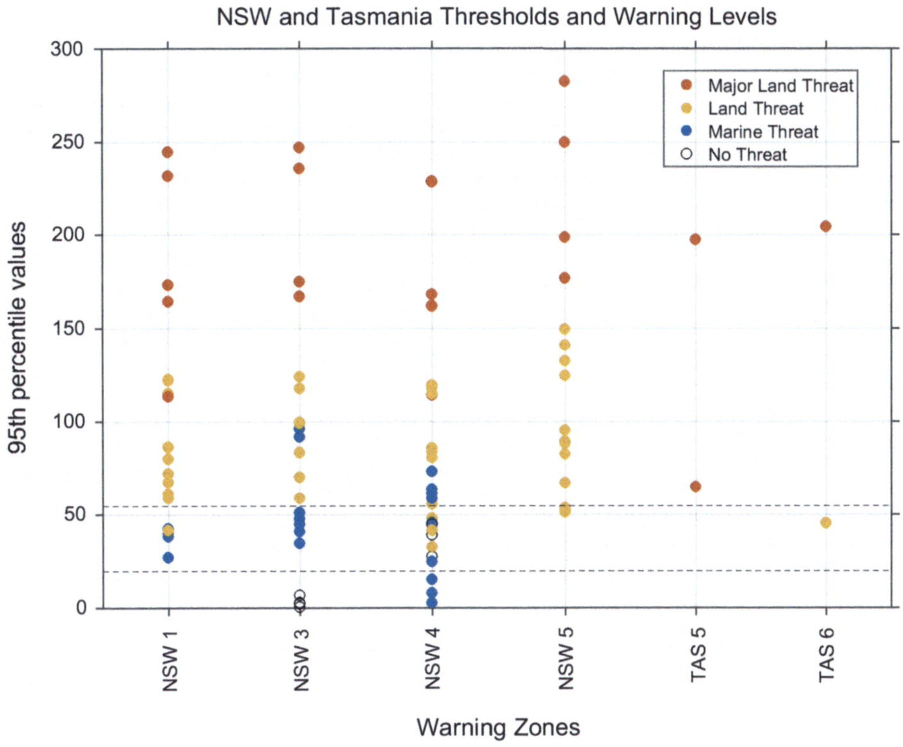

Figure 4
The 95th percentile values associated with each inundation scenario, with the colour designating the assigned threat levels from the inundation modelling results. Dashed horizontal lines indicate existing threat level thresholds (Land Threat threshold = 55 cm, Marine Threat threshold = 20 cm)

optimal threshold value is because there are very few 'No threat' examples in this dataset so there are few opportunities for a low threshold to create false alarms. An increased number of false alarms will decrease the Threat Score (see Eq. 1). The lack of 'No threat' examples is not surprising as the scenarios analysed in the studies used here were hazard assessments focussing on high-impact events. The conclusion we draw from this is that the underlying dataset used here is not fit for the purpose of analysing the Marine Threat threshold and will not be addressed any further in the present work. A comprehensive assessment of the Marine Threat threshold based on impacts experienced during actual and more suitable Marine Threat events can be found in Greenslade and Allen (2019).

The optimal Land Threat threshold for the whole dataset is found to be 48.5 cm and this results in a Threat Score of 0.86. This threshold compares reasonably well with the existing operational threshold of 55 cm if somewhat lower. The Threat Score for a threshold of 55 cm threshold is 0.79, so the benefit gained from lowering the threshold is not substantial. This is because by lowering the thresholds we gain some 'hits', but we also increase the number of 'false alarms'.

The optimal Major Land Threat threshold found here is between 150.5 cm and 162 cm. Any thresholds set within this range will result in a Threat Score of 0.83. Operationally the lower bound of this range would be used, as the general principle of the JATWC is to remain conservative where possible.

The results shown in Fig. 4 suggest that one-size-fits-all thresholds for all coastal zones may not be appropriate—implementing different thresholds for different zones may be a better option. The optimal thresholds for the NSW zones treated individually are shown in the lower 4 rows of Table 2. Tasmanian zones are not addressed here as there is insufficient data, although as discussed in the previous section, a

Table 2

Resulting optimal threshold values and Threat Scores for the overall dataset and for each zone individually

	Marine Threat		Land Threat		Major Land Threat	
	Threat score	Threshold range (cm)	Threat score	Threshold range (cm)	Threat score	Threshold range (cm)
All zones	0.95 (0.93)	7–8	0.86 (0.79)	48.5	0.86	150.5–162
NSW zone 1	1.0 (1.0)	0–27.5	0.94 (0.93)	40.5–41.5	0.80	123.5–164.5
NSW zone 3	1.0 (1.0)	7–35	0.92 (0.92)	52–59	1.0	125–167.5
NSW zone 4	0.89 (0.79)	0–3	0.75 (0.65)	74–81	0.83	120–162
NSW zone 5	n/a	n/a	1.0 (0.76)	0–51	1.0	150–177

Threat Scores using the operational thresholds of 20 cm (Marine Threat) and 55 cm (Land Threat) are shown in parentheses

different approach for Tasmania may be beneficial. For the Land Threat thresholds, these results suggest that zone 1 should have a lower threshold, and zones 3 and 4 should have higher thresholds, although the Threat Scores are not significantly improved for Zones 1 and 3. Zone 5 is already optimised for a threshold of 48.5 cm, although it should be noted that Zone 5 did not have any events for which the impacts were designated to be lower than Land Threat so firm conclusions cannot be drawn here. For the Major Land threat threshold, the proposed value of 150.5 cm is already within the optimal range for each zone so no benefit would be gained through changing this threshold. Deriving tailored thresholds could be a reasonable approach for the zones considered here, but the required information to apply this to all zones for the Australian coastline does not exist. Further modelling studies and considerable resources would be required to develop that information.

The results found here suggest that even tailored thresholds would not be ideal. As discussed in the previous section, while there is a general correlation between P95 and the amount of inundation, it is not clear-cut. Within an individual zone, some events associated with higher offshore amplitudes do not produce inundation while some events associated with lower offshore values do. This could be due to aspects such as different geographic locations of the sources and resulting varying directionality near the coast. This suggests that there may be a better 'predictor' than P95 within a coastal zone. Greenslade and Allen (2019) explored several other options for predictors, including other percentile values (50th, 90th, 100th). They also investigated factors relating to the assumed tsunami wavelength and reducing the number of grid points contributing to the percentile values. Optimal thresholds were determined for each of these alternate predictors using the Threat Score technique described earlier, but they did not find a predictor that performed substantially better than P95. One way to address this would be to provide a real-time forecast of inundation during an event (Wei et al. 2008; Allen and Greenslade 2016).

There are a number of assumptions and issues involved in this work that should be considered. One of the key assumptions is that the inundation modelling results are accurate—any interpretation of the results depends heavily on this assumption. Each of the prior studies on which the present study is based (i.e. NSW State Emergency Service and Office of Environment and Heritage 2012; Wilson et al. 2018; Kain et al. 2018; Allen and Greenslade 2016) undertook extensive validation of their modelling systems and results, so we can be reasonably confident of the results found here. A related assumption is that each of the inundation models used in the various studies would produce the same results (or at least the same threat level) if run with the same input data (boundary conditions and DEM). This was partially addressed in Uslu and Greenslade (2013) where the same scenario was run for MOST and for DELFT3D, with the results found to be in general agreement. In addition, as noted previously, there were a few cases where two separate studies considered the same scenario for a different site within the same tsunami warning zone. In these cases, identical threat levels had been independently assigned for the duplicate cases.

As noted earlier, it has been difficult to assign the threat levels in some cases. Those cases with significant inundation, or with negligible wave activity are straightforward, but there are a number of borderline cases, which raise some questions. For example, should an event for which the only inundation that occurs is on an uninhabited island receive a Land Threat? There is clearly inundation in a case such as this but it is not threatening to life or property. If a Land Threat is issued, and the entire coastal zone is evacuated on this basis, it could be unnecessarily costly. On the other hand, if a Marine Threat is issued and significant inundation occurs, then it could be argued that the threat level was not severe enough. Particularly if there is a possibility that the area in question may become populated at some stage in the future. In this work, we have assigned the Threat level on a case-by-case basis after careful inspection of the potential impacts.

Another issue to consider is that this study has derived the threshold values for each coastal zone based on the inundation modelling results at just one (or two) location(s) within the zone (with the exception of the two Tasmanian zones). The problem with this approach is that, for example, a Marine Threat may be assigned based on the results at one site within the zone, but there may be other sites within the zone that receive significant inundation, and would have required a Land Threat. This should not be a major issue here as the sites examined in each study were either those that were designated the highest risk (e.g. NSW State Emergency Service and Office of Environment and Heritage 2012) or the entire coastline within a zone was modelled (e.g. Kain et al. 2018). So we can be reasonably confident that the assigned threat levels represent the most significant impacts for each event.

This study has mostly focussed on flow depths to assign threat levels, although inundation extents and runups have also been taken into account. In relation to the issuing of warnings, it can be useful to provide an action statement relating to the distance one needs to go inland to avoid danger, in addition to guidance as to the safe elevations above sea level. This might be considered a more general approach than using inundation maps to design specific evacuation routes for a location. While an action statement could potentially be deduced from the inundation extents examined here, a Land Threat event with a maximum flow depth of, say, 1.5 m at one location, may have a very different inundation extent to an event with a maximum flow depth of 1.5 m at another location, depending on the local topography. Given this, it is not possible to determine a typical inundation extent for a Land Threat or Major Land Threat in the present study. Future work could involve examining inundation extents for each threat level on a site-by-site basis, in order to determine the relevant action statements. An alternative approach to that followed here could be to assign the threat levels according to inundation extents, provide action statements based on inundation extent, and then assess the safe elevation levels for each site or zone based on the flow depths. Both approaches are equally valid.

7. *Conclusions*

Tsunami warnings issued by the Joint Australian Tsunami Warning Centre (JATWC) have been assessed using inundation modelling results from pre-existing hazard assessment studies. In addition, a possible new tier 'Major Land Threat' is explored. The dataset was found to be not suitable for the purpose of assessing the Marine Threat threshold. In summary, it is recommended that the Land Threat threshold for the 95th percentile value of tsunami amplitude within each coastal zone is kept at 55 cm, and the recommended threshold for the possible new Major Land threat is 150 cm. Tailored thresholds for individual zones could be considered, but there is insufficient data at present to do this for any more than 3 or 4 of the existing 67 national coastal zones.

Acknowledgements

The authors would like to thank the NSW Office of Environment and Heritage (OEH), the NSW State Emergency Service (NSW SES), and Cardno for providing the inundation modelling results used in this study. This work was partly funded by the Natural Disaster Mitigation Program through NSW SES. This work was also funded through an award from the NSW Government under the State Emergency Management Projects 2014–15 program to HEP. KMW is supported by a University of Newcastle Faculty of Science Strategic Scholarship (50:50). The authors would like to thank Robert Greenwood and Eric Schulz for useful comments on the text.

References

Allen, S. C. R., & Greenslade, D. J. M. (2010). Model-based tsunami warnings derived from observed impacts. *Natural Hazards and Earth System Science, 10*(12), 2631–2642.

Allen, S. C. R., & Greenslade, D. J. M. (2016). A pilot tsunami inundation forecast system for Australia. *Pure and Applied Geophysics, 173*(12), 3955–3971. https://doi.org/10.1007/s00024-016-1392-y

Greenslade, D. J. M., & Allen, S. C. R. (2019) On the optimal amplitude thresholds for tsunami warning, Bureau Research Report no. 039. Australia: Bur. Met.

Greenslade, D. J. M., Allen, S. C. R., & Simanjuntak, M. A. (2010). An evaluation of tsunami forecasts from the T2 scenario database. *Pure and Applied Geophysics, 168*(6–7), 1137–1151. https://doi.org/10.1007/s00024-010-0229-3.

Greenslade, D. J. M., Simanjuntak, M. A., & Allen, S. C. R. (2009). An enhanced tsunami scenario database: T2. BMRC Research Report no. 014. Australia: Bur. Met.

Kain, C. L., Mazengarb, C., Rigby, E. H., Cohen, W., Simard, G., & Lewarn, B. (2018). Technical report on tsunami inundation modelling in South East Tasmania, Tasmanian Geological Survey Record UR2018_02.

Nielsen, O., Roberts, S., Gray, D., McPherson, A., & Hitchman, A. (2005). Hydrodynamic modelling of coastal inundation. MODSIM 2005 International Congress on Modelling and Simulation Modelling and Simulation Society of Australia and New Zealand. pp. 519–523.

NSW State Emergency Service and Office of Environment and Heritage (2012). Final Draft NSW Tsunami Inundation Modelling and Risk Assessment, Cardno Technical report LJ2874/Rep2703

Okada, Y. (1985). Surface deformation due to shear and tensile faults in a half-space. *Bulletin of the Seismological Society of America, 75,* 1135–1154.

Simanjuntak, M. A., Greenslade, D. J. M., & Allen, S. C. R. (2011). Extensions to the T2 tsunami scenario database, CAWCR Research Letters, Issue 7. Australia: Bur. Met.

Somerville, P., Hanslow, D. J., & Gissing, A. (2009). NSW Tsunami Risk–An Overview of the NSW Risk Assessment Scoping Study. In *Joint NSW and Victorian Flood Management Conference*, Albury-Wodonga, February 2009.

Stelling, G.S. (1984). On the construction of computational methods for shallow water flow problems. Tech. Rep. Rijkswaterstaat No. 35, Rijkswaterstaat. The Netherlands: The Hague.

Stelling, G. S., & Duinmeijer, S. (2003). A staggered conservative scheme for every Froude number in rapidly varied shallow water flows. *International Journal Numerical Methods In Fluids*. https://doi.org/10.1002/fld.537

Synolakis, C. E., Bernard, E. N., Titov, V. V., Kanoğlu, U., & Gonzalez, F. I. (2008). Validation and verification of tsunami numerical models. *Pure Applied Geophysics, 165*(11–12), 2197–2228.

Titov, V., Kanoglu, U., & Synolakis, C. (2016). Development of MOST for Real-Time Tsunami Forecasting. *Journal of Waterway Port Coastal and Ocean Engineering,*. https://doi.org/10.1061/(ASCE)WW.1943-5460.000035703116004.

Titov, V. V., & Synolakis, C. E. (1998). Numerical modeling of tidal wave runup. *Journal of Waterway Port Coastal and Ocean Engineering. 124*(4), 157–171.

Uslu, B., & Greenslade, D. J. M. (2013). Validation of tsunami warning thresholds using inundation modelling, CAWCR Technical Report No. 062. Australia: Bur. Met.

Wei, Y., Bernard, E. N., Tang, L., Weiss, R., Titov, V. V., Moore, C., et al. (2008). Real-time experimental forecast of the Peruvian tsunami of August 2007 for U.S. coastlines. *Geophysical Research Letters, 35,* 1–7. https://doi.org/10.1029/2007GL032250.

Wilson, K. M., Allen, S. C. R., & Power, H. E. (2018). The tsunami threat to Sydney Harbour, Australia: Modelling potential and historic events. *Scientific Reports, 8*(1), 15045.

(Received August 13, 2019, revised November 21, 2019, accepted November 22, 2019, Published online December 4, 2019)

Pure Appl. Geophys.177 (2020), 1437–1450

https://doi.org/10.1007/s00024-019-02364-4

Pure and Applied Geophysics

Machine Learning Algorithms for Real-time Tsunami Inundation Forecasting: A Case Study in Nankai Region

Ardiansyah Fauzi[1,2] and Norimi Mizutani[1]

Abstract—The most acceptable method to estimate tsunami inundation caused by a submarine earthquake is by conducting a nonlinear tsunami simulation. However, this method has the disadvantages of a relatively high computational cost and the necessity for immediate warning announcements when a tsunami is imminent. In this study, to overcome this problem, we apply two machine learning models, a convolutional neural network and a multilayer perceptron, to estimate tsunami inundation in real time. We run multiple fault scenarios and store the result of the maximum tsunami amplitude in a low-resolution grid and the associated tsunami inundation in a high-resolution grid in the database. The convolutional neural network selects tsunami inundation in the high-resolution grid as the forecast based on pattern similarity between the input, which is the results of linear forward modeling in the low-resolution grid, and the precomputed patterns in the database. Slightly different from the convolutional neural network, instead of selecting the best-fit scenario in the database, the multilayer perceptron directly generates the inundation forecast based on knowledge acquired during the training process. We conduct an experiment using the hypothetical future Nankai megathrust earthquake with Atashika and Owase Bays in Japan as the study cases. The results show that our proposed methods are extremely fast (less than 1 s) and comparable with nonlinear forward modeling. Therefore, the proposed methods can be used as a deterministic model for real-time simulation.

Keywords: Tsunami warning, tsunami inundation forecast, convolutional neural network, multilayer perceptron, Nankai megathrust earthquake.

1. Introduction

The Nankai Trough is one of the most active subduction zones in the world. Many massive submarine earthquakes have been generated and well recorded in this region. Based on historical records, Ishibashi (2004) showed that earthquakes in this zone have a recurrence interval between 100 and 150 years. Five major earthquakes in the Nankai Trough have been identified: the 1707 Hoei earthquake, the 1854 Ansei Tokai and Ansei Nankai earthquakes, the 1944 Tonankai earthquake, and the 1946 Nankai earthquake. Among those earthquakes, the 1707 Hoei event marked the strongest earthquake in the preindustrial history of Japan, and a giant tsunami followed it. Three main fault segments are in this region: the Tokai, the Tonankai, and the Nankai. In those historical earthquakes, the three segments ruptured individually or simultaneously. In the future, it is expected that the three segments will rupture simultaneously and generate a massive earthquake. Because the Nankai Trough is relatively close to the coast of southeastern Japan (less than 150 km away), if the earthquake induces a tsunami, it will reach the coast in a very short time. Therefore, a fast and robust tsunami early warning system is required to prevent large casualties caused by Nankai Trough earthquakes in the future.

Generally, the objective of tsunami forecasting is mainly to predict the tsunami waveform and tsunami inundation. Both are vital information during a tsunami event: tsunami waveform prediction is essential for early warning, while the inundation prediction is crucial for the evacuation process. In this study, we focus on the tsunami inundation prediction. Most of the challenges in predicting tsunami inundation in real time is dealing with the computational load. Running a nonlinear tsunami model is the most accurate method to predict tsunami inundation. However, with a very short effective period for announcing tsunami information, particularly in the

[1] Graduate School of Civil Engineering, Nagoya University, Nagoya, Japan. E-mail: fauzi.ardiansyah@k.mbox.nagoya-u.ac.jp; arfa.fauzi@gmail.com
[2] Department of Civil Engineering, Institut Teknologi Kalimantan, Kalimantan, Indonesia.

Nankai Trough region, it is not suitable to run a forward model that requires a simulation time of tens of minutes or even hours on a regular computer. Oishi et al. (2015) used a supercomputer to simulate tsunami inundation in real time and was able to generate tsunami inundation for Sendai City with only 1.5 min of computational time. Similarly, Musa et al. (2018) equipped a supercomputer simulation with a delivery/ mapping server to send the expected inundation map to the local government. Even so, the use of supercomputers, which are usually owned by the government, has limitations and difficulties, especially for countries or institutions that do not have such facilities. A real-time tsunami inundation forecasting ystem based on a database for far-field tsunamis has been developed by the National Oceanic and Atmospheric Administration (Tang et al. 2008; Titov et al. 2005). However, it may not be suitable for near-field tsunami events because it requires direct forward simulation in the system. A more comprehensive tsunami inundation forecasting system for near-field tsunamis has been developed by Gusman et al. (2014), which they called "Near-field Tsunami Inundation Forecasting" (NearTIF). This system finds the best tsunami inundation scenario in the database by matching precomputed and real-time computed tsunami waveforms at virtual observation stations in the coastal region. Recently, there have been further improvements in the use of offshore bottom pressure data for rapid coastal tsunami estimation (e.g., Igarashi et al. 2016; Maeda et al. 2015; Tanioka 2018; Tsushima et al. 2014; Yamamoto et al. 2016). Although such methods are useful for early tsunami warnings, additional procedures are still required to estimate tsunami inundation, as most of these methods are based on a linear long-wave model.

Here we improve the searching algorithm proposed by Gusman et al. (2014), which uses the tsunami waveform at virtual observation points as the input to find the best-fit scenario in the database. To speed up the matching algorithm, we use a convolutional neural network (CNN). Similar to the study conducted by Mulia et al. (2018), we use maximum tsunami amplitude in the low-resolution grid as the model input instead of tsunami waveforms at virtual observation points. Because the tsunami inundation database is usually numerically computed from many simple earthquake scenarios, we believe that simply selecting tsunami inundation from the best scenario in the database is not sufficient to represent the uniqueness and the complexity of an actual earthquake source. Therefore, we also use another machine learning framework, a multilayer perceptron (MLP), to forecast tsunami inundation directly based on the pattern of maximum tsunami amplitude in the low-resolution database.

2. *Database Development*

2.1. *Study Site*

In this experiment, we use the Nankai megathrust earthquake to test the performance of our proposed method. We choose two locations for study sites: Atashika and Owase Bays (Fig. 1a). Atashika Bay directly faces the Pacific Ocean and is approximately 1–3 km wide, while Owase Bay is a U-shaped bay approximately 7 km wide and 10 km long. Both areas are tsunami-prone areas that have experienced many destructive tsunamis in the past, such as the 1707 Hoei tsunami (Furumura et al. 2011), the 1944 Tonankai tsunami (Kikuchi et al. 2003), and the 1960 Chilean tsunami (Hatori et al. 1981). Because the location of both areas is close to the earthquake source in Nankai Trough, a tsunami generated in Nankai can reach those areas within a very short time (less than 20 min), as shown in Fauzi and Mizutani (2019). The unique U shape of Owase Bay may also generate tsunami amplification as a result of soliton fission due to resonance in the bay (Yamanaka et al. 2016).

2.2. *Fault Model Scenarios*

In this study, the same fault model scenarios as in Fauzi and Mizutani (2019) are used. Those scenarios are assumed as three types of simple rectangular faults: thrust (reverse fault with low dip angle), tsunami earthquake, and reverse. We set earthquake magnitudes for the thrust fault ranging from 8.0 to 9.0 (0.2 interval), and 7.0 to 8.4 (0.2 interval) for the tsunami earthquake and reverse fault; in total, there are 328 fault scenarios. The strike angle is set to 225°

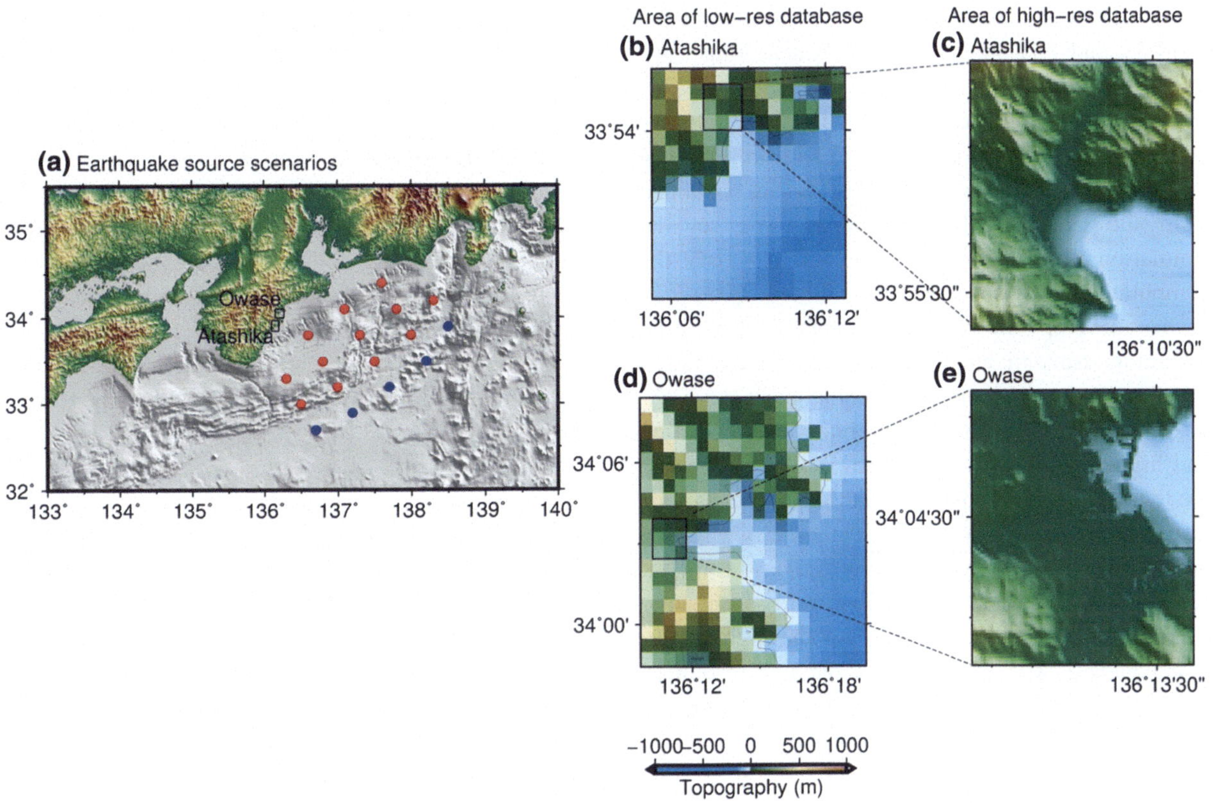

Figure 1

a Epicenter of fault scenarios used to develop the database. Red circles indicate the coordinate of the top center of the fault for thrust fault and tsunami earthquake scenarios. Blue circles indicate the coordinate of the top center of the fault for reverse fault scenarios. **b** Domain of low-resolution database at Atashika Bay. **c** Domain of high-resolution database at Atashika. **d** Domain of low-resolution database at Owase Bay. **e** Domain of high-resolution database at Owase

as an average orientation of the Nankai Trough subduction zone, while the rake angle is set to 90°. The dip angle for the thrust fault scenarios is determined according to SLAB 1.0 (Hayes et al. 2012), while for the tsunami earthquake and reverse fault scenarios, it is set to 10° and 45°, respectively. The depths of the fault for the thrust and tsunami earthquake scenarios are also determined from SLAB 1.0. For the reverse fault scenarios, the fault depth ranges from 5 to 20 km (5-km interval) according to the outer-rise seismicity depth in Nankai studied by Craig et al. (2014). A magnitude scaling relation for the plate-boundary earthquake proposed by Murotani et al. (2013) (Eq. 1) is used to calculate the area of the fault.

$$S = 1.34 \times 10^{(-10)} M_o^{(2/3)}, \qquad (1)$$

where S is the fault rupture area (km^2) and M_o is the seismic moment (Nm). The fault length (L) and width (W) can be derived from a simple relationship of $L = 2W$. Because the fault width for a tsunami earthquake is much shorter than the length, the fault width is set to be less than 50 km for tsunami earthquake scenarios (Tanioka and Satake 1996). To compute the initial sea surface for tsunami simulation, we use a coseismic deformation in an elastic half-space model (Okada 1985) to calculate seafloor deformation, and only the vertical component is used. The computed seafloor deformation is assumed to be the initial sea surface.

2.3. *Precomputed Tsunami Database*

We use a well-verified tsunami numerical model, JAGURS (Baba et al. 2015, 2016), to simulate 2-h tsunami propagation and inundation from the fault scenarios. The simulation is conducted in the spherical coordinate system. The computational domain consists of four nested grids with resolutions of 30, 10, 3.33, and 1.11 arcsec. We use a linear long-wave model for the low- and intermediate-resolution grids (30, 10, and 3.33 arcsec) and a nonlinear long-wave model for the high-resolution grid (1.11 arcsec). The topography and bathymetry datasets are obtained from the Cabinet Office of Japan. We set the simulation time step of $\Delta t = 0.25$ s to satisfy the computational stability, and a rise time of 60 s.

To develop the tsunami database, we store maximum tsunami amplitude in the low-resolution grid (30 arcsec) and tsunami inundation in the high-resolution grid (1.11 arcsec) to create low-resolution and high-resolution databases, respectively. Because the limit of inundation for each earthquake scenario is different in the high-resolution database, we fix the grid points that satisfy inundation coverage for all scenarios. We consider only the wet grid for the low-resolution database and dry grid for the high-resolution database. The coverages of the low-resolution and high-resolution domains are defined in the area of interest (Fig. 1b, e).

3. *Machine Learning*

In this study, we use two types of machine learning algorithms for two different tasks: CNN for classification and MLP for regression. Here, the CNN is proposed as an alternative to the matching algorithm suggested by Gusman et al. (2014) and Mulia et al. (2018), while the MLP is proposed to directly reproduce tsunami inundation in the high-resolution grid. Both proposed frameworks are implemented using an open-source and popular machine learning library, Tensorflow (Abadi et al. 2016).

3.1. *CNN Model*

The CNN is a machine learning algorithm that is usually used for image or pattern recognition problems (e.g., Ciregan et al. 2012; Garcia and Delakis 2004; Krizhevsky et al. 2012; Sermanet et al. 2013; Tompson et al. 2015), in which it is considered as the best among the other neural network (NN) methods (Krizhevsky et al. 2012). One of the advantages of a CNN over a regular NN is the capability to capture local patterns. Local patterns, which are obtained by down-sampling the input data into small fragments through the convolutional and pooling operator, improve the accuracy of global pattern recognition if they are examined from a large-scale view. A regular CNN typically consists of three layers: the convolutional layer, the pooling layer, and the fully connected layer.

Within the convolutional layer, the input data are filtered and transversed onto the feature map by applying a set of the convolutional kernel. The output of a convolutional layer can be analytically formulated as

$$f = \sigma(Wx + b). \tag{2}$$

where W are the weights of the kernel with kernel size $m \times n$, b is the bias of the kernel, x is the input kernel, and σ is the transfer function. For the transfer function σ, we select a rectified linear unit (ReLU) transfer function (Nair and Hinton 2010), $f(x) = \max(0, x)$, to introduce the nonlinearity of the network. The pooling layer compresses the size of the feature map and merges the semantically similar features into one feature (Lecun et al. 2015; LeCun et al. 1998). Typically, the pooling unit computes the maximum of a local unit kernel in one feature map. Then, all of the features are sent to the output layer through fully connected layer (conventional NN layer). The distribution over classes is mapped in the output layer by adding a classifier, that is, a softmax activation function.

We carefully investigate the layer configuration, and we finally use only one stage each of the convolutional layer, nonlinearity, and pooling layer, which are followed by two fully connected layers with 128 and 64 nodes. We set the kernel size as 3×3 and 2×2 for the convolutional and pooling layers, respectively. As typical deep-learning networks, a CNN may be subjected to overfitting caused by model complexity. It is usually typified by excellent training performance but unreliable testing performance. To avoid this problem, we apply a new kind of regularization term called dropout (Sutskever et al. 2014). The dropout algorithm randomly removes or ignores the hidden nodes and their incoming and outgoing connections with a given dropout probability p ($p = 0.2$ in this study). This technique reduces the complex co-adaptation in the hidden nodes and improves the model performance (Sutskever et al. 2014). To fit the network parameters to the training dataset, which is obtained by minimizing the categorical cross-entropy loss function, we use a back-propagation algorithm with an Adam optimizer (Kingma and Ba 2015).

3.2. MLP Model

The MLP is a popular and most-often-used artificial neural network. It is well known for its outstanding performance because of its ability for universal approximation. The mathematical formulation of the MLP is inspired by the structure of the human brain to deal with nonlinear tasks (Haykin 2009). Therefore, it is an effective method to solve regression problems (e.g., Bienvenido-Huertas et al. 2019), such as the problems in this study. The network structure of the MLP consists of three parts: an input layer, one or more hidden layers, and an output layer. The output value of a hidden node n can be expressed as

$$f = \sigma(W^l\sigma(W^{(l-1)} \cdots (W^0x + b^0) + b^{(l-1)}) + b^l), \tag{3}$$

where l indicates the final layer. For the activation function σ, we select the ReLU activation function (Nair and Hinton 2010), similar to the CNN model. We apply a loss function based on mean squared error and minimize it by using the Adam optimizer (Kingma and Ba 2015). Dropout and L2 norm regularizations with regularization parameter γ ($\gamma = 0.001$ in this study) are used to avoid over-fitting. We investigate the appropriate number of hidden layers and nodes and find that a relatively deep and wide network architecture is necessary to capture the pattern of nonlinear characteristics of tsunami inundation. An MLP network that consists of five hidden layers with 128 nodes is sufficient to produce a reliable prediction.

3.3. Training Datasets

Two types of databases are used in this study, of low resolution and high resolution, as explained in Sect. 2. The input data for training is the same for both the CNN and MLP models, that is, the low-resolution database, with a slightly different array configuration. For the low-resolution model, the array size of each scenario is 15×15 (Atashika) and 20×20 (Owase) for the CNN model, while for the MLP model, the array is reshaped into a one-dimensional array with sizes of 225 (Atashika) and 400 (Owase). The training target of the CNN model is an array of integers ranging from 1 to 328 to represent the scenario ranking, while the MLP model uses the high-resolution database as the training target. In the high-resolution model, the total number of grid points for each scenario are 6055 and 5826 for Atashika and Owase, respectively. We then stack the array from 328 scenarios in row vectors to create a low-resolution database with array sizes of $328 \times 15 \times 15$ (Atashika) and $328 \times 20 \times 20$ (Owase) for the CNN model and array sizes of 328×225 (Atashika) and 328×400 (Owase) for the MLP model. In a similar manner, the array sizes of the high-resolution database become 328×6055 (Atashika) and 328×5826 (Owase).

3.4. Testing Dataset

We apply our proposed method to a hypothetical future scenario of the Nankai megathrust earthquake (*M*8.7) (Central Disaster Management Council 2003) where the Tokai, Tonankai, and Nankai segments rupture simultaneously. The initial coseismic deformation of the hypothetical Nankai megathrust is shown in Fig. 2a. The hypothetical Nankai megathrust tsunami source is used by the Japanese government to formally estimate the tsunami risk and hazard caused by such a disastrous event. In a real event, once a reliable tsunami source is obtained, we then compute tsunami propagation using a linear tsunami model in the low-resolution grid to obtain the maximum tsunami amplitude. The maximum tsunami amplitude generated by the Nankai megathrust earthquake at the selected area (Fig. 2b, c) is then is used as input for our proposed methods to generate tsunami inundation forecasts. In addition, the simulated tsunami inundation in the high-resolution grid from a nonlinear tsunami model is considered as a pseudo observation to validate our proposed method.

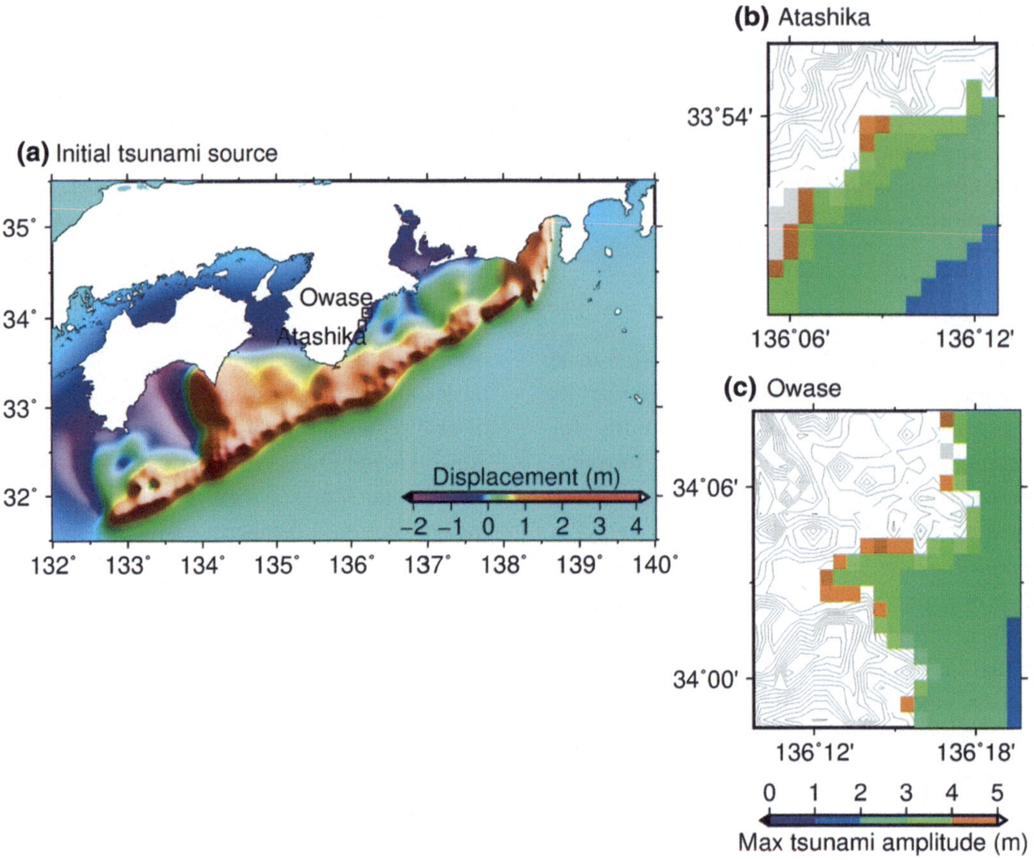

Figure 2
a Initial tsunami source of the hypothetical Nankai megathrust earthquake. **b** Maximum tsunami amplitude at Atashika as input to the networks. **c** Maximum tsunami amplitude at Owase as input to the networks

4. Tsunami Inundation Results

To quantify the prediction accuracy, the forecasted inundation is compared with the inundation from the forward modeling (depth and coverage). Because we use the pseudo observation, and because the tsunami inundation forecasts in our study may have inundation coverages different from that obtained by forward modeling, we select a relative error (d) to measure the model performance and avoid the errors caused by the logarithm of zero values

$$d = \frac{1}{N}\sum_{i=1}^{N}\left(2 \times \frac{\left(Y_i - \widehat{Y}_i\right)}{|Y_i| + \left|\widehat{Y}_i\right|}\right), \qquad (4)$$

where Y_i and $\widehat{Y}_i$ indicate the tsunami inundation depth from the forward modeling and prediction at point i, respectively, and d is a signed expression ranging from -2 to 2 that is positive for underestimation and negative for overestimation of the prediction over the observation. When d is closer to zero, it indicates better forecast accuracy. In addition, we also use a standard statistical measure, root mean square error (RMSE), which measures the average magnitude of the error.

4.1. CNN Model Results

Conventionally, comparing the result of direct linear forward modeling with the precomputed result in the database is the easiest method to find the best-fit scenario. This assumption was used by Gusman et al. (2014) to compare the tsunami waveforms from direct forward modeling with those in the database. However, when utilizing an extensive database (e.g., Baba et al. 2014), such a method is not practical because it may require a relatively long computational time. Mulia et al. (2018) also used a similar method by comparing maximum tsunami amplitudes in the low-resolution grid with those precomputed in the database. To simplify the searching process, the input and the database should go through a principal component analysis (PCA) to reduce the dimensionality and select the scenario that gives the smallest Euclidian distance relative to the direct forward modeling as the forecast.

Once it is trained, the CNN is able to find the best-fit scenario in near real time (Kolsch et al. 2018; Wang et al. 2019). Furthermore, the convolutional and pooling layers, which give the CNN an outstanding pattern recognition, may reduce the misrecognition caused by the complexity of the data. The comparison between CNN-estimated tsunami inundation and the tsunami inundation from the forward modeling at Atashika and Owase are shown in Figs. 3b and 4b, respectively, while the statistical measures are shown in Table 1. In Atashika, the CNN-estimated tsunami inundation is very similar to the tsunami inundation from the forward modeling having similar coverage, with $d = -0.003$ and RMSE = 1.882 m (Fig. 3b). The maximum inundation depth is slightly larger, by 0.53 m, than the maximum inundation depth from the forward modeling. In contrast, the tsunami inundation forecast in Owase shows underestimation of the tsunami inundation from the forward modeling, with $d = 0.548$ and RMSE = 2.117 m (Fig. 4b). Moreover, the forecast has a limit of inundation that is significantly lower than that of the forward modeling in the southern part of Owase (area below latitude 34°04'30"), with a total underestimation of 405 grid points. The forecasted maximum inundation depth in Owase also has a slightly larger error than that at Atashika with a discrepancy of 0.63 m.

In this study, we apply a *k*-fold cross-validation technique in training the network. The *k*-fold cross-validation is necessary to reduce the variance of the model and the bias, and thus allow us to estimate the generalization capability of the machine learning (Jiang and Wang 2017). We randomly divide the dataset into tenfolds, in which each fold consists of nine subsets for training the model, and the remaining subset for the testing. This folding procedure is repeated 10 times. The average value of the tenfolds is generally accepted to quantify the capability of the model against an independent dataset. Implementing tenfolds of *k*-fold, we find out that our model correctly recognizes the appropriate scenario for a training dataset with accuracy of 98.80% and 98.21% in Atashika and Owase, respectively.

4.2. MLP Model Result

Because every submarine earthquake has unique characteristics (i.e., coseismic deformation), the forecasted tsunami inundation from the best-fit scenario selected by the CNN may not closely represent the real tsunami inundation caused by the simple fault assumption used to develop the database. Unlike a CNN purposed for classification tasks, in which the algorithm will select the best-fit scenario from the database, the MLP has the ability to produce a unique output based on the knowledge acquired during the learning process. In other words, the MLP acts like tsunami inundation modeling based on the pattern of the training samples.

The comparison between MLP-estimated tsunami inundation and the forward modeling for Atashika and Owase are shown in Figs. 3c and 4c, respectively. The MLP-estimated tsunami inundation in Atashika underestimates the forward modeling with $d = 0.007$ and RMSE = 2.190 m, as shown in Table 1. The forecast poses inundation coverage similar to that of the forward modeling, even though it has a lower accuracy compared with the CNN, which is indicated by the larger RMSE value. The maximum inundation depth forecasted by the MLP is very similar to the forward modeling, with a discrepancy of 0.25 m. In Owase, the MLP provides a better result compared with the CNN, with $d = -0.061$ and RMSE = 1.142 m. The MLP-estimated maximum inundation depth is also almost identical to that of the forward modeling. An obvious difference between the tsunami inundation forecasts from the CNN and MLP in Owase is the limit of inundation. The forecast from the MLP successfully provided acceptable inundation coverage, with total overestimation of 104 grid points, whereas the CNN has an underestimation with a greater magnitude. Applying *k*-fold cross-validation with tenfolds produced an

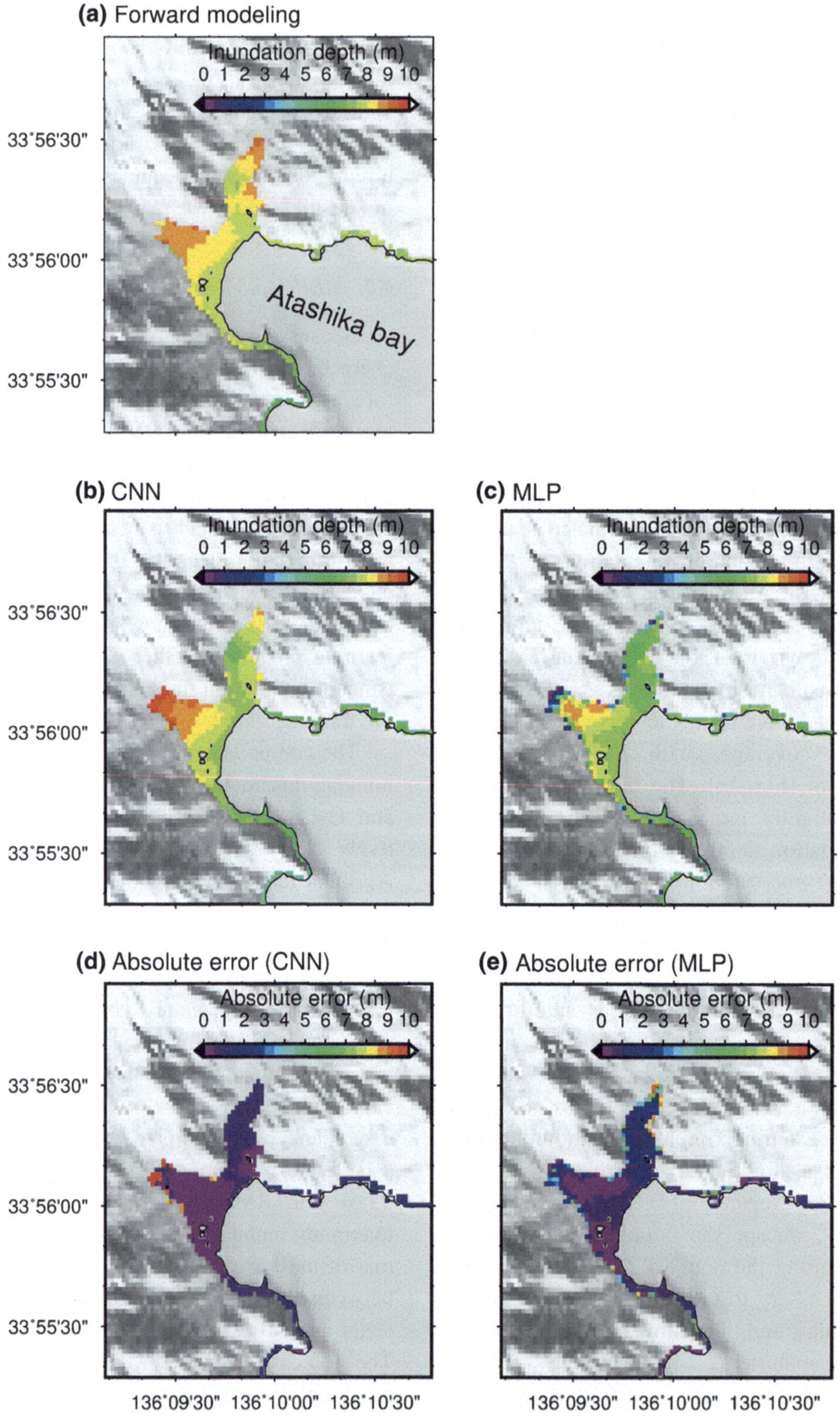

Figure 3

a Reference tsunami inundation at Atashika obtained from forward modeling. **b**, **c** Forecasts obtained from the proposed methods. **d**, **e** Absolute error between forecasts and forward modeling

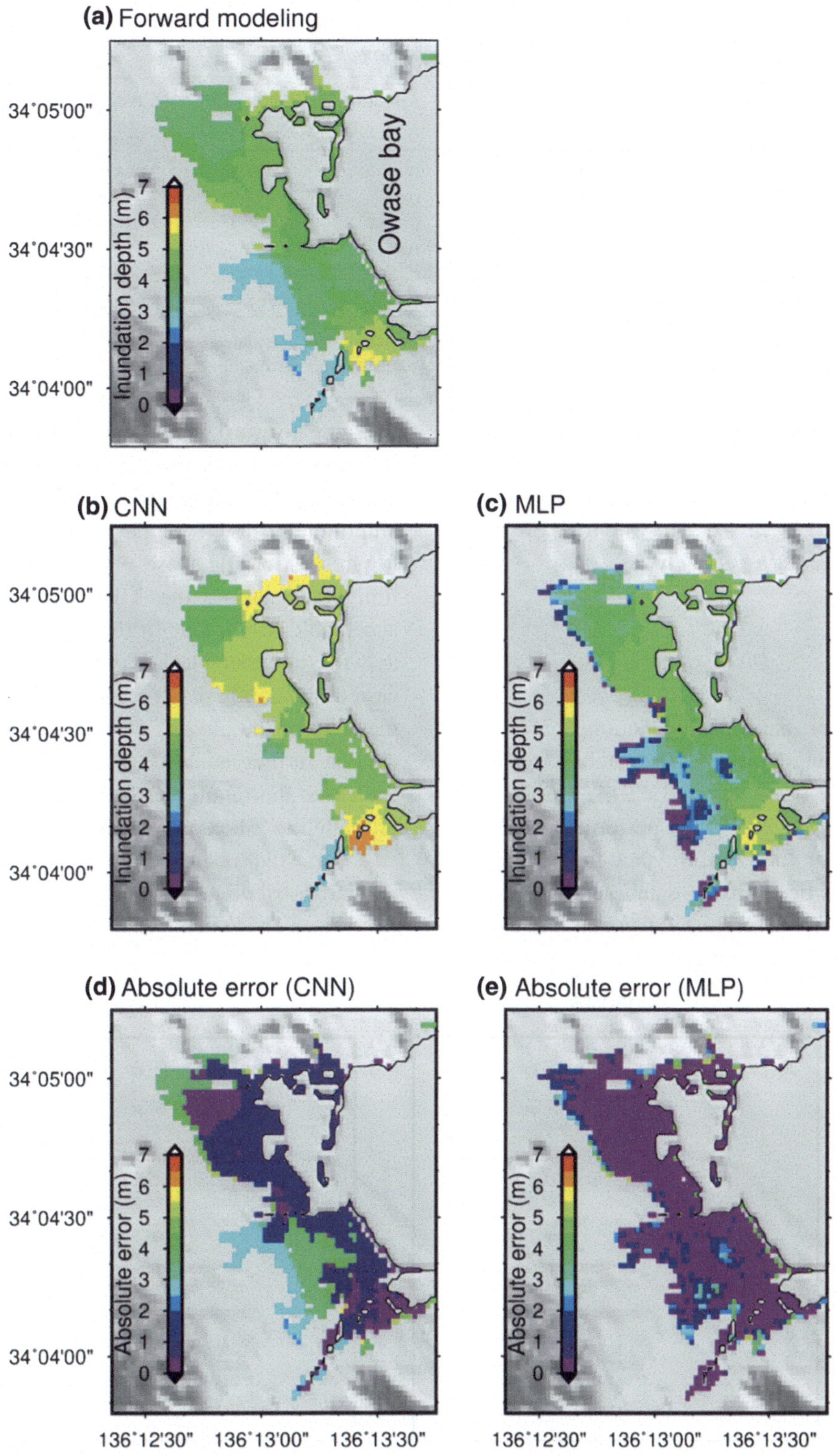

Figure 4
Reference tsunami inundation at Owase obtained from forward modeling. **b**, **c** Forecasts obtained from the proposed methods. **d**, **e** Absolute error between forecasts and forward modeling

Table 1

Statistical measures of the results

Method	Site	*d*	RMSE (m)	Min. (m)	Max. (m)	Mean (m)	Number of inundated grid points
Forward modeling	Atashika	–	–	0	9.378	7.735	527
	Owase	–	–	0	5.754	3.761	1279
CNN	Atashika	−0.003	1.882	0	9.904	7.588	545 (18)
	Owase	0.548	2.117	0	6.378	3.410	874 (−405)
MLP	Atashika	0.073	2.190	0	9.123	6.505	546 (19)
	Owase	−0.061	1.142	0	5.768	3.656	1383 (104)

Values in parentheses in the last column indicate the differences in the number of inundated grid points between forecasts and forward modeling. Positive values indicate overestimation, and negative values indicate underestimation

RMSE ranging from 0.35 m to 0.60 m and 0.32 m to 0.85 m for the testing dataset at Atashika and Owase, respectively. For the training dataset, the error ranges from 0.28 m to 0.75 m and 0.35 m to 0.50 m at Atashika and Owase, respectively.

5. Discussion

Like the other tsunami inundation forecasting systems that use an algorithm that matches scenarios against a database (Gusman et al. 2014; Mulia et al. 2018; Setiyono et al. 2017), the performance of the forecast depends on the variability and the number of scenarios used to develop the database. The underestimation of the CNN-forecasted tsunami inundation over the forward modeling at Owase indicates that the number of scenarios in our database is not sufficient, and thus the pattern of the maximum tsunami amplitude of the new input cannot fit a pattern in the low-resolution database. However, even with the limited scenarios, the MLP can provide a more acceptable result. One of the drawbacks of the MLP, as with the other black-box models, is that the forecasted inundation depth cannot represent the physical meaning of the physics of tsunami inundation.

For further analysis of the performance of the forecasts, the relative error at the grid points is also presented as a box and whiskers diagram (Fig. 5). At Atashika, the results from both methods show a similar distribution. For the CNN, the median of the relative error is slightly closer to zero than the MLP

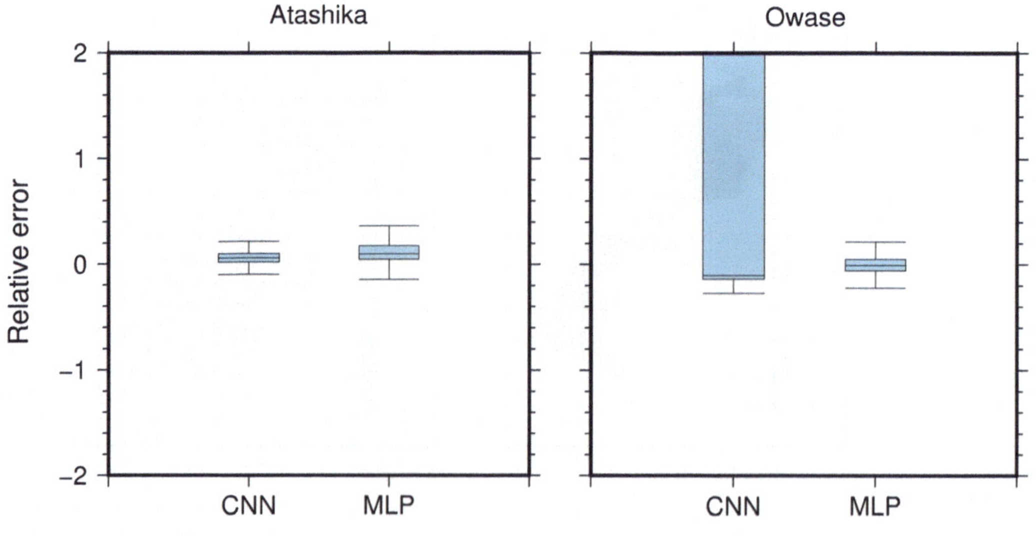

Figure 5
Box and whisker plots indicating the relative error between forward modeling and the forecasts at Atashika and Owase

result. The forecast from the CNN also possesses better accuracy than that of the MLP, as indicated by the range and length of the box and whiskers, though the differences are not significant. In contrast, the distribution of the relative error of the forecast at Owase exhibits characteristics different from that at Atashika. Most of the relative errors of the forecast from the CNN are distributed within a range of 0 to 2, which indicates a strong underestimation caused by a significant difference in the limit of inundation between the forecast and forward modeling. In contrast, the MLP presents a more satisfactory result, in which the relative errors are mostly distributed near zero. Additionally, in Figs. 3e and 4e, it is observed that the overestimation and underestimation of the forecasted inundation depth are mostly located near the edge of the limit of inundation. To suppress this effect, which is probably caused by the model complexity, the MLP requires an additional regularizer (L2 norm regularizer) in addition to the dropout regularizer. This may also explain why the MLP can capture the pattern of the inundation limit more easily than it can capture the inundation depth. A further study is being conducted to address this issue by using a stacked generalization ensemble method, and the results will be presented in the future.

The CNN and MLP models learn the pattern of the maximum tsunami amplitude during the training process. After the network has been trained, the prediction can be made quickly by performing a matrix multiplication procedure. We use a desktop computer equipped with an Intel i7 processor to conduct forward modeling, while a notebook computer with an i5 processor performed the machine learning simulation. Computationally, the nonlinear forward model requires about 40 min (on eight cores) for a 2-h simulation, while CNN and MLP require less than 1 s (0.069 and 0.084 s for CNN and MLP, respectively, on a single processor) to produce a tsunami inundation forecast. In a real event, although the time required for the prediction process can be ignored, the proposed methods depend heavily on a reliable tsunami source model. Considering that a tsunami caused by a hypothetical future Nankai megathrust earthquake can quickly reach the region used in this study, a robust and fast method to obtain a tsunami source model is needed. Seismic waves, which propagate faster than tsunamis, are promising for use in estimating the tsunami source about 5 to 10 min after the earthquake, as presented by Gusman and Tanioka (2013) and Kanamori (2015). Furthermore, computing the maximum tsunami amplitude using a linear model in the low-resolution grid takes less than 2 min on a regular computer (on eight cores). In total, to generate a tsunami inundation forecast, it only takes approximately 7–12 min. By using our proposed methods, we can provide more lead time, which is very important to give sufficient warning to coastal residents who need to evacuate the inundation area.

The advantages of our proposed methods over the NearTIF method (Gusman et al. 2014) are that our proposed methods can provide better and faster tsunami inundation prediction because the MLP can predict tsunami inundation using the learned scenarios and they have no waveform time-shifting procedure, as in NearTIF which may take some time. The ability of our proposed methods is similar to that of the method proposed by Mulia et al. (2018), which is able to provide a unique tsunami inundation forecast. However, in the method of Mulia et al. (2018), applying the interpolation method produces a tsunami inundation prediction that is exactly half of the Euclidian distance between the two closest samples, whereas the MLP produces inundation prediction based on the patterns learned during the training process. A problem arises when the number of scenarios is limited, so that the tsunami inundation from the two closest samples may be significantly different. Thus, the interpolation method may produce inundation prediction with significant overestimation or underestimation. Therefore, we would expect that the MLP provides a better tsunami inundation result compared with the method of Mulia et al. (2018).

For future development, adding more fault scenarios is essential to improve the prediction capability of the models. This can be accomplished by incorporating unique features from well-verified earthquakes or tsunami source models of previous events in the region, such as the 1707 Hoei (Furumura et al. 2011), the 1944 Tonankai (Baba et al. 2006), and the 1946 Nankai (Murotani et al. 2015) events. To consider the uncertainty of source properties of future tsunamigenic earthquakes, stochastic earthquake sources (Goda et al. 2018) are suggested for inclusion. Incorporating the influence of tides in shallow waters within the tsunami simulation could provide substantial results for tsunami hazard preparedness (Lee et al. 2015), and thus this effect should be considered in a future study. Additionally, although it requires high computational cost, incorporating a dispersive effect in the nonlinear tsunami model successfully reproduces soliton fission along the Sendai coast during the 2011 Tohoku tsunami (Baba et al. 2015). With U-shaped topography, Owase is also subjected to soliton fission effects (Yamanaka et al. 2016); therefore, applying dispersive effects may provide a more comprehensive database.

It is straightforward to apply the proposed methods in this study to other locations. However, because the model parameters (weight and bias) are particularly purposed for Atashika and Owase, it is necessary to re-train the model to fit each new specific location. Of course, the low-resolution and high-resolution domains need to be carefully adjusted to give a reliable result. In the near future, it is expected that these data-driven models will become a standard method for solving various problems, especially early warning tasks.

6. Conclusion

Two machine learning models (CNN and MLP) have been developed in this study. The concept of the CNN is similar to the method proposed by Mulia et al. (2018), which selects the best-fit scenario in the database based on similarity pattern between computed and precomputed maximum tsunami amplitude in the low-resolution grid. The difference is that the pattern recognition in the previous study is conducted by using a PCA, a common dimensionality reduction algorithm, while the dimensionality reduction is performed within the convolutional and pooling layers of a CNN in this study. Because the database is developed from limited simple fault scenarios, there is a possibility that the CNN is unable to find the appropriate best-fit scenario, as shown for Owase. However, the MLP, which is purposed for the regression task, is able to yield a more acceptable forecast than CNN when there is no scenario in the database that has a pattern similar to that produced by direct forward modeling. With a quick calculation time, both methods are strong candidates for a future tsunami inundation forecasting system.

Our proposed methods rely on accurate and reliable tsunami source models to generate inundation forecasts. Due to the short distance between the Nankai Trough and the nearby Japanese coast, a fast and reliable tsunami source model is needed. A tsunami source model based on W-phase inversion (Gusman and Tanioka 2013), which only requires about 10 min, is a promising method for integration with the proposed methods. In the future, we should focus on improving the reliability of the database by incorporating more scenarios. The network architecture may also need to be improved to accommodate more input variables, such as tsunami height at observation stations, to produce a better forecast.

Acknowledgements

Figures in this paper were made using GMT software (Wessel and Luis, 2017), which can be downloaded at http://gmt.soest.hawaii.edu/. We used the JAGURS tsunami simulation code (Baba et al. 2016, 2015), which is available at https://github.com/jagurs-admin/jagurs.

Publisher's Note Springer Nature remains neutral with regard to jurisdictional claims in published maps and institutional affiliations.

References

Abadi, M., Barham, P., Chen, J., Chen, Z., Davis, A., Dean, J., Devin, M., Ghemawat, S., Irving, G., Isard, M., Kudlur, M., Levenberg, J., Monga, R., Moore, S., Murray, D.G., Steiner, B., Tucker, P., Vasudevan, V., Warden, P., Wicke, M., Yu, Y., Zheng, X., Brain, G., Osdi, I., Barham, P., Chen, J., Chen, Z., Davis, A., Dean, J., Devin, M., Ghemawat, S., Irving, G., Isard, M., Kudlur, M., Levenberg, J., Monga, R., Moore, S., Murray, D.G., Steiner, B., Tucker, P., Vasudevan, V., Warden, P., Wicke, M., Yu, Y., Zheng, X. (2016). TensorFlow: A System for Large-Scale Machine Learning. In 12th USENIX Conference on Operating Systems Design and Implementation (pp. 265–283).

Baba, T., Ando, K., Matsuoka, D., Hyodo, M., Hori, T., Takahashi, N., et al. (2016). Large-scale, high-speed tsunami prediction for the Great Nankai Trough Earthquake on the K computer. *International Journal of High Performance Computing Applications, 30,* 71–84. https://doi.org/10.1177/1094342015584090.

Baba, T., Cummins, P. R., Hori, T., & Kaneda, Y. (2006). High precision slip distribution of the 1944 Tonankai earthquake inferred from tsunami waveforms: Possible slip on a splay fault. *Tectonophysics, 426,* 119–134. https://doi.org/10.1016/j.tecto.2006.02.015.

Baba, T., Takahashi, N., & Kaneda, Y. (2014). Near-field tsunami amplification factors in the Kii Peninsula, Japan for Dense Oceanfloor Network for Earthquakes and Tsunamis (DONET). *Marine Geophysical Researches, 35,* 319–325. https://doi.org/10.1007/s11001-013-9189-1.

Baba, T., Takahashi, N., Kaneda, Y., Ando, K., Matsuoka, D., & Kato, T. (2015). Parallel implementation of dispersive tsunami wave modeling with a nesting algorithm for the 2011 Tohoku tsunami. *Pure and Applied Geophysics, 172,* 3455–3472. https://doi.org/10.1007/s00024-015-1049-2.

Bienvenido-Huertas, D., Rubio-Bellido, C., Pérez-Ordóñez, J. L., & Moyano, J. (2019). Optimizing the evaluation of thermal transmittance with the thermometric method using multilayer perceptrons. *Energy Build., 198,* 395–411. https://doi.org/10.1016/j.enbuild.2019.06.040.

Central Disaster Management Council. (2003). Risk assessment results of Tokai-Tonankai-Nankai earthquake disaster. Tokyo, Japan: Cabinet Office, Government of Japan. http://www.bousai.go.jp/kaigirep/chuobou/9/pdf/zuhyou_2-2.pdf.

Ciregan, D., Meier, U., & Schmidhuber, J. (2012). Multi-column deep neural networks for image classification. In *Proceedings of the IEEE computer society conference on computer vision and pattern recognition*, pp. 333–338. Retrieved Aug 1, 2019, from https://doi.org/10.1109/CVPR.2012.6248110.

Craig, T. J., Copley, A., & Jackson, J. (2014). A reassessment of outer-rise seismicity and its implications for the mechanics of oceanic lithosphere. *Geophysical Journal International, 197,* 63–89. https://doi.org/10.1093/gji/ggu013.

Fauzi, A., & Mizutani, N. (2019). Tsunami inundation forecasting system based on database: a case study in Owase City, Japan. *International Journal of Offshore and Polar Engineering.*. https://doi.org/10.17736/ijope.2019.jc743. **(in press)**.

Furumura, T., Imai, K., & Maeda, T. (2011). A revised tsunami source model for the 1707 Hoei earthquake and simulation of tsunami inundation of Ryujin Lake, Kyushu, Japan. *Journal of Geophysical Research: Solid Earth.* https://doi.org/10.1029/2010JB007918.

Garcia, C., & Delakis, M. (2004). Convolutional face finder: A neural architecture for fast and robust face detection. *IEEE Transactions on Pattern Analysis and Machine Intelligence, 56,* 1408–1423. https://doi.org/10.1109/TPAMI.2004.97.

Goda, K., Yasuda, T., Mai, P. M., Maruyama, T., & Mori, N. (2018). Tsunami simulations of mega-thrust earthquakes in the Nankai–Tonankai Trough (Japan) based on stochastic rupture scenarios. *Geological Society, London, Special Publications.* https://doi.org/10.1144/sp456.1.

Gusman, A. R., & Tanioka, Y. (2013). W phase inversion and tsunami inundation modeling for tsunami early warning: Case Study for the 2011 Tohoku Event. *Pure and Applied Geophysics, 171,* 1409–1422. https://doi.org/10.1007/s00024-013-0680-z.

Gusman, A. R., Tanioka, Y., Macinnes, B. T., & Tsushima, H. (2014). A methodology for near-field tsunami inundation forecasting: Application to the 2011 Tohoku tsunami. *Journal of Geophysical Research Solid Earth, 119,* 8186–8206. https://doi.org/10.1002/2014JB010958.

Hatori, T., Aida, I., Iwasaki, S., & Hibiya, T. (1981). Field Survey of the Tsunamis Inundating Owase City-The 1944 Tonankai, 1960 Chile, and 1854 Ansei Tsunamis. *Bulletin. Earthquake Research Institute, University of Tokyo, 56,* 245–263.

Hayes, G. P., Wald, D. J., & Johnson, R. L. (2012). Slab1.0: A three-dimensional model of global subduction zone geometries. *J. Geophys. Res. Solid Earth, 117,* 01302. https://doi.org/10.1029/2011JB008524.

Haykin, S. S. (2009). *Neural networks and learning machines.* Upper Saddle River, NJ: Pearson Education.

Igarashi, Y., Hori, T., Murata, S., Sato, K., Baba, T., & Okada, M. (2016). Maximum tsunami height prediction using pressure gauge data by a Gaussian process at Owase in the Kii Peninsula, Japan. *Marine Geophysical Research, 37,* 361–370. https://doi.org/10.1007/s11001-016-9286-z.

Ishibashi, K. (2004). Status of historical seismology in Japan. *Annales Geophysicae, 47,* 339–368. https://doi.org/10.4401/ag-3305.

Jiang, G., & Wang, W. (2017). Error estimation based on variance analysis of k-fold cross-validation. *Pattern Recognition., 69,* 94–106. https://doi.org/10.1016/j.patcog.2017.03.025.

Kanamori, H. (2015). Earthquake hazard mitigation and real-time warnings of tsunamis and earthquakes. *Pure and Applied Geophysics, 172,* 2335–2341. https://doi.org/10.1007/s00024-014-0964-y.

Kikuchi, M., Nakamura, M., & Yoshikawa, K. (2003). Source rupture processes of the 1944 Tonankai earthquake and the 1945 Mikawa earthquake derived from low-gain seismograms. *Earth, Planets and Space., 55,* 159–172. https://doi.org/10.1186/BF03351745.

Kingma, D.P., Ba, J.L. (2015). Adam: A method for stochastic gradient descent. ICLR Int. Conf. Learn. Represent.

Kolsch, A., Afzal, M.Z., Ebbecke, M., Liwicki, M. (2018). Real-Time Document Image Classification Using Deep CNN and Extreme Learning Machines. In Proceedings of the International Conference on Document Analysis and Recognition, ICDAR. https://doi.org/10.1109/ICDAR.2017.217.

Krizhevsky, A., Sutskever, I., Hinton, G.E. (2012). ImageNet Classification with Deep Convolutional Neural Networks. In ImageNet Classification with Deep Convolutional Neural Networks. pp. 1090–1098.

Lecun, Y., Bengio, Y., & Hinton, G. (2015). Deep learning. *Nature, 521,* 436–444. https://doi.org/10.1038/nature14539.

LeCun, Y., Bottou, L., Bengio, Y., & Haffner, P. (1998). Gradient-based learning applied to document recognition. *Proceedings of the IEEE.* https://doi.org/10.1109/5.726791.

Lee, H. S., Shimoyama, T., & Popinet, S. (2015). Impacts of tides on tsunami propagation due to potential Nankai Trough earthquakes in the Seto Inland Sea, Japan. *Journal of Geophysical Research Oceans., 120,* 6865–6883. https://doi.org/10.1002/2015JC010995.

Maeda, T., Obara, K., Shinohara, M., Kanazawa, T., & Uehira, K. (2015). Successive estimation of a tsunami wavefield without earthquake source data: A data assimilation approach toward real-time tsunami forecasting. *Geophysical Research Letters, 42,* 7923–7932. https://doi.org/10.1002/2015GL065588.

Mulia, I. E., Gusman, A. R., & Satake, K. (2018). Alternative to non-linear model for simulating tsunami inundation in real-time.

Geophysical Journal International, 214, 2002–2013. https://doi.org/10.1093/GJI/GGY238.

Murotani, S., Satake, K., & Fujii, Y. (2013). Scaling relations of seismic moment, rupture area, average slip, and asperity size for M-9 subduction-zone earthquakes. *Geophysical Research Letters, 40,* 5070–5074. https://doi.org/10.1002/grl.50976.

Murotani, S., Shimazaki, K., & Koketsu, K. (2015). Rupture process of the 1946 Nankai earthquake estimated using seismic waveforms and geodetic data. *Journal of Geophysical Research: Solid Earth, 120,* 5677–5692. https://doi.org/10.1002/2014JB011676.

Musa, A., Watanabe, O., Matsuoka, H., Hokari, H., Inoue, T., Murashima, Y., et al. (2018). Real-time tsunami inundation forecast system for tsunami disaster prevention and mitigation. *The Journal of Supercomputing, 74,* 3093–3113. https://doi.org/10.1007/s11227-018-2363-0.

Nair, V., Hinton, G. (2010). Rectified Linear Units Improve Restricted Boltzmann Machines, in: Proceedings of the 27th International Conference on Machine Learning.

Oishi, Y., Imamura, F., & Sugawara, D. (2015). Near-field tsunami inundation forecast using the parallel TUNAMI-N2 model: Application to the 2011 Tohoku-Oki earthquake combined with source inversions. *Geophysical Research Letters, 42,* 1083–1091. https://doi.org/10.1002/2014GL062577.

Okada, Y. (1985). Surface deformation due to shear and tensile faults in a half-space. *Bulletin of the Seismological Society of America, 75,* 1135–1154.

Sermanet, P., Kavukcuoglu, K., Chintala, S., Lecun, Y. (2013). Pedestrian detection with unsupervised multi-stage feature learning. In Proceedings of the IEEE Computer Society Conference on Computer Vision and Pattern Recognition. https://doi.org/10.1109/CVPR.2013.465.

Setiyono, U., Gusman, A. R., Satake, K., & Fujii, Y. (2017). Pre-computed tsunami inundation database and forecast simulation in Pelabuhan Ratu, Indonesia. *Pure and Applied Geophysics, 178,* 3219–3235. https://doi.org/10.1007/s00024-017-1633-8.

Sutskever, I., Hinton, G., Krizhevsky, A., & Salakhutdinov, R. R. (2014). Dropout: A simple way to prevent neural networks from overfitting. *The Journal of Machine Learning Research, 15,* 1929–1958.

Tang, L. J., Titov, V. V., Wei, Y., Mofjeld, H. O., Spillane, M., Arcas, D., et al. (2008). Tsunami forecast analysis for the May 2006 Tonga tsunami. *Journal of Geophysical Research: Oceans., 113,* C12015. https://doi.org/10.1029/2008JC004922.

Tanioka, Y. (2018). Tsunami simulation method assimilating ocean bottom pressure data near a tsunami source region. *Pure and Applied Geophysics, 175,* 721–729. https://doi.org/10.1007/s00024-017-1697-5.

Tanioka, Y., & Satake, K. (1996). Tsunami generation by horizontal displacement of ocean bottom. *Geophysical Research Letters, 23,* 861–864. https://doi.org/10.1029/96GL00736.

Titov, V. V., González, F.I., Bernard, E.N., Eble, M.C., Mofjeld, H.O., Newman, J.C., Venturato, A.J. (2005). Real-time tsunami forecasting: Challenges and solutions. In Developing Tsunami-Resilient Communities: The National Tsunami Hazard Mitigation Program. pp. 41–58. https://doi.org/10.1007/1-4020-3607-8_3.

Tompson, J., Goroshin, R., Jain, A., LeCun, Y., Bregler, C. (2015). Efficient object localization using Convolutional Networks. In Proceedings of the IEEE Computer Society Conference on Computer Vision and Pattern Recognition. https://doi.org/10.1109/CVPR.2015.7298664.

Tsushima, H., Hino, R., Ohta, Y., Iinuma, T., & Miura, S. (2014). TFISH/RAPiD: Rapid improvement of near-field tsunami forecasting based on offshore tsunami data by incorporating onshore GNSS data. *Geophysical Research Letters, 41,* 3390–3397. https://doi.org/10.1002/2014GL059863.

Wang, X., Zhang, W., Wu, X., Xiao, L., Qian, Y., & Fang, Z. (2019). Real-time vehicle type classification with deep convolutional neural networks. *Journal of Real-Time Image Processing, 16,* 5–14. https://doi.org/10.1007/s11554-017-0712-5.

Wessel, P., & Luis, J. F. (2017). The GMT/MATLAB Toolbox. *Geochemistry Geophysics, Geosystems, 18,* 811–823. https://doi.org/10.1002/2016GC006723.

Yamamoto, N., Aoi, S., Hirata, K., Suzuki, W., Kunugi, T., & Nakamura, H. (2016). Multi-index method using offshore ocean-bottom pressure data for real-time tsunami forecast 4 Seismology. *Earth, Planets and Space.* https://doi.org/10.1186/s40623-016-0500-7.

Yamanaka, Y., Sato, S., Tajima, Y., & Shimozono, T. (2016). A study on soliton fission of Nankai trough tsunami. *Journal of Japan Society Civil Engineers Series B2, 72,* I_403–I_408. https://doi.org/10.2208/kaigan.72.i_403.

(Received August 21, 2019, revised November 5, 2019, accepted November 6, 2019, Published online November 11, 2019)

Pure Appl. Geophys. 177 (2020), 1451–1470

https://doi.org/10.1007/s00024-019-02381-3

Pure and Applied Geophysics

A Multiple-Parameter Methodology for Placement of Tsunami Sensor Networks

Joaquín Meza,[1] Patricio A. Catalán,[1,2,3] and Hiroaki Tsushima[4]

Abstract—A methodology to optimize the design of an offshore tsunami network array is presented, allowing determination of the placement of sensors to be used in a tsunami early warning system framework. The method improves on previous sensor location methods by integrating three commonly used tsunami forecast performance indicators as a measure of the predictive accuracy through a single cost function. The joint use of different tsunami parameters allows for a network that is less subject to bias found when using a single parameter. The resulting network performance was tested using a set of synthetic target scenarios and also verified against a model of the 2014 Pisagua event, suggesting that having such a network in place could have provided meaningful information for the hazard assessment. The small number of sensors required (three spanning nearly 700 km of the Northern Chile coast) may be useful in implementing such networks in places where funding of denser arrays is difficult.

Keywords: Tsunami, tsunami early warning, sensor network.

1. Introduction

Over the last decade, several tsunami events, such as those of Maule in 2010 and Tohoku in 2011, have further demonstrated the catastrophic and widespread potential for death and destruction inherent in tsunami waves and, consequently, the need to improve the reliability of tsunami early warning systems. The 2010 Maule earthquake (M_w8.8) generated a tsunami that caused severe damage and loss of life in coastal communities (e.g., Fritz et al. 2011) and highlighted the consequences of an ineffective alert (Soulé 2014). Events such as this have reaffirmed the importance of improving timely evacuation warnings, which are considered to be one of the most effective ways to reduce loss of human life and damage to coastal communities (Okal 2015).

The goal of a tsunami warning system is to deliver a timely and meaningful evaluation of the hazard to authorities and to the population at large, with the main objective of triggering evacuation. While the role of education is usually considered the cornerstone for successful responses, the role of accurate information regarding the actual hazard is also relevant (Okal 2015; Bernard and Titov 2015). Over the years, improving tsunami hazard assessment has followed different approaches, all attempting to maximize the lead time of the warning relative to tsunami arrival.

For far-field tsunami forecasting, where the coastal tsunami impact can be quantitatively evaluated well in advance of tsunami arrival, the approach is to combine monitoring of the actual tsunami along its propagation path with numerical simulations, and take advantage of this information to estimate the hazard; For instance, the USA has a worldwide network of offshore tsunami observatories, which are located near several subduction zones at distances equivalent to 30–60 min tsunami travel time from expected tsunamigenic earthquake sources. Data are used as input to inversion procedures to obtain an estimate of the earthquake or tsunami source (Percival et al. 2011), which allows forecasting a set of tsunami hazard products (Bernard and Titov 2015).

For near-field tsunami forecasting, on the other hand, the lapse between tsunami generation and arrival may be too short to perform tsunami source evaluations and forward simulations. Williamson and Newman (2019) show that, in some places, first

[1] Departamento de Obras Civiles, Universidad Técnica Federico Santa María, Valparaíso, Chile. E-mail: joaquin.meza@usm.cl; patricio.catalan@usm.cl

[2] Centro de Investigación para la Gestión Integrada del Riesgo de Desastres (CIGIDEN), Santiago, Chile.

[3] Centro Científico Tecnológico de Valparaíso-CCTVal, Universidad Técnica Federico Santa María, Valparaíso, Chile.

[4] Meteorological Research Institute, Japan Meteorological Agency, Tsukuba, Japan. E-mail: tsushima@mri-jma.go.jp

arrival can occur in as little as 5 min. In the field, Aránguiz et al. (2016) reported arrivals in less than 8 min for the M_w 8.4 Illapel event, based on anecdotal evidence. A relevant aspect of the problem is the finite time required to acquire enough data to allow for source inversion. Owing to the significantly different propagation speeds of tsunami and seismic waves, the preferred approach is to rely on seismic data alone. Although some seismic source solutions can be obtained in relatively short time, they are subject to uncertainties that can affect the accuracy of tsunami estimates, hence limiting their applicability for tsunami early warning purposes (Cienfuegos et al. 2018). In addition, computing tsunami propagation and inundation has a high computational cost. Consequently, to date, most operating tsunami early warning systems use table look-up procedures on datasets of simplified seismic scenarios previously calculated and stored in a database (e.g. Gica et al. 2008; Kamigaichi et al. 2009). This approach is used in countries such as Japan, Australia, and Indonesia, and recently, in Chile. This approach is affected by the limited accuracy of such assessments owing to the likely mismatch between the simple stored scenarios and the actual event (Behrens et al. 2010)

It has been noted that including tsunami data in the inversion process usually leads to improved results in estimating the tsunami hazard (e.g., Behrens et al. 2010; Gusman et al. 2014; Cienfuegos et al. 2018). Hence, it is highly desirable to incorporate tsunami data as early as possible. These observations can be acquired from tide gage data (Satake 1987; Satake and Kanamori 1991), satellite altimetry (Arcas and Titov 2006; Hirata et al. 2006), deep-ocean tsunameters (Titov 2009; Percival et al. 2011; Rabinovich and Eblé 2015), and cabled ocean-bottom pressure sensors (Baba et al. 2004; Tsushima et al. 2009, 2012). Among these, the use of coastal stations (tide gages) is typically not considered for early warning purposes, owing to the nil lead time and possible influence of coastal bathymetry on the hydrodynamics. Hence, offshore tsunami data from tsunameters such as OBPSs seem to be most appropriate for tsunami early warning systems.

Williamson and Newman (2019) analyzed the possible coastal locations where sensors can provide a meaningful lead time for near-field tsunamis. These locations concentrate along narrow bands that run roughly parallel to subduction zones. In addition, one of the most important factors for tsunami forecasting based on offshore tsunami data is the configuration of the array of tsunami stations. The accuracy of the reconstruction of the tsunami source strongly depends on the azimuthal coverage of observation stations with respect to the source area, which is improved when the sensors are located close to the main beam of tsunami energy (e.g., Pires and Miranda 2001; Piatanesi et al. 2001; Bernard et al. 2001). In addition, a large number of sensors enables better resolution of the finer structure of the tsunami source. For instance, Japan has a few submarine cabled seafloor observatory networks (e.g., the S-net, DONET1, and DONET2 systems) that provide data in real time. Tens of bottom pressure sensors have been installed or are planned to be installed (e.g., Kaneda et al. 2015; Kawaguchi et al. 2015) at an estimated cost of US$ 500M, according to Bernard and Titov (2015). Maintenance and operational costs also need to be accounted for, in addition to installation costs. This high cost may pose a significant hurdle for developing countries along subduction zones, such as Chile or other countries on the eastern Pacific seaboard, where there is an extensive seismogenic zone. To attempt to overcome this, it is valuable to study whether arrays with fewer sensors could provide a working solution at lower cost.

The number and placement of tsunameters have been based on expert judgment, considering technical aspects such as the variability and location of seismic sources (Schindelé et al. 2008), travel time (Schindelé et al. 2008; Omira et al. 2009; Williamson and Newman 2019), financial factors, for instance pertaining to installation or operation costs (Araki et al. 2008), and legal aspects, such as geographic boundaries or legal jurisdictions (Abe and Imamura 2013), among others. Regarding the construction of a sensor array, prior research indicates that two to four observation stations are capable of constraining the source parameters quite well if the stations are optimally located relative to the main tsunami energy beam, whereas adding more data does not significantly improve the inversion results (e.g., Percival et al. 2011; An et al. 2018). For example, Bernard et al. (2001) suggested that sensor spacing of about

175–700 km in the along-strike direction is required to characterize the main energy beam of $M_w \sim 8.0$ events using just three sensors. On the other hand, the relative distance between sensors and between sensors and the source will be constrained by the tsunami travel time and the duration of the tsunami record to be used in the inversion (e.g., Bernard et al. 2001). Hence, given an earthquake, it is possible to define an area (henceforth termed the influence area A_I) over which the tsunami waves have already propagated away from the source and allow for sufficient data to be collected (e.g., Williamson and Newman 2019). A minimum of two sensors must be considered inside this area. This is equivalent to defining a data observation time T_0, if the tsunami propagation speed is known. For the present application, T_0 is user defined, thereby limiting the areal coverage where sensors can be placed and how much data is used in the inversion.

Once possible locations of the sensors have been established, evaluations of the best configuration have adopted different approaches. Both Schindelé et al. (2008) and Omira et al. (2009) took into account local seismicity to identify possible sources, and designed the optimal placement of sensors using the travel time and a set of delays as the target function. They differ in that Schindelé et al. (2008) used 16 evenly spaced tsunami sources spanning a long stretch of coast of the Western Mediterranean Basin, which were used to test the efficacy of two predefined arrays. Thirteen tsunameters spaced at about 50–90 km yielded the best results. Instead, Omira et al. (2009) used just a single scenario in each of five tsunamigenic zones, though all were encompassed within a small domain, resulting in a common array of just three sensors. On the other hand, Spillane et al. (2008) applied an optimization approach to place Deep-ocean Assessment and Reporting of Tsunami (DART) buoys in the Aleutians Islands and Alaska regions. Results suggested that arrival time is the main restriction on tsunameter placement; it was found that adding more than three sensors did not improve the results significantly. Mulia et al. (2017) also used optimization methods, specifically a dimensionality reduction approach, to initiate the process. Unlike the previous cases where the goal was to address the performance for a wide range of sources, the goal of Mulia et al. (2017a, 2017b) was to identify the best placing of sensors to characterize a specific, large-magnitude, target scenario. Their focus was to resolve in great detail the characteristics of nonuniform slip by maximizing the accuracy of inverting a set of stochastic scenarios on a predefined spatial domain. Inverted sources were compared against each stochastic source, under the assumption that the tsunami would be well determined if the earthquake source was well retrieved. Hence, no evaluation of the tsunamis were performed. An initial set of 30 sensors was obtained, which was reduced to 23 sensors at specific locations after optimization. This highlights that, in pursuing the detailed spatial distribution of slip, a large number of sensors is required. Recently, Saunders (2018) proposed an improvement in slip characterization for the Cascadia Subduction Zone, testing five different dense sensor arrays, which coupled Global Navigation Satellite System (GNSS) sensors with tsunameters. To assess accuracy, they compared the root mean square of the difference between the inverted and input fault slip, maximum fault slip, tsunami amplitude at the coast, and percentage of coastline hit by high-amplitudes waves between the results recovered from the inversion and the input data.

Note that the performance of sensor networks has typically been quantified by analyzing their accuracy in predicting the arrival time (Bernard et al. 2001; Schindelé et al. 2008; Omira et al. 2009), the source slip (Mulia et al. 2017; Saunders 2018), or amplitude at the coast (Mulia et al. 2017b; Saunders 2018), among others. Typically, these parameters are analyzed independently. In this work, a methodology for estimating the optimal placement of a small numbers of sensors is used and tested. While similar in scope to previous studies (e.g., Schindelé et al. 2008; Omira et al. 2009), here other tsunami parameters are included in the assessment and combined into a single cost function to enable an objective comparison. In particular, three different tsunami parameters are considered in unison to find the optimal configuration.

2. Methodology

The overall objective of this work is to determine an optimal array configuration of offshore tsunami sensors for near-field tsunami forecasting, to be used as a data source for a tsunami inversion technique. To this end, a three-step process is applied. The first step consists of the implementation of the inversion procedure. Next, a number of different sensor arrays are used to invert a set of two target scenarios. Finally, the results of these source inversion tests are objectively compared by defining performance indicators. The analysis focuses on maximizing the forecast accuracy of three relevant tsunami parameters: arrival time, maximum tsunami amplitude, and forecast skill.

2.1. Implementation of Inversion Algorithm

The method is built on the premise that, to determine the tsunami source, an inversion procedure must be implemented. Here, the tsunami Forecasting based on Inversion for initial sea-Surface Height (tFISH) method (see Tsushima et al. 2009 for details) is used, by which tsunami waveform data are inverted (in real time in an operational setting) to estimate the initial distribution of sea-surface displacement. To this end, a set of unit tsunami sources are propagated to observation points to obtain their Green's functions, which are later used to perform the inversion and forecast the tsunami time series at target locations.

The initial sea surface model for each unit source corresponds to a Gaussian function. Here, a set of nearly 1000 unit sources is considered, each having dimensions of 700×700 arcsec, with their centers spaced by 0.15 arcdeg, thereby overlapping to allow for smooth variation of the sea-surface displacement using a finite number of discrete elements (Aida 1972). They cover an area spanning about $7° \times 3°$ (latitude, longitude, approximately 680×160 km) that coincides with the Northern Chile Gap (Comte and Pardo 1991; Metois et al. 2013), as shown in Fig. 1a, b. In what follows, other data sources such as DART buoys are not considered in the analysis, under the premise that a completely new system is being developed in the area of interest, even though they may be of benefit to the inversion process. Additionally, for the particular case of Chile, the location of existing DART buoys in the area of interest requires observation times longer than those studied herein (Williamson and Newman 2019).

All Green's functions were estimated in advance considering a prescribed duration of the event, T, using the tsunami model JAGURS (Baba et al. 2016). This is a parallelized numerical model that can solve the nonlinear Boussinesq dispersive equations in spherical coordinates with a leapfrog staggered-grid, finite-difference calculation scheme. However, for the present case, nonlinear terms were not considered. Bathymetry data were obtained from the GEBCO_2014 grid (version 20150318, www.gebco.net data) with spatial resolution of 30 arcsec (Weatherall et al. 2015). Data were stored in a database, from which appropriate tsunami free surface time series, $\eta(t)$, were extracted and stored for use in the inversion and forward calculations. These were collected at 212 offshore observation points, identified in Fig. 1a, b by black squares. Each of these denotes the possible location of a single observing station or sensor, and a collection of such sensors is termed a sensor array. Time series were also stored at seven coastal observation points located close to existing tide gages operated by the National Hydrographic and Oceanic Service of the Chilean Navy (SHOA, from its acronym in Spanish). The latter are used to evaluate the predictive performance at the coast (identified in Fig. 1a, b by red triangles). Typically, these are located at absolute depths greater than 200 m such than nonlinear effects can be neglected and linear superposition of tsunami time series can be performed with minimal errors. It is worth noting that tFISH considers the coseismic displacement in the inversion procedure, thereby allowing placement of sensors in the seismogenic zone. More details about tFISH can be found in Tsushima et al. (2009, 2012).

2.2. Numerical Experiments

For the purpose of testing and evaluating the performance of different sensor arrays, the accuracy of the inversion was evaluated using a set of prescribed tsunami sources, termed design scenarios. While it is possible to test a large set of sources over a

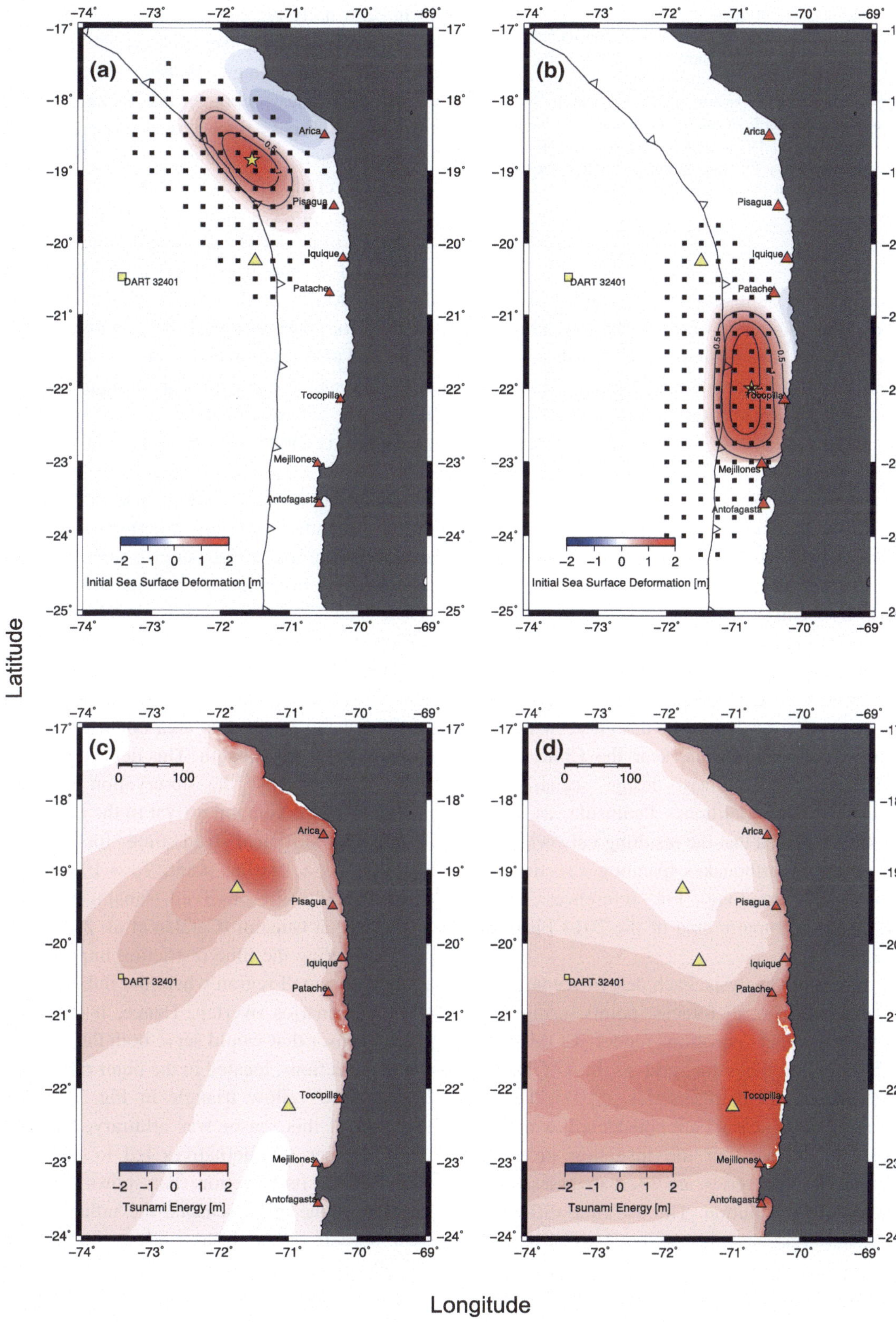
(a)
(b)
(c)
(d)
Arica
Pisagua
Iquique
Patache
Tocopilla
Mejillones
Antofagasta
DART 32401
Initial Sea Surface Deformation [m]
Tsunami Energy [m]
0 100
Latitude
Longitude

◀Figure 1
a, **b** Sea surface deformation for the scenarios considered. Stars indicate epicenters. Black squares show virtual observation points offshore, red triangles indicate forecasting points, and the yellow triangle represents the fixed sensor. **a** M_w 8.3 and **b** M_w 8.5 scenarios, located at the northern and southern end of the Northern Chile Gap, respectively. **c**, **d** Corresponding maps of tsunami wave height, considered a proxy for tsunami energy. **c**, **d** Final sensor configuration shown by yellow triangles

large domain, as in Schindelé et al. (2008), here the focus is on using sources located at the extremes of the area of interest. The assumption is that these will correspond to the most demanding configuration for the sensor arrays owing to their relative location to the main energy beam. A large number of sensor arrays are compared to identify the array that provides the best performance.

Two synthetic tsunamigenic earthquakes, with magnitudes M_w 8.3 and 8.5, respectively, are considered as the design scenarios. These were determined by Cienfuegos et al. (2014) as representative events for the Northern Chile Gap, based on the interseismic coupling model of Chlieh et al. (2011), information on interseismic slip rates, and convergence models. The northernmost design scenario is located just offshore of Arica, near the Chile–Peru border; whereas the southern design scenario is located just north of Mejillones Peninsula, as seen in Fig. 1. It is expected that the resulting network will be able to identify earthquakes spanning a section of coast about 680 km long. For reference, these scenarios flank the rupture area of the 2014 Pisagua earthquake.

To carry out the analysis, each design scenario is propagated forward to all forecast points, using the linear shallow water equation model COMCOT (Wang 2009) with the same grid as in JAGURS, thus providing observed time series, $\eta_{obs}(t)$. Those recorded at offshore sensors are considered as target time series to be used in the inversion process, whereas coastal time series are used to assess the performance of the solution. The use of a different tsunami model in propagating the target scenario and in preparing the database of Green's functions is implemented to reduce possible overfitting in the inversion. In addition, it is assumed that other sources of noise usually present in deep-ocean data, such as tides and seismic noise such as the recording of Rayleigh waves (e.g., Webb 1998), have been removed during preprocessing using detiding, or low- and high-pass filters, as described for instance in Rabinovich and Eblé (2015); For example, Tsushima et al. (2012) consider a 60-s moving average to process data for tFISH. Here, since synthetic time series are used, these effects are considered to have been filtered out already, as also done by Mulia et al. (2017). However, it is still possible for spurious signals to contaminate the time series and affect the inversion procedure. To simulate this, Gaussian noise with a maximum amplitude of 10% of the variance of the noise-free target tsunami time series is added (e.g., Romano et al. 2016; Mulia et al. 2017).

The relative position of each sensor with respect to the tsunami scenarios determines the tsunami arrival time at the sensor but also the amount of data usable in the inversion. In existing inversion procedures, each sensor uses a different observation time, sufficient to gather at least a quarter of the initial tsunami waveform. Here, the observation time T_0 is defined so as to determine an area such that any sensor located inside this area could record at least half a tsunami wavelength. This defines the influence area, A_l. Note that the observation time is also restricted by the tsunami arrival to the coast, in order to provide sufficient lead time for an eventual warning. As a starting point, $T_0 = 10$ min is used, consistent with the observed tsunami arrival of the 2014 Pisagua tsunami (Catalán et al. 2015).

Considering the time restriction imposed, there is a relatively small region where the influence areas A_l of both scenarios overlap. Hence, it is possible to place a sensor that would serve both the northern and southern sections, located in the outer rise offshore of Iquique (see yellow triangle in Fig. 1). While the selection of this sensor was arbitrary, the relatively small number of alternatives led to no significant sensitivity to its selection (not shown). To find the location of the other sensors, the influence area for each scenario was discretized every 0.25 arcdeg ($\sim$ 30 km) in both latitude and longitude to define a grid of possible sensor locations. This spacing is significantly smaller than the tsunami wavelength in the area. In addition, these offshore observation points are located at depths large enough for tsunami nonlinear effects to be considered negligible, thereby ensuring consistency with the assumptions of the inversion algorithm. As a result, 99 and 113 possible offshore observation points were defined in the northern and southern parts, respectively. A sensor array was defined by pairing the common sensor and each of the possible observation points. Note that potential observation points are located onshore and offshore of the trench. While this placement could be subject to further operational restrictions such as deployment cost, these are not considered herein, since the aim is to assess the methodology to select the optimal array. The difference in the number of observation points among the north and south regions is due to the difference in the source dimensions, and tsunami celerity arising from the bathymetry, which control the extent of A_l.

2.3. *Performance Assessment*

For each of the 212 sensor arrays (99 and 113, for each scenario), the observed tsunami time series are used to invert the tsunami source. Once a source is determined, coastal tsunami time series are obtained by linearly combining precomputed tsunami Green's functions weighted by the resulting initial distribution, and identified as forecasted data, $\eta_{for}(t)$. The use of Green's functions is preferred over directly modeling the inverted events (as done by Schindelé et al. 2008, for instance) because it allows for testing and comparing the solutions from a large number of sensor array configurations at low computational cost.

To assess the performance of each sensor array, the observed (from the full tsunami simulations using COMCOT) and the forecasted (from the linear combination using inverted source solutions) data are compared. Specifically, three tsunami parameters are considered. The first is the tsunami arrival time (e.g., Schindelé et al. 2008; Omira et al. 2009). This is considered to be an essential parameter for tsunami warning system frameworks to provide a timely hazard assessment. However, the definition of arrival time is relatively loose and could refer to different stages of the tsunami, such as the first exceedance of a threshold, the first local crest, first initial trough of N-waves, and others (e.g., Hayashi et al. 2011). Here, two different definitions are considered in order to make the analysis more robust. The first defines arrival time as the time T_1 at which the free surface first exceeds a certain arbitrary threshold, l_1

$$T_1(\eta) = \min(t \in (0, T) \mid \eta(t) > l_1). \qquad (1)$$

This arrival time definition is also used by the German–Indonesian Tsunami Early Warning System (GITEWS, Rakowsky et al. 2013), and for the present implementation a threshold of $l_1 = 0.05$ m is defined. However, considering that in some cases the tsunami time series might not exceed this prescribed threshold, a second arrival time is defined as the time when a proxy for the slope of the free surface exceeds the threshold l_2

$$T_2(\eta) = \min\left(t \in (0, T) \mid \frac{\partial \eta}{\partial t} > l_2\right). \qquad (2)$$

The aim is to establish a measure of the rate of change of the tsunami signal as an early proxy for the first local maximum. In this case, the value chosen was $l_2 = 0.002$. In doing this, it is assumed that the actual free surface slope is proportional to the rate of change as per the long-wave approximation

$$\frac{\partial \eta}{\partial x} \propto \frac{\partial \eta}{\partial t}. \qquad (3)$$

The accuracy in predicting the possible impact and magnitude of the tsunami is also relevant for a tsunami warning system. Another relevant parameter is the maximum tsunami amplitude (here denoted as H), which is the parameter used to categorize the hazard in most existing tsunami early warning systems. The maximum tsunami amplitude is estimated as

$$H = \max(\eta(t) - \eta(0) \mid t \in (0, T)). \qquad (4)$$

Both the arrival time and amplitude are statistical parameters that are estimated independently for observed and forecasted free surface time series, then compared. However, neither considers the accuracy in retrieving the shape of the waveform. Hence, a skill estimator (S_k) is computed, as this index is commonly used to evaluate model accuracy (e.g., Hampson et al. 2011)

$$S_k = \sqrt{\frac{\sum_{i=1}^{n}(\eta_{\text{for}} - \eta_{\text{obs}})^2}{\sum_{i=1}^{n}(\eta_{\text{obs}})^2}}, \quad (5)$$

where n corresponds to the number of time steps of the time series.

In assessing the accuracy, the error between observed and forecasted quantities is estimated for arrival times and maximum tsunami amplitudes. However, one possible difficulty in establishing a standard metric is that each of these parameters has its own scale with significantly different ranges of values; For example, while an error in arrival time of a few minutes can be considered acceptable (for instance, less than 2 min), a variation in height of more than 1 m can signify a large error and change the hazard category. To provide a common basis of comparison for all possible sensor configurations, the error of each parameter is nondimensionalized by dividing by the reference provided by the observed data. In addition, it is possible that one parameter having a large error could bias the combined assessment to be implemented. Therefore, the error estimated is capped under the assumption that errors larger than the observed value will be treated as equally significant. This is implemented as follows:

$$\Delta T_{\text{m}} = \min\left\{\left|\frac{T_{\text{m}}(\eta_{\text{obs}}) - T_{\text{m}}(\eta_{\text{for}})}{T_{\text{m}}(\eta_{\text{obs}})}\right|; 1\right\}, \quad (6)$$

$$\Delta H = \min\left\{\left|\frac{H(\eta_{\text{obs}}) - H(\eta_{\text{for}})}{H(\eta_{\text{obs}}) - \eta_{\text{obs}}(0)}\right|; 1\right\}, \quad (7)$$

$$S_k = \min\{S_k; 1\}, \quad (8)$$

where $m = 1, 2$ applies to the different time travel parameters ΔT_1 and ΔT_2. In order to couple both definitions of arrival times, the error estimator associated with this parameter was considered as the average of each percentage error, given by

$$\Delta T = \frac{\Delta T_1 + \Delta T_2}{2}. \quad (9)$$

By introducing the saturation value, errors ΔT_{m} or ΔH exceeding 100% are not penalized in excess and do not bias the overall error, allowing identification of the importance of the other parameters in the comparison. In the case of the skill, $S_k = 1$ indicates that the magnitude of the error is comparable to or greater than the observed values and zero values mean a perfect fit for the indicator. In addition, whether each quantity is under- or overestimated is not considered relevant, and absolute values are used instead.

It is proposed that, instead of analyzing these metrics independently, it is desirable to identify the sensor configuration that yields the minimum combined error. However, it might be relevant for the user to give preference to one of the metrics above the others. Hence, to combine all estimators into a single quantitative estimate, a forecasting accuracy function is introduced

$$F_{i,j}(\Delta T, \Delta H, S_k) = \alpha \Delta T_{i,j} + \beta \Delta H_{i,j} + \gamma S_k i, j, \quad (10)$$

which quantifies the total error of the estimation at any coastal forecasting point j given an offshore array i. Here, α, β and γ are weights that allow for user-defined tuning of the relative importance of each parameter. The sum of the weights should be 1, in order to preserve the comparison basis. In this way, it is possible to quantitatively compare the performance of all sensor arrays at any given forecast point. A conceptually similar approach was used by Behrens et al. (2010), although their objective was to evaluate the benefit of incorporating different sensors in an inversion procedure, rather than finding the best placement of them.

It is possible that comparisons at a single coastal point might also be subject to bias. To assess the overall predictive capability, the aggregated performance is computed by adding up the individual results at forecast points of interest, given a sensor array i

$$\text{EG}_i = \frac{1}{N}\sum_{j=1}^{N} F_{i,j}(\Delta T, \Delta H, S_k), \quad (11)$$

where N is the number of forecast points chosen. Upon calculating the global error by means of Eq. 11 considering all possible offshore arrays, the candidate array is selected as that having the minimal global error ($\min(\text{EG}_i)$). As in the case of Eq. 10, it is possible for Eq. 11 to be biased by a few forecast points. To this end, is required that each forecasting point j of the array should have an accuracy function value smaller than a threshold ($F_{i,j}(\Delta T, \Delta H, S_k) < \mu, \ \forall j$). This filter is equivalent to ensuring a minimal forecast capacity at each forecasting point j, and is applied after all sensor arrays have been compared.

3. Results

In Figs. 2 and 3, the accuracy of the estimation is presented as a space map of the value of each of the error estimates defined by Eqs. 6–10, presented as a color scale, where smaller values (red colors) indicate better accuracy. Each grid point corresponds to the location of the second sensor of the array for each scenario, which is defined as the "tested sensor," while the first is kept fixed at the location of the yellow triangle. The results shown correspond to the assessment obtained when using the observed time

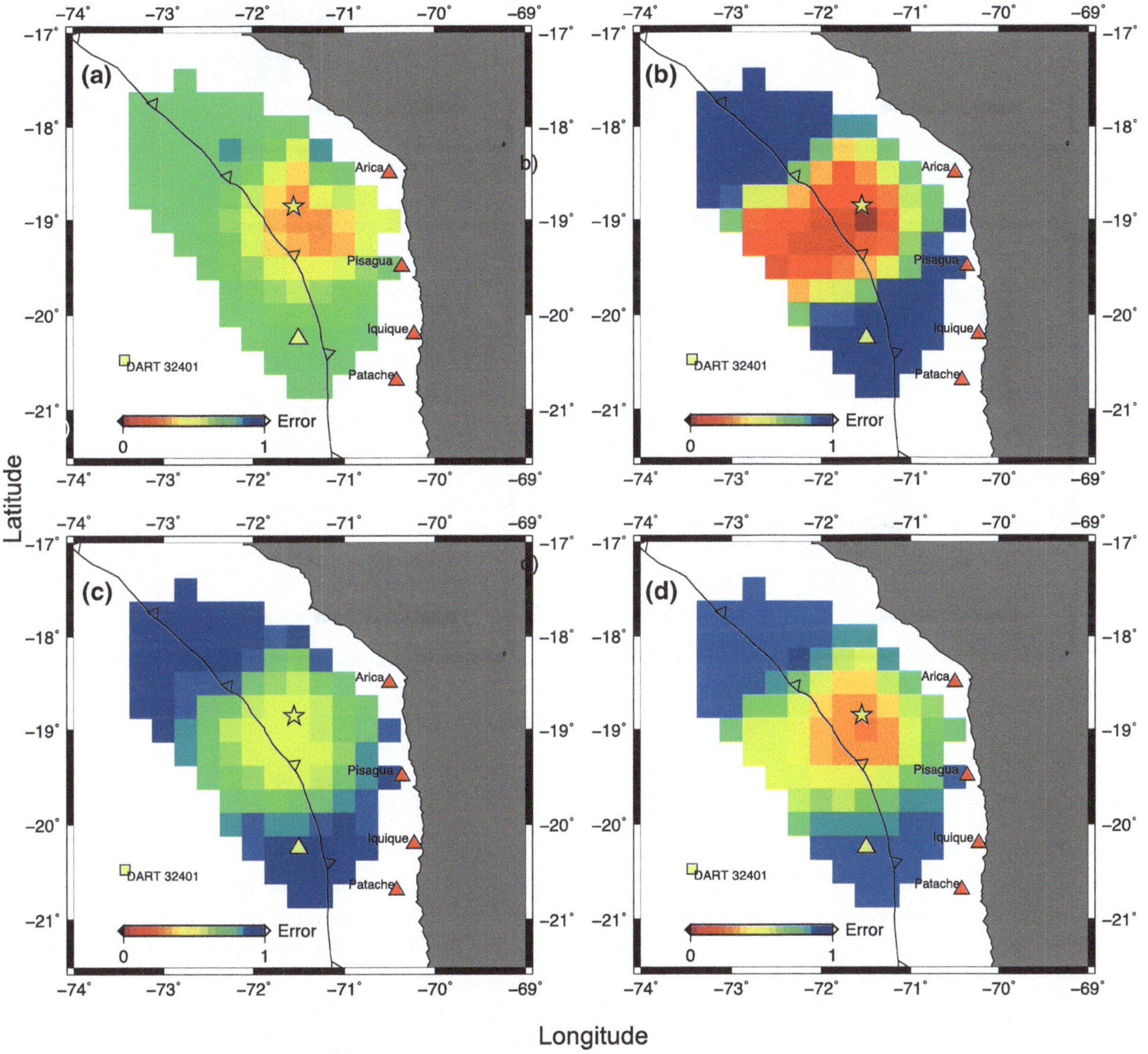

Figure 2
Spatial distribution of individual error estimates as a function of sensor arrays in the northern section. **a** ΔT (Eq. 9); **b** ΔH (Eq. 7); **c** S_k (Eq. 8); and **d** total forecasting accuracy function, $F_{i,j}$ (Eq. 10) estimated using as reference the coastal forecast point at Arica. Star indicates the scenario epicenter, the thin black line is the location of the trench, the yellow triangle denotes the location of the fixed sensor, and each grid point in the color map corresponds to a tested sensor

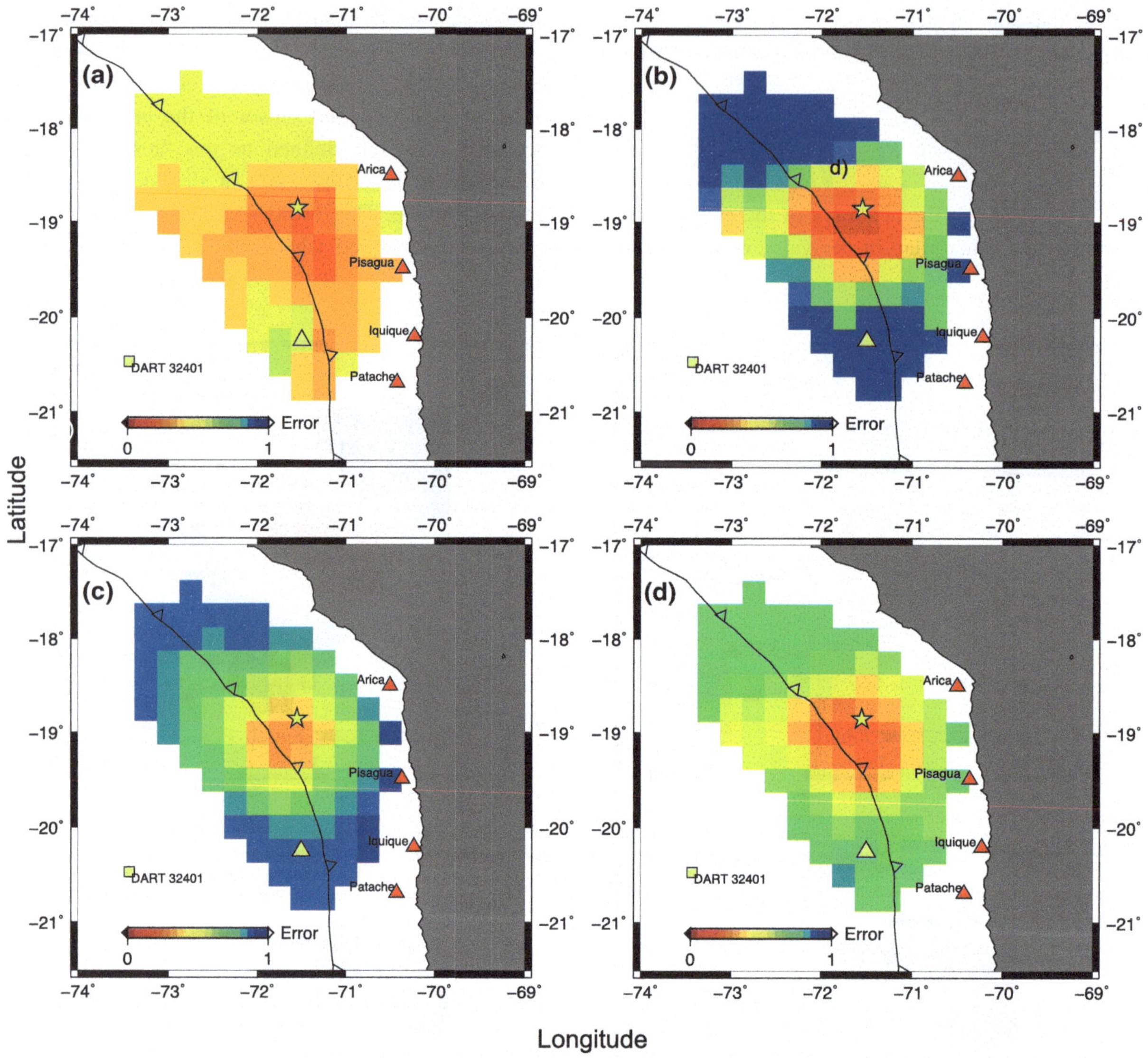

Figure 3
Same key as Fig. 2, using as reference the coastal forecast point at Patache

series at Arica and Patache as the coastal forecast point j of interest, respectively. For the purposes of demonstration, the results analyzed correspond to the case when $\alpha = \beta = 0.4$ and $\gamma = 0.2$.

The errors in arrival time, ΔT (Figs. 2a, 3a), are lower near the main energy beam, as shown by the warmer colors. Sensor locations close to the coast and the forecast point also show good performance. Arica and Patache show a distinct behavior, where Arica is more sensitive to the sensor location, with an error of $\Delta T \sim 0.36$ on average. Patache shows less sensitivity and better accuracy overall, partly because it benefits from the fixed sensor being located in front of it, making the results less sensitive to the placement of the secondary sensor.

The error in amplitude, ΔH, is shown in Figs. 2b and 3b. Unlike the case of the arrival time, the performance at Arica and Patache is similar, suggesting that the tested sensor dominates over the fixed one. The spatial distribution of the error ΔH is related to the directivity of the tsunami energy radiation. When the tsunami source has an aspect ratio (length to width) greater than 1, most of the tsunami energy is radiated perpendicular to the major axis of the source (Kajiura 1970, 1972). Consequently, errors smaller than 20% are found in some cases where the sensor is located close to the main energy beam (see Fig. 1c, d for reference on the energy beams). In contrast, when the sensor is located parallel to the major axis, where weaker amplitudes are radiated, the observed tsunami time series carries less information about the maximum tsunami energy, resulting in the initial sea surface being underestimated. This also leads to underestimation of the amplitude at forecast points. Thus, when sensors are located outside the main energy beam, accuracy decreases rapidly and reaches the saturation limit. The location of the fixed sensor does not allow for improved inversions, as it lies outside the main energy beam, explaining the similar performance at both sites. This marked difference in performance between the amplitude and arrival time errors highlights the need to also include the amplitude as a relevant tsunami parameter.

Similar conclusions are obtained from the skill indicator S_k (Figs. 2c, 3c), where higher forecasting accuracy is obtained for sensor arrays with at least one sensor located near the main energy beam. The loss of accuracy is not as well defined as with the amplitude, yet the minimum skill is close to $S_k \sim 0.35$ for Patache. Therefore, the skill is a more demanding parameter overall, but does not lead to a clear distinction among sensor arrays. The combination of these individual errors in the forecasting accuracy function, $F_{i,j}$ follows the same trend (Figs. 2d, 3d). The minimum global error is found in the area offshore of Arica and is influenced by the amplitude error. It aggregates the structure of the error in arrival time, arguably owing to the choice of weights being analyzed. On the other hand, at Patache, the global error distribution shows less contrast between locations than at Arica. These results reinforce the idea that a single error estimate such as arrival time does not suffice to identify the best placement, but it also shows that using a single forecast point as a reference can be affected by local dependencies. It is important to note that there is a smooth transition from lesser-quality results (cool colors) to good-quality results (warm colors) for all estimators, which means that the spatial discretization used in the sensor placement suffices to capture the error dependencies.

The aggregate of the forecasts at coastal points is estimated using Eq. 11, considering four coastal points ($N = 4$ in Eq. 11), namely Arica, Pisagua, Iquique, and Patache, for the northern design scenario, and Patache, Tocopilla, Mejillones, and Antofagasta for the southern case. The choice is due to their proximity to either source and because they have similar arrival times. Figure 4 shows the distribution of global error EG_i for both events. As before, the spatial maps show a relatively small area of better accuracy near the epicenter of the earthquakes, but the variability is reduced owing to the averaging effect of considering several forecast coastal points in unison in the evaluation. Despite this, it is still possible to identify locations where a sensor could be deployed to yield the best overall performance. Therefore, the final array configuration is determined by identifying the configuration with minimum global error EG_i, independently for each scenario, and for which the total error for each coastal forecast point satisfies $F_{i,j} \leq \mu = 0.55$, to ensure good quality at each forecast point. The coordinates of the selected sensor array are presented in Table 1. For reference, array number 62 out of 99 is obtained for the northern scenario, and array number 66 out of 113 for the southern one. In both cases, array configurations are numbered in accordance with the grid location, from west to east, then north to south.

4. Discussion

The methodology presented herein allows for objective comparison among several array configurations by using three relevant tsunami parameters. The method was tested by defining a minimum of two sensors to form a sensor array, under the premise that this represents the least expensive implementation. Testing for arrays comprising a larger number of sensors, and/or tsunami parameters, could be implemented in a similar manner, albeit at greater computational cost, but is not considered further in this work.

It is possible that, for actual implementations, the selection of a sensor may be subject to additional

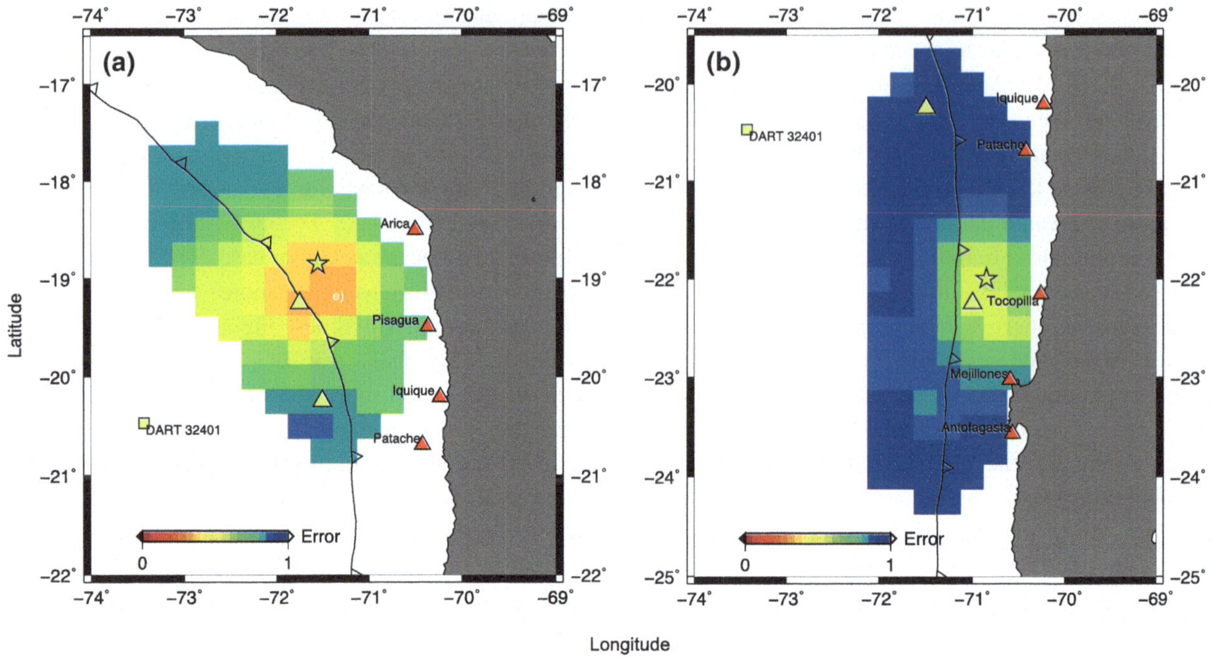

Figure 4
Space maps of global error, EG_i. Star indicates scenario epicenters, and yellow triangles show selected sensor array. **a** EG_i for the northern scenario, estimated using Arica, Pisagua, Iquique, and Patache. **b** EG_i for the southern scenario, estimated using Patache, Tocopilla, Mejillones, and Antofagasta

Table 1

Proposed sensor array configuration

Name	Latitude	Longitude
North	19.25° S	71.75° W
Fixed	20.25° S	71.50° W
South	22.25° S	71.00° W

restrictions. For example, the ability to transmit data in real time constrains to line-of-sight placement, or communications coverage; or deployment away from the trench to reduce the effect of seismic noise and coseismic signals, among many possible restrictions. Such restrictions are not considered here as they can be sensor specific, although they could be easily incorporated into the method by simply restricting the locations at which a sensor can be deployed.

The methodology expands on previous research by quantitatively incorporating different tsunami parameters into a cost function which can be minimized. However, the weights (α, β, and γ) in the cost function $F_{i,j}$ (Eq. 10) were chosen based on the relative importance of each parameter in previous studies and on the hazard categorization in tsunami early warning systems. Hence, both arrival time and tsunami amplitude were given a greater weight than skill. To evaluate the influence of this selection of weights in the assessment, a sensitivity analysis was carried out. Each weight was modified by up to ± 0.10 in steps of 0.05. In addition, cases when only one parameter is used were also considered, yielding 23 weight combinations as presented in Table 2.

This analysis is used to evaluate whether modifying the weights induces a change in the selection of the sensor array. In Fig. 5a–d, sample spatial maps of the global error EG_i are shown for some of the weight combinations (baseline combination, and combinations 10, 12, and 20; see Table 2 for details). It can be seen that, although there is a variation of the value of the global error, and also some variation of its spatial distribution, the overall structure remains consistent. The notable exception is combination 20, which only considers $\alpha = 1.0$, i.e., only the error in arrival times.

Table 2

Parameter space of weight values used

Weights	Combination number																						
	0	1	2	3	4	5	6	7	8	9	10	11	12	13	14	15	16	17	18	19	20	21	22
α	0.4	0.45	0.45	0.5	0.5	0.5	0.35	0.35	0.3	0.3	0.3	0.4	0.35	0.4	0.45	0.4	0.45	0.5	0.4	0.35	1.0	0	0
β	0.4	0.35	0.4	0.35	0.3	0.4	0.45	0.4	0.45	0.5	0.4	0.45	0.5	0.5	0.3	0.3	0.45	0.4	0.35	0.35	0	1.0	0
γ	0.2	0.2	0.15	0.15	0.2	0.1	0.2	0.25	0.25	0.2	0.3	0.15	0.15	0.1	0.25	0.3	0.1	0.1	0.25	0.3	0	0	1.0

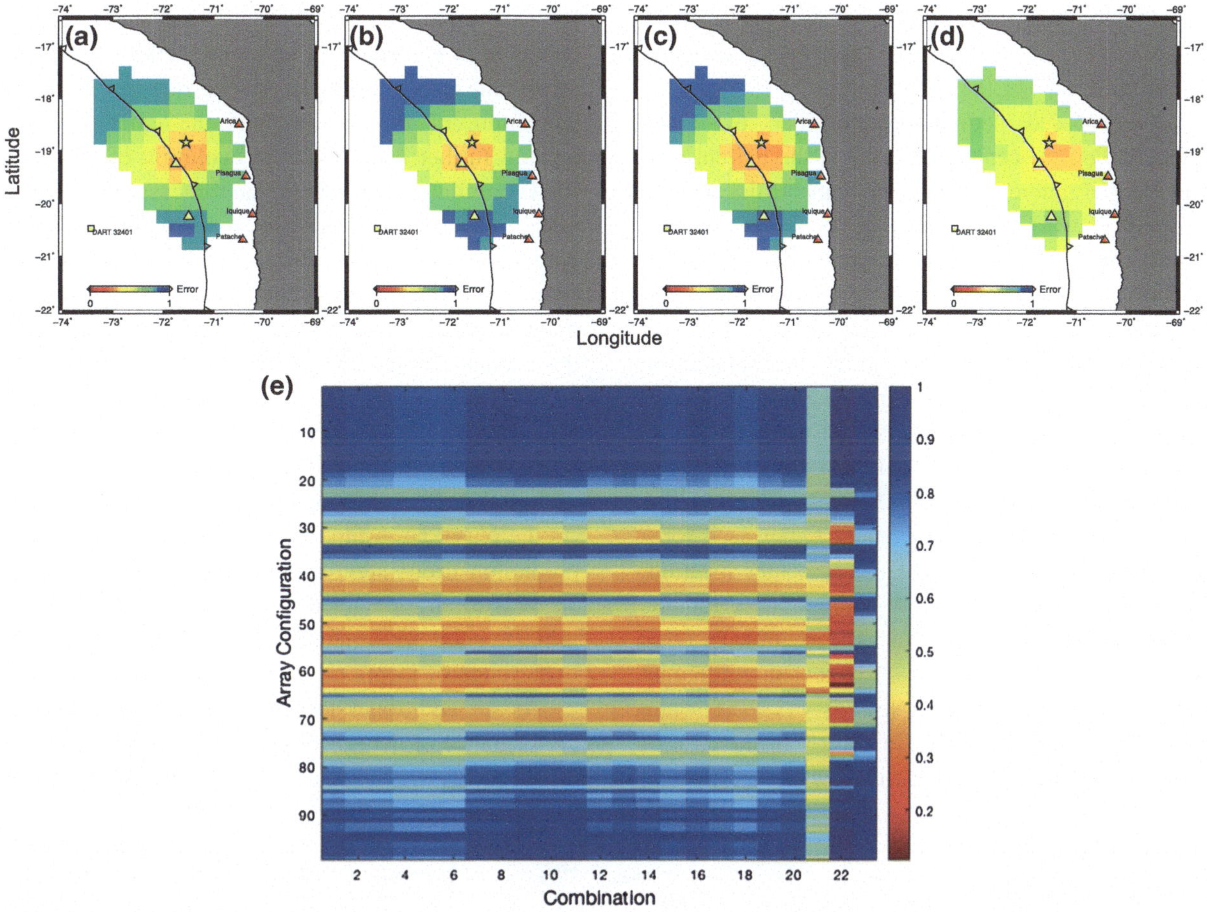

Figure 5

a–d Sample spatial maps of the global error, EG, as a function of the combination of weights. From left to right, combinations 1, 10, 12, and 20, respectively. **e** Matrix of the error EG (northern scenario) for each sensor array as a function of the combination of weights. Array configurations are numbered in accordance with the grid location, from west to east, then north to south

Figure 5e shows the value of the global error EG_i for each array configuration for the northern scenario, as a function of the weight combination. For all combinations, the minimum error is obtained for array 62, thereby showing no sensitivity to the distribution of weights. This could be due to several factors. For instance, perhaps the sensitivity range used in this test was not large enough to alter the result. However, even when only one parameter is considered, the solution remained unaltered. It could be argued that the solution is controlled by only one parameter. For instance, the error in amplitude ΔH yields the minimum errors in several arrays (see red colors for combination 21 in Fig. 5e). However, there are some other instances where the arrival time is the parameter yielding the minimum error (see combination 20). It is concluded that the use of a cost function combining both parameters maximizes the ability to capture well the overall structure of the tsunami signal. Moreover, the choice of weights initially proposed ($\alpha = \beta = 0.4, \gamma = 0.2$) appears to be a good compromise among them.

The proposed methodology considers two characteristic design scenarios located at the extremes of the area of interest. Here, the capabilities of the selected network are evaluated in other cases to ensure that the proposed configurations offer good performance not only for the design scenarios. To this end, the southern event (M_{w} 8.5) was modeled with different epicentral locations along the subduction zone, every 0.5 arcdeg along strike (as shown in Fig. 6). In addition, the observation time was changed ($T_0 = 10$, 15, and 20 min) to investigate the effect of record length on the forecast.

Figure 7 shows the matrices of $F_{i,j}$ at each forecast point j as a function of the epicenter location, considering the selected array configuration, for different observation times, T_0. The aggregate error decreases as T_0 is increased, as expected. Aggregate errors can reach values as low as $F_{i,j} \sim 0.1$ when $T_0 = 20$ min (compare Fig. 7a and c). In the case of Chile, where the seismic zone is located very close to the coast, this long data acquisition time may exceed the arrival time, thus preventing the use of this information as a trigger for early warnings. However, it could be considered in the later stages of the emergency cycle. For instance, this information can be used to refine initial hazard assessments derived by other means, such as the existing database of precomputed scenarios.

It is possible to observe a decrease in performance as the scenario approaches the reference forecast point. This results in poor accuracy at the northern forecast points (Arica, Pisagua, and Iquique) for events with epicenters at latitudes $-19.0°$ and $-20°$ (dark-blue data in upper-left corner of Fig. 7a), and a worsening of the accuracy as the epicenters are located further south (see the evolution of the error for Tocopilla). This is due to the forecast points being placed close to or inside the zone of the predicted coseismic deformation for the actual event, which prompts an inferred early arrival due to its displacement. Moreover, owing to the short observation time, the tsunami source solution results in a source with small spatial dimensions (see, for example, Fig. 8b, albeit for a different scenario). As a result of both situations, the arrival time yields a large error, $F_{i,j} \sim 0.9$–1.0. In the other cases, $F_{i,j}$ averages 0.64. As the observation time is increased, the error in the prediction drops significantly and values as low as $F_{i,j} \approx 0.40$ are obtained, which are considered to indicate good performance of the selected array for a wide range of locations.

As an additional test to evaluate the potential performance of the sensor array, the Pisagua tsunami on 1 April 2014 (M_{w} 8.1) was used as a source

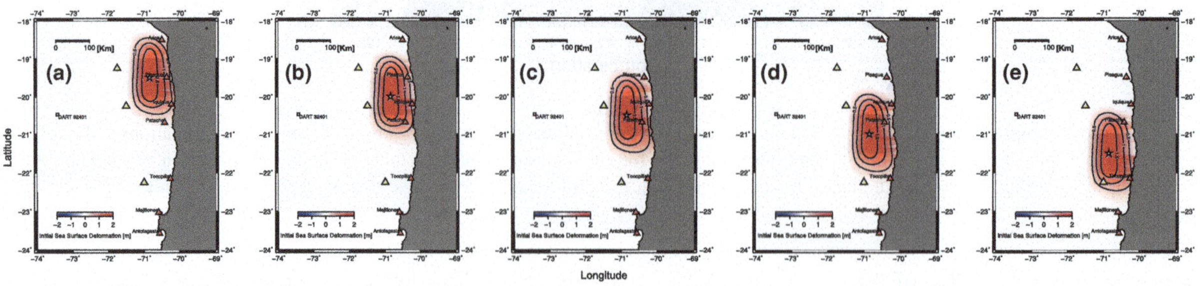

Figure 6
Initial sea-surface displacements used in evaluating the chosen sensor array. Red and yellow triangles indicate forecasting points and sensors, respectively. Epicenter is located at latitude **a** −19.5°; Parameter space of −20.0°, **c** −20.5°, **d** −21.0°, **e** −21.5°

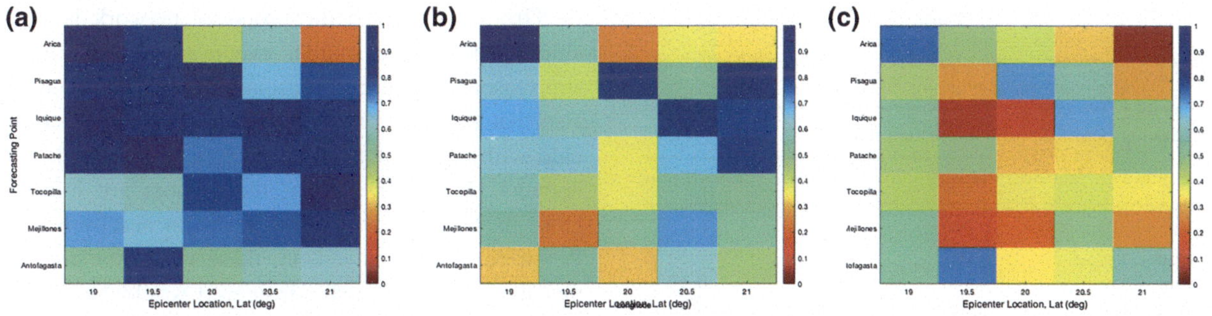

Figure 7

Forecast accuracy function $F_{62,j}$ as a function of reference forecast point j (vertical axis) for the final array configuration, considering different target scenario locations (horizontal axis) and observation time, T_0 of **a** 10 min, **b** 15 min, and **c** 20 min

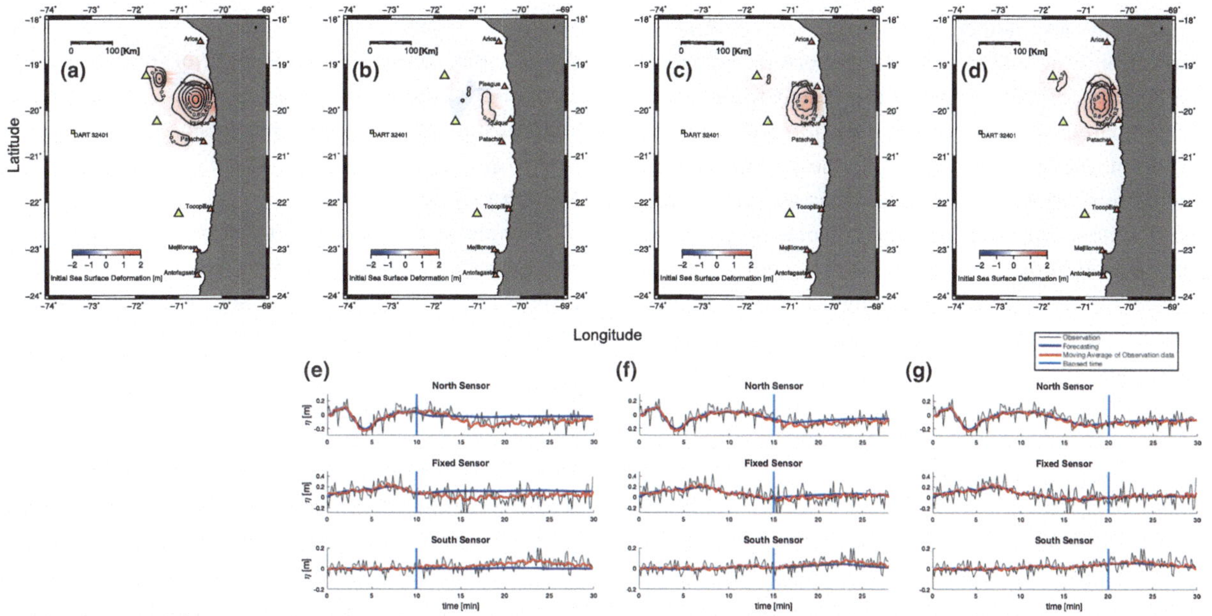

Figure 8

Initial sea-surface displacement obtained from different observation times for the 2014 Pisagua tsunami. **a** Target result resulting from using the slip distribution of Hayes et al. (2014) as source scenario. **b–d** Distribution of the initial sea-surface displacement estimated by tsunami waveform inversion considering $T_0 = 10$, 15, and 20 min. **e–g** Comparison of observed (black), observed data filtered with a nine-point moving average (red), and forecasted (blue) waveforms at the three sensors. Vertical light-blue lines in **e**, **f**, and **g** denote the observation time

scenario. Note that this event has a smaller magnitude than the scenarios used in designing the network, and also that it has a nonuniform slip distribution, with two main patches of slip. Synthetic tsunami waveforms at the location of the sensor array were obtained by a numerical simulation with COMCOT of the tsunami generated by the initial rupture model proposed by Hayes et al. (2014). As before, the accuracy of the assessment is evaluated using the coastal forecast points and different observation times. Gaussian noise with a maximum amplitude of 10% of the variance of the clean tsunami waveforms was added in the numerical experiments.

The results are summarized in Table 3, the initial surface solutions are shown in Fig. 8a–d, and observed (black) and forecasted (blue) time series are shown in Fig. 8e, f, for different observation times T_0. Within 10 min, the performance of the method is

Table 3

Summary of tsunami forecasting error $F_{i,j}$ considering different data observation times for 2014 Pisagua event

Forecasting point	Observation time		
	10 min	15 min	20 min
Arica	0.98	0.91	0.71
Pisagua	0.99	0.49	0.20
Iquique	0.97	0.62	0.33
Patache	0.95	0.55	0.35
Tocopilla	0.95	0.15	0.07
Mejillones	0.91	0.61	0.08
Antofagasta	0.79	0.52	0.40

inadequate, with large errors and an inverted initial surface that only detects a localized source. As the observation time is increased, the errors decrease significantly, especially when forecasted time series are compared with observed data filtered using a nine-point moving average (red lines). The reconstructed sea surface condition now has an appropriate spatial extent but smaller peak displacement. The solution is improved for $T_0 = 20$ min, and includes traces of the secondary peak northwest of the rupture. However, as mentioned above, such a sensor array would only have provided timely assessment for locations outside the main rupture zone, since the observed tsunami arrival at the tide gages at Iquique and Pisagua was less than 15 min (An et al. 2014; Catalán et al. 2015).

The effect of weighting was tested by comparing the accuracy of the estimates for the Pisagua tsunami for different combinations of weights. In general, most forecast points have a relatively stable assessment, typically varying by less than ±0.10 in the error estimate (Fig. 9). However, the situation changes significantly if only one parameter is used. For instance, if only time is considered (combination 20), the assessment for some forecast points is worse, e.g., Iquique and Arica. On the other hand, Patache shows a significant improvement, possibly influenced by its proximity to the fixed sensor. The situation reverses when only the skill is considered (combination 22); Arica and Iquique shows improved estimation but Patache worse. Therefore, the use of multiple parameters and multiple forecast locations in unison is essential for producing a more complete estimation of the error for determining the optimal configuration.

This suggests that the proposed network is capable of identifying smaller events with nonuniform slip reasonably well, in reasonable time. It is of note also that the location of this scenario does not coincide with either of the scenarios used to design the network. However, two of the sensors are located close to, but not directly in, the main energy beam. When considered in unison, these two tests suggest that the sensor network considered would be appropriate for tsunamis generated by earthquakes of magnitude similar and larger than that of the initial design scenarios. It is possible that a different network would be obtained if different design scenarios were used. Nevertheless, the methodology provides a means for identifying a suitable sensor array. This stresses that one relevant step prior to the implementation of the methodology would be to determine these design scenarios by other means. In the present case, the choice was based on data available for the Northern Chile Gap.

The above results suggest that the methodology is capable of delivering a working tsunami observation system comprising just three sensors. Although the general rule that the spacing between sensors should be in the range $\sim$ 200–400 km (Bernard et al. 2001) is somewhat preserved, the spacing between sensors differs, between 110 and 225 km, approximately. This yields a less dense network than the one proposed by either Schindelé et al. (2008) or Omira et al. (2009). The methodology also allows for the optimal placement of the sensors in this range of distances. On the other hand, the results show that, similar to what is assumed by Gusman et al. (2014), the optimal placement corresponds to the area where the maximum displacement occurs. Considering the inherent uncertainty in predicting the slip distribution of a future earthquake, stochastic methods could be incorporated into this methodology to further refine the results. However, with the simulations performed here, it is possible to see that just three sensors could provide a baseline solution that would have been appropriate for an event such as Pisagua 2014.

It must be noted, however, that several other factors could be considered in the analysis. Financial considerations such as implementation or operational costs, or technical constraints on sensor placement could be incorporated by limiting the placement of

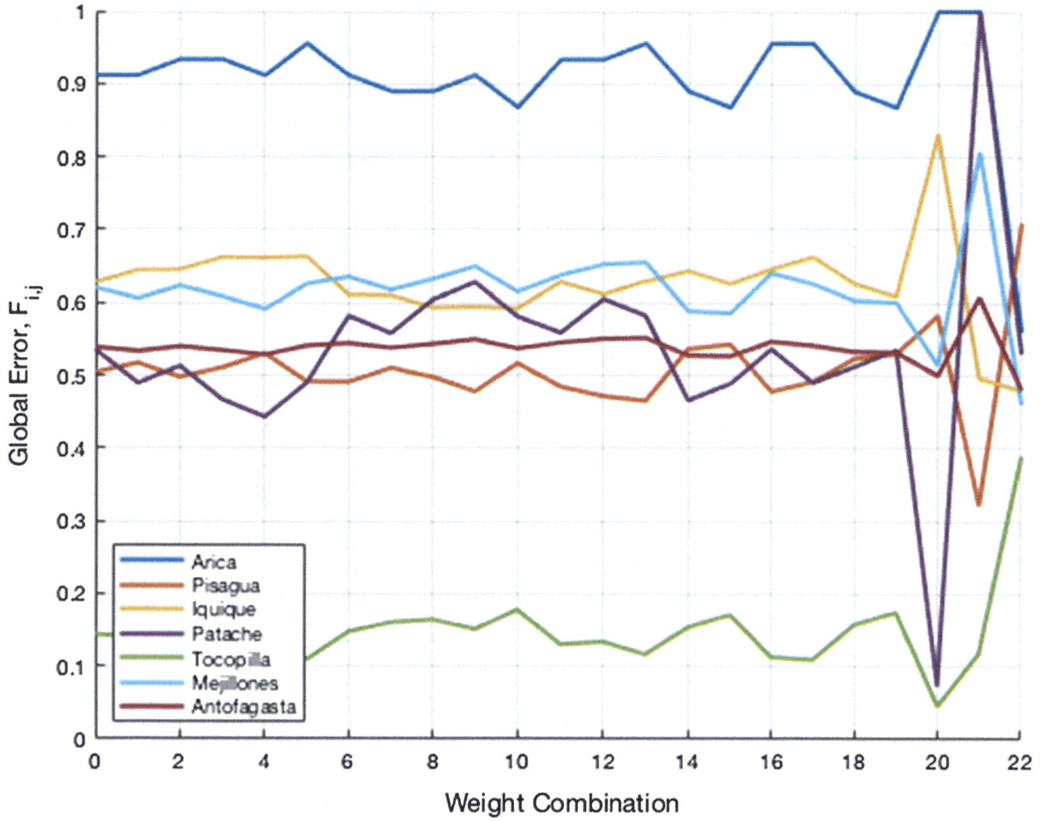

Figure 9
Forecasting accuracy from Eq. 10 considering different weight combinations in Table 2 for the Pisagua 2014 tsunami

the sensors during the definition of the possible sensor locations. These technical constraints can arise from a wide variety of situations that could affect the accuracy of the sensors. For example, seismic noise and Rayleigh waves can contaminate the signal for sensors near the source. However, it is noted that, in some cases, pressure sensors have been installed successfully near the source, for example during the 2016 Mie-oki earthquake (Wallace et al. 2016). The results shown herein are based on the assumption that the sensors are capable of providing filtered free surface data at sufficient temporal resolution, independent of the type of sensor. While an inversion algorithm can be subject to a wide range of errors and uncertainties on its own, for the purpose of the present work, it is hypothesized that using a common inversion procedure will weigh equally those errors, allowing comparison among different sensor configurations. The main goal of the presented procedure is the objective assessment of the best configuration, and it can be applied independently of the inversion method or nature of the sensors.

The implementation of the procedure is computationally intensive in two of its steps. First, establishing the database of Green's functions can be time consuming. The other aspect is the need to carry out inversions for a large number of sensor configurations. While in this case they were few (212 cases) because only pairs of sensors were considered, if a larger number of sensors is used, the processing cost could increase significantly. In such cases, it would be possible to apply optimization approaches to reduce the computational burden. Finally, to reduce the computational cost, forecasts at coastal stations were obtained by linear superposition. The use of this approach at shallow coastal points might be subject to inaccuracies arising from neglecting nonlinear interactions. Additionally, bathymetry-induced effects such as resonance were reduced here by placing these sensors at depths of about 200 m.

The use of tsunami data for the very first early warning is constrained by the distance to the source. At locations where arrival times are very short, it is also important to consider the actual benefit of incorporating such an evaluation for early warning. The present results reaffirm that short observation times yield less accurate results, highlighting the trade-off problem between a quick assessment and accuracy. The problem is further compounded by inherent uncertainties in inverting both earthquake and tsunami data (Cienfuegos et al. 2018). It seems reasonable to propose that, for cases such as this, higher-quality information obtained from inversion methods should be used for hazard assessment at later times in the emergency cycle, for instance to refine fast hazard assessments estimated by other means, and not necessarily as the method for first evaluation.

5. Conclusions

A methodology based on numerical simulations of near-field tsunami forecasting is presented to determine an optimal array configuration of offshore tsunami sensors. To provide an objective basis for the comparison, three parameters are considered simultaneously to assess the accuracy, thus expanding on previous methodologies that rely solely on a single parameter (e.g., arrival time). The joint use of the three estimators, viz. arrival time ΔT, tsunami amplitude ΔH, and model skill S_k, was robust when compared against a single parameter.

The methodology was tested in Northern Chile. Results showed that a configuration comprising just three sensors is capable of providing accurate estimations of tsunami arrival time and peak amplitude of the first wave. In this way, three sensors suffice to cover a $\sim$ 700 km stretch of coast when earthquakes in the M_w 8.0 range are considered.

Results show that there is a strong dependence between the location of the sensors and error estimators; arrival times are accurately predicted with sensors located offshore from coastal point of interest. In addition, better results for tsunami amplitude and skill are obtained when sensors are located inside the uplifted area or in front of it, near the main energy beam.

The proposed methodology shows potential for use as an operational tool in defining the location of possible tsunameters, especially for sparse configurations in countries where the financial cost of implementing and maintaining dense networks could be a hurdle.

Acknowledgements

This work was developed within the Research Project on Enhancement of Technology to Develop Tsunami-Resilient Community, sponsored by Science and Technology Research Partnership for Sustainable Development, SATREPS Program by Japan Science and Technology (JST), the Japan International Cooperation Agency (JICA). P.A.C. would also like to thank CONICYT for support through grants FONDEF D11I1119 and IT15I10001, FONDAP 1511017 (CIGIDEN), and PIA-Basal FB0821 (CCTVal). J.M. has been supported by CONICYT through the M.Sc. Scholarship CONICYT-PCHA/Magister Nacional/2015-22150620 and by Dirección General de Investigación, Innovación y Postgrado, UTFSM.

References

Abe, I., & Imamura, F. (2013). Problems and effects of a tsunami inundation forecast system during the 2011 Tohoku earthquake.

Journal of JSCE, *1*(1), 516–520. https://doi.org/10.2208/journalofjsce.1.1_516.

Aida, (1972). Numerical estimation of a tsunami source. *Zisin (Journal of the Seismological Society of Japan 2nd ser)*, *25*(4), 343–352. https://doi.org/10.4294/zisin1948.25.4_343.

An, C., Liu, P. L. F., & Meng, L. (2018). A sensitivity analysis of tsunami inversions on the number of stations. *Geophysical Journal International*, *214*(2), 1313–1323. https://doi.org/10.1093/gji/ggy212. http://oup.prod.sis.lan/gji/article-pdf/214/2/1313/25076307/ggy212.pdf.

An, C., Sepúlveda, I., & Liu, P. L. F. (2014). Tsunami source and its validation of the 2014 Iquique, Chile earthquake. *Geophysical Research Letters*, *41*(11), 3988–3994. https://doi.org/10.1002/2014GL060567.

Araki, E., Kawaguchi, K., Kaneko, S., & Kaneda, Y. (2008). Design of deep ocean submarine cable observation network for earthquakes and tsunamis. In *OCEANS 2008—MTS/IEEE Kobe Techno-Ocean* (pp. 1–4). https://doi.org/10.1109/OCEANSKOBE.2008.4531071.

Aránguiz, R., González, G., González, J., Catalán, P. A., Cienfuegos, R., Yagi, Y., et al. (2016). The 16 September 2015 Chile tsunami from the post-tsunami survey and numerical modeling perspectives. *Pure and Applied Geophysics*, *173*(2), 333–348. https://doi.org/10.1007/s00024-015-1225-4.

Arcas, D., & Titov, V. (2006). Sumatra tsunami: lessons from modeling. *Surveys in Geophysics*, *27*(6), 679–705. https://doi.org/10.1007/s10712-006-9012-5.

Baba, T., Ando, K., Matsuoka, D., Hyodo, M., Hori, T., Takahashi, N., et al. (2016). Large-scale, high-speed tsunami prediction for the Great Nankai Trough Earthquake on the K computer. *The International Journal of High Performance Computing Applications*, *30*(1), 71–84. https://doi.org/10.1177/1094342015584090.

Baba, T., Hirata, K., & Kaneda, Y. (2004). Tsunami magnitudes determined from ocean-bottom pressure gauge data around Japan. *Geophysical Research Letters*, *31*(8), L08303. https://doi.org/10.1029/2003GL019397.

Behrens, J., Androsov, A., Babeyko, A. Y., Harig, S., Klaschka, F., & Mentrup, L. (2010). A new multi-sensor approach to simulation assisted tsunami early warning. *Natural Hazards and Earth System Sciences*, *10*(6), 1085–1100. https://doi.org/10.5194/nhess-10-1085-2010.

Bernard, E., & Titov, V. (2015). Evolution of tsunami warning systems and products. *Philosophical Transactions of the Royal Society of London A: Mathematical, Physical and Engineering Sciences*, *373*(2053). https://doi.org/10.1098/rsta.2014.0371.

Bernard, E. N., González, F. I., Meinig, C., & Milburn, H. B. (2001). Early detection and real-time reporting of deep-ocean tsunamis. In *International Tsunami Symposium, Paper R-6* (pp. 7–10).

Catalán, P. A., Aránguiz, R., González, G., Tomita, T., Cienfuegos, R., González, J., et al. (2015). The 1 April 2014 Pisagua tsunami: Observations and modeling. *Geophysical Research Letters*, *42*(8), 2918–2925. https://doi.org/10.1002/2015GL063333.

Chlieh, M., Perfettini, H., Tavera, H., Avouac, J. P., Remy, D., Nocquet, J. M., et al. (2011). Interseismic coupling and seismic potential along the Central Andes subduction zone. *Journal of Geophysical Research*, *116*(B12), B12405. https://doi.org/10.1029/2010JB008166.

Cienfuegos, R., Catalán, P. A., Urrutia, A., Benavente, R., Aránguiz, R., & González, G. (2018). What can we do to forecast tsunami hazards in the near field given large epistemic uncertainty in rapid seismic source inversions? *Geophysical Research Letters*, *45*(10), 4944–4955. https://doi.org/10.1029/2018GL076998. https://agupubs.onlinelibrary.wiley.com/doi/pdf/10.1029/2018GL076998.

Cienfuegos, R., Suarez, L., Aránguiz, R., Gonzalez, G., González-Carrasco, J. F., Catalan, P. A., Dominguez, J. C., & Tomita, T. (2014). Reassessment of tsunami hazard in the City of Iquique, Chile, after the Pisagua Earthquake of April 2014. AGU Fall Meeting Abstracts NH13A-3727.

Comte, D., & Pardo, M. (1991). Reappraisal of great historical earthquakes in the northern Chile and southern Peru seismic gaps. *Natural Hazards*, *4*(1), 23–44. https://doi.org/10.1007/BF00126557.

Fritz, H., Petroff, C., Catalán, P. A., Cienfuegos, R., Winckler, P., Kalligeris, N., et al. (2011). Field survey of the 27 February 2010 Chile tsunami. *Pure and Applied Geophysics*, *168*, 1989–2010. https://doi.org/10.1007/s00024-011-0283-5.

Gica, E., Spillane, M. C., Titov, V. V., Chamberlin, C. D., & Newman, J. C. (2008). Development of the forecast propagation database tor NOAAs short-term inundation forecast for tsunamis (SIFT). Tech. Rep. NOAA Technical Memorandum OAR PMEL-139, National Oceanic and Atmospheric Administration.

Gusman, A. R., Tanioka, Y., MacInnes, B. T., & Tsushima, H. (2014). A methodology for near-field tsunami inundation forecasting: Application to the 2011 Tohoku tsunami. *Journal of Geophysical Research: Solid Earth*, *119*(11), 8186–8206. https://doi.org/10.1002/2014JB010958.

Hampson, R., MacMahan, J., & Kirby, J. T. (2011). A low-cost hydrographic kayak surveying system. *Journal of Coastal Research*, *27*, 600–603. https://doi.org/10.2112/JCOASTRES-D-09-00108.1.

Hayashi, Y., Tsushima, H., Hirata, K., Kimura, K., & Maeda, K. (2011). Tsunami source area of the 2011 off the Pacific coast of Tohoku Earthquake determined from tsunami arrival times at offshore observation stations. *Earth, Planets and Space*, *63*(7), 54. https://doi.org/10.5047/eps.2011.06.042.

Hayes, G. P., Herman, M. W., Barnhart, W. D., Furlong, K. P., Riquelme, S., Benz, H. M., et al. (2014). Continuing megathrust earthquake potential in Chile after the 2014 Iquique earthquake. *Nature*, *512*(7514), 295–298. https://doi.org/10.1038/nature13677.

Hirata, K., Satake, K., Tanioka, Y., Kuragano, T., Hasegawa, Y., Hayashi, Y., et al. (2006). The 2004 Indian Ocean tsunami: Tsunami source model from satellite altimetry. *Earth, Planets and Space*, *58*(2), 195–201. https://doi.org/10.1186/BF03353378.

Kajiura, K. (1970). Tsunami source, energy and the directivity of wave radiation. *Bulletin of the Earthquake Research Institute*, *48*, 835–869. https://doi.org/10.1007/BF02109296.

Kajiura, K. (1972). The directivity of energy radiation of the tsunami generated in the vicinity of a continental shelf. *Journal of the Oceanographical Society of Japan*, *28*(6), 260–277. https://doi.org/10.1007/BF02109296.

Kamigaichi, O., Saito, M., Doi, K., Matsumori, T., Tsukada, S., Takeda, K., et al. (2009). Earthquake early warning in japan: Warning the general public and future prospects. *Seismological Research Letters*, *80*(5), 717. https://doi.org/10.1785/gssrl.80.5.717.

Kaneda, Y., Kawaguchi, K., Araki, E., Matsumoto, H., Nakamura, T., Kamiya, S., et al. (2015). *Development and application of an advanced ocean floor network system for megathrust earthquakes and tsunamis* (chapter 25, pp. 643–662). Berlin: Springer. https://doi.org/10.1007/978-3-642-11374-1_25.

Kawaguchi, K., Kaneko, S., Nishida, T., & Komine, T. (2015). *Construction of the DONET real-time seafloor observatory for earthquakes and tsunami monitoring* (chapter 10, pp. 211–228). Berlin: Springer. https://doi.org/10.1007/978-3-642-11374-1_10.

Metois, M., Socquet, A., Vigny, C., Carrizo, D., Peyrat, S., Delorme, A., et al. (2013). Revisiting the north Chile seismic gap segmentation using GPS-derived interseismic coupling. *Geophysical Journal International*, *194*(3), 1283–1294. https://doi.org/10.1093/gji/ggt183.

Mulia, I. E., Gusman, A. R., & Satake, K. (2017a). Optimal design for placements of tsunami observing systems to accurately characterize the inducing earthquake. *Geophysical Research Letters*, *44*(24), 12,106–12,115. https://doi.org/10.1002/2017GL075791.

Mulia, I. E., Inazu, D., Waseda, T., & Gusman, A. R. (2017b). Preparing for the future Nankai Trough tsunami: A data assimilation and inversion analysis from various observational systems. *Journal of Geophysical Research: Oceans*, *122*(10), 7924–7937. https://doi.org/10.1002/2017JC012695.

Okal, E. A. (2015). The quest for wisdom: lessons from 17 tsunamis, 2004–2014. *Philosophical Transactions of the Royal Society of London A: Mathematical, Physical and Engineering Sciences*, *373*(2053). https://doi.org/10.1098/rsta.2014.0370.

Omira, R., Baptista, M. A., Matias, L., Miranda, J. M., Catita, C., Carrilho, F., et al. (2009). Design of a sea-level tsunami detection network for the Gulf of Cadiz. *Natural Hazards and Earth System Sciences*, *9*(4), 1327–1338. https://doi.org/10.5194/nhess-9-1327-2009.

Percival, D. B., Denbo, D. W., Eblé, M. C., Gica, E., Mofjeld, H. O., Spillane, M. C., et al. (2011). Extraction of tsunami source coefficients via inversion of DART buoy data. *Natural Hazards*, *58*(1), 567–590. https://doi.org/10.1007/s11069-010-9688-1.

Piatanesi, A., Tinti, S., & Pagnoni, G. (2001). Tsunami waveform inversion by numerical finite-elements Green's functions. *Natural Hazards and Earth System Sciences*, *1*(4), 187–194. https://doi.org/10.5194/nhess-1-187-2001.

Pires, M. (2001). Tsunami waveform inversion by adjoint methods. *Journal of Geophysical Research: Oceans*, *106*(C9), 19773–19796. https://doi.org/10.1029/2000JC000334. https://agupubs.onlinelibrary.wiley.com/doi/pdf/10.1029/2000JC000334.

Rabinovich, A. B., & Eblé, M. C. (2015). Deep-ocean measurements of tsunami waves. *Pure and Applied Geophysics*, *172*(12), 3281–3312. https://doi.org/10.1007/s00024-015-1058-1.

Rakowsky, N., Androsov, A., Fuchs, A., Harig, S., Immerz, A., Danilov, S., et al. (2013). Operational tsunami modelling with tsunami—Recent developments and applications. *Natural Hazards and Earth System Science*, *13*, 1629–1642. https://doi.org/10.5194/nhess-13-1629-2013.

Romano, F., Piatanesi, A., Lorito, S., Tolomei, C., Atzori, S., & Murphy, S. (2016). Optimal time alignment of tide-gauge tsunami waveforms in nonlinear inversions: Application to the 2015 Illapel (Chile) earthquake. *Geophysical Research Letters*, *43*(21), 11226–11235. https://doi.org/10.1002/2016GL071310.

Satake, K. (1987). Inversion of tsunami waveforms for the estimation of a fault heterogeneity: Method and numerical experiments. *Journal of Physics of the Earth*, *35*(3), 241–254. https://doi.org/10.4294/jpe1952.35.241.

Satake, K., & Kanamori, H. (1991). Use of tsunami waveforms for earthquake source study. *Natural Hazards*, *4*(2), 193–208. https://doi.org/10.1007/BF00162787.

Saunders, H. (2018) Augmenting onshore GNSS displacements with offshore observations to improve slip characterization for Cascadia subduction zone earthquakes. *Geophysical Research Letters*. https://doi.org/10.1029/2018GL078233. https://agupubs.onlinelibrary.wiley.com/doi/pdf/10.1029/2018GL078233.

Schindelé, F., Loevenbruck, A., & Hébert, H. (2008). Strategy to design the sea-level monitoring networks for small tsunamigenic oceanic basins: The western Mediterranean case. *Natural Hazards and Earth System Sciences*, *8*(5), 1019–1027. https://doi.org/10.5194/nhess-8-1019-2008.

Soulé, B. (2014). Post-crisis analysis of an ineffective tsunami alert: The 2010 earthquake in Maule, Chile. *Disasters*, *38*(2), 375–397. https://doi.org/10.1111/disa.12045.

Spillane, M. C., Gica, E., Titov, V. V., & Mofjeld, H. O. (2008). Tsunameter network design for the U.S. DART arrays in the Pacific and Atlantic oceans. NOAA Technical Memorandum OAR PMEL-143, National Oceanic and Atmospheric Administration.

Titov, V. (2009). *Tsunami forecasting, chapter 12* (pp. 371–400). Cambridge: Harvard University Press.

Tsushima, H., Hino, R., Fujimoto, H., Tanioka, Y., & Imamura, F. (2009). Near-field tsunami forecasting from cabled ocean bottom pressure data. *Journal of Geophysical Research*, *114*(B6), B06309. https://doi.org/10.1029/2008JB005988.

Tsushima, H., Hino, R., Tanioka, Y., Imamura, F., & Fujimoto, H. (2012). Tsunami waveform inversion incorporating permanent seafloor deformation and its application to tsunami forecasting. *Journal of Geophysical Research*, *117*(B3), B03311. https://doi.org/10.1029/2011JB008877.

Wallace, L. M., Araki, E., Saffer, D., Wang, X., Roesner, A., Kopf, A., Nakanishi, A., Power, W., Kobayashi, R., Kinoshita, C., Toczko, S., Kimura, T., Machida, Y., & Carr, S. (2016). Near-field observations of an offshore MW 6.0 earthquake from an integrated seafloor and subseafloor monitoring network at the Nankai trough, southwest Japan. *Journal of Geophysical Research: Solid Earth*, *121*(11), 8338–8351. https://doi.org/10.1002/2016JB013417. https://agupubs.onlinelibrary.wiley.com/doi/pdf/10.1002/2016JB013417.

Wang, X. (2009). COMCOT User's Manual.

Weatherall, P., Marks, K. M., Jakobsson, M., Schmitt, T., Tani, S., Arndt, J. E., et al. (2015). A new digital bathymetric model of the world's oceans. *Earth and Space Science*, *2*(8), 331–345. https://doi.org/10.1002/2015EA000107.

Webb, S. C. (1998). Broadband seismology and noise under the ocean. *Reviews of Geophysics*, *36*(1), 105–142. https://doi.org/10.1029/97RG02287.

Williamson, A. L., & Newman, A. V. (2019). Suitability of open-ocean instrumentation for use in near-field tsunami early warning along seismically active subduction zones. *Pure and Applied Geophysics*, *176*(7), 3247–3262. https://doi.org/10.1007/s00024-018-1898-6.

(Received June 28, 2019, revised November 25, 2019, accepted November 26, 2019, Published online December 6, 2019)

Pure Appl. Geophys. 177 (2020), 1471–1495
© 2019 Springer Nature Switzerland AG
https://doi.org/10.1007/s00024-019-02372-4

Pure and Applied Geophysics

Probabilistic Tsunami Hazard Assessment for Local and Regional Seismic Sources Along the Pacific Coast of Central America with Emphasis on the Role of Selected Uncertainties

Natalia Zamora[1,2] and Andrey Y. Babeyko[3]

Abstract—Historical data indicate that the Middle America subduction zone represents the primary tsunamigenic source that affects the Central American coastal areas. In recent years, the tsunami potential in the region has mainly been assessed using maximum credible earthquakes or historical events showing moderate tsunami potential. However, such deterministic scenarios are not provided with their adequate probability of occurrence. In this study, earthquake rates have been combined with tsunami numerical modeling in order to assess probabilistic tsunami hazard posed by local and regional seismic sources. The common conceptual framework for the probabilistic seismic hazard assessment has been adapted to estimate the probabilities of exceeding certain tsunami amplitudes along the Central American Pacific coast. The study area encompasses seismic sources related to the Central America, Colombia and Ecuador subduction zones. In addition to the classical subduction inter-plate events, this study also incorporates sources at the outer rise, within the Caribbean crust as well as intra-slab sources. The study yields conclusive remarks showing that the highest hazard is posed to northwestern Costa Rica, El Salvador and the Nicaraguan coast, southern Colombia and northern Ecuador. In most of the region it is 50 to 80% likely that the tsunami heights will exceed 2 m for the 500 year time exposure (T). The lowest hazard appears to be in the inner part of the Fonseca Gulf, Honduras. We also show the large dependence of PTHA on model assumptions. While the approach taken in this study represents a thorough step forward in tsunami hazard assessment in the region, we also highlight that the integration of all possible uncertainties will be necessary to generate rigorous hazard models required for risk planning.

Keywords: Seismic segmentation, earthquake rates, tsunami hazard, Central America, Colombia, Ecuador, subduction zones.

Electronic supplementary material The online version of this article (https://doi.org/10.1007/s00024-019-02372-4) contains supplementary material, which is available to authorized users.

[1] Research Center for Integrated Disaster Risk Management (CIGIDEN), CONICYT/FONDAP/15110017, Santiago, Chile. E-mail: natalia.zamora@cigiden.cl

[2] CYCLO Millennium Nucleus The Seismic Cycle Along Subduction Zones, Valdivia, Chile.

[3] GFZ German Research Centre for Geosciences, Telegrafenberg, 14473 Potsdam, Germany.

1. Introduction

There is an increasing awareness of tsunami threat worldwide and a need for preparing for these infrequent yet destructive events. Lately, more concern has arisen in regions like Central America where probabilities of large tsunamis are uncertain. Central America (CAM) lies in a region of high seismic rates where moderate to large earthquakes are common, and at least forty tsunamis (Fig. 1) have reached the coastal regions in the last 500 years (NGDC/WDS 2017). The likelihood of mega-earthquakes along CAM is unclear (Ye et al. 2013), although moderate earthquakes such as the 'tsunami earthquakes' caused by slow ruptures have triggered large tsunamis despite their moderate seismic moment (Kanamori 1972; Satake 1994; Borrero et al. 2014). The latter has raised interest in assessing the tsunamigenic potential of local sources by using different methods to account for the hazard in the region (Zamora and Babeyko 2015; Parsons and Geist 2009a).

There are two main approaches to estimating tsunami hazard. The most common approach is known as scenario-based, where numerical simulations are performed with credible worst-case scenarios (Lorito et al. 2008; Løvholt et al. 2012). Those scenarios are either designed by assuming synthetic events similar to historical ones, or are based on geological constraints to estimate maximum credible earthquake scenarios (MCE). Most of the tsunami studies in the CAM region have assessed tsunami threat using the scenario-based approach. Hence, in this approach neither the return periods of tsunamis, their probabilities of occurrence, nor their plausibility have been estimated. The second approach, called the Probabilistic Tsunami Hazard Assessment (PTHA),

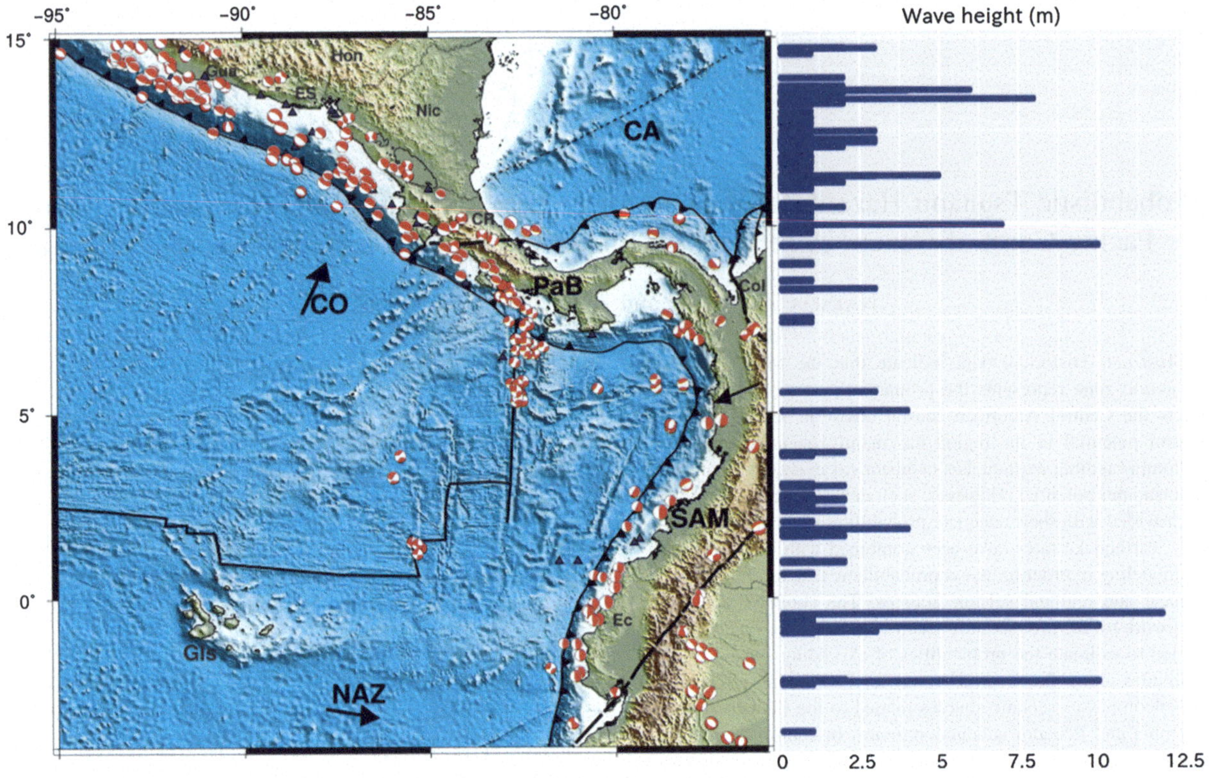

Figure 1
Tectonic setting of study area (left) and bar plot showing tsunami wave heights registered along the Pacific coast of Central America since 1900 (right). Runup data from Mexico to Ecuador are based on the National Geophysical Data Center (NGDC/WDS 2017) that covers a time of 500 years in this region. Note that some tsunami events that appear in the NGDC catalog do not have runup or wave height data but only limited information such as the year of occurrence

has become a necessary framework for dealing with multiple sources and accounting for their uncertainties (Grezio et al. 2017).

PTHA was established following procedures similar to those used for probabilistic seismic hazard assessment (Cornell 1968; McGuire 1976). The purpose of the seismic PTHA is to assess all possible earthquake sources that may contribute to the tsunami hazard and to statistically evaluate their corresponding tsunami impact in a probabilistic sense. PTHA results can be then used in risk assessment (Løvholt et al. 2012; Schaefer et al. 2015). For this, tsunami sources (e.g. earthquakes) and their probabilistic recurrence models are integrated with numerical simulations of tsunami propagation, or, alternatively, with empirical relationships between magnitude and wave runup to find tsunami rates of occurrence and return periods, as the probabilities of exceedance of wave height thresholds at particular coastal sites (Burroughs and Tebbens 2005; Geist and Parsons 2006; Thio et al. 2007; Sørensen et al. 2012). In addition to the seismically triggered tsunamis, PTHA has been also applied to other tsunami sources such as submarine landslides (Ward 2001) and asteroid impacts (Ward and Asphaug 2000).

The availability of tsunami runup records in tsunami catalogs for CAM might not be sufficient to make confident hazard estimates with only empirical relationships between earthquake parameters and local runup. To our knowledge, only two studies in Central America assess tsunami hazard in an empirical way. One covers the entire Caribbean basin (Parsons and Geist 2009a), while the other assessed tsunami hazard along the Pacific coast of Central America, taking into account hybrid methods by combining probability of earthquake occurrence and empirical relations to get tsunami amplitudes from Mexico to Ecuador (Brizuela et al. 2014). At the national level, some reports have been prepared under the Central American Probabilistic Risk Assessment (CAPRA) program (http://www.ecapra.org), nonetheless these models have not considered the associated uncertainties in the tsunami hazard modeling.

Therefore, while only two studies exist that provide probabilistic hazard estimates, most of the tsunami hazard research for this region has used a scenario-based deterministic approach (Zamora and Babeyko 2015; Fernández et al. 2004; Ortiz et al. 2001; Álvarez-Gómez et al. 2013). These studies considered historical scenarios to simulate tsunami threat to specific regions of Central America, for instance, El Salvador or the Puntarenas port, Costa Rica, and could provide a significant inputs for evacuation maps. Despite the large uncertainties drawn in PTHA, this approach is important since it allows to quantify probabilities of occurrence which are needed in many aspects of coastal risk planning.

The aim of this study is to assess the probability of exceedance of certain wave heights (tsunamis) along the Pacific coast of Central America by integrating local and regional tectonic sources along the Pacific coast from Guatemala to Ecuador, also highlighting the dependency of hazard estimates on model assumptions. The Monte Carlo technique has been implemented to account for seismic source uncertainties. Our study is based on a similar approach to the well-known PSHA in which the whole source region is divided into several seismo-tectonic zones characterized with constant seismic patterns (Cornell 1968).

The seismo-tectonic segmentation model is a crucial assumption for this PTHA. Several sources of information have been integrated to assess the possible asperities that could show different seismic patterns that allow to separate the region into seismo-tectonic segments. One of these constraints is the fact that strong seismic coupling is unevenly distributed, and mainly low coupling prevails along the Central America convergent margin (Alvarado et al. 2010; Álvarez-Gómez et al. 2008; Correa-Mora et al. 2009; Lyon-Caen et al. 2006; Kobayashi et al. 2014; LaFemina et al. 2009) which might partly explain the unlikely occurrence of megathrust earthquakes (Ye et al. 2013). The geodetic studies, and the geological and geomorphic imprints along Central America lead to a regional seismo-tectonic model (Alvarado et al. 2017) that has been adapted in this study to assess the tsunami hazard posed from the outer rise, inter-plate, intra-slab and crustal seismicity.

Alternatively, to explore the limits of PTHA uncertainty in regard to zonation models, we include hazard estimates that resulted from the premise that the CAM megathrust could rupture up to Mw 9.0 without any segmentation along strike (McCaffrey 2008). Another factor of PTHA uncertainty that we would like to test in our study is a concept of depth-dependent crustal rigidity (Bilek and Lay, 1999) which is believed to be a potential cause for the so-called 'tsunami earthquakes' (Kanamori 1972) known to occur in the Central American region.

2. *Methods: Steps in PTHA and Input Data*

The probabilistic tsunami hazard is assessed here by combining the seismic source rates with tsunami numerical modeling. The seismic source recurrence rate is used to sample magnitudes (Mi) from the Gutenberg-Richter type magnitude-frequency distribution with Monte Carlo techniques. The latter is an efficient tool to account for randomness of seismic occurrence as well as the fault characterization.

The method used in this study follows a similar approach as previous studies that assess the seismic hazard (Wiemer et al. 2008) and the tsunami hazard (Geist 2009; Sørensen et al. 2012). The first step is to determine earthquake or seismo-tectonic source zones and to estimate their statistical seismic parameters. This so called zone-based approach has also been the frame in regional probabilistic seismic hazard assessment in the region (Benito et al. 2012; Salazar et al. 2013). A sketch illustrating our approach is provided in Fig. 2.

2.1. *Regional Seismic Catalogs*

The regional seismic catalog (herein, CAT2011) resulted from an extensive compilation of several historical earthquakes and national catalogs (Rojas

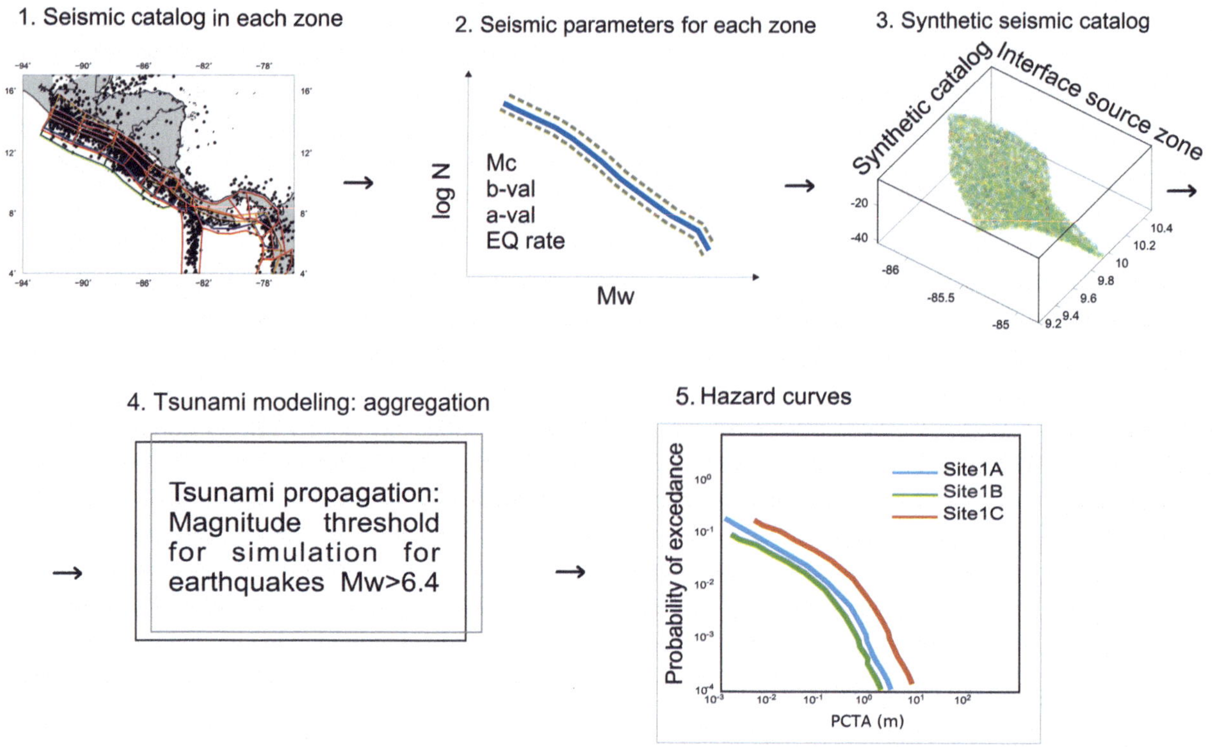

Figure 2
Schema that shows the steps of the probabilistic tsunami hazard assessment (PTHA) used in this study. *PCTA* peak coastal tsunami amplitudes

et al. 2013). The seismic databases of each country were homogenized with moment magnitude Mw using various empirical relationships, some of them defined for a specific country. In case of Panama, all events had Mw already calculated (Alvarado et al. 2017). Central American earthquake hypocenters have been collected through the Central America Seismological Center (CASC) since 1991. The CASC database is a complement to the global earthquake data since it covers earthquakes down to smaller magnitudes (Lindholm et al. 2004). The CASC was an initiative funded by the Norwegian Agency for International Cooperation, through the Center for the Prevention of Natural Disaster in Central America, and was created to improve national seismological networks and data exchange within Central American countries that could aid seismic zonation and hazard analysis among others.

CAT2011 has been generated in three main updating steps. The basis of this regional catalog was first described in Rojas et al. (1993), then updated during the RESISII project for the year 2007. Data of the Central American Seismological Centre (CASC) and the national catalogs have been integrated. CAT2011 has been updated until June 2011 as an integrated effort between the Central America national entities and the Polytechnic University of Madrid. Currently, it includes events starting from 1522. An extensive revision of the data and homogenization of magnitudes into Mw for all events of Mw $>$ 3.5 has been conducted by W. Rojas (persCom).

The CASC had contributed until 2011 as a continuous effort of the national seismological centers to assess seismicity patterns and to estimate seismic hazard. Constant improvements have been done to create a regional seismic catalog (Lindholm et al. 2004; Benito et al. 2012; Alvarado et al. 2017; Rojas et al. 1993). The 2011 catalog (CAT2011) initially contained about 55250 events, which was reduced to about 23000 earthquakes with magnitudes $>$ Mw 3.5 after using the de-clustering method proposed by Knopoff et al. (1982),

Additionally, for this study we obtained Colombian seismicity from the Comcat composite catalog (USGS 2017) by selecting earthquakes above Mw 4. For Ecuador, we employed the seismic catalog used in Beauval et al. (2013), which accounts for a minimum magnitude of Mw 3.5. For the entire catalog we also add information from different local studies.

We also retrieved the focal mechanisms from the GCMT website (Ekström et al. 2012) to get the main characterizations of the different tectonic regimes. This information was used to generate synthetic catalogs, as will be described later. A total of 2584 events for the period 1934–2013 have focal mechanism information, covering the geographic window delimited by the coordinates Lon 100° W–82° W and Lat 4° N–17° N, including the moment magnitudes above Mw 4.5.

2.2. *Seismo-Tectonic Zonation*

In this study, a zonation-based approach has been carried out, which means that geological and geomorphic imprints as well as geodetic and seismological characterization has been considered to differentiate segments in a similar way as in previous tsunami hazard assessment in other regions (Matias et al. 2013; Basili et al. 2013). It is assumed that within each seismic source zone, seismicity rates are constant and their parameters can be assessed by fitting the Gutenberg-Richter distribution (Cornell 1968).

The seismo-tectonic zonation model for Central America, Colombia and Ecuador (CAM-CE) is presented in Fig. 3. This model includes separate zones regarding their typical fault mechanisms which is not always undertaken in PSHA (e.g. Benito et al. 2012). Following Alvarado et al. (2017), we implemented a modified model that integrates such available data adapted for the PTHA (Zamora 2016). Particularly, several considerations are taken to separate these zones: characteristic focal mechanisms, seismic patterns, geomorphic imprints, coupling data, historical large ruptures (asperities) and gravimetry data available in several studies (refer to Alvarado et al. 2017).

It is important to mention that Alvarado et al. (2017) proposed an agreed regional seismo-tectonic zonation based on the integration of extensive research available for the Central American subduction. Therefore, only slight deviations from this model are seen in the model we used, in which e.g. outer rise seismicity has different rates than those of the interplate seismogenic zone (Álvarez-Gómez et al. 2012) that were not considered in previous regional seismic hazard studies (Benito et al. 2012), is treated separately. Thus, we estimate seismic rates of one of the most seismic active regions along the CAM, the inter-plate seismogenic zone, that has triggered most tsunamis in the region and separate the outer rise seismicity that could also generate tsunamis but with different rupture mechanism.

The tectonic segmentation of the Pacific side of Central America (CAM) is closely linked to the structures generated by the interaction of the Cocos, Caribbean, Nazca plates and the Panama block. The northern limit of the CAM is marked by the Motagua-Polochic fault system (PoF), a left-lateral shear zone that continues offshore along the Swan Islands transform fault. The PoF fault is a transform fault that accommodates deformation between the North American and the Caribbean plate. This deformation zone is a continuation of the Cayman tectonic system throughout the western Caribbean basin (Fig. S2). The transform plate boundary between the Caribbean and North American plates averages 18–20 $\pm$ 3 mm/yr of boundary motion (DeMets 2001). On the other side, along the southeastern end of the Cocos Plate the aseismic Cocos Ridge, with 20 km thick buoyant crust, could have been underthrusting for $\sim$ 1–5 Ma (Lonsdale and Klitgord 1978; de Boer et al. 1988, 1995; Sallarés et al. 2003). The convergence rate of the Cocos plate varies from 7.5 cm/year in northern CAM increasing the velocity to 9.3 cm/year towards southeast Costa Rica (DeMets 2001; DeMets et al. 2010).

The characterization of the CAM inter-plate seismogenic zone varies from fairly high coupling ($\sim$ 0.6) offshore Chiapas towards low-coupling ($\sim$ 0.25) offshore Guatemala and El Salvador (Franco et al. 2012), providing one of the criteria for defining seismo-tectonic zones. For example, GPS studies indicate very low coupling off the Nicaragua

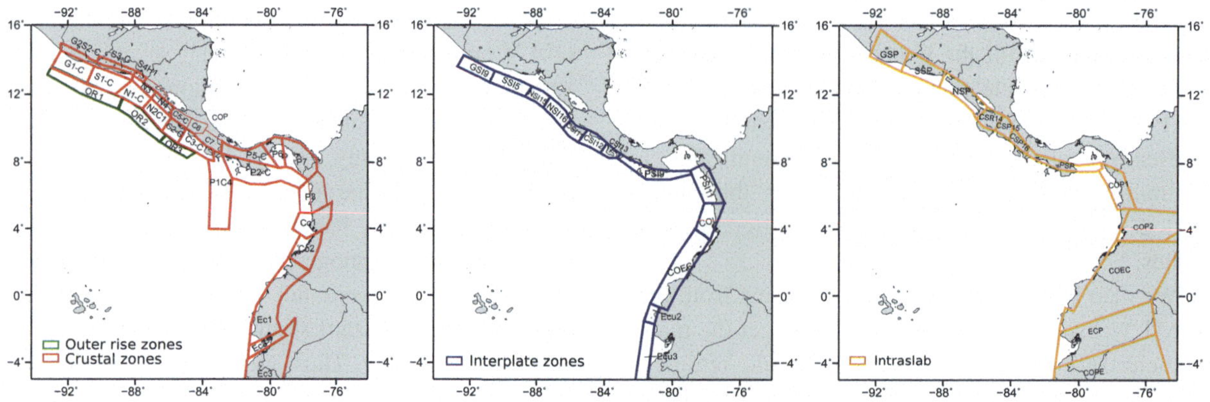

Figure 3
Delimitation of seismo-tectonic zones proposed for Central America, Colombia and Ecuador. The left panel shows outer rise zones (green) along the CAM together with the upper crustal zones along the Caribbean, the Panama block, the Andean block and the South American plate (red). Inter-plate zonation (subduction zone thrust earthquakes) is shown in blue at the central panel. The right panel presents zonation model for the intra-slab sources

and El Salvador segments (Alvarado et al. 2010; Correa-Mora et al. 2009; LaFemina et al. 2009). The low coupling has been explained by a lower interface normal stress due to the contact with the upper plate (Scholz and Small 1997; Protti et al. 1995). In contrast, the southern Central America margin has been proposed as a collision zone due to the Cocos Ridge subduction beneath the Panama block acting as an indenter (LaFemina et al. 2009). This feature could explain trench parallel velocities that dominate in Nicaragua (LaFemina et al. 2002). Based on the size of these segments, the maximum magnitude has been estimated from the scaling relations (Blaser et al. 2010). The latter scaling relations associate an area A to the magnitude moment Mw, and thus, one can estimate the seismic moment M_o. For that, we used the area of the segment following Aki's formula for the seismic moment (Aki 1966). The seismic moment (Eq. (1) of each segment of the megathrust is calculated based on its seismogenic area (A in km^2), mean slip δ and rigidity (μ) (here assumed as 30 GPa).

$$M_o = \mu A \delta, \tag{1}$$

The recurrence rates for the CAM-CE and the corresponding seismic parameters for each zone have been used as input data to the Monte-Carlo synthetic catalog generation technique. The seismic parameters for the inter-plate, crustal and intra-slab source zones are shown in Table 1. The seismic parameters such as b-value, earthquake annual rate and maximum magnitude have been estimated from the CAT2011 as explain in the next section. For the Colombian-Ecuador trench, seismic parameters within different zones were retrieved from the previous work by Beauval et al. (2013). A general geologic overview of the major seismic zones is given in the supplementary material.

2.2.1 Seismic Parameters

CAT2011 contains almost five hundred years of seismic records, but features a minimum magnitude Mw = 3.5 only after 1970. For the entire catalog, the magnitude of completeness (Mc) has been estimated as Mc = 4.0 after 1984 and Mc = 6.0 after 1821 according to the Stepp method (Stepp 1972). For each zone (Fig. 3), statistical seismic parameters lambda, beta, Mmin from the Gutenberg-Richter distribution (GRD) (Eq. 2) can be retrieved from the catalog by fitting the functional form of the frequency-magnitude distribution (FMD). In the present study, we employ the double truncated FMD as follows:

$$\lambda_{(Mi)} = e^{-\beta(M_i)} - e^{-\beta(M_{\max})} / e^{-\beta(M_{\min})} - e^{-\beta(M_{\max})} \tag{2}$$

where λ is the mean annual seismic rate, and $\beta = b \ln(10)$. The modified Gutenberg-Richter double truncated relation considers an upper magnitude bound, $M_{\max}$, since the rupture area of earthquakes

Table 1

Input data for Monte Carlo sampling for all seismo-tectonic zones considered in this study

Id	Depth (km)	Mw (min)	Mw (max)	Mw (max)	β	λ (4.5)	Rake	MFD
Outer rise seismic sources								
OR1	< 35	4.5	7.7	7.9	2.645	0.372	Normal	GR-T
OR2	< 35	4.5	7.7	7.9	2.64	1.431	Normal	GR-T
OR3	< 35	4.5	7.4	7.6	2.64	0.123	Normal	GR-T
Crustal seismic sources								
G1-C	< 15	4.5	7.5	7.7	2.42	1.115	Multi	GR-T
G2S2-C	< 20	4.5	6.7	6.9	2.107	0.28	Multi	GR-T
S1-C	< 25	4.5	5.8	6.1	3.694	0.601	Multi	GR-T
S3-C	< 15	4.5	6.8	7.1	2.105	0.974	Multi	GR-T
S4N5	< 20	4.5	7	7.3	1.95	0.992	Multi	GR-T
H1N4-C	< 20	4.5	6.3	6.5	2.429	0.315	Multi	GR-T
N1-C	< 15	4.5	7.8	8.1	2.96	3.584	Multi	GR-T
N3-C	< 25	4.5	6.8	7	2.033	0.291	Multi	GR-T
N4-C	< 25	4.5	6.5	6.7	3.172	0.178	Multi	GR-T
N2C1-C	< 15	4.5	7.3	7.5	3.006	4.524	Multi	GR-T
C2-C	< 15	4.5	7	7.1	2.091	0.352	Multi	GR-T
C3-C	< 15	4.5	7.5	7.7	2.309	1.098	Multi	GR-T
C4P1-C	< 30	4.5	7.7	7.9	2.615	16.206	Multi	GR-T
P2-C	< 20	4.5	7.3	7.5	1.992	2.267	Multi	GR-T
P3CO1	< 20	4.5	6.8	7	1.996	0.689	Multi	GR-T
P5-C	< 20	4.5	6.5	6.8	2.889	0.51	Multi	GR-T
P6-C	< 20	4.5	7.3	7.5	1.808	0.07	Multi	GR-T
P7-C	< 20	4.5	7.3	7.6	2.006	1.02	Multi	GR-T
CO2	< 20	4.5	7.1	7.3	2.07	0.11	Multi	GR-T
EC1	< 25	4.5	7.3	7.5	1.909	0.9	Multi	GR-T
EC2	< 25	4.5	7.5	7.7	2.208	0.34	Multi	GR-T
EC3	< 25	4.5	7.3	7.5	2.553	0.67	Multi	GR-T
Interplate source zones								
GSI9	8-40	4.5	7.9	8.1	2.383	5.32	Thrust	GR-T
SSI5	6-40	4.5	8.7	9	2.569	10.6	Thrust	GR-T
NSI15	6-40	4.5	7.9	8.1	2.834	8.99	Thrust	GR-T
NSI16	6-40	4.5	7.9	8.2	2.523	9	Thrust	GR-T
CSI11	6-40	4.5	8.1	8.3	1.914	2.24	Thrust	GR-T
CSI12	6-40	4.5	7.8	8.4	2.139	3.14	Thrust	GR-T
CSI13	6-40	4.5	7.6	7.8	1.9734	1.55	Thrust	GR-T
PSI9	20-40	4.5	7.7	7.9	2.086	1.155	Multi	GR-T
PSI11	20-100	4.5	7.1	7.3	1.971	1.008	Multi	GR-T
COL	5-40	4.5	7.9	8.1	2.3	3.02	Multi	GR-T
COEG	4-50	4.5	8.8	9	1.54	2.17	Thrust	GR-T
ECU2	4–50	4.5	7.2	7.5	2.346	0.86	Thrust	GR-T
ECU3	4–50	4.5	7.8	8	2.012	2.21	Thrust	GR-T
Intraplate sources								
GSP10	40–220	4.5	8.2	8.4	2.44	13.97	Multi	GR-T
SSP6	40–220	4.5	8	8.2	2.555	10.65	Multi	GR-T
NSP17	40–220	4.5	7.9	8	2.17	15.45	Multi	GR-T
CSP14	40–160	4.5	7.5	7.7	2.525	1.583	Multi	GR-T
CSP15	40–125	4.5	7.8	7.9	1.932	1.25	Multi	GR-T
CSP16	40–60	4.5	7.5	7.7	1.819	0.629	Multi	GR-T
PSP11	20–100	4.5	7.5	7.7	1.838	0.641	Multi	GR-T
COP1	> 100	4.5	7.5	7.7	1.02	0.502	Multi	GR-T
COP2	> 40	4.5	7.5	7.7	1.01	0.408	Multi	GR-T
COEC	50–150	4.5	7.5	7.7	2.78	2.17	Multi	GR-T
ECP	50–200	4.5	7.2	7.4	2.3	7.51	Multi	GR-T
COPE	50–150	4.5	7.5	7.7	1.9	3.53	Multi	GR-T
Interplate 1 zone CAM								
M9.0-1z	4–40	5.6	9.0	–	2.29	3.0	Thrust	GR-T

The only parameter varied within 2 models MCE and MEE was *Mmax* (truncation to Gutenberg-Richter relation). The upper bound *Mmax* for MEE model (left) and MCE (right). The λ is the annual rate of occurrence. Standard deviation for the depth sampling: 0.5. Multi refers to the variability of earthquake types assumed, this means a range of rakes used in each zone, are selected from the predominant rupture mechanism of past events. The rake range used for interplate zones varies between 75°–105°

and their slip are physically constrained. Further details of the estimation of maximum magnitudes are available in the supplementary material to this article. Earthquake occurrences (starting from the minimum magnitude) have been assumed to follow a Poisson process in time (Eq. 3). To be consistent with this premise, the seismic catalog has been de-clustered in order to avoid aftershocks in the seismic database, leaving only main or independent shocks.

$$P(H(x) > H_o, T) = 1 - e^{-\lambda(H(x) > H_o)(T)} \quad (3)$$

where P is the probability of exceeding a peak coastal amplitude (PCTA) threshold (H_o), λ is the mean annual rate of tsunami hits on a selected POI and T is the time window for the assessment.

In practice, seismic parameters for each zone were estimated from the regional earthquake catalogs using - Cosentino et al. (1977) and Kijko and Smit (2012). Based on the derived seismic zone parameters (Table 1), different synthetic seismic catalogs have been generated for each area of the fifty seismic zones ranging from Guatemala to Ecuador, and integrated into one single aggregated seismic catalog for the subsequent tsunami modeling. In fact, two comprehensive synthetic catalog were generated with the main difference in the upper bound ($M_{\max}$) of the GRD.

2.3. *Synthetic Catalog Generation Procedure*

To account for the randomness of the tsunami generation, the Monte-Carlo (MC) sampling for the fault parametrization was implemented to get the magnitude distribution as well as the multi-dimensional space of fault parameters and multiple sources of uncertainties (Rubinstein and Kroese 2008). The MC technique is an efficient tool to account for aleatory uncertainties by sampling different parameters (Ebel and Kafka 1999; Geist and Parsons 2006). The earthquake parameterization is thoroughly accounted for by assigning different ranges of focal mechanisms according to the characterization of the seismic sources.

In order to account for these uncertainties, we employed our code, *seisCat*. This code generates a synthetic catalog based on the frequency-magnitude distribution as well as distributions of other fault parameters (e.g., focal mechanism, geometry, depth distribution). Since source zones are determined in map-view (Fig. 3), the generation of a synthetic catalog additionally needs the distribution of earthquakes with depth within the given geographical limits. *seisCat* can be used as a means to distribute hypocenters uniformly inside a polygon with minimum and maximum depth levels (Sørensen et al. 2012), or project seismicity onto the predefined slab geometry uniformly or normally distributed.

Two reference synthetic catalogs have been developed, whose main difference lies in the upper truncation of the frequency-magnitude distribution. Whereas the first catalog corresponds to the expected earthquake (MEE) based on the expert opinion, the second catalog was compiled following the maximum credible earthquake (MCE) concept based on geological constraints. The maximum magnitudes in MCE appear to be larger particularly for some segments of the interplate seismogenic zones.

A third synthetic catalog has been included to test the effect of depth-dependent rigidity on hazard estimates. This catalog (MEE-r) is based on MEE and implements a rigidity profile as interpolated from estimations by Bilek and Lay (1999). Finally, a fourth synthetic catalog (M9.0-1z) considers the whole CAM megathrust as a single seismogenic zone capable to host up to M9 earthquakes (McCaffrey 2008). These two PTHA models will be presented in the Discussion section.

This synthetic seismicity has been generated based on the seismic parameters listed in Table 1. Sub-catalogs were developed employing the following criteria:

- Define the segment or polygon coordinates (Fig. 3) and the time window of 100 kyear to sample the synthetic seismicity. Within each segment we distribute epicenters randomly and uniformly
- The hypocenter of each earthquake is sampled from a uniform distribution. It is important to mention that this follows a three dimensional perspective (Fig. 4). In this study, the Slab1.0 slab geometry model (Hayes et al. 2012) is used to distribute seismicity within the different source zones by depth (Fig. 4). For instance, in case of inter-plate zones, earthquakes are distributed along the SLAB1.0 surface starting from 5–10 km depth toward to the down-dip limit, which follows the 40

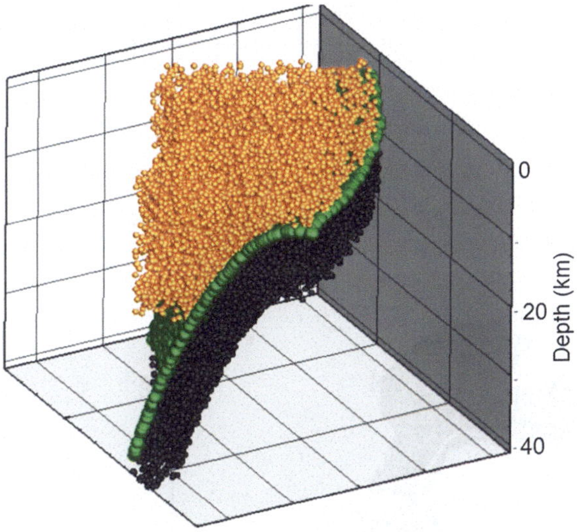

Figure 4
Synthetic earthquakes for the crustal, interplate and intraslab seismicity that has been sampled with Monte Carlo methods

km slab contour (50 km depth in Ecuador). For the upper crustal seismicity, scenarios are uniformly distributed from the surface down to 35 km depth. Similarly, the upper bound of the intra-slab synthetic events is provided by the Slab1.0 geometry.

- The magnitude (Mi) is sampled from the GRD's shown in Eq. (2). Figure 5 shows an example of the synthetic seismicity for all seismogenic zones.
- The fault size is obtained according to scaling relations to derive length (L) and width (W) for each sampled magnitude. These are estimated based on scaling relations for length and width in dependence of magnitude (Blaser et al. 2010). In addition, the latter provides a probability distribution function, which relates magnitude of the earthquake (Mi) to a corresponding width and length. This particular relation provides the variability given to the length that has important consequences in augmenting the variability of the rupture areas derived for each magnitude Mi.
- The slip is estimated with the seismic moment relation (Aki 1966; Hanks and Kanamori) as shown in Eq. (1).
- The strike for the inter-plate seismicity is sampled following the Slab1.0 model. For the cases of the upper crustal and intra-slab seismicity, we have applied a number of distribution intervals and gave weights for the distinctive ranges according to the prevailing azimuths of previous earthquakes in the GCMT catalog (Ekström et al. 2012), that are given at the bottom of Table 1.
- The dip angle variability follows a normal distribution based on the Slab1.0 model, or multi-ranges for the cases of upper crustal and intra-slab seismicity based on focal mechanisms of the GCMT catalog.
- The rake of each earthquake is sampled for different ranges using a multi-range choice, sampling from uniform and normal distributions (plus standard deviations) to increase variability. These ranges have been mostly used for upper crustal and intra-slab seismicity where earthquakes have larger variability of mechanisms. We have applied number of distribution intervals and gave weights for the different ranges according to the observed prevailing rake values of previous earthquakes. In the case of the outer rise, pure normal events were considered. Contrary, for the inter-plate seismogenic zones, a range of thrust events were assumed.

2.4. *Tsunami Modeling*

The set of multiple synthetic seismic catalogs allows for longer seismicity time span and a larger variability of fault parameters. These seismic catalogs corresponding to a window of time of 100 kyear are used to first simulate initial conditions for tsunami modeling. A threshold moment magnitude Mw = 6.4 is used. Each of these seismic events in the catalog has been simulated using the *easyWave* code that solves the linear long shallow water equations (LSWE) on a finite-difference staggered grid. *easyWave* employs Okada's (Okada 1985) dislocation model to calculate initial conditions for tsunami propagation from a set of fault parameters. The bathymetry used in this model is Gebco with one minute resolution (Becker et al. 2009).

easyWave employs fully reflecting boundary conditions along the coast. This means the code does not compute inundation and runup in an explicit way. Instead, we used indirect metric in order to assess the tsunami impact at the coast. We followed Kamigaichi

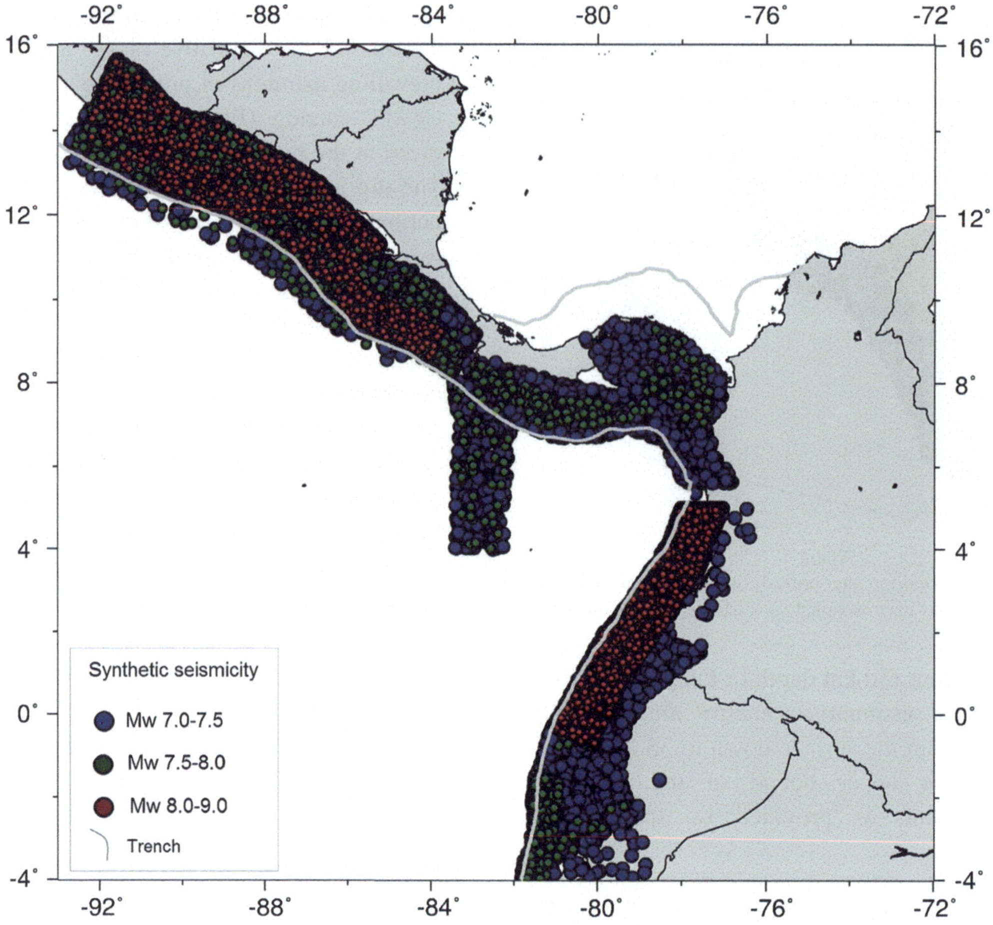

Figure 5
Map of the synthetic catalog used for model MEE containing about 184 k earthquakes Mw > 6.4

(2009) who suggested to record the wave height at some off-shore position, typically at 100 m depth, and then project this value to the coastline using the Green's Law (GL), as implemented for the Japanese Tsunami Early Warning System. In our computations, we saved maximum wave heights at water nodes located within 0–200 m depth interval. The post-processing procedure allowed us to select corresponding water nodes from the thousands of points along area of interest and to project their off-shore wave heights to the points of interest (POI) along the coast. This projection gave us our main metric: peak coastal tsunami amplitude (PCTA). Coastal tectonic correction (for the co-seismic uplift or subsidence) was additionally applied to the coastal POIs. Figure 6 shows the distribution of POIs used in this study, representing the selected coastal cities and harbors from Guatemala to southern Ecuador.

For this study, we have considered the Green's Law approach since nowadays precise modeling of the tsunami propagation and coastal impact over high-resolution nearshore bathymetry is still not feasible for regional-scale PTHA studies due to a lack of suitable high-resolution digital elevation models for the Pacific coast of Central America. Besides, even if high-resolution ($\sim$ 10–30 m) bathymetry models would be available, it is still computationally infeasible to model nearshore wave propagation for hundreds of thousands or millions of seismic scenarios needed for a full PTHA. There are examples of several local PTHA studies (e.g., González et al. (2009) for Seaside, Oregon; Lorito

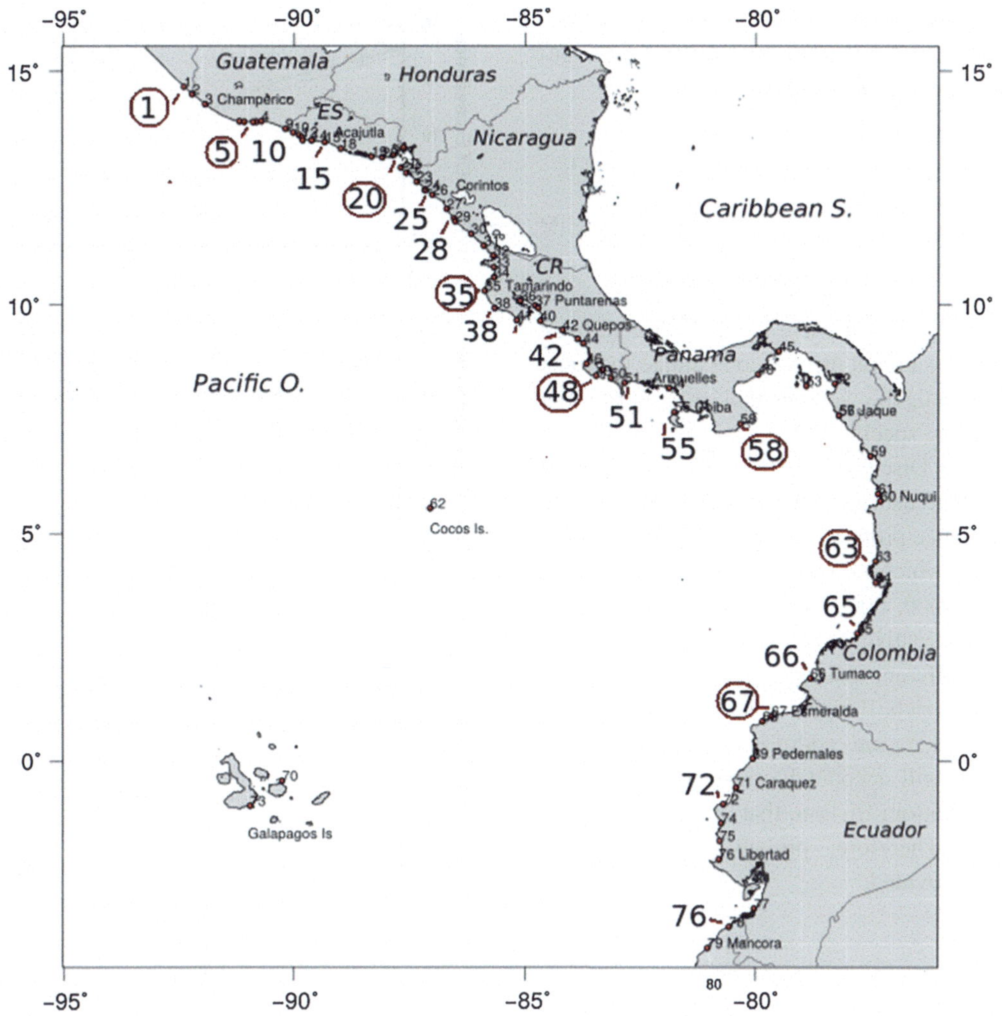

Figure 6
Map showing the location points of interest (POI) for the hazard estimates. Numbers refers to the POI number and selected coastal cities are shown as reference. *ES* El Salvador, *CR* Costa Rica

et al. (2015) for East Sicily), but the authors are not aware of any country- or regional- scale PTHA studies operating with high-resolution bathymetry models. Therefore, a first necessary step is to acquire reasonable amplification factors on the region as discussed by Hébert and Schindelé (2015). Due above reasons, regional-scale PTHA have to employ various approximations of the coastal impact from the coarse-resolution offshore propagation. Green's law is the most used and the most straightforward approximation accounting for the 1D wave shoaling. Because it has a singularity at zero water depth, wave height is usually projected to 1 m depth. Recently, several studies (Gailler et al. 2017; Glimsdal et al. 2019) proposed alternative methods of local amplification factors, which may better account for the particular nearshore bathymetry than the simple Green's law. The method suggested by Gailler et al. (2017), including the original scheme by Jamelot and Reymond (2015), requires either calibration to historical cases, or, again, simulations over very high-resolution bathymetry not available for Pacific Central America, and hence cannot be implemented in our study. The new method suggested by Glimsdal et al. (2019), in turn, does not necessarily require high-resolution bathymetry which makes it interesting for large-scale PTHA studies, on one side, but leaves open the question about its accuracy. In any case, the method of local amplification factors should be considered as a valid alternative for future large- and regional-scale PTHA studies.

2.5. *Hazard Estimates*

The resulting PCTA's are aggregated to provide hazard curves and hazard maps where the average return times of i.e. 100, 500, 1000 and 5000 years can be retrieved for different wave heights. Probabilistic analysis provides a good method to evaluate the most important tsunamigenic sources that are contributing to hazard for any specific POI.

Thus, in the Discussion section we used probabilistic maps that also show the expected PCTA with annual exceedance probabilities for 0.02, 0.01, 0.002, 0.001, 0.0002, which corresponds to the inverse of 50, 100, 500, 1000, 5000 years return time, respectively. The alternative tsunami hazard models are derived from different assumptions within a sensitivity analysis approach. If high resolution bathymetry is available for the region, additional tsunami inundation modelling will account for nonlinear processes and effects of bottom friction, that are not accounted in this study. Therefore, probabilistic inundation maps are not provided.

3. *Results*

PTHA results are presented for the two different synthetic catalogs. These results are shown as hazard curves for selected sites as well as probability maps. To facilitate analysis, most of the results will be given for the MEE model. A comparison between the two models will be provided in the discussion section.

A hazard curve represents annual probabilities of exceedance of different tsunami wave heights (or herein PCTA) at specific locations. It integrates wave heights resulting from all the sources coupled with their probabilities. Calculated hazard curves for selected POIs are shown in Fig. 7. Refer to Fig. 6 for the POIs location. Overall, these hazard curves illustrate that significant tsunamis display very low annual probabilities of exceedance in the whole region.

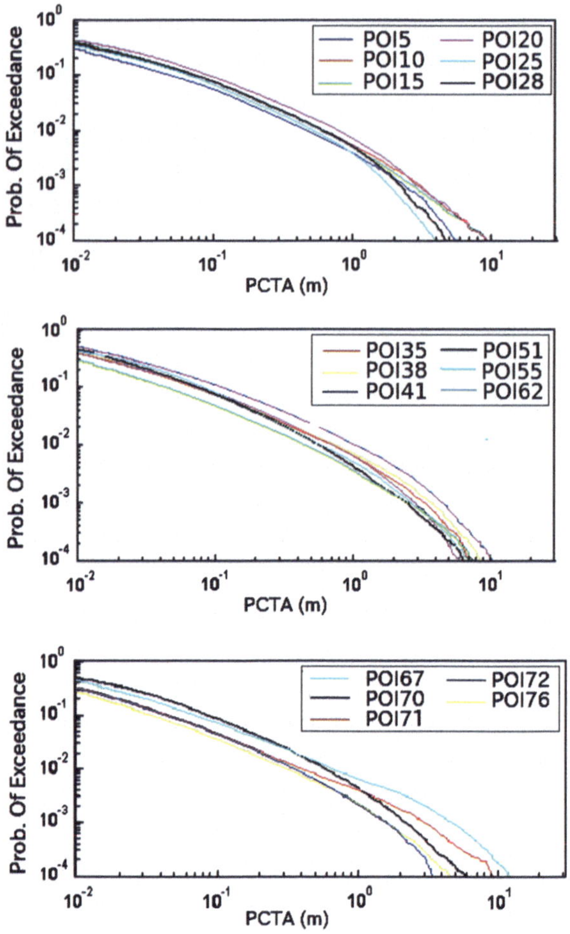

Figure 7
Upper, Hazard curves show the annual probability of exceeding a certain threshold for selected POIs from Guatemala to Nicaragua. Middle panel, hazard curves for selected POIs from south Nicaragua to Panama. Below, hazard curves for selected POIs from Colombia-Ecuador

The hazard curves for central and southern Central America have a distinct tail, which denotes a higher hazard with respect to the hazard in northern Central America (e.g. Guatemala). Notably, at shorter time windows of 50 year hazard appears very low with annual exceedance rates that are higher for southern Nicaragua, Costa Rica and Ecuador. Another observation from Fig. 7 is that overall in South America, annual probabilities of exceedance for larger PCTA's at 100 average return period (or i.e. the 1% probabilities on any given year) are higher than in Central America, except for southern Nicaragua, Costa Rica and specific coastal areas in Panama such as POI51 Armuelles and POI55 Coiba Island).

These tails in the hazard curves varies mainly in the coastal region of South America (Fig. 7). This can be related to the likelihood of larger events along the Colombia-Ecuador subduction zone or to the directivity of tsunami waves from sources of CAM that will not affect this particular sites, only in the long term where larger earthquakes are expected. This highlights the fact that very large tsunamis are expected with relatively larger return periods, which still not reconcile the 1992 and 2012 runups seen in a local scale (Fig. S3).

In summary, the examination of the hazard curves reveals that at the northern region of Central America tsunami probability is lower. This might be explained partially due to the wave directivity from southern seismic sources affecting mostly Nicaragua and Costa Rica.

An alternative way to show annual probabilities is presented on a map view showing the expected tsunami wave heights for specific mean return periods (times) or rates of exceedance. Here, the expected PCTA has been estimated for the 50, 100, 500, 1000, 5000 years.

In Fig. 8, we present the expected PCTA at selected locations for mean return periods according to the MEE model. The 50 years PCTA becomes higher towards the south, but generally does not exceed 1.0 m.

The largest PCTA for the 100 year mean return period are expected in Colombia-Ecuador (MEE model), while one of the maximum expected values corresponds to POI63. Furthermore, for the 1% annual exceedance probabilities, the expected PCTA lies within 1.5 m (MEE model). In this time frame, the most affected regions are still southern Nicaragua (POI26–POI30) and central-south Colombia (POI63–POI66). Along Panama, the highest hazard is posed by the South American seismic sources. Mainly along the coastal areas of Armuelles (POI51) and Chiriquí (POI54), expected PCTA of 1.5 m could be exceeded with a 1% annual probability. Similarly maximum tsunamis of about 1.5 m are expected in south Nicaragua.

In this case, with the current model, the 1% annual probability PCTA's values are not expected to overcome 2 m in the region as resulted in this model. The latter could highlight that larger recurrence rates are estimated from the frequency-magnitude distributions than might be expected, specially from the the Colombia-Ecuador interplate earthquakes where large earthquakes are known to have occurred in the last century.

These events, like the 1906 Colombia earthquake, appear to have long-term recurrence times. This means that in the longer return periods the hazard for Central America could be also dominated by large events expected from this subduction zone.

For the 500-year mean return period, tsunami hazard is higher in regions of South America characterized with PCTA values of 4 m (POI63). The estimated PCTA for the 500 year event is 3 m for both southern Nicaragua (POI25 to POI30) and Esmeralda, in Ecuador. Moreover, the tsunami hazard in Cocos island (Costa Rica) for the 500 year return period is about 2 m, and at the Galapagos Island PCTAs are expected to have about 2.5 m.

The PCTA for 1000 year event is estimated to reach 5 m in Esmeralda and surrounding coastal regions, 4.5 m in southern Nicaragua, and 4 m in the central and southern part of Costa Rica. Lowest hazard is expected in the regions of the Galapagos islands and Carate, southern Costa Rica as well as Cocos island, Here, PCTA is estimated to be 3 m.

In Central America, PCTAs are not expected to exceed 12 m. The hazard map shows that events corresponding to the 5000 year average return period have the highest hazard with 10 m as shown for reference in POI67 (Esmeralda). The run-up distribution from historical records showed in the last panel of Fig. 8 and Fig. S2 can provide an idea of the most affected coastal regions, and our results do not fully agree with that distribution. In the latter, the annual exceedance probability has been given for different return periods, showing only an agreement with larger tsunamis occurring along the southern Nicaragua coast and in Colombia-Ecuador. The coastal points (POI55 and POI58) are showing higher hazard. These coastal cities in Panama appear to be largely affected by tsunami directivity from the Colombia-Ecuador Trench, considering that the Panama seismic sources have smaller rates.

We further provide the hazard expressed as the probability that PCTA (assessed for a threshold of 2 m) will be exceeded at least once in a certain period

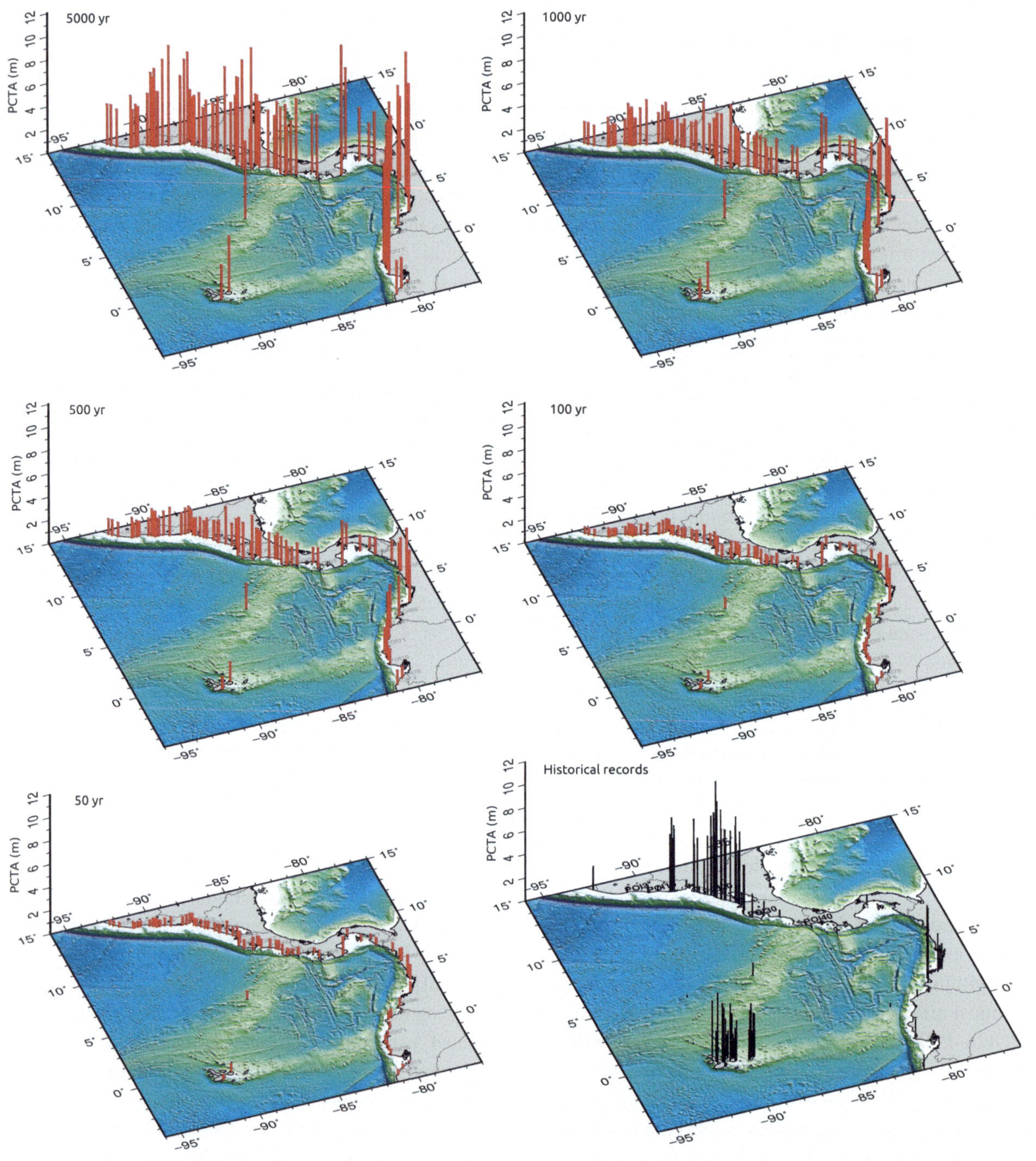

Figure 8
Hazard map showing the PCTA expected at mean return periods of 50, 100, 500, 1000 and 5000 years. The tsunami records shown in the last map (bars in black) are based on NGDC/WDS (2017). These results correspond to the MEE model

of time, as shown by the Eq. 3. In this case, the Fig. 9 shows probabilities of PCTA to reach specific values during the next 100 years.

Figure 10 shows aggregated wave heights exceeding certain thresholds for selected POIs. The hazard estimates have been grouped for northern

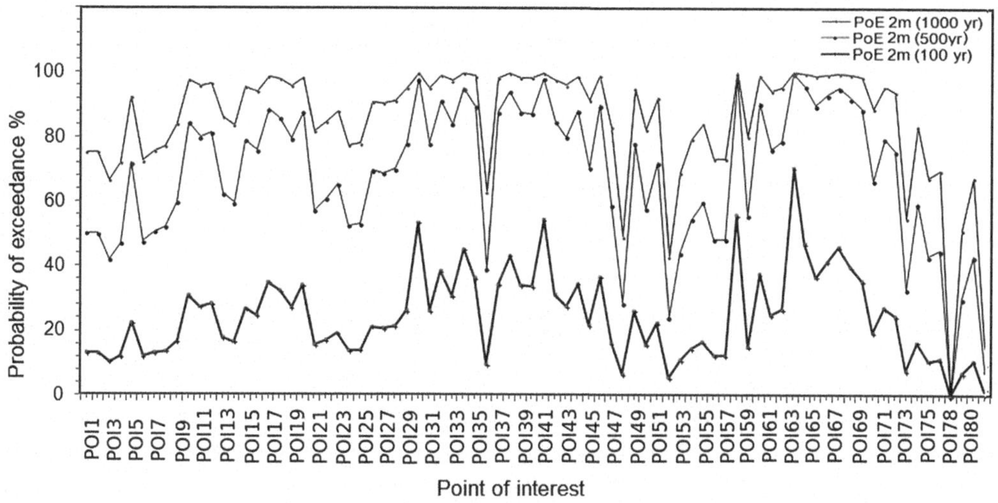

Figure 9
Probability of exceeding 2 m in for different times at selected POIs given in percentage. Refer to Fig. 6 for the POI location

CAM (upper panel), southern CAM (mid panel), and South America (lower panel). The tails of these curves varied significantly as seen in the plots of the upper panel. These curves correspond to three POIs in Nicaragua where maximum magnitude expected for the 100 year. The hazard in the POIs in Nicaragua decreases from north to south. Particularly, POI25 to POI29 show the higher hazard in these selected POIs (Fig. 10).

The hazard posed to Costa Rica is higher along the central Pacific coast POI42-POI44 (e.g. Quepos) than the POI48 (Carate) located about 140 km south of Quepos. The difference in the tail of the hazard curve for POI51 (Puerto Armuelles) could be related to a greater influence of several tsunamigenic sources from South America due to its location if compared to the other sites in Central America.

The results shown here are consistent with the fact that a variation of PCTA at different return periods is related to longer occurrence rates of large events mainly at the Ecuador-Colombia trench and Nicaragua. Also, we show in the following figure that large tsunamis could strike Central America with PCTA of more than 30 m (Fig. 11). Here, we show the maximum tsunamis extracted from the 100 kyear of simulations. However, as expected they appear to have very long recurrence times and are not well represented in the hazard curves. From Fig. 11, we can conclude that the influence of the maximum magnitudes, or upper bound that has been considered in the second model (MCE) leads to larger PCTAs in most of the POIs. We suggest that the local sources along northern Central America are increasing the hazard values, which are the ones whose upper bound was assumed larger.

Overall, these results show that local tsunamigenic sources have a higher influence in the hazard at e.g. the 500 year tsunami event, which is partially consistent with findings in other regions as suggested in other studies (González et al. 2009) but not necessarily in the regional sources. Also, variations in the hazard are reflected in the comparison of the two models with different maximum magnitudes (Fig. 12). The plot shows larger differences for the longer return periods comparing the two models based on the probability of exceeding a PCTA threshold of 2 m, where the maximum Mw is the only assumption that was changed to generate a second set of synthetic catalogs.

4. Discussion

4.1. Comparison of results with previous studies

The results for our two main models appear to differ with other regional studies. Figure 8 shows the PCTA expected for the mean return periods of

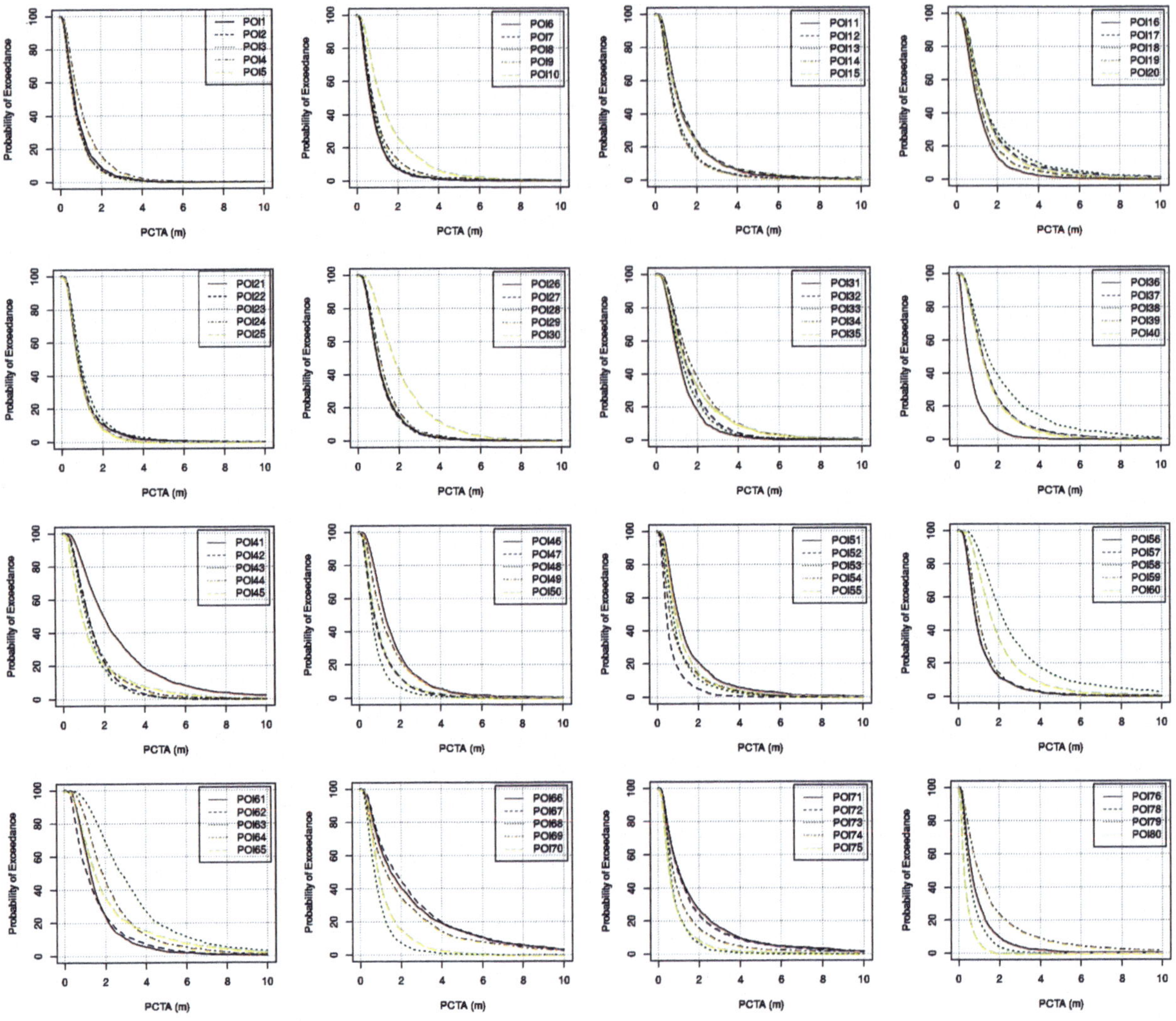

Figure 10
Probability of exceedance in 100 years in different POIS along CAM-CE. The probability is given as percentage. Refer to Fig. 6 for the POIs location

50–5000 years for the model MEE. The expected PCTA indicates rather low hazard level along most of the study area from Guatemala to Ecuador for the 50- and 100-year return periods. These results differ from those presented recently by Brizuela et al. (2014).

One reason for this difference could be the influence of b-values in the hazard curves. In their study, a seismic b-value of 0.67 is used to estimate earthquake recurrence rates for the entire region between south Mexico and Costa Rica (Brizuela et al. 2014), which not necessarily characterizes each zone. As shown by Hoechner et al. (2016), the b-value can have a large effect on the hazard estimates, therefore this could be one of the reasons of the highly different hazard estimates of these two studies.

Another explanation may be that since Brizuela et al. (2014) did not employ tsunami propagation simulations but instead used simplified source-to-coast tsunami run-up relations; our PTHA results differ from this due to the propagation component of the assessment chain. It would be very important to compare the two studies using the same input data in order to define where these differences arise and how to overcome them. Since results appear to be so

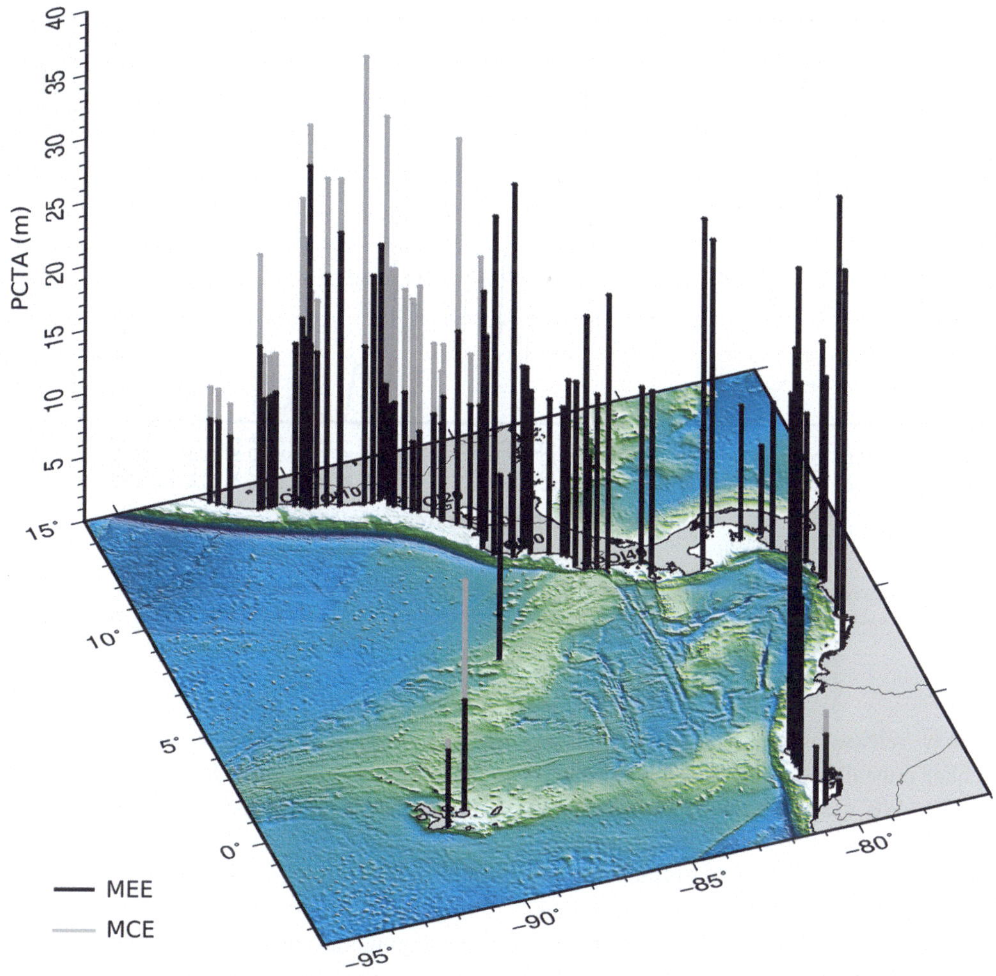

Figure 11
Maximum PCTA values extracted from all tsunami simulations. The black bars show the MEE model, and the MCE model is shown in gray. These large values in the map represents PCTA with larger return periods of those assessed in this study

different, it is necessary that future studies assess what causes these large discrepancies.

In this sense, Stein et al. (2017) have described that in principle the data and time windows used to describe the past seismicity of any region, as well as the models that are implemented to forecast future earthquakes will greatly effect the hazard estimates. The latter has been illustrated in regions with extensive paleotsunamis that have shown repeating large earthquakes and tsunamis (e.g. Cascadia, Chile), where a difference in seismic rates could be seen between assuming clustering of events different to the mean estimates (Stein et al. 2017). Thus, in this study we partially approach these uncertainties, and more models might be needed. Unfortunately, paleotsunami data showing longer recurrence rates that could be included in the hazard estimates is lacking in the CAM region.

Besides, Fig. 1 and Fig. S3 provide evidence of existing patterns of 500 years of tsunami data. The remarkable fact that the 1992 Nicaragua and the 2012 El Salvador tsunamis have been caused by moderate magnitude tsunami earthquakes, posing a hazard that could be inferred within the 1000 and 5000 year return periods. Both earthquakes generated larger tsunamis that could be expected from their moment magnitudes and hence could be characterized as 'tsunami earthquakes' (Kanamori 1972). Such

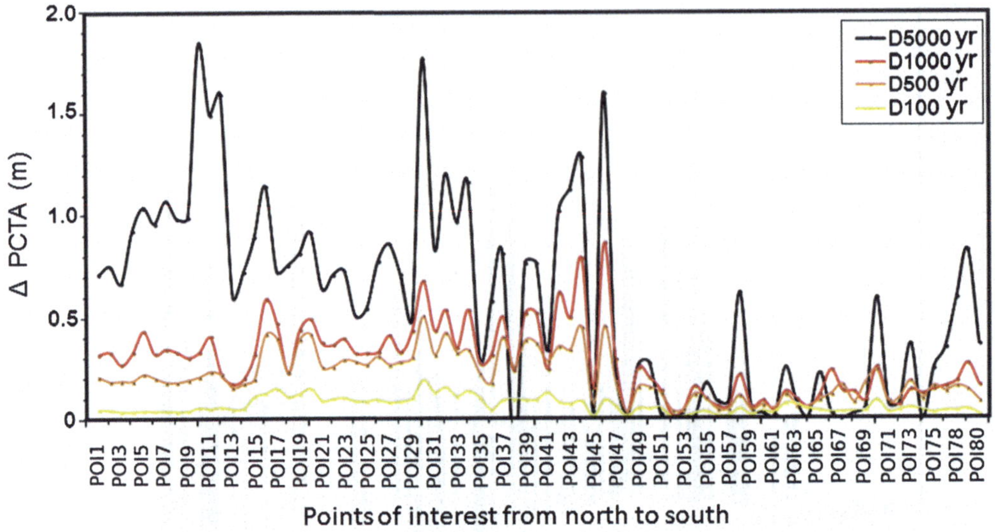

Figure 12
Difference in the PCTA between the two models (MCE and MEE) considered in this study to compare effect of Mmax. These differences are based on the threshold of exceeding 2 m wave height in the time (i.e. difference D5000 shown for the 5000 year time exposure is shown in gray)

earthquakes may release more co-seismic slip by rupturing weaker crustal material close to the trench (e.g., sediments). We consider lower rigidity estimates or concentration of shallow slip in our models. Thus, including more physical assumptions to the models and not only maximum magnitudes has highlighted the large dependency that PTHA has on the fundamental assumptions. Further studies comparing different methods will shed light on this aspect that is crucial for the tsunami hazard assessment because this dependency on the choice of the probability models (Stein et al. 2017).

4.2. Sensitivity analysis

To further understand the susceptibility of hazard estimates to major model assumptions, we considered two additional synthetic catalogs (ref. to Sect. 2.3 above). One of them (MEE-r) is based on our MEE reference catalog but implies depth-dependent rigidity in accord with Bilek and Lay (1999). This rigidity model (applied only to the CAM inter-plate seismogenic zone) accounts for the phenomenon of 'tsunami earthquakes'. Since in the present approach we model ruptures with the classical single 'Okada' fault model, we had to assume an effective rigidity value for the whole fault plane (e.g., no rigidity variation along the rupture width). This effective rigidity value was, in turn, picked from the depth-dependent rigidity profile according to the depth of the center of the fault. For ruptures with large width, this approach may result in the overestimation of their tsunamigenic potential, therefore we may consider this model as a 'rigidity end-member' model in our sensitivity analysis. Figure 13 compares PTHA results between the constant and depth-dependent rigidity models.

In this sense, there is relative agreement with the section of Nicaragua to Costa Rica in terms of the size of tsunami, showing that local segments prone to tsunami earthquakes could greatly contribute to the tsunami hazard posed in Central America and appear to be underestimated in our reference models MEE and MCE.

Thus, the distribution of run-ups clustered in three segments in all models through the area of this study, that corresponds with few events (Fig. 1 and S3). In Nicaragua, for example, most of the run-up values are related to the 1992 earthquake, and in Colombia-Ecuador, most of run-ups were generated from the 1906 and 1979 Colombia-Ecuador earthquakes. What seems to be consistent in the shorter term is that tsunami hazard is higher in Nicaragua and southern

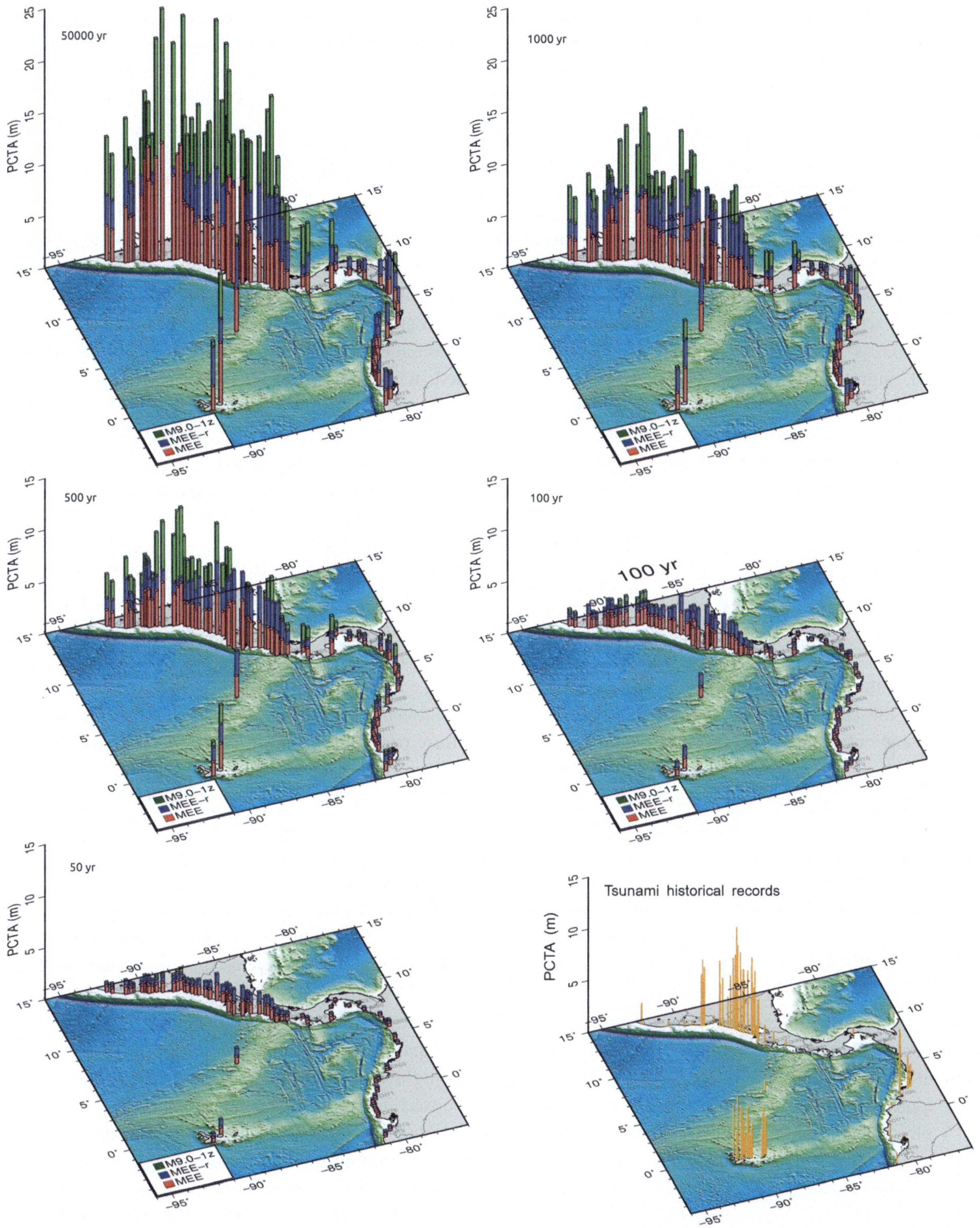

Figure 13
Each map shows the PCTA expected for several return periods for three models based only on the CAM interplate seismogenic zone. MEE stands for the same model shown in Fig. 8. MEE-r model includes the rigidity depth variation. M9.0-1z stands for the CAM single segment with uniform rigidity similar as it was used for MEE and MCE models. Refer to Table 1 to see the seismic parameters

Colombia to Ecuador. Particularly for Central America, the hazard curves show higher hazard in southern Nicaragua and Costa Rica in the long term (e.g POI31, POI35, POI41).

Additionally, we explore another source of uncertainty which is related to the seismo-tectonic segmentation and is associated to the earlier idea of McCaffrey (2008) that we cannot rule out that any subduction zone may produce a magnitude 9 or larger earthquake. Following this idea, we show hazard estimates by assuming the whole CAM megathrust as one single segment with Mmax 9.0 (model called 'M9.0-1z'), as shown in Fig. 13. As might be expected, significantly higher PCTA refer to the *large return periods* (> 500 years) due to the long recurrence rates of giant earthquakes. In the shorter perspective (return periods of 50 and 100 years), the effect of low shallow rigidity (i.e., 'tsunami earthquakes') becomes dominant. Also, it is observed from the hazard curves that the MAT poses a low tsunami hazard to coastal towns in South America.

4.3. Limitations of the Models

The limitations of the carried out model are related to the large spectrum of uncertainties which have been only partly accounted. Some of the assumptions such as Poisson process could be treated as epistemic uncertainty when alternative models such as Brownian Passage Time process (Matthews et al. 2002) are considered. This could hold e.g. for the short term forecasting that accounts for the fault interaction or aftershocks. However, for the mid-to-long-term assessment, the Poisson process is realistic and valid in PSHA as well as PTHA (Horspool et al. 2014; Parsons and Geist 2009b). Yet, a remaining point is the need to integrate more assumptions and spatio-temporal statistic techniques, that could constrain if the earthquake rates are solely controlling what appears to be low hazard estimates.

A physical limitation in PTHA derives from the simple rupture scenarios based on uniform slip, neglecting critical aspects such as the slip distribution influence on tsunami inundation (Li et al. 2016). In the case of the slab model, slip should be distributed within sub-fault models, but we didn't perform this in a stochastic way. Randomness is in turn given by the stochastic location of earthquake sources and the focal mechanisms, which have a high influence in local tsunami PCTA. More physical constraints have to be considered in PTHA models (Murphy et al. 2016). For example, in most of the models discussed here, we assumed uniform slip and a constant shear modulus of 30 GPa. Alternatively, we considered variations of the shear modulus along-dip. Allowing depth-varying shear modulus greatly influenced the hazard estimates. Such assumptions could provide different branches in the construction of a logic tree to integrate more uncertainties.

We identify three major limitations of our model. First, while we compared models for two maximum magnitudes truncating the GRD based on a regional model (Alvarado et al. 2017), we did not account for many alternative segmentation models that allow different maximum magnitudes nor the zone-less approach except for one case where the CAM is assumed to rupture free of segments with the possibility of earthquakes up to Mw 9.0. This could be critical since large earthquakes could rupture beyond the proposed segment limits (Okal et al. 2006) as shown in our model when considering mega-earthquakes in the one-segment interplate seismogenic zone that lead to larger PCTA in long return periods. Second, we restrict the assessment only to earthquake sources. An aspect that should be included in future work is to account for the tsunami hazard due to submarine landslides. Particularly, in Central America, significant slide imprints resembling large submarine landslides have been imaged (VonHuene et al. 2004), but their probability of occurrence is difficult to constrain. Third, we did not consider trans-Pacific events that are known to affect Central America (Rabinovich et al. 2013; Chacón-Barrantes and Gutiérrez-Echeverría 2017). In Fig. S3, we separated the recorded runups in the study area to highlight that the Galapagos and Cocos islands are exposed to long distance tsunamis as has been recorded by the 2011 Tohoku earthquake (yellow bars). Including this multi-hazard approach will give a more complete probabilistic assessment for this region. Another aspect that could affect the hazard variations in short distances are the bathymetric features in Central America. These differ in very clear segments, since the basin offshore Central

America is prone to edge waves. Those particular types of waves occur due to refraction and travel parallel to the coast. They can be typical of the shallow continental shelf as seen from numerical simulations (Zamora and Babeyko 2015).

4.4. *Validity and Reliability of the PTHA*

The main approach proposed for probabilistic seismic hazard assessment testing is addressed by comparing the overall forecast with ground motion data (Stirling and Gerstenberger 2010). Similar to that approach, the PTHA has to be tested. Nonetheless, the historical observations of wave heights along the coast of Central America, as in many other regions in the world is too short in duration, which disables the ability to make a robust testing. In this study, a qualitative comparison has been done by comparing wave height distribution given by tsunami catalog and the results for 500 and 1000 year return period. It has to be kept in mind that some tsunami records are given as run-up values and others as wave height (Fig. 1b).

Finally, the ultimate goal of the PTHA is to provide a forecast for the frequency of future tsunamis at selected coastal regions. This approach uses statistical analysis, which should be tested rigorously. However, unlike in probabilistic seismic hazard assessment (PSHA), where a set of algorithms have been implemented to test seismic hazard assessment (van Stiphout et al. 2011) based on short term observations, such testing is not realistic for tsunami hazard. However, future work should be planned to test for more tsunami hazard assessment techniques in this region, where 500 years of data might not be sufficient since the coastal areas were not always populated, thus the tsunami records could be incomplete to compare with the mean recurrence times of tsunamis in this region.

5. *Conclusions*

This study has embraced a comprehensive probabilistic tsunami approach to estimate hazard at specific POIs from Guatemala to Ecuador. Our approach is based on combining the assessment of seismic sources with numerical simulations of tsunamis. This method has the advantage of providing tsunami frequency based on exhaustive seismic data along Central America.

After establishing the seismic zonation based on a total of 50 individual seismic area sources, the Monte-Carlo approach was used to simulate a series of synthetic regional earthquake catalogs for the next 100 kyear. Monte-Carlo sampling from the multi-dimensional fault parameter space allows random uncertainties in the rupture process to be taken into account. In addition, the numerical simulation of tsunami wave propagation also considers bathymetric imprints that could affect the waveforms and additionally explain the variability of PCTA. The time span is something that has to be tested, for such studies larger time span (e.g. 500 kyear–1 myear) could be advisable to cover more seismic cycles, especially for the segments with longer recurrence times.

This numerical approach has an advantage over other studies, as it combines a great variability of parameters with realistic tsunami propagation, yet is too computationally expensive. Therefore, in a following paper, Tsunami Green Functions are used to facilitate this assessment and the sensitivity analysis. Two synthetic seismic catalogs resulted from joining fifty seismic zones, from Guatemala to Ecuador. Each catalog contains about 185,000 earthquakes larger than Mw 6.4. The two catalogs differ in the assumptions on their maximum magnitude. For each earthquake tsunami generation and propagation for 6 h were simulated and maximum wave heights saved at the thousands of near-coastal computational nodes, located within the depth interval from 0 to 500 m. In the post-processing phase, these wave heights were projected onto the selected 80 coastal POI's by means of the Green's law. To facilitate the analysis, hazard curves were presented for fewer POIs.

Our analysis shows that tsunami hazard is higher in Nicaragua and Colombia, where the probability of exceeding 2 m in 500 years is 50–70%, respectively. For 50 and 100 years return periods, expected tsunami wave heights are rather small in the whole study area, and hazard appears to be low. In general, wave heights up to 1 m are expected for the 100 years return periods. The latter could be expected from the historical tsunami catalogs, except for the 1992 and 2012 tsunami earthquake events that dominate the runup distributions along Central America if compared with all events occurred during the last 500 years.

The assessment with different assumptions such as different zonation for the interplate seismogenic zone and lower rigidity values shows a high variability of hazard estimates that cannot be neglected. The latter raise the question of the validity of results where only a reduced range of uncertainties are incorporated.

It is shown in this study that even when large magnitudes for Central America have been estimated for some segments, the hazard becomes significant only for higher return periods, which appears to be consistent with what has been argued from seismological and geodetic studies.

Ideally, after the various limitations described here have been addressed, this type of model that can cover various mean return periods could provide a foundation for engineers to modify tsunami building codes. In addition, evacuation and coastal planning have to be further developed in order to take into account future models of this type that cover different return periods, rather than simply basing them on maximum credible earthquake estimates.

Acknowledgements

This study was possible thanks to the financial support to NZ from the Helmholtz Association and the GeoSim Program and the GeoForschungsZentrum (GFZ-Potsdam). Deep thanks to B. Benito, Y. Torres, A. Rivas and R. García at the Politechnic of Madrid (UPM) for hosting NZ during the Central American PSHA Workshop held in 2011 and for the discussions on the probabilistic framework. Special acknowledge to W. Rojas, E. Molina, E. Camacho for discussions related to Central American seismic catalog and seismic rates of the region. M. Fernández is acknowledged for discussions of the Central America tsunami catalog. A. Hoechner is acknowledged for the fruitful discussions about PTHA uncertainties. M.Sørensen and L. Matias are acknowledged for comments on a first stage of the study. Most figures were drawn using the GMT software Wessel et al. (2013) and R-project (R Core Team 2014). We strongly appreciate the comments and suggestions of two anonymous reviewers and the guest editor of the special volume E. Okal.

References

Aki, K. (1966). Generation and propagation of G waves from the Niigata earthquake of June 16, 1964. *Bulletin of the Earthquake Research Institute*, *44*, 73–88.

Alvarado, D., DeMets, C., Tikoff, B., Hernandez, D., Wawrzyniec, T., Pullinger, C., et al. (2010). Forearc motion and deformation between El Salvador and Nicaragua: GPS, seismic, structural, and paleomagnetic observations. *Lithosphere*, *3*(1), 3–21.

Alvarado, G. E., Benito, B., Staller, A., Climent, Á., Camacho, E., Rojas, W., et al. (2017). The new Central American seismic hazard zonation: Mutual consensus based on up to day seismotectonic framework. *Tectonophysics*, *721*, 462–476.

Álvarez-Gómez, J. A., Meijer, P. T., Martínez-Díaz, J. J., & Capote, R. (2008). Constraints from finite element modeling on the active tectonics of northern Central America and the Middle America Trench. *Tectonics*, *27*(1), TC1008.

Álvarez-Gómez, J. A., Gutiérrez-Gutiérrez, O. Q., Aniel-Quiroga, I., & Gonzlez, M. (2012). Tsunamigenic potential of outer-rise normal faults at the Middle America trench in Central America. *Tectonophysics*, *574–575*, 133–143.

Álvarez-Gómez, J. A., Aniel-Quiroga, I., Gutiérrez-Gutiérrez, O. Q., Larreynaga, J., González, M., Castro, M., et al. (2013). Tsunami hazard assessment in El Salvador, Central America, from seismic sources through flooding numerical models. *Natural Hazards and Earth System Sciences*, *13*(11), 2927–2939.

Basili, R., Tiberti, M. M., Kastelic, V., Romano, F., Piatanesi, A., Selva, J., et al. (2013). Integrating geologic fault data into tsunami hazard studies. *Natural Hazards and Earth System Sciences*, *13*(4), 1025–1050.

Beauval, C., Yepes, H., Palacios, P., Segovia, M., Alvarado, A., Font, Y., et al. (2013). An earthquake catalog for seismic hazard assessment in Ecuador. *Bulletin of the Seismological Society of America*, *103*(2A), 773–786.

Becker, J., Sandwell, D., Smith, W., Braud, J., Binder, B., Depner, J., et al. (2009). Global bathymetry and elevation data at 30 Arc Seconds Resolution: SRTM plus. *Marine Geodesy*, *32*(4), 355–371.

Benito, M. B., Lindholm, C., Camacho, E., Climent, A., Marroquin, G., Molina, E., et al. (2012). A new evaluation of seismic hazard for the Central America region. *Bulletin of the Seismological Society of America*, *102*(2), 504–523.

Bilek, S. L., & Lay, T. (1999). Rigidity variations with depth along interplate megathrust faults in subduction zones. *Nature*, *400*, 443–446.

Blaser, L., Kruger, F., Ohrnberger, M., & Scherbaum, F. (2010). Scaling relations of earthquake source parameter estimates with special focus on subduction environment. *Bulletin of the Seismological Society of America*, *100*(6), 2914–2926.

Borrero, J. C., Kalligeris, N., Lynett, P. J., Fritz, H. M., Newman, A. V., & Convers, J. A. (2014). Observations and modeling of the August 27, 2012 earthquake and tsunami affecting El Salvador and Nicaragua. *Pure and Applied Geophysics*, *171*(12), 3421–3435.

Brizuela, B., Armigliato, A., & Tinti, S. (2014). Assessment of tsunami hazards for the Central American Pacific coast from southern Mexico to northern Peru. *Natural Hazards and Earth System Science*, *14*(7), 1889–1903.

Burroughs, S. M., & Tebbens, S. F. (2005). Power-law scaling and probabilistic forecasting of tsunami runup heights. *Pure and Applied Geophysics*, *162*(2), 331–342.

Chacón-Barrantes, S., & Gutiérrez-Echeverría, A. (2017). Tsunamis recorded in tide gauges at Costa Rica Pacific coast and their numerical modeling. *Natural Hazards*, *89*(1), 295–311.

Cornell, C. A. (1968). Engineering seismic risk analysis. *Bulletin of the Seismological Society of America*, *58*(5), 1583–1606.

Correa-Mora, F., DeMets, C., Alvarado, D., Turner, H. L., Mattioli, G., Hernandez, D., et al. (2009). GPS-derived coupling estimates for the Central America subduction zone and volcanic arc faults: El Salvador. *Honduras and Nicaragua, Geophysical Journal International*, *179*, 1279–1291.

Cosentino, P., Ficarra, V., & Luzio, D. (1977). Truncated exponential frequency-magnitude relationship in earthquake statistics. *Bulletin of the Seismological Society of America*, *67*(6), 1615–1623.

de Boer, J. Z., Defant, M. J., Stewart, R. H., Restrepo, J. F., Clark, L. F., & Ramirez, A. H. (1988). Quaternary calc-alkaline volcanism in western Panama: Regional variation and implication for the plate tectonic framework. *Journal of South American Earth Sciences*, *1*(3), 275–293.

de Boer, J. Z., Drummond, M. S., Bordelon, M. J., Defant, M. J., Bellon, H., & Maury, R. C. (1995). Cenozoic magmatic phases of the Costa Rican island arc (Cordillera de Talamanca). *Geological Society of America Special Papers*, *295*, 35–56.

DeMets, C. (2001). A new estimate for present-day Cocos-Caribbean plate motion: Implications for slip along the Central American volcanic arc. *Geophysical Journal International*, *28*(21), 4043–4046.

DeMets, C., Gordon, R. G., & Argus, D. F. (2010). Geologically current plate motions. *Geophysical Journal International*, *181*(1), 1–80.

Ebel, J., & Kafka, A. (1999). A Monte Carlo Approach to Seismic Hazard Analysis. *Bulletin of the Seismological Society of America*, *89*(4), 854–866.

Ekström, G., Nettles, M., & Dziewoński, A. (2012). The global CMT project 2004–2010: Centroid-moment tensors for 13,017 earthquakes. *Physics of the Earth and Planetary Interiors*, *200*, 1–9.

Fernández, M., Ortiz-Figueroa, M., & Mora, R. (2004). Tsunami hazards in El Salvador. *Geological Society of America Special Paper*, *375*, 435–444.

Franco, A., Lasserre, C., Lyon-Caen, H., Kostoglodov, V., Molina, E., Guzman-Speziale, M., et al. (2012). Fault kinematics in northern Central America and coupling along the subduction interface of the Cocos Plate, from GPS data in Chiapas (Mexico), Guatemala and El Salvador. *Geophysical Journal International*, *189*(3), 1223–1236.

Gailler, A., Hbert, H., Schindel, F., & Reymond, D. (2017). Coastal amplification laws for the French Tsunami warning center: Numerical modeling and fast estimate of tsunami wave heights along the French Riviera. *Pure and Applied Geophysics*, *175*, 1–16.

Geist, E. L. (2009). Phenomenology of tsunamis: Statistical properties from generation to runup. *Advances in Geophysics*, *51*, 107–169.

Geist, E. L., & Parsons, T. (2006). Probabilistic analysis of tsunami hazards. *Natural Hazards*, *37*(3), 277–314.

Glimsdal, S., Løvholt, F., Harbitz, C. B., Romano, F., Lorito, S., Orefice, S., et al. (2019). A new approximate method for quantifying tsunami maximum inundation height probability. *Pure and Applied Geophysics*, *176*, 3227–3246.

González, F. I., Geist, E. L., Jaffe, B., Kânolu, U., Mofjeld, H., Synolakis, C. E., et al. (2009). Probabilistic tsunami hazard assessment at Seaside, Oregon, for near- and far-field seismic sources. *Journal of Geophysical Research*, *114*(C11), C11023.

Grezio, A., Babeyko, A., Baptista, M. A., Behrens, J., Costa, A., Davies, G., et al. (2017). Probabilistic tsunami hazard analysis: Multiple sources and global applications. *Reviews of Geophysics*, *55*(4), 1158–1198.

Hayes, G. P., Wald, D. J., & Johnson, R. L. (2012). Slab1. 0: A three-dimensional model of global subduction zone geometries. *Journal of Geophysical Research: Solid Earth*, *117*(B1), B01302.

Hébert, H., & Schindelé, F. (2015). Tsunami impact computed from offshore modeling and coastal amplification laws: Insights from the 2004 Indian Ocean tsunami. *Pure and Applied Geophysics*, *172*(12), 3385.

Hoechner, A., Babeyko, A., & Zamora, N. (2016). Probabilistic tsunami hazard assessment for the Makran region with focus on maximum magnitude assumption. *Natural Hazards and Earth System Science*, *16*, 1339–1350.

Horspool, N., Pranantyo, I., Griffin, J., Latief, H., Natawidjaja, D. H., Kongko, W., et al. (2014). A probabilistic tsunami hazard assessment for Indonesia. *Natural Hazards and Earth System Sciences*, *14*(11), 3105–3122.

Jamelot, A., & Reymond, D. (2015). New tsunami forecast tools for the French polynesia tsunami warning system. *Pure and Applied Geophysics*, *172*, 791–804.

Kamigaichi, O. (2009). Tsunami forecasting and warning. In R. A. Meyers (Ed.), *Encyclopedia of Complexity and Systems Science SE-568* (pp. 9592–9618). New York: Springer.

Kanamori, H. (1972). Mechanism of tsunami earthquakes. *Physics of the Earth and Planetary Interiors*, *6*(5), 346–359.

Kijko, A., & Smit, A. (2012). Extension of the Aki-Utsu b-value estimator for incomplete catalogs. *Bulletin of the Seismological Society of America*, *102*(3), 1283–1287.

Knopoff, L., Kagan, Y. Y., & Knopoff, R. (1982). b Values for foreshocks and aftershocks in real and simulated earthquake sequences. *Bulletin of the Seismological Society of America*, *72*(5), 1663–1676.

Kobayashi, D., LaFemina, P., Geirsson, H., Chichaco, E., Abrego, A. A., Mora, H., et al. (2014). Kinematics of the western Caribbean: Collision of the Cocos Ridge and upper plate deformation. *Geochemistry, Geophysics, Geosystems*, *15*, 1671–1683.

LaFemina, P. C., Dixon, T. H., & Strauch, W. (2002). Bookshelf faulting in Nicaragua. *Geology*, *30*(8), 751–754.

LaFemina, P., Dixon, T. H., Govers, R., Norabuena, E., Turner, H., Saballos, A., et al. (2009). Fore-arc motion and Cocos Ridge collision in Central America. *Geochemistry, Geophysics, Geosystems*, *10*(5), 1–21.

Li, L., Switzer, A. D., Chan, C.-H., Wang, Y., Weiss, R., & Qiu, Q. (2016). How heterogeneous coseismic slip affects regional probabilistic tsunami hazard assessment: A case study in the South China Sea. *Journal of Geophysical Research: Solid Earth*, *121*, 6250–6272.

Lindholm, C., Redondo, C. A., & Bungum, H. (2004). Two earthquake databases for Central America. *Geological Society of America Special Papers*, *375*, 357–362.

Lonsdale, P., & Klitgord, J. (1978). Structure and tectonic history of the eastern Panama Basin. *Geological Society of America Bulletin*, *89*(7), 981–999.

Lorito, S., Tiberti, M. M., Basili, R., Piatanesi, A., & Valensise, G. (2008). Earthquake-generated tsunamis in the Mediterranean Sea: Scenarios of potential threats to Southern Italy. *Journal of Geophysical Research: Solid Earth*, *113*(B1), B1301.

Lorito, S., Selva, J., Basili, R., Romano, F., Tiberti, M. M., & Piatanesi, A. (2015). Probabilistic hazard for seismically induced tsunamis: Accuracy and feasibility of inundation maps. *Geophysical Journal International*, *200*(1), 574–588.

Løvholt, F., Glimsdal, S., Harbitz, C. B., Zamora, N., Nadim, F., Peduzzi, P., et al. (2012). Tsunami hazard and exposure on the global scale. *Earth-Science Reviews*, *110*, 58–73.

Løvholt, F., Khn, D., Bungum, H., Harbitz, C. B., & Glimsdal, S. (2012). Historical tsunamis and present tsunami hazard in eastern Indonesia and the southern Philippines. *Journal of Geophysical Research: Solid Earth*, *117*(B9), B09310.

Lyon-Caen, H., Barrier, E., Lasserre, C., a. Franco, I., Arzu, L., Chiquin, M., et al. (2006). Kinematics of the North AmericanCaribbean-Cocos plates in Central America from new GPS measurements across the Polochic-Motagua fault system. *Geophysical Research Letters*, *33*(19), L19309.

Matias, L. M., Cunha, T., Annunziato, A., Baptista, M. a, & Carrilho, F. (2013). Tsunamigenic earthquakes in the Gulf of Cadiz: Fault model and recurrence. *Natural Hazards and Earth System Science*, *13*(1), 1–13.

Matthews, M. V., Ellsworth, W. L., & Reasenberg, P. A. (2002). A Brownian model for recurrent earthquakes. *Bulletin of the Seismological Society of America*, *92*(6), 2233–2250.

McCaffrey, R. (2008). Global frequency of magnitude 9 earthquakes. *Geology*, *36*(3), 263. https://doi.org/10.1130/G24402A.1.

McGuire, R. (1976). FORTRAN computer program for seismic risk analysis. In Technical report, U.S. Geol. Surv., Open File Rep. No. 76–67.

Murphy, S., Scala, A., Herrero, A., Lorito, S., Festa, G., Trasatti, E., et al. (2016). Shallow slip amplification and enhanced tsunami hazard unravelled by dynamic simulations of mega-thrust earthquakes. *Science Reports*, *6*, 1–12.

NGDC/WDS. (2017). National geophysical data center, world data service, global historical tsunami database, national geophysical data center, NOAA. http://www.ngdc.noaa.gov/hazard/tsu_db.shtml/. Accessed 31 Dec 2017.

Okada, Y. (1985). Surface deformation due to shear and tensile faults in a half-space. *Bulletin of the Seismological Society of America*, *75*(4), 1135–1154.

Okal, E. A., Borrero, J. C., & Synolakis, C. E. (2006). Evaluation of tsunami risk from regional earthquakes at Pisco, Peru. *Bulletin of the Seismological Society of America*, *96*(5), 1634–1648.

Ortiz, M., Fernández Arce, M., & Rojas, W. (2001). Análisis de riesgo de inundación por tsunamis en Puntarenas, Costa Rica. *GEOS*, *21*(2), 108–113.

Parsons, T., & Geist, E. (2009a). Tsunami probability in the Caribbean region. *Pure and Applied Geophysics*, *165*(11–12), 2089–2116.

Parsons, T., & Geist, E. L. (2009b). Is there a basis for preferring characteristic earthquakes over a Gutenberg-Richter distribution in probabilistic earthquake forecasting? *Bulletin of the Seismological Society of America*, *99*(3), 2012–2019.

Protti, M., McNally, K., Pacheco, J., Gonzlez, V., Montero, C., Segura, J., et al. (1995). The March 25, 1990 (Mw = 7.0, ML = 6.8), earthquake at the entrance of the Nicoya Gulf, Costa Rica: Its prior activity, foreshocks, aftershocks, and triggered seismicity. *Journal of Geophysical Research: Solid Earth*, *100*(B10), 20345–20358.

R Core Team. (2014). *R: A Language and environment for statistical computing*. Vienna, Austria: R Foundation for Statistical Computing. http://www.R-project.org/.

Rabinovich, A. B., Candella, R. N., & Thomson, R. E. (2013). The open ocean energy decay of three recent trans-Pacific tsunamis. *Geophysical Research Letters*, *40*, 3157–3162.

Rojas, W., Camacho, E., Marroquin, G., Molina, E., & Benito, M. B. (2013). Evolution of the Earthquake Catalog in Central America. In *Meeting of the Americas, AGU, Mexico. S52A08*, Number AGU Meeting of the Americas.

Rojas, W., Bungum, H., & Lindholm, C. (1993). Historical and recent earthquakes in Central America. *Revista Geologica de America Central*, *16*, 5–22.

Rubinstein, R. Y. & Kroese, D. P. (2008). *Simulation and the Monte Carlo Method* (2nd editio ed.). Wiley.

Salazar, W., Brown, L., Hernández, W., & Guerra, J. (2013). An earthquake catalogue for El Salvador and neighboring Central American countries (1528–2009) and its implication in the seismic Hazard Assessment. *Journal of Civil Engineering and Architecture*, *7*(8), 1018–1045.

Sallarés, V., Flueh, E., Charvis, P., & Bialas, J. (2003). Seismic structure of Cocos and Malpelo Volcanic ridges and implications for hot spot-ridge interaction. *Journal of Geophysical Research*, *108*(B12), 2564. https://doi.org/10.1029/2003JB002431.

Satake, K. (1994). Mechanism of the 1992 Nicaragua tsunami earthquake. *Geophysical Research Letters*, *21*(23), 2519–2522.

Schaefer, A., Daniell, J., & Wenzel, F. (2015). State-of-the-Art in Tsunami Risk Modelling for a global perspective. In *EGU General Assembly Conference Abstracts*, Volume 17 of *EGU General Assembly Conference Abstracts*, pp. 5239.

Scholz, C. H., & Small, C. (1997). The effect of seamount subduction on seismic coupling. *Geology*, *25*, 487–490. https://doi.org/10.1130/0091-7613(1997)025<0487:TEOSSO>2.3.CO;2.

Sørensen, M. B., Spada, M., Babeyko, A., Wiemer, S., & Grünthal, G. (2012). Probabilistic tsunami hazard in the Mediterranean Sea. *Journal of Geophysical Research*, *117*(B1), B01305.

Stein, S., Salditch, L., Brooks, E., Spencer, B., & Campbell, M. (2017). Is the Coast Toast? Exploring Cascadia earthquake probabilities. *GSA Today*, *27*, 6–7.

Stepp, J.C. (1972). Analysis of completeness of the earthquake sample in the Puget Sound area and its effect on statistical estimates of earthquake hazard. In *Proc. of the 1st Int. Conf. on Microzonazion, Seattle*, Volume 2, pp. 897–910.

Stirling, M., & Gerstenberger, M. (2010). Ground motion-based testing of seismic hazard models in New Zealand. *Bulletin of the Seismological Society of America, 100*(4), 1407–1414.

Thio, H. K., Somerville, P., & Ichinose, G. (2007). Probabilistic analysis of strong ground motion and tsunami hazards in Southeast Asia. *Journal of Earthquake and Tsunami, 01*(02), 1–19.

USGS (2017). *USGS ComCat catalog*. United States Geological Survey. https://earthquake.usgs.gov/earthquakes/search/. Accessed 20 Jan 2017.

van Stiphout, T., Schorlemmer, D., & Wiemer, S. (2011). The Effect of Uncertainties on Estimates of Background Seismicity Rate. *Bulletin of the Seismological Society of America, 101*(2), 482–494.

VonHuene, R., Ranero, C. R., & Watts, P. (2004). Tsunamigenic slope failure along the Middle America Trench in two tectonic settings. *Marine Geology, 203*, 303–317.

Ward, S. N. (2001). Landslide tsunami. *Journal of Geophysical Research, 106*, 11201–11216.

Ward, S. N., & Asphaug, E. (2000). Asteroid impact tsunami: A probabilistic hazard assessment. *Icarus, 145*, 64–78.

Wessel, P., Smith, W. H. F., Scharroo, R., Luis, J., & Wobbe, F. (2013). Generic mapping tools: Improved version released. *Eos, Transactions American Geophysical Union, 94*, 45.

Wiemer, S., Giardini, D., Fäh, D., Deichmann, N., & Sellami, S. (2008). Probabilistic seismic hazard assessment of Switzerland: Best estimates and uncertainties. *Journal of Seismology, 13*(4), 449–478.

Ye, L., Lay, T., & Kanamori, H. (2013). Large earthquake rupture process variations on the Middle America megathrust. *Earth and Planetary Science Letters, 381*, 147–155. https://doi.org/10.1016/j.epsl.2013.08.042.

Zamora, N. (2016). *Probabilistic tsunami hazard analysis for Central America with focus on uncertainties*. Phd dissertation, Postdam University.

Zamora, N., & Babeyko, A. Y. (2015). Tsunami potential from local seismic sources along the southern Middle America Trench. *Natural Hazards, 80*(2), 901–934. https://doi.org/10.1007/s11069-015-2004-3.

(Received December 31, 2018, revised October 4, 2019, accepted November 14, 2019, Published online November 28, 2019)

Pure Appl. Geophys. 177 (2020), 1497–1520
© 2019 The Author(s)
https://doi.org/10.1007/s00024-019-02260-x

Pure and Applied Geophysics

Effect of Shallow Slip Amplification Uncertainty on Probabilistic Tsunami Hazard Analysis in Subduction Zones: Use of Long-Term Balanced Stochastic Slip Models

A. Scala,[1,2] S. Lorito,[2] F. Romano,[2] S. Murphy,[3] J. Selva,[4] R. Basili,[2] A. Babeyko,[5] A. Herrero,[2] A. Hoechner,[5] F. Løvholt,[6] F. E. Maesano,[2] P. Perfetti,[4] M. M. Tiberti,[2] R. Tonini,[2] M. Volpe,[2] G. Davies,[7] G. Festa,[1] W. Power,[8] A. Piatanesi,[2] and A. Cirella[2]

Abstract—The complexity of coseismic slip distributions influences the tsunami hazard posed by local and, to a certain extent, distant tsunami sources. Large slip concentrated in shallow patches was observed in recent tsunamigenic earthquakes, possibly due to dynamic amplification near the free surface, variable frictional conditions or other factors. We propose a method for incorporating enhanced shallow slip for subduction earthquakes while preventing systematic slip excess at shallow depths over one or more seismic cycles. The method uses the classic k^{-2} stochastic slip distributions, augmented by shallow slip amplification. It is necessary for deep events with lower slip to occur more often than shallow ones with amplified slip to balance the long-term cumulative slip. We evaluate the impact of this approach on tsunami hazard in the central and eastern Mediterranean Sea adopting a realistic 3D geometry for three subduction zones, by using it to model ~ 150,000 earthquakes with M_w from 6.0 to 9.0. We combine earthquake rates, depth-dependent slip distributions, tsunami modeling, and epistemic uncertainty through an ensemble modeling technique. We found that the mean hazard curves obtained with our method show enhanced probabilities for larger inundation heights as compared to the curves derived from depth-independent slip distributions. Our approach is completely general and can be applied to any subduction zone in the world.

Key words: Tsunamis, seismic-probabilistic tsunami hazard assessment, tsunami source models, stochastic seismic slip distributions.

Electronic supplementary material The online version of this article (https://doi.org/10.1007/s00024-019-02260-x) contains supplementary material, which is available to authorized users.

[1] Department of Physics "Ettore Pancini", University of Naples, Naples, Italy. E-mail: scala@fisica.unina.it; antonio.scala@ingv.it
[2] Istituto Nazionale di Geofisica e Vulcanologia, Sezione di Roma 1, Rome, Italy.
[3] Ifremer, Plouzané, France.
[4] Istituto Nazionale di Geofisica e Vulcanologia, Sezione di Bologna, Bologna, Italy.
[5] GFZ, Potsdam, Germany.
[6] NGI, Oslo, Norway.
[7] Geoscience Australia, Canberra, Australia.
[8] GNS Science, Lower Hutt, New Zealand.

1. Introduction

A relatively high rate of great seismic events ($M_w \geq 8.0$) characterized the last two decades. Most of these events occurred along subduction zones and triggered some of the strongest ever-recorded tsunamis (e.g., 2004 M_w 9.2 Sumatra–Andaman, 2010 M_w 8.8 Maule, and 2011 M_w 9.1 Tohoku). Some of these great earthquakes revealed unprecedented rupture features, for example the Tohoku earthquake that produced an unexpectedly large amount of slip (~ 50 m) just at the trench, resulting in a huge tsunami (e.g., Romano et al. 2014; Lorito et al. 2016; Lay 2018). Before this earthquake, it was commonly stated that the accretionary sedimentary wedges could not accumulate sufficient strain to produce a large co-seismic slip (e.g., Hyndman et al. 1997; Moore and Saffer 2001). Even smaller events, such as the 2010 M_w 7.8 Mentawai earthquake (classified as a tsunami earthquake, e.g., Yue et al. 2014), produced larger than expected tsunami waves due to relatively large slip at shallow depths. It was also observed that shallow subduction earthquakes tend to have a longer normalized source duration than deeper ones, which was explained by inverse dependence of the rigidity and/or stress drop with depth (Bilek and Lay 1999; Geist and Bilek 2001). Moreover, the depth-dependent frequency radiation recorded during great earthquakes (Wang and Mori 2011; Lay et al. 2012), featuring generally higher-frequency seismic radiation zones at depth, has been interpreted in the framework of geometrical and structural segmentation of the slab and variation of thermal properties with depth (Satriano et al. 2014). Some numerically simulated dynamic effects may indeed favor the up-dip propagation of subduction rupture. These include the bi-material effect (Rubin and Ampuero 2007; Ma and Beroza 2008; Scala et al. 2017), the interaction with radiation reflected and/or converted from the free surface (Nielsen 1998; Lotto et al. 2017; Scala et al. 2019), and the variability of frictional properties (Kozdon and Dunham 2013; Murphy et al. 2018).

All the above observations and their interpretation are relevant for tsunami hazard, as it is well-established that tsunamis and tsunami hazard are sensitive to slip complexity, including shallow slip features, not only in the near-field of the source, but also at regional distances (e.g., Geist 2002; McCloskey et al. 2007; González et al. 2009; Li et al. 2016).

Several strategies have been proposed to produce heterogeneous slip distributions for tsunami hazard calculations. The stochastic slip distributions are either computed from pre-defined statistical distributions (LeVeque et al. 2016; Sepúlveda et al. 2017) or constrained by models of real earthquakes (Mori et al. 2017). These strategies allow for the definition of large slip distribution ensembles to explore the slip distribution uncertainty. This variability in the source models can be propagated in the computation of tsunami (inundation) scenarios and hazard assessment.

Nevertheless, to our knowledge, a shallow slip amplification has not yet been included in tsunami hazard models (Grezio et al. 2017). Some techniques which use sets of data-driven slip distributions (Davies and Griffin 2018; Goda et al. 2014) or pseudo-dynamic kinematic seismic source descriptions (Song and Somerville 2010; Song et al. 2013) may intrinsically address the issue since they are based on seismic or tsunami observations. Moreover, dynamically modified k^{-2} kinematic source models have been proposed to account for shallow slip amplification and provided enhanced tsunamigenic potential (Murphy et al. 2016). While accounting for enhanced shallow slip, these approaches do not deal with the frequency of occurrence of shallow slip amplification. The enhanced shallow slip should be imposed in a manner such that the time-integrated rate of slip across the whole fault plane reflects the convergence rate and plate coupling over long time periods, that is, among other things, avoiding an unrealistic and unjustified slip accumulation at shallow depths (e.g., Wang and Dixon 2004).

In this work, we propose a methodology to produce sets of kinematic k^{-2} slip distributions considering slip amplification at shallow depths on the plate interface, while re-balancing the spatial distribution of the consequent long-term slip. For individual earthquakes, we consider systematically enhanced shallow slip. That is, the slip amplitude increases as the rigidity decreases, reaching its maximum at relatively shallow depths, apart from a less strongly coupled zone at the very shallowest part. However, since the long-term spatial slip rate needs to be compatible with the convergence rate and coupling along the interface (e.g., Nalbant et al. 2013), the enhanced slip for a single shallow earthquake should be balanced by an overall lower rate of occurrence for such events. In this way, the earthquakes simulated by this model produce reasonably uniform (hence realistic) long-term slip accumulation, enabling their incorporation in long-term hazard models.

We address the importance of such a model through a long-term seismic—probabilistic tsunami hazard assessment (S-PTHA; hazard for tsunamis generated by earthquakes only). The sensitivity of S-PTHA is demonstrated by using either our method featuring enhanced shallow slip, or the usual depth-independent heterogeneous slip distributions. To perform this sensitivity analysis we use the Mediterranean basin as a case study, this basin contains several subduction structures and has hosted several destructive tsunamis in the past (e.g., the 365 Crete tsunami). We employ a simplified version of the TSUMAPS-NEAM (http://www.tsumaps-neam.eu/) seismicity model. The model is simplified since we consider only subduction events while neglecting all crustal seismicity. In this area, the hazard due to crustal seismicity represents an important component of the total S-PTHA (Sørensen et al. 2012; Selva et al. 2016). One should eventually consider also hazard from non-seismic sources (e.g., Tonini et al. 2011; Grezio et al. 2012, 2015; Urlaub et al. 2018; Paris et al. 2019). As a further simplification in comparison to the TSUMAPS-NEAM model, we consider a narrower range of alternative models to describe the epistemic uncertainty, both as far as the seismicity rates and the inundation models are concerned.

In the following sections, we first introduce our assumptions on the depth-dependence of rigidity and seismic coupling. Then, we synthetically describe the subduction zones in the Mediterranean, which are needed to illustrate the procedure for constructing the seismic slip distributions. Finally, we conclude by explaining the different steps involved in the long-term S-PTHA case study in the Mediterranean Sea.

2. Method

2.1. Seismic Moment and Rigidity/Coupling Variation with Depth

The seismic moment of an earthquake can be defined as:

$$M_0 = \int_A \mu \cdot \delta \cdot dS = \langle \mu \cdot \delta \rangle \cdot A, \qquad (1)$$

where δ is the slip amplitude and μ the local rigidity over fault area A, whereas $\langle \mu \cdot \delta \rangle$ is the spatial average of slip times rigidity.

When the continuous quantity defined by Eq. (1) is modeled on a discretized mesh, the seismic moment can be approximated by the following summation of discrete quantities:

$$M_0 = \sum_{n=1}^{N} \mu_n \delta_n A_n, \qquad (2)$$

where N is the number of mesh cells, whereas μ_n and A_n represent the local rigidity and the surface size of the nth cell, respectively.

For a given value of the seismic moment M_0, the rigidity variations at different locations on the fault constrain the slip value at the same location; for example, the slip is larger in (shallow) low-rigidity areas. We here assume that the rigidity only varies with depth, while we neglect lateral (i.e., along-strike) variations.

Assuming constant stress drop, to fit the observation that the normalized source duration generally decreases with depth (Bilek and Lay 1999), the rigidity has been proposed to exponentially vary with depth as well, as shown in Fig. 1a, according to:

$$\mu(z) = 10^{a+bz}, \qquad (3)$$

where μ is rigidity in *GPa*, z is depth in *km*, and from the regression, the parameters a and b take the values of 0.5631 and 0.0437, respectively. A rigidity-depth dependence is also featured by the preliminary reference earth model (PREM, Dziewonski and Anderson 1981; Fig. 1a). However, as pointed out by Geist and Bilek (2001), to match observed tsunami amplitudes generated by several tsunami earthquakes (M_w 7.7, 1992 Nicaragua earthquake, Satake 1995; M_w 7.5, 1996 Peru earthquake, Heinrich et al. 1998; Tanioka and Satake 1996), the average rigidity of the shallow portion of the subduction zone needs to be somewhat higher than in the Bilek and Lay (1999) end-member case, yet still lower than implied by the PREM. In our analysis, we adopt an intermediate rigidity profile (Fig. 1a), computed as the average between the two aforementioned profiles.

To evaluate the long-term slip accumulation (in multiple seismic cycles), slip on the subduction faults should also account for the aseismic contribution produced by creeping and other non-tsunamigenic, non-radiative episodic slip events (e.g., Scholz 1998), including the contribution of after slip and viscoelastic relaxation phenomena (e.g., Sun and Wang 2015). The balance between the seismic and aseismic slip can be described as controlled by the seismic coupling factor, that is the seismic fraction of total plate convergence in a given time period, from now on called "coupling" for simplicity. The total long-term slip D_T released at the position x can be modelled as:

$$D_T = \delta_T(\boldsymbol{x}) + d_T(\boldsymbol{x}) = k(\boldsymbol{x}) \cdot D_T + [1 - k(\boldsymbol{x})] \cdot D_T, \qquad (4)$$

where $\delta_T(\boldsymbol{x})$ and $d_T(\boldsymbol{x})$ are the seismic and aseismic contributions to the slip at $\boldsymbol{x}$, respectively, and $k(x) = \tilde{K} \cdot K(x)$ is the coupling factor, modelled as the product between $\tilde{K}$, the average absolute coupling value of the specific subduction zone, and $K(\boldsymbol{x})$, the relative coupling variation at position $\boldsymbol{x}$. Whatever arbitrary choice we make about the coupling variation, it affects both the slip distribution of single events and the expected cumulative long-term slip. In this work, as a first approximation, a variation of the relative coupling with depth has been imposed: $K(z)$ is assumed to be a constant function along most of the

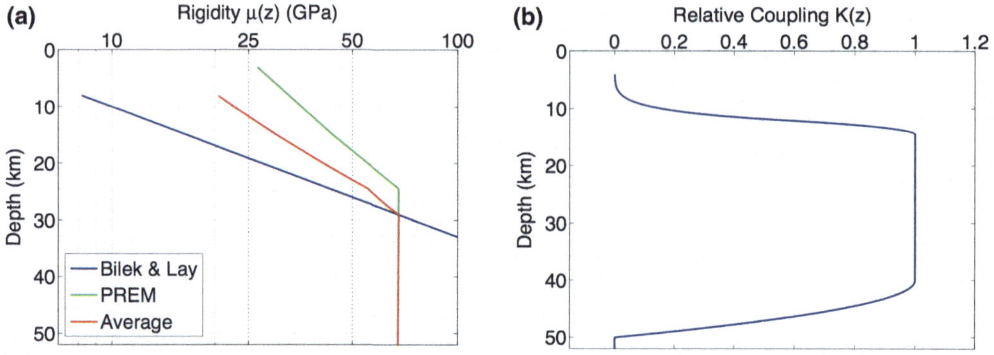

Figure 1

a Rigidity profiles as a function of the depth computed from the end-member case of Bilek and Lay (1999) (blue curve) and interpolating the PREM (Dziewonski and Anderson 1981) values (green line). The average between the two profiles is shown in red. **b** A priori hypothesis of relative coupling as a function of depth. We impose a homogeneous coupling decreasing toward the upper and deeper limits of the seismogenic depth interval. Symbols in the diagram titles for rigidity and relative coupling are the same as in Eq. (3) and in the description of the Eq. (4), respectively

subduction interface, while it monotonically decreases towards the shallowest and deepest boundaries of the seismogenic zone. The decrease at depth is modelled starting from 40 km depth (Tichelaar and Ruff 1993), possibly corresponding to the 350° isotherm, which may be among the factors controlling seismicity cut-off at depth (e.g., Hyndman 2013). Note, however, that the exact down-dip seismogenic limit is relatively unimportant as far as tsunamigenesis is concerned. At shallow depths, the decrease of the coupling occurs where the average dip angle decreases below 10°, generating an almost flat sedimentary wedge (Polonia et al. 2011; Gallais et al. 2012; Maesano et al. 2017), which makes it difficult to define an exact trench location in the specific tectonic context used here as a case study, as discussed in the next section.

This depth-dependent coupling function produces a maximum coupling up to the shallowest seismogenic portion of the subduction, possibly allowing for shallow slip amplification. Figure 1b illustrates the function $K(z)$ used in this work, with which we mimic a gradual coupling decrease toward the boundaries of the seismogenic zone.

2.2. *Subduction in the Mediterranean*

The case study focuses on three subduction zones within the Mediterranean basin, namely the Calabrian Arc, the Hellenic Arc, and the Cyprus Arc (Fig. 2).

The Calabrian Arc seismogenic and tsunamigenic potential is much debated; some authors suggested that the Calabrian Arc hosted the tsunamigenic earthquakes of 1693, M7.3, or of 1905, M7.0. For the source of the tsunami associated to the 1693 earthquake, however, there is no consensus on whether it was caused by the subduction interface, by a crustal fault, or by a landslide (Piatanesi and Tinti 1998; Tinti et al. 2001; Gutscher et al. 2006; Gerardi et al. 2008; Argnani et al. 2012). The Hellenic Arc subduction is considered by some authors as the locus of at least two M8 + earthquakes in 365 and 1303, which generated destructive tsunamis (Guidoboni et al. 1994; Papazachos and Papazachou 1997; Tinti et al. 2005; Lorito et al. 2008; Papadimitriou and Karakostas 2008; Ganas and Parsons 2009; De Martini et al. 2010; Maramai et al. 2014). Some others claim that these events were generated on shallower crustal faults embedded in the accretionary wedge (Shaw et al. 2008; Stiros 2010; England et al. 2015).

Available estimates of the convergence rates in these subduction zones are very variable, and we recall just some of them here. The Calabrian Arc convergence rate is estimated at 5 mm/year by Devoti et al. (2008) and recently at 1.5–1.6 mm/year if creeping, or 2.7 – 3.0 if temporarily locked by Carafa et al. (2018). The Hellenic Arc convergence rate is estimated at 23 mm/year in the Ionian Islands (Hollenstein et al. 2008), 35 mm/year in the western

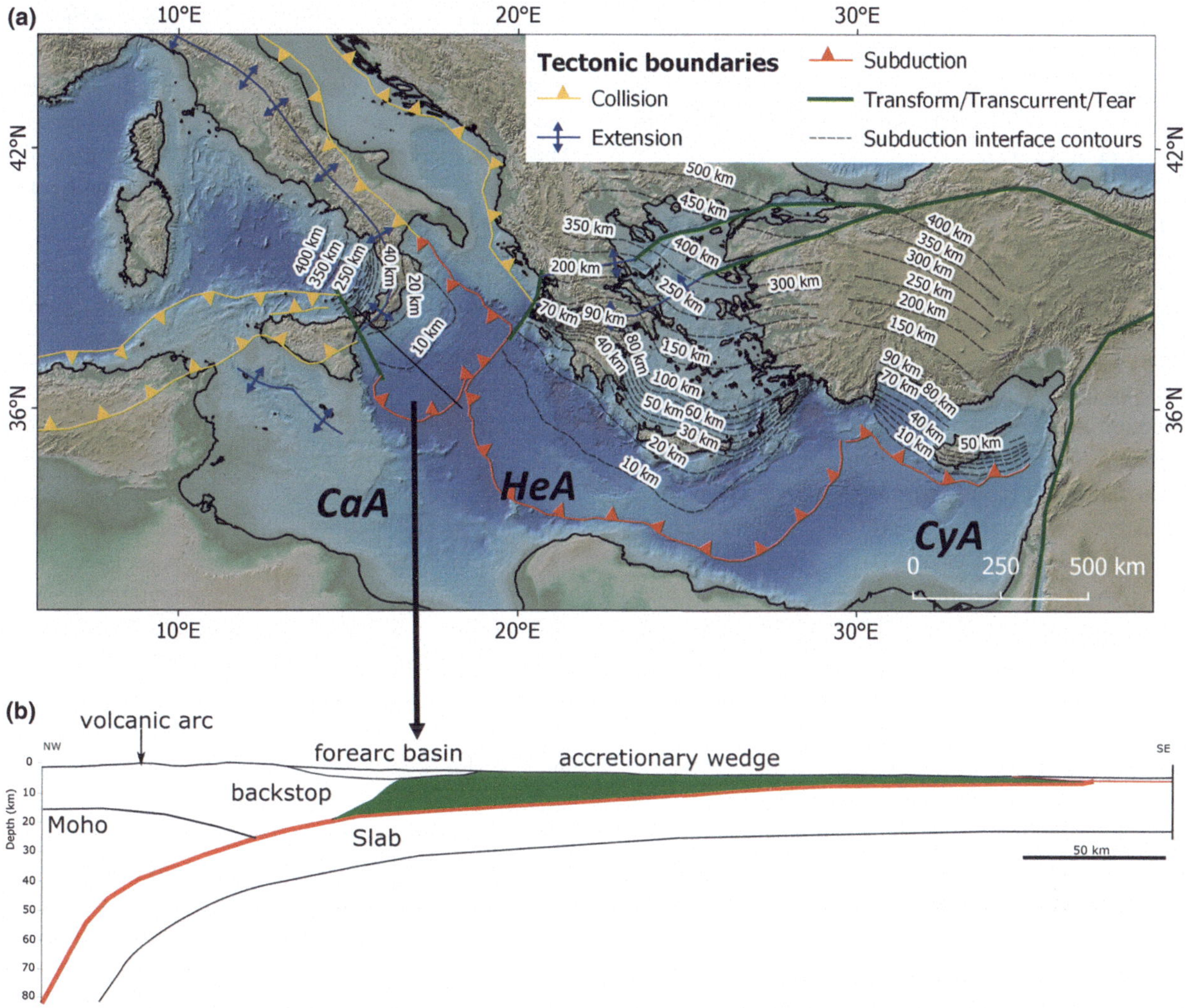

Figure 2

a Tectonic sketch of the Eastern Mediterranean region (CaA, Calabrian Arc; HeA, Hellenic Arc; CyA, Cyprus Arc). **b** Schematic profile across the Calabrian subduction zone. Note the wide and shallow subduction interface covered by the accretionary prism

part (Reilinger et al. 2006; Noquet 2012), and 10 mm/year in the eastern part (Reilinger et al. 2006). The Cyprus Arc convergence rate is estimated at 18 mm/year (Reilinger et al. 2006), 8–9 mm/year (Wdowinski et al. 2006), 12 mm/year (Howell et al. 2017) in the western part, and at 7–8 mm/year (Wdowinski et al. 2006), 5–8 mm/year (Noquet 2012) in the eastern part.

The values of the coupling coefficients in these three subduction zones are also highly debated. In the Calabrian Arc two competing interpretations—ranging from partially-locked to unlocked or inactive—were recently proposed (Carafa et al. 2018; Nijholt et al. 2018). In the Hellenic Arc interpretations range from full locking (Ganas and Parsons 2009) to low coupling (Shaw and Jackson 2010; Vernant et al. 2014), but also the presence of important along-strike coupling variations was proposed (Laigle et al. 2004).

Rates used afterward for S-PTHA are inherited from the TSUMAPS-NEAM Project. We refer to the Project documentation for further details. We just list for completeness here that, depending on the alternative TSUMAPS-NEAM modelling strategy, either the SHEEC-EMEC seismic catalog (Stucchi et al.

2013; Grünthal and Wahlström 2012), whose time span ranges from 1000 to 2006, was used; or, the b-values, convergence and coupling data from GEM Faulted Earth (Christophersen et al. 2015) and from Davies et al. (2017).

The geometry of the three slabs was initially derived from the European Database of Seismogenic Faults (EDSF; Basili et al. 2013) and then modified according to newer data where available. In particular, the Calabrian Arc is replaced by the more recent model by Maesano et al. (2017) which is derived from the interpretation of a dense network of seismic reflection profiles integrated with the analysis of the seismicity distribution with depth. The Hellenic Arc is the same as that in the EDSF, but we verified its consistency with recent works by Sodoudi et al. (2015) and Sachpazi et al. (2016). The Cyprus Arc was slightly modified in consideration of the results of recent works (Bakirci et al. 2012; Salaün et al. 2012; Howell et al. 2017; Sellier et al. 2013a, b) that are based on seismic reflection profiles and tomographic and seismological data and constrain the geometry of the western part of the slab.

According to the classification by Clift and Vannucchi (2004), these three subduction systems are of the accretionary type (e.g., Barbados, Nankai, or Makran), as opposed to erosional type (e.g., Mexico, Tonga, or Kermadec). Due to their relatively slow convergence rate, old age, and presence of a thick sedimentary cover onto the lower plate, they show huge accretionary wedges and no clear evidence of a trench, as it is filled with the accreted deformed sediments. In this configuration, there is a large portion of the shallow part of the lower plate in contact with highly-heterogeneous rocks along a low-angle interface, where one can expect large earthquake ruptures to propagate (e.g., Lallemand et al. 1994; Gutscher and Westbrook 2009). Such configuration is sometimes also considered to favor the occurrence of very shallow slip (e.g., Bilek and Lay 2018, and references therein). Due to the large extension of the accretionary prisms and the gentle dip of the interface (see Fig. 2b) our depth-dependent coupling profile determines that high-amplitude coseismic slip is released just below the sea bottom.

Starting from these subduction geometries, we built 3D triangular meshes with $\sim$ 15-km element size for all of them, using the Cubit mesh generator (http://cubit.sandia.gov). These geometries and their modeling should be compared to the Slab2 model that was very recently published by Hayes et al. (2018). The three meshes are illustrated in Fig. ESM1 in the Electronic Supplementary Material highlighting the depth variation over the modeled interfaces.

2.3. *Rupture Areas and Stochastic Slip Distributions*

We model tsunamigenic earthquakes in a range from $M_w = 6.0$ to $M_w = 9.0$ subdivided into 18 bins. Some previous sensitivity analysis showed that the lower limit of this range might still have a significant effect on the tsunami hazard depending on some particular local conditions (Selva et al. 2016).

We used the TSUMAPS-NEAM parameterization (http://www.tsumaps-neam.eu/documentation, see also Lorito et al. 2015; Selva et al. 2016) and a similar approach for hazard computation, though we considered fewer epistemic alternatives. Here, for example, we consider only one seismogenic depth interval for each subduction zone, and only one earthquake scaling relation (Strasser et al. 2010). The surface extension A is a priori constrained to the expected value of the respective magnitude-size scaling empirical relation, neglecting any predictive uncertainty regarding the fault length and width. Note that the surface extension is assumed independent of the average depth of the scenario, and therefore compatible with the hypothesis of uniform stress drop (Bilek and Lay 1999).

For each modeled magnitude and each possible position in the seismogenic domain of the mesh, a ruptured surface is built by starting from a geometrical center and then by iteratively adding more and more neighbor mesh cells. This procedure is arrested when the selected area exceeds the expected value from the selected scaling relation. Duplicated surfaces may be generated and they are considered only once. The surfaces whose centroid is farther than $0.1 \cdot \sqrt{A}$ from the initial cell centroid are discarded. This selection inevitably leads to less numerous surface sets as the earthquake magnitude increases. No constraint is imposed on the length L (distance along strike) and width W (distance along dip), and thus the set of ruptures explores a wide range of aspect ratios

L/W. It is generally observed that for most of the selected areas $L > W$ due to the larger extension along strike of the subduction zones.

We verified that, for each magnitude bin, the selected set of surfaces covers the irregular seismogenic surface rather homogeneously, apart from some tapering towards the edges of the seismogenic domain. Figures ESM2(a-b-c) in the Electronic Supplementary Material show the number of events generating slip within each cell of the Calabrian Arc mesh at three magnitude bins.

Within each of the identified rupture areas, five slip distributions are computed to explore the earthquake aleatory variability using a stochastic composite source model (Zheng et al. 1994; Ruiz et al. 2011). This model is based on the random spatial distribution of overlapping circular dislocations of different sizes over the pre-defined slipping surface. These individual dislocations will henceforth be referred to as "sub-asperities". The number of sub-asperities for a given size is defined to ensure that the slip spectral amplitude decays as k^2 (where k represents the radial wavenumber). The number of asperities of a given size is given by a power law relationship such that the cumulative distribution of sub-asperities against radius is:

$$N(r > R) = pR^{-2}, \tag{5}$$

where p is a fractal dimension constrained by the imposed seismic moment and stress drop (Zheng et al. 1994). In Eq. (5) the fractal dimension is 2 ensuring the k^{-2} decay of the slip spectral amplitude. In our model, the sub-asperities have radii ranging between $R_{min} \sim 5\Delta x$ and $R_{max} \sim 0.35W$, with Δx being the average of the mesh-cell linear sizes and W the width as inferred from the selected scaling relation. Each sub-asperity contains an individual slip distribution based on the Eshelby's (1957) circular crack slip function (Ruiz et al. 2011). The distribution of circular sub-asperities over non-planar faults is ensured by the implementation of a multi-lateration scheme that allows for the distance across non-planar surfaces to be accurately calculated (Herrero and Murphy 2018). Once all the sub-asperities have been placed on the fault surface, they are summed together producing a slip distribution that has the expected spectral amplitude k^{-2} decay.

The location of each sub-asperity is randomly chosen according to a Probability Density Function (PDF). In this approach, this PDF is in turn imposed as a combination of two PDFs. The first PDF is depth-independent and is either a Gaussian or a sum of several Gaussian functions. Both the number of Gaussian functions (from 1 to 4) and their centers are randomly drawn from a uniform distribution. The Gaussian function(s) provide a slight focusing of the slip, allowing exploration of the variability of the size and slip amplitude of the main sub-asperity. The second PDF is based on the distribution of rigidity and coupling with depth. The role of this second PDF is central since it is used to include the shallow slip amplification.

So, to obtain the five slip distributions previously mentioned, the first PDF is calculated five times, for all the rupture areas defined at all available positions for all earthquake magnitudes on each considered subduction zone. The detailed description of the features of the second PDF will be provided in the next sub-section.

2.4. *Slip Weight Function*

The final depth-dependent PDF is built by combining the Gaussian PDF with a Slip Weight Function (SWF), which is a function of rigidity and coupling. The average rigidity profile (Fig. 1a) allows us to define a rigidity value as $\mu_n = \mu(\bar{z}_n)$, where the subscript n refers to the n-th cell and $\bar{z}_n$ represents the average depth of the n-th cell. Similarly, the coupling associated with each cell can be defined as $K_n = K(\bar{z}_n)$. Figure 3 shows the assumed distributions of rigidity μ_n (panel a) and coupling K_{n} (panel b) for the Calabrian Arc. For a single earthquake, it is reasonable to expect the slip to be larger where the rigidity is smaller, and the coupling is larger. Therefore, we defined:

$$SWF_n = C_f \frac{K_n}{\mu_n}, \tag{6}$$

with SWF_n representing the cell-discretized Slip Weight Function and C_f is a normalization factor defined such that $\sum_{n=1}^{N} SWF_n = 1$, where N is the total number of cells on the seismogenic portion of the subduction interface. Once a specific rupture

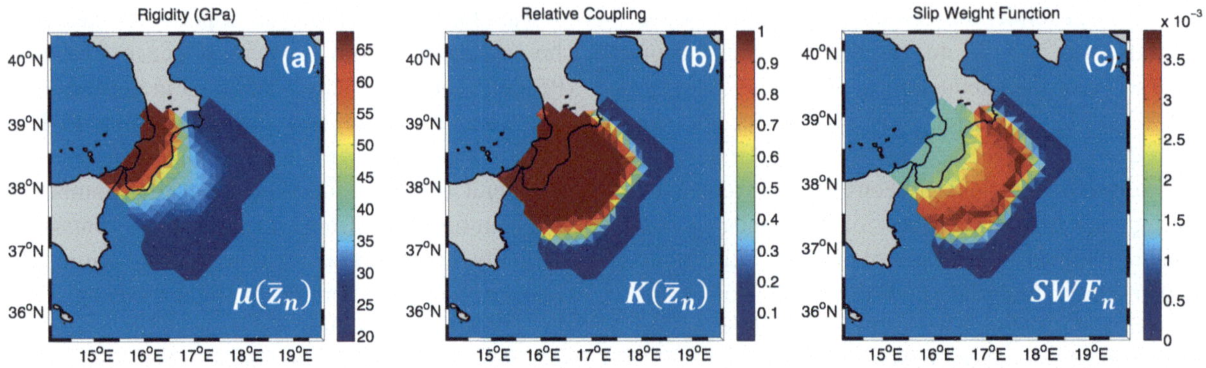

Figure 3

a Rigidity distribution expressed in GPa, **b** relative coupling, **c** slip weight function assumed for the Calabrian Arc. $\mu(\bar{z}_n)$ and $K(\bar{z}_n)$ are functions of the average depth $\bar{z}_n$ of the n-th cell (see the text before Eq. (6) for details)

surface is extracted, the restricted SWF_n is normalized and hence it is the second depth-dependent PDF. The SWF_n for the Calabrian Arc is shown in Fig. 3c.

In Fig. 4, a scheme for the k^{-2} slip distribution computation is presented for a $M_w = 8.6$ event on the Calabrian Arc. Figure 4a is an example of random multiple Gaussian PDF extraction, whose features were described in Sect. 2.2. The left-hand side of Fig. 4 summarizes the steps leading to the definition of one of the slip distributions for the case with depth-dependent rigidity and coupling. Hereafter, we refer to the set of slip distributions generated in this way as the "depth-dependent set". Figure 4b shows the SWF_n defined within the ruptured area. Figure 4c is the normalized product between the random Gaussian PDF (panel a) and the SWF_n (panel b). This PDF is used to modulate the distribution of the sub-asperity centers that represents the phase of the k^{-2} distribution.

For comparison, for each slip distribution in this set, we also compute a corresponding depth-independent k^{-2} slip map by considering uniform rigidity ($\mu = 33$ GPa) and coupling on the fault. In this case, the slip distribution depends only on the Gaussian PDF. Hereafter, we refer to this set of slip distributions generated in this simpler way as the "depth-independent set". The right-hand side of Fig. 4 shows that in this case the sub-asperity location is modulated only by the random Gaussian PDF.

Figure 4e, f show the slip distributions computed starting from the two different schemes. For the depth-dependent set, the effect of the variable SWF_n included in the k^{-2} PDF is to enlarge the shallow high-amplitude patch of slip along the strike direction. Moreover, the smaller value of the shallow rigidity with respect to the reference one (i.e., $\mu = 33$ GPa) contributes to the amplification of the maximum values within the patch. Since homogeneous coupling is imposed for the depth-independent set, it is worth noting that the slip decrease toward the shallower boundary is only due to the tapering effect of the Eshelby's (1957) slip distributions (Fig. 4f).

For lower magnitudes, due to the smaller rupture area in comparison to the mesh size, it is difficult to define the k^{-2} sub-asperities distribution properly. In the configuration presented in this work, it is not possible for $M_w < 8.5$. Hence, for smaller magnitudes, no stochastic selection of slip distribution parameter is performed.

Figure 4 ▶

Sketch of the steps for the definition of the slip distributions. Left-hand column. **a** A random Gaussian function is combined with **b** the SWF_n to define **c** a depth-dependent PDF controlling the location of the sub-asperities over the mesh. Right-hand column. **d** The PDF coincides with the random Gaussian function due to homogeneous rigidity and coupling. **e** Sample slip distribution belonging to the depth-dependent set. **f** Sample slip distribution belonging to the depth-independent set. For the same stochastic slip distribution, the depth-dependent SWF_n leads to a wider along-strike extension of the shallow slip asperity. The absolute maximum slip value is larger in panel **e** due to the smaller rigidity at shallow depths

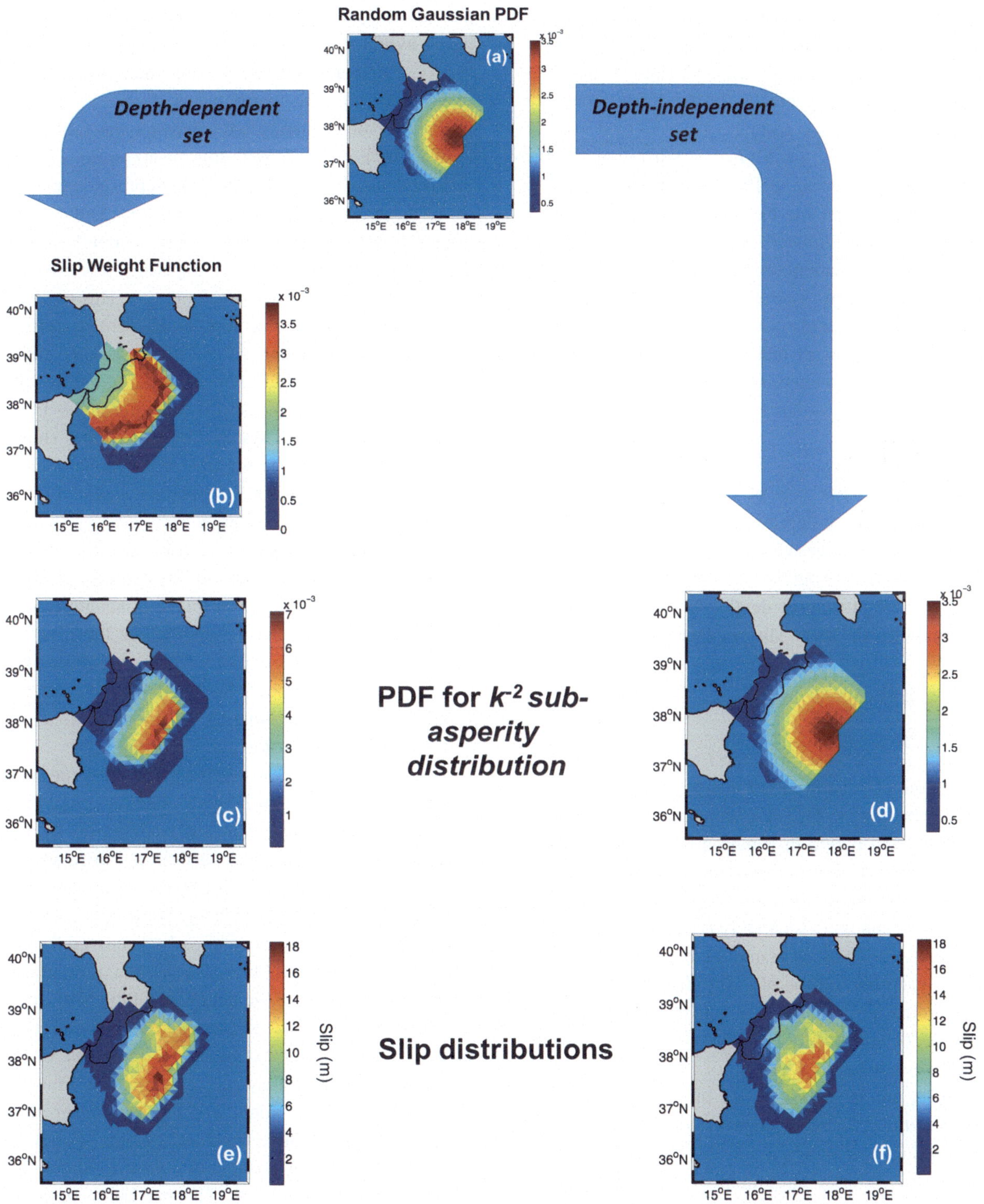
Random Gaussian PDF
Depth-dependent set
Depth-independent set
Slip Weight Function
PDF for k^{-2} sub-asperity distribution
Slip distributions
Slip (m)
(a)
(b)
(c)
(d)
(e)
(f)

Hence, for the depth-independent set, a uniform slip is imposed for the smaller magnitudes as: $\bar{\delta} = M_o/(\mu \cdot A)$, where μ is the uniform rigidity, and A is the rupture area. Conversely, for the depth-dependent set, for $M_w < 8.5$, we compute a normalized seismic moment $\tilde{M}_o = \sum_{n=1}^{\tilde{N}} \mu_n SWF_n A_n$, where $\tilde{N}$ is the number of rupturing cells. Considering the real seismic moment of the event M_0, the slip δ_n within each cell n is estimated as: $\delta_n = \frac{M_o}{\tilde{M}_o} SWF_n$. In the Electronic Supplementary Material, Figs. ESM3(a) and ESM3(b) show examples of a $M_w = 7.5$ earthquake with homogeneous slip distribution for the depth-independent set and a SWF_n-derived distribution for depth-dependent set, respectively.

2.5. Balancing Slip Probability

We defined two sets of slip distributions (either depth-dependent or depth-independent). To check the cumulative slip over the long term, that is to verify whether there is progressively larger unrealistic slip accumulation at shallow depths over multiple events, we computed the mean slip per earthquake $\hat{\delta}_n$, that is:

$$\hat{\delta}_n = \sum_{M_{w\min}}^{M_{w\max}} \sum_{i=1}^{N_{M_w}} \delta_{ni} \cdot P(M_w) \cdot P(Sl_i|M_w) \tag{7}$$

where δ_{ni} is the slip (in meters) generated in the nth cell by the i-th distribution for a given magnitude M_w. The probability $P(M_w)$ is computed from the cumulative tapered Pareto distribution (Kagan et al. 2010, Eq. 2). $P(Sl_i|M_w)$ represents the conditional probability for the slip distribution Sl_i, given the magnitude M_w. The mean $\hat{\delta}_n$ is separately computed for the two sets, considering all the sampled magnitude bins and, within each bin, all the N_{M_w} slip distributions.

Cumulating this mean slip over a large enough number of earthquakes, we obtain, up to a multiplicative constant, a proxy of the slip rate. For example, multiplying $\hat{\delta}_n$ by the product $\lambda \cdot year$, with λ representing the mean annual rate of the considered events (larger than $M_w = 6.0$ in our case) and *year* being an arbitrarily large number of years (e.g., the number of years after which we could expect at least one event at the maximum magnitude), a spatial pattern of the released co-seismic slip over the long term can be computed. Therefore, Eq. (7) must provide a $\hat{\delta}_n$ having a pattern compatible with the a priori hypothesis made on the coupling.

As a first attempt, we assume that, for a given magnitude, all the earthquakes have a uniform probability of occurring anywhere on the fault, that is $P(Sl_i|M_w) = \frac{1}{N_{M_w}} \forall i$. With this ansatz in Eq. (7), the resulting long-term slip from the depth-independent set turns out to be approximately uniform and $\hat{\delta}_n$ does not show any particular zone of slip accumulation on the subduction interface. A tapering towards the edges of the seismogenic zone emerges. This tapering is due both to the smaller number of events rupturing close to the boundaries as compared to those rupturing in the middle of the fault, and to the intrinsic tapering of the slip distribution. The Electronic Supplementary Material (Fig. ESM4), shows $\hat{\delta}_n$, normalized over the multiple n locations considered, from the depth-independent set of slip distributions, for the Calabrian Arc.

When the same ansatz is used for the depth-dependent distributions set, the systematic shallow slip amplification generates a spatial concentration of accumulated slip around the area where the SWF_n is maximum. This concentration of slip is highlighted in Fig. 5a, where we show the normalized $\hat{\delta}_n$ for the Calabrian Arc. Subdividing the seismogenic area into a series of along-strike sections (e.g., the black rectangle in Fig. 5a) we compute $\bar{\hat{\delta}}_n$ as the mean of the normalized $\hat{\delta}_n$ within each section. Figure 5b shows the variability of $\bar{\hat{\delta}}_n$ as a function of the average rigidity of the strike section, $\bar{\mu}$. For relatively small rigidity values ($\bar{\mu} < 30$ GPa), corresponding to the shallower depths, there is a drop in $\bar{\hat{\delta}}_n(\bar{\mu})$. This is caused by the same reasons already discussed for the depth-independent set, but also by the near-trench decreasing coupling in the definition of the SWF_n (see Eq. (6)). Hence, the shallow coseismic slip is depleted to a certain extent at the locations where the slip is being partially accommodated aseismically in the less coupled zone. For rigidity larger than 30 GPa, a systematic decrease of the $\bar{\hat{\delta}}_n(\bar{\mu})$ is observed that is approximately linear in the semi-logarithmic plot of Fig. 5b. The maximum $\bar{\hat{\delta}}_n(\bar{\mu})$ in Fig. 5b corresponds to the along-strike section highlighted by a black rectangle in Fig. 5a, b.

However, in the zone where a relative coupling $K(z) = 1$ is imposed, the total long-term slip should

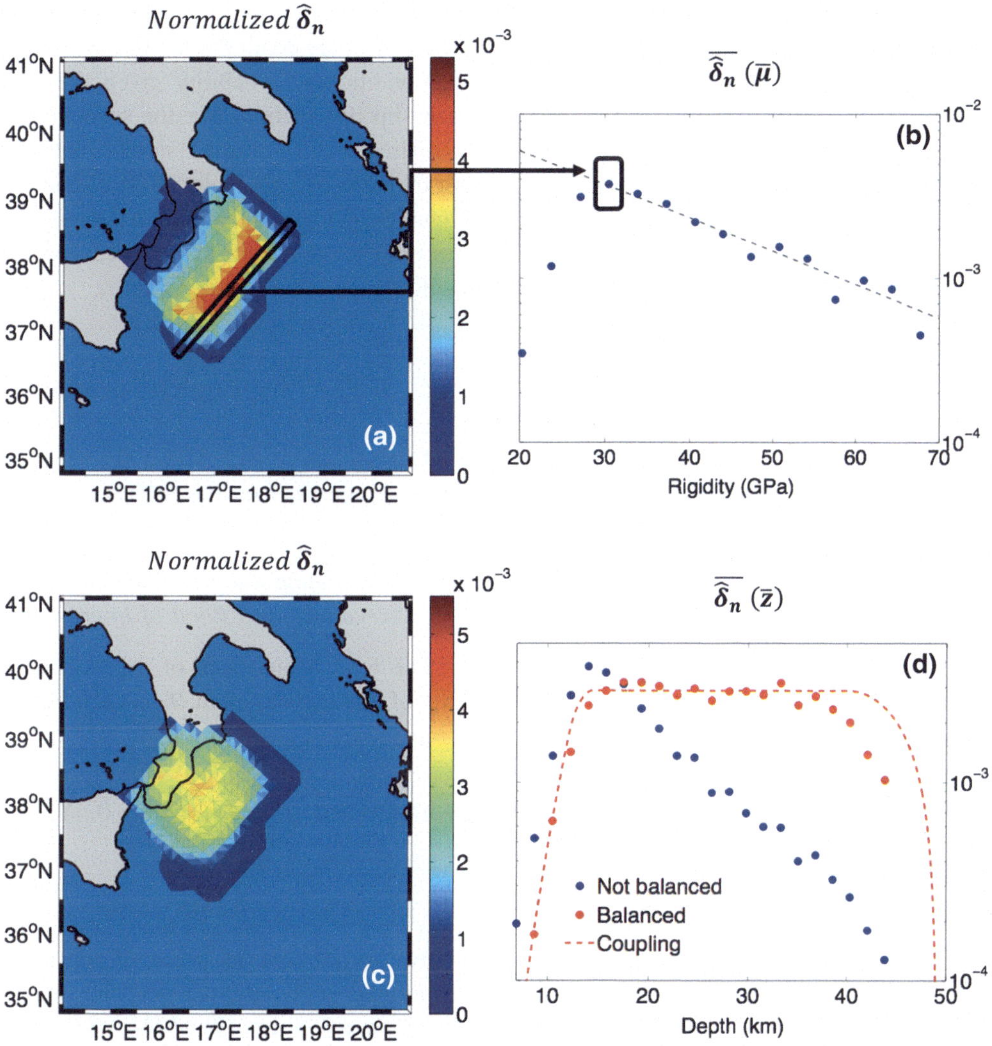

Figure 5

a Normalized stack of the slip computed from Eq. (7), used as an estimate of the total long-term slip. The domain is subdivided into along-strike sections. An example of such section is the black rectangle in panel **(a)**. From each of these along-strike sections the mean of the $\hat{\delta}_n$ is obtained. **b** $\overline{\hat{\delta}_n}(\bar{\mu})$, that is along-strike mean of $\hat{\delta}_n$, plotted versus the average rigidity within each section. Each point refers then to a different along-strike section and the black rectangle in panel **(b)** highlights the value corresponding to the section enclosed by the black rectangle in panel **(a)**. The black dashed line is the best-fit solution from which the parameter β is extracted (see text for details). **c** Here the $\hat{\delta}_n$ is computed when the $P(Sl_i|M_w)$ of Eq. (9) is used: no clear trend against the depth emerges as an effect of the imposed long-term slip balancing. **d** $\overline{\hat{\delta}_n}(\bar{z})$ is plotted for non-balanced (blue dots) and balanced (red dots) long-term stack of the slip. To compare the results with the a priori coupling hypothesis, the function $\frac{K(z)}{\int_0^\infty K(z')dz'}$ is also plotted (red dashed line)

be uniform. In other words, the quantity $\overline{\hat{\delta}_n}$ should track the behavior of the a priori imposed coupling. Instead, what we observe is a decrease in the amount of accumulated slip with depth.

To correct this unwanted feature, the parameter β is extracted from a linear regression $\log_{10}\overline{\hat{\delta}_n}(\bar{\mu}) = \alpha - \beta \cdot \bar{\mu}$. We compute these parameters considering only the points at those depths where $K(z) = 1$ (and $\beta > 0$). The best-fit solution is shown in Fig. 5b. From this regression, we can determine a "correction factor" to our initial ansatz, which is applied to make the mean in the Eq. (12) approximately uniform with depth. From simple geometrical considerations, we can define a horizontal line as: $\log_{10}\overline{\hat{\delta}_N^H}(\bar{\mu}) = \log_{10}\overline{\hat{\delta}_n}(\bar{\mu}) + \beta \cdot \bar{\mu}$. This latter equation can be re-arranged as:

$$\overline{\hat{\delta}_N^H}(\bar{\mu}) = \overline{\hat{\delta}_n}(\bar{\mu}) \cdot 10^{\beta\bar{\mu}} \quad (8)$$

Incorporating this correction into the definition of $P(Sl_i|M_w)$, Eq. (8) can be used to re-normalize the conditional probability of each slip distribution as a function of the average rigidity of the scenario itself:

$$P(Sl_i|M_w) = \frac{\frac{1}{N_{M_w}} \cdot 10^{\beta\bar{\mu}_i}}{\sum_{i=1}^{N(M_w)} \frac{1}{N_{M_w}} \cdot 10^{\beta\bar{\mu}_i}} = \frac{10^{\beta\bar{\mu}_i}}{\sum_{i=1}^{N(M_w)} 10^{\beta\bar{\mu}_i}}, \quad (9)$$

where now $\bar{\mu}_i$ is the average rigidity of the ith scenario defined for the magnitude M_w, and N_{M_w} is still the number of slip distributions defined for that magnitude. It is straightforward to verify that the discrete distribution of Eq. (9) is normalized and can be regarded as a PDF.

We further observe that this procedure generates a new distribution $\bar{\delta}_n(\bar{\mu})$ that under-corrects the decreasing trend with a non-zero parameter β, that is, this trend is still present after the correction. This occurs because the regression (black dashed line in Fig. 5b) is based on the local rigidity value of each cell, whereas the balancing can be performed only on a non-local property, that is the probability of occurrence of a particular slip distribution (computed from the average rigidity of the slip distribution). However, if we estimate the angular coefficients β_1 and β_2 for the first two iterations, we have $\beta_2 < \beta_1$, meaning that the remaining unwanted trend tends to be attenuated. Iterating the procedure and replacing the parameter β in Eq. (9) by $\beta_1 + \beta_2$ we get a new log-linear behavior having more gently steeping $\beta_3 < \beta_2$. Finally, imposing a tolerance, a limited number of iterations m is always found such that $\beta_m \sim 0$. Therefore, replacing β in Eq. (9) by $\sum_{l=1}^{m-1} \beta_l$ balances the mean of the slip defined in Eq. (7). Within the presented scheme, and for all the three subduction-zones, it was verified that after two iterations the value of β is reduced by at least an order of magnitude.

The stack computed with Eqs. (7) and (9) and using $\beta = \sum_{l=1}^{m-1} \beta_l$ is shown in Fig. 5c for the Calabrian Arc. The long-term seismic slip now quite satisfactorily matches the desired coupling. Figure 5d shows the final $\overline{\hat{\delta}_n^H}$ as a function of the along-strike section average depth $\bar{z}$ when Eq. (9) is used to define the conditional probability given the magnitude of each slip distribution. For the sake of clarity, $\overline{\hat{\delta}_n^H}$ is compared with the same quantity plotted in Fig. 5b, but they are now both plotted as a function of depth. Figure 5d evidences that the total balanced long-term slip matches the a priori imposed coupling both at shallower and intermediate depths.

3. *From Slip Distributions to S-PTHA*

This section describes how the balanced slip distributions are used for the S-PTHA.

3.1. *Mean Annual Rates of Tsunami Hazard Intensity Exceedance at a Point of Interest (POI)*

The total annual rate of exceedance of a given level of inundation height H_0 at each POI can be now computed as (e.g., Lorito et al. 2015):

$$\lambda_{POI}(H > H_0) = \sum_{i=1}^{N_e} P_{POI}[(H > H_0|Sl_i)] \cdot \lambda_j \cdot P(M_w) \cdot P(Sl_i|M_w) \quad (10)$$

where $P(Sl_i|M_w)$ is the balanced slip distribution Sl_i conditional probability, given the magnitude M_w, computed as in Eq. (9); $P(M_w)$ is the cumulative tapered Pareto as in Eq. (7); λ_j is the mean annual rate for earthquakes with $M_w \geq 6$, for the j-th subduction zone ($j = 1, 2, 3$); finally, $P_{POI}[(H > H_0|Sl_i)]$ is the conditional probability of exceedance of the tsunami intensity threshold H_0 at a given POI.

The mean annual rates λ_j are inherited from TSUMAPS-NEAM. The TSUMAPS-NEAM model considers epistemic uncertainty and the uncertainty on hazard curves is quantified as an ensemble distribution (Marzocchi et al. 2015; Selva et al. 2016). Here, for the sake of simplicity, we always consider only the mean of the epistemic uncertainty.

$P_{POI}[(H > H_0|Sl_i)]$ is evaluated starting with the computation of each individual slip distribution (see the example in Fig. 6a). The sea-bottom coseismic displacement generated from a slip distribution is computed by using dislocations on triangular subfaults in a homogeneous Poisson's solid half space (Meade 2007). The water column acts as a low-pass filter when the sea-bottom displacement is transferred to the sea-surface. This attenuation is considered by applying a two-dimensional filter of the form $1/\cosh(kH)$, where k is the wavenumber and H the effective height of the water column (Kajiura 1963). The greater importance of the filtering with respect to other approximations, like the linear combination described below, was shown for example by Løvholt et al. (2012). The sea surface displacement obtained in this way from the slip distribution of Fig. 6a is shown in Fig. 6b.

To produce virtual mareograms at the POIs (which lie approximately on the 50 m isobath), the sea surface elevation is used as the initial condition for pre-computed tsunami Green's functions. The Green's functions are the elementary mareograms produced at the POIs by Gaussian-shaped, of $\sim$ 4 km standard deviation ($\sim$ 20 km base width) and spacing $\sim$ 7 km, elementary sea surface elevations (Molinari et al. 2016). These mareograms were simulated with Tsunami-HySEA, which is a non-linear shallow water GPU-optimized and NTHMP benchmarked code (de la Asunción et al. 2013; Macías et al. 2016, 2017). The simulation time was 8 h on a spatial domain enclosing the entire Mediterranean from the Gibraltar Strait (with a small buffer in the Atlantic) to the Eastern Mediterranean including the Aegean and Marmara Seas. The topo-bathymetry employed is SRTM30 + , which has a resolution of 30 arc-seconds ($\sim$ 900 m) and is available at http://topex.ucsd.edu/WWW_html/srtm30_plus.html. The coefficients for linearly combining the mareograms produced by the elementary sources are those allowing the optimal reconstruction of the initial sea level displacement as a linear combination of Gaussian elementary displacements (Fig. 6c, which is the reconstruction of the displacement in Fig. 6b). This reconstruction is based on the potential energy of the displacement field, following Molinari et al. (2016). An example of the virtual mareogram obtained as a linear combination using the coefficients determined in this way is presented in Fig. 6d. The approximations introduced by this

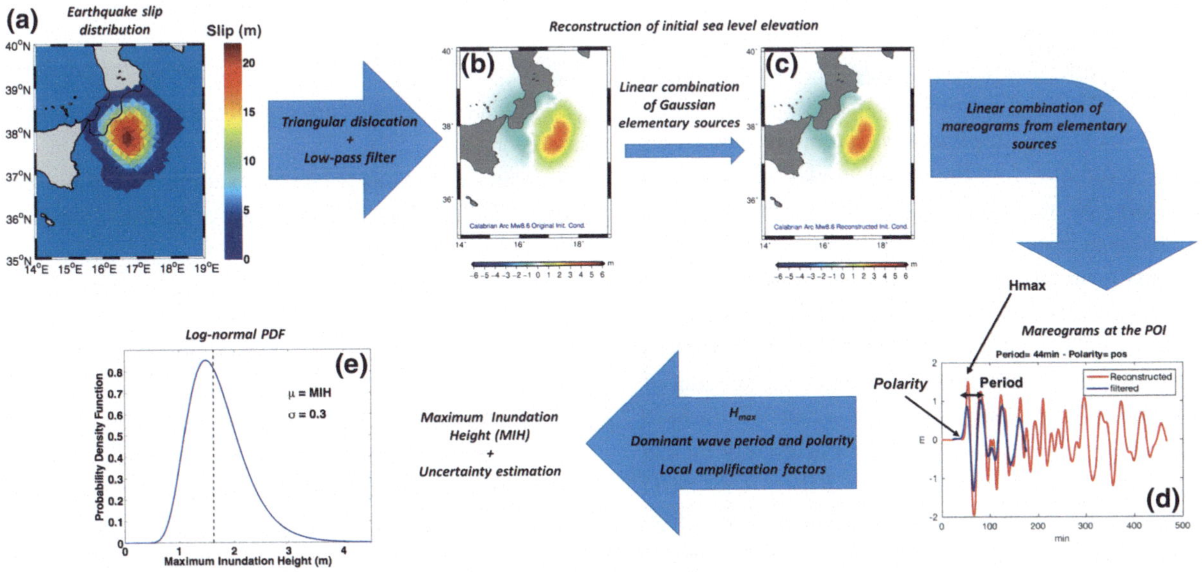

Figure 6

From the slip distributions to the tsunami probability $P_{POI}[(H > H_0|Sl_i)]$. **a** A slip distribution on the Calabrian Arc. **b** The corresponding initial sea level elevation by using the dislocations on triangular subfaults and the low-pass filter. **c** The reconstructed initial sea level elevation from the linear combination of Gaussian elementary sources **d** The synthetic mareogram at the POI. **e** From the analysis of the dominant wave period and polarity and the application of the corresponding local amplification factor, the tsunami log-normal PDF is computed technique were addressed by Molinari et al. (2016); in particular, it was noted that the non-linearity of tsunami propagation, in the framework of the present linear combination scheme, did not introduce a significant bias but just some dispersion of the residuals of the reconstructed mareograms. Noting this is important since unwanted dispersion or non-linearities related to the numerical scheme might be potential issues arising from using small elementary sources for approximating large-scale tsunamis, which the superposition may indeed practically fix (e.g., Baba and Cummins, 2005). Here, we illustrate this approximation for two earthquakes on the Calabrian Arc (the same as in Fig. 6 and Fig. ESM5) and on the Hellenic Arc (Fig. ESM6), by showing the residuals of the reconstructed field, and of several mareograms at different POIs. We also point out that dispersion of the residuals of the linear combinations with respect to directly simulated mareograms (Molinari et al. 2016) is addressed by an error propagation technique, within the Glimsdal et al. (2019) scheme described here below.

Further, and finally, we need to convey simulated offshore tsunami heights into amplified heights at the coastline using approximated amplification factors (Løvholt et al. 2013, 2015; Glimsdal et al. 2019). For a specific point on the coastline, the amplified height acts as a proxy for the Maximum Inundation Height (MIH) on the coast beyond. As discussed by Løvholt et al. (2013), the amplification factor method gives exact estimates for the MIH under the special condition of non-breaking plane waves, but this method also assumes incident plane waves and hence neglects local effects such as focusing and refraction. It was recently shown that the amplification factor applied to the offshore value at the POI provides a good and almost unbiased estimator of the median of the whole MIH distribution for a set of onshore transects over a coastline stretch of a few kilometers behind the POI (Glimsdal et al. 2019). This MIH distribution is generally well approximated by a log-normal distribution (see also Davies et al. 2017). Glimsdal et al. (2019) estimated the amplification factors as a function of the offshore wave period and polarity for the TSUMAPS-NEAM POIs in the Mediterranean. They also took into account the coastal bathymetry around each POI. In this study, we use their amplification factors. To estimate the median MIH throughout the coast behind a POI, we take the product of the offshore maximum height at the POI, as estimated from the virtual mareogram, with the specific local amplification factor from Glimsdal et al. (2019). In doing this, we extract for each scenario the period and the polarity of the leading wave at the POI estimated, as illustrated in the example of Fig. 6d, from the reconstructed mareogram. Further examples of maxima, period and polarity extraction, following Glimsdal et al. (2019), are given in Figs. ESM5 and ESM6 in the Electronic Supplementary Material. Finally, following Davies et al. (2017), for a given slip distribution Sl_i, we compute the conditional probability $P_{POI}[(H > H_0|Sl_i)]$ from a log-normal distribution having the computed MIH as median and a standard deviation $\sigma = 0.3$.

3.2. S-PTHA

In the classical hypothesis of earthquake occurrence as a Poissonian arrival time process, the probability of at least one exceedance of a threshold level of inundation height H_0 over an exposure time T is given by (e.g., Geist and Parsons 2006):

$$p(H > H_0) = 1 - e^{-\lambda(H > H_0) \cdot T}. \tag{11}$$

The hazard curves are computed through Eq. (11) for each POI.

In the following S-PTHA examples for the case-study in the Mediterranean, an exposure time $T = 50$ year is adopted.

4. S-PTHA Sensitivity

In this section, we show a sensitivity analysis performed using the case study which considers three subduction zones in the Mediterranean as potential tsunami sources. In Fig. 7, we show the hazard curves at three POIs each one located nearby one of the subduction zones.

We compare the hazard curves obtained from the depth-independent and the depth-dependent slip distribution sets. In this latter case, we present the tsunami hazard obtained by either using or not using

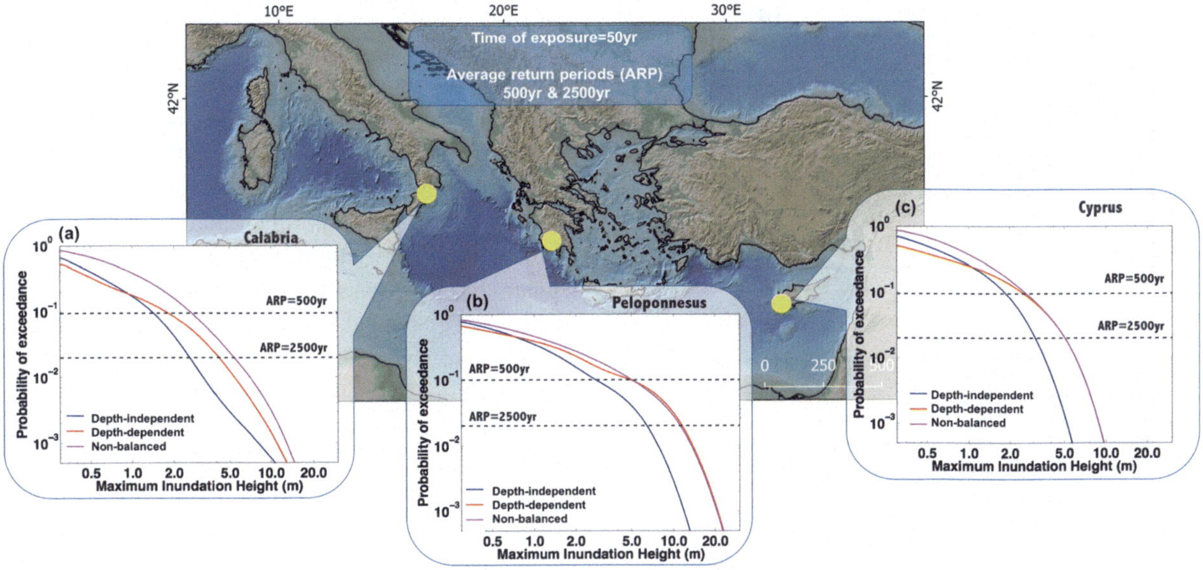

Figure 7
Hazard curves at three reference POIs on **a** the Calabrian coast, close to the Calabrian Arc subduction zone; **b** the Peloponnesus peninsula, close to the Hellenic Arc subduction zone; and **c** the Cyprus Island, close to the Cyprus Arc subduction zone. Blue and red lines are the hazard curves from the depth-independent and the balanced depth-dependent slip distributions, respectively. Magenta lines depict the hazard curves from unbalanced depth-dependent slip distributions that are obtained without imposing a spatially uniform slip rate

the balancing of the total long-term slip presented in Sect. 2.5. The probabilities of exceedance corresponding to the average return periods (ARP) of 500 year ($\sim 10\% \in 50yr$) and 2500 year ($\sim 2\% \in 50yr$) are also highlighted.

We observe a repeating pattern in the hazard curves for the three models. This similarity at the three locations could perhaps have been expected since each of the three sites is relatively close and landward of the neighboring subduction zone.

Our model from the balanced depth-dependent set, as compared to the classically-used depth-independent case, features a lower probability of smaller intensities, to which a decreased probability of occurrence of the shallow lower magnitude events may contribute, and exhibits a larger probability for higher intensities, likely due to the shallow slip amplification associated with the largest events rupturing almost everywhere over the subduction interface. The cross-over point between the two hazard curves slightly oscillates between MIH = 0.75–1.2 m; it also always occurs for ARPs shorter than 500 year.

It is also worth noting that the unbalanced model would overestimate the tsunami hazard at all the evaluated ARPs for the Calabrian site, or at least up to the 500 year ARP in the other cases, due to the accumulated shallow slip excess. The balanced and unbalanced models tend to provide more and more similar results for longer ARPs/larger intensities. Again, this is likely because larger magnitude events, producing larger slip and overall larger tsunamis, also feature wider ruptures along-dip, on which the balancing effect is less pronounced.

The S-PTHA sensitivity can also be illustrated by directly comparing the tsunami hazard maps. Tsunami hazard maps are obtained from hazard curves by plotting on a map view the MIH values corresponding to a fixed probability/ARP level. Defining MIH_{D-D} and MIH_{D-I} as the balanced depth-dependent and depth-independent MIH at each POI corresponding to a given ARP, respectively, we show in Fig. 8 the difference $MIH_{D-D} - MIH_{D-I}$ for the two ARPs of 500 year (Fig. 8a) and 2500 year (Fig. 8b) mentioned above. For the first case, on the south-west coasts of the Peloponnesus, on the western coasts of Libya and Cyprus, and on the Ionian coast

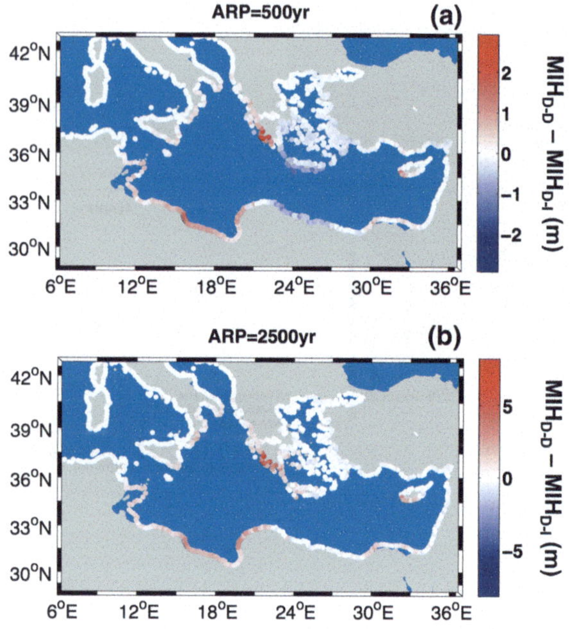

Figure 8
Difference $MIH_{D-D} - MIH_{D-I}$ for two different average return periods (ARPs). Panel (**a**) ARP = 500 year corresponding to MIH having $\sim 10\%$ to be overcome in 50 year. Panel (**b**) ARP = 2500 year corresponding to MIH having $\sim 2\%$ to be overcome in 50 year

of Calabria, the depth-dependent case provides larger MIH estimates compared to the depth-independent case. Elsewhere, we found $\mathrm{MIH_{D-D}} < \mathrm{MIH_{D-I}}$. However, as the ARP increases (corresponding to smaller probability of exceedance and larger expected maximum inundation heights), the balanced depth-dependent set tends to provide larger MIH estimates over all the coastlines in the vicinity of the subduction zones, as well as relatively far and perpendicularly to the source (ideally along the main tsunami energy propagation direction). As an example, for ARP = 2500 year, we found that the depth-independent slip distributions may lead to MIH underestimation from $\sim$ 0.5 to $\sim$ 6 m compared to the depth-dependent distributions case (Fig. 8b).

Two further sensitivity tests were performed to address how the S-PTHA depends on the slip distribution features.

The first one is a sensitivity test to the variation of the rigidity profile. As shown in Sect. 2, the earthquake occurrence probability and the slip distributions are constrained by the choice on the rigidity/coupling profiles. Hence, we repeated the analysis for the two extreme rigidity profiles of Fig. 1a (Bilek and Lay 1999; and the PREM model, Dziewonski and Anderson 1981). We found that, even for these end-member cases, the depth-dependent probability of occurrence (Eq. (9)) ensures estimates of the balanced mean slip per earthquake $\hat{\delta}_n$ similar to the one shown in Fig. 5c. The results of this sensitivity test are summarized in Fig. 9, where the tsunami hazard curves are computed at the same POIs of Fig. 7. For the sake of comparison, the depth-dependent and the depth-independent tsunami hazard curves of Fig. 7 are also plotted in Fig. 9. At all three POIs, the expected overall inverse dependence of the tsunami hazard with the rigidity is obtained. Regardless, these two extreme cases still feature a smaller hazard at lower levels of inundation and a larger hazard for the higher MIH as compared to the depth-independent case.

The second and final sensitivity concerns the minimum magnitude for which the stochastic slip is modelled. To extend stochastic k^{-2} slip distributions to the events with smaller magnitude ($7.9 \leq M_w > 8.6$) we reduced the minimum size R_{min} of the stochastic slip asperities from $5\Delta x$ to Δx (See Sect. 2.3). It is worth stressing that in all the cases, below the imposed limit magnitude, for the depth-dependent distributions a slip value proportional to the SWF_n is assigned to each mesh cell (see Eq. (6) and Sect. 2.4), while above this limit the stochastic slip also allows spatial slip heterogeneity (e.g., to have small concentrated patches of large slip). The results of this sensitivity test are shown in Fig. 10 where the hazard curves of Fig. 7 are again also reported. Compared to the original results, a crossover between the two sets is still present, while an overall increase of the hazard occurs. This means that in a real application and for coastlines in the near-field of the tsunami source, the stochastic slip should be applied as much as possible even at relatively low magnitudes, for which a finer discretization might be necessary.

The last two analyses also evidenced a higher hazard sensitivity at the Cyprus POI with respect to that at the other POIs. This might be due to the shallower minimum seismogenic depth combined with the vicinity of the Cyprus coast to the trench. Such configurations appear to be characterized by a

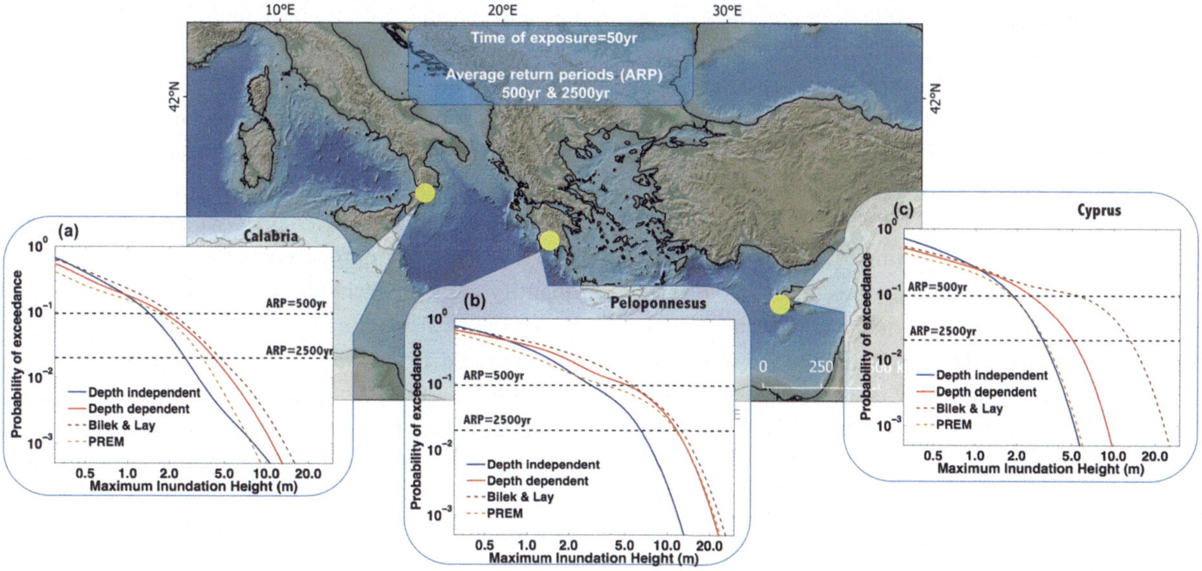

Figure 9
Hazard curves at the same reference POIs of Fig. 7 for the sensitivity test against different rigidity profiles. The hazard curves obtained from Bilek and Lay (brown dashed lines) and PREM (orange dashed lines) rigidity profiles are compared with the "depth-dependent" and the "depth-independent" cases of Fig. 7 (red and blue solid lines respectively)

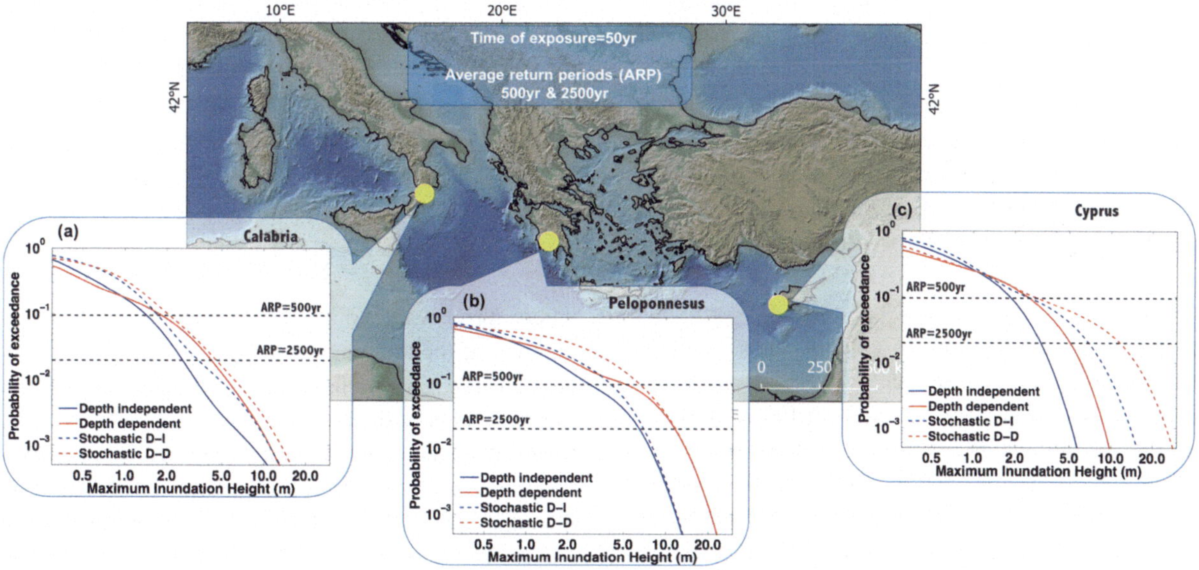

Figure 10
Hazard curves at the same reference POIs of Fig. 7 for the sensitivity test extending stochastic slip distributions down to $M_w = 7.9$. The hazard curves obtained for "depth-dependent" and "depth-independent" case (red and blue dashed lines respectively) are compared with the similar cases already shown in the Fig. 7 (red and blue solid lines respectively)

larger model (epistemic) uncertainty, therefore at least a more detailed description (i.e., a finer discretization) of the seismic scenarios is recommended to partly reduce this uncertainty.

5. Discussion

We proposed a method for S-PTHA that allows for the exploration of the expected natural variability of the seismic slip on a subduction interface to a significant extent. Future possible improvements may include further aspects that are not fully addressed here, which we nevertheless discuss in this section.

For example, here the rupture area was limited quite tightly around the expected (best guess) value from the earthquake scaling relations. Although some along-strike variability was imparted, more systematic sampling of different aspect ratios would be perhaps desirable (e.g., Davies and Griffin 2018). Moreover, we quite subjectively judged the sampling of slip variability (five slip distributions per each rupture area) to sufficiently represents the tsunami hazard of our case study, also given the very dense spatial sampling of the rupture positions, which effectively makes the number of slip samples larger than five. However, this is known to be a challenging issue to manage (e.g., LeVeque et al. 2016; Sepúlveda et al. 2017) and a quantitative hazard convergence testing with respect to the sample dimension could be performed. Note that the proposed approach is in principle suitable for including a larger variability of all the seismic parameters, certainly including rupture size and slip distributions.

Moreover, our model considers only the end-member case of depth-dependent rigidity and uniform stress drop, whereas a more realistic model explaining observed earthquake durations should also include a variable stress drop with depth (Bilek and Lay 1999; Saloor and Okal 2018). In fact, the rigidity values implied by the constant stress drop end-member case are too low to explain some tsunami observations (Geist and Bilek 2001). Hence, we employed higher rigidity values than those of Bilek and Lay (1999), which however slightly underpredict the observed durations. To avoid this inconsistency, in a future update, we might then impart a decrease of the stress drop toward the surface, along with an increase of the rupture length, to compensate for the shorter duration, while preserving the rigidity value at a given depth. This addition would imply a modification of the SWF_n definition, which would then become: $SWF_n = CK_n/\left(\mu_n L_n^2\right)$, with L_n depending on the average depth of the cell. The consequent combined effect of a reduced shallow slip amplification and of a narrower rupture aspect ratio would have a complex impact on the resulting tsunami and on the tsunami hazard in the near field, which certainly deserves further studies.

In our model the shallow slip amplifications are all only roughly represented through the rigidity variability as a proxy for the fault conditions in the broader sense. As a consequence, our model generates systematic larger slip where the rigidity decreases. Other improvements should be oriented to quantify and consider shallow slip amplifications due to geometrical, frictional, and structural features, as they emerge from rupture dynamics modelling (Ma and Beroza 2008; Murphy et al. 2016, 2018; Scala et al. 2017, 2019), in a more thorough way. This would address for which tectonic settings and to what extent the modelled seismogenic zones are expected to feature shallow co-seismic slip amplification.

We employed a simplified 1-D coupling model that is considered an acceptable assumption in the absence of specific local coupling models. However, the procedure can be readily extended, possibly including a lateral variation of the coupling, if such a model becomes available. It is also worth noting that the seismic coupling only appears in the definition of the SWF_n. Its net effect is to reduce the slip amount at very shallow depths for a single event. Alternatively, the coupling could be included in the definition of the conditional probability of occurrence as:

$$P(Sl_i|M_w) = \frac{10^{b\bar{\mu}_i} \cdot K(\bar{z}_i)}{\sum_{i=1}^{N(M_w)}\left[10^{b\bar{\mu}_i} \cdot K(\bar{z}_i)\right]}, \quad (12)$$

where $K(\bar{z}_i)$ is the coupling computed at the average depth $\bar{z}_i$ of the slipping area. The introduction of this term would ensure that an earthquake breaking a limited portion of the less-coupled zone is more likely than an event slipping only within the less-coupled zone. We verified through preliminary tests that this approach only slightly modifies the hazard curves without changing the qualitative comparison between depth-independent and depth-dependent curves of Figs. 7,9 and 10.

The seismic slip distributions here presented are based on the k^{-2} paradigm that is widely used by the seismological community due to its ability to reproduce several macroscale direct observations. However, more efforts are needed to compare our slip distributions with real observations systematically.

Recently, the analysis of Source Time Functions is giving important answers about some macroscopic source properties, such as stress drop, duration, and rupture velocity for subduction events as compared to crustal earthquakes and when different subduction zones are considered (Chounet et al. 2018; Chounet and Vallée 2018). Provided enough resolution, a similar approach could be attempted to look for evidence of macroscopic differences between subduction events occurring at different depths, as already highlighted in terms of moment normalized radiated energy (Newman and Okal 1998; Saloor and Okal 2018). Seismic inversion catalogs (e.g., SRCMOD; USGS and others; Mai and Thingbaijam 2014; Ye et al. 2016) could be used to build "data-driven" SWF_n as a basis for stochastic slip distributions (Mai and Beroza 2002; Goda et al. 2014). Finally, further consistency testing, such as with mareographic and runup tsunami data (Davies and Griffin 2018), would be also desirable.

We also point out that given the, on the average, limited sea depths in the Mediterranean and the relatively short distances and propagation times, the dispersion may be considered perhaps negligible (e.g. Glimsdal et al. 2013). Hence, the results produced by the Tsunami-HySEA code used here may be considered accurate enough. Otherwise, the dispersion would influence the final hazard results, combining with the effects introduced by our depth-dependent model.

6. Conclusions

In this work, we proposed a methodology to define stochastic slip distributions for moderate-to-large magnitude earthquakes in a subduction zone, accounting for possible shallow slip amplification. These sets of events are made compatible with the convergence rate and depth-dependent coupling along the subduction interface. Depth-dependent seismicity features have been already investigated or reviewed as described in several recent papers (see Lay et al. 2012 or discussions in Lay 2018 and references therein). Such features were for example interpreted as controlled by the variability of either geometrical or structural and thermal factors (e.g., Satriano et al. 2014; Bletery et al. 2016) and deserve further investigation.

For illustrative and sensitivity testing purposes, we performed a simplified S-PTHA using the proposed approach for exploring the earthquake slip aleatory variability of three subduction zones in the Mediterranean. The proposed method, however, is completely general and it can thus be applied to any other subduction zone.

The shallow slip amplification is included in the single-event distributions through the definition of a depth-dependent Slip Weight Function, directly proportional to the coupling and inversely proportional to the rigidity variation. The k^{-2} slip distributions thus obtained are characterized by larger patches of higher amplitude slip as compared to the depth-independent case.

To make these single-event slip distributions compatible with the expected long-term slip reflecting convergence rate and the coupling, a depth-dependent probability of occurrence must be defined. Since this probability of occurrence is imposed to increase with increasing average rigidity of the rupture area, it is evident that the largest magnitude seismic events, rupturing almost everywhere on the surface fault, can be considered spatially equiprobable for a fixed magnitude value. Therefore, to balance the systematic slip amplification due to those largest events, the probability of occurrence of relatively smaller events ($6.0 \leq M_w \leq 8.5$) needs to be changed accordingly, with the deeper ones more probable than the shallower ones. By comparing our depth-dependent S-PTHA approach for subduction zone earthquakes to the more common depth-independent one, we found a lower probability for smaller hazard intensities and a higher probability for larger hazard intensities.

Some possible improvements of this approach have been extensively discussed, including consideration of variable stress drop together with more systematic exploration of the earthquake magnitude-size relations, and lateral along-strike seismic coupling variations.

Acknowledgements

The approach proposed here was used in the computation of a probabilistic tsunami hazard map for the North-East Atlantic, Mediterranean and connected seas (NEAM) area in the framework of the European Project TSUMAPS-NEAM (http://www.tsumaps-neam.eu/), co-financed by the European Union Civil Protection Mechanism, Agreement Number: ECHO/SUB/2015/718568/PREV26. The authors acknowledge Tom Parsons and Kenneth Ryan, for the fruitful scientific discussions about the topics of this work. The authors also sincerely thank the Editor, Yuichiro Tanioka, and two anonymous Reviewers for their very useful comments which led to a substantial revision and improvement of this manuscript.

References

Argnani, A., Armigliato, A., Pagnoni, G., Zaniboni, F., Tinti, S., & Bonazzi, C. (2012). Active tectonics along the submarine slope of south-eastern Sicily and the source of the 11 January 1693 earthquake and tsunami. *Natural Hazards and Earth System Sciences, 12,* 1311–1319. https://doi.org/10.5194/nhess-12-1311-2012.

Baba, T., & Cummins, P. R. (2005). Contiguous rupture areas of two Nankai Trough earthquakes revealed by high-resolution tsunami waveform inversion. *Geophysical Research Letters, 32,* L08305. https://doi.org/10.1029/2004GL022320.

Bakırcı, T., Yoshizawa, K., & Özer, M. F. (2012). Three-dimensional S-wave structure of the upper mantle beneath Turkey from surface wave tomography. *Geophysical Journal International, 190*(2), 1058–1076. https://doi.org/10.1111/j.1365-246X.2012.05526.x.

Basili, R., Kastelic, V., Demircioglu, M. B., Garcia Moreno, D., Nemser, E. S., & Petricca, P., et al. (2013). The European Database of Seismogenic Faults (EDSF) compiled in the framework of the Project SHARE. http://diss.rm.ingv.it/share-edsf/. https://doi.org/10.6092/ingv.it-share-edsf.

Bilek, S. L., & Lay, T. (1999). Rigidity variations with depth along interpolate megathrust faults in subduction zones. *Nature, 400,* 443–446. https://doi.org/10.1038/22739.

Bilek, S. L., & Lay, T. (2018). Subduction zone megathrust earthquakes. *Geosphere, 14*(4), 1468–1500. https://doi.org/10.1130/GES01608.1.

Bletery, Q., Thomas, A. M., Rempel, A. W., Karlstrom, L., Sladen, A., & De Barros, L. (2016). Mega-earthquakes rupture flat megathrusts. *Science, 354*(6315), 1027–1031. https://doi.org/10.1126/science.aag0482.

Carafa, M. M. C., Kastelic, V., Bird, P., Maesano, F. E., & Valensise, G. (2018). A "Geodetic Gap" in the Calabrian Arc: Evidence for a locked subduction megathrust? *Geophysical Research Letters, 45*(4), 1794–1804. https://doi.org/10.1002/2017GL076554.

Chounet, A., & Vallée, M. (2018). Global and interregion characterization of subduction interface earthquake derived from source time functions properties. *Journal of Geophysical Reasearch: Solid Earth, 123,* 5831–5852. https://doi.org/10.1029/2018JB015932.

Chounet, A., Vallée, M., Causse, M., & Courboulex, F. (2018). Global catalog of earthquake rupture velocities shows anticorrelation between stress drop and rupture velocity. *Tectonophysics, 733,* 148–158. https://doi.org/10.1016/j.tecto.2017.11.005.

Christophersen, A., Berryman, K., & Litchfield, N. (2015) The GEM Faulted Earth Project, Version 1.0, April 2015, GEM Faulted Earth Project. https://doi.org/10.13117/GEM.GEGD.TR2015.02.

Clift, P., & Vannucchi, P. (2004). Controls on tectonic accretion versus erosion in subduction zones: Implications for the origin and recycling of the continental crust. *Review of Geophysics, 42,* RG2001. https://doi.org/10.1029/2003rg000127.

Davies, G., & Griffin, J. (2018). The 2018 Australian probabilistic tsunami hazard assessment: Hazard from earthquake generated tsunamis. Record 2018/41. Geoscience Australia, Canberra. https://doi.org/10.11636/Record.2018.041.

Davies, G., Griffin, J., Løvholt, F., Glimsdal, S., Harbitz, C., Thio, H. K., et al. (2017). A global probabilistic tsunami hazard assessment from earthquake sources. *Geological Society, London, Special Publications.* https://doi.org/10.1144/SP456.5.

de la Asunción, M., Castro, M. J., Fernández-Nieto, E. D., Mantas, J. M., Ortega, S., & González-Vida, J. M. (2013). Efficient GPU implementation of a two waves TVD-WAF method for the two-dimensional one layer shallow water system on structured meshes. *Computers & Fluids, 80,* 441–452. https://doi.org/10.1016/j.compfluid.2012.01.012.

De Martini, P. M., Barbano, M. S., Smedile, A., Gerardi, F., Pantosti, D., Del Carlo, P., et al. (2010). A unique 4000-year long geological record of multiple tsunami inundations in the Augusta Bay (eastern Sicily, Italy). *Marine Geology, 276,* 42–57. https://doi.org/10.1016/j.margeo.2010.07.005.

Devoti, R., Riguzzi, F., Cuffaro, M., & Doglioni, C. (2008). New GPS constraints on the kinematics of the Apennines subduction. *Earth and Planetary Science Letters, 273,* 163–174. https://doi.org/10.1016/j.epsl.2008.06.031.

Dziewonski, A. M., & Anderson, D. L. (1981). Preliminary reference earth model. *Physics of the Earth and Planetary Interiors, 25*(4), 297–356. https://doi.org/10.1016/0031-9201(81)90046-7.

England, P., Howell, A., Jackson, J., & Synolakis, C. (2015). Palaeotsunamis and tsunami hazards in the Eastern Mediterranean. *Philosophical Transaction of the Royal Society A, 373,* 20140374. https://doi.org/10.1098/rsta.2014.0374.

Eshelby, J. D. (1957). The determination of the elastic field of a ellipsoidal inclusion and related problems. *Proceedings of the Royal Society of London, A241,* 376–396. https://doi.org/10.1098/rspa.1957.0133.

Gallais, F., Gutscher, M.-A., Klaeschen, D., & Graindorge, D. (2012). Two-stage growth of the Calabrian accretionary wedge in the Ionian Sea (Central Mediterraean): Constraints from depth-migrated multichannel seismic data. *Marine Geology, 326–328,* 28–45. https://doi.org/10.1016/j.margeo.2012.08.006.

Ganas, A., & Parsons, T. (2009). Three-dimensional model of Hellenic Arc deformation and origin of the Cretan uplift. *Journal of Geophysical Research, 114,* B06404. https://doi.org/10.1029/2008JB005599.

Geist, E. L. (2002). Complex earthquake rupture and local tsunamis. *Journal of Geophysical Research, 107*(B5), ESE2-1–ESE2-16. https://doi.org/10.1029/2000JB000139.

Geist, E. L., & Bilek, S. L. (2001). Effect of depth-dependent shear modulus on tsunami generation along subduction zones. *Geophysical Research Letters, 28*(7), 1315–1318. https://doi.org/10.1029/2000GL012385.

Geist, E. L., & Parsons, T. (2006). Probabilistic analysis of tsunami hazards. *Natural Hazards, 37*(3), 277–314. https://doi.org/10.1007/s11069-005-4646-z.

Gerardi, F., Barbano, M. S., De Martini, P. M., & Pantosti, D. (2008). Discrimination of tsunami sources (earthquake versus landslide) on the basis of historical data in eastern sicily and Southern Calabria. *Bulletin of the Seismological Society of America, 98,* 2795–2805. https://doi.org/10.1785/0120070192.

Glimsdal, S., Løvholt, F., Harbitz, C., Romano, F., Lorito, S., Orefice, S., et al. (2019). A new approximate method for quantifying tsunami maximum inundation height probability. *Pure and Applied Geophysics, 10,* 1–20. https://doi.org/10.1007/s00024-019-02091-w.

Glimsdal, S., Pedersen, G. K., Harbitz, C. B., & Løvholt, F. (2013). Dispersion of tsunamis: Does it really matter? *Natural Hazards and Earth System Sciences, 13,* 1507–1526. https://doi.org/10.5194/nhess-13-1507-2013.

Goda, K., Mai, P. M., Yasuda, T., & Mori, N. (2014). Sensitivity of tsunami wave profiles and inundation simulations to earthquake slip and fault geometry for the 2011 Tohoku earthquake. *Earth, Planets and Space, 66,* 105. https://doi.org/10.1186/1880-5981-66-105.

González, F. I., Geist, E. L., Jaffe, B., Kânoğlu, U., Mofjeld, H., Synolakis, C. E., et al. (2009). Probabilistic tsunami hazard assessment at seaside, Oregon, for near- and far-field seismic sources. *Journal of Geophysical Research: Oceans, 114,* C11023. https://doi.org/10.1029/2008JC005132.

Grezio, A., Babeyko, A., Baptista, M. A., Behrens, J., Costa, A., Davies, G., et al. (2017). Probabilistic tsunami hazard analysis: multiple sources and global applications. *Review of Geophysics, 55*(4), 1158–1198. https://doi.org/10.1002/2017RG000579.

Grezio, A., Sandri, L., Marzocchi, W., Argnani, A., Gasparini, P., & Selva, J. (2012). Probabilistic tsunami hazard assessment for Messina Strait Area (Sicily, Italy). *Natural Hazards, 64,* 329–358. https://doi.org/10.1007/s11069-012-0246-x.

Grezio, A., Tonini, R., Sandri, L., Pierdominici, S., & Selva, J. (2015). A methodology for a comprehensive probabilistic tsunami hazard assessment: Multiple sources and short-term interactions. *Journal of Marine Science and Engineering, 3,* 23–51. https://doi.org/10.3390/jmse3010023.

Grünthal, G., & Wahlström, R. (2012). The European-Mediterranean earthquake catalogue (EMEC) for the last millennium. *Journal of Seismology, 16,* 535–570. https://doi.org/10.1007/s10950-012-9302-y.

Guidoboni, E., Comastri, A., & Traina, G. (1994). *Catalogue of ancient earthquakes in the mediterranean area up to the 10th century,* Vol. 1, ING-SGA, pp. 504. Update catalogues available on-line at http://storing.ingv.it/cfti/cfti4/#.

Gutscher, M.-A., Roger, J., Baptista, M. A., Miranda, J. M., & Tinti, S. (2006). The source of the 1693 Catania earthquake and tsunami (Southern Italy): New evidence from tsunami modeling of a locked subduction fault plane. *Geophysical Research Letters, 33,* L08309. https://doi.org/10.1029/2005GL025442.

Gutscher, M.-A., & Westbrook, G. K. (2009). Great earthquakes in slow subduction, low taper margins. In S. Lallemand, F. Funiciello, (Eds.,) Subduction zone dynamics, pp. 119–134. Springer, Berlin, https://doi.org/10.1007/978-3-540-87974-9.

Hayes, G. P., Moore, G. L., Portner, D. E., Hearne, M., Flamme, H., Furtney, M., et al. (2018). Slab2, a comprehensive subduction zone geometry model. *Science.* https://doi.org/10.1126/science.aat4723.

Heinrich, P., Schindele, F., & Guibourg, S. (1998). Modeling of the February 1996 Peruvian tsunami. *Geophysical Reasearch Letters, 25,* 2687–2690. https://doi.org/10.1029/98GL01780.

Herrero, A., & Murphy, S. (2018). Self-similar slip distributions on irregular shaped faults. *Geophysical Journal Inetrnational, 213*(3), 2060–2070. https://doi.org/10.1093/gji/ggy104.

Hollenstein, C., Müller, M. D., Geiger, A., & Kahle, H. G. (2008). Crustal motion and deformation in Greece from a decade of GPS measurements, 1993–2003. *Tectonophysics, 449,* 17–40. https://doi.org/10.1016/j.tecto.2007.12.006.

Howell, A., Jackson, J., Copley, A., McKenzie, D., & Nissen, E. (2017). Subduction and vertical coastal motions in the eastern Mediterranean. *Geophysical Journal International, 211*(1), 593–620. https://doi.org/10.1093/gji/ggx307.

Hyndman, R. D. (2013). Downdip landward limit of Cascadia great earthquake rupture. *Journal of Geophysical Research, 118,* 5530–5549. https://doi.org/10.1002/jgrb.50390.

Hyndman, R. D., Yamano, M., & Oleskevich, D. A. (1997). The seismogenic zone of subduction thrust faults. *Island Arc, 6*(3), 244–260. https://doi.org/10.1111/j.1440-1738.1997.tb00175.x.

Kagan, Y. Y., Bird, P., & Jackson, D. D. (2010). Earthquake patterns in diverse tectonic zones of the globe. *Pure and Applied Geophysics, 167,* 721–741. https://doi.org/10.1007/s00024-010-0075-3.

Kajiura, K. (1963). The leading wave of a tsunami. *Bulletin of Earthquake Research Institute, University of Tokyo, 41,* 535–571.

Kozdon, J. E., & Dunham, E. M. (2013). Rupture to the trench: Dynamic rupture simulations of the 11 March 2011 Tohoku earthquake. *Bulletin of the Seismological Society of America, 103*(2B), 1275–1289. https://doi.org/10.1785/0120120136.

Laigle, M., Sachpazi, M., & Hirn, A. (2004). Variation of seismic coupling with slab detachment and upper plate structure along the western Hellenic subduction zone. *Tectonophysics, 391*(1–4), 85–95. https://doi.org/10.1016/j.tecto.2004.07.009.

Lallemand, S., Schnürle, P., & Malavieille, J. (1994). Coulomb theory applied to accretionary and nonaccretionary wedges:

Possible causes for tectonic erosion and/or frontal accretion. *Journal of Geophysical Research, 99,* 12033–12055. https://doi.org/10.1029/94JB00124.

Lay, T. (2018). A review of the rupture characteristics of the 2011 Tohohu-oki Mw 9.1 earthquake. *Tectonophysics, 733,* 4–36. https://doi.org/10.1016/j.tecto.2017.09.022.

Lay, T., Kanamori, H., Ammon, C. J., Koper, K. D., Hutko, A. R., Ye, L., et al. (2012). Depth-varying rupture properties of subduction zone megathrust faults. *Journal of Geophysical Research, 117,* B04311. https://doi.org/10.1029/2011JB009133.

LeVeque, R. J., Waagan, K., González, F. I., Rim, D., & Lin, G. (2016). Generating random earthquake events for probabilistic tsunami hazard assessment. *Pure and Applied Geophysics, 173*(12), 3671–3692.

Li, L., Switzer, A. D., Chan, C.-H., Wang, Y., Weiss, R., & Qiu, Q. (2016). How heterogeneous coseismic slip affects regional probabilistic tsunami hazard assessment: A case study in the South China Sea. *Journal of Geophysical Research, 121*(8), 6250–6272. https://doi.org/10.1002/2016JB013111.

Lorito, S., Romano, F., & Lay, T. (2016). Tsunamigenic major and great Earthquakes (2004–2013): Source processes inverted from seismic, geodetic, and sea-level data. In: Meyers, R. A. (Eds.), Encyclopedia of complexity and systems science, Springer, New York, ISBN 13:978-0-387-30440-3.

Lorito, S., Selva, J., Basili, R., Romano, F., Tiberti, M. M., & Piatanesi, A. (2015). Probabilistic hazard for seismically induced tsunamis: Accuracy and feasibility of inundation maps. *Geophysical Journal International, 200*(1), 574–588. https://doi.org/10.1093/gji/ggu408.

Lorito, S., Tiberti, M. M., Basili, R., Piatanesi, A., & Valensise, G. (2008). Earthquake-generated tsunamis in the Mediterranean Sea: Scenarios of potential threats to Southern Italy. *Journal of Geophysical Research: Solid Earth, 113,* B01301. https://doi.org/10.1029/2007JB004943.

Lotto, G. C., Dunham, E. M., Jeppson, T. N., & Tobin, H. J. (2017). The effect of compliant prisms on subduction zone earthquakes and tsunamis. *Earth and Planetary Science Letters, 458,* 213–222.

Løvholt, F., Glimsdal, S., Lynett, P., & Pedersen, G. (2015). Simulating tsunami propagation in fjords with long-wave models. *Natural Hazards and Earth System Sciences, 15,* 657–669. https://doi.org/10.5194/nhess-15-657-2015.

Løvholt, F., Lynett, P., & Pedersen, G. (2013). Simulating run-up on steep slopes with operational Boussinesq models; capabilities, spurious effects and instabilities. *Nonlinear Processes in Geophysics, 20,* 379–395. https://doi.org/10.5194/npg-20-379-2013.

Løvholt, F., Pedersen, G., Bazin, S., Kühn, D., Bredesen, R. E., & Harbitz, C. (2012). Stochastic analysis of tsunamirunup due to heterogeneous coseismic slip and dispersion. *Journal of Geophysical Research, 117,* C03047. https://doi.org/10.1029/2011JC007616.

Ma, S., & Beroza, G. C. (2008). Rupture dynamics on a bimaterial interface for dipping faults. *Bulletin of the Seismological Society of America, 98*(4), 1642–1658. https://doi.org/10.1785/0120070201.

Macías, J., Castro, M. J., Ortega, S., Escalante, C., & González-Vida, J. M. (2017). Performance benchmarking of tsunami-HySEA model for NTHMP's inundation mapping activities. *Pure and Applied Geophysics, 174*(8), 3147–3183. https://doi.org/10.1007/s00024-017-1583-1.

Macías, J., Mercado, A., González-Vida, J. M., Ortega, S., & Castro, M. J. (2016). Comparison and computational performance of Tsunami-HySEA and MOST models for LANTEX 2013 scenario: Impact assessment on Puerto Rico coasts. *Pure and Applied Geophysics, 173*(12), 3973–3997. https://doi.org/10.1007/s00024-016-1387-8.

Maesano, F. E., Tiberti, M. M., & Basili, R. (2017). The Calabrian Arc: Three-dimensional modelling of the subduction interface. *Scientific Reports, 7,* 8887. https://doi.org/10.1038/s41598-017-09074-8.

Mai, P. M., & Beroza, G. C. (2002). A spatial random field model to characterize complexity in earthquake slip. *Journal of Geophysical Research, 107*(B11), 2308. https://doi.org/10.1029/2001JB000588.

Mai, P. M., & Thingbaijam, K. K. S. (2014). SRCMOD: An online database of finite fault–fault rupture models. *Seismological Research Letters, 85*(6), 1348–1357. https://doi.org/10.1785/0220140077.

Maramai, A., Brizuela, B., & Graziani, L. (2014). The Euro-Mediterranean Tsunami Catalogue. *Annals of Geophysics, 57*(4), S0435. https://doi.org/10.4401/ag-6437.

Marzocchi, W., Taroni, M., & Selva, J. (2015). Accounting for epistemic uncertainty in PSHA: Logic Tree and ensemble modeling. *Bulletin of the Seismological Society of America, 105*(4), 2151–2159. https://doi.org/10.1785/0120140131.

McCloskey, J., Antonioli, A., Piatanesi, A., Sieh, K., Steacy, S., Nalbant, S. S., et al. (2007). Near-field propagation of tsunamis from megathrust earthquakes. *Geophysical Research Letters, 34,* L14316. https://doi.org/10.1029/2007GL030494.

Meade, B. J. (2007). Algorithms for the calculation of exact displacements, strains, and stresses for triangular dislocation elements in a uniform elastic half space. *Computation Geosciences, 33*(8), 1064–1075. https://doi.org/10.1016/j.cageo.2006.12.003.

Molinari, I., Tonini, R., Lorito, S., Piatanesi, A., Romano, F., Melini, D., et al. (2016). Fast evaluation of tsunami scenarios: Uncertainty assessment for a Mediterranean Sea database. *Natural Hazards and Earth System Sciences, 16,* 2593–2602. https://doi.org/10.5194/nhess-16-2593-2016.

Moore, J. C., & Saffer, D. (2001). Updip limit of the seismogenic zone beneath the accretionary prism of southwest Japan: An effect of diagenetic to low-grade metamorphic processes and increasing effective stress. *Geology, 29*(2), 183–186. https://doi.org/10.1130/0091-7613(2001)029%3c0183.

Mori, N., Muhammad, A., Goda, K., Yasuda, T., & Ruiz-Angulo, A. (2017). Probabilistic tsunami hazard analysis of the pacific coast of Mexico: Case study based on the 1995 Colima earthquake tsunami. *Frontiers in Built Environment, 3*(34), 1–16. https://doi.org/10.3389/fbuil.2017.00034.

Murphy, S., Di Toro, G., Romano, R., Scala, A., Lorito, S., Spagnuolo, E., et al. (2018). Tsunamigenic Earthquake simulations using experimentally derived friction laws. *Earth and Planetary Science Letters, 486,* 155–165. https://doi.org/10.1016/j.epsl.2018.01.011.

Murphy, S., Scala, A., Herrero, A., Lorito, S., Festa, G., Trasatti, E., et al. (2016). Shallow slip amplification and enhanced tsunami hazard unravelled by dynamic simulations of mega-thrust earthquakes. *Scientific Reports, 6,* 35007. https://doi.org/10.1038/srep35007.

Nalbant, S., McCloskey, J., Steacy, S., NicBhloscaidh, M., & Murphy, S. (2013). Interseismic coupling, stress evolution, and earthquake slip on the Sunda megathrust. *Geophysical Research Letters, 40*(16), 4204–4208. https://doi.org/10.1002/grl.50776.

Newman, A. V., & Okal, E. (1998). Teleseismic estimates of radiated seismic energy: The E/MO discriminant for tsunami earthquakes. *Journal of Geophysical Research Atmospheres, 1032*(B11), 26885–26898. https://doi.org/10.1029/98JB02236.

Nielsen, S. B. (1998). Free surface effects on the propagation of dynamic rupture. *Geophysical Research Letters, 25*(1), 125–128. https://doi.org/10.1029/97GL03445.

Nijholt, N., Govers, R., & Wortel, R. (2018). On the forces that drive and resist deformation of the south-central Mediterranean: A mechanical model study. *Geophysical Journal International, 214*(2), 876–894. https://doi.org/10.1093/gji/ggy144.

Noquet, J.-M. (2012). Present-day kinematics of the Mediterranean: A comprehensive overview of GPS results. *Tectonophysics, 579,* 220–242. https://doi.org/10.1016/j.tecto.2012.03.037.

Papadimitriou, E., & Karakostas, V. (2008). Rupture model of the great AD 365 Crete earthquake in the southwestern part of the Hellenic Arc. *Acta Geophysics, 56,* 293–312. https://doi.org/10.2478/s11600-008-0001-6.

Papazachos, B., & Papazachou, C. (1997). *The earthquakes of Greece* (pp. 304). Thessaloniki: Editions Ziti.

Paris, R., Ulvrova, M., Selva, J., Brizuela, B., Costa, A., Grezio, A., et al. (2019). Probabilistic hazard analysis for tsunamis generated by subaqueous volcanic explosions in the Campi Flegrei caldera, Italy. *Journal of Volcanology and Geothermal Research, 379,* 106–116. https://doi.org/10.1016/j.jvolgeores.2019.05.010.

Piatanesi, A., & Tinti, S. (1998). A revision of the 1693 eastern Sicily earthquake and tsunami. *Journal of Geophysical Research, 103*(B2), 2749–2758. https://doi.org/10.1029/97JB03403.

Polonia, A., Torelli, L., Mussoni, P., Gasperini, L., Artoni, A., & Klaeschen, D. (2011). The Calabrian Arc subduction complex in the Ionian Sea: Regional architecture, active deformation, and seismic hazard. *Tectonics, 30,* TC5018. https://doi.org/10.1029/2010tc002821.

Reilinger, R., McClusky, S., Vernant, P., Lawrence, S., Ergintav, S., Cakmak, R., et al. (2006). GPS constraints on continental deformation in the Africa-Arabia-Eurasia continental collision zone and implications for the dynamics of plate interactions. *Journal of Geophysical Research, 111*(B5), B05411. https://doi.org/10.1029/2005JB004051.

Romano, F., Trasatti, E., Lorito, S., Piromallo, C., Piatanesi, A., Ito, Y., et al. (2014). Structural control on the Tohoku earthquake rupture process investigated by 3D FEM, Tsunami and Geodetic Data. *Scientific Reports, 4,* 5631. https://doi.org/10.1038/srep05631.

Rubin, A. M., & Ampuero, J.-P. (2007). Aftershock asymmetry on a bimaterial interface. *Journal of Geophysical Research, 112,* B05307. https://doi.org/10.1029/2006JB004337.

Ruiz, J. A., Baumont, D., Bernard, P., & Berge-Thierry, C. (2011). Modelling directivity of strong ground motion with a fractal, k $^{-2}$, kinematic source model. *Geophysical Journal International, 186,* 226–244. https://doi.org/10.1111/j.1365-246X.2011.05000.x.

Sachpazi, M., Laigle, M., Charalampakis, M., Diaz, J., Kissling, E., Gesret, A., et al. (2016). Segmented Hellenic slab rollback driving Aegean deformation and seismicity. *Geophysical Research Letters, 43*(2), 651–658. https://doi.org/10.1002/2015GL066818.

Salaün, G., Pedersen, H. A., Paul, A., Farra, V., Karabulut, H., Hatzfeld, D., et al. (2012). High-resolution surface wave tomography beneath the Aegean-Anatolia region: Constraints on upper-mantle structure. *Geophysical Journal International, 190,* 406–420. https://doi.org/10.1111/j.1365-246X.2012.05483.x.

Saloor, N., & Okal, E. A. (2018). Extension of the energy-to-moment parameter Θ to intermediate and deep earthquakes. *Physics of the Earth and Planetary Interiors, 274,* 37–48. https://doi.org/10.1016/j.pepi.2017.10.006.

Satake, K. (1995). Linear and nonlinear computation of the 1992 Nicaragua earthquake tsunami. *Pure and Applied Geophysics, 144,* 455–470. https://doi.org/10.1007/BF00874378.

Satriano, C., Dionicio, V., Miyake, H., Uchida, N., Vilotte, J.-P., & Bernard, P. (2014). Structural and thermal control of seismic activity and Megathrust rupture dynamics in subduction zones: Lessons from the Mw 9.0, 2011 Tohoku earthquake. *Earth and Planetary Science Letters, 403,* 287–298. https://doi.org/10.1016/j.epsl.2014.06.037.Satrianoetal.2014.

Scala, A., Festa, G., & Vilotte, J.-P. (2017). Rupture dynamics along bimaterial interfaces: A parametric study of the shear-normal traction coupling. *Geophysical Journal International, 209*(1), 48–67. https://doi.org/10.1093/gji/ggw489.

Scala, A., Festa, G., Vilotte, J.-P., Lorito, S., & Romano, F. (2019). Wave interaction of reverse-fault rupture with free surface: Numerical analysis of the dynamic effects and fault opening induced by symmetry breaking. *Journal of Geophysical Research: Solid Earth, 124,* 1743–1758. https://doi.org/10.1029/2018JB016512.

Scholz, C. H. (1998). Earthquakes and friction laws. *Nature, 391*(6662), 37–42. https://doi.org/10.1038/34097.

Sellier, N.C., Loncke, L., Vendeville, B.C., Mascle, J., Zitter, T., Woodside, J. & Loubrieu, B. (2013a). Post-Messinian evolution of the Florence Ridge area (Western cyprus Arc), Part I: Morphostructural analysis. *Tectonophysics.* https://doi.org/10.1016/j.tecto.2012.04.001.

Sellier, N. C., Vendeville, B. C., & Loncke, L. (2013b). Post-Messinian evolution of the Florence Rise area (*Western Cyprus* Arc) Part II: Experimental modelling. *Tectonophysics, 591,* 143–151. https://doi.org/10.1016/j.tecto.2011.07.003.

Selva, J., Tonini, R., Molinari, I., Tiberti, M. M., Romano, F., & Grezio, A. (2016). Quantification of source uncertainties in Seismic Probabilistic Tsunami Hazard Analysis (SPTHA). *Geophysical Journal International, 205*(3), 1780–1803. https://doi.org/10.1093/gji/ggw107.

Sepúlveda, I., Liu, P. L.-F., Grigoriu, M., & Pritchard, M. (2017). Tsunami hazard assessments with consideration of uncertain earthquake slip distribution and location. *Journal of Geophysical Research: Solid Earth, 122,* 7252–7271. https://doi.org/10.1002/2017JB014430.

Shaw, B., Ambraseys, N. N., England, P. C., Floyd, M. A., Gorman, G. J., Higham, T. F. G., et al. (2008). Eastern Mediterranean tectonics and tsunami hazard inferred from the AD 365 earthquake. *Nature Geoscience, 1,* 268–276. https://doi.org/10.1038/ngeo151.

Shaw, B., & Jackson, J. (2010). Earthquake mechanisms and active tectonics of the Hellenic subduction zone. *Geophysical Journal International, 181*(2), 966–984. https://doi.org/10.1111/j.1365-246X.2010.04551.x.

Sodoudi, F., Brüstle, A., Meier, T., Kind, R., Friederich, W., & Egelados Working Group. (2015). Receiver function images of the Hellenic subduction zone and comparison to microseismicity.

Solid Earth, 6(1), 135–151. https://doi.org/10.5194/se-6-135-2015.

Song, S. G., & Somerville, P. (2010). Physics-based earthquake source characterization and modeling with geostatistics. *Bulletin of the Seismological Society of America, 100*(2), 482–496. https://doi.org/10.1785/0120090134.

Song, S. G., Dalguer, L. A., & Mai, P. M. (2013). Pseudo-dynamic source modelling with 1-point and 2-point statistics of earthquake source parameters. *Geophysical Journal International, 196*(3), 1770–1786. https://doi.org/10.1093/gji/ggt479.

Sørensen, M. B., Spada, M., Babeyko, A., Wiemer, S., & Grünthal, G. (2012). Probabilistic tsunami hazard in the Mediterranean Sea. *Journal of Geophysical Research, 117,* B01305. https://doi.org/10.1029/2010JB008169.

Stiros, S. C. (2010). The 8.51 magnitude, AD365 earthquake in Crete: Coastal uplift, topography changes, archaeological and historical signature. *Quaternary International, 216,* 54–63. https://doi.org/10.1016/j.quaint.2009.05.005.

Strasser, F. O., Arango, M. C., & Bommer, J. J. (2010). Scaling of the source dimensions of interface and intraslab subduction-zone earthquakes with moment magnitude. *Seismological Research Letters, 81*(6), 941–950. https://doi.org/10.1785/gssrl.81.6.941.

Stucchi, M., Rovida, A., Gomez Capera, A. A., Alexandre, P., Camelbeeck, T., Demircioglu, M. B., et al. (2013). The SHARE European Earthquake Catalogue (SHEEC) 1000-1899. *Journal of Seismology, 17,* 523–544. https://doi.org/10.1007/s10950-012-9335-2.

Sun, T., & Wang, K. (2015). Viscoelastic relaxation following subduction earthquakes and its effects on afterslip determination. *Journal of Geophysical Research: Solid Earth, 120,* 1329–1344. https://doi.org/10.1002/2014JB011707.

Tanioka, Y., & Satake, K. (1996). Fault parameters of the 1896 Sanriku tsunami earthquake estimated from tsunami numerical modeling. *Geophysical Research Letters, 23,* 1549–1552. https://doi.org/10.1029/96GL01479.

Tichelaar, B. W., & Ruff, L. J. (1993). Depth of seismic coupling along subduction zones. *Journal of Geophysical Research, 98*(B2), 2017–2037. https://doi.org/10.1029/92JB02045.

Tinti, S., Armigliato, A., & Bortolucci, E. (2001). Contribution of tsunami data analysis to constrain the seismic source: The case of the 1693 eastern Sicily earthquake. *Journal of Seismology, 5,* 41–61. https://doi.org/10.1023/A:1009817601760.

Tinti, S., Armigliato, A., Pagnoni, G., & Zaniboni, F. (2005). Scenarios of giant tsunamis of tectonic origin in the Mediterranean. *ISET Journal of Earthquake Technology, Paper n. 464, 42*(4), 171–188.

Tonini, R., Armigliato, A., Pagnoni, G., Zaniboni, F., & Tinti, S. (2011). Tsunami hazard for the city of Catania, eastern Sicily, Italy, assessed by means of Worst-case Credible Tsunami Scenario Analysis (WCTSA). *Natural Hazards and Earth System Sciences, 11,* 1217–1232. https://doi.org/10.5194/nhess-11-1217-2011.

Urlaub, M., Petersen, F., Gross, F., Bonforte, A., Puglisi, G., Guglielmino, F., et al. (2018). Gravitational collapse of Mount Etna's southeastern flank. *Science Advances, 4,* eaat 9700. https://doi.org/10.1126/sciadv.aat9700.

Vernant, P., Reilinger, R., & McClusky, S. (2014). Geodetic evidence for low coupling on the Hellenic subduction plate interface. *Earth and Planetary Science Letters, 385,* 122–129. https://doi.org/10.1016/j.epsl.2013.10.018.

Wang, K., & Dixon, T. (2004). "Coupling" semantics and science in earthquake research. *Eos (Transactions, American Geophysical Union), 85,* 180. https://doi.org/10.1029/2004EO180005.

Wang, D., & Mori, J. (2011). Rupture process of the 2011 off the Pacific coast of Tohoku Earthquake (Mw 9.0) as imaged with back-projection of teleseismic P-waves. *Earth Planets Space, 63,* 603–607. https://doi.org/10.5047/eps.2011.05.029.

Wdowinski, S., Ben-Avraham, Z., Arvidsson, R., & Ekström, G. (2006). Seismotectonics of the Cyprian Arc. *Geophysical Journal International, 164*(1), 176–181. https://doi.org/10.1111/j.1365-246X.2005.02737.x.

Ye, L., Lay, T., Kanamori, H., & Rivera, L. (2016). Rupture characteristics of major and great (Mw $\geq$ 7.0) megathrust earthquakes from 1990 to 2015: 1. Source parameter scaling relationships. *Journal of Geophysical Research: Solid Earth, 121,* 826–844. https://doi.org/10.1002/2015JB012426.

Yue, H., Lay, T., Rivera, L., Bau, Y., Yamazaki, Y., Cheung, K. F., et al. (2014). Rupture process of the 2010 Mw 7.8 Mentawai tsunami earthquake from joint inversion of near-field hr-GPS and teleseismic body wave recordings constrained by tsunami observations. *Journal of Geophysical Research: Solid Earth, 119,* 5574–5593. https://doi.org/10.1002/2014JB011082.

Zheng, Y., Anderson, J. G., & Yu, G. (1994). A composite source model for computing realistic synthetic strong ground motions. *Geophysical Research Letters, 21*(8), 725–728. https://doi.org/10.1029/94GL00367.

(Received January 6, 2019, revised June 10, 2019, accepted June 14, 2019, Published online June 28, 2019)

Pure Appl. Geophys. 177 (2020), 1521–1548

https://doi.org/10.1007/s00024-019-02299-w

Pure and Applied Geophysics

Sensitivity of Probabilistic Tsunami Hazard Assessment to Far-Field Earthquake Slip Complexity and Rigidity Depth-Dependence: Case Study of Australia

GARETH DAVIES[1] and JONATHAN GRIFFIN[1,2]

Abstract—Probabilistic Tsunami Hazard Assessment (PTHA) often proceeds by constructing a suite of hypothetical earthquake scenarios, and modelling their tsunamis and occurrence-rates. Both tsunami and occurrence-rate models are affected by the representation of earthquake slip and rigidity, but the overall importance of these factors for far-field PTHA is unclear. We study the sensitivity of an Australia-wide PTHA to six different far-field earthquake scenario representations, including two rigidity models (constant and depth-varying) combined with three slip models: fixed-area-uniform-slip (with rupture area deterministically related to magnitude); variable-area-uniform-slip; and spatially heterogeneous-slip. Earthquake-tsunami scenarios are tested by comparison with DART-buoy tsunami observations, demonstrating biases in some slip models. Scenario occurrence-rates are modelled using Bayesian techniques to account for uncertainties in seismic coupling, maximum-magnitudes and Gutenberg-Richter b-values. The approach maintains reasonable consistency with the historical earthquake record and spatially variable plate convergence rates for all slip/rigidity model combinations, and facilitates partial correction of model-specific biases (identified via DART-buoy testing). The modelled magnitude exceedance-rates are tested by comparison with rates derived from long-term historical and paleoseismic data and alternative moment-conservation techniques, demonstrating the robustness of our approach. The tsunami hazard offshore of Australia is found to be insensitive to the choice of rigidity model, but significantly affected by the choice of slip model. The fixed-area-uniform-slip model produces lower hazard than the other slip models. Bias adjustment of the variable-area-uniform-slip model produces a strong preference for 'compact' scenarios, which compensates for a lack of slip heterogeneity. Thus, both heterogeneous-slip and variable-area-uniform-slip models induce similar far-field tsunami hazard.

Key words: Probabilistic tsunami hazard assessment, sensitivity analysis.

[1] Positioning and Community Safety Division, Geoscience Australia, Canberra, Australia. E-mail: gareth.davies@ga.gov.au

[2] *Present Address*: Department of Geology, University of Otago, PO Box 56, Dunedin 9054, New Zealand.

1. Introduction

Destructive tsunamis are most often generated by large subduction zone earthquakes (Grezio et al. 2017). Although the highest runup usually occurs near to the source, earthquake-generated tsunamis show strong directivity and can remain hazardous at trans-oceanic distances (Ben-Menahem and Rosenman 1972). This was illustrated by the far-field impacts of the 2004 Sumatra-Andaman tsunami (300 deaths in Somalia), the 1960 Chile tsunami (203 deaths in Hawaii and Japan), and the 1946 Aleutian tsunami (162 deaths in California, the Marquesas and Hawaii) (Okal et al. 2002; Fritz and Borrero 2006; Okal 2011). The latter sites range between 4000 and 17,000 km from the tsunami source. Probabilistic Tsunami Hazard Assessments (PTHAs) suggest far-field subduction earthquakes can contribute at first-order to the hazard even for sites exposed to near-field subduction sources, such as Crescent City (near Cascadia) and Napier (near the Hikurangi trench) (Gonzalez et al. 2009; Geist and Parsons 2016; Power et al. 2017).

A key challenge for earthquake-generated tsunami hazard assessments concerns representing earthquakes and their occurrence-rates given substantial uncertainties in the underlying science (Selva et al. 2016; Davies et al. 2017; Power et al. 2017). Most studies follow a computational approach which requires specifying each earthquake scenario's location, moment magnitude M_w, focal mechanism, rupture extent, rigidity and spatial distribution of slip, and subsequently modelling the resulting tsunami (Grezio et al. 2017). The plausible variation of earthquake parameters and occurrence-rates is often very large; e.g., on the Kermadec-Tonga trench

Berryman et al. (2015) suggested the maximum earthquake magnitude $M_{w,max}$ could be anywhere between [8.1–9.6], implying large uncertainty in potential far-field tsunami impacts. The representation of rupture area, spatial slip variability and rigidity is also not standardized, with different approaches potentially leading to first-order differences in the modelled tsunami size when M_w is fixed (Geist and Bilek 2001; Gica et al. 2007; Davies et al. 2015; Mueller et al. 2015; Li et al. 2016; Butler et al. 2017; Mori et al. 2017). Compared with scenario based hazard assessments, Probabilistic Tsunami Hazard Assessment (PTHA) methodologies have the great advantage that such uncertainties can be explicitly integrated into the analysis, but the hazard calculations remain sensitive to choice of models and representation of uncertainties (Grezio et al. 2017). Competing models/parameters must be weighted appropriately to ensure limited weight is placed on unlikely models (which is often nontrivial to ensure in practice, as for Probabilistic Seismic Hazard Assessment, Bommer and Scherbaum 2008). Bayesian methods offer a useful approach to this problem, as initial weights can be updated based on the model's consistency with data (Parsons and Geist 2009; Grezio et al. 2010, 2017; Selva et al. 2016; Davies et al. 2017). However testing and sensitivity analyses remain critical for informing PTHA modelling decisions, by focussing attention on the most significant model weaknesses and the most influential sources of uncertainty (Li et al. 2016; Sepúlveda et al. 2019; Volpe et al. 2019).

Currently there is no consensus regarding how earthquake rupture complexity (i.e. variability of fault dimensions and spatial slip distribution) should be represented for PTHA. Often rupture complexity is considered most important for near-field tsunamis and of limited significance at far-field sites (e.g. Geist 2002; Okal and Synolakis 2008), whereas other studies suggest it is important also for far-field tsunamis (i.e. more than 1000 km from the source, Gica et al. 2007; Li et al. 2016; Butler et al. 2017). In the near-field case, it is clear that if M_w and the rupture-area are fixed then slip heterogeneity substantially influences modelled tsunami wave heights and inundation (Geist 2002; Mueller et al. 2015; Ruiz et al. 2015). However this does not necessarily imply that near-field tsunami hazard assessments must use heterogeneous-slip. An et al. (2018) found that if the rupture area and location were calibrated, then optimal uniform-slip scenarios matched near-field tsunami observations nearly as well as optimal heterogeneous-slip scenarios. Fewer studies have focussed on the far-field case. Okal and Synolakis (2008) compared the far-field tsunami radiation pattern of a uniform-slip scenario with a weakly heterogeneous scenario (slip within 80–120% of the mean). The far-field tsunami radiation pattern was similar in both cases; however, finite-fault inversions suggest historical earthquakes have much more than 20% slip variation (e.g. Poisson et al. 2011; Lay 2018) which may induce greater far-field effects. Li et al. (2016) compared PTHA calculations derived from uniform and heterogeneous-slip earthquakes on the Manila trench, finding slip heterogeneity substantially increased the peak nearshore tsunami amplitude at both near-field and far-field sites (e.g. $\simeq 35\%$ at 500 year average return interval (ARI)). Butler et al. (2017) modelled tsunamis in Hawaii due to a range of uniform-slip $M_w8.6$ Aleutian Island earthquakes with varying length and width, finding a factor $\simeq 2$ variation in the modelled far-field tsunami size. Butler et al. (2017) also found $M_w9.25$ Aleutian earthquakes with higher near-trench slip (and reduced deep slip) produced more inundation in Hawaii than similar uniform-slip scenarios, with some locations being more sensitive to shallow slip than others. Gica et al. (2007) studied the sensitivity of tsunami wave heights in Hawaii to variations in the dimensions of uniform-slip earthquakes in Chile, Japan, and the Aleutians. Their modelled wave heights varied by a factor $\simeq 2$ due to doubling/halving the rupture length, width and slip (while preserving M_w). They also report that, along the main beam of tsunami energy, the sensitivity of the wave to the rupture dimensions did not decrease with increasing distance from the source, suggesting that rupture complexity influences even the far-field tsunami behaviour.

In addition to rupture complexity, PTHAs may be affected by the representation of the fault rigidity μ. The rigidity mediates the relationship between M_w and slip (and thus the tsunami size); all else being equal, lower μ implies higher slip for fixed M_w and produces a larger offshore tsunami. In practice the rigidity of subduction zones is not very well constrained (Bilek and Lay 1999; Geist and Bilek 2001), but tsunami hazard studies often assume constant $\mu \simeq 3 - 6 \times 10^{10}$ Pa (e.g. Butler et al. 2017; Fukutani et al. 2018; Kalligeris et al. 2017). Despite this common practice, the use of depth-varying μ (with low near-trench values) appears necessary to simulate the large tsunamis observed historically following shallow 'tsunami-earthquake' events (e.g. Geist and Bilek 2001; Newman et al. 2011b; Hébert et al. 2012), albeit not in every such case (Newman et al. 2011a). For example, low rigidity was used to simulate near-field runup of up to 10 m resulting from the 2006 M_w7.7 Java tsunami-earthquake (Hébert et al. 2012), an event which also produced the highest historically observed tsunami runup in Australia [7.9 m at Steep Point $\simeq$ 1800 km from the source, Prendergast and Brown (2012)].

Even though the rigidity model is highly significant for interpreting such events, it remains unclear whether the overall tsunami hazard is sensitive to the choice of constant or depth-varying rigidity, assuming the modelled earthquake occurrence-rates are constrained with a combination of earthquake catalogue data and moment conservation methodologies (e.g. as in Kagan and Jackson 2013; Rong et al. 2014; Davies et al. 2017). The combination of these methodologies is desirable because earthquake catalogue data alone has limited power to constrain high M_w exceedance-rates (Zöller 2013, 2017). Importantly, moment conservation arguments imply the time-integrated earthquake slip-rate should balance the seismically coupled fraction of the tectonic convergence-rate (e.g. Bird and Kagan 2004; Bird and Liu 2007), suggesting that for fixed M_w, low-rigidity high-slip earthquakes should occur less often than those with higher-rigidity and lower-slip. This would reduce the effect of low-rigidity tsunami-earthquake type events on the overall hazard. Thus to understand how the hazard is affected by the rigidity model, it is necessary to apply moment-conservation approaches in conjunction with both constant and depth-varying rigidity models. Compared with the constant-rigidity case, moment-conservation approaches are more complex to apply with depth-varying rigidity because there is no longer a one-to-one relation between the scenario M_w and its spatially integrated slip.

This study considers the sensitivity of a PTHA to the representation of far-field earthquake scenarios and their occurrence-rates. The work was conducted as part of the 2018 Australian PTHA and the broader study is described in two detailed technical reports (Davies and Griffin 2018; Griffin and Davies 2018). In addition Davies (2019) tests the modelled tsunami scenarios against DART-buoy observations without consideration of scenario frequencies or hazard. The current study focusses on the hazard calculation for offshore sites, and its sensitivity to key modelling assumptions. Inundation hazard assessments can be developed on a site-by-site basis by combining an offshore PTHA with high-resolution inundation models (e.g. Lane et al. 2012), but for simplicity that step is not undertaken herein. Because the importance of earthquake rupture complexity for far-field tsunami hazard is unclear, the tests feature three alternative slip models with varying degrees of complexity. For all slip models, the sensitivity of the hazard to the rigidity representation is also examined by re-interpreting the scenario M_w on the basis of either constant or depth-varying rigidity. To enable this the current study develops scenario occurrence-rate models which are applicable to multiple slip and rigidity representations. The approach generalises the global-scale methodology of Davies et al. (2017) which constrained scenario rates using earthquake catalogues and plate convergence rates, but was limited to scenarios with constant rigidity, uniform slip and a deterministic M_w-vs-area relation. As well as generalising to a range of slip and rigidity models, the new occurrence-rate methodology makes more efficient use of earthquake catalogue data, and leads to a better match between the modelled long-term earthquake-slip rates and spatial variations of tectonic convergence.

2. *Earthquake-Tsunami Scenario Database*

Below we present key features of the 2018 Australian PTHA earthquake-tsunami scenario database used in this study. Further details are provided in Davies and Griffin (2018).

2.1. Source-Zone Discretization

The database includes major Pacific and Indian Ocean earthquake source-zones, and minor sources close to Australia (Fig. 1a). Fault geometries were defined using SLAB 2.0 or SLAB 1.0 where possible (Hayes et al. 2012, 2018), and elsewhere were schematized using linear or parabolic profiles (Griffin and Davies 2018). The fault geometries extend between the trench and a maximum seismogenic depth estimated from Berryman et al. (2015) and Griffin and Davies (2018). Most source-zones were modelled with only thrust earthquake scenarios, which are by far the largest contributor to the computed hazard in Australia, and for brevity the treatment of non-thrust sources is not presented herein.

Two alternative rigidity models are tested, featuring constant-rigidity ($\mu = 30$ GPa) and depth-varying rigidity (Fig. 1b). The latter was fit to rigidity estimates for subduction earthquakes by Bilek and Lay (1999) (Fig. 1b). It exhibits low rigidities for shallow earthquakes (but always ≥ 10 GPa), and transitions to the preliminary reference earth model (PREM) at greater depths (Dziewonski and Anderson 1981; Geist and Bilek 2001).

Each source-zone was discretized into unit-sources with dimensions of $\simeq 50 \times 50$ km^2, although this varies to match the source-geometry (Fig. 1c–h). To better represent non-planar fault geometries the unit-sources were subdivided into a large number of rectangular ‘sub-unit-sources’ (Davies et al. 2017). The vertical co-seismic seabed deformation associated with 1 m of slip on each unit-source was computed using the homogeneous elastic half-space model (Okada 1985) integrated over sub-unit-sources. The ocean surface deformation was derived from this using a Kajiura filter (Kajiura 1963;

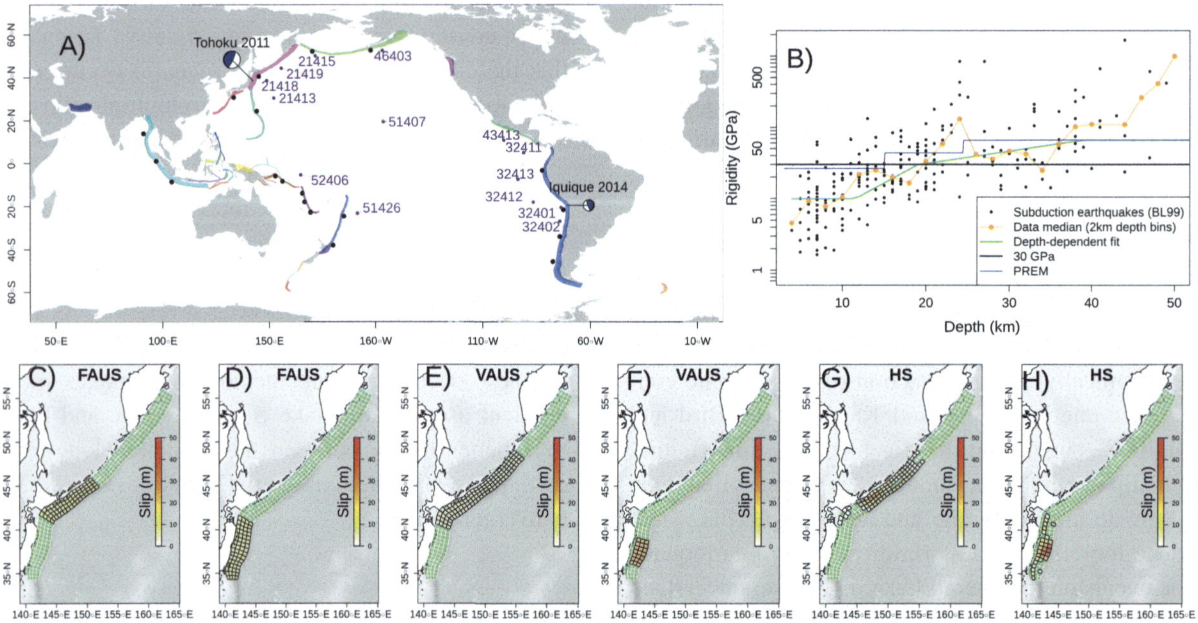

Figure 1
Overview of the earthquake-tsunami scenario database: **a** earthquake source-zones used in this study. Numbers show DART buoy locations. Black points along large source-zones denote segment boundary locations based on Berryman et al. (2015), which are given 50% weight in the scenario rate calculation (with 50% on an unsegmented model, Sect. 3.1); **b** Rigidity-vs-depth on subduction-zones using data from Bilek and Lay (1999), with the depth-dependent and constant-rigidity (30 GPa) models used here for thrust scenarios. The preliminary reference earth model (PREM) is also shown (Dziewonski and Anderson 1981); **c**, **d** example fixed-area-uniform-slip (FAUS) scenarios.; **e**, **f** example variable-area-uniform-slip (VAUS) scenarios; **g**, **h** example heterogeneous-slip (HS) scenarios. All examples **c–h** are on the Kurils–Japan source-zone

Glimsdal et al. 2013). The resulting unit-source tsunami was modelled for 36 h with the linear shallow water equations on a 1-arc-min grid with elevation based on GEBCO14 and GA250 (Whiteway 2009; Weatherall et al. 2015). The tsunami model domain includes global longitudes $[-40, 320]$, latitudes from $[-72, 65]$, with boundary conditions being periodic (east–west) and reflective (north–south) (Davies and Griffin 2018).

2.2. *Earthquake-tsunami scenarios*

Earthquake scenarios consist of linear combinations of unit-sources (Fig. 1c–h), with the associated tsunami being a linear combination of the unit-source tsunamis (Thio et al. 2007; Burbidge et al. 2008). On each source-zone the constant-rigidity earthquake scenarios were initially generated as detailed below, with magnitudes ranging from $7.2, 7.3, \ldots, 9.7, 9.8$ for computational convenience. Magnitudes above a source-zone specific $M_{w,max} < 9.8$ will later be assigned a rate of zero (Sect. 3.1), while in practice $M_w7.2$ earthquakes will only generate small waves near Australia.

To understand how the hazard is affected by the earthquake representation, three different kinds of constant-rigidity scenarios were created (Fig. 1c–h):

- **Fixed-area-uniform-slip (FAUS)** scenarios have uniform-slip, with magnitude-dependent length and width based on the scaling relations of Strasser et al. (2010) ignoring the predictive uncertainty terms (Fig. 1c, d). For each magnitude the FAUS scenarios include a fixed number of unit-sources along-strike and down-dip (e.g. 16×4 for the examples in Fig. 1c, d, determined following Davies et al. (2017)) and the set of unit-sources is moved through all possible source-zone locations. By iterating over all magnitudes, the full set of FAUS scenarios is produced.
- **Variable-area-uniform-slip (VAUS)** scenarios also have uniform-slip, but account for $\pm 2\sigma$ predictive uncertainties in the rupture length and width using the scaling relations of Strasser et al. (2010) (Fig. 1e, f). At least 15 VAUS scenarios were generated for each FAUS scenario, all having the same magnitude and a random length and width (with independent residuals). The VAUS scenario location is also random, but at least partially overlaps the 'parent' FAUS scenario. The number of 'child' VAUS scenarios per 'parent' FAUS scenario was increased to more than 15 if necessary, to ensure at least 200 VAUS scenarios on the source-zone at each magnitude. These numbers were found sufficient to obtain convergent hazard results at a range of sites around Australia, which was tested by splitting the scenario set into two equal groups and graphically comparing the tsunami maximum-stage vs return-period curves (Davies and Griffin 2018).
- **Heterogeneous-slip (HS)** scenarios are generated using the VAUS scenario's rupture dimensions but have spatially non-uniform slip (Fig. 1g, h). The slip is randomly generated using a k^{-2} type model, specifically the S_{NCF} model of Davies et al. (2015). The S_{NCF} model was chosen because it had the best performance among eight k^{-2} type models tested by comparison with 66 finite-fault inversions (Davies et al. 2015). One HS scenario is generated per VAUS scenario. This implies a similar parent-child relationship exists between the FAUS (parent) scenarios and HS (child) scenarios, with 15 or more 'children' for each parent. Multi-site convergence tests indicate there are enough scenarios to support our hazard calculations (Davies and Griffin 2018).

Further implementation details are provided elsewhere (Davies and Griffin 2018; Davies 2019). A key concept is the parent-child relation between FAUS scenarios and the other scenario types, which will be exploited when assigning occurrence-rates to scenarios (Sect. 3.2) and for model testing (Sect. 2.3).

The effect of depth-varying rigidity is simulated by re-labelling the constant-rigidity scenario magnitudes, without otherwise changing their slip or area. Each unit-source is assigned a depth-dependent rigidity (Fig. 1b), assuming for simplicity the trench is 4 km below mean-sea-level, and these rigidities are used to re-compute each scenario's magnitude. Scenarios in shallow, low-rigidity regions thus behave like constant-rigidity earthquakes with higher magnitude, and conversely for high-rigidity regions. It is necessary to adjust the scenario rate computations with depth-varying rigidity to maintain consistency with instrumental magnitude observations and tectonic constraints (Sect. 3.3), which affects the resulting hazard. This is not specific to our magnitude-relabelling approach; for example, Scala et al. (2019) propose a completely different treatment of depth-varying rigidity for PTHA which also requires scenario rate modifications.

2.3. Biases in Tsunami Scenarios

Scenario generation methods were tested by comparing random database scenarios with 18 earthquake-generated tsunamis observed at DART buoys in 2006–2016 (see two examples in Fig. 2a, b). Full details are reported elsewhere (Davies 2019); below we summarise key results that affect the scenario occurrence-rate modelling in this study. For each

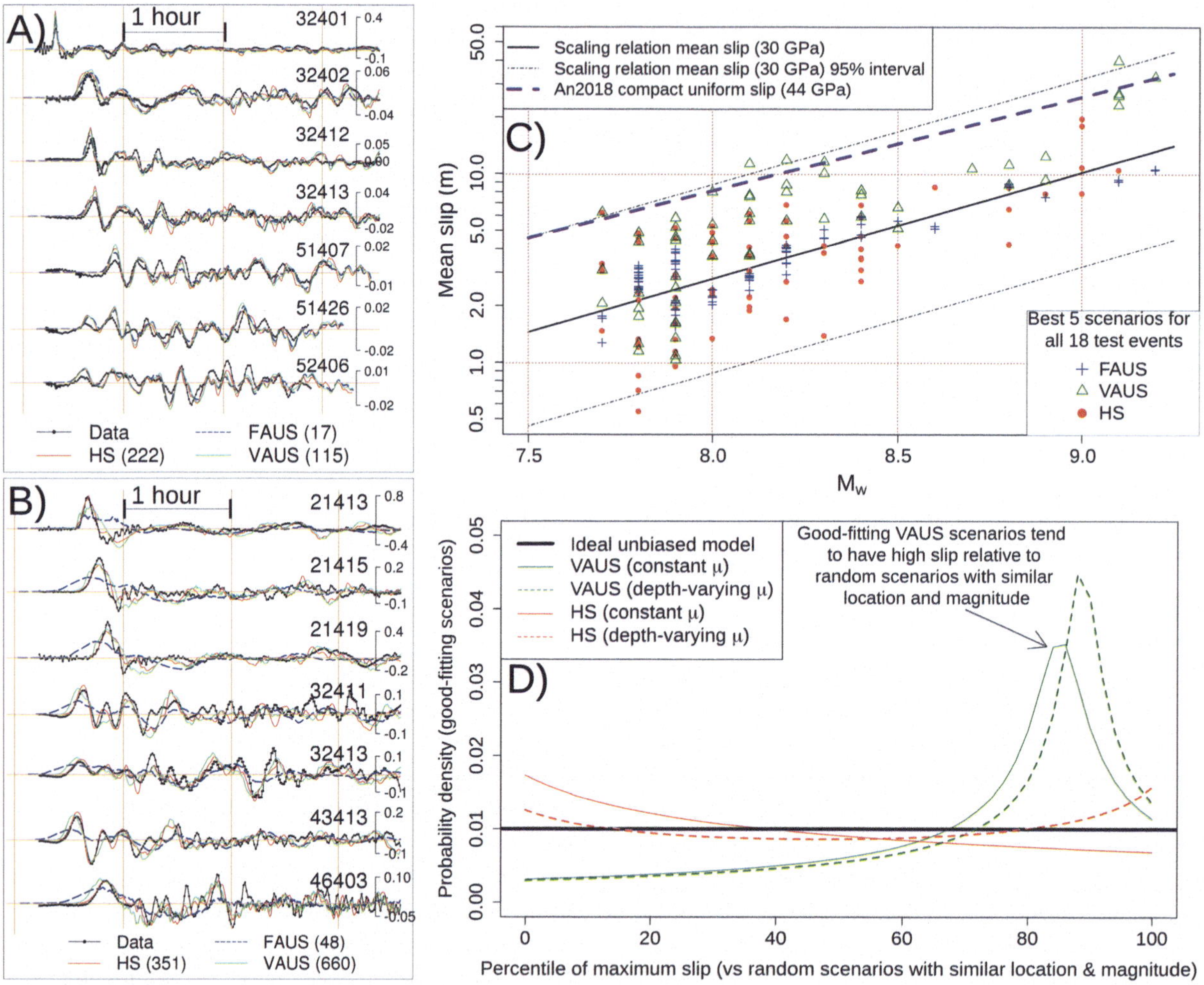

Figure 2

a Example good-fitting database scenarios for the 2014/04/01 M_w8.2 Iquique (Chile) earthquake-generated tsunami using the constant μ FAUS, VAUS and HS models. Each panel compares the tsunami observed at a DART buoy with the same three database scenarios (see Fig. 1 for DART locations). Vertical scale in meters. Modelled time-series are temporally offset by the optimum value determined in the goodness-of-fit calculation; **b** same as panel-A for the 2011/03/11 M_w9.1 Tohoku tsunami, using the depth-varying μ scenarios. For brevity only 7 of the DART observations are shown; **c** magnitude vs mean-slip for the 5 best-fitting database scenarios for all 18 test events (constant μ case). The scaling relation mean-slip is inferred from the Strasser et al. (2010) length and width scaling relations assuming uncorrelated residuals. The 'compact-uniform-slip' curve was derived from the area scaling relation of An et al. (2018), $A = 2.89 \times 10^{-11} M_0^{2/3}$, assuming $\mu = 44$ GPa because they used PREM rigidities; **d** probability density for the maximum-slip percentile of good-fitting events. This is not applied to FAUS scenarios because by construction they have little variability scenario generation method, each of the 18 observed tsunamis was compared with all database scenarios having 'similar earthquake location and magnitude'. Scenarios are defined as having 'similar earthquake location and magnitude' as an observed event if their M_w is within ± 0.15 of the GCMT catalogue value, and their parent FAUS scenario includes unit-sources within half a scaling-relation length and width of the GCMT hypocenter (Davies 2019). This approach to testing was applied because in PTHA, scenario occurrence-rates are generally modelled as a function of the earthquake location and magnitude (e.g. Selva et al. 2016; Power et al. 2017; Davies et al. 2017); thus to avoid biases in the hazard, the 'randomly generated tsunami waveforms' should represent real tsunamis generated by earthquakes with similar location and magnitude. For each observed event, a weighted least-squares goodness-of-fit statistic was used to measure the agreement between the observed tsunami at DART buoys and all aforementioned scenarios (Davies 2019). The goodness-of-fit statistic includes an optimal time-offset (Lorito et al. 2008; Romano et al. 2016; Ho et al. 2019) to account for processes that can delay wave-propagation, but are not treated in our linear shallow water model (Watada et al. 2014; Allgeyer and Cummins 2014; Baba et al. 2017).

Figure 2a, b illustrates the best-fit FAUS, VAUS and HS database scenarios identified with the above procedure for 2 of the 18 test events; the 2014 M_w8.2 Iquique (Chile) tsunami and the 2011 M_w9.1 Tohoku tsunami (see Davies and Griffin 2018; Davies 2019 for other examples). For the 2014 event a reasonable agreement is obtained between observations and the best database scenario for every model type (FAUS, VAUS, HS) (Fig. 2a). In contrast, for the Tohoku event all FAUS scenarios produced long-period low-amplitude waves which poorly matched observations both near and far from the earthquake source, whereas the best VAUS and HS scenarios performed well (Fig. 2b). These visual observations are consistent with quantitative goodness-of-fit results (Davies 2019).

Analysis of the good-fitting earthquake scenarios was used to further assess model biases (Fig. 2c, d; Davies (2019)). If a model gives an unbiased representation of random tsunamis, then the properties of good-fitting earthquake scenarios should not differ systematically from random scenarios with similar location and magnitude, when considered over all 18 test events. Conversely if a model has some bias (e.g. producing too many low-slip, high-area scenarios), we may see statistical differences between good-fitting and random earthquake scenarios. Figure 2c, d suggests the good-fitting VAUS model scenarios most often have high slip relative to the scaling-relation used to construct them, which indicates bias in the VAUS model. In contrast, the good-fitting HS scenarios exhibit mean-slip variability within the expected $\pm 2\sigma$ range of random scenarios, without any strong preference for high or low values (Fig. 2c, d). The FAUS scenarios have little variability by construction and so their biases cannot be usefully quantified with this approach, but we emphasise that some historical events are not well modelled with FAUS scenarios (e.g. Tohoku, Fig. 2b).

The VAUS model biases are qualitatively consistent with the results of An et al. (2018). They simulated near-field tidal gauge and DART buoy observations for six tsunamis using both heterogeneous and uniform slip models, and found near-optimal results could consistently be obtained using compact uniform-slip earthquakes with a low aspect ratio ($L = W$) and high-slip (Fig. 2c). We infer the compact nature of good-fitting VAUS scenarios allows representation of rupture asperities, which have a dominant influence on the resulting tsunami. The HS model can simulate those asperities directly and so avoids similar biases (Davies 2019).

3. Scenario Rates

The scenario rate modelling methodology was designed to meet the following objectives:

1. Applicable to all earthquake-slip and rigidity models in Sect. 2.2.
2. On each source-zone the modelled magnitude-frequency distribution should be reasonably consistent with plate convergence rates and instrumental seismicity.

3. The method should treat uncertainties in the seismic coupling, maximum magnitudes and the Gutenberg-Richter b value.
4. Spatial variations in the scenario rates should reflect variations in plate convergence.

Our approach generalises that of Davies et al. (2017), because the latter is only applicable to FAUS type scenarios with constant rigidity (violating objective 1) and has some weaknesses regarding spatial variations in scenario rates (objective 4) that are addressed herein. Our new approach also makes more efficient use of earthquake catalogue data to constrain uncertainties (objective 3). The approach involves firstly modelling each source-zone's magnitude-frequency distribution (Sect. 3.1), and secondly partitioning these rates among individual scenarios (Sect. 3.2). Possible segmentation of large source-zones (boundaries in Fig. 1) is treated by applying the model separately to: (A) the unsegmented source-zone, and; (B) the union of individual segments; with 50% weight on each interpretation. Ruptures are allowed to cross segment boundaries in any case (as occurred e.g. for the 2007 Solomons earthquake, Lorito et al. (2015a)) so the primary effect of segmentation is to enhance spatial variations in the source-zone's magnitude-frequency distribution. The use of depth-varying-rigidity introduces additional complications which are addressed in Sect. 3.3. The results are tested in Sect. 3.4. The calculations which convert magnitude exceedance-rates to tsunami hazard metrics are described in Sect. 3.5.

3.1. Source-Zone Integrated Magnitude Exceedance-Rates

Seismicity on each source-zone is assumed to follow one of two Gutenberg-Richter (GR) type relations with different tail behaviour; a characteristic GR model (Kagan 2002):

$$\begin{aligned} GR(x) &= 10^{a-bx} \quad \text{for} \quad x \leq M_{w,max} \\ &= 0 \quad \text{for} \quad x > M_{w,max} \end{aligned} \tag{1}$$

and a truncated GR model (Kagan 2002):

$$\begin{aligned} GR(x) &= 10^{a-bx} - 10^{a-bM_{w,max}} \quad \text{for} \quad x \leq M_{w,max} \\ &= 0 \quad \text{for} \quad x > M_{w,max} \end{aligned} \tag{2}$$

For both models $GR(x)$ gives the rate of earthquakes (events/year) with magnitude $\geq x$ as a function of three unknown parameters: a, b, and the maximum magnitude $M_{w,max}$.

Uncertainty in each source-zone's magnitude-frequency distribution is represented using a logic-tree (Annaka et al. 2007). The logic-tree includes every combination of the GR model type and the parameters b, $M_{w,max}$, and the seismic coupling fraction c. The Gutenberg-Richter a value is derived from these via moment conservation arguments (Bird and Kagan 2004; Bird and Liu 2007; Davies et al. 2017). The individual parameters vary through a source-zone specific set of values with 'prior weights' described below. The prior weight of each parameter combination is defined as the product of its individual parameter prior weights. Some parameter combinations will predict unrealistic magnitude-frequency distributions compared with historical seismicity and this is dealt with by using earthquake catalogue data to update the weights via Bayes theorem (Davies et al. 2017).

The two GR models (Eqs. 1, 2) are assigned prior weights of 30% (characteristic) and 70% (truncated). The b parameter is assigned twenty equally spaced values between 0.7 and 1.2 (Berryman et al. 2015; Davies et al. 2017) with uniform prior weights. The coupling prior weights are a 50:50 combination of two different distributions. The first takes the lower, preferred and upper coupling values from Berryman et al. (2015) (with default values of 0.3, 0.5, 0.7 where no information is available) and assumes they define the lower limit, median and upper-limit of the prior coupling cumulative distribution function, with linear interpolation between these values. The second is a uniform distribution assigned to 20 coupling values in [0.1, 1.3]. This latter distribution is deliberately uninformative. The lower coupling limit ($c = 0.1$) is conservative, to prevent source-zones with rapid convergence and limited historical seismicity from being assigned a very low coupling after the Bayesian weight-update. Considering the model herein cannot treat non-stationary seismicity, which could allow for longer time intervals between large earthquakes than expected under stationarity, such conservatism is warranted (Stirling and Gerstenberger 2018). The upper coupling limit ($c = 1.3$) is also conservative and deliberately exceeds the physical limit ($c \leq 1$). This was done because modelled seismicity depends on the product of c, the fault area A_T, and the mean horizontal tectonic convergence rate $\dot{s}$, but uncertainties in the geometry and convergence are not directly treated in our methodology. The $\dot{s}$ values are defined using the models of Bird (2003), Koulali et al. (2015, 2016) and Griffin and Davies (2018), averaged over the source-zone. If the convergence direction is oblique to the trench at any point then the component of convergence that maximises $\dot{s}$ is used, up to a maximum of 50° away from pure thrust. This allows oblique subduction as suggested for the Puysegur source-zone (Hayes and Furlong 2010). Although potentially enabling overestimation of seismicity, this is counteracted by the Bayesian update which will down-weight high c values when inconsistent with observed seismicity (e.g. on 'quiet' source-zones with substantial convergence). By partially weighting the Berryman et al. (2015) coupling values the model can reflect the results of paleoseismic and geodetic analyses, which are informative for some source-zones (e.g. Cascadia) but not otherwise straightforward to include.

The maximum magnitude $M_{w,max}$ ranges between lower and upper limits. The lower limit reflects the largest earthquake thought to have occurred on the source-zone (based on Ekstrom et al. 2012; Storchak et al. 2012; Berryman et al. 2015), plus a small perturbation (0.05) to ensure it has non-zero rate according to the truncated *GR* model (Eq. 2). On the South-American trench, a lower limit $M_{w,max} = 9.2$ is used to represent the 1960 Chile earthquake for consistency with tsunami and geodetic inversions (Moreno et al. 2009; Fujii and Satake 2013), even though seismic wave inversions suggest a higher magnitude ($\simeq 9.5$; Kanamori 1977; Engdahl and Villasenor 2002). The upper limit $M_{w,max}$ is defined using the minimum of two scaling-relation based constraints. Firstly $M_{w,max}$ is less than the magnitude of a 'compact' earthquake that fills the source zone according to the M_w-vs-area scaling relation of Strasser et al. (2010), where the area is evaluated at − 1 prediction standard-deviation to represent a 'compact' event. Secondly $M_{w,max}$ is less than the magnitude of a 'narrow' earthquake which fills the down-dip width of the source-zone, using the M_w-vs-width scaling relation of Strasser et al. (2010) with width evaluated at − 2 prediction standard deviations (representing a 'narrow' earthquake). Forty $M_{w,max}$ values are initially created with equal spacing between these lower and upper limits, and uniform prior weights. Finally all $M_{w,max}$ values are subsequently clipped to a maximum of 9.6, because scaling-relation based $M_{w,max}$ limits can be very high on large source-zones (Berryman et al. 2015; Davies et al. 2017).

The *GR* parameter a is derived from the above parameters using a fault-based seismic-moment conservation approach (Bird and Kagan 2004; Bird and Liu 2007; Davies et al. 2017). The relation between earthquake scenarios and tectonic convergence is:

$$\sum_{e \in E} r_e (SA)_e = \frac{A_T (c\dot{s}) \xi}{\cos(\overline{\delta})} \tag{3}$$

Here e denotes an individual earthquake scenario within the set of modelled scenarios E, r_e is its rate (events/year), and $(SA)_e$ is its spatially integrated slip (m^3). The full source-zone fault area is A_T (m^2) with average coupled horizontal convergence rate $(c\dot{s})$ (m/year) and cosine of mean dip $\cos(\overline{\delta})$. See Meade and Loveless (2009) for justification of the latter factor in Eq. 3, which is not universally applied in moment balance approaches (e.g. Kagan and Jackson 2013); irrespective for our analysis its influence is largely offset when constraining the coupling coefficient (below). The factor $\xi \in [0 - 1]$ accounts for the convergence due to earthquakes with magnitude $< M_{w,min}$, which are not represented among the modelled events E. For the current study $M_{w,min} = 7.2$, but given our scenario discretization this represents a bin of magnitudes $7.2 \pm \Delta/2$ where $\Delta = 0.1$ (Sect. 2.2). Assuming constant rigidity, the proportion of integrated slip due to the modelled scenarios is:

$$\xi = \frac{\int_{(M_{w,min} - \Delta/2)}^{M_{w,max}} gr(x) M_0(x) dx}{\int_{-\infty}^{M_{w,max}} gr(x) M_0(x) dx} \tag{4}$$

where $M_0(x)$ is the seismic moment as a function of the magnitude x, and $gr(x) = -\frac{dGR}{dx}$ (technically we assume infinitesimal smoothing is applied to the GR models near $x = M_{w,max}$ so the derivative is well defined). Due to cancellation Eq. 4 is independent of a. Assuming constant rigidity, the a parameter can be derived from Eqs. 3 and 4 using $c, \dot{s}, A_T, \overline{\delta}, b$ and $M_{w,max}$ (known for each logic-tree branch). It is not necessary to know the individual scenario rates r_e because with constant rigidity the spatially integrated slip $(SA)_e$ is a function of magnitude alone. This reasoning fails with depth-varying rigidity, necessitating a special treatment of that case (Sect. 3.3).

An extremely wide range of Gutenberg–Richter type models GR_i result from these priors, as illustrated on the unsegmented Kermadec–Tonga source-zone (Fig. 3a). Both individual logic-tree branches and the prior-mean curve itself may deviate substantially from the observed seismicity rates. On all source-zones the observed seismicity is taken from the GCMT catalogue including events with $M_w > (M_{w,min} - \Delta/2) = 7.15$, hypocenter within 0.4° of the unit-sources, depth <71 km, having at least one nodal plane with rake within 50° of pure thrust and strike within 50° of the nearest unit-source's strike.

To obtain better agreement between the rate models and aforementioned data, the prior weights are updated using Bayes theorem:

$$w_i = \frac{w_i' L(\text{data}|i)}{\sum_i w_i' L(\text{data}|i)} \tag{5}$$

Here w_i is the posterior weight of the i'th logic-tree branch, w_i' is the prior weight, and $L(\text{data}|i)$ is the likelihood of the data if the i'th logic-tree branch is true. The likelihood is calculated from the rate model and the assumption of stationary seismicity:

$$L(\text{data}|i) = L_{\text{count}}(\text{data}|i) L_{\text{magnitude}}(\text{data}|i) \tag{6}$$

$$L_{\text{count}}(\text{data}|i) = \frac{\exp(-\lambda_i)(\lambda_i)^n}{n!} \tag{7}$$

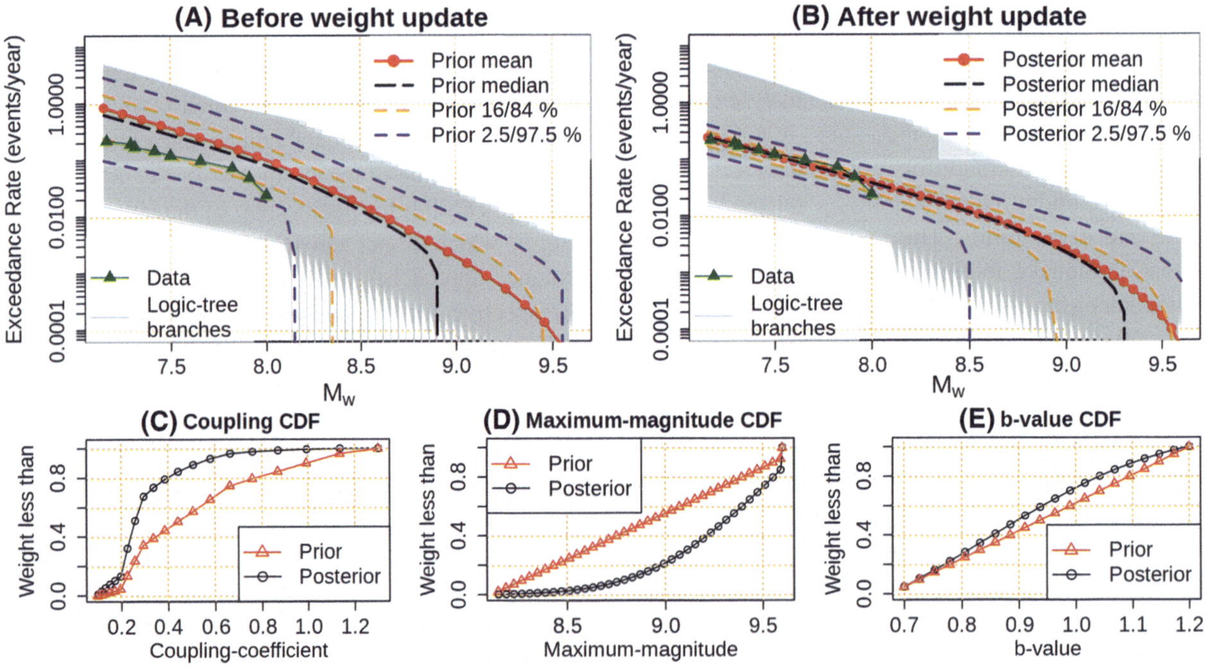

Figure 3

Source-zone magnitude-frequency modelling, using the unsegmented Kermadec-Tonga source as an example. **a** Magnitude-frequency curves prior to weight update. Note the individual logic-tree branches (thin grey curves) appear as a grey shading in much of the plot because there are many; **b** magnitude-frequency curves after weight update; **c** prior and posterior c weights; **d** prior and posterior $M_{w,max}$ weights; **e** prior and posterior b-value weights

$$L_{\text{magnitude}}(\text{data}|i) = \left(\frac{T}{\lambda_i}\right)^n \prod_{l=1}^{n} \left(-\frac{dGR_i}{dx}(M_{w,l})\right) \quad (8)$$

Equation 7 is a standard model for observed counts from a Poisson process, where n is the number of events in the data which spans an observation time T (years), and $\lambda_i = GR_i(M_{w,min} - \Delta/2)T$ gives the mean number of events predicted by the i'th logic-tree branch in time T. Equation 8 was derived by converting GR_i to a probability density for a random magnitude, with the likelihood being the product of the density at each observed magnitude $M_{w,l}$. The approach improves upon that of Davies et al. (2017) which ignored the observed magnitude distribution (i.e. $L_{\text{magnitude}} = 1$). In the special case with no observed events ($n = 0$), the current study also assumes $L_{\text{magnitude}} = 1$. The observation of zero events is nonetheless informative because it suggests logic-tree branches which predict frequent events are unlikely to be correct, and Eq. 7 will down-weight such branches accordingly.

Bayesian updating has a dramatic impact on the modelled earthquake rates and their uncertainties (Fig. 3a, b). In general the mean rate becomes more consistent with historical observations, and uncertainties at low magnitudes are reduced (Fig. 3b). On the unsegmented Kermadec–Tonga source-zone, the results reflect that historical seismicity was relatively low compared to the prior model (Fig. 3a). In principle this could indicate low coupling, and/or that significant slip is released in higher magnitude earthquakes (in which case frequent low-magnitude earthquakes are not required to balance tectonic convergence). However the observations are unlikely if c is high and $M_{w,max}$ is low, because in this case more frequent low-magnitude earthquakes are required to balance tectonic convergence. As a result the Bayesian update shifts more weight onto lower c values, and higher $M_{w,max}$ values (Fig. 3c, d).

3.2. *Partitioning the Source-Zone Integrated Exceedance-Rates Among Scenarios*

For each logic-tree branch i, the individual scenario occurrence-rate $r_{e,i}$ (events/year) for any scenario e with magnitude $M_{w,e}$ must be consistent with the source-zone integrated exceedance-rate function GR_i. Supposing all scenarios have magnitudes arranged as $M_{w,min}, M_{w,min} + \Delta, M_{w,min}+2\Delta, \ldots$ representing bins of width Δ (as for our constant rigidity scenarios where $\Delta = 0.1$), this implies:

$$r_{e,i} = Pr(e|M_w = M_{w,e})\big(GR_i(M_{w,e} - \Delta/2) - GR_i(M_{w,e} + \Delta/2)\big) \quad (9)$$

Here $Pr(e|M_w = M_{w,e})$ is the conditional probability that scenario e occurs, assuming some random scenario with the same magnitude occurs on the same source-zone. This will be modelled as independent of the logic-tree branch i.

Many studies assume uniform scenario conditional probabilities (e.g. Horspool et al. 2014; Løvholt et al. 2014; Lorito et al. 2015b):

$$Pr(e|M_w = M_{w,e}) \propto 1. \quad (10)$$

Alternatively, the scenario conditional probability may be manipulated to make earthquakes more likely to occur on rapidly converging, wider parts of the source-zone. For the case of FAUS scenarios with constant rigidity Davies et al. (2017) proposed to represent this via:

$$Pr(e|M_w = M_{w,e}) \propto (\dot{s}_e/S_e) \quad (11)$$

where S_e is the scenario's mean slip, and $\dot{s}_e$ is the average horizontal convergence rate where e occurs on the source-zone.

A problem with both Eqs. 10 and 11 is they artificially concentrate the time-integrated slip in the middle of the source zone (along-strike), even when the tectonic convergence and source-geometry is uniform. For illustration, consider a FAUS earthquake with 3×2 unit-sources which occurs on a uniform source-zone with 7×2 unit-sources (Fig. 4a). An earthquake of this size can occur in five different positions on the source-zone. Among those five scenarios, unit-sources at the along-strike edges will be included in only one, whereas unit-sources in the middle of the source-zone will be included in three scenarios (Fig. 4a). If all five scenarios have the same slip and occurrence-rate, the time-integrated slip rate would be 3 times greater in the interior of the source-zone than at the along-strike edges. This conflicts with the assumed uniform

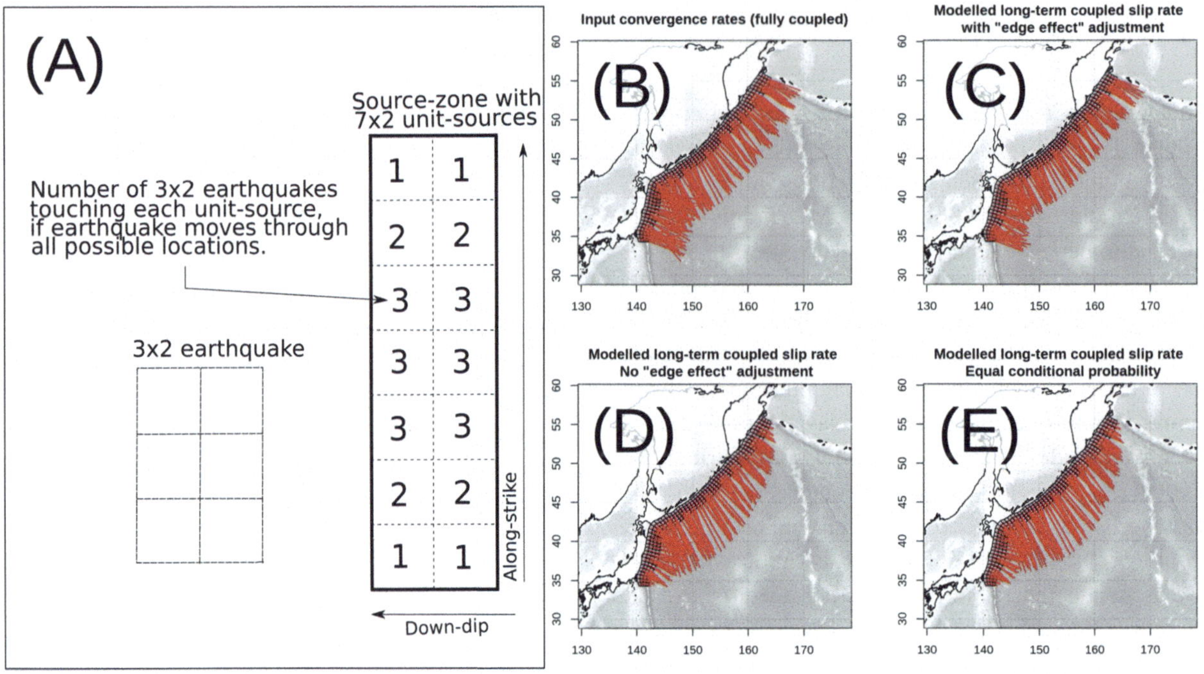

Figure 4

Edge-effect biases in time-integrated slip rates. **a** If a 3 × 2 earthquake is moved through all possible locations on a source-zone, then fewer scenarios will touch unit-sources at the along-strike extremities. If not explicitly treated, this concentrates time-integrated slip in the source-zone interior; **b** spatial distribution of the tectonic convergence rate prescribed for the Kurils–Japan source-zone. The modelled time-integrated slip rate should be similar to this; **c** modelled time-integrated slip rates using Eq. 12 to correct for edge-effects; **d** modelled time-integrated slip rates using Eq. 11, which does not include an edge-effect correction; **e** Modelled time-integrated slip rates using the 'equal conditional probability' approach without any edge-effect correction (Eq. 10)

horizontal convergence and illustrates the 'edge-effect' bias.

The same edge-effect bias occurs in realistic applications which integrate over all scenarios. Figure 4b–e demonstrates this on the Kurils-Japan subduction-zone. All examples use constant rigidity FAUS scenarios, and GR_i in Eq. 9 is replaced with the posterior mean magnitude exceedance-rate over all GR_i in the logic-tree. Although the input tectonic convergence does not vary strongly in space (Fig. 4b), the time-integrated slip rate is concentrated in the middle of the source-zone using either Eq. 10 (Fig. 4e) or Eq. 11 (Fig. 4d) as the scenario conditional probability model. The results of these two approaches would differ more substantially if the source-zone had greater spatial variations in convergence, but clearly both approaches suffer the edge-effect bias.

To approximately correct for the edge-effect, FAUS scenarios which touch an along-strike edge unit-source should have a higher occurrence-rate. Herein this is achieved by modifying Eq. 11 to:

$$Pr(e|M_w = M_{w,e}) \propto (\dot{s}_e/S_e)(1 + kI_e). \tag{12}$$

Here $I_e = 1$ for FAUS scenarios e which include unit-sources on the along-strike edge of the unsegmented source-zone (and $I_e = 0$ otherwise), and k is a source-zone specific constant. For each unsegmented source-zone, k is determined numerically to give the smallest least-squares difference between the spatial distribution of the horizontal tectonic convergence rate and the model's time-integrated slip rate (using the posterior mean over all logic-tree branches). To account for the seismic coupling, which may lead to modelled slip rates being a fraction of the convergence rate, the variables are normalised (divided by their mean over all unit-sources) before computing k; thus only the

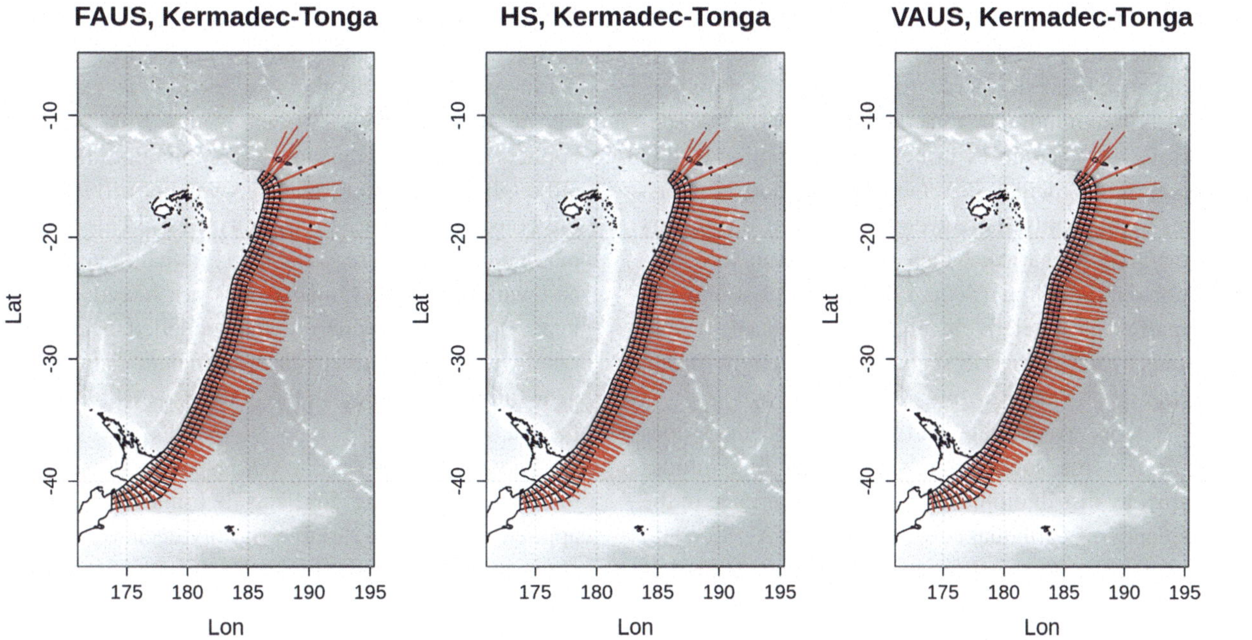

Figure 5
Comparison of modelled time-integrated slip rates (posterior mean over all logic-tree branches) on the Kermadec-Tonga source-zone, computed using the FAUS, VAUS and HS slip models

spatial pattern of slip is considered. This approximation implies the conditional probability does not vary among logic-tree branches, a property shared by Eqs. 10 and 11. This simplifies some calculations and reduces the file-storage required for our analysis, although in-principle a more complex treatment of k could be developed. When modelling segmented source-zones the corresponding unsegmented k value is used, because scenarios can span multiple segments and receive a partial weight from each based on the fraction of their moment that occurs there. Equation 12 leads to better agreement between the spatial patterns of convergence in the input data and the model (Fig. 4b, c), as compared with the use of Eq. 10 (Fig. 4e) or Eq. 11 (Fig. 4d).

When source-zones include three or more unit-sources down-dip, a related edge-effect occurs in the down-dip direction. This reduces time-integrated slip rates in the deepest and shallowest row of unit-sources relative to the mid-depth unit-sources, which are included in more scenarios. The down-dip edge-effect is relatively small compared with the along-strike variant discussed above; for instance in Fig. 4c the shallow and deep unit-sources have modelled time-integrated slip rate on average 81% of the mid-depth sources. An approximate correction could be developed by adapting the above approach, but no attempt was made to do this in the current study because subduction-zone coupling is likely to be genuinely higher at mid-depths (e.g. Bilek and Lay 2018).

The arguments in this section thus far apply only to FAUS scenarios, because justification of Eqs. 11 and 12 requires that rupture area is a deterministic function of magnitude (Davies et al. 2017). To extend the approach to non-FAUS scenarios (Fig. 5) the occurrence-rate of each individual FAUS scenario is partitioned among its ‘child’ scenarios. Recall each FAUS scenario has at least 15 ‘child’ HS scenarios (and similarly for VAUS), all with the same magnitude and partial location overlap with their parent (Sect. 2.2). The simplest occurrence-rate partitioning method is to equally split the parent FAUS scenario’s occurrence-rate among its children. However an unequal partition is preferable to partially correct known model biases, such as those established earlier (most significantly for the VAUS model, Sect. 2.3). Thus in the current study an unequal partition is applied to each group of N child scenarios ($N \geq 15$), separately for VAUS and HS. Each scenario is assigned a local slip percentile ($100r/(N+1)$ where r is its maximum-slip rank among the other N scenarios), and the curves in Fig. 2d are applied to this percentile to define the unequal partition weights. Thus significantly more weight is placed on higher-slip VAUS scenarios, which show good-fit to observations more often (Sect. 2.3). In contrast the HS scenarios have relatively uniform weights (Fig. 2d) because good-fitting HS scenarios do not show strong preference for high or low slip.

The occurrence-rate partitioning procedure does not affect the source-zone's integrated magnitude-frequency distribution, so there is no difference in the frequency of earthquakes with FAUS, VAUS or HS. However the use of different earthquake slip models can lead to slight changes to the modelled spatial distribution of time-integrated slip on unit-sources, because the slip on a single FAUS scenario will not be spatially identical to the integrated slip on its child HS or VAUS scenarios. In practice these differences average out and are small enough to be ignored (Fig. 5). For instance at 95% of unit-sources in Fig. 5 the time-integrated slip rate of the HS and VAUS models differs by $<10\%$ from the FAUS result.

3.3. Extension to Depth-Varying Rigidity

The arguments in Sects. 3.1 and 3.2 do not directly extend to the depth-varying rigidity case, because there is no longer a one-to-one relation between a scenario's magnitude and its spatially integrated slip $(SA)_e$. This complicates the previously simple relation between magnitude and convergence, invalidating Eq. 4 and the associated a parameter calculation. It is also not obvious how to apply the scenario conditional probability model (Eqs. 9 and 12) to scenarios with variable-rigidity magnitudes, considering the latter are not arranged in a small set of discrete values $(7.2, 7.3, \ldots)$.

A key component of our solution to these problems is to parameterise the source-zone's magnitude-frequency distribution as a function of the earthquake's 'constant rigidity magnitude', even when the rigidity is modelled as depth-varying. Obviously the constant-rigidity magnitude then differs from the 'true magnitude'. The difference between these quantities may be treated as a 'perturbation of the magnitude' (Fig. 6) and is explicitly treated below. The key point is that Eqs. 1, 2 are applied using the constant-rigidity magnitude for x, although the posterior logic-tree weights will differ in the constant and variable-rigidity cases. This GR re-parameterisation is reasonable given that the constant and variable rigidity magnitudes are heavily correlated (Fig. 6). Considering the modelled events are rare, we are unlikely to have sufficient data in the near future to distinguish the goodness-of-fit of either parameterisation.

The great benefit of the GR re-parameterisation is that most reasoning in Sects. 3.1 and 3.2 can be applied directly to variable-rigidity scenarios, using constant rigidity magnitudes in Eqs. 1, 2, 3, 4, 9 and 12. The prior logic-tree weights are applied without modification, noting they are in any case very diffuse. However, adjustments are necessary when using magnitude observations to update the logic-tree weights (Eqs. 7, 8), because the GR_i magnitude-frequency distribution now gives the exceedance-rate in terms of the 'constant-rigidity' magnitude M_w^c rather than the 'depth-varying-rigidity magnitude' M_w^v which is represented by observations. To enable the logic-tree weight update, the exceedance-rate function of M_w^v (denoted $\Lambda_i(M_w^v)$) is derived as detailed below, and then used in place of GR_i in Eqs. 7 and 8. This ensures the logic-tree weight update treats the observed magnitudes as representing depth-varying rigidity earthquakes.

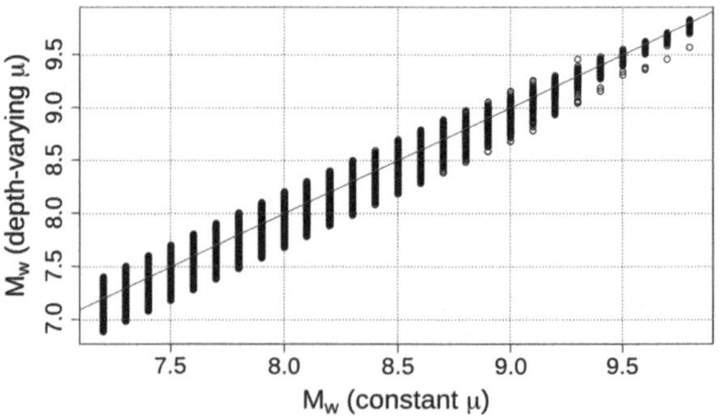

Figure 6
Constant rigidity magnitude vs depth-varying rigidity magnitude for HS scenarios on the Kermadec-Tonga source-zone

To derive Λ_i, consider the magnitude perturbation m' due to depth-varying rigidity:

$$m' = M_w^v - M_w^c. \quad (13)$$

For a random observed event e, we can consider m'_e as a random variable with distribution conditional on its unknown constant-rigidity magnitude (Fig. 6):

$$Pr(m'_e \leq X) = F(X|M_{w,e}^c) \quad (14)$$

where $F(X|M_{w,e}^c)$ is the cumulative distribution function of m'_e for a random scenario e with constant-rigidity magnitude $M_{w,e}^c$. Figure 6 highlights that for scenarios in our database, the variability of m' is magnitude dependent, with less variance at higher magnitudes because ruptures cover a larger area, averaging out the effect of rigidity variation. In the current study F is modelled empirically on each source-zone using the differences between M_w^v and M_w^c in the scenario database. It will be necessary to evaluate F at a continuous set of M_w^c values, so interpolation is used in between the discrete scenario values $(7.2, 7.3, \ldots 9.8)$, while extrapolation outside this range uses the nearest boundary M_w^c value (7.2 or 9.8). Finally, supposing that $GR_i(x)$ gives the source-zone's exceedance-rate for M_w^c on a particular logic-tree branch, the associated exceedance-rate for the variable-rigidity magnitudes is:

$$\Lambda_i(M_w^v) = \int_{-\infty}^{\infty} \left(1 - F(M_w^v - x|x)\right) gr_i(x) dx \quad (15)$$

where $gr_i = -\frac{dGR_i}{dx}$ (technically we assume infinitesimal smoothing of the GR models near $M_{w,max}$ so the derivative is well-defined). Notice the term in large parenthesis in the integrand gives the probability that the magnitude perturbation is large enough for the variable rigidity magnitude to exceed M_w^v when the constant rigidity magnitude is x.

The Λ_i exceedance-rate model is computed numerically from Eq. 15 and used in place of GR_i for all calculations associated with Eqs. 7 and 8, thus treating the observed magnitudes as having depth-varying rigidity when defining the logic-tree weights. The individual scenario occurrence-rates are then computed as in the constant-rigidity case (Sect. 3.2), but using revised logic-tree weights.

3.4. *Testing the Modelled Magnitude Exceedance-Rates*

As a first test the modelled and observed seismicity are compared at the global scale (Fig. 7). Although this data was used to update the logic-tree weights and is thus not independent of the model, the comparison is useful because it may make model biases more obvious than tests at the individual source-zone level, where data sparsity permits a wide range of plausible magnitude-frequency curves (e.g. Fig. 3). The modelled mean magnitude exceedance-rates are in reasonable agreement with the GCMT observations using both constant and depth-varying μ (Fig. 7). The models give slightly higher exceedance-rates than empirical estimates, but are within 95% confidence intervals for the true exceedance-rate inferred from the data (Garwood 1936). At magnitudes $\gtrsim 8.4$ the depth-varying μ model predicts slightly greater exceedance-rates than the constant μ model (Fig. 7) because large earthquakes occur predominantly on wide, deep source-zones, where depth-varying rigidities on average exceed the constant value (30 GPa). For a fixed magnitude, higher rigidity implies less average slip, thus exceedance-rates should increase to produce the same horizontal

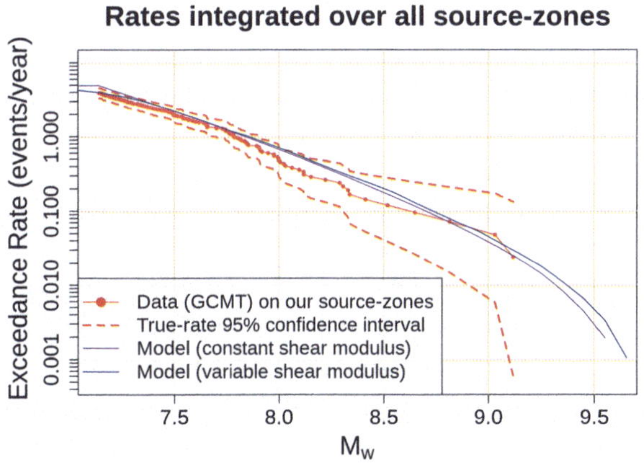

Figure 7

Comparison of globally integrated magnitude exceedance-rates model and the GCMT catalogue. The catalogue data combines all earthquakes used to update the source-zone logic-tree weights. The modelled exceedance-rates were derived by summing the mean scenario occurrence-rates. This leads to some artefacts in the depth-varying-rigidity curve at low magnitudes (around $M_{w,min}$), due to the use of a minimum magnitude combined magnitude-relabelling technique. However this is inconsequential for the hazard and is properly accounted for in the logic-tree weight update (Sect. 3.3) convergence. This effect is partially offset by the Bayesian update of logic-tree weights using earthquake catalogue data, which constrains the exceedance-rates at commonly occurring magnitudes irrespective of the rigidity and convergence. Thus the globally-integrated exceedance-rates do not differ much at magnitudes $\lesssim 8.4$, irrespective of the rigidity model (Fig. 7).

To compare the models herein with other published magnitude exceedance-rates it is useful to restrict the analysis to some prescribed region (i.e. a subset of unit-sources that matches the region considered in other studies, Table 1). To do this, the modelled occurrence-rates for HS scenarios inside the prescribed region are summed to create a 'local' magnitude exceedance-rate curve. To account for the magnitude discretization, the local exceedance-rates are computed directly at magnitude bin-boundaries (i.e. $M_w \geq 7.15, 7.25, \ldots 9.75$), and linear interpolation of M_w vs $\log_{10}(\text{exceedance-rate})$ is used to evaluate the 'local' exceedance-rate at other magnitudes. When HS scenarios have only part of their integrated slip in the prescribed region, they are included with proportionately down-weighted occurrence-rate. Credible intervals are derived by partitioning the full source-zones percentile uncertainties, using the same approach applied to the mean curve.

The constant and depth-varying μ models usually give similar magnitude exceedance-rates on subsets of subduction zones (Table 1). Differences reflect the source-zone's maximum depth below the trench, as specified using the upper estimates of Berryman et al. (2015). On source-zones with relatively deep seismicity (Alaska, Chile, Japan, Sumatra—with modelled depths extending 45–55 km below the trench) the constant rigidity model predicts slightly lower exceedance-rates, as was observed for global seismicity due to the trade-off between higher rigidities and exceedance-rates under moment conservation constraints (Fig. 7). The converse applies for relatively shallow source-zones (Nankai and Cascadia, with modelled depths extending 25–30 km below the trench).

The modelled ARIs are generally comparable to the range of estimates in other studies using longer-term historical or paleo data, and/or alternative moment conservation techniques (Table 1). The credible intervals are wide, but this is also true of 95% intervals reported elsewhere (Table 1, Rong et al. 2014; Butler et al. 2016). The uncertainty is further emphasised by comparing ARIs from different studies at the same site (Table 1). For example on the relatively well studied Alaska megathrust (Kodiak to Prince William Sound), Wesson et al. (2007, 2008) drew on multi-site Holocene stratigraphy to infer an ARI $\simeq$ 650 years for earthquakes like the 1964 event (which had M_w9.2), whereas Butler et al. (2016) estimated $M_w \geq 9$ ARI $\simeq$ 1403 (447–9800) years using an expanded set of paleo data. Recently Shennan et al. (2018) inferred the 1964 earthquake was the only event in the last 2000 years to rupture the entire region, by combining paleoseismic results at 22 sites. Our model's results near Alaska are most similar to Butler et al. (2016), although the Wesson et al. (2007, 2008) ARI's are within the 95% interval (Table 1). Wesson et al. (2007, 2008) assumed $M_w \geq 9$ earthquakes never occur further west (Semidi and Shumagin segments) whereas our model does not represent such fine-scale variations in seismicity. Thus if the latter segments are included in the prescribed region, our modelled $M_w \geq 9$ ARIs continue to reduce (Table 1).

Table 1 also highlights the diversity of ARIs estimated from moment-conservation type arguments (including our model). For example in the Tohoku region of Japan, Kagan and Jackson (2013) estimated that $M_w \geq 9$ events have ARI $\simeq$ 300–400 years by constraining the parameters of a tapered Gutenberg-Richter type model with tectonic convergence rates and earthquake catalogue data. In the same region Butler et al. (2016) obtained an ARI of 1148 (490–3448) using the 'regionally scaled global moment rate' (RSGR) method, whereas our method suggests ARIs intermediate between these two estimates (Table 1).

Systematic differences are expected between our model and the RSGR method used by Butler et al. (2016). The latter partitions global subduction seismicity among source-zones in proportion to their length and trench-normal convergence rate (Burbidge et al. 2008). It thus ignores variability in $M_{w,max}$, coupling, and the source-zone width, whereas these factors are included in our model, albeit with

Table 1

Magnitude exceedance average-recurrence-interval (ARI, units of years) for sub-regions of some subduction zones

Region	M_w	Range along-strike (lon,lat)	ARI (constant μ)	ARI (depth-varying μ)	ARI (other)	Other category	Other comment
Alaska near 1964 earthquake	9	(−145.3, 59.15) to (−153.97, 55.26)	1494 (670–11,978)	1319 (594–8071)	1403 (447–9804); B16	P	
"	9	"	"	"	1350 (593–3969); B16	M	RSGR method
"	9?	(−144.49, 59.21) to (−153.97, 55.26)	1399 (628–11244)	1230 (554–7499)	650; W08	P	Events similar to 1964 earthquake
Alaska extended to Shumagin	9	(−144.49, 59.21) to (−162.3, 53.21)	770 (339–6916)	707 (314–5074)			
Cascadia subduction zone	9	(−125.03, 40.64) to (−127.94, 49.61)	1586 (542–∞)	2174 (672-∞)	1000 (285–5000); R14	MPH	Generic b-value, their Fig. 5
"	"	"	"	"	1000 (222-∞); R14	MPH	Source-zone b-value, their Fig. 5
Chile near 1960 earthquake	8	(−75.86, −45.62) to (−74.57, −37.58)	60 (35–138)	59 (35–142)	65 ± 41; B08	H	
"	8.5	"	172 (103–501)	166 (98–505)	128 ± 46; B08	H	
"	8.6	"	216 (131–677)	206 (122–678)	292 ± 93; M18	P	Magnitude 8.6 is a lower bound
"	9	"	597 (350–5889)	532 (313–4593)	873 (283–6803); B16	P	
"	9	"	"	"	950 (421–2778); B16	M	RSGR method
"	9?	"	"	"	350; C17	P	Mean time between 4 events which may be similar to 1960 earthquake
Japan near 2011 earthquake	9	(144.28,40.09) to (142.28,35.02)	649 (366–1730)	552 (315–1501)	300–400; K13	MH	
"	9	"	"	"	1157 (405–7143); B16	P	
"	9	"	"	"	1148 (490–3438); B16	M	RSGR method
Kamchatka near 1952 earthquake	9	(162,52.48) to (157.23,48.8)	832 (465–2643)	717 (409–2205)	1203 (506–4950); B16	P	
"	"	"	"	"	1305 (554–3921); B16	M	RSGR method
Nankai offshore of Japan	8	(138.19,34.01) to (132.43,30.81)	130 (66–481)	163 (87—577)	124 ± 93; B08	H	
"	8.6	"	651 (268–∞)	1085 (380–164005)	661; B08	H	
Sumatra near 2004 earthquake	9	(96.12,1.57) to (92.68,14.45)	758 (353–5054)	728 (331–2910)	1653 (617–9174); B16	P	
"	9	"	"	"	1470 (662–4254); B16	M	RSGR method
"	9?	"	"	"	500; R13	P	Mean time between 3 events with similar extent to 2004 earthquake

ARIs for the constant μ and depth-varying μ models are compared with other published ARIs. The 'Other category' column indicates the published ARI methodology, involving paleo-data (P), long-term historical data (H), and/or moment-conservation arguments (M). ARIs in parenthesis are 95% credible intervals. Where credible intervals were not provided, the uncertainty is reported following the original study (but is not a 95% credible interval). Magnitudes followed by '?' were inferred when no M_w was provided; in these cases paleo events were identified as being 'similar' to some historical event as described in the comment. Reference abbreviations are: B08 (Burbidge et al. 2008), B16 (Butler et al. 2016), C17 (Cisternas et al. 2017), K13 (Kagan and Jackson 2013), M18 (Moernaut et al. 2018), R13 (Rajendran 2013), R14 (Rong et al. 2014), W08 (Wesson et al. 2008) significant uncertainty (e.g. Fig. 3). Therefore compared with our model the RSGR method should predict more frequent earthquakes on narrow source-zones with apparently low coupling (e.g. Marianas trench), and conversely on wide source-zones with apparently high coupling (e.g. South America). This seems consistent with predictions of each model near Chile, Japan, Kamchatka and Sumatra (Table 1). However the relationship does not extend to the wide Alaskan megathrust, where both approaches give similar results (Table 1). This is because the Alaskan segment of our Alaska–Aleutian source-zone does not feature GCMT thrust earthquakes with $M_w > 7.15$, causing the Bayesian weight update to prefer moderately low coupling (prior mean $c = 0.75$; posterior mean $c = 0.35$). The effect is moderated by the 50% weight assigned to the unsegmented Alaska-Aleutians source which exhibits more GCMT seismicity (prior mean $c = 0.64$; posterior mean $c = 0.49$). Nonetheless the lower overall coupling in our model effectively offsets the high width on the Alaskan megathrust, ultimately leading to results similar to the RSGR method.

Considering our model makes limited use of site-specific long-term data (to facilitate global application), our ARIs are not assumed to be more accurate than others in Table 1 at any particular site. However the fact that our results are comparable to a range of other studies, many of which employ much longer term observational data, gives some support to the method.

3.5. Maximum-Stage Exceedance-Rates at Hazard Points

For each earthquake-tsunami scenario, the maximum-stage (i.e. maximum simulated water-level above ambient sea level) is used to describe the tsunami size at any given location. In practice the tsunami model time-series are not stored at every grid cell due to file storage limitations; instead they are stored at a set of around 20,000 locations (termed 'hazard points') which are globally distributed but have much higher density near Australia (Davies and Griffin 2018). For each hazard point the tsunami maximum-stage exceedance-rates are derived from the earthquake magnitude-frequency models as described below.

First consider a single unsegmented source-zone (or an individual segment of a segmented source-zone). If the earthquake-slip and rigidity models are specified then each individual earthquake-tsunami scenario e on the source has an associated family of occurrence-rates $r_{e,i}$ which are derived by partitioning each logic-tree GR_i curve among all scenarios on the source (Eqs. 9, 12). Each GR_i and $r_{e,i}$ also have an associated Bayesian posterior weight w_i (Sects. 3.2 and 3.3). For a given hazard point p and scenario e, denote the tsunami maximum-stage as $\eta_{e,p}$. Then for each GR_i there is an associated maximum-stage exceedance-rate curve at p, given by:

$$\begin{aligned}\Upsilon_i^p(x) &= \sum_e r_{e,i}\mathcal{I}(\eta_{e,p} > x) \\ &= \sum_e \Bigg[\left(GR_i\left(M_{w,e} - \frac{\Delta}{2}\right) - GR_i\left(M_{w,e} + \frac{\Delta}{2}\right)\right) \\ &\qquad Pr(e|M_w = M_{w,e})\mathcal{I}(\eta_{e,p} > x)\Bigg]\end{aligned} \tag{16}$$

where the second step simply expands $r_{e,i}$ using Eq. 9 and is useful below. Here $\Upsilon_i^p(x)$ gives the rate (average number of events per year) of earthquake-tsunamis with maximum-stage greater than x at point p, assuming that GR_i is correct. The indicator function $\mathcal{I}()$ is defined to be unity if its argument is true, and zero otherwise. Equation 16 leads to a family of maximum-stage exceedance-rate curves at each hazard point for each unsegmented source-zone (or individual segment). To summarise the results it is natural to define the logic-tree-mean maximum-stage exceedance-rate curve $\Upsilon^p(x)$ (events/year) as:

$$\begin{aligned}\Upsilon^p(x) &= \sum_i w_i \Upsilon_i^p(x) \\ &= \sum_e \Bigg[\left(\overline{GR}(M_{w,e} - \frac{\Delta}{2}) - \overline{GR}(M_{w,e} + \frac{\Delta}{2})\right) \\ &\qquad Pr(e|M_w = M_{w,e})\mathcal{I}(\eta_{e,p} > x)\Bigg]\end{aligned} \tag{17}$$

where $\overline{GR}$ is the posterior-mean magnitude exceedance-rate curve over all logic-tree branches. The second step follows from Eq. 16 and highlights that individual $\Upsilon_i^p(x)$ do not need to be calculated. This facilitates efficient computation, and is a benefit of the scenario conditional probability model being independent of i (Sect. 3.2). This logic-tree-mean maximum-stage exceedance-rate can be straightforwardly generalised to multiple source-zones by summation of their $\Upsilon^p(x)$ values. On source-zones with segmentation 50% weight is placed on the union-of-segments interpretation, and the remainder on the unsegmented interpretation (as was done for the magnitude exceedance-rates).

Percentile maximum-stage exceedance-rate curves are also useful to indicate the uncertainty in the tsunami hazard (Power et al. 2017). For example, consider the 84th percentile maximum-stage exceedance-rate at point p, denoted $\Upsilon^{p,84}(x)$ with units (events/year) which is a function of the desired maximum-stage x. For the case of an unsegmented source-zone or single segment, $\Upsilon^{p,84}(x)$ is defined as the *smallest* number such that:

$$\sum_i \left(w_i \mathcal{I}\left(\Upsilon_i^p(x) \le \Upsilon^{p,84}(x)\right)\right) \ge 84/100 \qquad (18)$$

For a given maximum-stage x, this implies 84% of the logic-tree weight is assigned to maximum-stage exceedance-rates $\le \Upsilon^{p,84}(x)$. Eq. 18 directly generalises to other percentiles in the open interval (0, 100). Note a different definition was used in Davies and Griffin (2018) for computational expedience; in comparison Eq. 18 is more rigorous but relatively expensive to compute because all $\Upsilon_i^p(x)$ are required. Although Eq. 18 is used herein, the impact on our percentile uncertainty calculations is small; for instance the 84th percentile results in Sect. 4.2 differ from those of Davies and Griffin (2018) by less than 5% at 90% of hazard points.

The generalisation of exceedance-rate percentiles (Eq. 18) to multiple source-zones is more complex than for the logic-tree-mean (Eq. 17) because it depends on assumptions about the dependence of uncertainties between source-zones; we defer full discussion of this to Sect. 4.2. However note that for a possibly segmented source-zone, the maximum-stage exceedance-rate for a given percentile and maximum-stage is computed by summing the results of Eq. 18 on each individual segment (which prevents cancellation of uncertainties due to segmentation, and is consistent with the co-monotonic treatment discussed in Sect. 4.2). Given a 50% weight on both the 'union-of-segments' and 'unsegmented' interpretations, the combined distribution of exceedance-rates for a given maximum-stage is a 50:50 mixture of the 'union-of-segments' and 'unsegmented' exceedance-rate distributions. The latter two distributions can be computed individually from their percentiles; it is then straightforward to derive the full mixture distribution and compute any desired percentiles directly.

4. *Results*

4.1. *Sensitivity of Offshore Hazard to the Chosen Slip and Rigidity Model*

The sensitivity analysis focusses on the ARI = 500 year tsunami maximum-stage at sites offshore of Australia (Fig. 8a). To compute this, for every source-zone the logic-tree-mean maximum-stage exceedance-rate (Eq. 17) is calculated at 100 maximum-stage values logarithmically spaced from 0.02 to 20 m; then for each maximum-stage the exceedance-rates are summed over all source-zones, and finally the 'ARI = 500 year maximum-stage' is defined as the maximum-stage that results in this summed exceedance-rate being 1/500, interpolating as required. Irrespective of the chosen slip or rigidity model, wave shoaling over the continental shelf leads to strong shore-normal gradients in tsunami size (Fig. 8a). This complicates interpretation at the continental scale, and so for subsequent analysis the results are normalised to 100 m depth using Green's law (i.e. multiplied by $(\text{depth}/100)^{1/4}$). Normalisation greatly reduces the depth-dependence of the results and emphasises regional patterns in the offshore tsunami size (Fig. 8b).

For each combination of earthquake slip and rigidity model, the normalised ARI = 500 maximum-stage is depicted in Fig. 9. Comparison of models with constant μ (top row) and depth-varying μ (bottom row) indicates the choice of rigidity model has a minor effect on the results when the earthquake-slip model is fixed (Fig. 9). Switching between rigidity models typically leads to point-wise changes

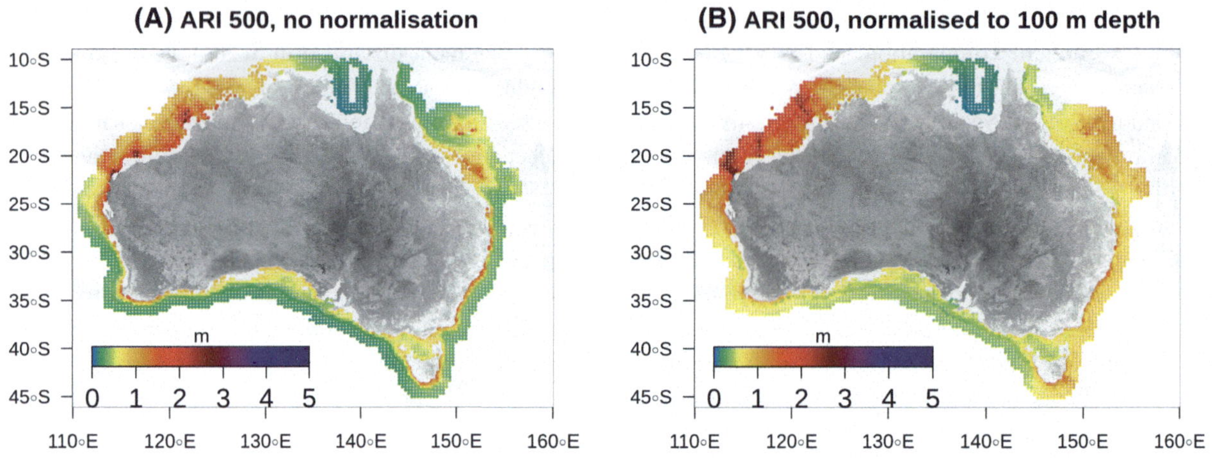

Figure 8

a Maximum-stage at ARI = 500 years, using the HS model with constant μ for illustration. **b** The same data normalised to a depth of 100 m using Green's law (i.e. multiplied by $(\text{depth}/100)^{0.25}$), to reduce the effect of depth variations

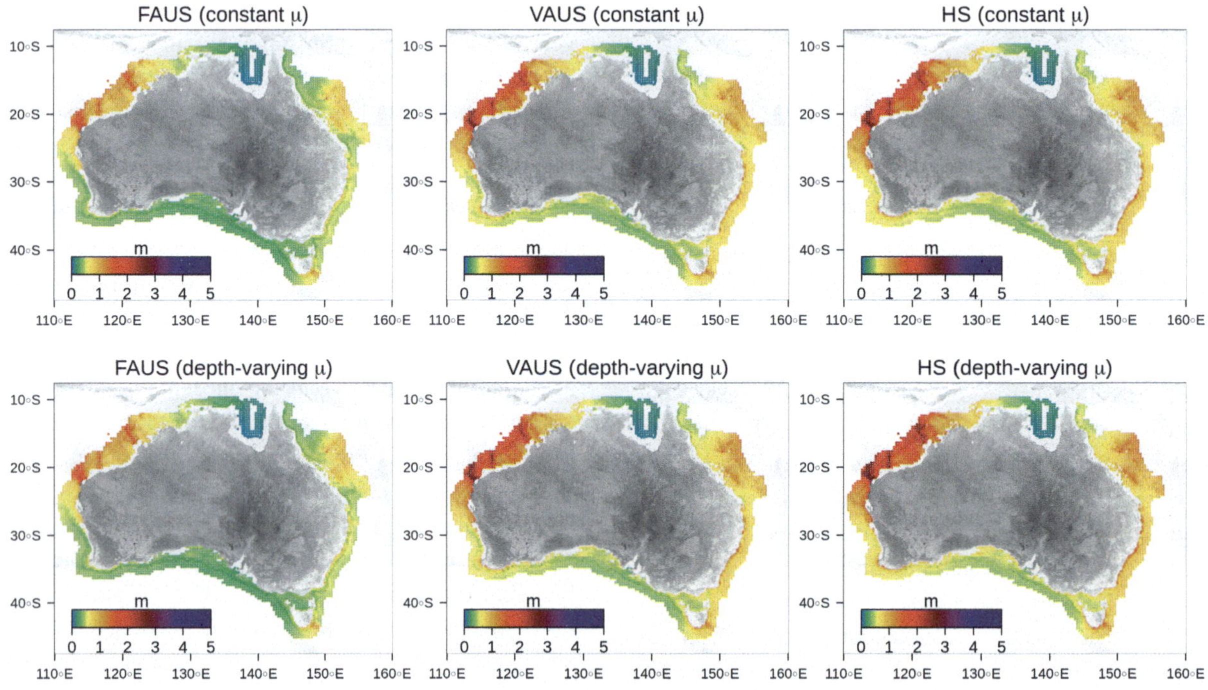

Figure 9

Maximum-stage at ARI = 500 years with Green's law normalisation to 100 m depth, for all slip and rigidity models

of a few percent, and while the difference varies from site to site it is always less than 10%.

The choice of earthquake slip model has a more substantial impact on the results (Fig. 9). The FAUS model produces smaller ARI = 500 tsunamis than both the VAUS and HS models (Fig. 9); for example the FAUS/VAUS ARI = 500 maximum-stage ratio is around 0.67 (median over all sites), with 90% of sites in (0.6–0.84) irrespective of the rigidity. Differences between the VAUS and HS models are much smaller

(Fig. 9), with the VAUS/HS ARI=500 maximum-stage ratio $\simeq 0.91$ (median over all sites), with 90% of sites in (0.85–0.96).

These differences reflect both the varying capacity of the earthquake slip models to produce large tsunamis, and our application of bias-adjustment via a non-uniform partition of the parent FAUS scenario occurrence-rates (Sect. 3.2). Recall bias-adjustment was not applied to the FAUS model because its scenarios have little variability by construction, so even though they poorly represent some observed tsunamis (Fig. 2b) no improvement can be obtained by preferentially weighting some subset of scenarios. Conversely, both the HS and VAUS models produce more variable scenarios that are amenable to bias adjustment. For the VAUS model this resulted in a strong preference for compact, high-slip scenarios (Fig. 2d), which tend to produce larger tsunamis than uniform-slip scenarios with similar magnitude but low or median slip. This is the key driver of differences between the VAUS and FAUS hazard results (Fig. 9). While the HS model was subject to a much smaller bias-adjustment, the similarity of the HS and VAUS results reflects that HS scenarios can simulate slip asperities directly without recourse to compact rupture area. Considering the VAUS model completely ignores earthquake slip heterogeneity, it is remarkable that the difference with a heterogeneous-slip model is only around 10% once a preference for compact ruptures is accounted for (Fig. 9).

4.2. *Sensitivity of Offshore Hazard to Epistemic Uncertainty in the Magnitude-Frequency Distributions*

Uncertainties in PTHA also result from the uncertain frequency of large-magnitude earthquakes (e.g. Fig. 3). It is important to understand the relative significance of this compared with the choice of slip and rigidity model, to help guide future improvements to PTHA methodologies (Sepúlveda et al. 2019). A complication arises because the site-specific hazard is often affected by multiple source-zones. Thus we must determine whether the uncertain earthquake frequencies on these source-zones are independent, or exhibit some kind of epistemic-uncertainty dependence. For example, the frequency of $M_w > 9$ earthquakes on the Kermadec-Tonga trench is highly uncertain due to poor constraints on $M_{w,max}$ (Fig. 3), but if future research demonstrated that $M_{w,max} > 9$ on the Tonga segment, we ask if this would influence our belief that $M_{w,max} > 9$ elsewhere (e.g. on the nearby Kermadec segment, or other source-zones)? If the answer to such questions is always 'no' then the uncertainties are independent, and otherwise they are dependent.

Dependence does not affect the mean hazard (because the mean of the sum of random variables is always equal to the sum of their means), but does affect percentile uncertainty calculations. If two source-zones show positive dependence in the magnitude-frequency epistemic-uncertainty, then the hazard uncertainty will increase at coastal sites affected by both. Although it is unclear how to best specify inter-source-zone epistemic-uncertainty dependence, in many situations independence seems unlikely. Source-zone parameters such as $M_{w,max}$, coupling coefficients and *GR* b-values are often hypothesised to be related to other physical properties of the source-zone (e.g. McCaffrey 1997; Scholz and Campos 2012; Nishikawa and Ide 2014; Bilek and Lay 2018), and if correct such theories imply epistemic-uncertainty dependence, because source-zones with similar properties will deviate similarly from the model (assuming those properties are not already accounted for). For example it has been hypothesised that subduction-zones with high down-dip curvature are less likely to host large earthquakes due to heterogeneities in shear strength (Bletery et al. 2016). This may or may not be correct, but if true it suggests the relatively high-curvature Solomon, New Hebrides, Kermadec-Tonga, Philippines, Marianas and Scotia subduction zones may host high-magnitude earthquakes infrequently compared with our model (which does not explicitly consider curvature). This possibility implies some epistemic-uncertainty dependence between these source-zones. A conflicting hypothesis is that $M_{w,max}$ is limited only by the source-zone size (McCaffrey 2008). This may or may not be correct, but if correct implies the true frequency of high magnitude earthquakes will be high relative to the model on most source-zones (i.e. wherever the model places significant weight on smaller $M_{w,max}$ values). The key point is that our model may have structural errors. This suggests some inter-source-zone epistemic-uncertainty dependence, albeit difficult to specify.

To robustly account for the unknown epistemic-uncertainty dependence structure, herein the multi-source-zone maximum-stage exceedance-rate percentile curves are computed assuming every source-zone simultaneously attains the same percentile (Fig. 10). This is termed 'co-monotonic' dependence (Deelstra et al. 2009). For instance the 16th-percentile panel in Fig. 10 is computed assuming that at each hazard point, the 16th percentile maximum-stage exceedance-rate curve is 'true' for all source-zones simultaneously, so for any maximum-stage these exceedance-rates may be summed to derive the multi-source-zone exceedance-rate (and similarly for the other percentiles). Co-monotonicity is widely used to robustly model dependence in economic theories of decision under risk and uncertainty (Deelstra et al. 2009) and prevents uncertainties on multiple source-zones from partially cancelling, as would occur under independence. Because most coastal sites are significantly affected by only a few source-zones, the site-specific maximum-stage exceedance-rate uncertainties are not reliant on the global correctness of the co-monotonic assumption, but rather that it describes the dependence of locally significant source-zones (Davies et al. 2017). The co-monotonic approach bypasses the need to fully describe this dependence structure, albeit at the expense of some conservatism.

Compared with the choice of slip and rigidity model, epistemic uncertainty in the magnitude-frequency distributions leads to large uncertainty in the ARI = 500 maximum-stage (Fig. 10). Taking the HS model with constant μ for illustration, the 16th, 50th and 84th percentile values in Fig. 10 are respectively $\simeq 50\%$, 87% and 126% of the mean discussed earlier (top right-panel of Fig. 9). The latter ratios vary spatially, but mostly within $\pm 10\%$. The dominance of magnitude-frequency related uncertainties in our model is qualitatively consistent with results of recent tsunami hazard assessments in the South China Sea (Li et al. 2016; Sepúlveda et al. 2019). They found the maximum-stage at ARIs $\simeq$ 100–1000 years changed by well over 100% due to the choice of magnitude-frequency model, as compared with changes $\simeq 20 - 60\%$ due to the earthquake slip representation.

4.3. Global Scale Results

Although our study is focussed on Australia, some results were stored globally to facilitate model testing and interpretation (Fig. 11). At the global level the model suggests large waves are most likely around major subduction zones. The South America and Kurils-Japan subduction-zones are particularly prominent because they are wide, converging relatively rapidly, definitely have $M_{w,max} > 9$, and their GCMT earthquake history leads the model to favour reasonably high coupling (posterior mean $c \simeq 0.77$ in both cases). The model also predicts substantial

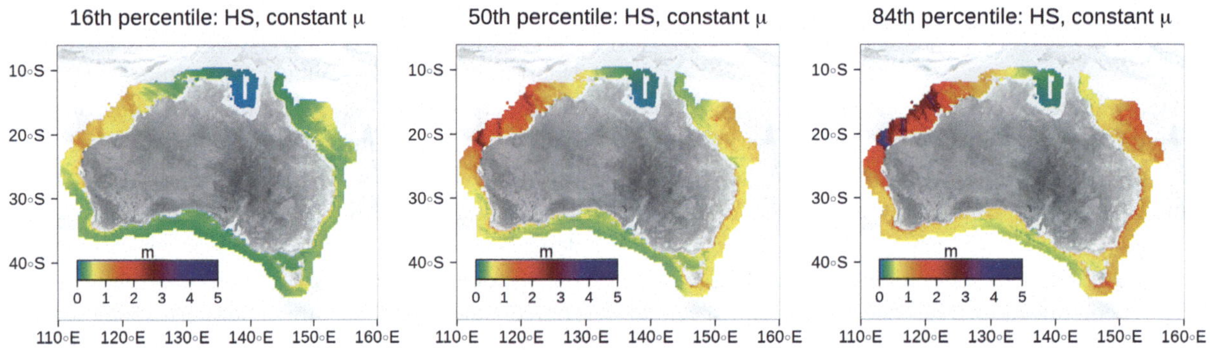

Figure 10
Percentile uncertainties in maximum-stage at ARI = 500 years with Green's law normalisation to 100 m depth, due to epistemic-uncertainty in magnitude-frequency distributions. Results use the HS model with constant μ

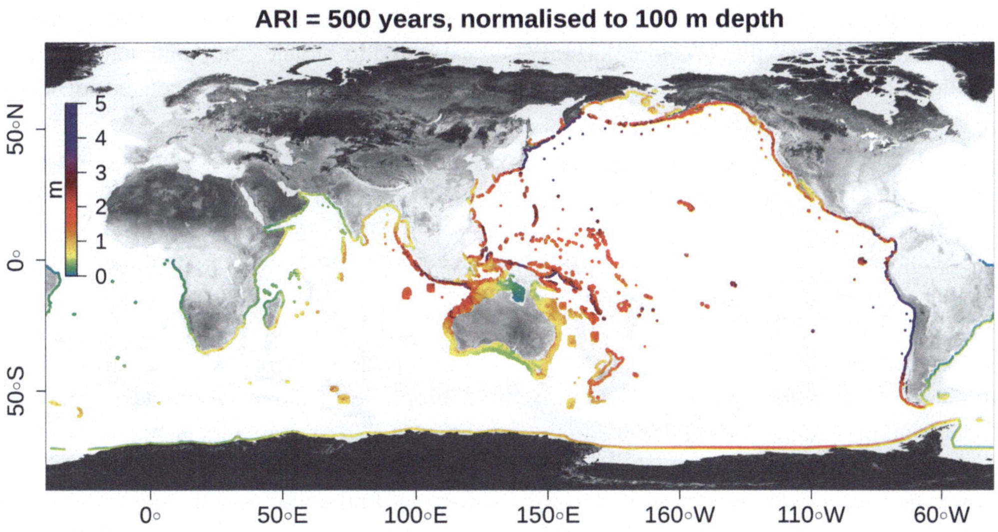

Figure 11
Maximum-stage (m) at ARI = 500 years, normalised to 100 m depth using a Green's law factor $((\text{depth}/100)^{0.25})$ to reduce the effect of depth variations. The HS model with constant μ was used

penetration of large waves into the central Pacific Ocean (Fig. 11). Noting that both the 1946 Aleutian and 1960 Chile earthquakes led to large far-field tsunami runup at central Pacific sites such as Hawaii, the Marquesas and Easter Island (NGDC 2018), this result seems qualitatively reasonable. Compared with the global results the hazard in Australia is moderate overall, except on the northwest coast which is directly exposed to tsunamis generated on the eastern Sunda Arc (Fig. 11).

It is illuminating to assess the FAUS, VAUS and HS model return periods corresponding to larger offshore waves observed near Japan during the 2011 Tohoku tsunami. During this event two GPS gauges near Iwate in Japan recorded maximum-stage values exceeding 6 m. They were located about 12 km offshore in 200 m depth and separated by 40 km (termed 'Iwate M' and 'Iwate S' in Satake et al. 2013, see the latter study for locations and observed time-series). Our nearest model point falls between these gauges but slightly further offshore (395 m depth). If the wave shoaling follows Green's law then the corresponding maximum-stage is around 5.1 m at the model point, which has a modelled ARI of 970 years (HS), 1208 years (VAUS), and never occurs with the FAUS model (ARI $= \infty$). This further highlights the potential for FAUS scenarios to underestimate offshore wave-heights, and the tendency for comparable results to be obtained using either the (bias-adjusted) VAUS or HS scenarios.

5. Conclusions

The framework in this paper facilitates a consistent treatment of earthquake-scenario rates for PTHA using different slip and rigidity models, while maintaining reasonable consistency with the historical earthquake record and tectonic constraints. It features a number of improvements relative to previous approaches for large-scale PTHA (e.g. Power et al. 2017; Davies et al. 2017):

- The edge-effect adjustment leads to a better match between modelled time-integrated slip rates and spatial variations in tectonic convergence
- Earthquake catalogue data is more efficiently used to control the logic-tree weights, in a manner that also accounts for the choice of rigidity model.
- Non-uniform scenario weights are used to partially offset biases in the earthquake slip models.

This provides a suitable basis to study the sensitivity of PTHA results to the choice of earthquake slip and rigidity model.

Within our framework the choice of rigidity model has a small effect on the offshore hazard. Although shallow low rigidity zones may permit surprisingly large tsunamis for their magnitude, tectonic constraints imply such events should occur less often than higher-rigidity events with similar magnitude and lower slip. A similar trade-off was identified by Scala et al. (2019) who used a completely different 'slip-amplification' approach to represent the effect of depth-varying rigidity on earthquake slip. Our 'magnitude-relabelling' approach implies the rigidity model affects the hazard by changing our interpretation of the integrated slip released by historical earthquakes (which in turn affects the logic-tree weights for the magnitude-frequency distributions). Compared with slip-amplification (Scala et al. 2019), the magnitude-relabelling approach has the practical advantage of not requiring any re-computation of tsunami waveforms. However it is not obvious whether one-or-other approach should be preferred on theoretical grounds, or the extent to which their results differ. This should be considered in future research.

Our results confirm that the representation of earthquake slip has a significant effect on PTHA, even in the far-field (Li et al. 2016). This is significant because PTHAs often employ FAUS-like uniform-slip scenarios with magnitude-dependent length and width based on a scaling relation (e.g. Løvholt et al. 2014; Roshan et al. 2016; Davies et al. 2017; Kalligeris et al. 2017). In our study some historical tsunamis were poorly represented with the FAUS approach, and it predicted significantly lower tsunami hazard in Australia than the other approaches. Thus we suggest the FAUS approach should not be used, even in the far-field. While the HS and VAUS scenarios showed better performance, there was a clear tendency for good-fitting VAUS scenarios to be 'compact' relative to the scaling-relation predictions. Once accounted for via bias-adjustment, the VAUS and HS scenarios produced similar ARI=500 maximum-stage estimates for Australia. This suggests both HS and VAUS can usefully represent earthquake-generated tsunamis for PTHA. For some hazard modelling applications the use of both scenario types could be beneficial to represent epistemic uncertainties in tsunami generation. In other cases the fewer degrees-of-freedom of VAUS scenarios may be exploited to reduce computational effort, as noted by An et al. (2018) in the context of tsunami early warning.

Although the representation of earthquake size and slip is important for producing realistic tsunami wave-forms, in this study the largest source of uncertainty remains the earthquake magnitude-frequency relations. Future work may reduce these uncertainties by more efficiently using paleo-seismic and long-term historical observations, coupled with appropriate treatment of the data uncertainties (Griffin et al. 2018). Refinements of the model's structure should simultaneously be considered (e.g. allowing for non-Poissonion event times, Geist 2014; Moernaut et al. 2018) and are likely to become more important when additional data is used to constrain the model. In addition, the development of more refined representations of inter-source-zone epistemic-uncertainty dependence may allow the tsunami hazard uncertainty to be reduced at some sites affected by multiple source-zones, as compared with the co-monotonic treatment applied herein.

Acknowledgements

This paper is published with the permission of the CEO of Geoscience Australia. This project was undertaken with the assistance of resources and services from the National Computational Infrastructure (NCI), which is supported by the Australian Government. Comments from Hadi Ghasemi and two anonymous reviewers improved the paper. The tsunami scenario database and source-code used to construct it are freely available, and may be accessed following instructions at https://github.com/GeoscienceAustralia/ptha/tree/master/ptha_access.

Publisher's Note Springer Nature remains neutral with regard to jurisdictional claims in published maps and institutional affiliations.

References

Allgeyer, S., & Cummins, P. (2014). Numerical tsunami simulation including elastic loading and seawater density stratification. *Geophysical Research Letters*, *41*(7), 2368–2375. https://doi.org/10.1002/2014GL059348.

An, C., Liu, H., Ren, Z., & Yuan, Y. (2018). Prediction of tsunami waves by uniform slip models. *Journal of Geophysical Research: Oceans*. https://doi.org/10.1029/2018jc014363.

Annaka, T., Satake, K., Sakakiyama, T., Yanagisawa, K., & Shuto, N. (2007). Logic-tree approach for probabilistic tsunami hazard analysis and its applications to the Japanese Coasts. *Pure and Applied Geophysics*, *164*, 577–592. https://doi.org/10.1007/s00024-006-0174-3.

Baba, T., Allgeyer, S., Hossen, J., Cummins, P. R., Tsushima, H., Imai, K., et al. (2017). Accurate numerical simulation of the far-field tsunami caused by the 2011 Tohoku earthquake, including the effects of boussinesq dispersion, seawater density stratification, elastic loading, and gravitational potential change. *Ocean Modelling*, *111*, 46–54. https://doi.org/10.1016/j.ocemod.2017.01.002.

Ben-Menahem, A., & Rosenman, M. (1972). Amplitude patterns of tsunami waves from submarine earthquakes. *Journal of Geophysical Research (1896–1977)*, *77*(17), 3097–3128. https://doi.org/10.1029/JB077i017p03097.

Berryman, K., Wallace, L., Hayes, G., Bird, P., Wang, K., Basili, R., Lay, T., Pagani, M., Stein, R., Sagiya, T., Rubin, C., Barreintos, S., Kreemer, C., Litchfield, N., Stirling, M., Gledhill, K., Haller, K., & Costa, C. (2015). The GEM Faulted Earth Subduction Interface Characterisation Project: Version 2.0 - April 2015. Tech. rep., GEM.

Bilek, S. L., & Lay, T. (1999). Rigidity variations with depth along interplate megathrust faults in subduction zones. *Nature*, *400*, 443–446.

Bilek, S.L., & Lay, T. (2018). Subduction zone megathrust earthquakes. *Geosphere*. https://doi.org/10.1130/GES01608.1

Bird, P. (2003). An updated digital model of plate boundaries. *Geochemistry Geophysics Geosystems*, *4*(3), 1–52.

Bird, P., & Kagan, Y. Y. (2004). Plate-tectonic analysis of shallow seismicity: Apparent boundary width, beta, corner magnitude, coupled lithosphere thickness, and coupling. *Bulletin of the Seismological Society of America*, *94*(6), 2380–2399.

Bird, P., & Liu, Z. (2007). Seismic hazard inferred from tectonics: California. *Seismological Research Letters*, *78*(1), 37–48.

Bletery, Q., Thomas, A. M., Rempel, A. W., Karlstrom, L., Sladen, A., & De Barros, L. (2016). Mega-earthquakes rupture flat megathrusts. *Science*, *354*(6315), 1027–1031. https://doi.org/10.1126/science.aag0482.

Bommer, J. J., & Scherbaum, F. (2008). The use and misuse of logic trees in probabilistic seismic hazard analysis. *Earthquake Spectra*, *24*(4), 997–1009. https://doi.org/10.1193/1.2977755.

Burbidge, D., Cummins, P., Mleczko, R., & Thio, H. (2008). A probabilistic tsunami hazard assessment for Western Australia. *Pure and Applied Geophysics*, *165*, 2059–2088. https://doi.org/10.1007/s00024-008-0421-x.

Butler, R., Frazer, L. N., & Templeton, W. J. (2016). Bayesian probabilities for Mw 9.0+ earthquakes in the Aleutian Islands from a regionally scaled global rate. *Journal of Geophysical Research: Solid Earth*. https://doi.org/10.1002/2016JB012861

Butler, R., Walsh, D., & Richards, K. (2017). Extreme tsunami inundation in Hawai'i from Aleutian–Alaska subduction zone earthquakes. *Natural Hazards*, *85*(3), 1591–1619. https://doi.org/10.1007/s11069-016-2650-0.

Cisternas, M., Garrett, E., Wesson, R., Dura, T., & Ely, L. (2017). Unusual geologic evidence of coeval seismic shaking and tsunamis shows variability in earthquake size and recurrence in the area of the giant 1960 Chile earthquake. *Marine Geology*, *385*, 101–113. https://doi.org/10.1016/j.margeo.2016.12.007.

Davies, G. (2019). Tsunami variability from uncalibrated stochastic earthquake models: Tests against deep ocean observations 2006–2016. *Geophysical Journal International*, *218*(3), 1939–1960. https://doi.org/10.1093/gji/ggz260.

Davies, G., & Griffin, J. (2018). The 2018 Austrailan probabilistic tsunami hazard assessment: Hazards from earthquake generated tsunamis. Tech. rep., Geoscience Australia Record 2018/41. https://doi.org/10.11636/Record.2018.041

Davies, G., Horspool, N., & Miller, V. (2015). Tsunami inundation from heterogeneous earthquake slip distributions: Evaluation of synthetic source models. *Journal of Geophysical Research: Solid Earth*, *120*(9), 6431–6451. https://doi.org/10.1002/2015JB012272.

Davies, G., Griffin, J., Løvholt, F., Glimsdal, S., Harbitz, C., Thio, H. K., et al. (2017). A global probabilistic tsunami hazard assessment from earthquake sources. *Geological Society, London, Special Publications*,. https://doi.org/10.1144/sp456.5.

Deelstra, G., Dhaene, J., & Vanmaele, M. (2009). An overview of comonotonicity and its applications in finance and insurance. In: Advanced Mathematical Methods for Finance, Springer, New York.

Dziewonski, A. M., & Anderson, D. L. (1981). Preliminary reference earth model. *Physics of the Earth and Planetary Interiors*, *25*, 297–356.

Ekstrom, G., Nettles, M., & Dziewonski, A. (2012). The global CMT project 2004–2010: Centroid-moment tensors for 13,017 earthquakes. *Physics of the Earth and Planetary Interiors*, *200–201*, 1–9. https://doi.org/10.1016/j.pepi.2012.04.002.

Engdahl, E., & Villasenor, A.(2002). Global seismicity: 1900–1999. International Handbook of Earthquake and Engineering Seismology 81A.

Fritz, H. M., & Borrero, J. C. (2006). Somalia field survey after the December 2004 Indian Ocean Tsunami. *Earthquake Spectra*, *22*(S3), 219–233. https://doi.org/10.1193/1.2201972.

Fujii, Y., & Satake, K. (2013). Slip distribution and seismic moment of the 2010 and 1960 Chilean Earthquakes inferred from tsunami waveforms and coastal geodetic data. *Pure and Applied Geophysics*, *170*, 1493–1509. https://doi.org/10.1007/s00024-012-0524-2.

Fukutani, Y., Suppasri, A., & Imamura, F. (2018). Quantitative assessment of epistemic uncertainties in tsunami hazard effects on building risk assessments. *Geosciences*. https://doi.org/10.3390/geosciences8010017.

Garwood, F. (1936). Fiducial limits for the poisson distribution. *Biometrika*, *28*(3–4), 437–442. https://doi.org/10.1093/biomet/28.3-4.437.

Geist, E. (2002). Complex earthquake rupture and local tsunamis. *Journal of Geophysical Research*. https://doi.org/10.1029/2000JB000139.

Geist, E., & Bilek, S. (2001). Effect of depth-dependent shear modulus on tsunami generation along subduction zones. *Geophysical Research Letters*, *28*(7), 1315–1318.

Geist, E., & Parsons, T. (2016). Reconstruction of far-field tsunami amplitude distributions from earthquake sources. *Pure and Applied Geophysics*,. https://doi.org/10.1007/s00024-016-1288-x.

Geist, E. L. (2014). Explanation of temporal clustering of tsunami sources using the epidemic-type aftershock sequence model. *Bulletin of the Seismological Society of America*, *104*(4), 2091–2103. https://doi.org/10.1785/0120130275.

Gica, E., Teng, M. H., Liu, P. L. F., Titov, V., & Zhou, H. (2007). Sensitivity analysis of source parameters for earthquake-generated distant tsunamis. *Journal of Waterway, Port, Coastal, and Ocean Engineering*, *133*(6), 429–441. https://doi.org/10.1061/(ASCE)0733-950X(2007)133:6(429).

Glimsdal, S., Pedersen, G., Harbitz, C., & Løvholt, F. (2013). Dispersion of tsunamis: does it really matter? *Natural Hazards and Earth System Sciences*, *13*, 1507–1526. https://doi.org/10.5194/nhess-13-1507-2013.

Gonzalez, F. I., Geist, E. L., Jaffe, B., Kanoglu, U., Mofjeld, H., Synolakis, C. E., et al. (2009). Probabilistic tsunami hazard assessment at Seaside, Oregon, for near- and far-field seismic sources. *Journal of Geophysical Research*, *114*(C11023), 1–19. https://doi.org/10.1029/2008JC005132.

Grezio, A., Marzocchi, W., Sandri, L., & Gasparini, P. (2010). A Bayesian procedure for probabilistic tsunami hazard assessment. *Natural Hazards*, *53*(1), 159–174. https://doi.org/10.1007/s11069-009-9418-8.

Grezio, A., Babeyko, A., Baptista, M. A., Behrens, J., Costa, A., Davies, G., et al. (2017). Probabilistic tsunami hazard analysis: Multiple sources and global applications. *Reviews of Geophysics*, *55*(4), 1158–1198. https://doi.org/10.1002/2017RG000579, 2017RG000579.

Griffin, J., & Davies, G. (2018). Earthquake sources of the Australian plate margin: Revised models for the 2018 national tsunami and earthquake hazard assessments. Tech. rep., Geoscience Australia Professional Opinion 2018/xx.

Griffin, J., Nguyen, N., Cummins, P., & Cipta, A. (2018). Historical earthquakes of the Eastern Sunda Arc: Source mechanisms and intensity-based testing of Indonesia's National seismic hazard assessment historical earthquakes of the Eastern Sunda Arc. *Bulletin of the Seismological Society of America*, *109*(1), 43–65. https://doi.org/10.1785/0120180085.

Hayes, G. P., & Furlong, K. P. (2010). Quantifying potential tsunami hazard in the Puysegur subduction zone, south of New Zealand. *Geophysical Journal International*, *183*, 1512–1524.

Hayes, G. P., Wald, D. J., & Johnson, R. L. (2012). Slab1.0: A three-dimensional model of global subduction zone geometries. *Journal of Geophysical Research*. https://doi.org/10.1029/2011JB008524

Hayes, G. P., Moore, G. L., Portner, D. E., Hearne, M., Flamme, H., Furtney, M., et al. (2018). Slab2, a comprehensive subduction zone geometry model. *Science*,. https://doi.org/10.1126/science.aat4723.

Hébert, H., Burg, P., Binet, R., Lavigne, F., Allgeyer, S., & Schindelé, F. (2012). The 2006 July 17 Java (Indonesia) tsunami from satellite imagery and numerical modelling: A single or complex source? *Geophysical Journal International*, *191*(3), 1255–1271. https://doi.org/10.1111/j.1365-246X.2012.05666.x.

Ho, T. C., Satake, K., Watada, S., & Fujii, Y. (2019). Source estimate for the 1960 Chile earthquake from joint inversion of geodetic and transoceanic tsunami data. *Journal of Geophysical Research: Solid Earth*, *124*(3), 2812–2828. https://doi.org/10.1029/2018JB016996.

Horspool, N., Pranantyo, I., Griffin, J., Latief, H., Natawidjaja, D. H., Kongko, W., et al. (2014). A probabilistic tsunami hazard assessment for Indonesia. *Natural Hazards and Earth System Sciences*, *14*, 3105–3122. https://doi.org/10.5194/nhessd-2-3423-2014.

Kagan, Y. Y. (2002). Seismic moment distribution revisited: 1 statistical results. *Geophysical Journal International*, *148*, 520–541.

Kagan, Y. Y., & Jackson, D. D. (2013). Tohoku earthquake: A surprise? *Bulletin of the Seismological Society of America*, *103*, 1181–1194. https://doi.org/10.1785/0120120110.

Kajiura, K. (1963). The leading wave of a tsunami. *Bulletin of the Earthquake Research Institute*, *41*, 535–571.

Kalligeris, N., Montoya, L., Ayca, A., & Lynett, P. (2017). An approach for estimating the largest probable tsunami from far-field subduction zone earthquakes. *Natural Hazards*, *89*, 233. https://doi.org/10.1007/s11069-017-2961-9.

Kanamori, H. (1977). The energy release in great earthquakes. *Journal of Geophysical Research*, *82*(20), 2981–2987.

Koulali, A., Tregoning, P., McClusky, S., Stanaway, R., Wallace, L., & Lister, G. (2015). New Insights into the present-day kinematics of the central and western Papua New Guinea from GPS. *Geophysical Journal International*, *202*(2), 993–1004. https://doi.org/10.1093/gji/ggv200.

Koulali, A., Susilo, S., McClusky, S., Meilano, I., Cummins, P., Tregoning, P., et al. (2016). Crustal strain partitioning and the associated earthquake hazard in the eastern Sunda-Banda Arc. *Geophysical Research Letters*, *43*(5), 1943–1949. https://doi.org/10.1002/2016GL067941,2016GL067941.

Lane, E. M., Gillibrand, P. A., Wang, X., & Power, W. (2012). A probabilistic tsunami hazard study of the Auckland Region, Part II: Inundation modelling and hazard assessment. *Pure and Applied Geophysics*, *170*(9–10), 1635–1646. https://doi.org/10.1007/s00024-012-0538-9.

Lay, T. (2018). A review of the rupture characteristics of the 2011 Tohoku-oki Mw 9.1 earthquake. *Tectonophysics*, *733*, 4–36. https://doi.org/10.1016/j.tecto.2017.09.022.

Li, L., Switzer, A. D., Chan, C. H., Wang, Y., Weiss, R., & Qiu, Q. (2016). How heterogeneous coseismic slip affects regional probabilistic tsunami hazard assessment: A case study in the South China Sea. *Journal of Geophysical Research: Solid Earth*, *121*(8), 6250–6272. https://doi.org/10.1002/2016JB013111, 2016JB013111.

Lorito, S., Piatanesi, A., & Lomax, A. (2008). Rupture process of the 18 April 1906 California earthquake from near-field tsunami waveform inversion. *Bulletin of the Seismological Society of America*, *98*, 832–845. https://doi.org/10.1785/0120060412.

Lorito, S., Romano, F., & Lay, T. (2015a). Tsunamigenic major and great earthquakes (2004–2013): Source processes inverted from seismic, geodetic, and sea-level data. *Encyclopedia of Complexity and Systems Science*. https://doi.org/10.1007/978-3-642-27737-5_641-1.

Lorito, S., Selva, J., Basili, R., Romano, F., Tiberti, M., & Piatanesi, A. (2015b). Probabilistic hazard for seismically induced tsunamis: Accuracy and feasibility of inundation maps. *Geophysical Journal International*, *200*, 574–588. https://doi.org/10.1093/gji/ggu408.

Løvholt, F., Glimsdal, S., Harbitz, C., Horspool, N., Smebye, H., de Bono, A., et al. (2014). Global tsunami hazard and exposure due to large co-seismic slip. *International Journal of Disaster Risk Reduction*, *10*, 406–418. https://doi.org/10.1016/j.ijdrr.2014.04.003.

McCaffrey, R. (1997). Influences of recurrence times and fault zone temperatures on the age-rate dependence of subduction zone seismicity. *Journal of Geophysical Research: Solid Earth*, *102*(B10), 22839–22854. https://doi.org/10.1029/97JB01827.

McCaffrey, R. (2008). Global frequency of magnitude 9 earthquakes. *Geology*, *36*(3), 263–266. https://doi.org/10.1130/G24402A.1.

Meade, B. J., & Loveless, J. P. (2009). Block modeling with connected fault-network geometries and a linear elastic coupling estimator in spherical coordinates. *Bulletin of the Seismological Society of America*, *99*(6), 3124–3139. https://doi.org/10.1785/0120090088.

Moernaut, J., Van Daele, M., Fontijn, K., Heirman, K., Kempf, P., Pino, M., et al. (2018). Larger earthquakes recur more periodically: New insights in the megathrust earthquake cycle from lacustrine turbidite records in south-central Chile. *Earth and Planetary Science Letters*, *481*, 9–19. https://doi.org/10.1016/j.epsl.2017.10.016.

Moreno, M., Bolte, J., Klotz, J., & Melnick, D. (2009). Impact of megathrust geometry on inversion of coseismic slip from geodetic data: Application to the 1960 Chile earthquake. *Geophysical Research Letters*, *36*, L16310. https://doi.org/10.1029/2009GL039276.

Mori, N., Mai, P. M., Goda, K., & Yasuda, T. (2017). Tsunami inundation variability from stochastic rupture scenarios: Application to multiple inversions of the 2011 Tohoku, Japan earthquake. *Coastal Engineering*, *127*, 88–105. https://doi.org/10.1016/j.coastaleng.2017.06.013.

Mueller, C., Power, W., Fraser, S., & Wang, X. (2015). Effects of rupture complexity on local tsunami inundation: Implications for probabilistic tsunami hazard assessment by example. *Journal of Geophysical Research (Solid Earth)*, *120*, 488–502. https://doi.org/10.1002/2014JB011301.

Newman, A. V., Feng, L., Fritz, H. M., Lifton, Z. M., Kalligeris, N., & Wei, Y. (2011a). The energetic 2010 M_W 7.1 Solomon Islands tsunami earthquake. *Geophysical Journal International*, *186*, 775–781. https://doi.org/10.1111/j.1365-246X.2011.05057.x.

Newman, A. V., Hayes, G., Wei, Y., & Convers, J. (2011b). The 25 October 2010 Mentawai tsunami earthquake, from real-time discriminants, finite-fault rupture, and tsunami excitation. *Geophysical Research Letters*, *38*(L05302), 1–7. https://doi.org/10.1029/2010GL046498.

NGDC (2018) National Geophysical Data Center/World Data Service Global Historical Tsunami Database. 10.7289/V5PN93H7. https://www.ngdc.noaa.gov/hazard/tsu_db.shtml. Last accessed 24 Sept 2015.

Nishikawa, T., & Ide, S. (2014). Earthquake size distribution in subduction zones linked to slab buoyancy. *Nature Geoscience*, *7*, 904–908. https://doi.org/10.1038/ngeo2279.

Okada, Y. (1985). Surface deformation due to shear and tensile faults in a half-space. *Bulletin of the Seismological Society of America*, *75*(4), 1135–1154.

Okal, E. A. (2011). Tsunamigenic earthquakes: Past and present milestones. *Pure and Applied Geophysics*, *168*, 969–995. https://doi.org/10.1007/s00024-010-0215-9.

Okal, E. A., & Synolakis, C. E. (2008). Far-field tsunami hazard from mega-thrust earthquakes in the Indian Ocean. *Geophysical Journal International*, *172*, 995–1015. https://doi.org/10.1111/j.1365-246X.2007.03674.x.

Okal, E. A., Synolakis, C. E., Fryer, G. J., Heinrich, P., Borrero, J. C., Ruscher, C., et al. (2002). A field survey of the 1946 Aleutian tsunami in the far field. *Seismological Research Letters*, *73*(4), 490–503. https://doi.org/10.1785/gssrl.73.4.490.

Parsons, T., & Geist, E. L. (2009). Tsunami probability in the Caribbean Region. *Pure and Applied Geophysics*, *165*, 2089–2116. https://doi.org/10.1007/s00024-008-0416-7.

Poisson, B., Oliveros, C., & Pedreros, R. (2011). Is there a best source model of the Sumatra 2004 earthquake for simulating the consecutive tsunami? *Geophysical Journal International*, *185*, 1365–1378. https://doi.org/10.1111/j.1365-246X.2011.05009.x.

Power, W., Wang, X., Wallace, L., Clark, K., & Mueller, C. (2017). The New Zealand Probabilistic Tsunami Hazard Model: development and implementation of a methodology for estimating tsunami hazard nationwide. *Geological Society, London, Special Publications*. https://doi.org/10.1144/SP456.6.

Prendergast, A., & Brown, N. (2012). Far-field impact and coastal sedimentation associated with the 2006 Java tsunami in West Australia. *Natural Hazards*, *60*, 69–79. https://doi.org/10.1007/s11069-011-9953-y.

Rajendran, K. (2013). On the recurrence of great subduction zone earthquakes. *Current Science*, *104*(7), 880–892.

Romano, F., Piatanesi, A., Lorito, S., Tolomei, C., Atzori, S., & Murphy, S. (2016). Optimal time alignment of tide-gauge tsunami waveforms in nonlinear inversions: Application to the 2015 Illapel (Chile) earthquake. *Geophysical Research Letters*, *43*(21), 11,226–11,235. https://doi.org/10.1002/2016GL071310,2016GL071310.

Rong, Y., Jackson, D. D., Magistrale, H., & Goldfinger, C. (2014). Magnitude limits of subduction zone earthquakes. *Bulletin of the Seismological Society of America*, *104*(5), 2359–2377. https://doi.org/10.1785/0120130287.

Roshan, A. D., Basu, P. C., & Jangid, R. S. (2016). Tsunami hazard assessment of Indian coast. *Natural Hazards*, *82*(2), 733–762. https://doi.org/10.1007/s11069-016-2216-1.

Ruiz, J. A., Fuentes, M., Riquelme, S., Campos, J., & Cisternas, A. (2015). Numerical simulation of tsunami runup in northern Chile based on non-uniform k^{-2} slip distributions. *Natural Hazards*.

Satake, K., Fujii, Y., Harada, T., & Namegaya, Y. (2013). Time and space distribution of coseismic slip of the 2011 Tohoku Earthquake as inferred from tsunami waveform data. *Bulletin of the Seismological Society of America*, *103*, 1473–1492. https://doi.org/10.1785/0120120122.

Scala, A., Lorito, S., Romano, F., Murphy, S., Selva, J., Basili, R., et al. (2019). Effect of shallow slip amplification uncertainty on probabilistic tsunami hazard analysis in subduction zones: Use of long-term balanced stochastic slip models. *Pure and Applied Geophysics*,. https://doi.org/10.1007/s00024-019-02260-x.

Scholz, C.H., & Campos, J. (2012). The seismic coupling of subduction zones revisited. *Journal of Geophysical Research*. https://doi.org/10.1029/2011JB009003

Selva, J., Tonini, R., Molinari, I., Tiberti, M., Romano, F., Grezio, A., et al. (2016). Quantification of source uncertainties in Seismic Probabilistic Tsunami Hazard Analysis (SPTHA). *Geophysical Journal International*, *205*, 1780–1803. https://doi.org/10.1093/gji/ggw107.

Sepúlveda, I., Liu, P. L. F., & Grigoriu, M. (2019). Probabilistic Tsunami Hazard Assessment in South China Sea with consideration of uncertain earthquake characteristics. *Journal of Geophysical Research: Solid Earth*. https://doi.org/10.1029/2018JB016620.

Shennan, I., Brader, M. D., Barlow, N. L., Davies, F. P., Longley, C., & Tunstall, N. (2018). Late Holocene paleoseismology of Shuyak Island, Alaska. *Quaternary Science Reviews, 201*, 380–395. https://doi.org/10.1016/j.quascirev.2018.10.028.

Stirling, M., & Gerstenberger, M. (2018). Applicability of the Gutenberg–Richter relation for major active faults in New Zealand. *Bulletin of the Seismological Society of America, 108*(2), 718–728. https://doi.org/10.1785/0120160257.

Storchak, D., Giacomo, D. D., Bondar, I., Harris, J., Engdahl, E., Lee, W., Villasenor, A., Bormann, P., & Ferrari, G. (2012). ISC-GEM global instrumental earthquake catalogue (1900–2009): GEM Technical Report 2012-01. Tech. rep., GEM. https://doi.org/10.13117/GEM.GEGD.TR2012.01.

Strasser, F., Arango, M., & Bommer, J. J. (2010). Scaling of the source dimensions of interface and intraslab subduction-zone earthquakes with moment magnitude. *Seismological Research Letters, 81*(6), 941–950. https://doi.org/10.1785/gssrl.81.6.941.

Thio, H. K., Somerville, P., & Ichinose, G. (2007). Probabilistic analysis of strong ground motion and tsunami hazards in Southeast Asia. In: Proceedings from 2007 NUS-TMSI Workshop, National University of Singapore.

Volpe, M., Lorito, S., Selva, J., Tonini, R., Romano, F., & Brizuela, B. (2019). From regional to local SPTHA: Efficient computation of probabilistic tsunami inundation maps addressing near-field sources. *Natural Hazards and Earth System Sciences, 19*(3), 455–469. https://doi.org/10.5194/nhess-19-455-2019.

Watada, S., Kusumoto, S., & Satake, K. (2014). Traveltime delay and initial phase reversal of distant tsunamis coupled with the self-gravitating elastic earth. *Journal of Geophysical Research: Solid Earth, 119*(5), 4287–4310. https://doi.org/10.1002/2013jb010841.

Weatherall, P., Marks, K. M., Jakobsson, M., Schmitt, T., Tani, S., Arndt, J. E., et al. (2015). A new digital bathymetric model of the world's oceans. *Earth and Space Science, 2*(8), 331–345. https://doi.org/10.1002/2015EA000107.

Wesson, R. L., Boyd, O. S., Mueller, C. S., Bufe, C. G., Frankel, A. D., & Petersen, M. D. (2007). Revision of Time-Independent Probabilistic Seismic Hazard Maps for Alaska, Open-File Report 2007–1043. United States Geological Survey: Tech. rep.

Wesson, R. L., Boyd, O. S., Mueller, C. S., & Frankel, A. D. (2008). Challenges in making a seismic hazard map for Alaska and the Aleutians. In: Freymueller, J. (ed), Active Tectonics and Seismic Potential of Alaska, American Geophysical Union. https://doi.org/10.1029/179GM22

Whiteway, T. (2009). Australian Bathymetry and Topography Grid, June 2009. Tech. rep., Geoscience Australia Record 2009/21.

Zöller, G. (2013). Convergence of the frequency-magnitude distribution of global earthquakes: Maybe in 200 years. *Geophysical Research Letters, 40*(15), 3873–3877. https://doi.org/10.1002/grl.50779.

Zöller, G. (2017). Comment on "Estimation of Earthquake Hazard Parameters from Incomplete Data Files. Part III. Incorporation of Uncertainty of Earthquake-Occurrence Model" by Andrzej Kijko, Ansie Smit, and Markvard A. Sellevoll. *Bulletin of the Seismological Society of America*, 107(4):1975. https://doi.org/10.1785/0120160193.

(Received June 5, 2019, revised July 31, 2019, accepted August 3, 2019, Published online August 9, 2019)

Pure Appl. Geophys. 177 (2020), 1549–1568

https://doi.org/10.1007/s00024-019-02384-0

Pure and Applied Geophysics

Tsunami Inundation and Maritime Hazard Modelling for a Maximum Credible Tsunami Scenario in Southeast Tasmania, Australia

Claire L. Kain,[1] Barrie Lewarn,[2] Edward H. Rigby,[3] and Colin Mazengarb[1]

Abstract—The east coast of Tasmania is directly exposed to tsunamis originating from the Puysegur subduction zone, off New Zealand's southwest coast. However, the potential impacts of tsunami inundation for coastal communities and Hobart Airport, and risks to maritime operations in Hobart Port, are poorly understood. The purpose of this project is to simulate a maximum credible earthquake/tsunami/high tide scenario (a Mw 8.7 rupture of the Puysegur subduction zone) across a 17,000 km^2 model area and provide outputs that can inform local hazard management plans. Modelling was performed using the ANUGA hydrodynamic library, along with a detailed elevation model constructed from LiDAR, photogrammetric contours and bathymetric datasets. The modelling strategy comprised three primary scenarios: a 13 h simulation of tsunami activity focusing on maritime hazard from waves and currents, with two further 4 h simulations focusing on coastal inundation for 71 coastal communities and Hobart Airport. The models were run on a variable triangulated mesh with a detailed Manning's *n* surface roughness model. Tsunami erosion of sand dunes was accounted for through the development of an erosion operator. Results predict severe inundation (> 4 m flow depth) in exposed east-facing areas and the maritime hazard assessment advises that water disturbance would pose a significant threat to marine craft. The assessment suggests that the feasibility of shipping evacuation from Hobart port is questionable, given the timeframes involved and the nature of simulated water disturbance. Modelling results suggest that the dune line in front of Hobart Airport would not be breached in this instance and so inundation would not occur.

Keywords: Tsunami, modelling, hazard, emergency management.

1. Introduction

The east coast of Australia, including Tasmania, is exposed to tsunamis originating from many source regions around the Pacific Ocean. Southeast Tasmania is particularly exposed to tsunamis originating along the Puysegur subduction zone, located off the southwest coast of New Zealand. Several small tsunamis have been recorded in Tasmania since 1858 (Goff and Chagué-Goff 2014; Morris and Mazengarb 2009; Uslu et al. 2011), but no large events have occurred in recorded history. However, geological evidence suggests the occurrence of several significant tsunami events within the last 4000 years (Clark et al. 2011).

In the absence of eyewitness records of damaging tsunami events, numerical modelling of tsunamis can provide a method of predicting potential impacts and investigating the risk to coastal communities. Initial tsunami modelling for southeast Tasmania was undertaken in 2009 (Van Putten et al. 2009) and results indicated that parts of the coastline could be significantly affected by a tsunami generated from a large rupture of the Puysegur subduction zone, off New Zealand's southwest coast. Emergency managers have since sought greater detail on potential impacts for coastal communities, infrastructure and operations in Hobart Port.

The impacts of tsunami waves and currents for maritime infrastructure and operations has long been apparent in historical records, yet this aspect of tsunami hazard is commonly neglected (Borrero et al. 2015; Lynett et al. 2014). Following the large tsunamis in recent decades, damage and disruptions were experienced in ports adjacent to the tsunami source region, but also in far-field transoceanic

[1] Mineral Resources Tasmania, PO Box 56, Rosny Park, Hobart 7018, Tasmania, Australia. E-mail: claire.kain@stategrowth.tas.gov.au; colin.mazengarb@stategrowth.tas.gov.au

[2] Australian Maritime College Search, University of Tasmania, Locked Bag 1400, Launceston 7248, Tasmania, Australia. E-mail: Barrie.lewarn@utas.edu.au

[3] Rienco Consulting, PO Box 602, Sorell 7172, Tasmania, Australia. E-mail: ted.rigby@rienco.com.au

locations. For example, the 1960 Chile tsunami caused surging, currents and damage in harbours as far afield as Australia, Japan, New Zealand and Hawaii (Beccari 2009; Cox and Mink 1963; Heath 1976; Kato et al. 1961). Following the 2004 Indian Ocean tsunami, port areas were completely destroyed in Banda Aceh, and tsunami currents caused mooring lines to break and ships to be cast adrift as far away as Madagascar and Reunion Island (Okal et al. 2006a, b). Similarly, the 2011 Tohoku tsunami destroyed port infrastructure along the adjacent coastline, and effects were felt across the Pacific Ocean with coastal erosion and strong currents causing disruption to harbour operations in California (Wisniewski and Wolski 2012). Research has begun to address this gap and numerical models have been developed that are capable of simulating tsunami current speeds, which in conjunction with modelled wave data and coastal inundation mapping can allow an educated assessment of the hazard to maritime operations and infrastructure (e.g. Hasan et al. 2017; Lynett et al. 2012, 2017; Muhari et al. 2015).

The purpose of this work is to model the impacts of a maximum credible earthquake/tsunami/high tide scenario in southeast Tasmania using new high resolution bathymetric and topographic data, and to explore the maritime hazard posed by such an event. The project scope includes an assessment of the risk to the Hobart airport runway and terminal from potential tsunami inundation.

2. Methods

The model scenario represents a maximum credible earthquake/tsunami/high tide combination, which corresponds to a Mw 8.7 earthquake on the Puysegur Subduction Zone. The rupture zone for this scenario covers a length of 425 km and a width of 125 km, with a dip of 15° and slip of 6.65 m. Tsunami source and propagation modelling were performed as part of the first probabilistic tsunami hazard assessment (PHTA) for Australia (Burbidge et al. 2008). This event represents an approximate annual exceedance probability (AEP) of one in 13,000 years. Initially, the AEP of this scenario was calculated as one in 10,000 years, but this has since been revised. Van Putten et al. (2009) modelled a number of Tasmanian tsunami inundation scenarios at a coarser resolution, in order to explore the effects of different rupture geometries and earthquake magnitudes, and we selected an event that corresponds with this study's 'worst case scenario'.

The model domain covers an area of approximately 17,000 km^2 (Fig. 1) and the resulting coastal inundation map series covers 71 communities. This area includes greater Hobart, which is Tasmania's largest city and has a population of approximately 232,000 (Australian Bureau of Statistics 2018).

The modelling strategy comprised three scenarios, as follows:

(A) Maritime hazard (13 h simulation).

(B) Coastal inundation (4 h simulation).

(C) Hobart airport hazard and dune erosion modelling (4 h simulation).

Note that the start time of the simulation was 1.4 h post-earthquake, to capture the moment the tsunami arrived at the model boundary and minimise computational time. The 13 h simulated outputs for Scenario A were inspected and a 4 h simulation was deemed sufficient for the higher resolution coastal inundation modelling in Scenarios B and C, as the maximum impacts occurred within this window.

2.1. Input Data Preparation

Four main sets of input data were required to accurately model nearshore tsunami propagation and inundation, including: Boundary condition hydrographs, mesh resolution zones, a digital elevation model and a surface roughness model (Manning's roughness coefficient: *n*). A polygon shape file was used to specify areas of potential dune erosion.

2.1.1 Boundary Conditions

The boundary condition hydrographs represent the incoming tsunami water level and momentum in deep water (100 m depth contour, Fig. 2). These data are freely available from Geoscience Australia for a range of tsunami rupture and deep water modelling scenarios (Burbidge et al. 2008) and we have taken the levels for the above mentioned event as the inputs to our nearshore modelling. We note that the PTHA

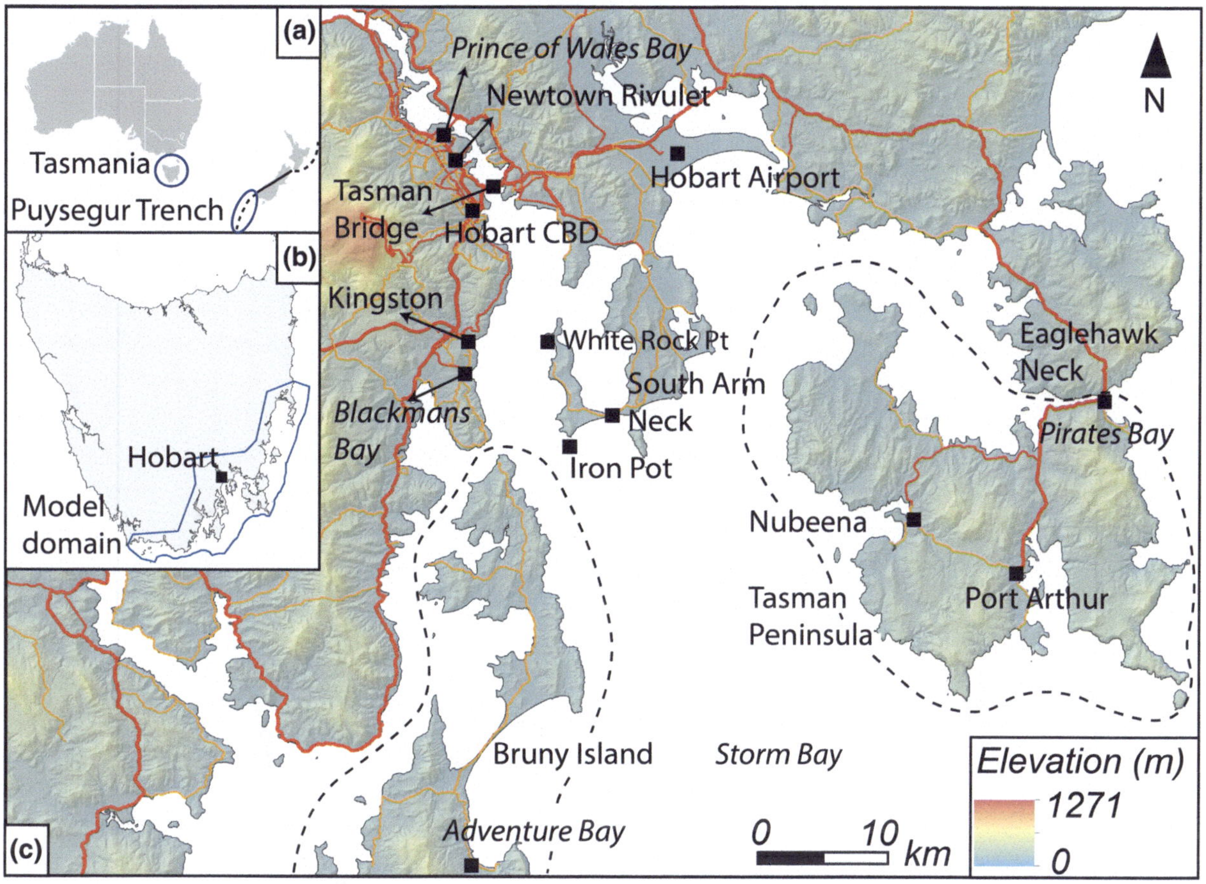

Figure 1
a Location of Tasmania with respect to the Puysegur subduction zone. **b** Coverage of the model domain. **c** Locations of key areas and places mentioned in the text

has been updated since this work was completed (Davies and Griffin 2018). The boundary condition dataset covered a 5 h period, beyond which time the seaward boundary reverted to an initial stage, zero momentum, non-reflective boundary. The top and bottom segments of the domain boundary (i.e. the northernmost and southernmost ocean segments not subject to the time series boundary conditions) were transmissive, thus allowing the waves to exit the domain.

2.1.2 Mesh Resolution

The resolution of the modelling mesh controls the level of detail of the modelling process and outputs; however, processing time increases substantially when a finer mesh is used. An unstructured triangular mesh was employed, to allow variation in mesh resolution across the irregularly shaped study area. Mesh resolution varies from a 400 m triangular mesh in deep ocean and high land areas to a 10 m triangular mesh in coastal areas of particular interest. Details of the mesh resolution zones are given in Table 1 and Fig. 3.

2.1.3 Elevation Model

A seamless elevation model was constructed by combining bathymetric and topographic data from a variety of sources (Table 2), using the terrain geoprocessing tools in ArcGIS. To avoid introducing artefacts due to overlapping datasets, the datasets were masked so that the newest and/or highest resolution data took precedence in any given location. In addition to the data listed in Table 2, additional elevation points were added at 20 m intervals along

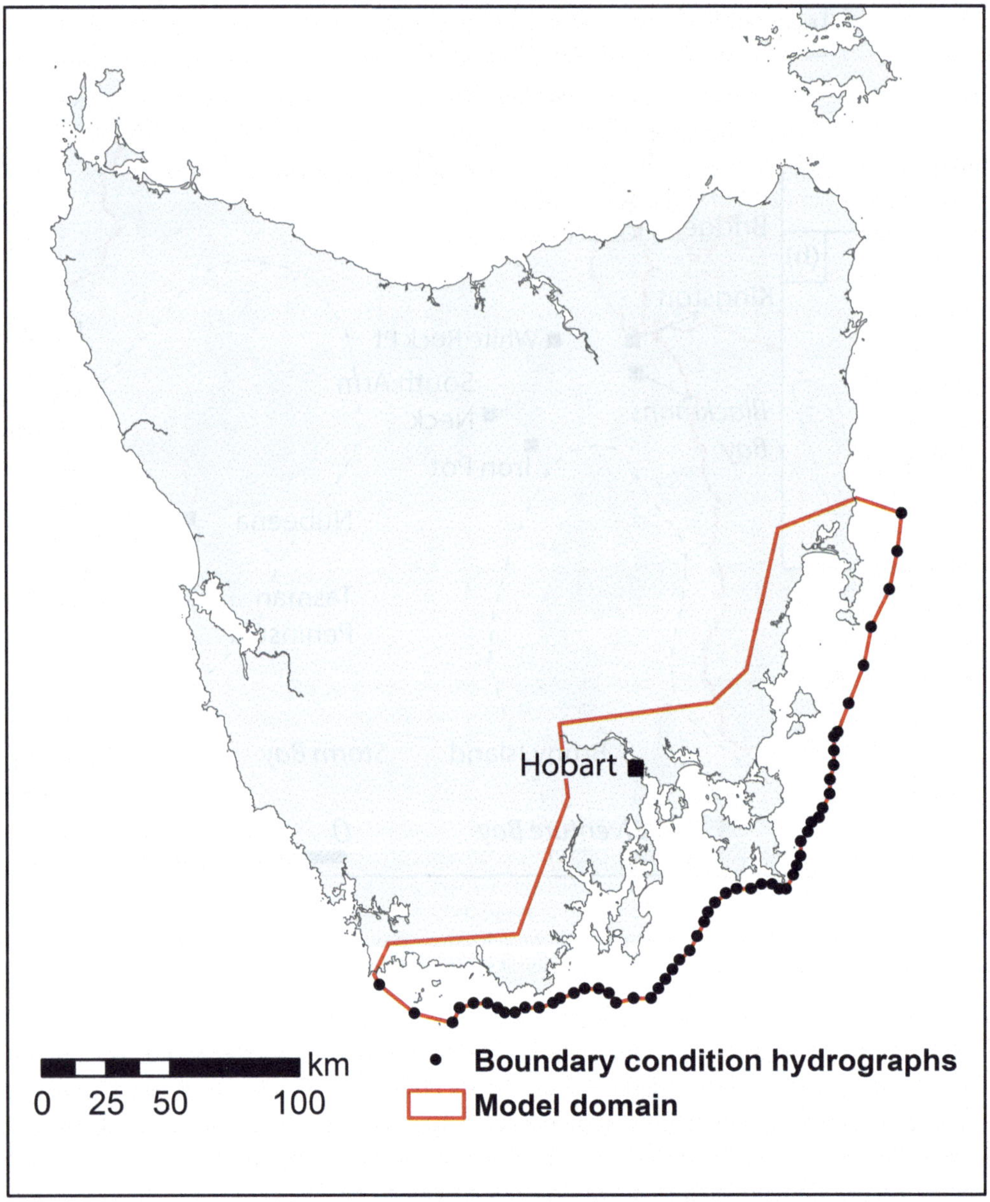

Figure 2
Location of the model domain and boundary condition hydrographs

the mapped coastline (at mean sea level and highest astronomical tide) to ensure the coastal margin was properly constrained.

2.1.4 Mannings n Surface Roughness Model

Surface roughness is an important control on wave attenuation and run-up distances. Factors such as vegetation cover, rivers and the presence or density of buildings can result in considerable variation in flooding patterns. Variations in surface cover were mapped in ArcGIS and accounted for in the model, using the Manning's n parameter (e.g. Bricker et al. 2015; Chow 1959). A number of publicly available datasets were used to generate the surface map, including the statewide road network, water course

Table 1

Mesh resolution zones and areas covered

Mesh resolution (triangle length, m)	Coverage
Coarse: 400 × 400	Open ocean and land > 15 m ASL
Medium: 200 × 200	Offshore areas − 30 > − 10 m ASL
Fine: 50 × 50	Rural areas − 10 m > 10 m ASL
Very fine: 20 × 20	Urban areas of interest
Extra fine: 10 × 10	Hobart Airport, Blackmans Bay

and cadastre layers (Land Information System Tasmania 2016) and the TASVEG vegetation map (DPIPWE 2013). Building footprints were extracted from the LiDAR data. The Manning's n coefficient for each surface type, as assigned based on the Australian Rainfall and Runoff Guidelines (Commonwealth of Australia—Geoscience Australia 2016), is listed in Table 3 and a spatial representation is given in Fig. 4.

2.2. Modelling Using ANUGA

Modelling was performed using the free and open source ANUGA hydrodynamic modelling library (Nielsen et al. 2005; Roberts et al. 2019). ANUGA was developed by the Australian National University (ANU) and Geoscience Australia (GA) and has been widely used to simulate tsunami inundation since its public release in 2006 (e.g. Baldock et al. 2007; Jakeman et al. 2010; Sterzai et al. 2015; Wilson et al. 2018). Several validation studies have compared ANUGA with other 2D hydrodynamic models or solutions and found the results aligned well (Fernandes 2009; Mungkasi and Roberts 2013), while modelling of historical events has also been undertaken for Sydney Harbour (Wilson et al. 2018) and the model solution found to be within an acceptable level of uncertainty. In addition, inundation

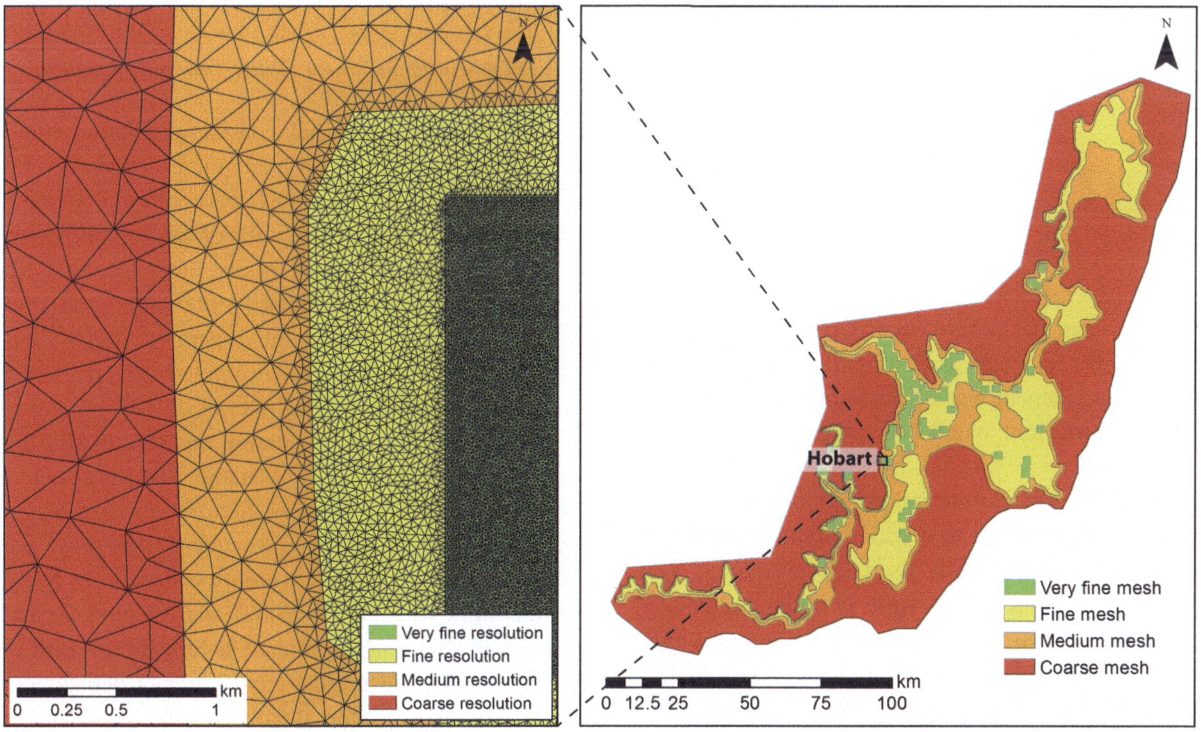

Figure 3
Details of mesh resolution zones across the study area

Table 2

Data sources used to construct the elevation model

Data source	Type	Year obtained	References
TasPorts bathymetry	Swath bathymetry	2010–2013	Unpublished
SeaMap: Bruny Bioregion Dataset	Sonar point data (Bathymetry)	2001	Barrett et al. (2001)
SeaMap: Derwent Estuary Programme	Sonar point data (Bathymetry)	2007	Lucieer et al. (2007)
SeaMap: Pittwater Estuary	Sonar point data (Bathymetry)	2002	Davies et al. (2002)
AusENC (Australian Hydrographic Office)	Depth soundings (Bathymetry)	Up to 2015	Licenced data
Private kayak soundings: Huon River	Depth soundings (Bathymetry)	Unknown	Unpublished
Australian Bathymetry and Topography Grid	National gridded bathymetry model (200 m resolution)	2009	Whiteway (2009)
Terrestrial LiDAR surveys: Coal Mines (Tasman Peninsula) 2015 DPAC Coastal 2014 Geoscience Australia, Greater Hobart 2013 Mt Wellington, 2011		2011–2015	Geoscience Australia, National Elevation Data Framework (NEDF)
Tas DEM 25 m	Photogrammetric (topographic) contours		DPIPWE (2007)

Table 3

Manning's n coefficients of roughness applied in the model, as assigned by surface type

Mannings *n*	Surface type
0.5	Solid buildings
0.071	Built-up areas
0.055	Vegetated areas
0.035	Land (default)
0.03	Bare ground
0.025	Water courses
0.018	Roads
0.01	Oceans and estuaries

resulting from the 2004 Indian Ocean tsunami was modelled by Jakeman et al. (2010) using ANUGA, and good agreement was obtained between modelled data and measured inundation extents, flow depths and velocities.

ANUGA uses a finite volume modelling method for solving shallow water wave equations and the DE0 flow algorithm was used in this study. The tsunami modelling scripts were built in Python 2.7 and major processing was performed through cloud hosted computing on Amazon Web Service. The processing time for the longest model run (maritime hazard scenario; a 13 h simulation) was approximately 80 h.

The initial water level was set at highest astronomical tide, to represent a worst case scenario of tsunami and tidal interaction. An assumed value of 0.8 m AHD was applied across the entire domain, which represents an average across the region (range 0.70–0.86 m AHD).

2.2.1 Development of the Dune Erosion Operator

In undertaking the initial coastal inundation modelling, it became clear that there were several locations where sand dunes were providing some level of protection for property or assets behind the dune line. Hobart Airport is one such location. To better understand and quantify the protection provided by these dunes, an operator script was developed in Python and incorporated into the model scripts for Scenario C (Rigby et al. 2017).

The purpose of this operator was to quantify the increase in risk from a tsunami scouring out a protective dune line, and so focuses on the erosion process rather than the subsequent transport and deposition process. The erosion mechanism is based on the work of Froelich (2002) and assumes a clear water scour with no significant sediment entrained in the approaching wave that could impact detachment rates. At each time step, the operator firstly simulates the detachment of sand by water flowing over the surface, by calculating the bed stress for each mesh

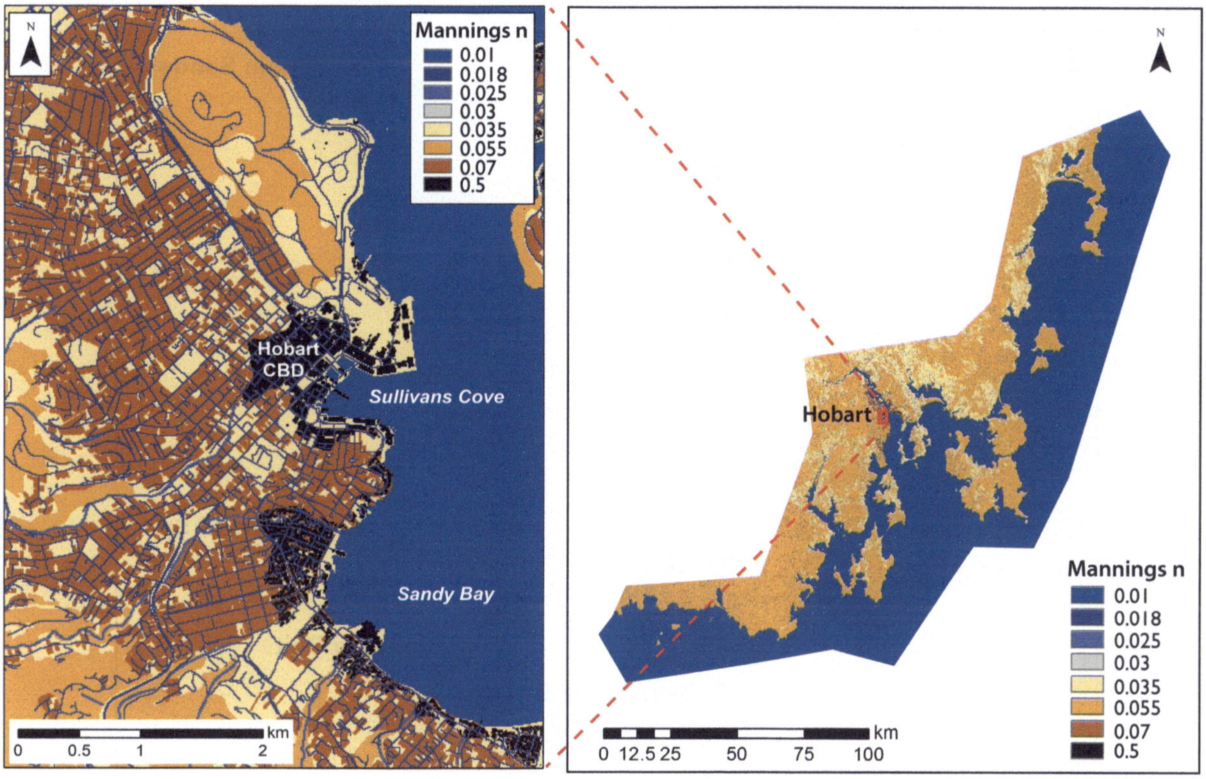

Figure 4
Manning's *n* zones as applied across the model domain

centroid (Eq. 1) and comparing this value with the detachment rate (the critical bed shear stress T_{crit} as per Froelich 2002; Eq. 2). The operator then simulates the collapse, fluidisation and removal of sand as a consequence of the erosion process creating face slopes that would be steeper than their angle of repose. Each triangle within the erosion zone is checked against neighbouring triangles with the aim of creating a stable surface that does not drop below a specified base level (1–3 m AHD depending on location). Further details of the development and testing of this operator script are provided by Rigby et al. (2017):

$$T_{bed} = \frac{Wd * g * n^2 * m^2}{d^{2.333}} \quad (1)$$

$$S_k = \frac{Kd * (T_{bed} - T_{crit})}{Sd} \quad (2)$$

where: Wd = water mass density (1000 kg m^3), g = acceleration due to gravity (9.8 ms^{-2}), n = Manning's n (sand n = 0.025), m = absolute momentum ($(mx^2 + my^2)^{0.5}$), d = water depth (ANUGA stage-elevation), Kd = the detachment rate (0.0250 kg s^{-1} m^{-2} Pa^{-1}), Sd = sediment mass density (1800 kg^{-3} for packed siliceous sand).

The primary objective of Scenario C was to explore the level of protection afforded by the present (2015) Seven Mile Beach dune line, which separates Hobart Airport from the waters of Frederick Henry Bay. However, the opportunity was also taken to investigate the possible loss of protection at twelve other sites that are potentially prone to erosion (Fig. 5).

2.3. *Data Processing*

Raster and time series outputs were extracted from the netcdf model files and analysed to map maximum flow depths and maximum velocity.

CSV files of time series data for water level and velocity were extracted at 187 point locations near

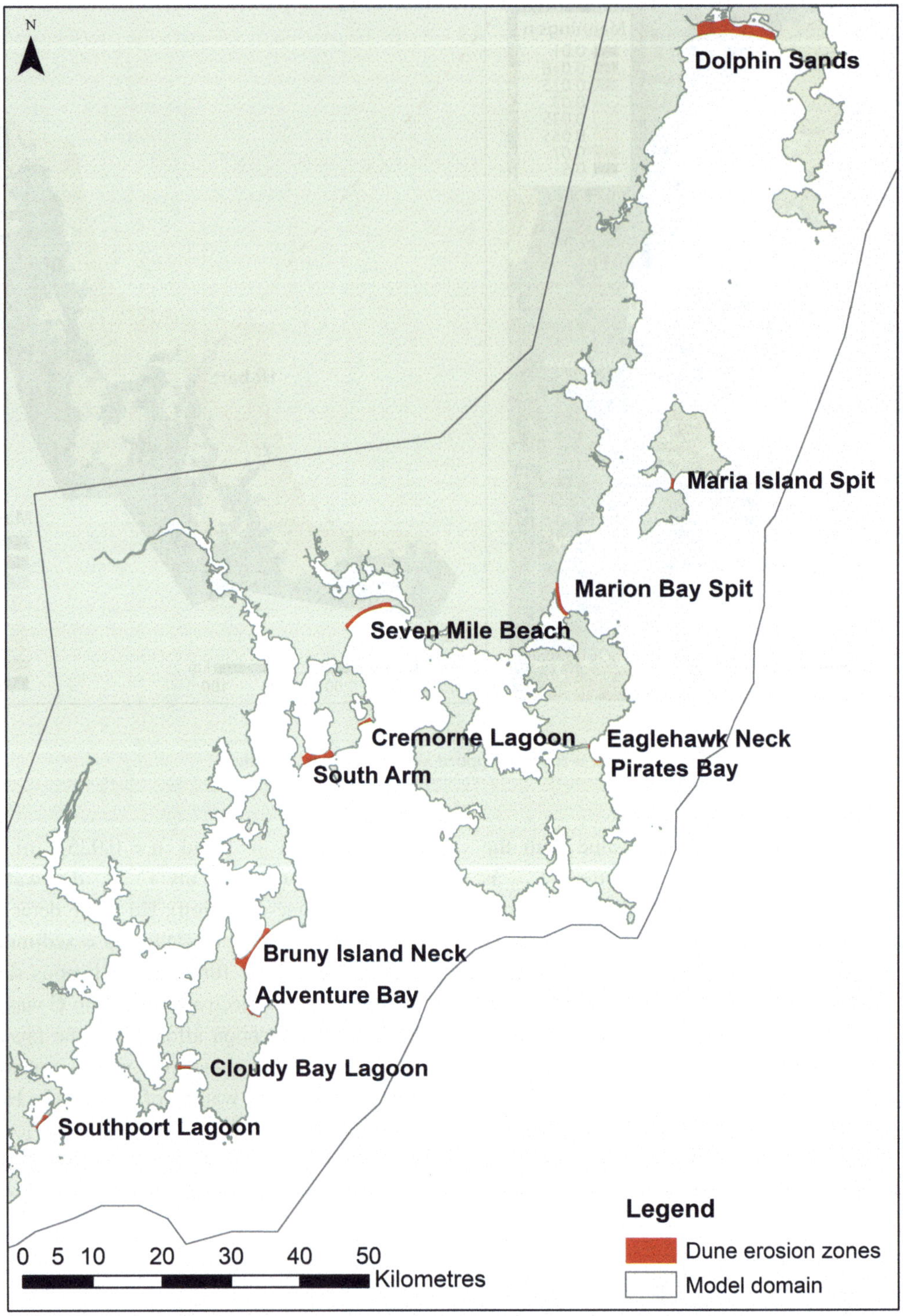

Figure 5
Location of the areas modelled with the active dune erosion operator

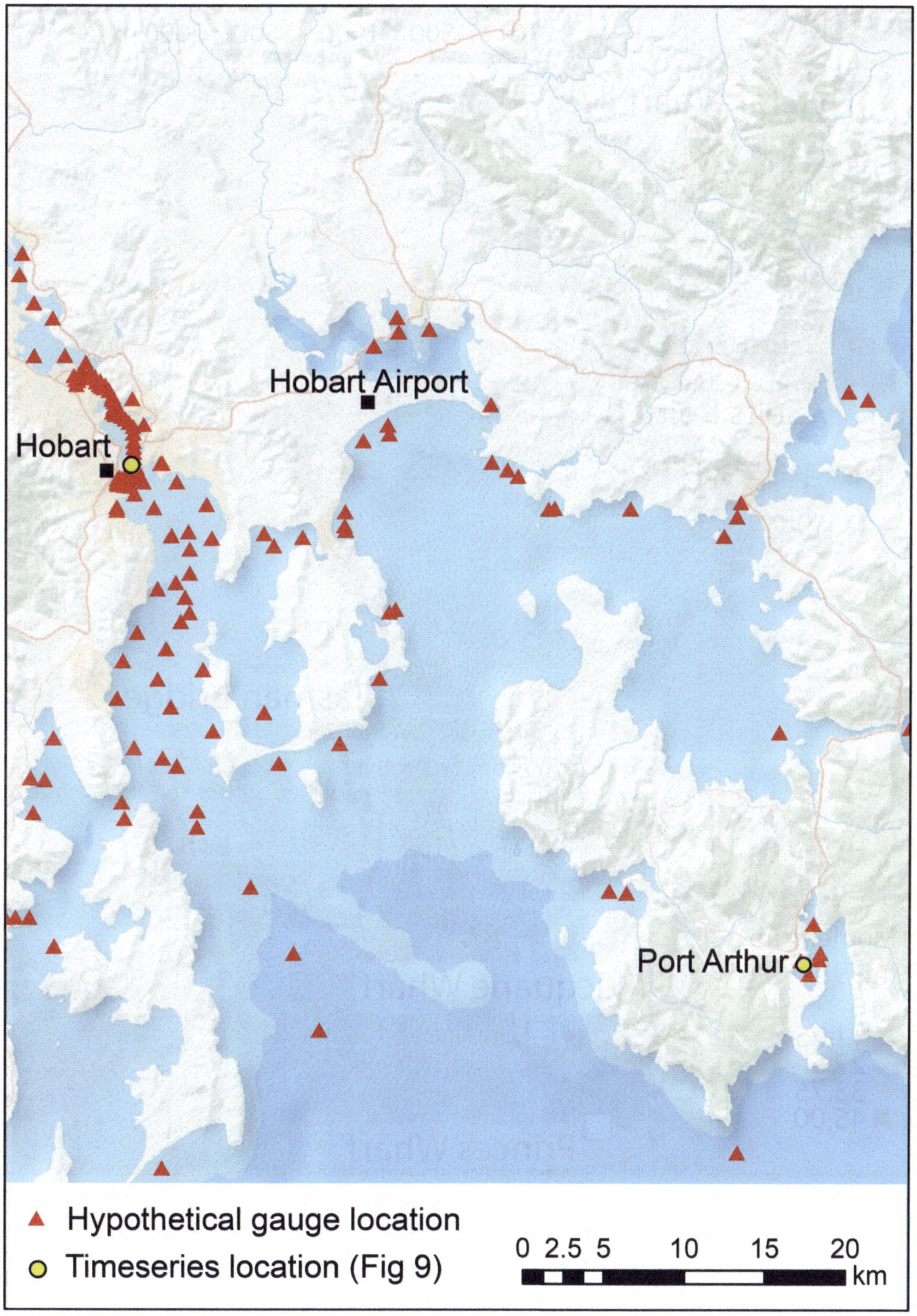

Figure 6
Gauge locations for the maritime hazard scenario

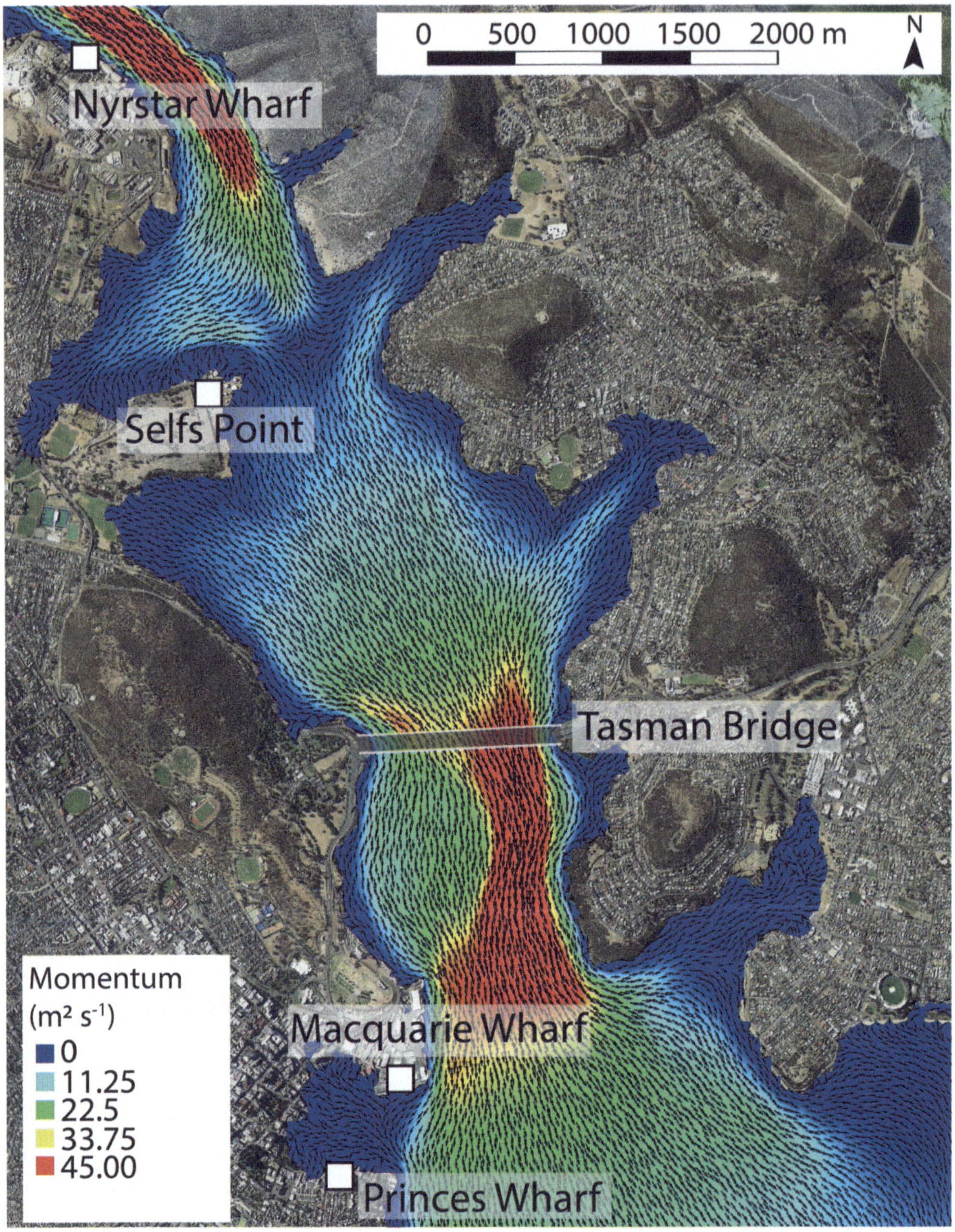

Figure 7
Snapshot of simulated tsunami momentum in the Derwent Estuary near Hobart city. Locations of port and dock areas mentioned in the maritime hazard results are also shown

coastal communities or where maritime hazard may be significant, such as shipping lanes, marinas and embayments (Fig. 6). The following parameters were calculated from the time series for the maritime hazard assessment: tsunami arrival time (arrival of the first peak post-earthquake, with a threshold of 0.1 m), maximum wave height (wave crest—trough), time of maximum wave height, period, average wavelength of the first two waves, maximum current speed, maximum celerity (wave speed), maximum

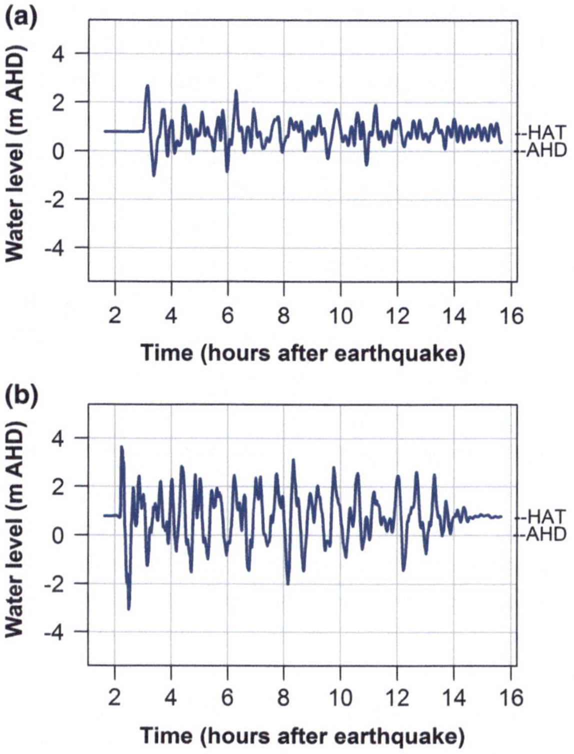

Figure 8
a Tsunami time series plot for the primary shipping channel near Tasman Bridge. **b** Tsunami time series plot for Port Arthur

possible instantaneous speed (current speed + celerity), maximum water level (m AHD), minimum water level (m AHD) and turbulence ratio (wave height: water depth). Note that wave heights are used for the maritime hazard analysis, rather than wave amplitudes, as the total variation from peak to trough is important for assessing shipping hazard and berthing requirements.

The final outputs include video animations of the modelled tsunami and a series of maps depicting maximum possible impacts around 71 coastal communities, which show coastal inundation, maximum offshore current speed and potentially turbulent areas (Kain et al. 2018).

2.4. *Field Validation*

Site visits were undertaken in order to visually assess the inundation extents at important sites or areas where modelled impacts were significant. Sites were photographed and key infrastructure at risk of flooding was identified. In some cases, features were identified that may increase or mitigate the mapped risk; for example, subsurface drains could allow greater tsunami penetration, whereas dense vegetation may limit it.

3. *Results*

The geometry of the subduction zone and rupture scenario is such that the wave would arrive as a leading peak (rising water level), rather than a trough (receding water). The modelled tsunami wavelengths are in the order of 500 m nearshore (increasing to > 5 km offshore) and, consequently, the tsunami manifests as a gradual rise and fall in water level that occurs over a period of 10–20 min. Significant seiching and wave reflections are predicted in the channel and embayments, and these would generate considerable water disturbance and localised amplification. The first tsunami wave would reach the exposed east coast approximately 2 h and 10 min post-earthquake, and around 3 h in the Hobart CBD area. A snapshot of the simulated momentum in the Derwent Estuary is given in Fig. 7.

3.1. *Maritime Hazard Analysis*

Within the principal shipping channel, modelled wave heights vary from 2.4 m at the southern end (Fig. 8a) to 5.1 m off Newtown Bay, with many values falling between 3.5 and 4.5 m. Maximum modelled wave speed falls between 8.0 ms^{-1} (mid channel at the entrance to Prince of Wales Bay) and 16.8 ms^{-1}. Maximum induced current speeds are in the order of 1.0–3.6 ms^{-1} and severe turbulence would not be expected in the shipping channel, as the water there is deep relative to the wave height.

The tsunami impacts would be variable in the port and dock areas. Maximum predicted wave heights range from 2.7 to 5.0 m and maximum simulated celerity varies between 2.5 and 15.9 ms^{-1}. In general, predicted wave speeds are < 5.1 ms^{-1} within the dock areas and along the shore, but higher values occur further out in the channel. Turbulence would be expected in most port and dock areas, with Prince of Wales Bay, Macquarie Wharf and Princes

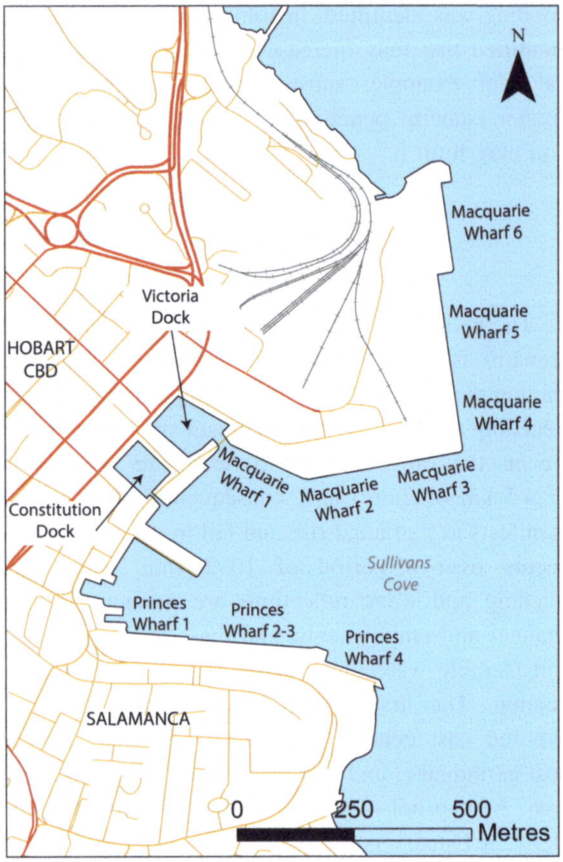

Figure 9
Locations of the berths at Macquarie Wharf and Princes Wharf

Wharf particularly affected. A backup of water is observed in places, most notably at Constitution Dock near Hobart CBD (Fig. 9). Following inundation by the first wave in the model, the water does not fully drain from the area before the arrival of the second wave and the water level remains at least 1.5 m higher than the starting level for the duration of the model run. However, at no point does the water height exceed that of the first wave peak level.

Potential consequences include damage to boats and infrastructure from collisions, breakage of moorings due to water level or currents, spills and pollution, and foundering or sinking of smaller craft. A summary of berth information and modelled tsunami wave parameters is given in Table 4. Because most of the commercial shipping around Hobart port have drafts greater than the inundation levels, there is little likelihood of these ships being lifted onto the berth. However, smaller vessels/craft with shallower drafts, e.g. less than 1.2 m at the CSIRO Wharf (Princes Wharf 4; Fig. 9), are at risk of being lifted on to the berth deck. There is also a significant risk of vessels breaking adrift if moorings are not tended to deal with the rising and falling levels. This could cause major damage to the drifting ship, other vessels and infrastructure.

For vessels and small craft underway at the time of the tsunami, loss of navaids and steering capacity due to the tsunami wave(s) and induced currents is expected. These issues could then lead to collisions and would be a particular concern for large ships around the Tasman Bridge (Fig. 1). The maximum modelled wave height in the main navigational channel at the bridge is 5.1 m, with a simulated maximum current speed of 2.5 ms^{-1}. Given that the channel is 100 m wide with clearance of 44 m and a maximum permitted vessel passage of 185 m (in length), consequences could be catastrophic.

Modelled tsunami arrival times range from 2.5 to 3.25 h post-earthquake, with a difference of 36 min between the outer limit of the shipping channel (off Iron Pot, to the anchorages at Prince of Wales Bay (Fig. 1). A timeline based on the modelled event and the Tasmanian State Tsunami Emergency Response Plan, is given in Table 5.

Given an evacuation warning time window of 1 h and sufficient crew, pilot boat availability and standby engine power, it may be possible for some large ships to evacuate from Hobart port and Nyrstar/Risdon wharves. However, the required passage through Tasman Bridge limits the feasibility of evacuation. The decision to evacuate to deeper water depends heavily on the ship's mobilisation status as well as distance to deeper water, and in many cases it may be safer to remain in port and evacuate crew to higher ground. Specific mobilisation and evacuation times have been calculated for vessels that frequent the Hobart port areas (Table 6). For smaller craft, the primary advice is to secure the craft and evacuate the crew to higher ground. Significant water disturbance and turbulence is expected at all marinas and anchorages around Hobart, making small craft manoeuvrability difficult and/or unsafe.

Table 4

Summary of berth data alongside modelled tsunami wave heights and induced current speeds for each locality (locations shown in Figs. 1, 7 and 9)

Location or berth	Declared depth (metres)	Height of berth above chart datum (metres)	Number of bollards	Usage	Tsunami arrival time (post earthquake)	Max wave height (m)	Inundation height above berth deck (m)	Induced current speed (ms^{-1})
Main navigational channel off Iron Pot	14.1 m (Iron Pot Bar)	–	–	Shipping channel	2 h 30 m	4.7	–	1.8
Main navigational channel, Tasman Bridge		–	–	Shipping channel	3 h	5.1	–	2.5
Main navigational channel off Newtown Bay				Shipping channel	3 h 10	5.1	–	3.2
Princes Wharf 1	7.3	3.31	13	Antarctic Supply, Small	3 h	3.4	0	0.3
Princes Wharf 2-3	9.0	3.25	8	Cruise, Naval	3 h	3.1	0	0.5
Princes Wharf 4	7.2	2.79		CSIRO	3 h	3.6	0	0.6
Macquarie 1	4.1	3.1 to 4.01		Small Fishing, Fish Unloading Facility	3 h	3.1	1.0	0.4
Macquarie 2	9.1	4.01	10	Larger Fishing, Caustic Acid	3 h	3.8	1.0	0.6
Macquarie 3	9.9	4.01	12	Car Carriers, Bulk Products, Cruise, Antarctic	3 h	3.5	1.0	0.7
Macquarie 4–5	13.0	4.01–2.79	15	Container Traffic, Break Bulk, Cruise, Antarctic	3 h	3.2	1.0	0.8
Macquarie 6	11.4	2.79–3.2	13	Lay-up berth	3 h	3.3	1.0	1.0
Self's Point	14.4	3.71	10 on wharf; 4 on	dolphins	Fuel Terminal, Bunkering Facility	3 h 10 m	4.9–5.0	0.5
2.1–2.4								
Risdon (Nyrstar)	10.2	3.33	19	Concentrates, Acid, Fertiliser	3 h 10 m	2.7–4.5	~ 0.1	2.6–2.8

A map of berth sites at Princes and Macquarie Wharves is provided in Fig. 9

3.2. Coastal Hazard and Inundation

This section is concerned primarily with results from Scenario B (as outlined in Sect. 2). Simulated coastal inundation is generally greatest in exposed eastern areas or where river channels act as a conduit for inland penetration. A funnelling of the tsunami energy is observed in the simulation as it passes through the narrow channel northwest of the Tasman Bridge, which may also be exacerbated by wave reflection from the steep bathymetry on the eastern shore.

Modelled flooding around the central Hobart waterfront (Fig. 10) extends up to 100 m inland in places, but depths are no more than 1 m with a maximum run-up elevation of 2.5 m AHD. The results indicate that the wharf areas and surrounding car parks would be inundated, and the buildings along

Table 5

Timeline of tsunami arrival and predicted warning times

Warnings and wave arrival summary (approximations)	
Event	Elapsed time
Earthquake occurrence	0
National Tsunami watch alert	30 min
Specific watches and warnings issued	1 h
Tasmania Police issue additional warnings	
Realistic notification of Hobart public	2 h
Wave reaches Tasmania	2 h
Wave reaches Iron Pot	2 h 35 min
Wave reaches Sullivans Cove	3 h
Wave reaches Selfs Point	3 h 05 min
Wave reaches Risdon (Nyrstar)	3 h 08 min
Wave reaches Prince of Wales Bay	3 h 10 min

Tsunami watch and warning estimates derived from the Tasmanian State Tsunami Response Plan

the waterfront would most likely be affected with potentially significant impacts for underground infrastructure. Modelling results also suggest that the tsunami would be funnelled into the narrow embayments on both sides of the Derwent Estuary, causing flooding at the head of these bays and impacting the nearshore marinas and recreational areas.

The most heavily affected areas are Tasman Peninsula and Bruny Island (Fig. 1), with flow depths predicted to reach > 4 m with an inundation limit of several hundred metres inland in many locations. Significant inundation is predicted at Port Arthur, Nubeena and Adventure Bay, with simulated flooding reaching several hundred metres inland and a maximum flow depth of 5 m near the coast. A time series graph for Port Arthur is given in Fig. 8b.

In particular, simulated water levels at Eaglehawk Neck reach 14 m AHD and severe flooding could occur through dune breaching and overtopping. This extreme modelled water level arises from the interaction of the tsunami wave with the shape of Pirates Bay in the simulation. We suggest this water level not be taken at face value, as the reality would vary with a small change in wave characteristics, timing or approach direction at this location. In this simulation, a combination of wave refraction from the neighbouring bay and wave reflection from the semicircular embayment results in the formation of large vortices at each end of the embayment, along with constructive wave interference and focusing that together cause a simulated wave amplitude of 14 m. Notably, the actual incoming wave crest elevation (prior to the amplification) is approximately 7 m, which is in line with other exposed coastal locations.

Regardless of any occurrence of tsunami amplification, the impact would remain severe. Any significant pre-existing low point in east-facing dunes would still be exploited as a conduit by an incoming tsunami of 7 m. This level would result in several state highways becoming compromised, which would isolate communities beyond the choke points of Eaglehawk Neck and Bruny Neck, and some coastal locations north of Tasman Peninsula.

Table 6

Estimated time to mobilise and evacuate to deeper water for ships that frequent Hobart Port

Estimated time to reach				Distance to White Rock Pt (km)	Distance to open ocean (km)
Location	Vessel type	Open ocean*	Deep water in shipping channel**		
Nyrstar Wharf	Bulk carrier	2 h 35 m	1 h 20 m	17.6	45.4
Selfs Point Wharf	Oil tanker	2 h 25 m	1 h 25 m	15.7	43.5
Princes Wharf	Research vessel	1 h 50 m	55 m	12.0	39.8
Macquarie Wharf	Cruise ship	1 h 35 m	50 m	12.0	39.8

*45 m depth, Storm Bay; **33 m depth, White Rock Pt

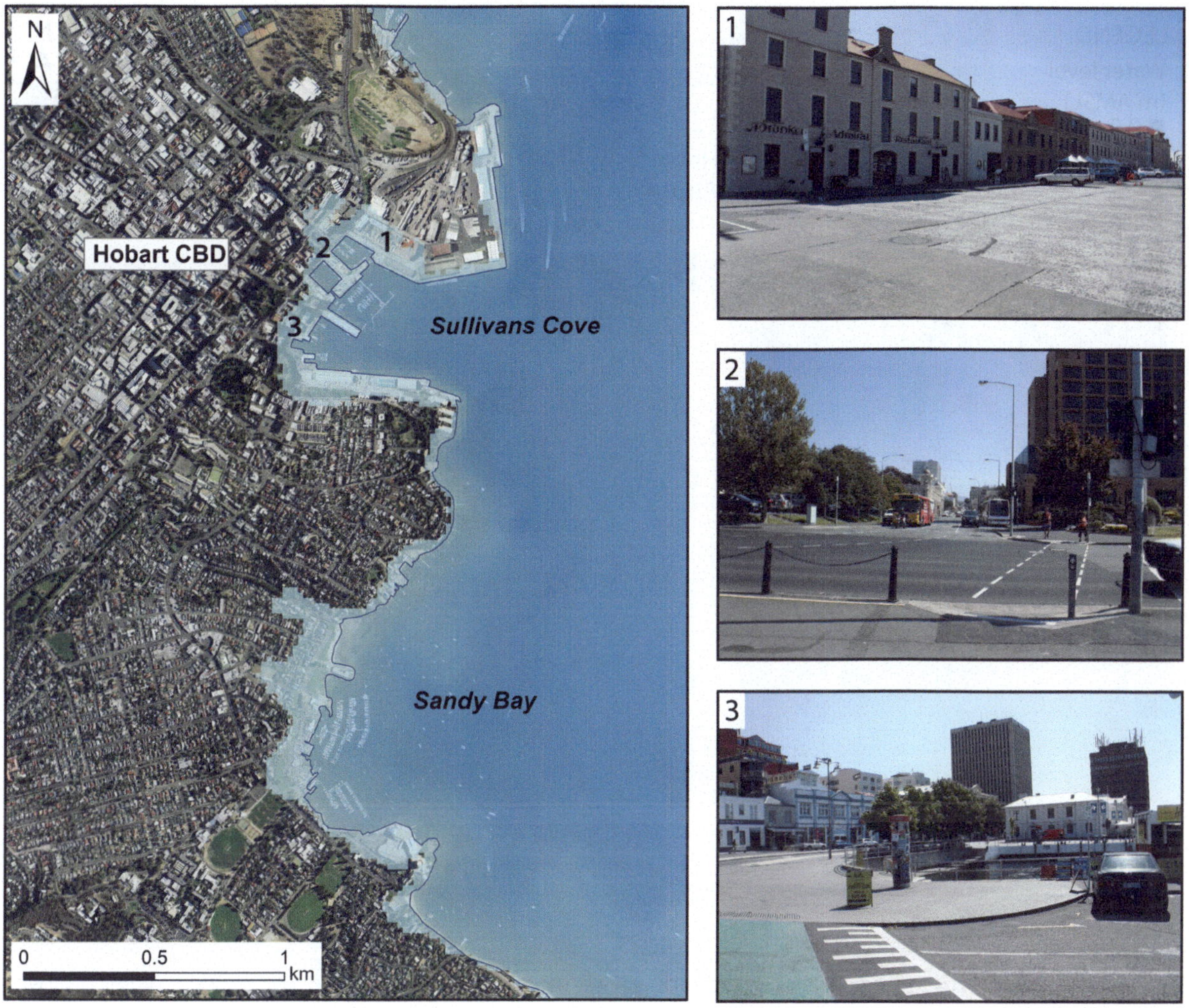

Figure 10
Modelled inundation in Hobart CBD and Sandy Bay, accompanied by field validation photos showing areas of predicted flooding around the city waterfront

Predicted inundation is less severe at Hobart city waterfront and in the bays inland from the CBD. However, significant flooding is predicted in Kingston and flooding via the Newtown Rivulet would compromise highway access between Moonah and Newtown (Fig. 1).

3.3. Dune Erosion and Risk to Hobart Airport

This section presents results from Scenario C (see Sect. 2). Modelling results show that the wave height would be significantly reduced as the tsunami travels through Frederick Henry Bay (Fig. 11). As such, the dune line (as present in 2016) is sufficient to protect the airport runway and infrastructure from inundation for the modelled event. Little dune erosion was apparent in the simulation at Seven Mile Beach, although this could change if the dune structure altered (for example, through a blowout or removal of part of the dune). It is also important to note that part of the dune line is only marginally higher than the approaching wave amplitude, so a relatively minor reduction in dune height could create significant local breaches. The dune line in front of the runway is well

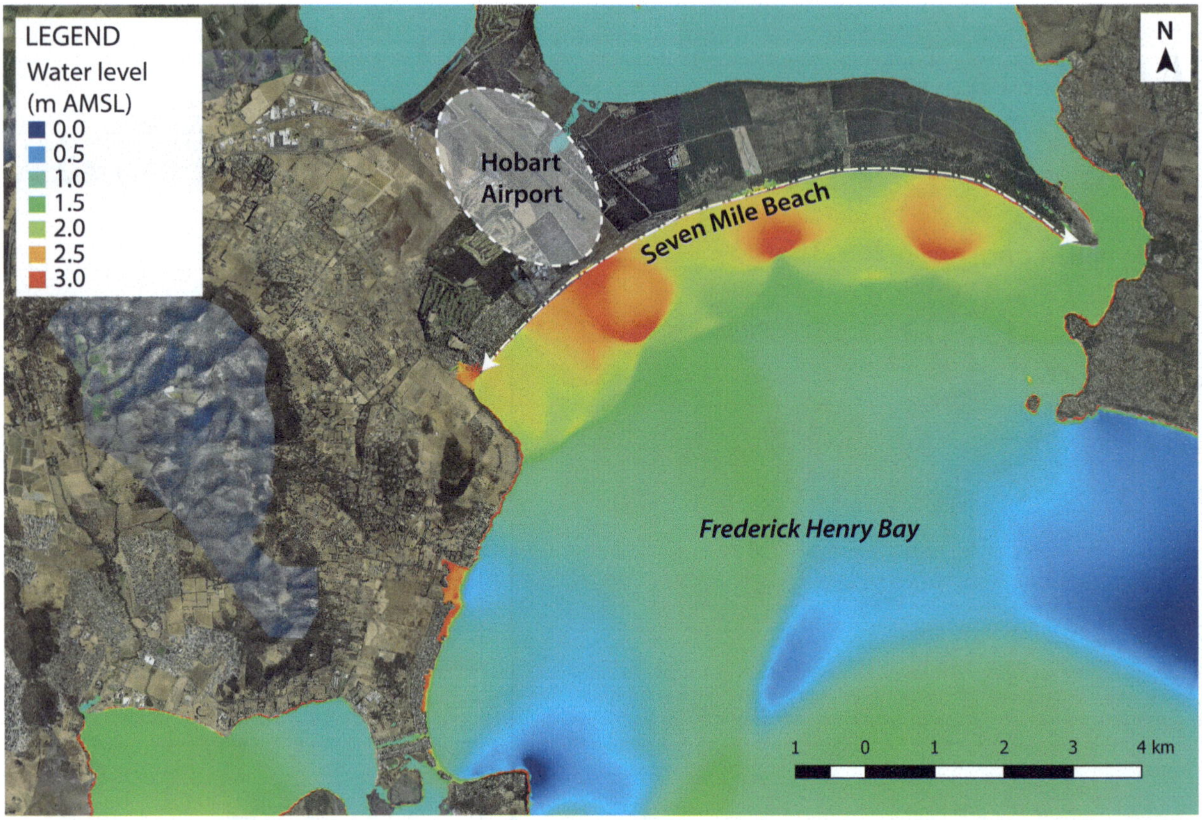

Figure 11
Snapshot of the modelled wave as it reaches Seven Mile Beach in front of Hobart Airport. The dune line is sufficiently high that the wave is largely reflected back from the beach (red pattern, nearshore) and does not penetrate the dunes in front of the runway

developed, but if it were to be removed, an incoming tsunami wave would likely breach the barrier at this location.

In addition to the detailed modelling of the airport site, the opportunity was taken to investigate the possible loss of dune protection at other erosion prone sites in southeast Tasmania. Erosion-enabled modelling suggests that significant erosion and inundation could potentially affect infrastructure and/or settlements at the following sites: Eaglehawk Neck, Bruny Island Neck, Adventure Bay Beach, South Arm Neck and Pirates Bay Beach (Fig. 5). A detailed analysis of the impact of the erosion operator was performed at Bruny Neck, where approximately 900,000 m^3 of sand was removed during the simulation. As shown in Fig. 12a, up to 2.5 m of sand was eroded in places. A comparison of modelled tsunami flow depths with and without the erosion operator showed a water level increase of 3 m where the eroded dunes allowed greater tsunami penetration (Fig. 12b). Conversely, the comparison showed a decrease in maximum stage immediately seaward of the eroded dune line, as a consequence of reduced wave reflection.

4. Discussion

Modelling is an excellent tool for estimating the impacts of a tsunami event, but some inherent limitations must be recognised. Firstly, ANUGA is a 2D model library, which means that vertical motion cannot be resolved in the simulation. As such, factors such as wave breaking and turbulence are not directly simulated.

The tidal level was set to HAT at the beginning of the simulation and the subsequent water level and

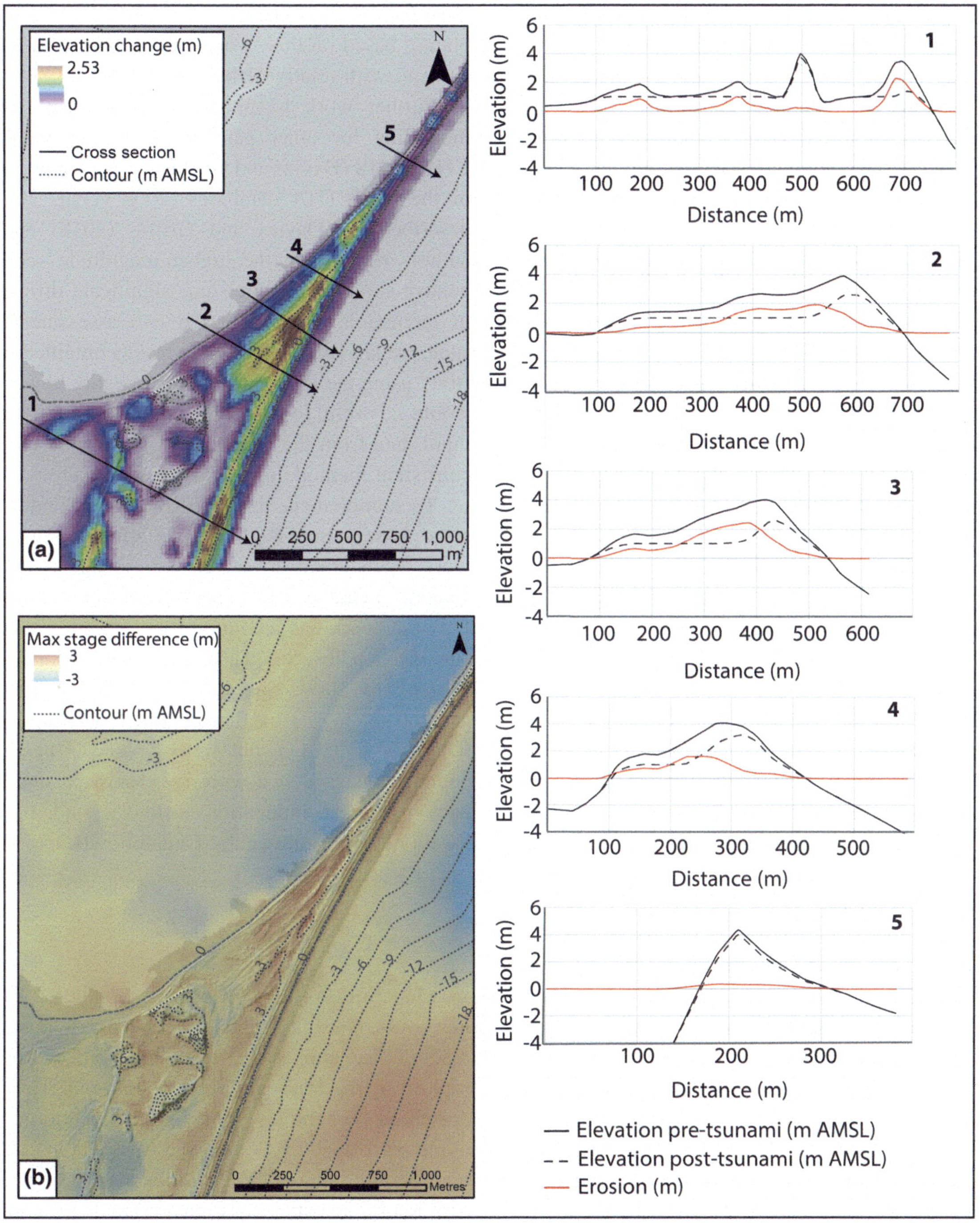

Figure 12
Comparison of the effects of the dune erosion operator at Bruny Neck. **a** Map of erosion patterns across the study area, with the location of cross section profiles 1–5. Plots of the pre and post-tsunami elevation along these profiles are presented to the right of the figure. **b** Map showing the difference in maximum stage when comparing model runs with and without the dune erosion operator enabled

current fluctuations do not account for ongoing tidal variation and currents. The interaction of the tsunami with tidal fluctuations was not feasible given the model size and computational constraints. However, the sensitivity of the modelled inundation to tidal phasing is an interesting avenue for further work. Additionally, it would be beneficial to model historical events and compare the results with tide gauge observations, in order to better understand and quantify the uncertainties within our model data.

Onshore tsunami flooding is controlled by coastal geometry, erosion and land use/cover. The event was modelled based on elevation data from a snapshot in time, so changes in dune form or land cover would influence actual tsunami behaviour. Every effort was made to accurately model the effects of vegetation and building cover through the Manning's roughness model input, but this is again a best approximation of reality. Similarly, it is important to note that the parameters used in the dune erosion operator do not distinguish between bare sand and dunes that are vegetated with grass and/or trees. We recommend that further work be done to test the variation in detachment rates for bare and stabilised dunes.

The resolution of the mesh with respect to topographic changes has caused anomalous run-up values in areas of steep relief. Because the mesh represents a modelled surface generated from an elevation model, it does not properly represent vertical or near-vertical surfaces such as coastal cliffs and artificial structures like wharf edges. It is possible to use break lines in ANUGA to account for buildings and vertical features, but due to the large spatial extent and complexity of our model this was not attempted. As a consequence, the maps may occasionally show flow depths that appear extreme (e.g. 10 m along the cliffs at Lindisfarne). The higher the mesh resolution, the less this problem occurs, as the closer the modelled surface reflects reality. These anomalies are relatively easy to identify in the maps and have little effect on the interpretation of the inundation extent.

These results do not constitute a tsunami risk assessment. This would require consideration of a variety of tsunami sources with a range of magnitudes and return periods. However, an analysis of vulnerable places, properties and infrastructure should be undertaken alongside a review of Southeast Tasmania's tsunami emergency response plan(s) based on this 'worst case scenario' of a large tsunami with relatively little warning time.

Further work is underway to perform tsunami modelling for other parts of Tasmania, using the PTHA 2018 (Davies and Griffin 2018). A comparison of the two PTHA databases (2008 versus 2018) is described by Davies and Griffin (2018) and the authors report that the higher magnitude scenarios behave similarly despite some significant differences in methodology between the two assessments. As such, we are confident that the results obtained in this study using the PTHA 2008 remain valid; however, it would be useful to re-run the inundation model using input data from the PTHA 2018 in order to ensure a consistent methodology across Tasmania.

To complement the modelling work and understand past tsunamis that have affected Tasmania, work is underway to search for geological palaeotsunami evidence in exposed eastern locations. Previous work has recognised the scope for palaeotsunami studies in Tasmania (e.g. Morris and Mazengarb 2009; Sharples 2006) and a previous palaeotsunami site has been studied at Bruny Island (Clark et al. 2011), but further work is needed to connect this evidence with a wider event and compare modelled inundation distances and flooding locations with sedimentary tsunami evidence.

5. *Conclusions*

Modelling results suggest that Southeast Tasmania could be significantly affected by a maximum credible tsunami event, resulting from a Mw 8.7 rupture of the Puysegur subduction zone. The work presented here builds upon initial inundation modelling performed by Geoscience Australia in 2009, by incorporating new high-resolution elevation data and addressing specific questions related to maritime hazard and risk to Hobart Airport. This approach allows a "whole-coast" hazard assessment for the region and highlights the differences in tsunami timing and coastal interactions across the entire area.

Tsunami arrival times range from approximately 2 h at the eastern coast, to 3 h at Hobart wharves. Significant water disturbance in the form of currents, extreme water level changes and turbulence is predicted in all coastal and nearshore environments, but does not generally extend to the main shipping channel in the Derwent estuary. Such disturbance would cause considerable risk to marine craft and could result in ships breaking adrift, damage to boats and infrastructure, and pollution. The maritime hazard assessment suggests that the feasibility of evacuation to deeper water is questionable, given the short timeframes involved, and securing/tending of vessels and evacuation of crew to land is generally recommended.

The simulations predict severe inundation levels (> 4 m flow depth) in exposed coastal communities on the east coast. Significant inundation (> 3 m flow depth) is predicted in the embayments along both shores of the Derwent estuary, including the Hobart city waterfront and wharves. The potential for dune erosion was included in the modelling for Hobart Airport, and results suggest that the current dune line would not be breached or overtopped by a tsunami of this magnitude at this location. However, inundation could occur if the dune was to be substantially lowered or partially removed in the future.

These results could be useful for an analysis of vulnerable communities and infrastructure for the case of a maximum credible tsunami, and feed into regional hazard management planning. In addition, palaeotsunami studies would help validate the modelling results and provide useful information regarding inundation extents and recurrence intervals of tsunamis that have previously affected Tasmania.

Acknowledgements

This project was funded by a grant from the Natural Disaster Resilience Grants Programme (NDRGP), in conjunction with Mineral Resources Tasmania. Assistance with ANUGA troubleshooting was kindly provided by Stephen Roberts of ANU. Claire Kain and Colin Mazengarb publish with permission of the Director of Mines, Tasmania.

Publisher's Note Springer Nature remains neutral with regard to jurisdictional claims in published maps and institutional affiliations.

References

Australian Bureau of Statistics. (2018). Regional Population Growth, Australia, 2017–2018. https://www.abs.gov.au/AUSSTATS/abs@.nsf/mf/3218.0. Accessed 23 Aug 2019.

Baldock, T. E., Barnes, M. P., Guard, P. A., Hie, T., Hanslow, D., Ranasinghe, R., Gray, D., & Nielsen, O. (2007). Modelling tsunami inundation on coastlines with characteristic form. In: Proceedings of the 16th Australasian Fluid Mechanics Conference: Gold Coast, Australia, pp. 939–942.

Barrett, N., Sanderson, J. C., Lawler, M., Halley, V., & Jordan, A. (2001). Mapping of inshore marine habitats in south-eastern Tasmania for marine protected area planning and marine management. Tasmanian Aquaculture and Fisheries Institute Technical Report Series 7.

Beccari, B. (2009). Measurements and impacts of the Chilean tsunami of May 1960 in New South Wales, Australia. New South Wales State Emergency Service Report 05/0576, p 42.

Borrero, J. C., Lynett, P. J., & Kalligeris, N. (2015). Tsunami currents in ports. *Philosophical Transactions of the Royal Society A*. https://doi.org/10.1098/rsta.2014.0372.

Bricker, J. D., Gibson, S., Takagi, H., & Imamura, F. (2015). On the need for larger Manning's roughness coefficients in depth-integrated tsunami inundation models. *Coastal Engineering Journal*. https://doi.org/10.1142/s0578563415500059.

Burbidge, D., Mleczko, R., Thomas, C., Cummins, P., Nielsen, O., & Dhu, T. (2008). A probabilistic tsunami hazard assessment for Australia. Geoscience Australia Professional Opinion 2008/04, Canberra.

Chow, V. T. (1959). *Open-channel hydraulics*. New York: McGraw-Hill Book Company.

Clark, K., Cochran, U., & Mazengarb, C. (2011). Holocene coastal evolution and evidence for paleotsunami from a tectonically stable region, Tasmania, Australia. *The Holocene, 21,* 883–895.

Commonwealth of Australia—Geoscience Australia. (2016). ARR: australian rainfall and runoff; a guide to flood estimation. http://arr.ga.gov.au/arr-guideline. Accessed 11 Dec 2016.

Cox, D. C., & Mink, J. F. (1963). The tsunami of 23 May 1960 in the Hawaiian Islands. *Bulletin of the Seismological Society of America, 53,* 1191–1209.

Davies, P. E., Crawford, C. M., Wells, F. A., Dunstan, P., & Mitchell, I. M. (2002). Environmental flow assessment of the lower coal river and pitt water estuary. Report to DPIPWE from Freshwater Systems and the Tasmanian Aquaculture and Fisheries Institute. p. 203.

Davies, G., & Griffin, J. (2018). The 2018 Australian probabilistic tsunami hazard assessment: hazard from earthquake generated tsunamis. Geoscience Australia, Record 2018/41. https://doi.org/10.11636/Record.2018.041

Department of Primary Industries, Parks, Water and Environment (DPIPWE). (2007). LIST Tasmania 25 metre Digital Elevation Model. http://listdata.thelist.tas.gov.au/opendata/index.html#LIST_Tasmania_25_Metre_Digital_Elevation_Model. Accessed 1 Aug 2016

Department of Primary Industries, Parks, Water and Environment (DPIPWE). (2013). TASVEG 3.0: Tasmanian vegetation monitoring and mapping program. Resource Management and Conservation Division

Fernandes, M. A. G. (2009). Systematic comparison on the inundation response of AnuGA and COMCOT tsunami modelling codes applied to the Boca do Rio and Alvor Bay area. Unpublished PhD Thesis, Univerisidade do Algarve. p. 75.

Froelich, D. (2002). IMPACT project field tests 1 and 2 "Blind" simulation. 2nd IMPACT Project Workshop, Mo-i-Rana Norway.

Goff, J., & Chagué-Goff, C. (2014). The Australian tsunami database: a review. *Progress in Physical Geography, 38,* 218–240.

Hasan, H., Lodhi, H. A., LeVeque, R. J., Lodi, S. H., & Ahmed, S. (2017). Assessing tsunami risk for Karachi Port through simulation of currents that were reportedly produced there by the 1945 Makran tsunami. In: Proceedings of the 16th World Conference on Earthquake, Santiago Chile, January 2017

Heath, R. A. (1976). The response of several New Zealand harbours to the 1960 Chilean tsunami. *Bulletin of the Royal Society of New Zealand, 15,* 71–82.

Jakeman, J. D., Nielsen, O. M., Van Putten, K., Mleczko, R., Burbidge, D., & Horspool, N. (2010). Towards spatially distributed quantitative assessment of tsunami inundation models. *Ocean Dynamics, 60,* 1115–1138.

Kain, C. L., Mazengarb, C., Rigby, E. H., Cohen, B., Simard, G., & Lewarn, B. (2018). Technical report on tsunami inundation modelling in South East Tasmania. Tasmanian Geological Survey Report UR2018_02. p. 160.

Kato, Y., Suzuki, Z., Nakamura, K., Takagi, A., Emura, M., Ito, M., et al. (1961). The Chile tsunami of 1960 observed along the Sanriku coast of Japan. Report series of Tohoku University, 5. *Geophysics, 13,* 107–125.

Land Information System Tasmania. (2016). ListData Open data. https://listdata.thelist.tas.gov.au/opendata/. Accessed 11 Dec 2016.

Lucieer, V. L., Lawler, M., Morffew, M., & Pender, A. (2007). Estuarine habitat mapping in the derwent—2007. A resurvey of marine habitats by seamap tasmania. Tasmanian Aquiculture and Fisheries Institute, University of Tasmania: Hobart.

Lynett, P. J., Borerro, J. C., Weiss, R., Son, S., Greer, D., & Renteria, W. (2012). Observations and modelling of tsunami-induced currents in ports and harbors. *Earth and Planetary Science Letters, 327–328,* 68–74.

Lynett, P. J., Borrero, J., Son, S., Wilson, R., & Miller, K. (2014). Assessment of the tsunami-induced current hazard. *Geophysical Research Letters, 41,* 2048–2055.

Lynett, P. J., Gately, K., Wilson, R., Montoya, L., Arcas, D., Aytore, B., et al. (2017). Inter-model analysis of tsunami-induced coastal currents. *Ocean Modelling, 114,* 14–32.

Morris, M. K., & Mazengarb, C. (2009). Historical accounts of tsunamis in Tasmania. Tasmanian Geological Survey Record 2009/04. Department of Infrastructure, Energy and Resources, Mineral Resources Tasmania.

Muhari, A., Charvet, I., Tsuyoshi, F., Suppasri, A., & Imamura, F. (2015). Assessment of tsunami hazards in ports and their impact on marine vessels derived from tsunami models and the observed damage data. *Natural Hazards.* https://doi.org/10.1007/s11069-015-1772-0.

Mungkasi, S., & Roberts, S. G. (2013). Validation of ANUGA hydraulic model using exact solutions to shallow water wave problems. *Journal of Physics: Conference Series, 423,* 012029. https://doi.org/10.1088/1742-6596/423/1/012029.

Nielsen, O., Roberts, S., Gray, D., McPherson, A., & Hitchman, A. (2005). Hydrodynamic modelling of coastal inundation. MODSIM 2005 In: International Congress on Modelling and Simulation, Modelling and Simulation Society of Australia & New Zealand, pp. 518–523.

Okal, E. A., Fritz, H. M., Raveloson, R., Joelson, G., Pancockova, P., & Rambolamanana, G. (2006a). Madagascar field survey after the December 2004 Indian Ocean tsunami. *Earthquake Spectra, 22,* 263–283.

Okal, E. A., Sladen, A., & Okal, E. A. S. (2006b). Rodrigues, Mauritius and Reunion islands field survey after the December 2004 Indian Ocean tsunami. *Earthquake Spectra, 22,* 241–261.

Rigby, E., Mazengarb, C., Barthelmess, A., & Kain, C. (2017). Experiences developing and applying an ANUGA dune erosion operator to assess risk from loss of dune protection during a tsunami. In: Proceedings of the 13th Hydraulics in Water Engineering Conference. pp. 8

Roberts, S., Nielsen, O., Gray, D., Sexton, J., & Davies, G. (2019). ANUGA user manual. Geoscience Australia. https://anuga.anu.edu.au/

Sharples, C. (2006). *Indicative mapping of Tasmanian coastal vulnerability to climate change and sea-level rise: explanatory report* (p. 173). Tasmania: Consultant Report to Department of Primary Industries & Water.

Sterzai, P., Coren, F., & Creati, N. (2015). The application of open-source software ANUGA installed on a BlueJ/Q computer for a tsunami modelling in coastal areas of city of Trieste using high resolution laser-scanning and multibeam data. *Planet-Risk, 3*(1), 121–125.

Uslu, B., Power, W., Greenslade, D., Eblé, M., & Titov, V. (2011). The July 15, 2009 Fiordland, New Zealand Tsunami: real-time assessment. *Pure and Applied Geophysics, 168*(11), 1963–1972.

Van Putten, K., Fountain, L., Griffin, J., Sexton, J., Nielsen, O., & Wilson, R. (2009). Capacity building for tsunami planning and preparation: inundation models for south east Tasmania. Geoscience Australia Professional Opinion. No.2009/07. Geoscience Australia: Canberra.

Whiteway, T. (2009). *Australian bathymetry and topography grid, June 2009. Scale 1:5 000 000.* Canberra: Geoscience.

Wilson, K. M., Allen, S. C. R., & Power, H. E. (2018). The tsunami threat to Sydney Harbour, Australia: modelling potential and historic events. *Scientific Reports.* https://doi.org/10.1038/s41598-018-33156-w.

Wiśniewski, B., & Wolski, T. (2012). The safety of the shipping and ports in the aspect of the tsunami events. *Scientific Journals, 30*(102), 150–157.

(Received September 4, 2019, revised November 25, 2019, accepted November 29, 2019, Published online December 13, 2019)

Pure Appl. Geophys. 177 (2020), 1569–1581

https://doi.org/10.1007/s00024-019-02319-9

Pure and Applied Geophysics

Far-Field Tsunami Hazard Assessment Along the Pacific Coast of Mexico by Historical Records and Numerical Simulation. Part II: Tsunami-Induced Currents in the Port of Ensenada, Baja California

LAURA G. ORTIZ-HUERTA[1]

Abstract—The time evolution of the velocity field in and around the Port of Ensenada, induced by large distant tsunamis as produced by Mw 9.3 hypothetical earthquakes around the Pacific Ocean, is analyzed through the numerical modeling of distant tsunamis. The results indicate that tsunami-induced currents are 4–6 knots (~ 2 to 3 m/s) at the harbor entrance and 2–4 knots (~ 1 to 2 m/s) inside and outside the harbor. Low amplitude tsunamis, as well as the reverberance or coda of large distant tsunamis, may produce currents of 2 knots (~ 1 m/s) along the harbor channel induced by the interaction of coastal and harbor seiches. Visual scrutiny of the resulting velocity field at time steps of 1 min, as well as the mathematical concept of residual velocity, reveal transient eddies fed by flood and ebb currents that produce transversal currents by the harbor entrance. Currents induced by large distant tsunamis are practically negligible at depths greater than 120 m (15 km offshore from the harbor).

Key words: Port of Ensenada, distant tsunamis, velocity field, harbor seiches, numerical modeling.

1. Introduction

The Port of Ensenada, Baja California, Mexico, located in the northwestern part of the Peninsula of Baja California (116.62°W; 31.85°N; Fig. 1), is considered to be one of the most important ports along the Pacific Coast of Mexico. This port is attending large cargo and touristic vessels (~ 300 m long; ~ 70,000 Tons), as well as fishing ships and small vessels (e.g., www.puertoensenada.com.mx). Despite being located in a seismically active region, no local or regional generated tsunamis have been noticed in the short time history of Ensenada (~ 137 years). In contrast, historical as well as recent distant tsunamis produced by large earthquakes around the Pacific Ocean have been well documented in the Port of Ensenada (Fig. 2; Table 1). Among them, the historic Chile (22 May 1960, M_w 9.5), and Alaska (27 March 1964, M_w 9.3) tsunamis produced the most significant water levels observed in the port. Since the Intergovernmental Coordination Group for the Pacific Tsunami Warning System (ICG/PTWS) started operations in 1965 (ITSU Master Plan 1999), no formal tsunami warnings were issued at that time. Nonetheless, for the Alaska 1964 tsunami, a particular telephone call was received from the Port of San Diego, California, telling that Kodiak, Alaska, was devastated by a strong earthquake, and that a seismic tidal wave was in progress in the Pacific. At that time, harbor operations in the Port of Ensenada consisted mainly in attending the tuna fishing fleet and the floating dock for their maintenance. As such, the primary objective during the arrival of the Alaska tsunami was to secure the floating dock by tightening and releasing its moorings from pier #1 according to the demand imposed by the tsunami-induced water level variations. Most of the inhabitants in the port neighborhood evacuated to the hills where they spent the night until the next morning after seeing for themselves that nothing happened in the harbor (Port Captain J. Prieto-Guzman, inedited narrative, 1964).

As for the recent Tohoku tsunami (11 March 2011, M_w 9.0), port authorities in Ensenada indicated that, since the tsunami would arrive during the neap tide period, it would be contained between high and low tide levels, thereby reducing the inundation impact. However, large cargo and tourism vessels evacuated to open sea as a precautionary measure after seen the striking images of the Tohoku tsunami

[1] Instituto de Investigaciones Oceanológicas, Universidad, Autónoma de Baja California (UABC), Ensenada, Baja California, Mexico. E-mail: laura_ortiz@uabc.edu.mx; lortiz.huerta@gmail.com

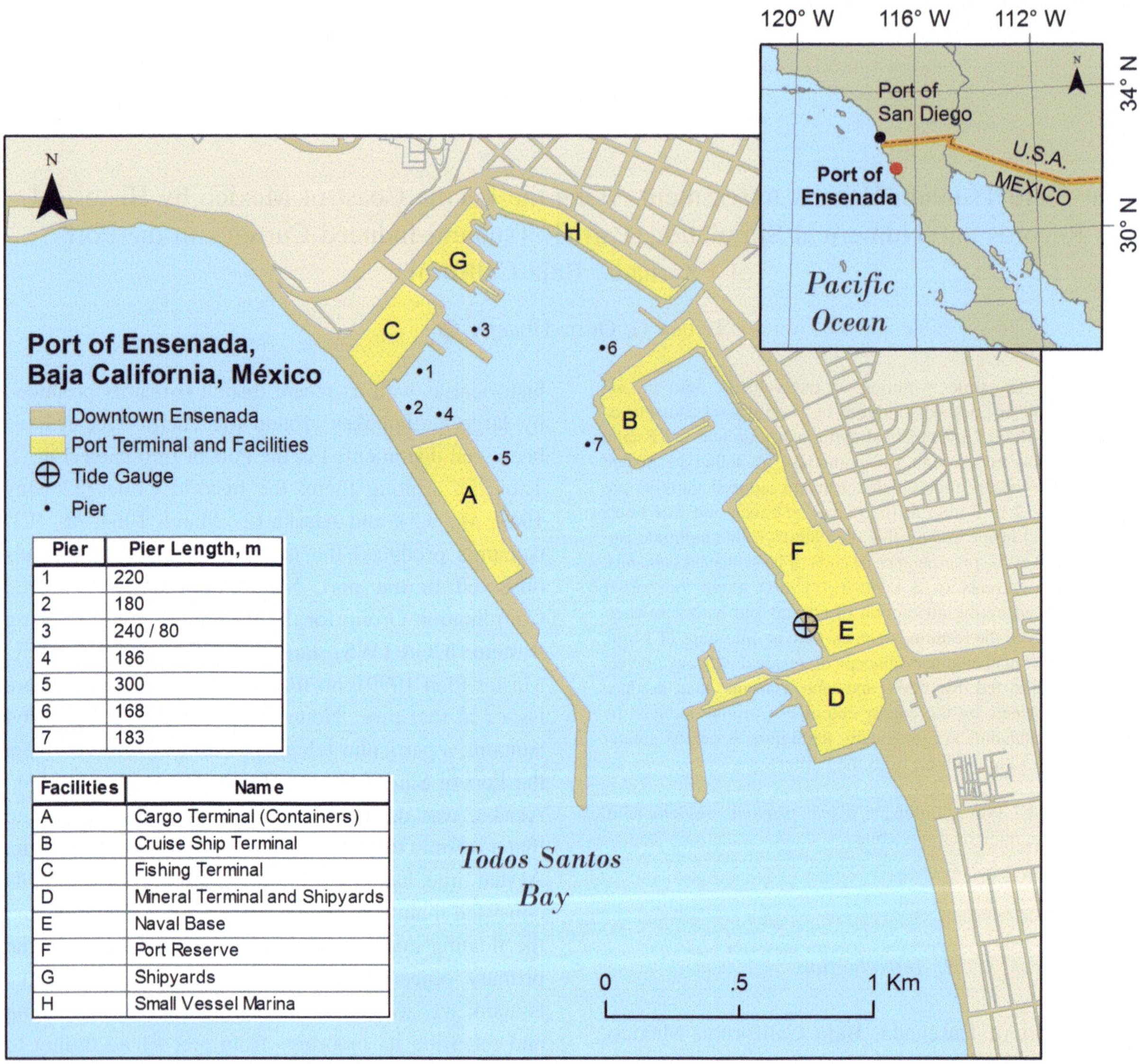

Figure 1
Map and facilities of the Port of Ensenada, Baja California, Mexico

broadcasted live by the NHK World-Japan television network. Because the tsunami arrived during the daylight hours, strong currents and eddies were observed for the first time by myself and other eyewitnesses for several hours, as traced by floating debris and suspended sediments at the port entrance. During the Chile tsunami (27 February 2010, M_w 8.8) the port was closed to navigation but reopened a few hours after the arrival of the tsunami, given that the effects within the port were considered practically unharmful. During the remaining tsunamis illustrated in Fig. 2, the Port of Ensenada remained open to navigation after the tsunami advisories were issued by the Pacific Tsunami Warning Center.

In addition to the tsunami inundation hazard, harbors are naturally vulnerable to tsunami-induced currents and, consequently, to the potential damage by drifting structures of varying sizes. As for the tsunami experiences in the Port of Ensenada, harbor operations have been guided solely by empirical knowledge. Therefore, an analysis of expected tsunami-induced currents is necessary to anticipate harbor operations as well as

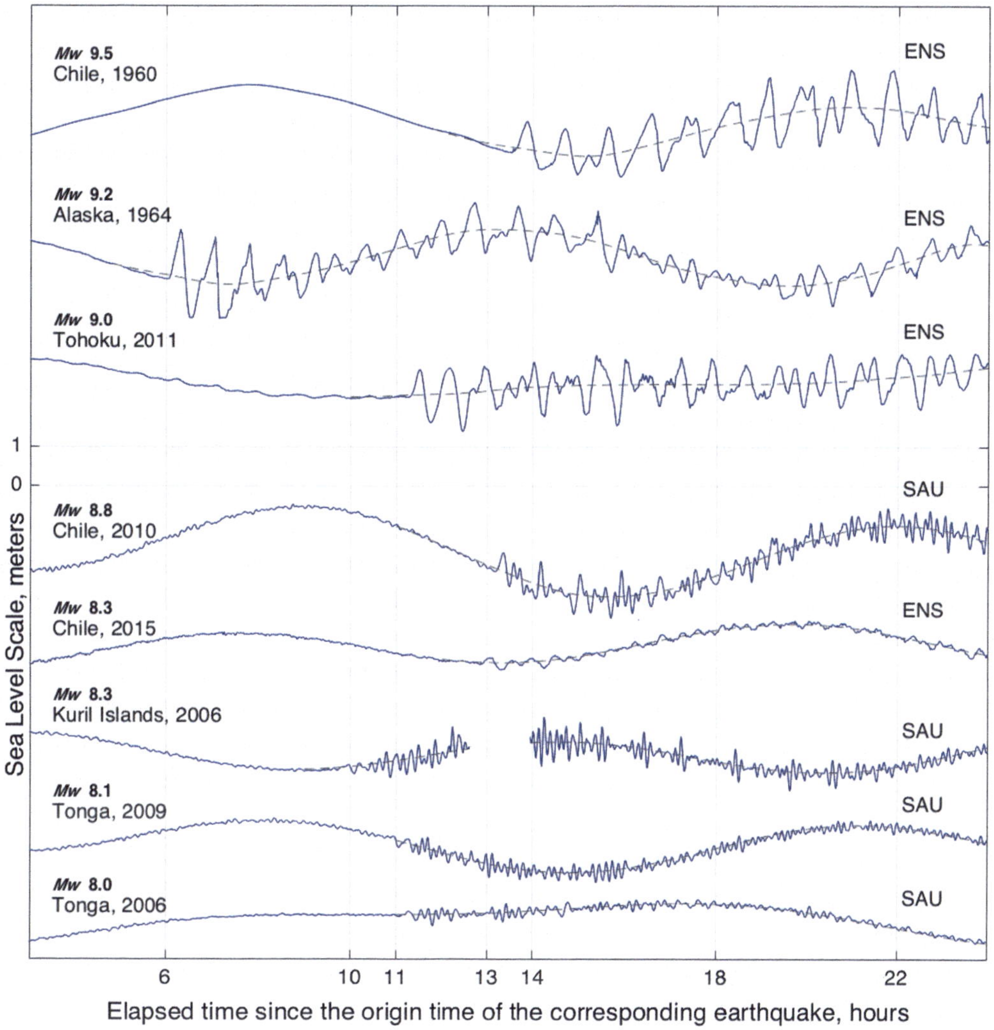

Figure 2
Historical and recent tsunami records recorded in Todos Santos Bay at the Ports of Ensenada (ENS), and Sauzal (SAU), Baja California, Mexico. Dashed lines indicate the corresponding predicted tide. The origin region and the magnitude of the earthquakes are indicated in the corresponding marigram or sea level record. See Fig. 5 for locations of Ensenada and Sauzal in Todos Santos Bay

Table 1

Basic statistics of the tsunami records illustrated in Fig. 2

	Chile 1960	Alaska 1964	Tohoku 2011	Chile 2010	Chile 2015	Kuril 2006	Tonga 2009	Tonga 2006
H_s (m)	1.57	1.28	1.26	0.79	0.22	0.69	0.39	0.28
T_s (min)	57	45	43	14	24	8	10	13
Max H (m)	1.91	2.21	1.82	0.97	0.31	1.00	0.60	0.44

H_s is the significant wave height (crest to trough) or the average height of the highest one-third waves during the first 12 h after the tsunami arrival and T_s is the significant wave period or the average zero-crossing wave period of the highest one-third waves

safety maritime evacuation zones as a guideline to the establishment of a tsunami response harbor operation plan.

Here after, tsunami-induced currents will be expressed in knots which are more familiar to navigators to whom this study is addressed.

2. *Numerical Modeling of Distant Tsunamis*

A comprehensive analysis of expected tsunami hazard along the Pacific Coast of Mexico for distant tsunamis, originated by $M_w \sim 9.3$ synthetic earthquakes around the Pacific Ocean, indicates that tsunami heights in the Port of Ensenada are expected to be 1–2.5 m above the still water level (Ortiz-Huerta et al. 2018). Therefore, two case scenarios of tsunami-induced currents in the Port of Ensenada are presented in this study as thresholds representatives of the 23 tsunami-case scenarios analyzed in Ortiz-Huerta et al. (2018). The upper threshold scenario corresponds to a tsunami originated by an earthquake in the Aleutian Trench aside of the western edge of the 27 March 1964 Alaska earthquake, while the lower threshold scenario corresponds a tsunami originated above the rupture area of the 11 March 2011 Tohoku earthquake (regions 3 and 11 respectively, in Ortiz-Huerta et al. 2018). The focal mechanism of the earthquakes in both case scenarios is prescribed here by the FOCAL-II dislocation model defined by a uniform slip distribution of 20 m on a rectangular large thrust fault ($600 \times 180\ \mathrm{km}^2$) dipping 15° at the depth of 20 km. The goodness of the FOCAL-II dislocation model was probed to be adequate for the numerical simulation of large distant tsunamis, such as the ones produced by the Chile 1960, Alaska 1964, and Tohoku 2011 earthquakes (Ortiz-Huerta et al. 2018).

The transoceanic propagation modeling of the tsunami in both case scenarios is computed with the numerical tsunami propagation model of Goto et al. (1997), which solves the vertically integrated (depth-averaged) linearized shallow water momentum and continuity equations or linear long wave equations in a rotating Earth (e.g., Dronkers 1964; Pedlosky 1979):

$$\frac{\partial \eta}{\partial t} + \nabla \cdot \mathrm{M} = 0, \tag{1}$$

$$\frac{\partial \mathrm{M}}{\partial \mathrm{t}} + 2\Omega \times M + gh\nabla\eta = 0. \tag{2}$$

In Eqs. (1, 2), t is time, η is the vertical displacement of the water surface above the still water level (the equipotential surface), h is the ocean depth, g is the gravitational acceleration and Ω is the angular velocity of the Earth. $\mathrm{M} = [U, V]$ is the horizontal depth-averaged volume flux vector, where $U = u(\eta + h)$ and $V = v(\eta + h)$ are the horizontal depth-averaged volume flux vectors in longitudinal and latitudinal spherical coordinates, and u and v are the corresponding water particle velocities.

The above set of Eqs. (1, 2) are solved by finite differences in an explicit staggered leap-frog scheme at time steps of 2 s in a spatial grid resolution of 2 min. The current bathymetry or domain of integration of the numerical model for the transoceanic tsunami propagation corresponds to the ETOPO2v2 dataset (National Geophysical Data Center 2006) in a matrix of 5761×3891 nodes.

In both case scenarios, the tsunami initial condition or sea surface deformation was taken as the vertical deformation of the seafloor produced by the FOCAL-II dislocation model, as computed by the expression given by Mansinha and Smylie (1971).

The modeling of the tsunamis in the target region is computed at time steps of 2 s in a set of interconnected nested grids within the transoceanic domain of integration (grid resolution 38 m, Fig. 3; grid resolution 353 m, Fig. 5), where full non-linear terms (advective and bottom friction) are included in the expressions (3, 4) of Eqs. (1, 2) in a rectangular coordinate system $(x,\ y)$, disregarding Coriolis terms:

$$\frac{\partial \eta}{\partial t} + \frac{\partial U}{\partial x} + \frac{\partial V}{\partial y} = 0 \tag{3}$$

$$\begin{aligned} &\frac{\partial U}{\partial t} + \frac{\partial}{\partial x}\left(\frac{U^2}{D}\right) + \frac{\partial}{\partial y}\left(\frac{UV}{D}\right) + gD\frac{\partial \eta}{\partial x} \\ &\quad + \frac{gn^2}{D^{7/3}} U\sqrt{U^2 + V^2} = 0, \\ &\frac{\partial V}{\partial t} + \frac{\partial}{\partial x}\left(\frac{UV}{D}\right) + \frac{\partial}{\partial y}\left(\frac{V^2}{D}\right) \\ &\quad + gD\frac{\partial \eta}{\partial y} + \frac{gn^2}{D^{7/3}} V\sqrt{U^2 + V^2} = 0. \end{aligned} \tag{4}$$

The Gauckler–Manning roughness coefficient n in the bottom friction terms (Eqs. 4) was set to 0.025 (cf., Titov et al. 2003).

The Shalow Water Model Equations (Eqs. 3, 4) has been widely used to adequately reproduce tsunami records recorded in tide gauges in bays and

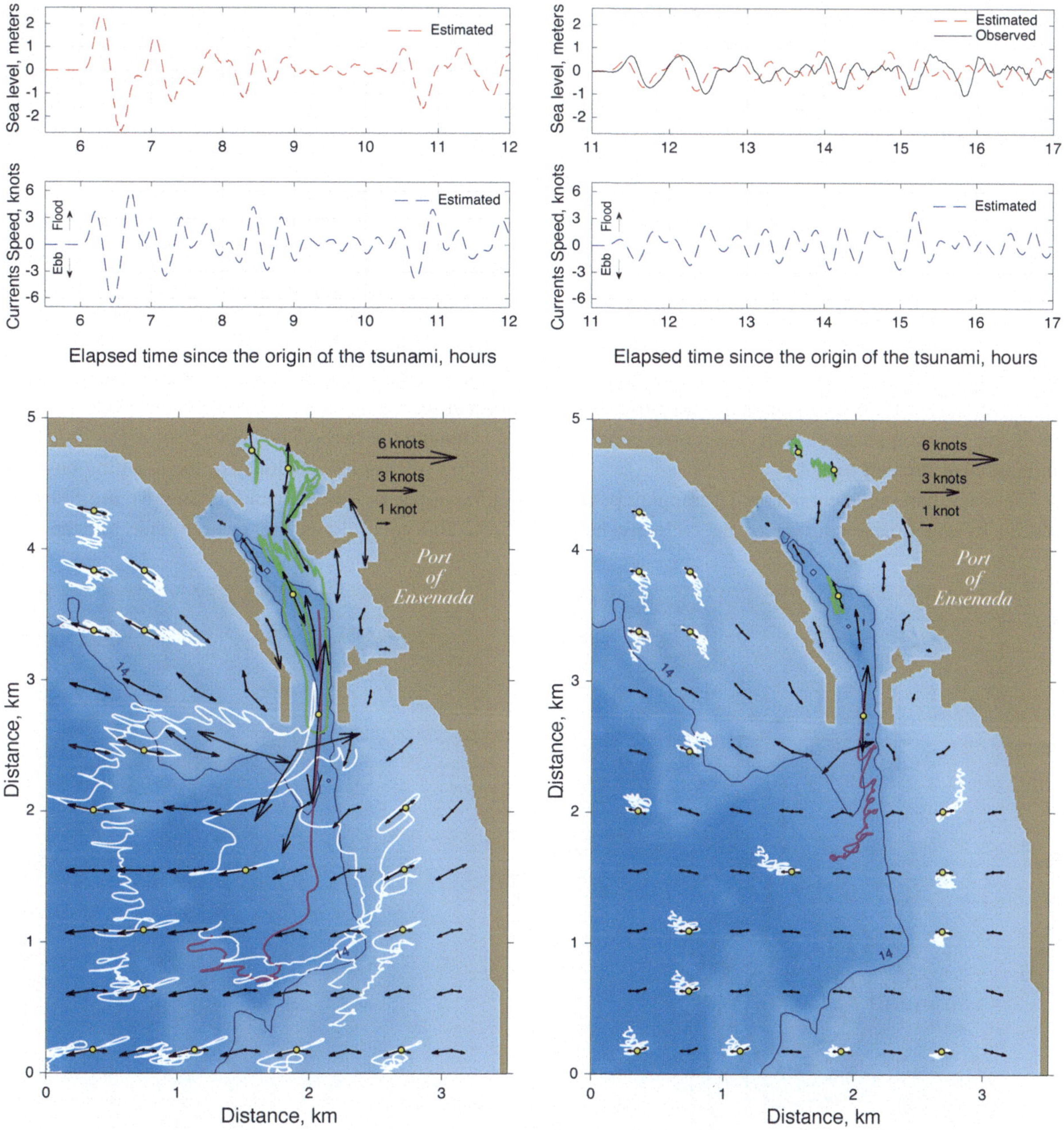

Figure 3
Left panels (upper threshold scenario); Right panels (lower threshold scenario). Upper panels: sea level and currents at the harbor entrance. Lower panels: maximum flood and ebb currents inside and around the harbor reached during the first 6 h after the tsunami arrival. Yellow dots at the base of velocity vectors indicate the initial position of water particle Lagrangian paths. White lines indicate Lagrangian paths outside the harbor; green inside the harbor and the magenta line indicate the path departing from the harbor entrance. The 14 m isobath is depicted in blue

harbors. Recently, a large of number of records of tsunami-induced currents by the Tohoku 2011 tsunami, recorded in bays and harbors, have allowed the validation of the deep-averaged equations to adequately reproduce the time evolution of velocity fields (e.g., Arcos and LeVeque 2015).

3. Tsunami-Induced Currents in the Port of Ensenada

The time evolution of the velocity field in and around the harbor resulting from the numerical model is examined here at time steps of 1 min. The resulting tsunami-induced currents in both case scenarios is illustrated in Fig. 3. Currents in the upper threshold scenario reached 6 knots at the harbor entrance and 2–4 knots inside and outside the harbor, while currents in the lower threshold scenario reached 4 knots at the harbor entrance and 1–3 knots inside and outside the harbor. Note that the observed marigram of the 11 March 2011 Tohoku tsunami is also traced in Fig. 3 for its comparison with the corresponding synthetic or estimated one.

The water particle excursion or Lagrangian paths, induced by the time evolution of the velocity field during the first 6 h after the arrival of the tsunami, ranged from 1 to 6 km by the harbor entrance in the upper threshold scenario, while excursion paths are in general less than 0.5 km in the lower threshold scenario. In particular, in the upper threshold scenario, Lagrangian paths are indicating a large eddy like circulation outside the harbor entrance (2 km in diameter, Fig. 3). Furthermore, typical of a simple model basin with narrow strait, two pair of transient vortices are revealed in both case-scenarios (Fig. 4) by employing the mathematical concept of residual velocity (the time-average of velocity field; e.g., Imasato 1983) despite of the complex geometry of the harbor.

Outside the harbor, in Todos Santos Bay in which the Port of Ensenada is located, the resulting currents in the upper threshold scenario (Fig. 5), are found to be 2–3 knots along the eastern side of the bay

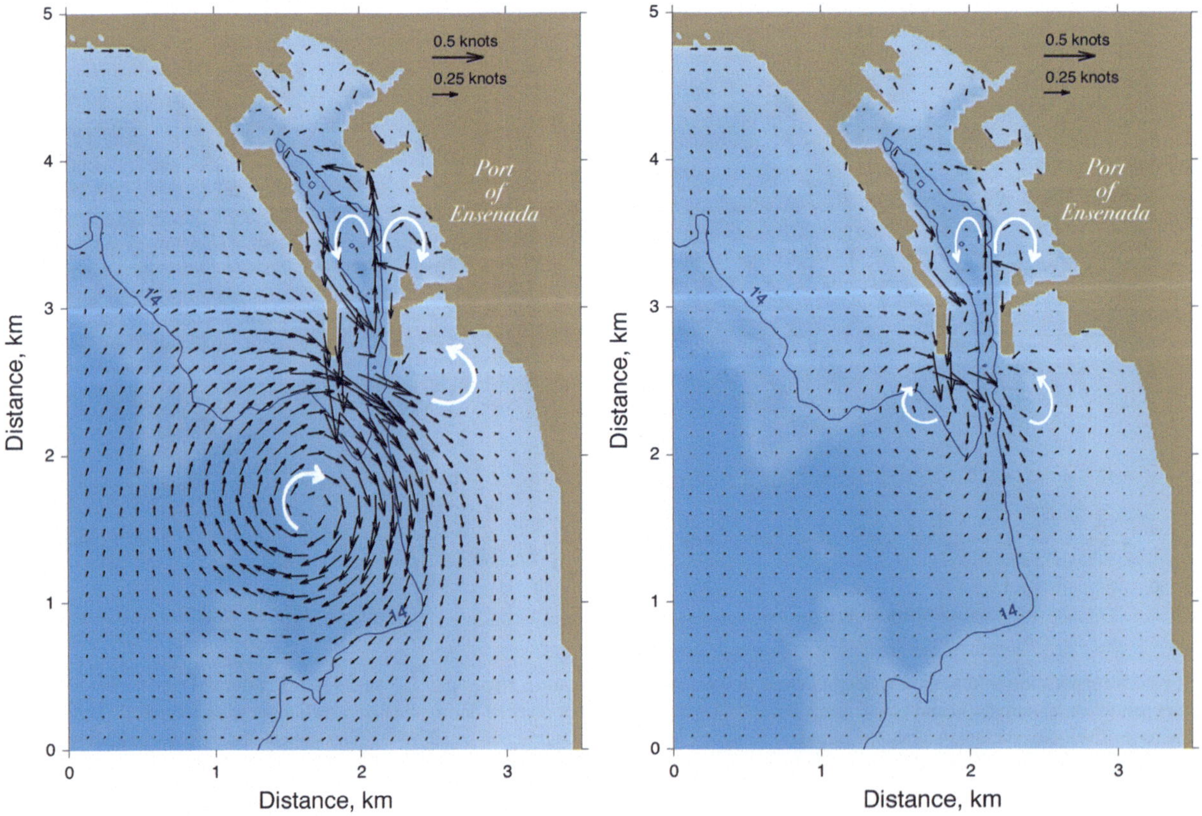

Figure 4
Residual velocity field inside and outside the harbor (left panel, upper threshold scenario; right panel, lower threshold scenario). White arrows highlight the two pair of transient vortices revealed by the residual velocity field. Residual velocity fields were computed by the time-average of the first 6 h of the velocity field after the tsunami arrival

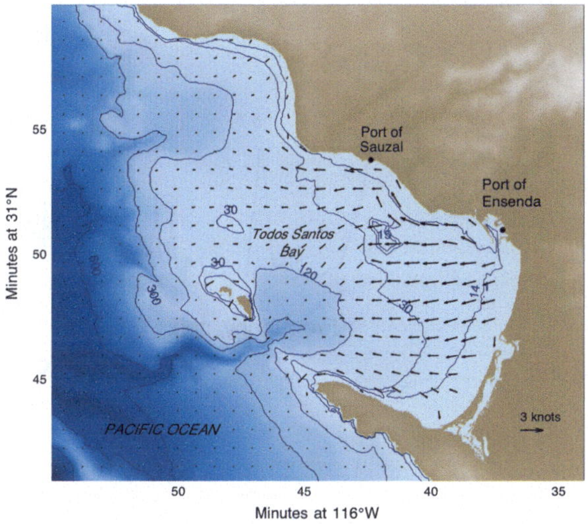

Figure 5
Upper threshold scenario: maximum tsunami-induced currents in Todos Santos Bay, whether it corresponds to flow or ebb. Blue lines indicate isobaths, depths in meters. Note that ebb currents exceed flood currents, as illustrated in Fig. 3

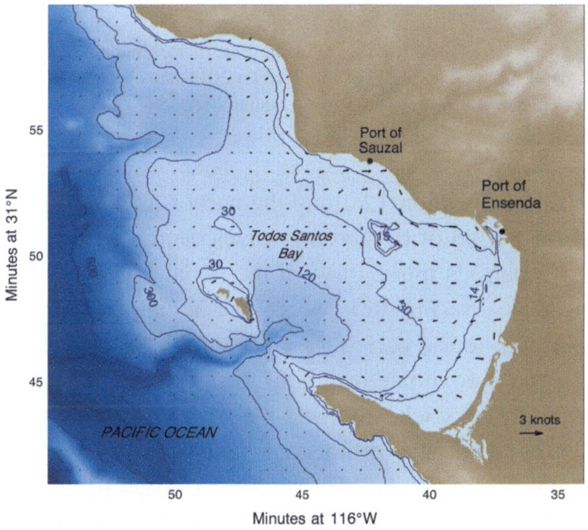

Figure 6
Lower threshold scenario: maximum tsunami-induced currents in Todos Santos Bay, whether it corresponds to flow or ebb. Blue lines indicate isobaths, depths in meters

between the 14 and 30 m isobaths, and are less than 2 knots in depths greater than 30 m. In contrast, the resulting currents in the lower threshold scenario (Fig. 6), are less than 2 knots between the 14 and 30 m isobaths, and practically negligible at depths greater than 30 m. In both case scenarios, currents are practically negligible at depths greater than 120 m (15 km offshore from the harbor).

4. Coastal Seiches and Currents in the Port of Ensenada

It is well known that tsunamis originated in different regions around the Pacific have the same sea level spectral shape in a specific bay or harbor due to the natural local resonant eigen-modes or coastal seiches (e.g., Rabinovich 1997; Zaytsev et al. 2017), i.e., in the Port of Ensenada, the sea level variance during the Tohoku 2011 tsunami was found largely amplified in the whole spectrum while keeping the spectral shape of the background sea level spectrum at the periods T $\sim$ [51, 29, 21, 6, and 3 min] (cf. Fig. 7, in Zaytsev et al. 2017). In particular, the harbor sea level frequency response in the Port of Ensenada, as computed by the spectral ratio (admittance and coherence; Fig. 7a–c) between the synthetic tsunami records, at the tide gauge location, and at 2 km outside the harbor (both records resulting from the Tohoku 2011 tsunami model, Eqs. 3, 4) reveals two resonant periods with an amplification factor of 3.5 centered at the period of 21 min, and an amplification factor of 2.5 at the period of 6.7 min.

The period $T_0 = 21$ min may correspond to the fundamental zeroth Helmholtz longitudinal mode, which is also called the pumping mode because it is related to periodic mass transport—pumping—through the harbor entrance (e.g., Rabinovich 2009); while the period at $T_1 = 6.7$ min $\cong T_0/3$ may correspond to the period of the first Helmholtz longitudinal mode according to the formula $T_n = 4L/(2n+1)/\sqrt{gh}$ for modes $n = 0, 1, \ldots$ in a rectangular open-mouth basin.

In order to illustrate the profiles of sea level and currents along the harbor axis, corresponding to both resonant periods, harmonic water-level variations with a moderate wave height ($H = 0.2$ m) were introduced as forced boundary condition in an artificial channel towards the harbor entrance in the hydrodynamic numerical model of Goto et al. (1997), applicable to the Port of Ensenada (Eqs. 3, 4; Fig. 8) in a matrix of 201 $\times$ 454 nodes with a grid resolution of 10 m. The resulting profiles of sea level and currents along the harbor axis corresponding to both resonant periods are illustrated in Fig. 9.

The high coherence level for periods longer than 19 min (Fig. 7c), indicate that these periods

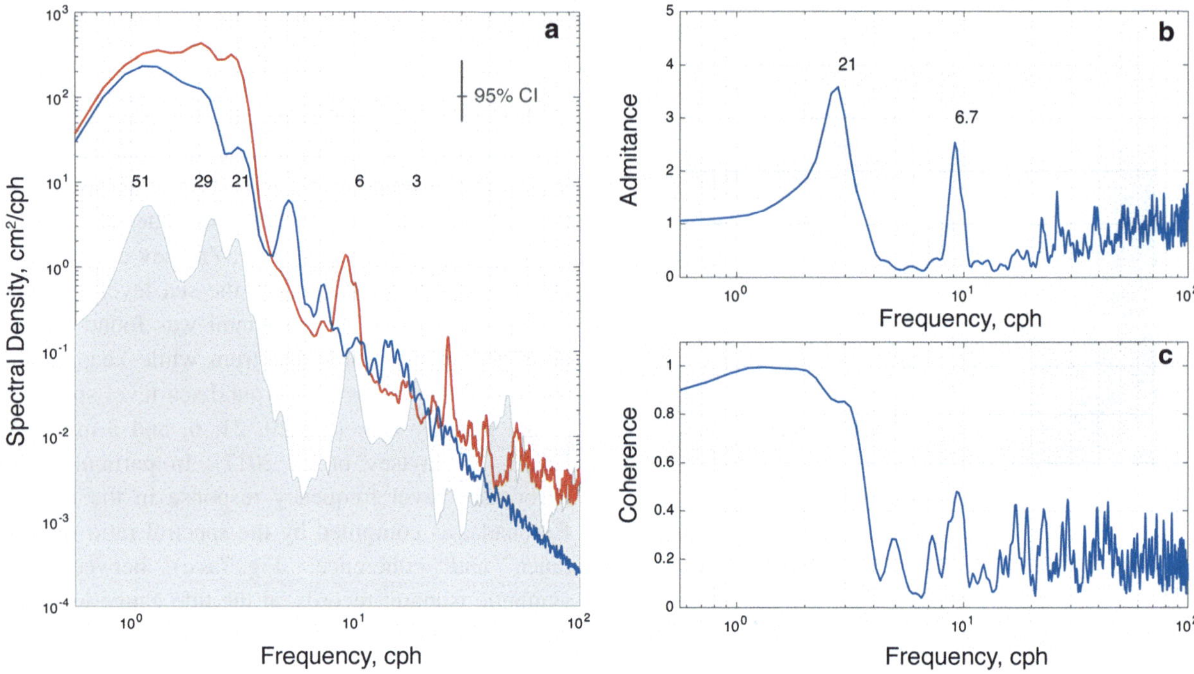

Figure 7

Sea level Spectral Density: **a** Background Spectrum (shaded area) obtained from the recorded sea level in the Port of Ensenada at the tide gauge location in an ordinary day, in comparison with the Spectra obtained from the model results of the Tohoku 2011 tsunami inside (red) and outside (blue) the port; **b**, **c** Admittance and coherence between the inside and outside tsunami records

correspond to coastal seiches originated outside the harbor where the resonant period $T_0 = 21$ min corresponds to both, a harbor seiche, as well as to a coastal seiche outside the harbor. The low coherence level at the resonant period $T_1 = 6.7$ min indicates that this period corresponds to a harbor seiche, that besides of being excited by sea level fluctuations outside the harbor, is also excited by the multiple wave reflections within the complex harbor geometry.

It is noticeable the lack of variance in the sea level spectrum at 6 cph ($T = 10$ min) in both, the background spectrum, and in the Tohoku 2011 tsunami spectrum at the tide gauge location in the Port of Ensenada (Fig. 7a; cf. Fig. 7, in Zaytsev et al. 2017). This is because the nodal line of the harbor resonant mode ($T_0 = 21$ min), which is located outside the harbor (Fig. 9a), migrates towards the tide gauge location (location 12 along the harbor axis in Fig. 8) as the period of the forcing waves decreases. Shorter wave periods move forward the nodal line up to the location 15 (Fig. 9c), where the nodal line corresponding to the period $T_1 = 6.7$ min is located.

An approximate insight of the spectral and spatial structure of seiches outside the harbor is given here by the empirical orthogonal functions (EOF; e.g., Tolkova and Power 2011; Aránguiz 2015) of the sea level time history resulting from the numerical modeling of the Tohoku 2011 tsunami. The first EOF, the one with the largest variance (35%), has a composed spectrum (Fig. 10a) with a dominant peak at the period 60 min, and a secondary peak centered at 20 min. By speculation, its spatial structure (Fig. 10b) resembles a fundamental zeroth Helmholtz longitudinal mode in Todos Santos Bay. The second EOF (variance 33%) has a composed spectrum

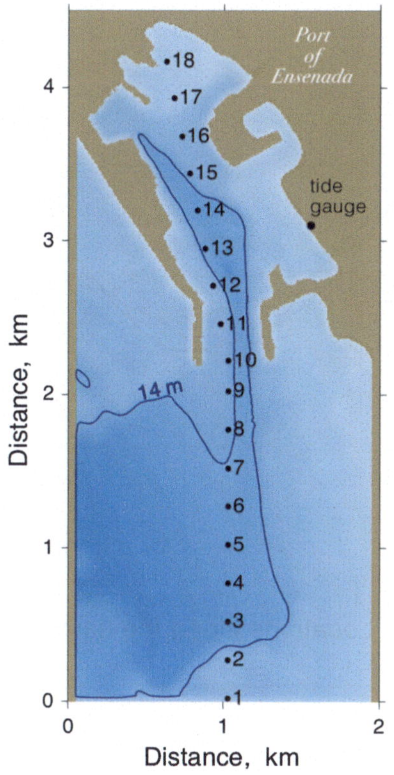

Figure 8
Artificial channel towards the Port of Ensenada with an open boundary at the channel mouth. Dots are indicating 18 locations along the channel and harbor axis. The 14 meters isobath is indicated in blue

(Fig. 10c) with two dominant peaks at 30 and 60 min, and a secondary peak centered at 20 min; its spatial structure (Fig. 10d) resembles a "continental shelf trapped tsunami mode". The variance (10%) of the third EOF is distributed in a narrow spectral band between 20 and 30 min (Fig. 11a), while its spatial structure (Fig. 11b) resembles a series of coastal trapped waves, which might be a result of transient "coastal trapped edge waves tsunami modes" (e.g., González et al. 1995), captured by the orthogonal decomposition method. The variance (6%) of the fourth EOF is distributed in a wide spectral band between 10 and 60 min, which makes difficult its physical interpretation. It is noticeable that a spectral peak appears in all of the EOFs near the fundamental harbor resonant period (T_0 = 21 min).

5. *Discussion and Conclusions*

As a result of coastal and harbor seiches, a large number of significant waves ($H > 1$ m) are observed in the tsunami records of the Chile 1960, Alaska 1964, and Tohoku 2011 tsunamis for more than 24 h after the tsunami arrivals (Fig. 2), consequently, large flood and ebb currents (4–6 knots, as the ones illustrated in Fig. 3) can be expected to occur by the harbor entrance during the same period of time that would make navigation difficult inside and around the harbor for several hours after the arrival of a tsunami. Moreover, tsunami heights greater than 1.5 m above the still water-level would represent a considerable hazard if the tsunami were to arrive during the spring tides, as the piers in the harbor are 1.5 meters above the Mean Higher High Water (MHHW) and 1.23 m above the Mean Spring Tide Level (MSTL, defined here as the average of the maximum heights observed during consecutive spring tide cycles).

Low amplitude tsunamis (H_s = 0.69–0.79 m), as the ones illustrated in Fig. 2; Table 1, may not represent inundation hazard, however, those may produce important currents of 2–3 knots at the harbor entrance as a result of the interaction of coastal and harbor seiches that may excite the natural harbor seiches at the resonant periods T_0 = 21 min, and T_1 = 6.7 min (e.g., Figs. 7, 8). Note that the lowest amplitude recorded tsunami (Chile 2015) has a significant wave height and period (H_s = 0.22 m; T_s = 24 min) similar to those of the forced boundary condition (H = 0.2 m; T = 21 min) at the mouth of the artificial channel in Fig. 9, that in turn produced currents of ~ 1.5 knots at the harbor entrance and wave heights of ~ 1 m at the harbor head (Figs. 8a, b).

Figures 3, 4, 5 and 6 of maximum expected tsunami-induced currents mapped inside and outside the Port of Ensenada, were carefully confectioned for practical use to anticipate harbor operations, as well as to identify safety maritime evacuation zones. However, whether estimated tsunami-induced currents may be considered a hazard to navigation or harbor operations will depend on the judgement of navigators and port authorities.

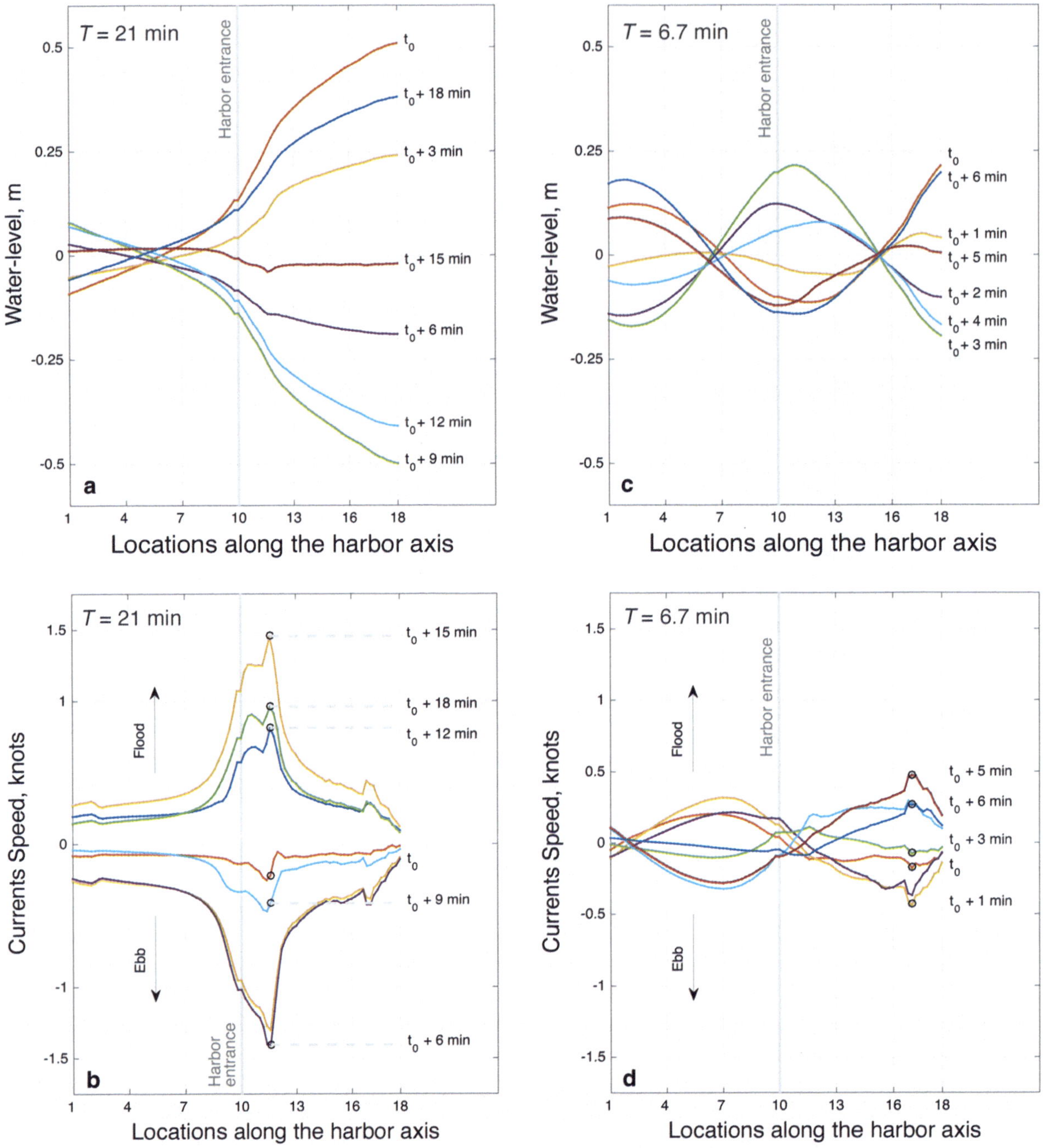

Figure 9
Profiles of water-level and currents along the harbor axis within one wave period: **a**, **b** for the resonant period T = 21 min; **c**, **d** for the resonant period T = 6.7 min. Colors to easily follow single profiles

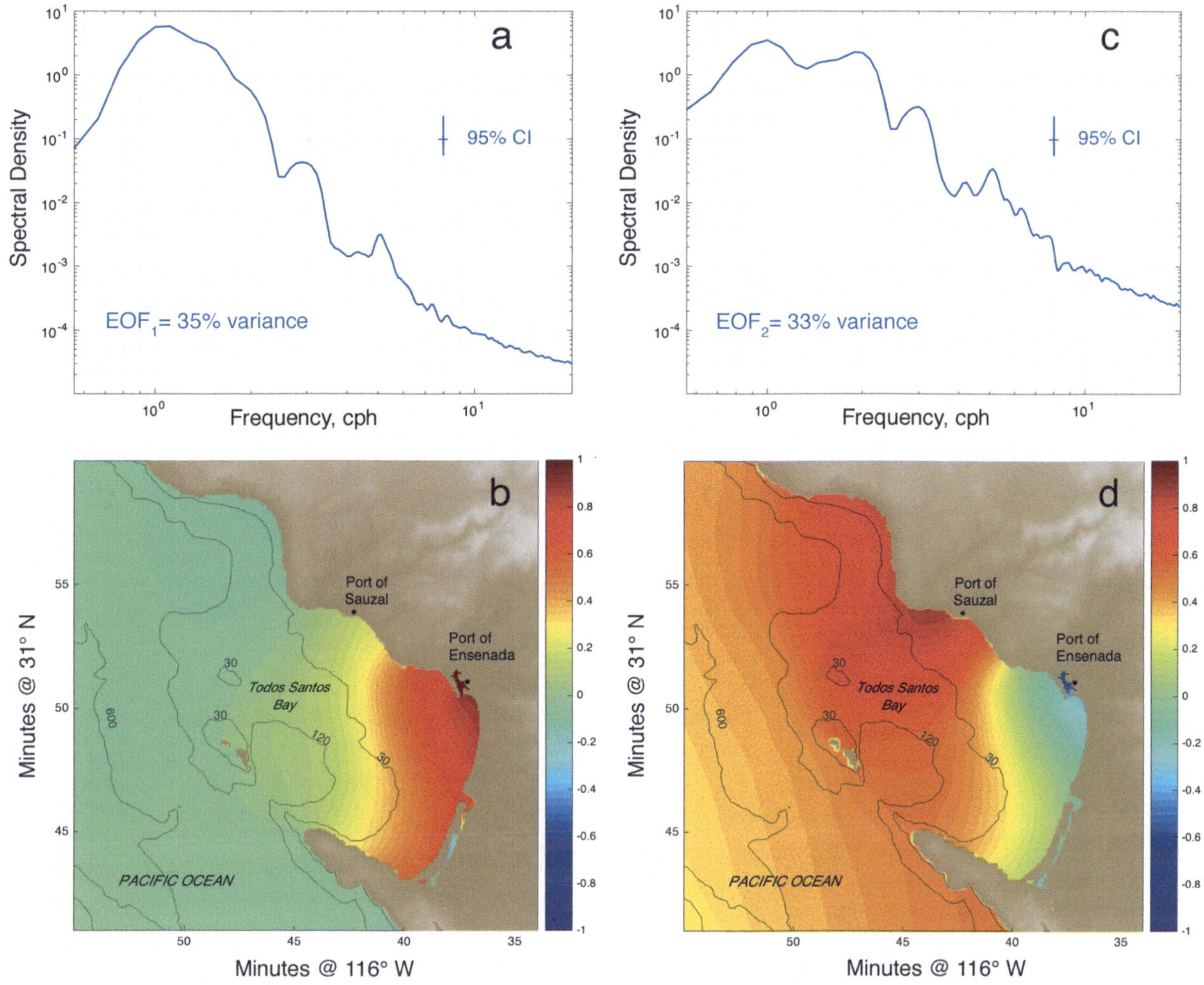

Figure 10
Spectral and spatial distribution of the first (**a**, **b**) and second (**c**, **d**) EOFs. Spectral density weighted by the corresponding percentage of variance. Color scale indicate the corresponding normalized amplitude of the EOFs spatial distribution. Contour lines indicate isobaths with depths annotated in meters

It is of a practical use to close a harbor to navigation after a tsunami warning is issued. However, reopen it to navigation would depend on the magnitude of the tsunami-induced currents and sea-level variations inside and around the harbor for the safety of navigation. Therefore, it would be desirable for a modern port to deploy an adequate array of pressure-gauges and current-meters operating in real time for marine traffic control to be able to close or reopen port operations to prevent damage and economic losses (e.g., https://tidesandcurrents.noaa.gov/ports).

In addition, tide-induced currents were computed by introducing the tide level of one tidal day during the spring tides (1.6 m peak to trough) at the harbor entrance in the numerical model (Eqs. 3, 4; Fig. 8). The resulting currents (0.2 knots) are representative of the natural background currents in the harbor, and served as a reference frame for the magnitude of the tsunami-induced currents estimated above.

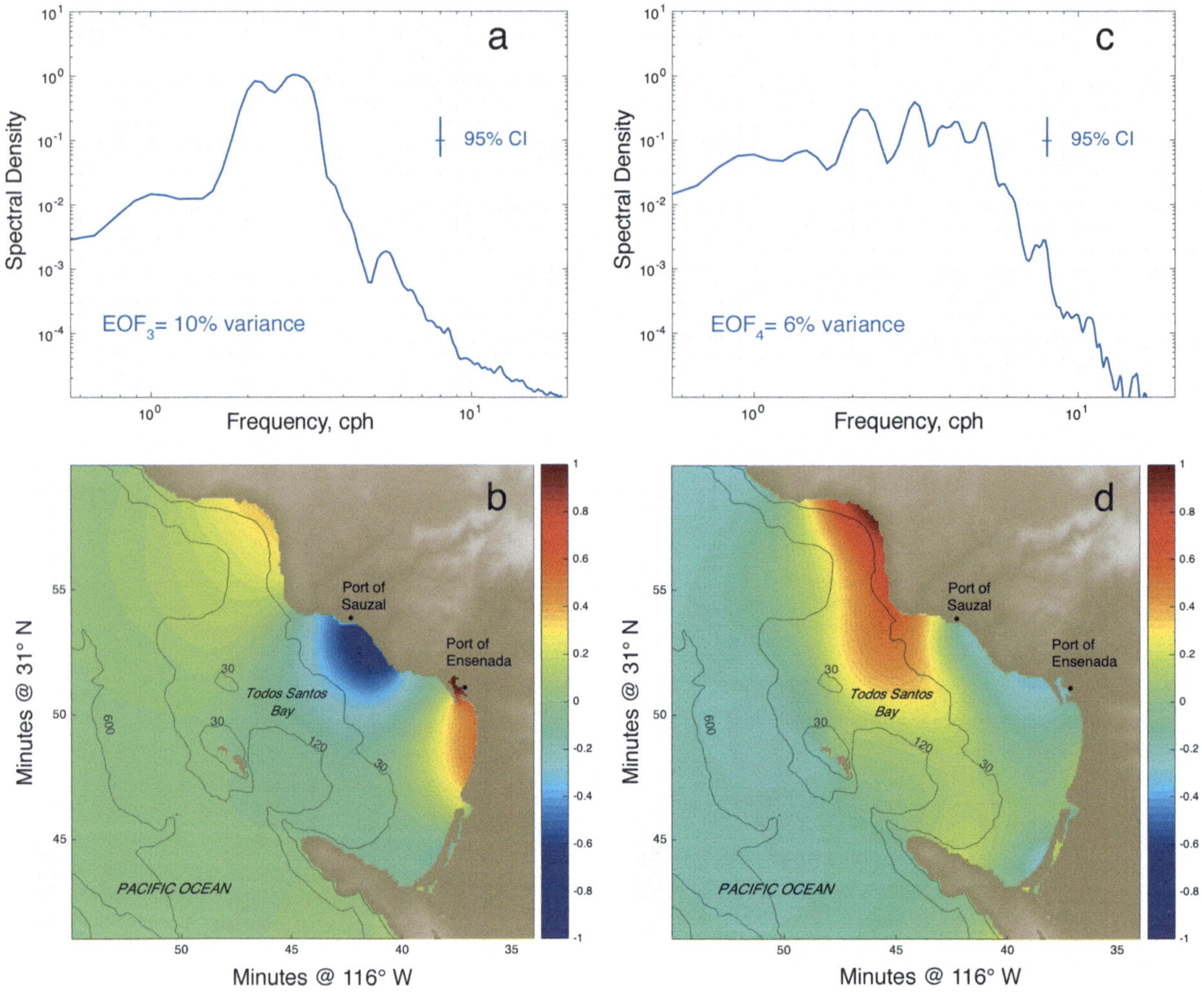

Figure 11
Spectral and spatial distribution of the third (**a**, **b**) and fourth (**c**, **d**) EOFs. Spectral density weighted by the corresponding percentage of variance. Color scale indicate the corresponding normalized amplitude of the EOFs spatial distribution. Contour lines indicate isobaths with depths annotated in meters

Acknowledgements

To CONACyT (México) through the Ph.D. program scholarship CVU221676 that made this study possible. Tsunami records courtesy of the Joint Sea Level Network Operations of Mexico, operated by CICESE, IMT, SEMAR, and UNAM. ETOPO2v2 bathymetric data courtesy of the National Geophysical Data Center (2006). High-resolution bathymetric data around the Port of Ensenada courtesy of the Division of Coasts and Harbors of the Mexican Institute of Transportation (IMT). Thanks to both anonymous reviewers and the Editor for their sound and constructive comments that enhanced the contents of the manuscript, and to Professor Modesto Ortiz for the fructiferous discussions throughout this study.

Funding

This study was funded by Consejo Nacional de Ciencia y Tecnología (Grant No. 221676).

Publisher's Note Springer Nature remains neutral with regard to jurisdictional claims in published maps and institutional affiliations.

References

Aránguiz, R., (2015). *Tsunami Resonance in the Bay of Concepción (Chile) and the effect of Future Event.* Handbook of Coastal Disaster Mitigation for Engineers and Planners. http://dx.doi.org/10.1016/B978-0-12-801060-0.00006-X.

Arcos, M. E. M., & LeVeque, R. J. (2015). Validating velocities in the GeoClaw tsunami model using observations near Hawaii from the 2011 Tohoku tsunami. *Pure and Applied Geophysics, 172*(3–4), 849–867. https://doi.org/10.1007/s00024-014-0980-y.

Dronkers, J. J. (1964). *Tidal computations in rivers and coastal waters.* Amsterdam: North-Holland Publishing Company.

González, F. I., Satake, K., Boss, E. F., & Mofjeld, H. O. (1995). Edge wave and non-trapped modes of the 25 April 1992 Cape Mendocino tsunami. *Pure and Applied Geophysics, 144*(3–4), 409–426. https://doi.org/10.1007/BF00874375.

Goto, C., Ogawa, Y., Shuto, N., and Imamura, F. (1997). IUGG/IOC Time Project, in IOC Manuals and Guides No. 35, UNESCO, Paris.

Imasato, N. (1983). What is tide-induced residual current? *Journal of Physical Oceanography, 13*(7), 1307–1317. https://doi.org/10.1175/1520-0485(1983)013%3c1307:WITIRC%3e2.0.CO;2.

ITSU Master Plan, (1999). *Tsunami Warning System in the Pacific Master Plan.* Intergovernmental Oceanographic Commission (IOC) of UNESCO.

Mansinha, L., & Smylie, D. E. (1971). The displacement fields of inclined faults. *Bulletin of the Seismological Society of America, 61*(5), 1433–1440.

National Geophysical Data Center. (2006). 2-minute Gridded Global Relief Data (ETOPO2) v2. *National Geophysical Data Center, NOAA.*. https://doi.org/10.7289/v5j1012q.

Ortiz-Huerta, L. G., Ortiz, M., & García-Gastélum, A. (2018). Far-field tsunami hazard assessment along the Pacific coast of Mexico by historical records and numerical simulation. *Pure and Applied Geophysics, 175*(4), 1305–1323. https://doi.org/10.1007/s00024-018-1816-y.

Pedlosky, J. (1979). *Geophysical fluid dynamics.* New York: Springer. https://doi.org/10.1007/978-1-4684-0071-7.

Rabinovich, A. B. (1997). Spectral analysis of tsunami waves: Separation of source and topography effects. *Journal of Geophysical Research: Oceans, 102*(C6), 12663–12676.

Rabinovich, A. B. (2009) "Seiches and harbor oscillations," Handbook of Coastal and Ocean Engineering, ed. Kim, Y. C. (World Scientific, Singapore), pp. 193–236. https://doi.org/10.1142/9789812819307_0009.

Titov, V.V., González, F.I., Mofjeld, H.O., & Venturato, A.J. (2003). NOAA time seattle tsunami mapping project: procedures, data sources, and products. US Department of Commerce, National Oceanic and Atmospheric Administration, Oceanic and Atmospheric Research Laboratories, Pacific Marine Environmental Laboratory.

Tolkova, E., & Power, W. (2011). Obtaining natural oscillatory modes of bays and harbors via Empirical Orthogonal Function analysis of tsunami wave fields. *Ocean Dynamics, 61*(6), 731–751. https://doi.org/10.1007/s10236-011-0388-5.

Zaytsev, O., Rabinovich, A. B., & Thomson, R. E. (2017). The 2011 Tohoku tsunami on the coast of Mexico: A case study. *Pure and Applied Geophysics, 174*(8), 2961–2986. https://doi.org/10.1007/s00024-017-1593-z.

(Received April 12, 2019, revised August 13, 2019, accepted August 14, 2019, Published online September 9, 2019)

Pure Appl. Geophys. 177 (2020), 1583–1593

https://doi.org/10.1007/s00024-019-02332-y

Pure and Applied Geophysics

Tsunami Run-Up on a Plane Beach in a Tidal Environment

IRA DIDENKULOVA[1,2] and EFIM PELINOVSKY[2,3,4,5]

Abstract—We analytically solve the nonlinear shallow water theory for the tsunami run-up on a plane beach in the presence of tide and show that over a plane beach the tide in the nearshore zone can be considered static (uniform in space and frozen in time). With this assumption, we find an exact analytical solution for the tsunami run-up height as a function of the amplitude of the incident wave and analyse the influence of the tide on the tsunami run-up characteristics. We confirm these results by calculation of the total tsunami and tidal run-up field using analytical solutions of the initial nonlinear shallow water equations.

Key words: Tsunami run-up, tide, plane beach, nonlinear shallow water theory, Carrier-Greenspan transformation.

1. Introduction

Tsunami wave run-up on a beach usually occurs in the presence of a tide, whose amplitude may be significant. If a tsunami occurs during the high tide, it penetrates much further inland and results in a larger hazard than the same tsunami during low tide. Furthermore, during the low tide, even a significant tsunami may remain unnoticed, and such situations have been observed. This is why it is common to discuss "tideless" characteristics of tsunamis, subtracting tide from the tide-gauge records or from observation data. Nowadays, the tide itself is pre-calculated with high accuracy and calculation errors are minimal. Usually, for the maximal sea level height, the tidal level with a tsunami or storm surge is summarized by a linear superposition; see, for instance, Shevchenko (2017).

However, a tsunami's impact on a beach in the presence of a tide is a highly nonlinear process, where simple superposition of the tsunami and tide is often insufficient. Kulikov et al. (1998) and Thomson et al. (2001) point out the significant role of the tide in tsunami generation by landslides. During high tide, the beach slope liquefies, and then during the following low tide it may lose stability and slide down into the water, generating a tsunami. Wave interaction with the tide also influences the coastal zone ecology, in particular, mixing of fresh groundwaters with the salted seawater (Xin et al. 2010). Numerical simulation of tsunami propagation at different tidal levels reveals the strong dependence of the maximal tsunami flow depth on the tidal level, whereas the maximal tsunami run-up height is not so impacted by the tide (Lima et al. 2010). Depending on the tidal phase, tidal oscillations may also reinforce or weaken the tsunami while the tsunami only approaches the bay (Behera et al. 2011). Tsunami penetration into rivers also depends on the tidal level, and numerical simulations underline the important role of tsunami interaction with the tide and river runoff (Chanson 2011; Tolkova 2013, 2018; Tolkova et al. 2015).

Most studies of tidal influence on tsunami run-up characteristics mentioned above are based on numerical simulations, while there is a clear lack of corresponding analytical solutions, except the works of Tolkova (2018) on tsunami-tide interaction in rivers. To the best of our knowledge, the only analytical work on an open beach is our recent work on tsunami-tide interaction on the vertical wall (Didenkulova and Pelinovsky 2018). Here, we analytically study the influence of tide on the tsunami run-up height for another typical case of a plane

[1] Department of Marine Systems, Tallinn University of Technology, Tallinn, Estonia. E-mail: ira.didenkulova@msi.ttu.ee
[2] Nizhny Novgorod State Technical University n.a. R.E. Alekseev, Nizhny Novgorod, Russia.
[3] Institute of Applied Physics, Nizhny Novgorod, Russia.
[4] National Research University, Higher School of Economics, Moscow, Russia.
[5] Special Research Bureau for Automation of Marine Researches, Yuzhno-Sakhalinsk, Russia.

beach. The study is performed in the framework of the 1D nonlinear shallow water theory. The article is organized as follows. The mathematical model is described in Sect. 2. The tidal oscillations on a plane beach are discussed in Sect. 3, while the Carrier and Greenspan transform for tsunami run-up in the presence of tidal currents is described in Sect. 4. The wave-breaking criterion is discussed in Sect. 5. Examples of the tsunami-tide interaction based on exact solutions of nonlinear shallow water equations are given in Sect. 6. The main results are summarized in the Conclusion.

2. Mathematical Model

Let us consider the classical problem of long wave run-up on a plane beach in the presence of tide. The study is performed in the framework of the nonlinear shallow water theory:

$$\frac{\partial \eta}{\partial t} + \frac{\partial}{\partial x}[(h+\eta)u] = 0, \tag{1}$$

$$\frac{\partial u}{\partial t} + u\frac{\partial u}{\partial x} + g\frac{\partial \eta}{\partial x} = 0, \tag{2}$$

where η is sea level displacement, u is flow velocity averaged over the water depth, g is gravitational acceleration, $h(x) = -\alpha x$ is the unperturbed water depth, and α is the tangent of the beach slope (Fig. 1). The x-axis is directed onshore, so that the shore is located at point $x = 0$ for the still water conditions. The boundary condition at the shore is determined by the total water depth equal to zero:

$$H(x,t) = h(x) + \eta(x,t) = 0. \tag{3}$$

Equation (3) also defines oscillations of the moving shoreline in time $x(t)$.

The solution of Eqs. (1)–(2) is usually found using Carrier-Greenspan transformation (Carrier and Greenspan 1958), which reduces the system of Eqs. (1)–(2) to the linear wave equation:

$$\frac{\partial^2 \Phi}{\partial \lambda^2} - \frac{\partial^2 \Phi}{\partial \sigma^2} - \frac{1}{\sigma}\frac{\partial \Phi}{\partial \sigma} = 0, \tag{4}$$

where all physical variables are expressed explicitly through the wave function $\Phi(\sigma, \lambda)$:

$$\eta = \frac{1}{2g}\left(\frac{\partial \Phi}{\partial \lambda} - u^2\right), \quad u = \frac{1}{\sigma}\frac{\partial \Phi}{\partial \sigma}, \tag{5}$$

$$x = \frac{1}{2\alpha g}\left(\frac{\partial \Phi}{\partial \lambda} - u^2 - \frac{\sigma^2}{2}\right), \quad t = \frac{1}{\alpha g}(\lambda - u). \tag{6}$$

Note that the variable σ is proportional to the total water depth of the basin:

$$\sigma = 2\sqrt{gH(x,t)}; \tag{7}$$

hence, the point $\sigma = 0$ corresponds to the moving shoreline. Thus, the wave equation, Eq. (4), is solved on the semi-axis $\sigma \geq 0$, and the boundary condition at

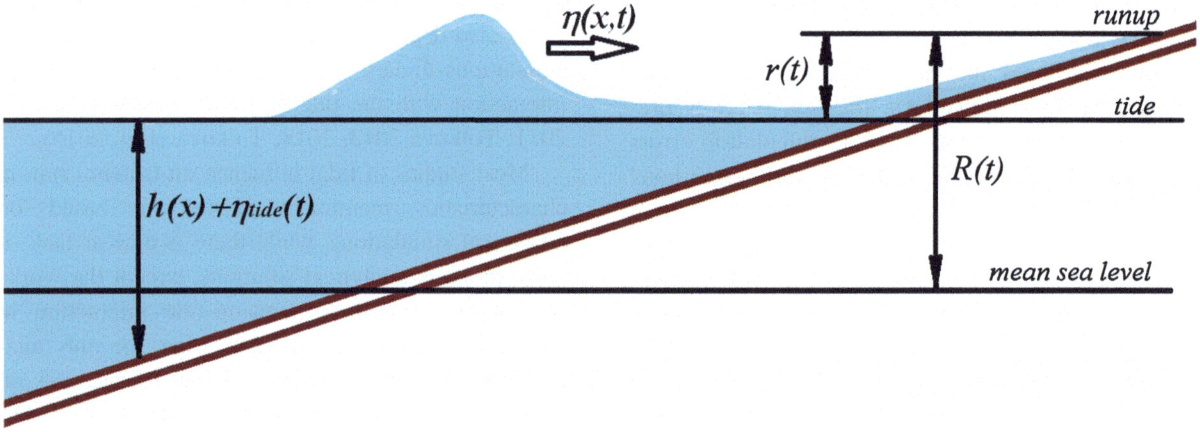

Figure 1
Basic geometry of the problem

$$\frac{\partial x}{\partial \sigma} = (\lambda - g\alpha t)\frac{\partial t}{\partial \sigma} - \frac{\sigma}{2}\frac{\partial t}{\partial \lambda}. \qquad (29)$$

After expressing the time t from Eq. (26) and substituting it into the last two equations, Eqs. (28)–(29), they are transformed to

$$\left(\frac{\partial^2 u}{\partial \lambda^2} - \frac{\partial^2 u}{\partial \sigma^2}\right) - \frac{3}{\sigma}\frac{\partial u}{\partial \sigma} = 0, \qquad (30)$$

$$g\alpha\frac{\partial x}{\partial \sigma} = -u\frac{\partial u}{\partial \sigma} - \frac{\sigma}{2} + \frac{\sigma}{2}\frac{\partial u}{\partial \lambda}. \qquad (31)$$

Now the total flow velocity can be presented as a sum of tidal flow u_{tide} and tsunami flow v:

$$u = u_{\text{tide}} + v, \quad v = \frac{1}{\sigma}\frac{\partial \Phi}{\partial \sigma}, \qquad (32)$$

After substitution of Eq. (32) into Eq. (31) and integration, we obtain:

$$2g\alpha x = -\frac{\sigma^2}{2} + \frac{\partial \Phi}{\partial \lambda} - v^2 - 2u_{tide}v + \Psi(\lambda), \qquad (33)$$

where the function $\Psi(\lambda)$ is not defined yet. Substituting Eq. (32) into Eq. (30), we obtain the wave equation for the function $\Phi(\sigma, \lambda)$

$$\frac{\partial^2 \Phi}{\partial \lambda^2} - \frac{\partial^2 \Phi}{\partial \sigma^2} - \frac{1}{\sigma}\frac{\partial \Phi}{\partial \sigma} = 0. \qquad (34)$$

Now from Eqs. (20) and (27) we find the displacement of the water surface:

$$\eta = \frac{1}{2g}\left(\frac{\partial \Phi}{\partial \lambda} - v^2 - 2u_{\text{tide}}v + \Psi\right) - \eta_{\text{tide}}. \qquad (35)$$

If there are no waves, there is no water displacement above the tidal level and, therefore, the function $\Psi(\lambda) = 2g\eta_{\text{tide}}$ is constant. Finally, Eqs. (33) and (35) can be re-written in the form:

$$2g\alpha x = -\frac{\sigma^2}{2} + \frac{\partial \Phi}{\partial \lambda} - v^2 - 2u_{\text{tide}}v + 2g\eta_{\text{tide}}, \qquad (36)$$

$$\eta = \frac{1}{2g}\left(\frac{\partial \Phi}{\partial \lambda} - v^2 - 2u_{\text{tide}}v\right). \qquad (37)$$

Therefore, the nonlinear problem of tsunami run-up on a beach in the presence of tide was reduced to the wave equation, Eq. (34), with subsequent calculation of the desired physical characteristics by Eqs. (32), (36), (37), and

$$\alpha g t = \lambda - u_{\text{tide}} - v. \qquad (38)$$

As Eqs. (32), (36), (37), and (38) show, the influence of the tidal flow is manifested not only by the additional changes in the shoreline oscillations and the tsunami travel time, but also by the additional term $u_{\text{tide}}v$ in Eqs. (36)–(37).

The boundary conditions at the shoreline for the function $\Phi(\sigma, \lambda)$ do not change and are determined by Eq. (8).

For correct definition of the wave field far offshore, let us consider asymptotic expressions of the time and coordinate

$$2g\alpha x = -\frac{\sigma^2}{2} + 2g\eta_{\text{tide}}, \qquad (39)$$

$$\alpha g t = \lambda - u_{\text{tide}}. \qquad (40)$$

The solution of Eq. (34) for a monochromatic wave with the frequency ω has a form similar to Eq. (9)

$$\Phi = AJ_0\left(\frac{\omega\sigma}{g\alpha}\right)\sin\left(\frac{\omega\lambda}{g\alpha} + \varphi\right). \qquad (41)$$

Far offshore, Eq. (41) can be asymptotically presented as the sum of two waves propagating in opposite directions. Let us consider the wave propagating onshore:

$$\Phi_{in} = A\sqrt{\frac{g\alpha}{2\pi\omega\sigma}}\sin\left[\frac{\omega}{g\alpha}(\lambda - \sigma)\right]. \qquad (42)$$

Here, the phase shift is eliminated by the transformation of time. The incident wave in physical variables can be found from Eq. (37) using Eqs. (39), (40), and (42).

$$\eta = H_0\sin\left[\omega\left(t + \frac{u_{\text{tide}}}{\alpha g} - \frac{2}{\alpha}\sqrt{g(h_0 + \eta_{\text{tide}})}\right)\right], \quad H_0 = \frac{A}{4g}\sqrt{\frac{\omega}{g\alpha\sqrt{h_0 + \eta_{\text{tide}}}}}. \qquad (43)$$

Here, h_0 is the water depth at the location offshore, where the incident wave with an amplitude of H_0 is defined. From Eq. (43), it follows that tsunami amplitude attenuates with the distance from the shore according to Green's law ($h^{-1/4}$). The time shift in Eq. (43) corresponds to the tsunami travel time to the shore taking into account the tidal flow.

The solution for the run-up of a sinusoidal wave on a plane beach is well known (Pelinovsky and Mazova 1992; Madsen and Fuhrman 2008); therefore, we can write it down also for the case with the tide

$$\frac{r}{H_0} = 2\sqrt{\frac{2\pi\omega c}{\alpha g}}, \quad c = \sqrt{g(h_0 + \eta_{\text{tide}})}. \tag{44}$$

Here, r is the maximal run-up height above the tidal level (Fig. 1), and H_0 is the initial tsunami amplitude. Equation (44) is universal in this form, and the influence of the tide is manifested by the change in celerity in the initial location. Finally, the tsunami run-up height above the still water level (Fig. 1) is:

$$R = \eta_{\text{tide}} + 2H_0\sqrt{\frac{2\pi\omega c}{\alpha g}}, \quad c = \sqrt{g(h_0 + \eta_{\text{tide}})}. \tag{45}$$

It follows from Eq. (45) that tide mostly affects the total water level rather than the run-up height above the tidal level. Figure 2 shows the changes of the maximal run-up height of a sinusoidal wave with an amplitude of H_0 = 1 m and a period of 10 min on a plane beach of 1:50 slope as a function of distance to the shore $L = h_0/$ for different tidal levels from − 2 to 2 m with a step of 0.5 m.

The explicit contribution from η_{tide}, which is the trivial effect of the tide, makes the implicit effect through the wave celerity c difficult to assess. To make it visible, in Fig. 2 we also plot Eq. (45) both keeping and deleting η_{tide} in the wave celerity c. It can be seen that this effect is small, which confirms the conclusions above tide mostly affecting the total water level.

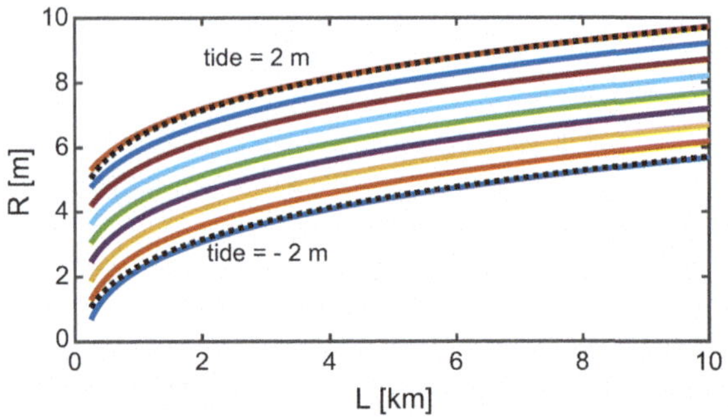

Figure 2
Maximal run-up height of a sinusoidal wave with an amplitude of H_0 = 1 m and a period of 10 min on a plane beach of 1:50 slope as a function of distance to the shore $L = h_0/$ for different tidal levels from − 2 m to 2 m with a step of 0.5 m. Black dotted lines for tidal levels of − 2 m and 2 m indicate the maximal run-up height not considering tidal oscillations η_{tide} in wave celerity c in Eq. (45)

5. Wave Breaking Criterion

As in our previous works (see, for example, Didenkulova 2009), the wave breaking criterion can be found by analysing the shoreline velocity. The moving shoreline corresponds to the point $\sigma = 0$, where $v = V(\lambda)$. Taking into account Eq. (38), we get

$$v(t) = V\left(t + \frac{u_{\text{tide}}}{\alpha g} + \frac{v}{\alpha g}\right), \tag{46}$$

where the function $V(t)$ corresponds to the shoreline velocity in linear approximation. The tidal flow velocity influences only the tsunami travel time and can be eliminated from Eq. (46) by re-defining the time. Then, the tsunami breaking condition, which is associated with the gradient catastrophe (time derivative dv/dt tends to infinity), has the following form:

$$Br = \frac{\max(\mathrm{d}V/\mathrm{d}t)}{g\alpha} = 1. \tag{47}$$

Since the tidal flow does not influence the wave breaking criterion, Eq. (47), i.e., the derivative from the constant tidal flow is zero, it can be eliminated, and the wave breaking criterion, Eq. (47), can be expressed through the vertical water oscillations on the shore:

$$Br = \frac{\max(d^2R/dt^2)}{g\alpha^2} = 1. \tag{48}$$

Equation (48) has the same form as in Didenkulova (2009), who derived it for a tideless sea; therefore this criterion, Eq. (48), is universal for both tidal and tideless seas. However, the water oscillations on the shore $R(t)$ depend on the tidal level; therefore, the wave breaking indirectly depends on the tide.

So, for the sinusoidal wave with an amplitude of H_0 = 1 m and a period of 10 min climbing a plane beach of 1:50 slope (conditions of Fig. 2), mentioned in Sect. 4, the wave breaking criterion is $Br = 0.03$, which corresponds to very gentle wave surging (weakly nonlinear case). In Sect. 6 we study two nonlinear cases: (1) for a tsunami wave with an amplitude of $H_0 = 15$ m and a period of 5 min climbing a plane beach of 1:20 slope, the wave breaking criterion is $Br = 0.27$; (2) for a tsunami wave with an amplitude of $H_0 = 8.5$ m and a period of 5 min climbing a plane beach of 1:50 slope, the wave breaking criterion is $Br = 0.95$ (strongly nonlinear case, just before the wave breaking).

6. *Examples of Tsunami-Tide Interaction*

In this section we study the dynamics of tsunami waves on a plane beach in the presence of tide by means of exact solutions of the nonlinear shallow water equations using the Carrier-Greenspan transform, i.e., Eqs. (4)–(6). This should also help to validate the approximate theory developed above. Tide in these calculations is not static. The function $\Phi(\sigma, \lambda)$ is taken as a sum of tidal Eq. (9) and tsunami Eq. (41) wave functions:

$$\Phi = A_0 J_0\left(\frac{\Omega\sigma}{g\alpha}\right)\sin\left(\frac{\Omega\lambda}{g\alpha}\right) + A J_0\left(\frac{\omega\sigma}{g\alpha}\right)\sin\left(\frac{\omega\lambda}{g\alpha} + \varphi\right), \tag{49}$$

where φ is the phase shift between them. So, Eq. (49) gives the explicit wave solution in the (λ, σ) parameter space, and Eqs. (5) and (6) give transformation formulas back to physical space. In that case the numerical calculations are not used to solve the differential equation, just to calculate the transformations.

For the first example, we consider a tsunami wave with an amplitude of 15 m and period of 5 min, while the tide has an amplitude of 4 m and period of 12 h. Hereafter, by tsunami and tide amplitudes we mean amplitudes of the vertical displacement of the moving shoreline (run-up amplitudes on the beach), which are related to amplitudes A and A_0 in Eq. (49):

$$A = \frac{2\alpha g^2}{\omega} R, \quad A_0 = \frac{2\alpha g^2}{\omega} R_0. \tag{50}$$

The corresponding beach slope is 1:20, and the phase shift is $\varphi = 0$. In this case the wave breaking parameter is $Br = 0.27$, which corresponds to moderate nonlinearity of tsunami waves. The wave field is shown in Fig. 3, while the velocity field is plotted in Fig. 4. This shows that the tsunami and tide interact almost linearly through superposition, while the velocity field is fully determined by tsunami waves.

For the second example, we consider a tsunami wave with an amplitude of 8.5 m and of period 5 min and tide with an amplitude of 4 m and period of 6 h. Such shorter period tides are often observed in the coastal zone (Munk and Cartwright 1966; Gallagher and Munk 1971). The beach slope is $\alpha = 1{:}50$ and phase shift $\varphi = 0$. The wave breaking parameter for this case is $Br = 0.95$, which corresponds to strongly nonlinear waves just before breaking. The wave field for this case is shown in Fig. 5, in which one can see the intensification of the wave motion, including tide, near the coastal zone. The velocity field is still mostly defined by the tsunami; however, the tidal oscillations are noticeable compared with the previous example (see Fig. 6).

Wave run-up oscillations and shoreline velocity are shown in Figs. 7, 8, 9, and 10. One can conclude from Fig. 7 that wave run-up oscillations represent a superposition of the tsunami and tide. This is also confirmed by Fig. 8, which shows the comparison of tsunami dynamics at maximum and minimum tidal levels. To make the coupling effects of a tsunami and tide more visible, we also plotted their linear superposition (computed separately and then added in the physical space). As Fig. 8 shows, the difference between them is minor and is manifested in nonlinear deformation of the tsunami wave shape.

As was already mentioned, the shoreline velocity is practically defined by the tsunami (see Fig. 9). Nonlinear effects related to the tidal phase are manifested in weak variations of the shoreline velocity (see Fig. 10). Figure 10 shows that during the low tide, the run-up velocities increase, while during the high tide the run-down velocities increase. The range of these variations at both high and low tide are about 0.1 m/s.

Note that we considered only the zero phase shift. Since here the tsunami is represented by a sine wave, due to a significant difference of the tsunami and tidal periods, there is always a moment when the tsunami crest coincides with the high and low tide for any

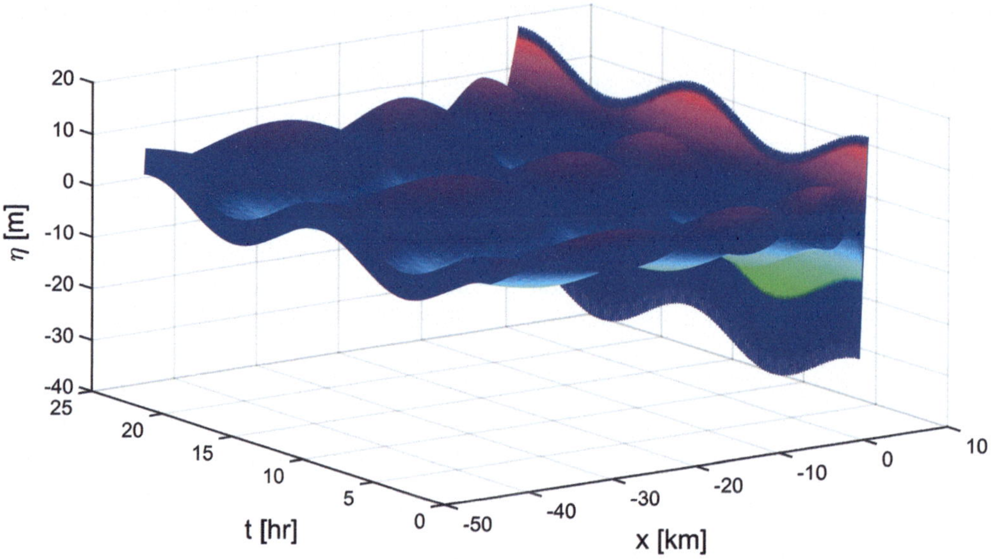

Figure 3
Dynamics of tsunami waves with amplitude of 15 m and period of 5 min in the presence of tide with an amplitude of 4 m and period of 12 h; the beach slope is $\alpha = 1{:}20$ and phase shift is $\varphi = 0$

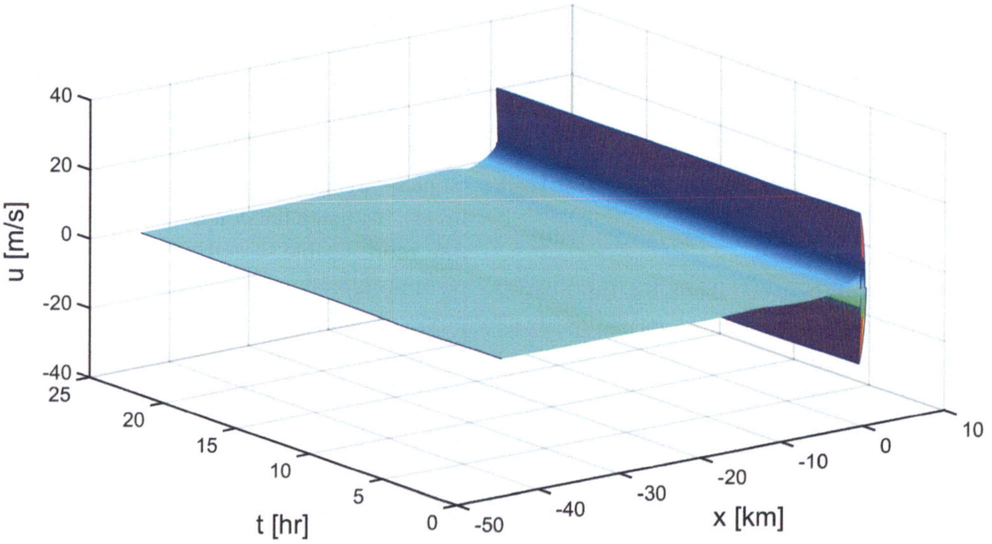

Figure 4
Velocity field for tsunami waves with an amplitude of 15 m and period of 5 min in the presence of tide with an amplitude of 4 m and period of 12 h; the beach slope is $\alpha = 1{:}20$ and phase shift is $\varphi = 0$

phase shift. This effect is illustrated in Figs. 7 and 8. Therefore, in the sine wave approximation the effect of the initial phase shift is not principal. However, it may play an important role in solitary tsunami waves.

7. Conclusion

In this article, we continue with our analytical study of tsunami interaction with the tide started in Didenkulova and Pelinovsky (2018) for tsunami run-

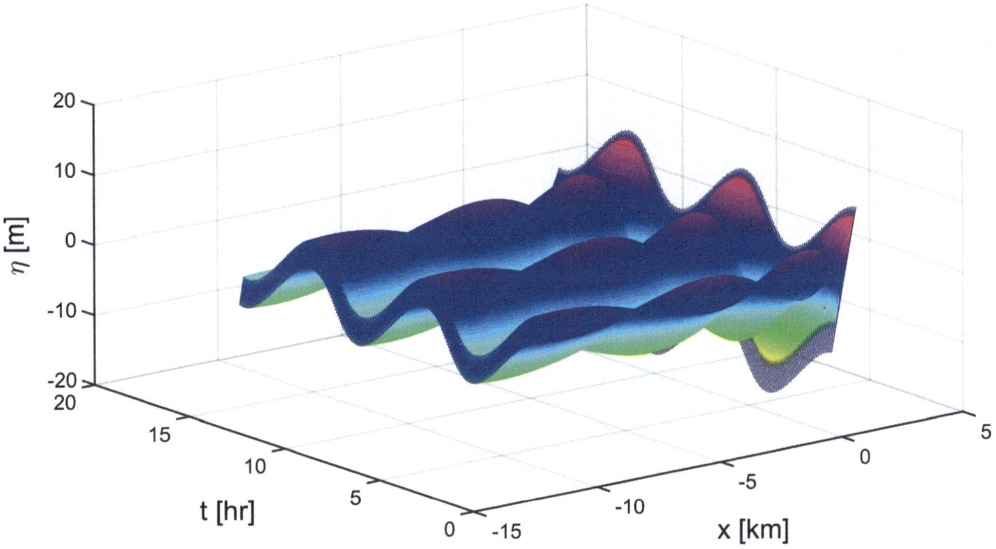

Figure 5
Dynamics of tsunami waves with an amplitude of 8.5 m and period of 5 min in the presence of tide with an amplitude of 4 m and period of 6 h; the beach slope is $\alpha = 1{:}50$ and phase shift is $\varphi = 0$

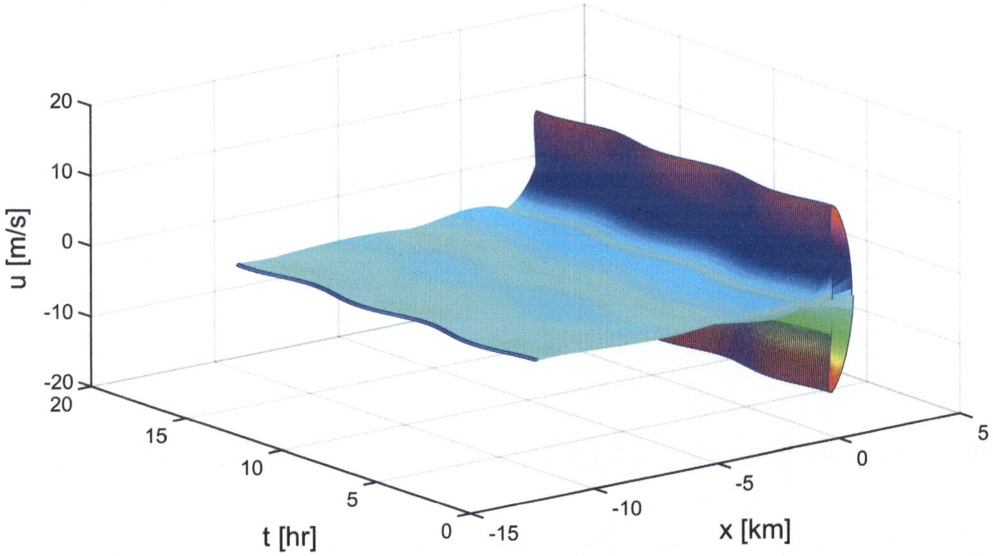

Figure 6
Velocity field for tsunami waves with an amplitude of 8.5 m and period of 5 min in the presence of tide with an amplitude of 4 m and period of 6 h; the beach slope is $\alpha = 1{:}50$ and phase shift is $\varphi = 0$

up on a vertical wall. Here, we study tsunami-tide interaction on a gentle plane beach in the framework of the nonlinear shallow water theory. First, we use an approximate theory where the tide in the nearshore zone is considered uniform in space during the tsunami travel time to the shore. For this case, by applying the classical Carrier-Greenspan transformation, we find an exact analytical solution for the tsunami run-up height as a function of the tsunami amplitude offshore and study tsunami run-up characteristics for different tidal levels. Second, we study the interaction of tsunami and tidal waves by means of analytical solutions of the initial nonlinear shallow water equations. In these calculations the tide

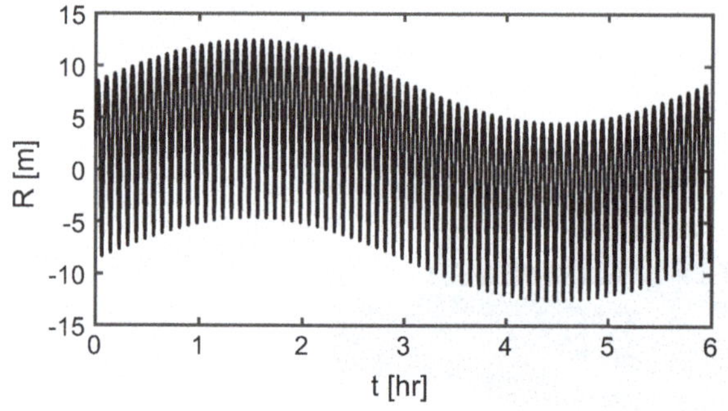

Figure 7
Run-up oscillations for tsunami waves with an amplitude of 8.5 m and period of 5 min in the presence of tide with an amplitude of 4 m and period of 6 h; the beach slope is α = 1:50 and phase shift is φ = 0

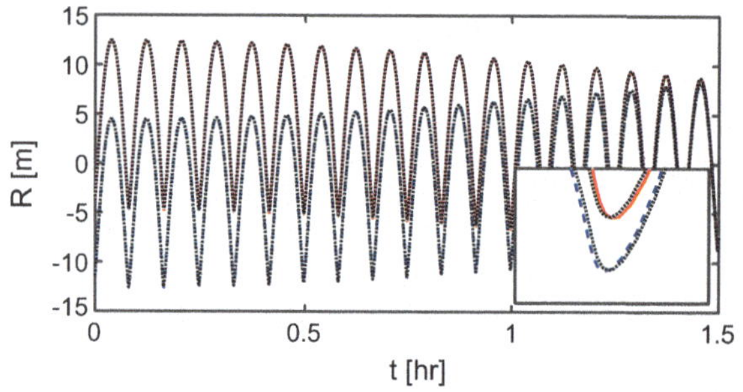

Figure 8
Tsunami run-up oscillations for full coupling of tsunami and tide during high tide (red solid line) and low tide (blue dashed line) for conditions in Fig. 7. Black dotted lines indicate run-up oscillations for linear superposition

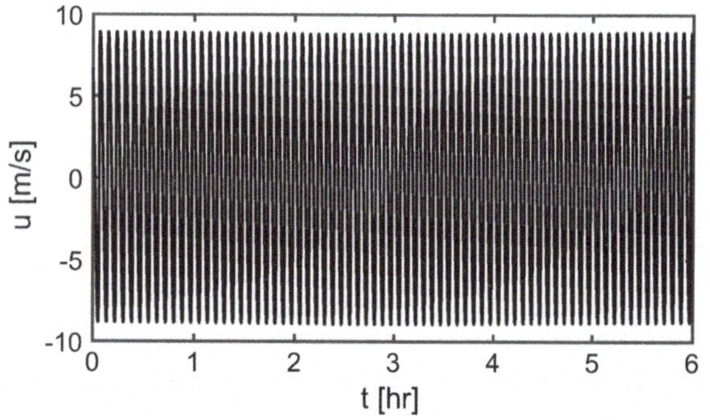

Figure 9
Shoreline velocity for tsunami waves with an amplitude of 8.5 m and period of 5 min in the presence of tide with an amplitude of 4 m and period of 6 h; the beach slope is α = 1:50 and phase shift is φ = 0

is not considered static. This showed that even for strongly nonlinear tsunami waves (i.e., just before breaking) the run-up oscillations on a beach can be

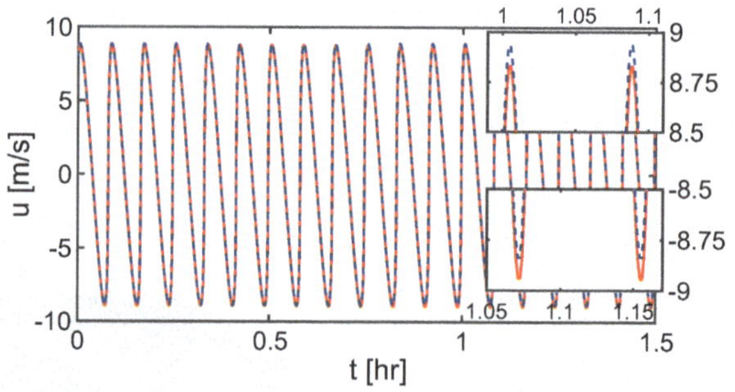

Figure 10
Shoreline velocity during high tide (red solid line) and low tide (blue dashed line) for conditions in Fig. 9

presented as a linear superposition of tsunami and tide. Nonlinear effects are manifested in weak variations of the shoreline velocity, so that during the low tide the run-up velocities increase, while during the high tide the run-down velocities increase.

This result is interesting and also unexpected. It is related to the special case of a plane beach, where tidal flows are weak, and the main effects are caused by sea level variations. In rivers, straits, and inlets, tidal flow may be extremely strong and may block tsunami wave propagation. Many such effects are considered in the book by Tolkova (2018). Another important effect that may influence our result is wave breaking, which is not taken into account in this article.

Our analysis of the simple interaction between tide and tsunami is important for understanding the applicability of analytical theory of long wave run-up, which is often used as a benchmark to test numerical codes.

The given study for a plane beach can be considered theoretical support for existing practical procedures to measure tsunami wave height above the tidal level.

Acknowledgements

This study was supported by ETAG project PUT1378, RFBR grants (17-05-00067, 18-05-80019, and 18-35-20026), and a grant from the President of the Russian Federation for state support of leading scientific schools of the Russian Federation (NSh-2685.2018.5).

Publisher's Note Springer Nature remains neutral with regard to jurisdictional claims in published maps and institutional affiliations.

References

Antuono, M., & Brocchini, M. (2007). The boundary value problem for the nonlinear shallow water equations. *Studies in Applied Mathematics, 119,* 73–93.

Antuono, M., & Brocchini, M. (2010). Solving the nonlinear shallow-water equations in physical space. *Journal of Fluid Mechanics, 643,* 207–232.

Behera, M. R., Murali, K., & Sundar, V. (2011). Effect of the tidal currents at the amphidromes on the characteristics of an N-wave-type tsunami. *Proc. Institution of Mechanical Engineers, Pt M – J Engineering for Marine Environment, 225,* 43–59.

Brocchini, M., & Gentile, R. (2001). Modelling the run-up of significant wave groups. *Continental Shelf Research, 21,* 1533–1550.

Carrier, G. F., & Greenspan, H. P. (1958). Water waves of finite amplitude on a sloping beach. *Journal of Fluid Mechanics, 4,* 97–109.

Carrier, G. F., Wu, T. T., & Yeh, H. (2003). Tsunami run-up and draw-down on a plane beach. *Journal of Fluid Mechanics, 475,* 79–99.

Chanson, H. (2011). *Tidal bores, Aegir, Eagre, Mascaret, Pororoca: theory and observations.* Singapore: World Scientific.

Didenkulova, I. (2009). New trends in the analytical theory of long sea wave runup. In E. Quak & T. Soomere (Eds.), *Applied wave mathematics: selected topics in solids, fluids, and mathematical methods* (pp. 265–296). New York: Springer.

Didenkulova, I. I., Kurkin, A. A., & Pelinovsky, E. N. (2007). Run-up of solitary waves on slopes with different profiles. *Izvestiya, Atmospheric and Oceanic Physics, 43*(3), 384–390.

Didenkulova, I., & Pelinovsky, E. (2011). Nonlinear wave evolution and runup in an inclined channel of a parabolic cross-section. *Physics of Fluids, 23*(8), 086602.

Didenkulova, I., & Pelinovsky, E. (2018). Tsunami wave run-up on a vertical wall in tidal environment. *Pure and Applied Geophysics, 175,* 1387–1391.

Dobrokhotov, S Yu., & Tirozzi, B. (2010). Localized solutions of one-dimensional non-linear shallow-water equations with velocity $c = \sqrt{x}$. *Russian Mathematical Surveys, 65*(1), 177–179.

Gallagher, B. S., & Munk, W. H. (1971). Tides in shallow water: spectroscopy. *Tellus, 23,* 346–363.

Kânoglu, U. (2004). Nonlinear evolution and run-up-rundown of long waves over a sloping beach. *Journal of Fluid Mechanics, 513,* 363–372.

Kânoglu, U., & Synolakis, C. (2006). Initial value problem solution of nonlinear shallow water-wave equations. *Physical Review Letters, 97,* 148–501.

Kulikov, E. A., Rabinovich, A. B., Fine, I. V., Bornhold, B. D., & Thomson, R. E. (1998). Tsunami generation by landslides at the Pacific coast of North America and the role of tides. *Oceanology, 38,* 323–328.

Lima, V. V., Miranda, J. M., Baptista, M. A., Catalao, J., Gonzalez, M., Otero, L., et al. (2010). Impact of a 1755-like tsunami in Huelva, Spain. *Natural Hazards and Earth System Sciences, 10,* 139–148.

Madsen, P. A., & Fuhrman, D. R. (2008). Run-up of tsunamis and long waves in terms of surf-similarity. *Coastal Engineering, 55,* 209–223.

Munk, W. H., & Cartwright, D. E. (1966). Tidal spectroscopy and prediction. *Philosophical Transactions of the Royal Society of London, A-Mathematical and Physical Sciences, 259,* 533–581.

Pedersen, G., & Gjevik, B. (1983). Run-up of solitary waves. *Journal of Fluid Mechanics, 142,* 283–299.

Pelinovsky, E., & Mazova, R. (1992). Exact analytical solutions of nonlinear problems of tsunami wave run-up on slopes with different profiles. *Natural Hazards, 6,* 227–249.

Raz, A., Nicolsky, D., Rybkin, A., & Pelinovsky, E. (2018). Long wave run-up in asymmetric bays and in fjords with two separate heads. *Journal of Geophysical Research Oceans, 123,* 2066–2080.

Rybkin, A., Pelinovsky, E., & Didenkulova, I. (2014). Nonlinear wave run-up in bays of arbitrary cross-section: Generalization of the Carrier-Greenspan approach. *Journal of Fluid Mechanics, 748,* 416–432.

Shevchenko, G. V. (2017). How the tide influences dangerous level rises on the coast of the Sea of Okhotsk and adjacent areas in cases of tsunami and storm surges. *Oceanology, 57*(5), 621–631.

Synolakis, C. E. (1987). The runup of solitary waves. *Journal of Fluid Mechanics, 185,* 523–545.

Thomson, R. E., Rabinovich, A. B., Kulikov, E. A., Fine, I. V., & Bornhold, B. D. (2001). On numerical simulation of the landslide-generated tsunami of November 3, 1994 in Skagway Harbor, Alaska. In G. T. Hebenstreit (Ed.), *tsunami research at the end of a critical decade, advances in natural and technological hazards research* (Vol. 18, pp. 243–282). Dordrecht: Kluwer.

Tinti, S., & Tonini, R. (2005). Analytical evolution of tsunamis induced by near-shore earthquakes on a constant-slope ocean. *Journal of Fluid Mechanics, 535,* 33–64.

Tolkova, E. (2013). Tide-tsunami interaction in Columbia River, as implied by historical data and numerical simulations. *Pure and Applied Geophysics, 170,* 1115–1126.

Tolkova, E. (2018). *Tsunami propagation in tidal rivers.* New York: Springer.

Tolkova, E., Tanaka, H., & Roh, M. (2015). Tsunami observations in rivers from a perspective of tsunami interaction with tide and riverine flow. *Pure and Applied Geophysics, 172,* 953–968.

Xin, P., Robinson, C., Li, L., Barry, D., & Bakhtyar, R. (2010). Effects of wave forcing on a subterranean estuary. *Water Resources Research, 46,* W12505.

(Received August 1, 2018, revised September 1, 2019, accepted September 19, 2019, Published online October 9, 2019)

Pure Appl. Geophys. 177 (2020), 1595–1616

https://doi.org/10.1007/s00024-019-02411-0

Pure and Applied Geophysics

Tsunami Modelling with Static and Dynamic Tides in Drowned River Valleys with Morphological Constrictions

KAYA M. WILSON[1] and HANNAH E. POWER[1]

Abstract—Tsunami modelling is widely used to estimate the potential impacts of tsunamis. Models require a tide input, which can be either static, representing a specific tide level, such as Highest Astronomic Tide or dynamic, which represents a moving tide level. Although commonly used, static tide inputs do not account for tsunami–tide interactions, which are known to be non-linear and more significant in estuaries when compared to the open coast. To demonstrate the differences between tsunami models using static or dynamic tide inputs, a series of models were carried out for two New South Wales estuaries, Sydney harbour and port hacking. Model boundary conditions phased a M_W 9.0 Puysegur source tsunami with multiple tide scenarios. Fourteen distinct scenarios with dynamic tides were created by phasing the largest tsunami wave peak at regular intervals across the tidal range. For comparison, static tide models were run using equivalent tide levels. The situations where static tide models provide results comparable or more conservative than dynamic tide models are for the first 1–2 h after tsunami arrival, at high tides, and when compared to dynamic falling tides at the same tide level. Differences are most apparent upriver of geomorphological constrictions. The effects of geomorphological constrictions were further examined using idealised model setups with a constriction variable. Results show that constrictions affect downriver maximum water levels, tsunami wave heights, upriver water accumulation and inundation maxima and distributions. These results have implications for estuaries vulnerable to erosion at constriction sites during a tsunami event.

Keywords: Tsunami modelling, tide–tsunami interactions, tsunami propagation, Sydney harbour, port hacking, tsunami erosion.

1. Introduction

Tsunamis are a global natural hazard and Australian coastal communities have differing levels of exposure (Davies 2018). In New South Wales (NSW), exposure around estuaries is particularly high, where reduced tidal ranges and protection from wind waves have allowed low-lying development to occur (Hanslow et al. 2018). Geological and historic evidence show that tsunamis have affected NSW in the past (Dominey-Howes 2007). Accounts from the Sydney harbour impacts of the May 1960 Chilean tsunami describe dragged and broken moorings, vessels being swept into bridges by large waves, and extreme currents. Significant scouring and erosion was also reported (Beccari 2009). The 1960 Chilean tsunami was generated by a subduction zone earthquake, the dominant cause of tsunamis worldwide (UNESCO 2015). Despite the rare occurrence of damaging tsunamis when compared to other hazards, the state of New South Wales (NSW), Australia, identifies tsunamis as a 'high' risk hazard due to the potential catastrophic consequences (NSW 2017).

Tsunami modelling is used in emergency management to provide a best estimate of potential tsunami impacts. The modelling process typically uses a digital representation of the environment and applies the non-linear shallow water wave equations that have been shown to best represent tsunami behaviour (Synolakis et al. 2008; Titov and Synolakis 1988). Tides can be applied to the models as either a static water level, which represents a constant tide level, such as HAT or Mean Sea Level (MSL) (e.g. Dall'Osso et al. 2014) or as a dynamic rising and falling water level (e.g. Cardno 2013; Wilson et al. 2018). Static tides provide a simpler and easier approach to tsunami modelling and are commonly used. Dynamic tides however, provide a more accurate representation of the natural environment and can reduce the limitations of the model (AIDR 2018).

When addressing the limitations of tsunami modelling with a static tide, the propagation

[1] University of Newcastle, University Drive, Callaghan, NSW 2308, Australia. E-mail: kaya.wilson@newcastle.edu.au

behaviour of both tides and tsunamis must be considered. Tide and tsunami propagation in long, shallow estuaries show distinct characteristics when compared to the open coast:

The propagation of tides in estuaries is more complex than just a rising and falling of the water level (Roy et al. 2001). Depending on estuary length, tides in estuary channels may behave like a progressive or standing wave or some combination of the two. At any one time, it may be simultaneously high tide at one location in an estuary and low tide at another (e.g. Wright et al. 1973).

Tidal range is also affected. As tides propagate up long estuaries, the influences of shoaling and friction have competing effects. As the tide propagates along an increasingly shallow estuary channel, a reduction of water depth induces shoaling, causing tidal range to increase with distance from the coast. Friction however, between the tide and estuary channel, dissipates energy and reduces the tidal range, eventually overwhelming shoaling. For long estuaries, tidal distortion may also occur, where the duration of the falling tide is relatively longer than the rising tide (e.g. Aubrey and Spear 1985).

The work of Tolkova (2018) used both observations and analytical solutions to show that three key phenomena occur when tsunamis propagate in rivers and long estuaries: tsunami–tide interaction, water accumulation, and tsunami waves travelling as bores.

Tsunami–tide interaction has been shown, through both observation and numerical simulation, to occur in a non-linear manner in estuaries and may exacerbate tsunami impacts (Kowalik and Proshutinsky 2010; Myers and Baptista 2001; Zhang et al. 2011). Tsunami modelling of historic events using a static tide input have shown good agreement with observed impacts at the open coast. However, the same comparisons were shown to underestimate inundation extents around riverine estuaries (Zhang et al. 2011).

In agreement with the open coast observations, analytical solutions of the nonlinear shallow water wave equations have shown that tsunami run-up on a plane beach follows a linear relationship between tsunami waves and tide. Weak, non-linear effects are only apparent in shoreline run-up and run-down velocities; during low tide the run-up velocities increase, while during high tide, the run-down velocities increase (Didenkulova and Pelinovsky 2019). Non-linear tsunami–tide interaction was however shown for run-up on a vertical wall, whereby wave breaking effects reduced tsunami run-up height to a greater extent at low tide, than at high tide (Didenkulova and Pelinovsky 2018).

Tsunami–tide interactions are most apparent in shallow water. Comparisons between the sum of separate tide and tsunami models with combined tide and tsunami models have shown that the extent of tsunami–tide interaction depends on tidal amplitude and phase (Kowalik and Proshutinsky 2010). Tidal current is also important: numerical modelling (Tolkova 2013; Tolkova et al. 2015) and observations have shown that tsunamis in rivers dissipate most with a receding tide, less as the tide starts to rise and least with a high tide (Yeh et al. 2012; Wilson and Torum 1972; Tanaka and Tinh 2012).

Upriver water accumulation has also been observed during tsunami events in rivers and may occur further upriver than wave motion is visible. At these upriver locations beyond visible wave motion, tsunamis may appear as a slow rise and fall in water level with no apparent negative amplitudes (Abe 1986). Water can also accumulate in a rivers' floodplain and even cause greater damage on the return flow to the ocean when compared to tsunami arrival (Tanaka et al. 2008). Although these observations have generally occurred some distance upriver, it has also been shown that a strong, near-field tsunami can cause water accumulation in the lower river, whereby the tsunami restricts river drainage (Tolkova et al. 2015).

The propagation of tsunamis as bores up rivers has been observed visually and in the case of the 2011 Tohoku tsunami entering the Kitakami River in Japan, has also been recorded by instruments. Records show the bores that formed in Kitakami to be a series of steady waveforms that propagated upriver until reaching a closed weir at 17.2 km from the river entrance (rkm). The greater part of the tsunami was reflected back, whilst a proportion overtopped the weir and a 1 m high wave continued upriver (Tolkova and Tanaka 2016).

The observations of tsunami propagation, tide propagation and tsunami–tide interactions in rivers and estuaries show that particular care must be taken when modelling tsunamis in estuarine locations.

Sydney harbour and port hacking in NSW are both long NSW estuaries classified as Drowned River Valleys (DRVs) under the Roy et al. (2001) estuary classification system (Fig. 1). DRVs are named as such because they are river valleys that were submerged by rising sea-levels during the last glacial melting (Roy et al. 2001). Both estuaries are referred to as tide-dominated estuaries as the tide is the dominant energy input into the system. Accumulated sediment deposited over the years by the flood tide, have created large flood-tide deltas of marine sands. Central basins upriver of the deltas are mostly made up of clay (Roy 1998; Troedson 2015; Masselink and Hughes 2003). Due to their small catchment, Sydney harbour and port hacking experience relatively little river input.

The unconsolidated sediments of the flood tide deltas are vulnerable to erosion in the event of a tsunami. Tsunamis are known to cause major morphological changes to the coastline and in particular to estuaries. Extensive erosion, sediment transport and deposition may take place over a huge area in a very short period of time (e.g. Mori and Takahashi 2012). In West Banda Aceh, the erosional imprints of the 2004 Indian Ocean Tsunami extend to 500 m from the shoreline and exceed 2 rkm along riverbed

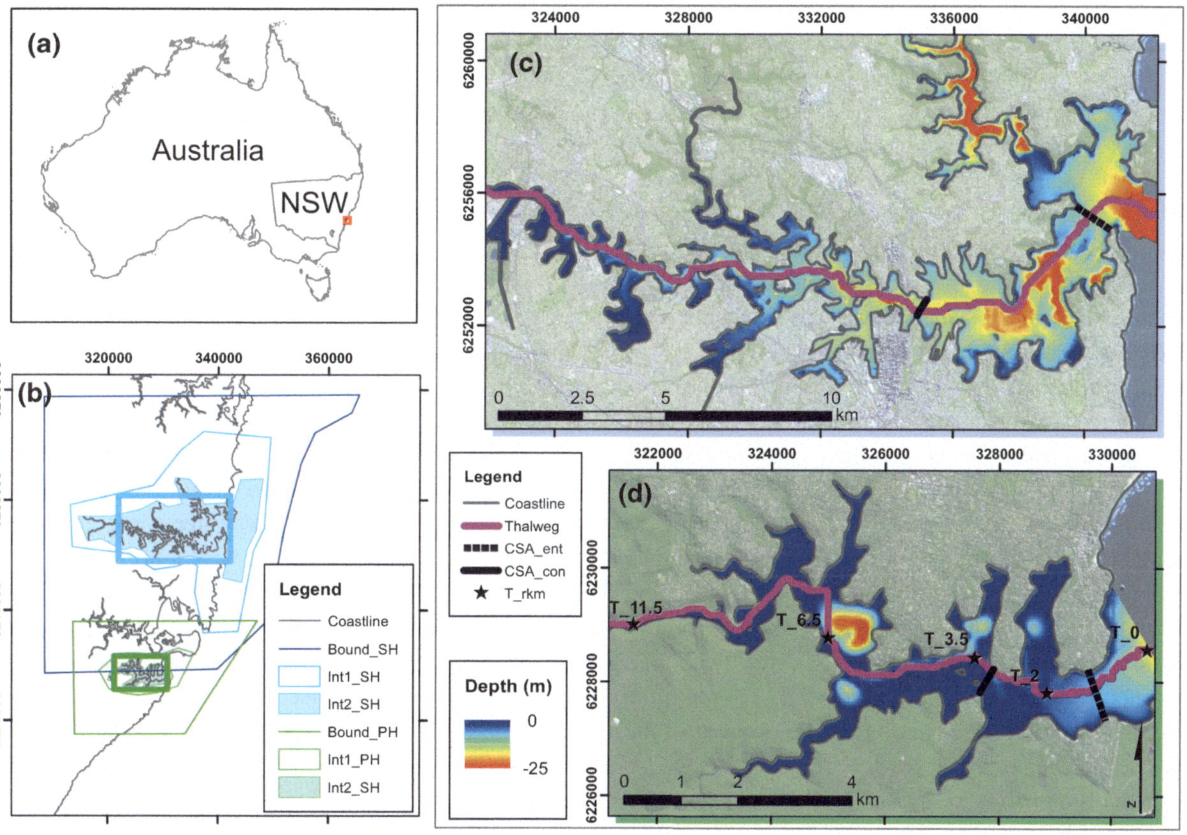

Figure 1

Study site location map: **a** sites in Australia; **b** tsunami model setups for Sydney harbour (SH) and port hacking (PH), where 'Bound' represents the relevant bounds of each model domain, and Int represents the internal model domains., where Int1 uses a mesh with maximum triangle areas of 1250 m^2 and Int2 areas of 450 m^2. Coloured boxes show the study site map locations of; **c** Sydney harbour site with bathymetry and; **d** port hacking site with bathymetry. 'CSA_ent' represents the entrance cross sectional area (CSA) and 'CSA_con' the constriction CSA as detailed in Sect. 2.1, 'T_rkm' denotes points along the thalweg, with river kilometres (rkm) provided. T_rkm locations are referred to throughout this paper. Maps **c** and **d** share a legend. All coordinates are provided in WGS84 UTM zone 56S consistent with the projection required by the hydrodynamic model used (ANUGA, see Sect. 2) estuaries (Paris et al. 2009). The trend in Khao Lak, Thailand during the same tsunami was for the widening of estuary mouths and the development of erosional channels (Scheucher and Vortisch 2010). These morphological changes alter the vulnerability to erosion that may occur in the future from both tsunamis and wind waves. In particular the deepening or widening of a river mouth may allow waves to propagate further inland, creating a higher energy environment upriver (Tanaka et al. 2012).

Numerical models have been developed to understand sediment transport due to tsunamis (e.g. Takahashi et al. 2000). The Takahashi et al. (2000) model includes the bed load rate, the suspended load rate and the exchange load rate in order to understand the process of bathymetric change. When wind wave effects are also considered, model results are comparable to measured bathymetric changes caused by tsunamis (Ranasinghe et al. 2013). Ranasinghe et al. (2013) assessed the bathymetric changes caused by the 2004 Indian Ocean Tsunami (IOT) at Kirinda Fishery Harbour, in Sri Lanka, and used sediment modelling to better understand the sediment transport that occurred during the event. Results showed that deposition dominates the inflow and erosion dominates the backflow of a tsunami. The majority of erosion measured was likely caused by the first backward (seaward) rushing wave, with sediment held in suspended form and then deposited in the nearshore area after water fluctuation had abated. It is notable that the largest wave is the leading wave in the IOT boundary conditions used by Ranasinghe et al. (2013). If waves with greater wave heights were to occur later in the wave sequence, the proportion of erosion caused by the first wave may differ. Although the majority of erosion may occur early on when the largest wave leads, deposition tends to occur after the event (Ranasinghe et al. 2013).

When compared to the open coast, tsunami-tide interaction in estuaries show cumulative effects that can exacerbate tsunami impacts (Kowalik and Proshutinsky 2010; Myers and Baptista 2001; Zhang et al. 2011). Estuaries may therefore be where the limitations of tsunami modelling with a static tide are greatest and so furthering understanding of these limitations is an asset to emergency management efforts. The erosional effects of tsunamis in estuaries must also be taken into account, as significantly altering the geomorphology of estuaries may create unanticipated exposure. The flood-tide deltas of accumulated marine sand in the port hacking and Sydney harbour estuaries may be locations particularly vulnerable to tsunami-induced erosion.

This paper compares tsunami modelling results using both static and dynamic tides in order to show where static tide models may underestimate or provide incorrect potential impacts in the case of drowned river valleys. Further exploration is given to the effects of morphological constrictions, which when present can cause a major disruption to water flow and when absent allow greater wave propagation. Thise paper also considers the possible effects of erosion, which may reduce such constrictions causing more extensive upriver impacts than those estimated by modelling with a static elevation dataset.

2. Methods

The open source, hydrodynamic model ANUGA was selected as suitable for tsunami inundation modelling. ANUGA uses the DE1 flow algorithm (Davies and Roberts 2015) and has been validated and used extensively to model the hydrodynamics of tsunamis (Jakeman et al. 2010; Jakir et al. 2015; Nielsen et al. 2005; Wilson et al. 2018). ANUGA solves the nonlinear shallow water wave equations, which have been shown to best represent tsunami behaviour (Titov and Synolakis 1988).

The model input requirements include elevation data, boundary conditions, tide data and friction values. A complete list of all model scenarios is provided in Appendix 1.

2.1. Model Input: Elevation Data

The elevation datasets developed by Wilson and Power (2018) were used for the study sites of Sydney harbour and port hacking with a gridded resolution of 10 m. Triangular mesh resolutions within the ANUGA model were selected by convergence testing. Within the primary area of interest, the triangular mesh generated by ANUGA had triangles with a maximum area of 450 m^2 for both study sites and idealised dataset models.

Idealised sites were created in order to simplify a drowned river valley estuary and isolate geomorphological constriction as a variable influencing tsunami propagation (Fig. 2). Idealised site elevation datasets were based on the dimensions of the port hacking estuary: the lengths of the idealised estuaries are 15 rkm, with a funnel-shaped entrance (2 km in width at the entrance funnelling to 0.85 km at 1.5 rkm) and convergence from estuary mouth to the furthest point inland or upriver. The coastline was smoothed using the PAEK Smoothing algorithm in the Smooth Line Cartography tool in ESRI ArcGIS 10.3™, with a smoothing tolerance of 200 m. Idealised elevation was generated using a combination of Python and ESRI ArcGIS 10.3™. Idealised estuary depths are a constant − 3 m at MSL. This extends to the estuary mouth, where the elevation slopes to a depth of − 100 m, 9 km seaward. Topography is constant at 15 m elevation. The slope from 15 m elevation, to the coastline at 0 m MSL, and the estuary depth at − 3 m was interpolated across 60 m with the coastline at a constant 0 m elevation, a slope angle of 16.7° (Fig. 2a). Interpolation was performed by the ESRI ArcGIS 10.3™ Topo to Raster tool.

A constriction variable was introduced to the idealised sites at 3 rkm (12 km from the model boundary). Three levels of constriction were created, which resulted in cross sectional area (CSA) reductions of 12%, 45% and 78% relative to the idealised dataset with no constriction, when measured at MSL. Representations of the idealised elevation datasets are shown in Fig. 2.

For the comparison of different morphological constrictions, the percentage reduction in cross-sectional area (CSA; %CSA reduction) was used as a normalised measure across all sites. %CSA reduction was calculated by measuring the CSA of the estuary entrance after initial funnelling, (CSA_ent; Figs. 1c, d, 2a) and the CSA of the constriction (CSA con; Figs. 1c, d, 2). %CSA reduction was calculated for all sites at all tide levels as:

$$\%CSA\ reduction = \left(\frac{CSA\ ent - CSA\ con}{CSA\ ent}\right) * 100$$

2.2. *Model Input: Tsunami Boundary Conditions*

Three different boundary conditions were used in this study: a scenario wave sequence was used for the site models, a single wave peak was used to compare rising and falling tide models for the port hacking site, and an idealised wave was used for the idealised elevation dataset models.

The tsunami time series used for the Sydney harbour and port hacking model boundary conditions was sourced from the T2 tsunami scenario database (Greenslade et al. 2009), which uses the MOST model (Titov and Synolakis 1988) to determine seafloor deformation and tsunami propagation. The tsunami time series was extracted from the MOST model at a location along the 100 m depth contour offshore Sydney harbour and represents a M_W 9.0 event originating at the Puysegur trench, near New Zealand (Fig. 3b, c). The largest wave leads in this wave sequence, with a positive amplitude of 1.3 m and a wave height from peak to trough (H) of 1.9 m. Positive amplitudes occurring later in the sequence remain below 0.6 m. An event of this magnitude was selected due to the potential significant impacts, which enable the differentiation of tested variables in this study.

Further details regarding the modelling of T2 tsunami waves and justification for the use of a M_W 9.0 event can be found in Wilson et al. (2018), see their 'Methods: Tsunami Wave Train Boundary Conditions.'

In order to test the differences between rising and falling dynamic tide models and the effects of later waves in the wave sequence, an adaptation of the T2 M_W 9.0 tsunami wave sequence was modelled in the port hacking model domain. Only the leading wave peak was used, with an amplitude of 1.3 m and no negative wave trough. Subsequent waves in the wave sequence were removed. Boundary conditions were formed by summing this single peak wave to rising and falling dynamic tides at the same tide level of 0.22 m (Fig. 3a).

For the idealised elevation dataset models, an idealised tsunami wave was used. This idealised wave has a single peak, with an amplitude of 0.72 m, a single trough at − 0.72 m (H = 1.44 m). and a

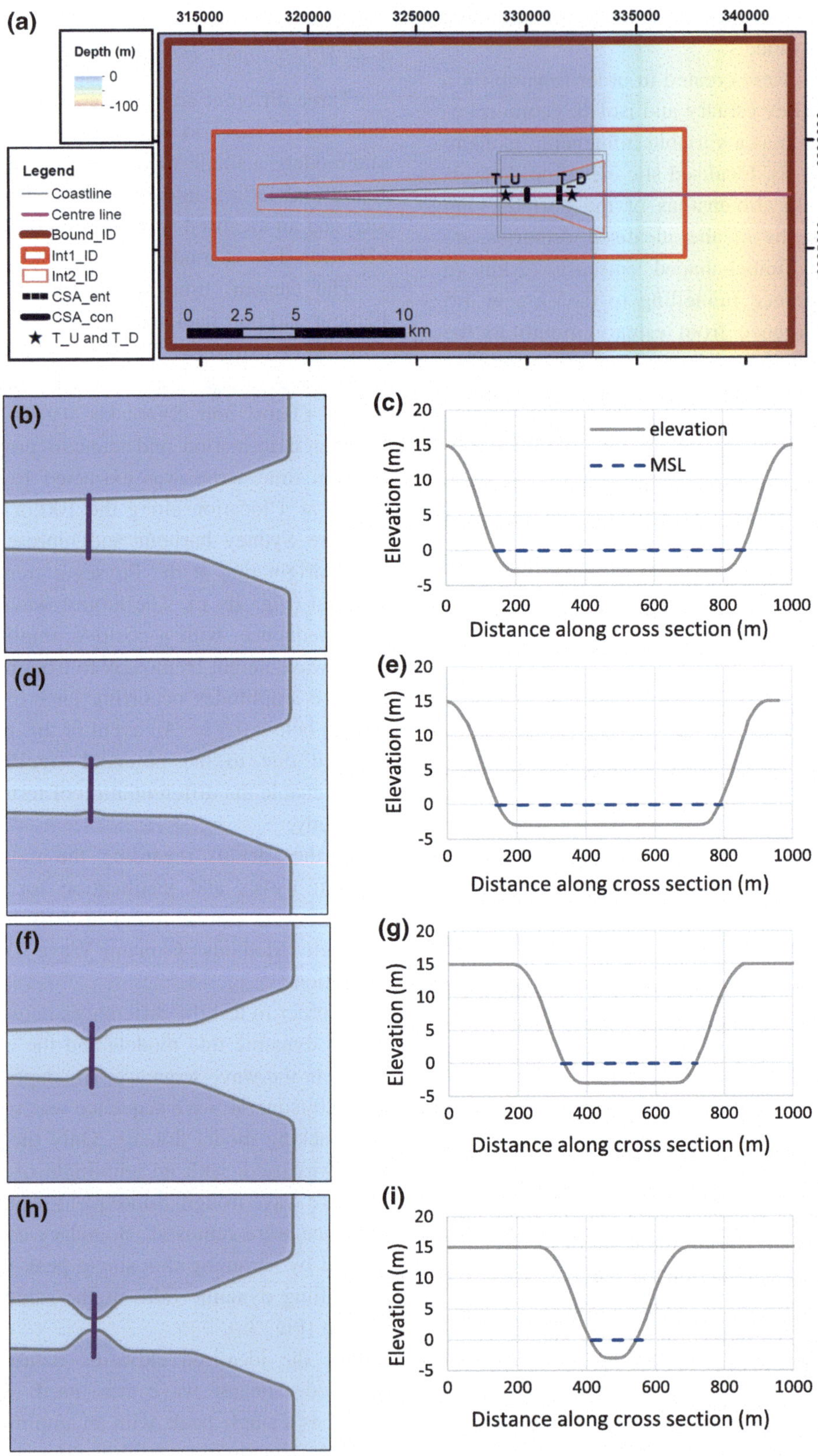
(a)
315000
320000
325000
330000
335000
340000
Depth (m)
0
-100
Legend
Coastline
Centre line
Bound_ID
Int1_ID
Int2_ID
CSA_ent
CSA_con
T_U and T_D
T_U
T_D
0
2.5
5
10
km
6230000
6225000
(b)
(c)
elevation
MSL
Elevation (m)
Distance along cross section (m)
(d)
(e)
(f)
(g)
(h)
(i)

◀Figure 2
Idealised elevations datasets: **a** model setup, where 'Bound_ID' represents the bounds of the model domain, Int1_ID represents the internal model domain using a mesh with maximum triangle areas of 1250 m^2, and Int2_ID represents the internal model domain using a mesh with maximum triangle areas of 450 m^2. 'CSA_ent' represents the location of the entrance CSA after initial funnelling and 'CSA_con' represents the location of the constriction CSA. T_U and T_D are thalweg positions representing positions upriver and downriver of the constriction position respectively and are referred to throughout this paper. The double-lined box is an extent indicator for **b**, **d**, **f**, and **h**; **b** idealised site with no constriction showing the location of the constriction cross section given in **c**; **d** idealised site with 12% constriction at MSL showing the location of the constriction cross section given in **e**; **f** idealised site with 45% constriction at MSL showing the location of the constriction cross section given in **g**; **h** idealised site with 78% constriction at MSL and location of the constriction cross section given in **i**

wave period of 26 min (Fig. 3a). This idealised wave differs from the scenario wave sequences in the peak/trough amplitude symmetry and the total wave height (1.44 m), which is mid-way between the heights of the first and second waves in the Puysegur 9.0 M_W sequence (Fig. 3).

2.3. *Model Input: Tide Data*

Tide data used were extracted from one tidal cycle, from the spring tides of January 2015 at the Middle Harbour Tide Gauge (151° 15′ 30.72″, − 33° 49′ 31.56″) in Sydney harbour. The tide cycle showing the largest peak to trough tidal range of those spring tides was selected, with a range of 1.76 m. Tide data were converted to Mean Sea Level (MSL) from the local datum, Zero Fort Denison.

T2 data was summed with this tide data in order to provide dynamic tide boundary conditions (Fig. 3c). The peak tsunami wave was phased with a range of tide levels: high tide (0.97 m), low tide (− 0.79 m), rising tide levels of − 0.64 m, − 0.44 m, − 0.14 m, 0.23 m, 0.6 m, 0.86 m and falling tide levels of 0.88 m, 0.61 m, 0.22 m, − 0.19 m, − 0.54 m, and − 0.75 m (see Appendix 1).

Static tide models were run by applying a single tide level to the entire model domain. The tsunami boundary condition was then run at this level. For example, for high tide models, the still water level across the model domain was + 0.97 m relative to MSL. The tsunami wave sequence was then summed to 0.97 m and run as a boundary condition (Fig. 3b).

If the static tide level was within a 0.02 m tolerance, some static tide level model results were used for comparison with both rising and falling dynamic tide models. For example the static tide level of 0.6 m was used for comparison with both the rising dynamic model with the tsunami phased at 0.6 m and the falling dynamic model with the tsunami phased at 0.61 m. This was done as a sensitivity analysis showed that differences in total inundation volume between static tide models at 0.6 m and 0.61 m tide levels for both Sydney harbour and port hacking were less than 1% for both sites; 0.6% for Sydney harbour and 0.85% for port hacking.

2.4. *Model Input: Friction*

A Manning's value of 0.02 was considered suitable for the entire model domain. Experimentation has shown this value to be applicable for seabed roughness, road surface and sand/gravel (Cardno 2013).

2.5. *Model Duration*

The model duration for dynamic tide models was 12 h. The peak tsunami wave was introduced at 2.5 h, allowing for tide-only model 'burn in' time prior to tsunami arrival. Static tide models were run for 6 h as tidal movement was not a factor and there was no concern that water level maxima could occur after the leading (and largest) wave had occurred. Tsunami impact maxima such as maximum inundation, which were calculated for results analysis were based on these model durations.

3. *Results*

Terrestrial inundation, maximum water levels, and water level time series were extracted from the tsunami models. Model scenarios are described according to the tide data used for modelling, i.e., static or dynamic tides, and the tide level at which the peak of the largest tsunami wave was coincident, *e.g.* 'low tide' or '0.23 m'. Idealised site scenarios are described according to the tide data used, the tide

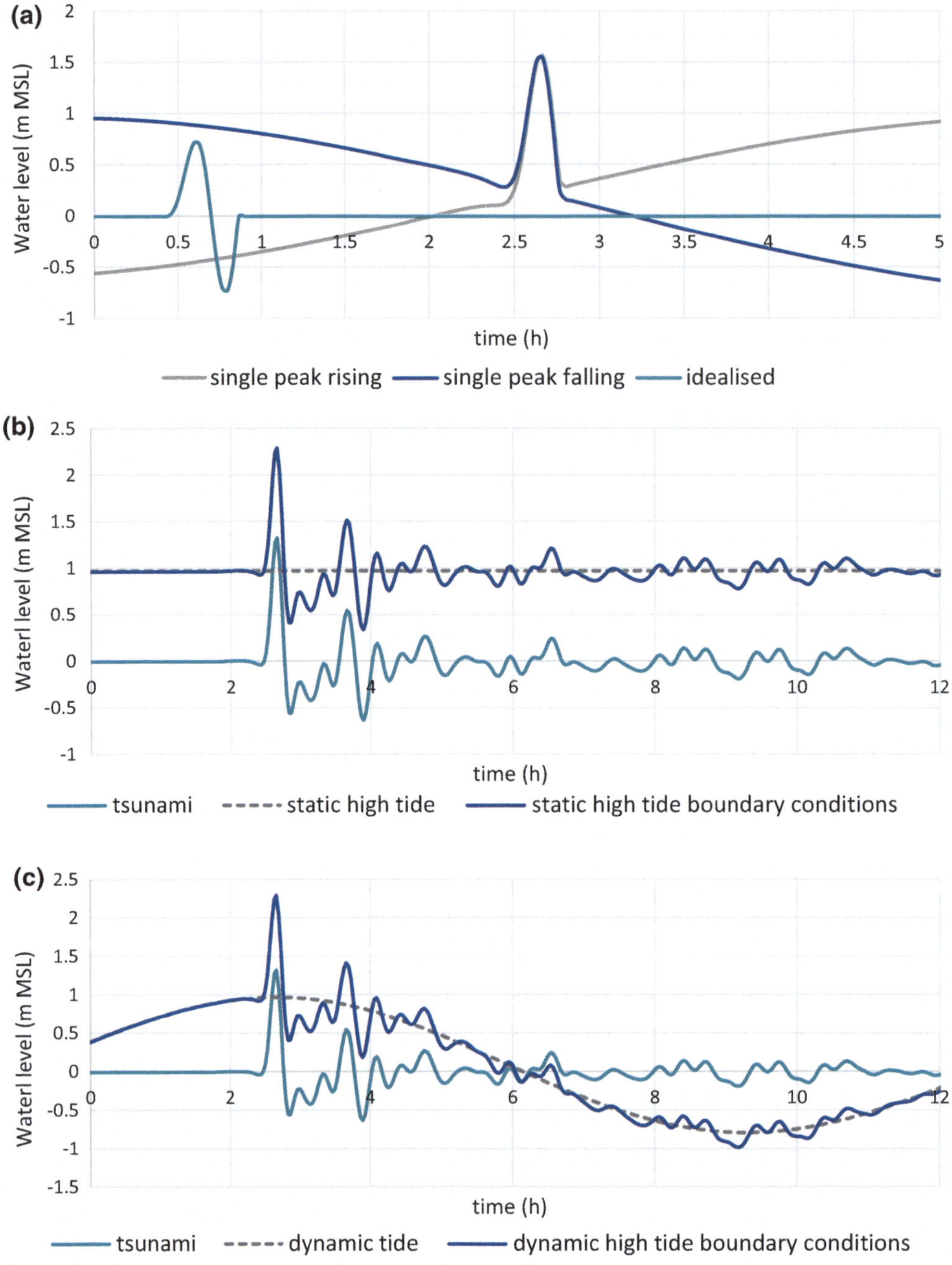

Figure 3

Tsunami and tide boundary conditions: **a** additional tsunami boundary conditions used: single peak rising and falling, and idealised tsunami used for idealised models; **b** static high tide boundary conditions and components as an example of summing a tsunami and static tide water level; **c** dynamic high tide boundary conditions and components as an example of summing a tsunami and dynamic tide water level

level and also the degree of constriction to the estuary geomorphology at 3 rkm.

Details of all model scenarios are provided in Appendix 1.

Results should be viewed in the context of the tidal propagation patterns for each estuary. Tidal propagation upriver for Sydney harbour, the longest estuary in this study showed a marginal

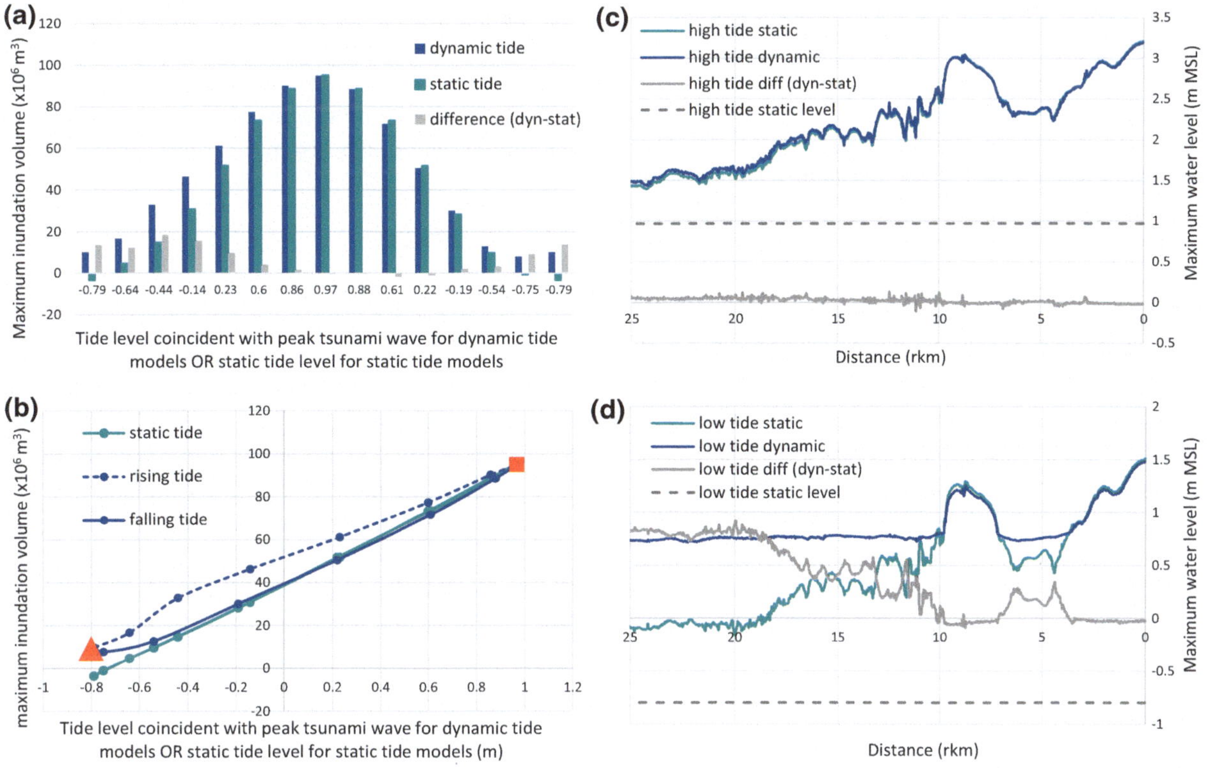

Figure 4

Sydney harbour tsunami model impact measures: **a** maximum inundation volume above high tide level for all models; **b** maximum inundation volume plotted against XYZ to show differences between models with rising and falling tides; **c** maximum water level along the thalweg for high tide models; and **d** maximum water level along the thalweg for low tide models

hypersynchronous pattern, whereby slight shoaling occurs in the lower estuary (tidal range increases from 1.80 to 1.87 m) before friction effects cause the tidal range to fall to 1.25 m in the furthest reaches of the estuary. Tidal propagation in port hacking tended more towards hyposynchronous behaviour where friction effects are greater than shoaling effects and tidal range falls as the tide propagates upriver (from 1.82 m at the mouth of the estuary to 0.78 m at the upper reaches). Idealised sites experience slight tidal shoaling with tidal range increasing from 1.8 to 1.97 m along the estuary.

3.1. Sydney Harbour

3.1.1 Maximum Inundation

Terrestrial inundation is dependent on tidal phase and inundation increases with increasing tide levels (Fig. 4a). For high tide models, maximum inundation volumes are marginally greater for static tide models than for dynamic tide models. The difference is less than 0.5% of the dynamic high tide model maximum inundation volume (95 million m^3) and can be considered negligible (Fig. 4a). The lowest inundation volumes occur at low tide and the differences between static and dynamic tide models are more significant. The dynamic low tide model shows 134% more inundation than the static low tide model (Fig. 4a).

Tide current also affects maximum inundation volume. Figure 4b illustrates the differences between dynamic tide models for rising and falling tides. Rising tide dynamic models show greater inundation volumes than falling tide dynamic models, for the same tide level. For example, at a tide level of 0.22 m, the difference is 21% of the falling dynamic tide volume (Fig. 4b). Falling tide dynamic models are more similar to static tide models than to rising tide dynamic models, showing mostly negligible differences (Fig. 4b).

3.1.2 Maximum Water Level

Maximum water levels along the southern thalweg in Sydney harbour are shown in Fig. 4c, d for high and low tide models respectively (see Fig. 1c for the path of the Sydney harbour southern thalweg).

Both high and low tide models show a general downward trend in water levels for the first 5 rkm and a water level peak between 7 and 10 rkm. The location of this peak is seaward of the constriction in the harbour near the Sydney Opera House (Fig. 4c, d). The small peaks shown between 1.5 and 3 rkm are caused by shoaling with rapid change in depth at that location (Fig. 1c).

High tide models show only negligible differences between static and dynamic tide models (Fig. 4c). Low tide models, however, show maximum water levels to be higher for dynamic tide models between 4 and 7 rkm, where differences remain under 0.5 m and also for 10–25 rkm, where differences increase to 0.8 m with increasing distance upriver (Fig. 4d).

3.2. Port Hacking

3.2.1 Maximum Inundation

Terrestrial inundation trends for port hacking models are consistent with inundation trends for Sydney harbour models; maximum inundation volumes are dependent on tidal phase and current (Fig. 5a, b).

For high tide models, maximum inundation volumes are 6.5% greater for static tide models than dynamic tide models (Fig. 5a). Consistent with Sydney harbour models, the lowest inundation volumes occur at low tide and the differences are more

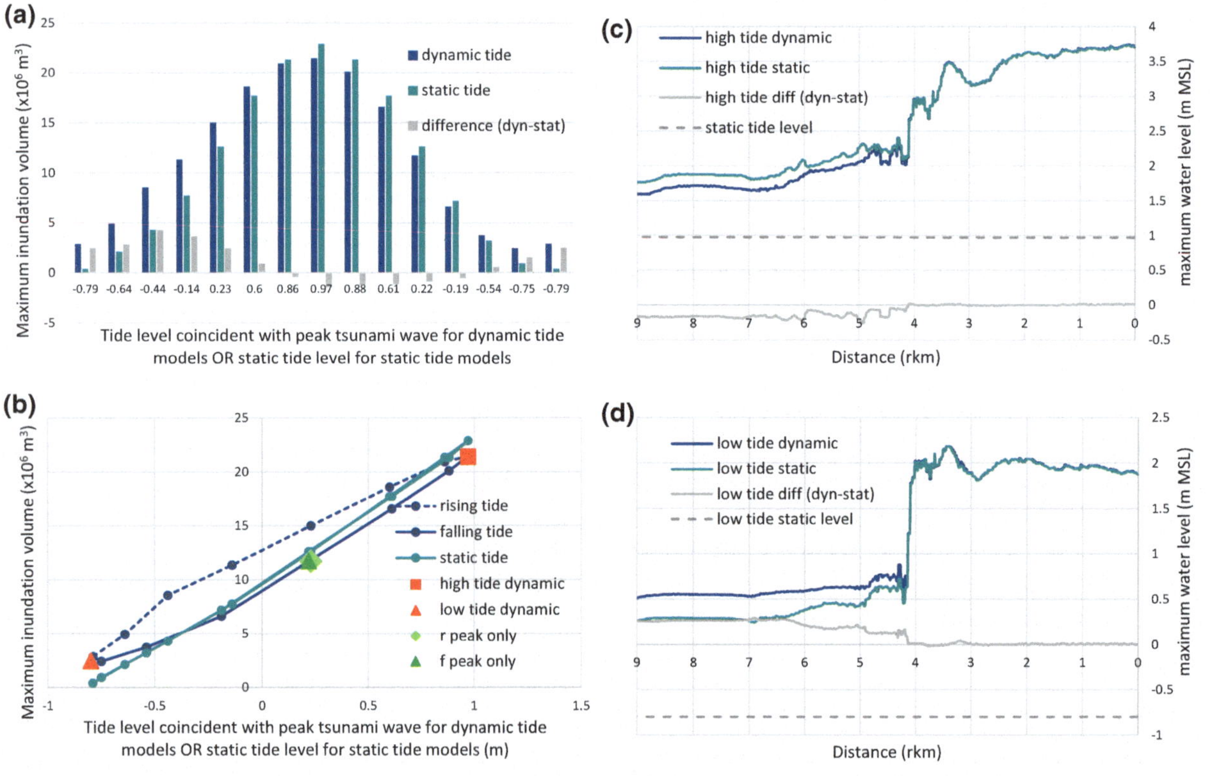

Figure 5

Port hacking tsunami model impact measures: **a** maximum inundation volume for all models; **b** maximum inundation volume plotted against XYZ to show differences between models with rising and falling tides. 'r peak only' represents the dynamic model with a single peak tsunami summed to a rising tide. 'f peak only' represents the dynamic model with a single peak tsunami summed to a falling tide; **c** maximum water level along thalweg for high tide models, **d** maximum water level along thalweg for low tide models significant. For low tide models, dynamic tide models show greater inundation than static tide models; 5.7 times that of static tide models (Fig. 5a).

Rising tide dynamic models show greater inundation than falling tide dynamic models, for the same tide levels (Fig. 5a, b). For tide levels below 0.6 m, static tide models show inundation volumes closer to falling tide dynamic models than rising tide models (Fig. 5b). For tide levels 0.6 m and above, however, the static tide models show inundation volumes closer to those of rising tide models (Fig. 5b).

Additional models were run substituting the tsunami wave sequence with a single peak tsunami wave and summing to both a rising and falling dynamic tide at a tidal level of 0.23 m (Fig. 3a). Dynamic, single peak wave models show minimal inundation difference between rising and falling tides (Fig. 5b).

3.2.2 *Maximum Water Level*

Maximum water levels along the port hacking thalweg (see Fig. 1d) are shown for high tide models (Fig. 5c) and for low tide models (Fig. 5d).

Both high and low tide models show negligible differences in maximum water levels between static and dynamic tide models for 0–4 rkm. Between 4 and 4.3 rkm, both high and low tide models show a sharp decrease in maximum water levels (Fig. 5c, d). This decrease in maximum water levels coincides with a major geomorphological constriction in port hacking (Fig. 1d). Upriver of this constriction, between 4 and 9 rkm, high tide models show maximum water levels 0.05–0.2 m higher for static tide models (Fig. 5c) and low tide models show maximum water levels 0.12–0.26 m higher for dynamic tide models (Fig. 5d).

3.2.3 *Wave Propagation*

Dynamic and static models of equivalent tide levels show negligible differences in water levels for 1–2 h after the arrival of the largest wave. After 2 h, the rising or falling tide of the dynamic models causes differentiation in water levels when compared to static tide models. These differences in water levels increase over time (Fig. 6).

Figure 6 shows water level time series for locations in port hacking at increasingly high rkm, from the mouth of the estuary, T_0 to location T_11.5 (numbers provide rkm, locations shown in Fig. 1d), which experiences drying at low tide (Fig. 6j). Tidal distortion is apparent (Fig. 6a–e) in the most upriver locations of the estuary. At the location T_11.5 (Fig. 1d), the tide rises over the course of 1 h and takes > 7 h to fall (Fig. 6e). At this location, low tide is dry, therefore negative tide amplitudes do not occur and the tide rises and falls above MSL. The water level time series of the high tide static model at the same location shows that the elevated upriver water level facilitates the speed of tsunami propagation and the tsunami arrival time is earlier for the static than the dynamic tide model (Fig. 6e). The negative amplitudes of the tsunami also show greater amplitude reduction than positive amplitudes as the tsunami propagates upriver (Fig. 6).

3.3. *Idealised Sites*

3.3.1 *Maximum Stage*

High tide models across all idealised sites, show little difference in maximum water levels between static and dynamic models along the thalweg (Fig. 7). Dynamic low tide models show higher maximum water levels than static low tide models upriver of the constriction (Fig. 7). This is most pronounced for the model with 78% CSA reduction, where differences are between 0.1 and 0.5 m (Fig. 7d). The peaks in maximum water levels that occur seaward/downriver of the constrictions, increase in amplitude with increasing degrees of constriction. High tide peaks for models with no constriction remain < 4 m in amplitude (Fig. 7a), whereas high tide peaks for 78% CSA constriction reach 5.6 m (Fig. 7d).

3.4. *Effects of Constriction*

Further examination of idealised model results shows how tsunamis propagate across geomorphological constrictions and how propagation differs when modelled with a static or dynamic tide. Low tide models are examined due to the differences observed in Fig. 7d. Figure 8a shows the initial

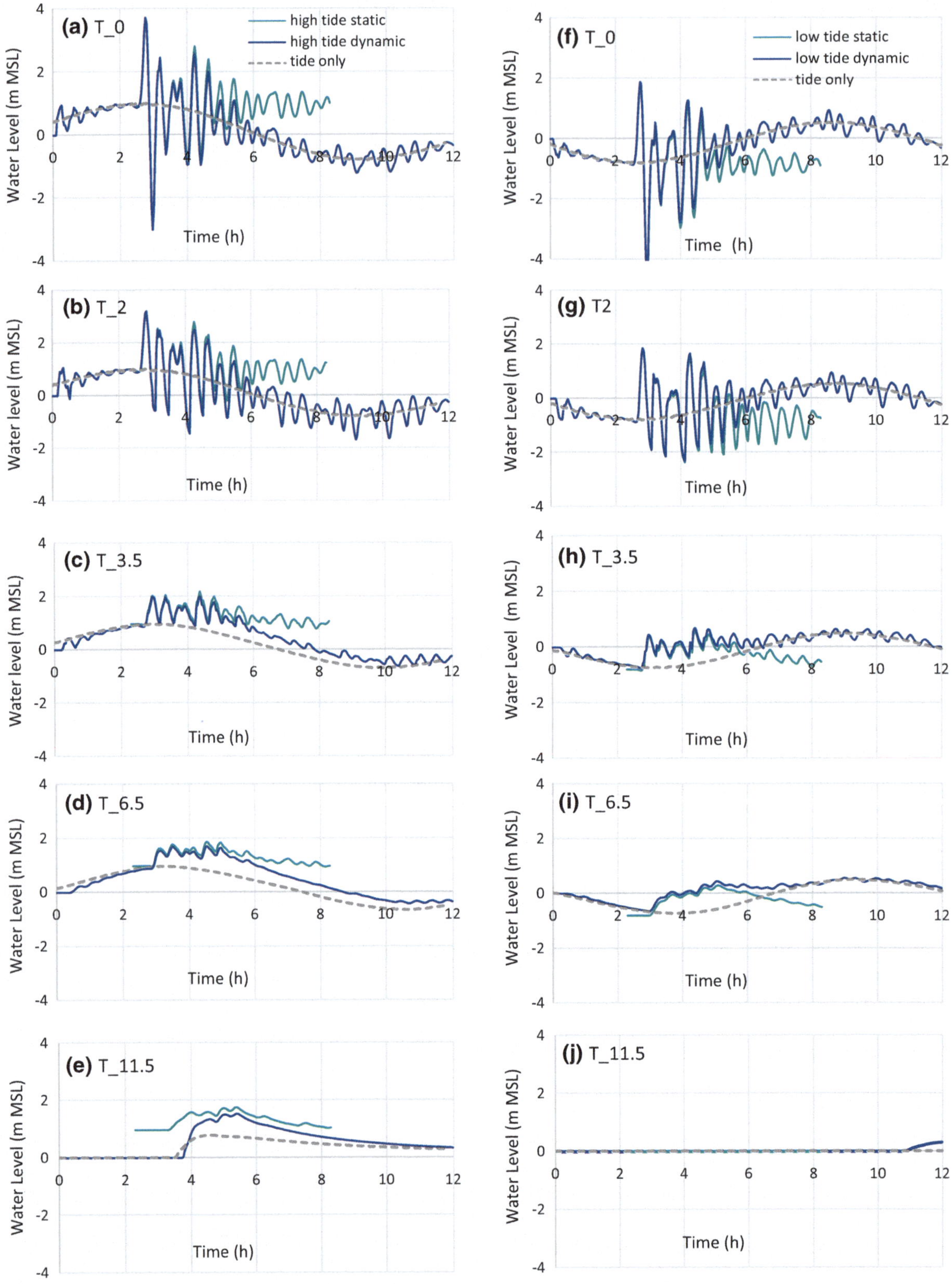
(a) T_0
high tide static
high tide dynamic
tide only
Water Level (m MSL)
Time (h)
(b) T_2
Water level (m MSL)
Time (h)
(c) T_3.5
Water level (m MSL)
Time (h)
(d) T_6.5
Water Level (m MSL)
Time (h)
(e) T_11.5
Water Level (m MSL)
Time (h)
(f) T_0
low tide static
low tide dynamic
tide only
Water Level (m MSL)
Time (h)
(g) T2
Water Level (m MSL)
Time (h)
(h) T_3.5
Water Level (m MSL)
Time (h)
(i) T_6.5
Water Level (m MSL)
Time (h)
(j) T_11.5
Water Level (m MSL)
Time (h)

◀Figure 6
Port hacking water level time series data showing tide and tsunami propagation at specified locations. Locations are denoted by T_rkm on the panels and provided on the map in Fig. 1d. **a–e** Show water level time series for high tide models. **f–j** Show water level time series for low tide models

tsunami wave propagating without a constriction at low tide, with some reduction in amplitude. Figure 8b shows deformation of the initial wave when encountering the largest CSA reduction (78%CSA at MSL) at low tide. The wave peaks, and a smaller wave is able to propagate through the constriction.

Figure 8d and e show the time series of the sequence of tsunami waves modelled at low tide at location T_D, seaward/downriver of the constriction location (Fig. 2a), for models with no constriction (Fig. 8d) and models with 78% constriction (when measured at MSL) (Fig. 8e). Figure 8f and 8g show the same models at location T_U, upriver of and therefore after propagating through the constriction.

3.4.1 Wave Height Reduction and Water Accumulation

When the tsunamis encounter a channel constriction, the positive wave amplitudes increase, some energy is reflected and some is able to propagate through the channel (Fig. 8b). Wave height reduction across the constriction is measured by comparing the wave height of the first wave in the tsunami sequence, both downstream (at location T_D) and upstream (T_U) of the constriction. Wave height H, is measured from peak to trough.

A reduction in wave heights across the constriction location (as measured from T_D to T_U) is evident for all idealised sites. At location T_U, for the idealised site with no constriction at low tide, the highest water levels occur at the 3rd and 4th wave peaks (Fig. 8f). Dynamic and static tide models reach similar water levels (1.5 m for the 3rd peak and 1.6 m for the 4th). Some water accumulation occurs upriver of the constriction: the water level of the trough of the 8th wave for the static tide model is − 0.6 m, a 0.2 increase from the initial low tide static level of − 0.8 m (Fig. 8f).

The reduction in wave heights across the constriction is more significant for a 78% constriction and is greater for waves occurring earlier in the wave sequence (Fig. 8g). The highest water levels of the dynamic tide model occur at the 7th wave peak (0.9 m). The highest water levels of the static tide model occur at the 6th wave peak (0.6 m). Water accumulation leads to a 0.8 m increase in wave trough levels for the static tide model (Fig. 8g).

Wave height reduction for the first tsunami wave across the constriction (as measured from T_D to T_U), is compared for all models in Fig. 8c. The data is normalised to a percentage reduction of wave height for comparison with all models.

The trend lines show that the wave height reduction across the constriction increases exponentially as the channel becomes more constricted. An increase in %CSA reduction from 15 to 20% leads to almost no change in %H reduction. An increase from 80 to 85 %CSA reduction, however, may result in a > 5% increase in %H reduction. Port hacking results fit well to the idealised site trend lines, whereas Sydney harbour results show a lower %H reduction than expected from the idealised site trend lines. Idealised sites are based on port hacking dimensions including the constriction rkm, so this is expected. Among other variables, the measured Sydney harbour constriction is much further upriver than the port hacking constriction.

3.4.2 Inundation Distribution

Wave changes over time are shown in Fig. 9a–c) for static low tide models with no constriction (idealised) (Fig. 9a), 78% constriction (idealised) (Fig. 9b), and port hacking (Fig. 9c). Sequential timesteps show how constrictions cause downriver/seaward wave amplification and reflection.

The wave reflection caused by the constrictions reduces tsunami impact upriver. Consequently, inundation distribution is dependent on the degree of constriction. Considering inundation as occurring either upriver or downriver of the constriction (see Fig. 10a) shows the effects of constriction on inundation distribution with upriver inundation decreasing with increasing %CSA reduction.

The trendline in Fig. 10b shows that differences in upriver inundation for sites with constrictions with less than a 60% reduction in CSA are small when compared to sites with a CSA reduction of > 60%: an

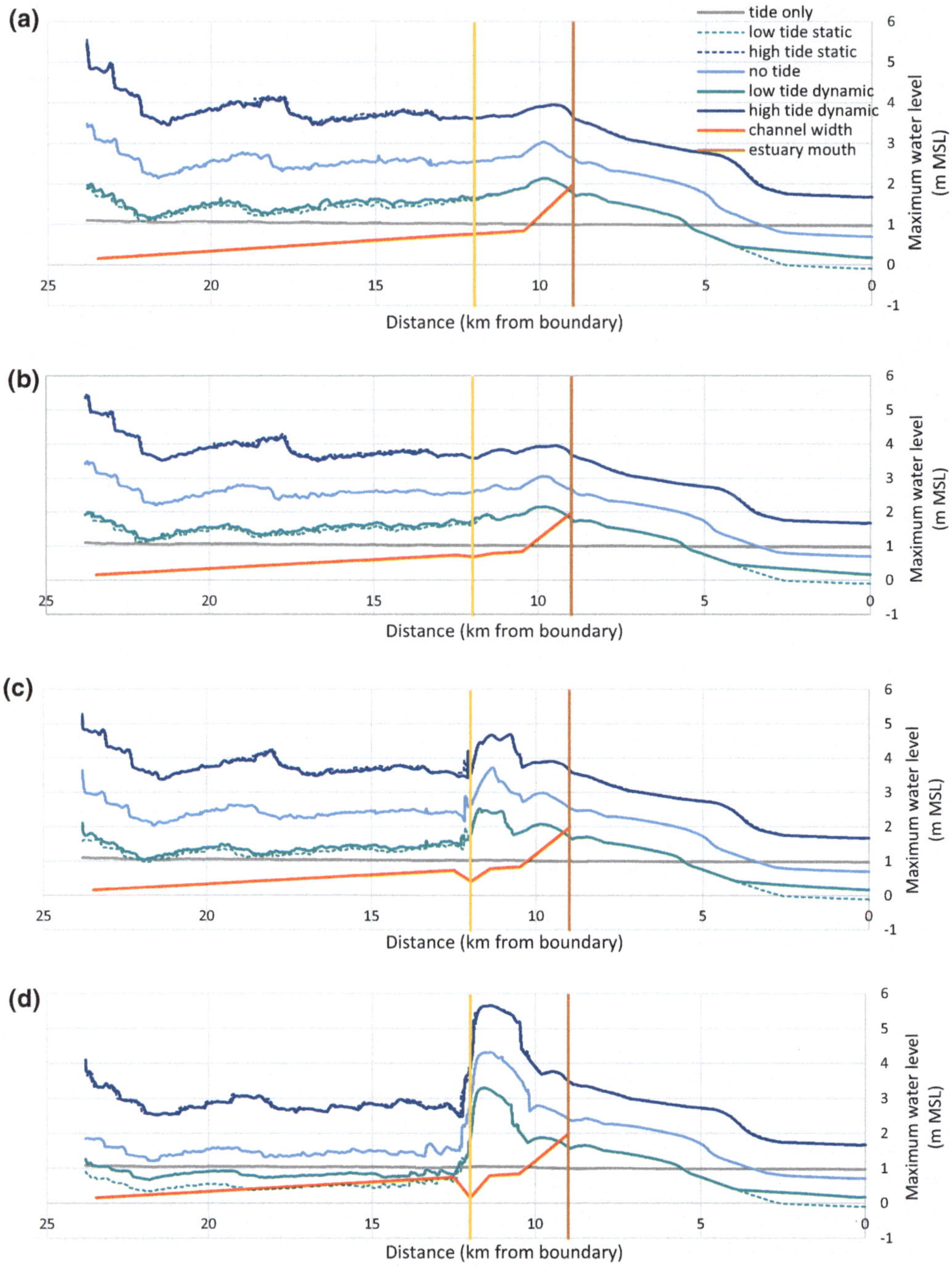

Figure 7

Maximum water levels along the thalweg and centre line extending to the boundary for the idealised site models: **a** maximum water level along the thalweg and centre line for idealised models with no constriction; **b** maximum water level along the thalweg and centre line for idealised models with 12% CSA reduction at MSL; **c** maximum water level along the thalweg and centre line for idealised models with 45% CSA reduction at MSL; **d** maximum water level along the thalweg and centre line for idealised models with 78% CSA reduction at MSL. The thalweg and centre line path is provided in Fig. 2 with details of idealised site model setup and elevation variables

increase in %CSA reduction from 23 to 57% resulted in an upriver inundation volume decrease of 6%, whereas an increase of %CSA reduction from 57 to 90% resulted in an upriver inundation volume decrease of 38%. For these same sites, downriver

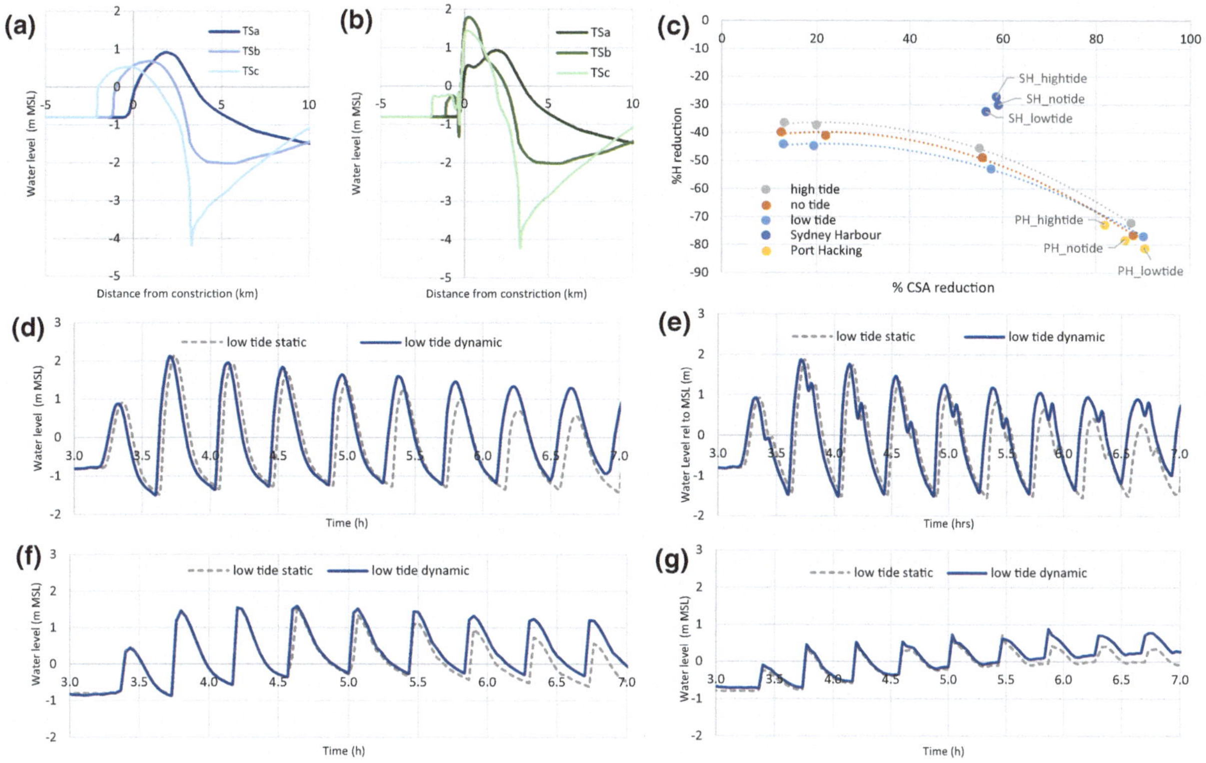

Figure 8
Tsunami wave propagation through geomorphological constriction: **a** tsunami wave propagation along the thalweg for the idealised low tide model with no constriction showing water levels for three time steps, TSa, TSb and TSc. TSa represents the tsunami at 21 min 20 s, TSb at 24 min 20 s, TSc at 27 min. The x axis shows distance from constriction, whereby positive values are distance downriver/seaward; **b** tsunami wave propagation along the thalweg for the idealised low tide model with 78% constriction showing the three consecutive timesteps, TSa, TSb and TSc as per **a**; **c** %H reduction across the constriction for all models plotted against %CSA reduction (see Sect. 2.1); **d** time series at T_D (immediately downriver/seaward of the constriction) for idealised models with no constriction; **e** time series at location 'T_D' for idealised models with 78% constriction; **f** time series at location 'T_U' (upriver of constriction location) for idealised models with no constriction; **g** time series at location 'T_U' for idealised models with 78% constriction

inundation remained at similar volumes, resulting in the total inundation volume decreases (Fig. 10b).

4. *Discussion*

The use of static tide levels in tsunami modelling is common practise and this paper shows that, in drowned river valley estuaries, static tides do not always provide the most conservative results when compared to models using dynamic tides.

At the estuary mouths of the two study sites examined in this paper, Sydney harbour and port hacking, differences between water level maxima for static and dynamic tide models are negligible, however, they become much more important within the estuaries (Figs. 4, 5). This is in agreement with other studies which show shallow channels to have more significant tsunami–tide interactions than the open coast (e.g., Kowalik and Proshutinsky 2010; Zhang et al. 2011)

Estuary channel constrictions due to narrowing geomorphology at the two study sites examined in this paper affect tide and tsunami propagation upriver where differences between models using static or dynamic tides increase. Differences are most relevant and pronounced for low tide models (Figs. 4d, 5d), when the cross-sectional area of the constriction is at its smallest. Wave breaking effects would also be greatest when estuary depths are at their shallowest

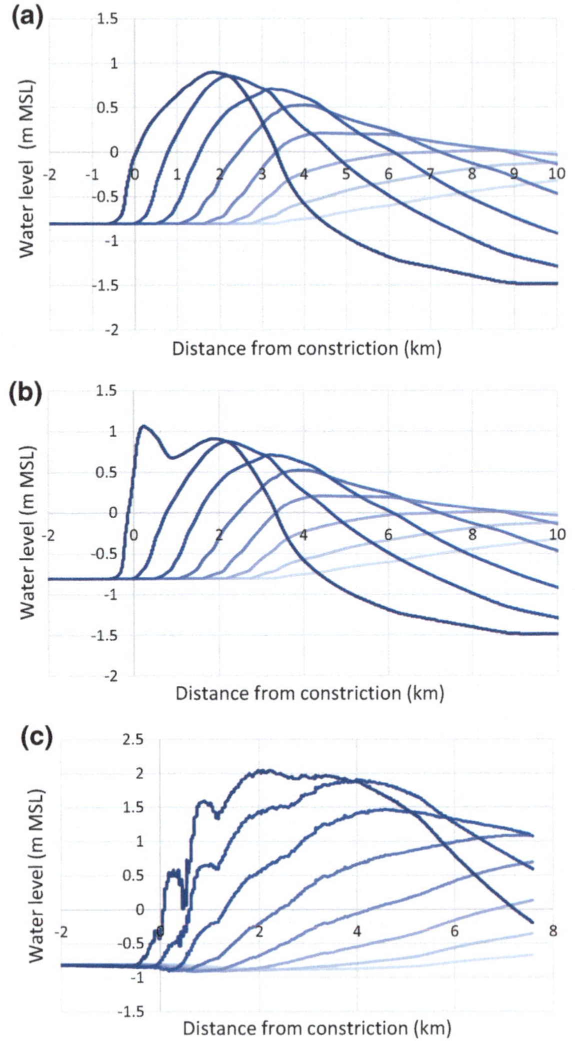

Figure 9
Wave amplification and reflection caused by constrictions: **a** water level along the thalweg for the idealised static low tide model with no constriction, for eight consecutive model timesteps, at time intervals of 120 s, with increasing line darkness; **b** water level along the thalweg for the static low tide idealised model with 78% constriction; **c** water level along the thalweg for static low tide port hacking model

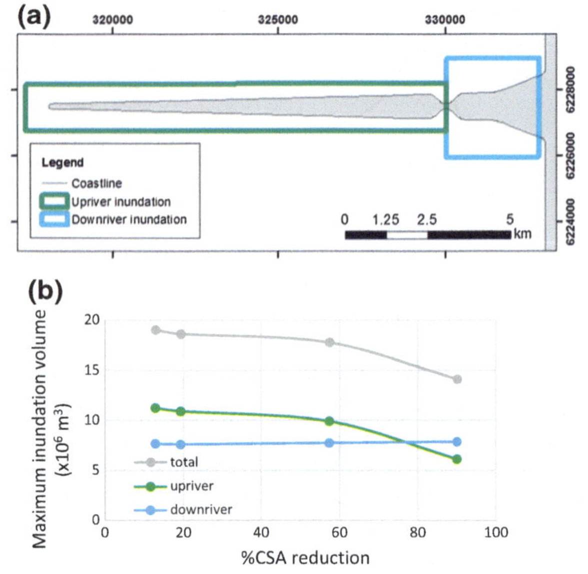

Figure 10
Inundation distribution of idealised models at port hacking: **a** inundation categorisation for idealised models into upriver and downriver inundation; **b** inundation distribution for idealised static tide models

and the region of the constriction may experience non-linear tsunami–tide interactions similar to those shown by Didenkulova and Pelinovsky (2019) for waves encountering a vertical wall.

For dynamic tide models where the tsunami coincides with low tide, the rising water levels allow the latter waves of the tsunami to propagate through the constriction with reduced energy loss. Thus, dynamic low tide models generally show higher maximum water levels upriver than static low tide models (Figs. 4d, 5d). This effect of a rising tide is apparent at all tide levels other than the highest tides, with dynamic rising tide models showing greater maximum inundation volumes than equivalent static and dynamic falling tide models (Figs. 4a, 5a).

Modelling with a hypothetical single wave peak tsunami was completed to isolate the effects of tide-tsunami phasing on the leading and largest wave in the tsunami wave train. These models showed rising and falling dynamic tide port hacking models to have minimal differences in maximum inundation volume (Fig. 5b). This indicates that the differences in maximum inundation volumes that occur in this study between rising and falling dynamic tide models are largely due to the impacts of waves that follow the leading (largest) tsunami wave. For other tsunamis with the largest wave occurring later in the wave sequence, the differences that occur between dynamic rising and falling tide models may be greater. These differences between static and dynamic tide models that occur over time typically become apparent ~ 1–2 h after the arrival of the leading (largest) tsunami wave (Fig. 6).

The effects of bathymetric constrictions on tide and tsunami propagation in riverine estuaries are a major contributor to the differences observed between static and dynamic tide models. To further investigate the effects of bathymetric constrictions, idealised sites with a constriction variable were modelled. This allowed insights into upriver wave propagation and provided a proxy for the potential sediment erosion that may occur during a tsunami. These models showed that bathymetric constrictions affect downriver maximum water levels, the degree of tsunami wave height reduction across the constriction, the degree of upriver water accumulation, as well as inundation volumes and distributions.

Wave height attenuation occurs with propagation upriver due to friction, and result show that this attenuation disproportionately affects wave trough amplitudes (Figs. 6, 8). When tsunami waves encounter a channel constriction, the wave shoals, some energy is reflected and some is able to propagate through the constriction (Fig. 8b). The degree of wave height attenuation (%H reduction) across the constriction is shown to be exponentially correlated to the degree of constriction as measured by the % reduction in estuary cross-sectional area (Fig. 8c).

The effects of constrictions are therefore most apparent when considering the differences between static and dynamic tide models for the idealised models with the most significant constriction (78% CSA constriction when measured at MSL). Differences between static and dynamic tide models are greatest at low tide, with dynamic tide models showing higher maximum water levels and maximum inundation volumes. This occurs because static low tide models show consistently high wave height reduction for the entire sequence of tsunami waves encountering the constriction. Dynamic low tide models show high wave height reduction for tsunami waves arriving earlier in the sequence, however, wave height reduction is reduced for waves occurring later in the wave sequence as the tide rises and constriction effects are reduced (Fig. 8e, g).

The tsunamis modelled in this study also cause upriver water accumulation, which is consistent with real world observations (e.g. Abe 1986). All models from this study show that water accumulates upriver throughout the duration of the tsunami. Constriction effects enhance upriver water accumulation with upriver water levels increasing to higher levels for models with greater channel constrictions. Static and dynamic tide models show similar rates of upriver water accumulation, however in the case of low tide models, the rising tide of dynamic models increases upriver water levels (Fig. 8e, g).

The effects of both water accumulation and wave height reduction are most apparent for the idealised models with the most significant constriction at low tide. For these models, the cumulative effects of a higher rate of water accumulation when compared to models with less constriction cause maximum upriver water levels to occur later in the model duration (Fig. 8g). By this time in the model duration, the tide in dynamic tide models has risen and larger waves are able to propagate through the constriction accentuating the differences between static and dynamic low tide model results (Fig. 8g). Therefore dynamic low tide models show higher maximum water levels upriver of constrictions than static low tide models (Figs. 4a, 5a, 7d).

The presence of constrictions influence both total inundation and inundation distribution. By reducing the energy that can propagate through an estuary, constrictions decrease the degree of upriver inundation (Fig. 10b). With the idealised elevation datasets modelled, constrictions also decrease the total inundation volumes. It is notable that significant reflection occurs at the constriction sites (Fig. 9a–c). At locations with more complex geomorphology, and different angles of incidence, this reflection may have additional effects on downriver inundation.

The results from idealised models with a constriction variable also provide a proxy for erosion. The location of the dominant constriction in port hacking (Fig. 1d) coincides with the position of the flood tide delta and is primarily composed of marine sands (Roy 1998; Troedson 2015; Masselink and Hughes 2003). Should this constriction be eroded and the channel widened by a tsunami, upriver inundation would increase.

Numerous studies have shown erosion by tsunamis to be substantial and extensive (Mori and Takahashi 2012; Paris et al. 2009; Scheucher and Vortisch 2010; Tanaka et al. 2012). Validated models using tsunamis with the first wave as a largest, as is the case with the tsunami used in this study, show that sediment may be suspended with the first tsunami wave and not deposited until after the tsunami has ceased (Ranasinghe et al. 2013). The marine sand of the flood tide deltas of Sydney harbour and port hacking would be vulnerable to erosion if exposed to a tsunami. Elevation datasets show that the flood tide deltas of Sydney harbour and port hacking create some degree of constriction to the estuary channels. In particular, the shallow, marine sand flood tide delta of port hacking coincides with the dominant constriction examined by this study. If port hacking's flood tide delta were severely eroded by a tsunami, the results of this study would suggest that the constriction effects caused by the flood tide delta would be reduced, causing tsunami impacts to increase upriver.

In contrast to port hacking, the Sydney harbour flood tide delta is situated at 5 rkm and, if eroded, would also increase the cross-sectional area of the channel, however due to the wide channel at this location it is not the most impactful constriction (Fig. 1c). Effects of erosion and the increase to river cross-sectional area at either the flood tide delta or the constriction at 10 rkm would also potentially increase upriver tsunami impacts. Since the constrictions in Sydney harbour are not as significant as those in port hacking, the potential increased tsunami impact due to erosion of estuarine constrictions not anticipated to be as significant in Sydney harbour when compared to port hacking.

A major limitation of the majority of tsunami modelling is that a static elevation dataset is assumed and tsunami induced erosion and deposition is unaccounted for. This study shows clearly that erosion to morphological constrictions, which causes an increase in river cross-sectional area may have a significant impact on upriver tsunami impacts. The implications of the static elevation assumption may therefore be particularly consequential for drowned river valleys with channel constrictions vulnerable to erosion. In these cases, modelling results may significantly underestimate potential inundation upriver of constrictions. For emergency management where conservative results are most necessary, this is a particularly important finding.

One limitation of the model used in this study is the use of the non-linear shallow water wave equations. These equations have been widely used and are validated for tsunami modelling (Jakeman et al. 2010; Jakir et al. 2015; Nielsen et al. 2005; Wilson et al. 2018), however, they best represent an unbroken wave (Horrillo et al. 2015). Results from this study show that tsunami waves would likely break and propagate upriver as a bore (Fig. 8f, g). Modelling the propagation of tsunamis that have broken and evolved into bores is challenging and requires the consideration of dispersive effects and energy dissipation in addition to nonlinear effects (e.g. Tissier et al. 2011; Grue et al. 2008). Comparison of numerically modelled tsunami waves with laboratory standards have shown that, in the case of breaking waves, the maximum wave amplitude model errors for a selection of non-linear shallow water wave equation models, are between 5 and 12%. Model errors for non-breaking maximum wave amplitudes were between 2 and 10%. The limitations of using the non-linear shallow water wave equations are therefore greater upriver, once the tsunami waves have broken (Horrillo et al. 2015).

A further limitation to this study is the interpolation conducted to compile the underlying elevation datasets used for Sydney harbour and port hacking as the data coverage was not complete. Thus, there may be influential bathymetric features that were excluded from the elevation datasets, however, interpolation was kept to a minimum and the Wilson and Power (2018) datasets used in this study were the best available at the time of research.

The idealised models are limited by the simplification of estuary depths, the waveforms of the tsunamis modelled, and the absence of river flow. The constant estuary depth of the idealised datasets, removes the bed slope variable known to affect phenomenon such as the location of the highest maximum water levels and the reach of upriver tsunami propagation (Tolkova 2018; Tanaka et al. 2008).

This study did not consider a range of tsunamis. Variations in the tsunami waveforms could therefore influence results, in particular, wave sequences where the largest wave does not occur as the first wave.

The absence of river flow in these models may be acceptable for Sydney harbour and port hacking, which have small catchment areas (Masselink and Hughes 2003), however, care should be taken when extrapolating to river-type estuaries in general, which may have more influential river flow. Further, as this study only involved two drowned river valley study sites, extrapolation to other estuary types may not hold.

This study supports existing literature that emphasises the importance of tsunami–tide interactions in shallow, riverine estuaries. The limitations of static tide models are most significant upriver of major morphological constrictions that affect tsunami and tide propagation, and > 1–2 h after tsunami arrival. Considering a decrease in an estuary constriction as a proxy for tsunami-induced erosion, for example, representing erosion of the flood tide deltas present in most drowned river valleys, this paper has shown that static elevation datasets have the potential underestimate tsunami impacts upriver. These results are consequential for tsunami modelling used for emergency management and should be considered in future research.

5. Conclusions

Static tide models offer a simpler tsunami model than dynamic tide models, whilst still providing usable results. Here we show that, for high tides, static tide models can be considered suitable, with only marginal differences in upriver impacts and more conservative results over time, thus making them effective for emergency management purposes. For tidal stages other than high tide, static tide models should not be considered as necessarily more conservative than dynamic tide models. Static tide models for tide levels other than high tide are shown to be reliable for estuary mouths prior to wave propagation across major morphological constrictions, and only for the first 1–2 h after tsunami arrival.

The effects of morphological constrictions on tsunami impacts in riverine estuaries are significant. Morphological constrictions are shown to affect wave heights, water accumulation upstream of the constriction, and inundation extents. The more significant the constrictions, the greater the impacts on tsunami behaviour. Constrictions that are vulnerable to tsunami induced erosion, such as the flood tide deltas of drowned river valleys, should be considered as potential gateways to increased upriver inundation should they be significantly eroded during a major tsunami event. With typical tsunami models using a static elevation dataset, erosion is unaccounted for and thus care should be taken when interpreting results for locations with constrictions comprised of sediment vulnerable to erosion.

Acknowledgements

We have gratefully received data and assistance from Geoscience Australia (GA), the Bureau of Meteorology (BOM), the NSW Department of Primary Industries and Environment (DPIE), the NSW Manly Hydraulics Laboratory and the Land Processes Distributed Active Archive Center (LP DAAC). In particular, we thank Jane Sexton and Gareth Davies of GA and Diana Greenslade and Stewart Allen of BOM. Aaron Scott and the Academic and Research Computing Services team at the University of Newcastle have been instrumental with IT and model setup. The authors also thank the editor and two reviewers for their helpful and constructive feedback. KMW is supported by a University of Newcastle Faculty of Science Strategic Scholarship (50:50).

Appendix 1

See Table 1.

Table 1

Model scenarios

Model	Tsunami	Tide	Elevation dataset
SH_d_neg079	T2 P9.0 M_W	Dynamic − 0.79 m low	Sydney harbour
SH_d_neg064	T2 P9.0 M_W	Dynamic − 0.64 m rising	Sydney harbour
SH_d_neg044	T2 P9.0 M_W	Dynamic − 0.44 m rising	Sydney harbour
SH_d_neg014	T2 P9.0 M_W	Dynamic − 0.14 m rising	Sydney harbour
SH_d_023	T2 P9.0 M_W	Dynamic 0.23 m rising	Sydney harbour
SH_d_060	T2 P9.0 M_W	Dynamic 0.60 m rising	Sydney harbour
SH_d_086	T2 P9.0 M_W	Dynamic 0.86 m rising	Sydney harbour
SH_d_097	T2 P9.0 M_W	Dynamic 0.97 m high	Sydney harbour
SH_d_088	T2 P9.0 M_W	Dynamic 0.88 m low	Sydney harbour
SH_d_061	T2 P9.0 M_W	Dynamic 0.61 m falling	Sydney harbour
SH_d_022	T2 P9.0 M_W	Dynamic 0.22 m falling	Sydney harbour
SH_d_neg019	T2 P9.0 M_W	Dynamic − 0.19 m falling	Sydney harbour
SH_d_neg054	T2 P9.0 M_W	Dynamic − 0.54 m falling	Sydney harbour
SH_d_neg075	T2 P9.0 M_W	Dynamic − 0.75 m low	Sydney harbour
SH_s_neg079	T2 P9.0 M_W	Static − 0.79 m low	Sydney harbour
SH_s_neg064	T2 P9.0 M_W	Static − 0.64 m	Sydney harbour
SH_s_neg044	T2 P9.0 M_W	Static − 0.44 m	Sydney harbour
SH_s_neg014	T2 P9.0 M_W	Static − 0.14 m	Sydney harbour
SH_s_023	T2 P9.0 M_W	Static 0.23 m	Sydney harbour
SH_s_060	T2 P9.0 M_W	Static 0.60 m	Sydney harbour
SH_s_061	T2 P9.0 M_W	Static 0.61 m	Sydney harbour
SH_s_086	T2 P9.0 M_W	Static 0.86 m	Sydney harbour
SH_s_097	T2 P9.0 M_W	Static 0.97 m high	Sydney harbour
SH_s_neg019	T2 P9.0 M_W	Static − 0.19 m	Sydney harbour
SH_s_neg054	T2 P9.0 M_W	Static − 0.54 m	Sydney harbour
SH_s_neg075	T2 P9.0 M_W	Static − 0.75 m	Sydney harbour
SH_tideonly	None	Dynamic tide cycle	Sydney harbour
SH_waveonly	T2 P9.0 M_W	None	Sydney harbour
PH_d_neg079	T2 P9.0 M_W	Dynamic − 0.79 m low	Port hacking
PH_d_neg064	T2 P9.0 M_W	Dynamic − 0.64 m rising	Port hacking
PH_d_neg044	T2 P9.0 M_W	Dynamic − 0.44 m rising	Port hacking
PH_d_neg014	T2 P9.0 M_W	Dynamic − 0.14 m rising	Port hacking
PH_d_023	T2 P9.0 M_W	Dynamic 0.23 m rising	Port hacking
PH_d_060	T2 P9.0 M_W	Dynamic 0.60 m rising	Port hacking
PH_d_086	T2 P9.0 M_W	Dynamic 0.86 m rising	Port hacking
PH_d_097	T2 P9.0 M_W	Dynamic 0.97 m high	Port hacking
PH_d_088	T2 P9.0 M_W	Dynamic 0.88 m low	Port hacking
PH_d_061	T2 P9.0 M_W	Dynamic 0.61 m falling	Port hacking
PH_d_022	T2 P9.0 M_W	Dynamic 0.22 m falling	Port hacking
PH_d_neg019	T2 P9.0 M_W	Dynamic − 0.19 m falling	Port hacking
PH_d_neg054	T2 P9.0 M_W	Dynamic − 0.54 m falling	Port hacking
PH_d_neg075	T2 P9.0 M_W	Dynamic − 0.75 m low	Port hacking
PH_s_neg079	T2 P9.0 M_W	Static − 0.79 m low	Port hacking
PH_s_neg064	T2 P9.0 M_W	Static − 0.64 m	Port hacking
PH_s_neg044	T2 P9.0 M_W	Static − 0.44 m	Port hacking
PH_s_neg014	T2 P9.0 M_W	Static − 0.14 m	Port hacking
PH_s_023	T2 P9.0 M_W	Static 0.23 m	Port hacking
PH_s_060	T2 P9.0 M_W	Static 0.60 m	Port hacking
PH_s_061	T2 P9.0 M_W	Static 0.61 m	Port hacking
PH_s_086	T2 P9.0 M_W	Static 0.86 m	Port hacking
PH_s_097	T2 P9.0 M_W	Static 0.97 m high	Port hacking
PH_s_neg019	T2 P9.0 M_W	Static − 0.19 m	Port hacking
PH_s_neg054	T2 P9.0 M_W	Static − 0.54 m	Port hacking
PH_s_neg075	T2 P9.0 M_W	Static − 0.75 m	Port hacking
PH_tideonly_petr	None	Dynamic tide cycle leading peak	Port hacking

Table 1 *continued*

Model	Tsunami	Tide	Elevation dataset
PH_tideonly_trpe	None	Dynamic tide cycle leading trough	Port hacking
PH_notide	T2 P9.0 M_W	None	Port hacking
ID_d_097_nocon	Idealised peak trough	Dynamic high tide level 0.97 m	Idealised no constriction
ID_d_neg080_nocon	Idealised peak trough	Dynamic low tide level − 0.8 m	Idealised no constriction
ID_s_097_nocon	Idealised peak trough	Static high tide level 0.97 m	Idealised no constriction
ID_s_neg080_nocon	Idealised peak trough	Static low tide level − 0.8 m	Idealised no constriction
ID_s_notide_nocon	Idealised peak trough	None	Idealised no constriction
ID_s_tideonly_nocon	None	Dynamic tide cycle	Idealised no constriction
ID_d_097_con12	Idealised peak trough	Dynamic high tide level 0.97 m	Idealised 12% constriction
ID_d_neg080_con12	Idealised peak trough	Dynamic low tide level − 0.8 m	Idealised 12% constriction
ID_s_097_con12	Idealised peak trough	Static high tide level 0.97 m	Idealised 12% constriction
ID_s_neg080_con12	Idealised peak trough	Static low tide level − 0.8 m	Idealised 12% constriction
ID_s_notide_con12	Idealised peak trough	None	Idealised 12% constriction
ID_s_tideonly_con12	None	Dynamic tide cycle	Idealised 12% constriction
ID_d_097_con45	Idealised peak trough	Dynamic high tide level 0.97 m	Idealised 45% constriction
ID_d_neg080_con45	Idealised peak trough	Dynamic low tide level − 0.8 m	Idealised 45% constriction
ID_s_097_con45	Idealised peak trough	Static high tide level 0.97 m	Idealised 45% constriction
ID_s_neg080_con45	Idealised peak trough	Static low tide level − 0.8 m	Idealised 45% constriction
ID_s_notide_con45	Idealised peak trough	None	Idealised 45% constriction
ID_s_tideonly_con45	None	Dynamic tide cycle	Idealised 45% constriction
ID_d_097_con78	Idealised peak trough	Dynamic high tide level 0.97 m	Idealised 78% constriction
ID_d_neg080_con78	Idealised peak trough	Dynamic low tide level − 0.8 m	Idealised 78% constriction
ID_s_097_con78	Idealised peak trough	Static high tide level 0.97 m	Idealised 78% constriction
ID_s_neg080_con78	Idealised peak trough	Static low tide level − 0.8 m	Idealised 78% constriction
ID_s_notide_con78	Idealised peak trough	None	Idealised 78% constriction
ID_s_tideonly_con78	None	Dynamic tide cycle	Idealised 78% constriction

References

Abe, K. (1986). Tsunami propagation in rivers of the Japanese Islands. *Continental Shelf Research, 5*(6), 655–677.

AIDR: Australian Institute of Disaster Resilience. (2018). Australian Disaster Resilience Tsunami hazard modelling guidelines.

Aubrey, D. G., & Spear, P. E. (1985). A study of non-linear tidal propagation in shallow inlet/estuarine systems Part I: Observations. *Estuarine, Coastal and Shelf Science, 21*(2), 185–205.

Beccari, B. (2009). Measurements and Impacts of the Chilean Tsunami of May 1960 in New South Wales, Australia. Report by NSW SES, State Headquarters.

Cardno. (2013). NSW Tsunami Inundation Modelling and Risk Assessment. NSW State Emergency Service and Office of Environment and Heritage Report. https://www.ses.nsw.gov.au/media/2593/nswtsunamiinundationmodellingriskassessment-report.pdf. Accessed December, 2019.

Dall'Osso, F., Dominey-Howes, D., Moore, C., Summerhayes, S., & Withycombe, G. (2014). The exposure of Sydney (Australia) to earthquake-generated tsunamis, storms and sea-level rise: A probabilistic multi-hazard approach. *Scientific Reports, 4,* 7401.

Davies, G. (2018). The 2018 Australian probabilistic tsunami hazard assessment. *Geoscience Australia Record.* https://doi.org/10.11636/Record.2018.041.

Davies, G. & Roberts, S. (2015). Open source flood simulation with a 2D discontinuous-elevation hydrodynamic model. In *Proceedings of MODSIM 2015.* 21st International Congress on Modelling and Simulation, Gold Coast, Australia, 29 Nov to 4 Dec 2015.

Didenkulova, I., & Pelinovsky, E. (2018). Tsunami wave run-up on a vertical wall in tidal environment. *Pure and Applied Geophysics, 175,* 1387–1391. https://doi.org/10.1007/s00024-017-1744-2.

Didenkulova, I., & Pelinovsky, E. (2019). Tsunami run-up on a plane beach in a tidal environment. *Pure and Applied Geophysics.* https://doi.org/10.1007/s00024-019-02332-y.

Dominey-Howes, D. (2007). Geological and historical records of tsunami in Australia. *Marine Geology, 239*(1–2), 99–123.

Greenslade, D. J. M., Simanjuntak, M. A. & Allen, S. C. R. (2009). An Enhanced Tsunami Scenario Database: T2. CAWCR Technical Report No. 014.

Grue, J., Pelinovsky, E. N., Fructus, D., Talipova, T., & Kharif, C. (2008). Formation of undular bores and solitary waves in the Strait of Malacca caused by the 26 December 2004 Indian Ocean tsunami. *Journal of Geophysical Research.* https://doi.org/10.1029/2007JC004343.

Hanslow, D. J., Morris, B. D., Foulsham, E., & Kinsela, M. (2018). A regional scale approach to assessing current and potential future exposure to tidal inundation in different types of estuaries. *Scientific Reports, 8,* 7065.

Horrillo, J., Grilli, S. T., Nicolsky, D., Roeber, V., & Zhang, J. (2015). Performance benchmarking tsunami models for

NTHMP's inundation mapping activities. *Pure and Applied Geophysics, 172,* 869–884. https://doi.org/10.1007/s00024-014-0891-y.

Jakeman, J. D., et al. (2010). Towards spatially distributed quantitative assessment of tsunami inundation models. *Ocean Dynamics, 60*(5), 1115–1138.

Jakir, M., Cummins, P., Roberts, S., & Allgeyer, S. (2015). Time reversal imaging of the tsunami source. *Pure and Applied Geophysics, 172,* 969–984.

Kowalik, Z., & Proshutinsky, A. (2010). Tsunami–tide interactions: A cook inlet case study. *Continental Shelf Research, 30*(6), 633–642.

Masselink, G., & Hughes, M. G. (2003). *Introduction to coastal processes and geomorphology.* London: Hodder Arnold.

Mori, N., & Takahashi, T. (2012). Nationwide post event survey and analysis of the 2011 Tohoku earthquake tsunami. *Coastal Engineering Journal, 54*(1), 1250001.

Myers, E. P., & Baptista, A. M. (2001). Analysis of factors influencing simulations of the 1993 Hokkaido Nansei-Oki and 1964 Alaska tsunamis. *Natural Hazards, 23,* 1–28.

Nielsen, O., Roberts, S., Gray, D., McPherson, A. & Hitchman, A. (2005). Hydrodynamic modelling of coastal inundation. In *MODSIM 2005 International Congress on Modelling and Simulation, Modelling and Simulation Society of Australia and New Zealand* (pp. 519–523). ISBN: 0-9758400-2-9.

NSW. (2017). 2017 State level emergency risk assessment. NSW Office of Emergency Management.

Paris, R., Wassmer, P., Sartohadi, J., Lavigne, F., Barthomeuf, B., Desgages, E., et al. (2009). Tsunamis as geomorphic crises: Lessons from the December 26, 2004 tsunami in Lhok Nga, West Banda Aceh (Sumatra, Indonesia). *Geomorphology, 104*(1–2), 59–72.

Ranasinghe, D., Goto, K., Takahashi, T., Takahashi, J., Wijetunge, J., Nishihata, T., et al. (2013). Numerical assessment of bathymetric changes caused by the 2004 Indian Ocean tsunami at Kirinda Fishery Harbor, Sri Lanka. *Coastal Engineering, 81,* 67–81.

Roy, P. S. (1998). Cenozoic geology of the coast and shelf (In: Scheibner E. & Basden H. (eds), Geology of New South Wales—Synthesis, Volume 2: Geological Evolution. Geological Survey of New South Wales). *Memoirs Geology, 13*(2), 385–961.

Roy, P. S., Jones, A. R., Yassini, I., Gibbs, P. J., Coates, B., West, R. J., et al. (2001). Structure and function of south-east Australian estuaries. *Estuarine, Coastal & Shelf Science, 53*(3), 351–384.

Scheucher, L., & Vortisch, W. (2010). Sedimentological and geomorphological effects of the Sumatra–Andaman Tsunami in the area of Khao Lak, southern Thailand. *Environmental Earth Sciences, 63*(4), 785–796.

Synolakis, C. E., Bernard, E. N., Titov, V. V., Kanoğlu, U., & Gonzalez, F. I. (2008). Validation and verification of tsunami numerical models. *Pure and Applied Geophysics, 65*(11–12), 2197–2228.

Takahashi, T., Shuto, N., Imamura, F., & Asai, D. (2000). Modelling sediment transport due to tsunamis with exchange rate between bed load layer and suspended load layer. In *Proceedings International Conference on Coastal Engineering 2000, Vol. 2 ASCE* (pp. 1508–1519).

Tanaka, H., Ishino, K., Nawarathna, B., Nakagawa, H., Yano, S., Yasuda, H., et al. (2008). Field investigation of disaster in Sri Lankan rivers caused by the 2004 Indian Ocean Tsunami. *Journal of Hydroscience and Hydraulic Engineering, 26*(1), 91–112.

Tanaka, H. & Tinh, N. X. (2012). The 2010 Chilean and the 2011 Tohoku tsunami waves impact to rivers in the Tohoku region, Japan. In *Proceedings of 33rd International Conference on Coastal Engineering.* 1(33).

Tanaka, H., Tinh, N., Umeda, M., Hirao, R., Pradjoko, E., Mano, A., et al. (2012). Coastal and estuarine morphology changes induced by the 2011 great east Japan earthquake tsunami. *Coastal Engineering Journal, 54*(1), 1250010.

Tissier, M., Bonneton, P., Marche, F., Chazel, F. & Lannes, D. (2011). Nearshore dynamics of tsunami-like undular bore using a fully nonlinear Boussinesq model. *Journal of Coastal Research,* SI 64 (Proceedings of the 11th International Coastal Symposium), 603-607 Szczecin, Poland, ISSN 0749-0208.

Titov, V. V., & Synolakis, C. E. (1988). Numerical modelling of tidal wave runup. *Journal of Waterway, Port, Coastal and Ocean Engineering, 124*(4), 157–171.

Tolkova, E. (2013). Tide–tsunami interaction in Columbia river, as implied by historical data and numerical simulations. *Pure and Applied Geophysics, 170*(6–8), 1115–1126.

Tolkova, T. (2018). *Tsunami propagation in tidal rivers.* Berlin: Springer. https://doi.org/10.1007/978-3-319-73287-9.

Tolkova, E., & Tanaka, H. (2016). Tsunami bores in Kitakami river. *Pure and Applied Geophysics, 173*(12), 4039–4054.

Tolkova, E., Tanaka, H., & Roh, M. (2015). Tsunami observations in rivers from a perspective of tsunami interaction with tide and riverine flow. *Pure and Applied Geophysics, 172*(3–4), 953–968.

Troedson, A. L. (2015). Coastal Quaternary geology mapping: Southern Hunter to northern Illawarra, NSW. Geological Survey of New South Wales Report GS2015/1188.

UNESCO. (2015). International Tsunami Information Centre. http://itic.ioc-unesco.org/index.php?option=com_content&view=category&layout=blog&id=1160&Itemid=1077&lang=en.

Wilson, K. M., Allen, S. C. R., & Power, H. E. (2018). The tsunami threat to Sydney Harbour, Australia: Modelling potential and historic events. *Scientific Reports, 8,* 15045.

Wilson, K. M., & Power, H. E. (2018). Seamless bathymetry and topography datasets for New South Wales, Australia. *Scientific Data, 5,* 1180115. https://doi.org/10.1038/sdata.2018.115.

Wilson, B. W., & Torum, A. (1972). *Effects of the tsunamis: An engineering study. In the great Alaska earthquake of 1964: Oceanography and coastal engineering (Committee on the Alaska Earthquake)* (pp. 361–526). Washington, DC: National Academy of Sciences.

Wright, L. D., Coleman, J. M., & Thom, B. G. (1973). Processes of channel development in a high-tide-range environment: Cambridge Gulf-Ord River delta, Western Australia. *Journal of Geology., 81,* 15–41.

Yeh, H., Tolkova, E., Jay, D., Talke, S., & Fritz, H. (2012). Tsunami hydrodynamics in the Columbia River. *Journal of Disaster Research, 7*(5), 604–608.

Zhang, Y. J., Witter, R. C., & Priest, G. R. (2011). Tsunami–tide interaction in 1964 Prince William Sound tsunami. *Ocean Modelling, 40,* 246–259.

(Received October 1, 2019, revised December 24, 2019, accepted December 27, 2019, Published online January 17, 2020)

Pure Appl. Geophys. 177 (2020), 1617–1638

https://doi.org/10.1007/s00024-019-02388-w

Pure and Applied Geophysics

The Tsunami Inundation Hazard of the Maltese Islands (Central Mediterranean Sea): A Submarine Landslide and Earthquake Tsunami Scenario Study

Christof Mueller,[1] Aaron Micallef,[2,3] Daniele Spatola,[2] and Xiaoming Wang[1]

Abstract—The tsunami hazard for the Maltese Islands is poorly defined, and historic records are available for only two recent events. Most of the population and touristic infrastructure of the archipelago is concentrated along the eastern low-lying coastline, which is exposed to tsunamis from near-field and far-field sources. In this study we present a scenario-based tsunami inundation study to assess the impact of potential significant cases. We simulated four scenarios—two submarine landslide sources (outer Malta Plateau slide and Gela Basin slide) and two earthquake sources mimicking events comparable to the 365 A.D. western Hellenic Arc event and the 1693 south-east of Sicily event. We find that all scenarios cause inundation in densely populated low-lying bays or rias of Mellieha Bay, Xemxija, Salini, Gzira, Msida, Marsaskala, St Thomas Bay, Marsaxlokk and Birzebbuga. The largest inundation extents and flow depths (> 10 m) are produced by the two landslide sources and the western Hellenic Arc earthquake. Of interest is the role of the Malta Escarpment and Sicily in amplifying and reflecting tsunami waves, and in generating consistent hot spots along the eastern coastline of Malta.

Keywords: Tsunami inundation, tsunami scenario study, earthquake, submarine landslide, Maltese Islands, Mediterranean Sea.

1. Introduction

The Maltese Islands are located in the central Mediterranean Sea (Fig. 1). The tsunami hazard for the Maltese Islands is poorly defined, and historic records are available for only two recent events (Mottershead et al. 2017). The first is the 11th January 1693 earthquake of south-east Sicily, which caused considerable damage to many structures on Malta (Galea 2007). It also generated a tsunami that was observed to have caused withdrawal of the sea and seafloor exposure at Xlendi in south Gozo (De Soldanis 1746). Mottershead et al. (2017) estimated run-up values of 1–4 m for this event. The second event is the 28th December 1908 earthquake of Messina. The impact of the tsunami generated by this earthquake was widely reported in the news (Borg et al. 2016), and witnesses from Gozo all the way to the south of Malta reported run-ups of up to 1.5 m (Baratta, 1910). At Marsaxlokk (south of Malta), the tsunami inundated the land up to 300 m inland. Two additional events of unusual sea level rise were reported on the 9th July 1973 and 26th March 1983, but these were likely generated by meteorological perturbations (Camilleri 2006).

Since no loss of life or major damage was reported during these tsunami events, a perception that the tsunami threat to Malta is insignificant has been encouraged. There is, in fact, no mention of tsunamis in Maltese law, and only reference to flooding, in relation to seasonal heavy rainfall, is made. Geomorphic evidence published in recent studies, however, is changing this perception, because it points to events of greater magnitude than any in recorded history. Evidence of extreme waves, in the form of depositional features (e.g. boulders, sediment sheets) and erosional features (e.g. sockets, scarps, scars, scoured terrain), has been documented along the eastern coast of the Maltese Islands (Mottershead et al. 2014, 2015; Biolchi et al. 2016). At a number of sites, the geomorphic features occur above the maximum storm wave threshold, up to a maximum of 22 m above sea level. The maximum storm wave height for the north-eastern Maltese Islands, based on wave records acquired between 2007 and

[1] GNS Science, Lower Hutt, New Zealand.
[2] Marine Geology and Seafloor Surveying, Department of Geosciences, University of Malta, Msida, Malta. E-mail: aaron.micallef@um.edu.mt
[3] Helmholtz Centre for Ocean Research, GEOMAR, Kiel, Germany.

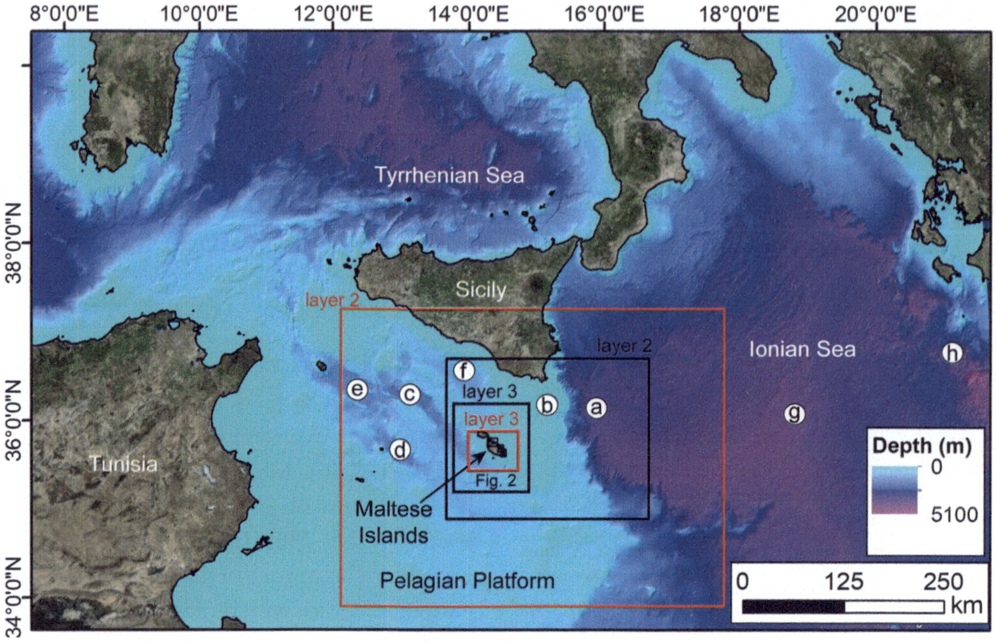

Figure 1
Location of the Maltese Islands in the central Mediterranean Sea. The boundaries for the computational domains of the DEM grids used for the earthquake (in black) and submarine landslide (in red) source scenarios are shown. **a** Malta Escarpment; **b** Malta Plateau; **c** Malta Graben; **d** Linosa Graben; **e** Pantelleria Graben; **f** Gela Basin; **g** Calabrian Arc; **h** Hellenic Arc and Trench System. Background bathymetry from EMODnet bathymetry (http://www.emodnet-bathymetry.eu)

2013, is approximately 7.5 m (Drago et al. 2013). Tsunami waves, with run-ups of up to 10 m above sea level, have thus been proposed to be responsible for the formation of these features (Mottershead et al. 2014, 2015).

Because of their location, the Maltese Islands are exposed to tsunamis from near-field and far-field sources across the entire Mediterranean region. The probability of a tsunami wave occurring in the Mediterranean Basin in the next 30 years that exceeds 1 m at the coast has been estimated to be close to certainty (Sørensen et al. 2012). Numerical modelling has shown that sources originating offshore Algeria, the western Hellenic Arc and the Calabrian Arc have the highest potential of generating a significant tsunami impacting the Maltese Islands (Tinti et al. 2005; Papadopoulos 2009). Field and marine indications of near-field and far-field tsunamis have been reported from nearby areas (e.g. Sicily) (De Martini et al. 2012; Gerardi et al. 2012; Polonia et al. 2016; San Pedro et al. 2017). The Maltese Islands have the 7th highest population density in the world and receive almost two million tourists per year. Since most of the population and touristic infrastructure are concentrated along the eastern low-lying coastline, the risk of a potential tsunami disaster is high. In view of all this, quantifying the tsunami hazard of the Maltese Islands is imperative.

In this study, we take a first step towards assessing part of the tsunami inundation hazard for Malta by simulating inundation for a small number of potential worst case scenarios, based on both earthquake and submarine landslide sources in the vicinity of the Maltese Islands. Our aim is to extend the tsunami catalogue of the Maltese Islands by incorporating extreme events that have a recurrence rate of > 300 years, and to ultimately reach a higher level of tsunami resilience in the Maltese Islands. It is important to note that the aim of this paper is to use proxies of historic sources to illustrate credible worst case scenarios rather than explain historic events. Thus, some of the source examples that we use are based on variations of parameters from published sources. This study does not intend to perform probabilistic tsunami inundation hazard assessment.

2. Regional Setting

2.1. The Maltese Islands

The Maltese Islands comprise one of the shallowest elements of the Pelagian Platform, a carbonate platform that extends from northern Africa to southern Sicily (Fig. 1). Located 95 km south of Sicily and 290 km east of Tunisia, the archipelago consists of three main islands—Malta, Gozo and Comino—and some smaller inhabited islands, for a total area of 316 km^2 and a population of $\sim$ 450,000. The Maltese Islands consist of a shallow water, Oligo-Miocene sedimentary succession (Pedley et al. 1976), which is disrupted by two generations of normal faults (Illies 1981). Tectonic activity has had a predominant control on the islands' geomorphology (Magri 2006). Because of uplift associated with the northern rim of Malta Graben to the south-west, the islands are tilted towards the north-east (Fig. 2). This explains why a low-lying rocky shoreline characterises the eastern coast, whereas high cliffs can be found along the west. The eastern coastline is interrupted by a number of inlets that coincide with either the drowned mouths of bedrock-incised river valleys (also called rias, e.g. Valletta harbours), or fault-controlled grabens (e.g. Xemxija Bay). The steeper, western coastline of the archipelago consists of either spreading slopes or vertical cliffs, which reach up to 200 m high in places.

The seafloor surrounding the Maltese Islands comprises an irregularly shaped continental shelf (Fig. 2). Towards the east, the shelf is gently sloping and its edge (at 130 m depth) is located at a distance of up to 9 km from the coast. In the northern half of the archipelago, the shelf edge is marked by a steep escarpment up to 70 m high. Towards the west, the shelf is considerably narrower and steeper, and at places the shelf edge is located a few tens of metres from the coastline. A more detailed description of the Maltese continental shelf is provided by Micallef et al. (2013) and Foglini et al. (2016).

Further offshore, the seafloor is characterised by four morphologic provinces (Fig. 1). The Maltese Islands are located at the south-western edge of the Malta Plateau, a 100–150 m deep, gently sloping asymmetric ridge (Micallef et al. 2011). The eastern boundary of the Malta Plateau and the Pelagian Platform is marked by the Malta Escarpment, which is a steep, 290 km long carbonate cliff with a relief in excess of 3 km (Micallef et al. 2019). To the west of the Maltese Islands, the Pelagian Platform is traversed by the Sicily Channel Rift Zone, which comprises three principal NW–SE trending grabens (Pantelleria, Malta, Linosa) that reach depths of 1700 m (Reuther and Eisbacher 1985; Civile et al. 2010). The Gela Basin, which is the foredeep of the Maghrebian fold-and-thrust belt, reaches depths of 900 m and is located to the north of the Maltese Islands.

2.2. Potential Tsunami Sources

2.2.1 Submarine Landslides

Known submarine landslides around the Maltese Islands include: (1) 160 small landslides in the drowned late Pleistocene palaeolandscape offshore the Maltese Islands (Micallef et al. 2013) (2) the Gela Slide in the northern Gela Basin (Trincardi and Argnani 1990); (3) 14 landslides in the eastern Gela Basin (Foglini et al. 2011; Minisini et al. 2007); (4) 48 landslides along the top of the Malta Escarpment (Micallef et al. 2016); (5) > 200 landslides across the Malta Escarpment (Micallef et al. 2019); (6) 4 landslides offshore eastern Sicily (Urgeles and Camerlenghi 2013); and (7) 7 landslides in the Ionian Basin (Urgeles and Camerlenghi 2013) (Fig. 3a). There are no historical or sedimentological records of tsunamis associated with any of these submarine landslides. Not all of these submarine landslides can cause tsunamis. A histogram of submarine landslide areas, presented in Fig. 3b, shows how > 80% of the landslides are relatively small, with an area < 30 km^2, and they are unlikely to generate regional/large-scale tsunamis. The histogram excludes the 160 small landslides in the first group because these are thought to be terrestrial landslides originated in a subaerial setting (Micallef et al. 2013).

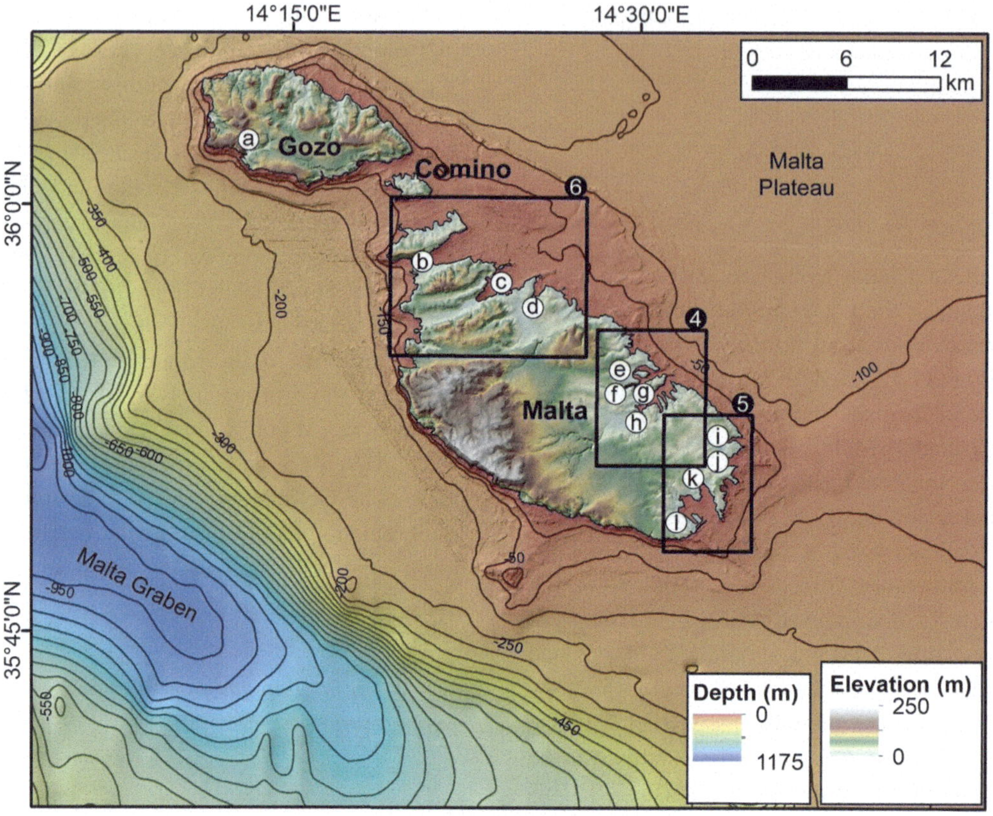

Figure 2
Topographic map of the Maltese Islands and bathymetric map of the surrounding seafloor (isobaths at 50 m intervals) for the DEM grid layer 3 used for the earthquake simulations. The boundaries of grids 4–6 are shown. Background bathymetry from EMODnet bathymetry (http://www.emodnet-bathymetry.eu). **a** Xlendi; **b** Mellieha Bay; **c** Xemxija; **d** Salini; **e** Gzira; **f** Msida; **g** Valletta; **h** Marsa; **i** Marsaskala; **j** St Thomas Bay; **k** Marsaxlokk; **l** Birzebbuga. **b–d** Computational domains 4–6

2.2.2 Earthquakes

The major earthquake hazard in the central Mediterranean region arises from the following structures (Figs. 1, 3a):

(i) The Malta Escarpment: The northern segment of the Malta Escarpment, north of Siracusa, is characterised by NNW–SSE, east dipping extensional faults and a lithospheric scale lateral slab tear between the Ionian oceanic basin and the Hyblean Plateau (Argnani and Bonazzi 2005; Gutscher et al. 2016; Polonia et al. 2016). These active faults are mechanically capable of generating 6–7 Mw earthquakes and may be responsible for some enigmatic historical earthquakes.

(ii) The Calabrian Arc: The Calabrian Arc has been characterised by subduction of remnants of the Mesozoic Tethys Ocean for the last 80 Ma (Faccenna et al. 2001). Subduction today is restricted to a 150 km zone between the Strait of Messina and Catanzaro (south Italy) and is associated with slab roll-back that started in the late Miocene (Goes et al. 2004). Although intraslab seismicity is known to occur below 40 km, regional shallow seismicity is limited in both frequency and magnitude (Maesano et al. 2017). Large historical earthquakes, such as the 1693 earthquake (Mw 7.3) in south-eastern Sicily (Gutscher et al. 2006) and the 1905 earthquake (Mw 7.0) in central Calabria (Galli 2005), have been associated with the shallow portion of the plate interface or slab.

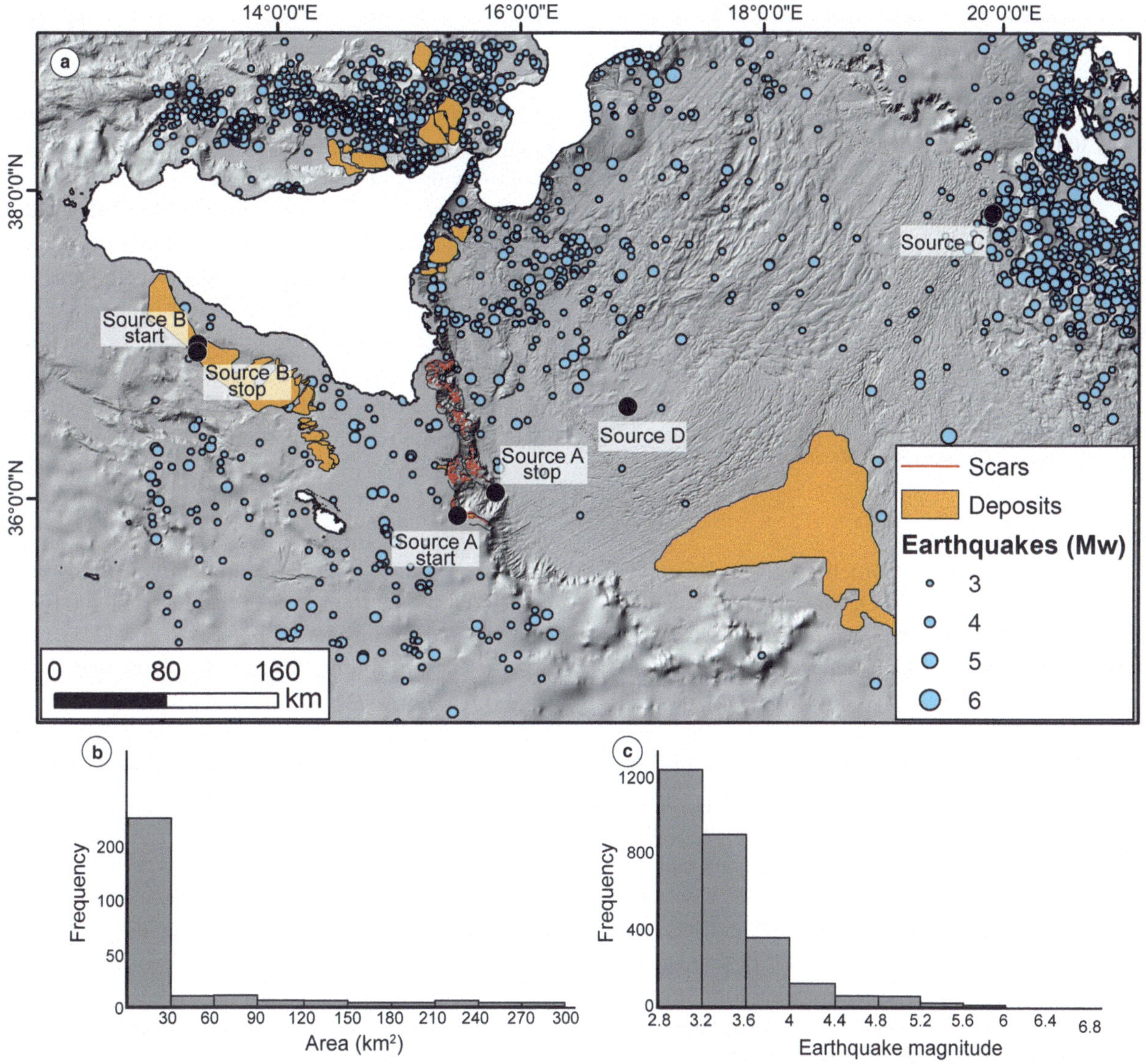

Figure 3

a Sources of tsunamis (submarine landslides and earthquakes) around the Maltese Islands. The sources used in this study are marked in black. We assume landslide mass as a semi-ellipsoid defined by its length L, width W, and thickness T. In the figure we indicate the start and end points of the sliding motion of these block models. Histograms for **b** submarine landslide area (up to 300 km^2) and **c** earthquake magnitude (source: http://cnt.rm.ingv.it)

(iii) The Hellenic Arc and Trench System: The Hellenic Arc and Trench System, and its back-arc region of the Aegean Sea, are the most seismically active region in the Mediterranean. The main ongoing processes include the westward extrusion of the Anatolia continental block along the dextral strike-slip North Anatolian Fault boundary, the NNE-ward subduction of the Eastern Mediterranean lithosphere beneath the Hellenic Arc, the subsequent SSW–NNE extension of the Aegean back-arc region, the collision of NW Greece with the Apulian block in the northern Ionian Sea, and the incipient collision with the Libyan promontory south of Crete (Papadopoulos et al. 2014). As a result of this stress build-up along the boundaries and in the interiors of the crustal blocks involved, shallow, interplate and intermediate-depth seismicity is extremely high (e.g. 365 A.D. earthquake (Mw 8.4) in the western segment of the Hellenic Arc; 1303 A.D. earthquake (Mw 8.0) in the eastern segment of the Hellenic Arc; 1956 earthquake (Mw 7.5) in the south Aegean Sea).

A histogram showing the frequency distribution of earthquake magnitude affecting the southern region of the central Mediterranean Sea is presented in Fig. 3c.

3. *Data and Methods*

3.1. Numerical Model

The tsunami model, Cornell Multi-grid Coupled Tsunami model (COMCOT), was used to calculate co-seismic displacements of earthquake sources, motions of landslides, and their resulting tsunamis. COMCOT was originally developed at Cornell University, USA in the 1990s (Liu et al. 1998) and, since 2009, it has been under development at GNS Science, New Zealand (Wang and Power 2011). COMCOT is publicly available, has been systematically validated against analytical, experimental and benchmark studies (Liu et al. 1994a, 1995; Cho 1995; Wang 2008 ; Wang and Power 2011), and has been widely used by researchers to study different aspects of tsunamis, including tsunami generation mechanism, propagation, run-up and coastal inundation (Liu et al. 1994b, 1998; Wang and Liu 2006, 2007; Wijetunge et al. 2008; Beavan et al. 2010; Mueller et al. 2015).

The model uses a modified staggered leap-frog finite difference scheme to solve linear/nonlinear shallow water equations in spherical and cartesian coordinates with improved dispersion relationship (Cho 1995; Wang 2008). To account for the shallowing of water depths and ensure enough spatial resolution in near-shore regions, a two-way nested grid configuration is implemented to use a relatively large grid spacing to efficiently simulate the propagation of tsunamis in the deep ocean and use refined grid spacing in coastal regions where high spatial resolutions are needed (Wang and Power 2011). This approach can well balance the computational efficiency and the numerical accuracy.

Multiple source mechanisms have been integrated in the tsunami simulation package of COMCOT, including subaerial/submarine landslides and earthquakes with transient rupture and/or variable slip distributions (Wang and Power 2011). Landslide tsunami is calculated through a dynamically coupled landslide-tsunami model that simulates both landslide motion and its dynamically generated tsunami (Wang et al. 2016). The landslide motion can be described by sequential snapshots of topographic variation, modelled as a cohesive block sliding with the sliding functions proposed by Watts et al. (2003) and Enet and Grilli (2007), or simulated as an avalanche-type failure (Wang et al. 2016; Mountjoy et al. 2018; Liu et al. 2018).

The use of a sliding block model in our study area is justified by the translational nature of the landslide sources under consideration (see Sect. 3.2.1). We have used the sliding function of Enet and Grilli (2007) built in COMCOT to describe the submarine landslide motions with a deceleration phase introduced at the locations where the slope merges with the flat floor. We assume landslide mass as a semi-ellipsoid defined by its length L, width W, and thickness T (Fig. 4). The edge area where the semi-ellipsoid approaches the sliding surface is smoothed through an exponential function to avoid singularities due to the discontinuity of spatial derivatives at the edge of the sliding mass. In Fig. 3 we give the start and end points of the sliding motion of these block models.

The earthquake sources considered in this study are modelled with simple rectangular geometries, with each having a length, width, dip, strike, rake and uniform slip assigned to them. These parameters have been extracted from the literature and are given in the following section. The co-seismic surface displacement is calculated using the elastic fault displacement theory of Okada (1985). The reader should note that this is the most basic geometry that can be assumed for earthquake sources. We are neglecting effects of non-uniform distribution of slip and complex geometries such as rupture surface curvatures. Non-uniform slip has been shown to have a first order effect on local tsunami wavefields (Mueller et al. 2015).

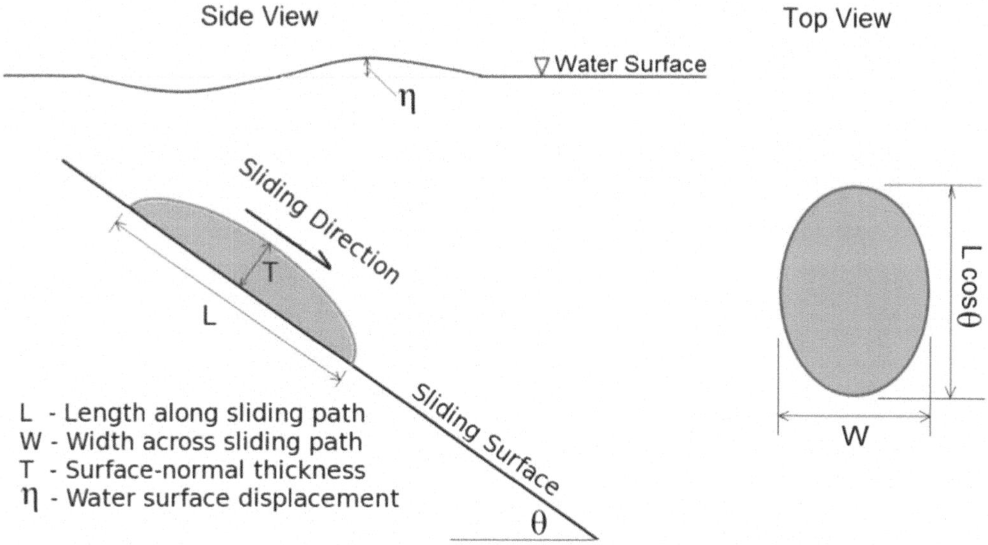

Figure 4
Sketch of the landslide block geometry used in COMCOT. We assume landslide mass as a semi-ellipsoid defined by its length L, width W, and thickness T

In this study, absorbing boundary condition (Wang and Power 2011) is applied to the four boundaries of the modelling domain to avoid potential reflections from the open sea boundaries. In linear model calculations, fully reflective wall boundaries are implemented along the coastal lines. In contrast, in non-linear model calculations, e.g. for land inundation simulations, a dynamically moving wet/dry boundary algorithm (Cho 1995; Wang and Power 2011) is applied to determine instantaneous fronts of tsunami inundation. Simulations were performed assuming current sea levels without any tide level adjustments.

3.2. Example Tsunami Source Scenarios

We have chosen four tsunami sources (A-D) to demonstrate the extent of inundation of worst case scenarios on Malta. The sources chosen here are modelled after submarine mass movement events and earthquakes that have been documented in the literature. We point out that the purpose here is not to accurately reconstruct these tsunami sources in order to explain historic or prehistoric events, but rather to use similar sources to these to give an idea of potential worst case tsunami scenarios on Malta. This approach is useful to assess potential maximum extent of inundation and define the upper end of the spectrum of possible events. We have not conducted an exhaustive search of the tsunami source parameter space to determine if our sources qualify as 'absolute' worst-cases. However, the tsunami generated by the sources chosen here are significant and are likely to fall into the upper part of the impact spectrum. Tsunami severity is very much dependent on the location where the tsunami is impacting. Thus, the definition of an 'absolute' worst-case also depends on the specific target location under consideration.

3.2.1 Submarine Landslide Sources

Source A is modelled after the largest submarine landslide reported along the outer Malta Plateau (Fig. 3; Table 1). Documented by Micallef et al. (2016) using high resolution geophysical data, the submarine landslide scar is located upslope of Cumecs Canyon. The crescent-shaped landslide scar has a length of 6.2 km, a width of 20 km, a thickness of 0.08 km, and a volume of 5.2 km^3. Because of the absence of debris within the landslide scar, all the material is assumed to have been evacuated from the scar during a single translational event. Loss of

Table 1

Landslide source attributes

	Source A	Source B
Location	Malta Plateau	Gela Basin
Longitude (°) (start–stop)	15.4639–15.7653	13.5701–13.5267
Latitude (°) (start–stop)	35.9247–36.0735	36.9790–36.9190
Average Slope (°)	9.0	0.9
Bulk Density (g/cm^3)	1.85	1.85
Length (km)	6.2	15.0
Width (km)	20	115
Thickness (km)	0.08	0.70
Volume (km^3)	5.2	632
Reference	Micallef et al. (2016)	Trincardi and Argnani (1990)

Landslides are modelled as ellipsoid block models on a sliding surface. Volumes have been approximated by calculating the volume of a half ellipsoid of the given dimensions

Table 2

Earthquake source attributes

	Source C	Source D
Location	Western Hellenic Arc	South-east of Sicily
Longitude (°)	19.84	16.83
Latitude (°)	37.79	36.62
Strike (°)	340	225
Dip (°)	35	5
Depth (km)	23.5	6.3
Width (km)	80	160
Length (km)	160	120
Slip (m)	8.9	2.0
Moment magnitude	8.4	8.1
Reference	Gailler et al. (2016)	Gutscher et al. (2006)

Earthquakes have been modelled assuming rectangular geometries with uniform distribution of slip. Depth values are given for the centre of the source plane. Rake angles are assumed to be uniform at 90 degrees (thrust fault subduction earthquake mechanism). Moment magnitude values assume a rigidity of 5×10^{11} dyne/cm^2

support at the base and seismicity are the most plausible triggers (Micallef et al. 2016).

Source B is modelled after the largest submarine landslide in the Gela Basin (Fig. 3; Table 1). First described by Trincardi and Argnani (1990) using geophysical data, the Gela Slide has a length of 15.0 km, a width of 115.0 km, a thickness of 0.7 km and a volume of 630 km^3. It has been interpreted as a slide that mobilised a sedimentary sequence within the Gela foredeep basin 600 ka BP.

3.2.2 Earthquake Sources

Source C is modelled after a strong earthquake (Mw 8.4) that occurred in 365 A.D. and that has been related to the subduction process along the Hellenic Trench system (Flemming 1978; Shaw et al. 2008) (Fig. 3; Table 2). The source of this earthquake is thought to be a reverse fault in the overriding plate (Stiros and Drakos 2006; Papadimitriou and Karakostas 2008; Shaw et al. 2008; Shaw and Jackson 2010; Stiros 2010). The earthquake generated a co-seismic displacement of shorelines in western Crete and Antikithira, reaching up to 9 m in south-west Crete (Pirazzoli et al. 1982). We have created an example source representing an earthquake in the northernmost part of the western Hellenic Arc and based on parameters taken from Gailler et al. (2016). We have chosen this source region because it generated the largest known tsunami in the eastern Mediterranean Sea and could potentially explain the geomorphic evidence of extreme waves along the eastern coast of the Maltese Islands.

Source D is modelled after the most powerful earthquake (Mw 7.4) that affected south-east Sicily in 1693, yet its source remains unresolved. The main hypotheses on the possible source fault can be summarised as follows: (1) the Scordia-Lentini graben depression between Catania and Augusta, based on morphological and geological considerations (D'Addezio and Valensise 1993); (2) the Malta Escarpment (offshore Augusta), according to seismic data interpretations (Piatanesi and Tinti 1998); (3) a subduction fault associated to a mean co-seismic slip of 2 m (Gutscher et al. 2006). We based the parameters for this source model on the description found in Gutscher et al. (2006) because the subduction zone source is described in enough detail to model a tsunami source based on it (Fig. 3; Table 2). We have chosen this source because it generated the largest tsunami documented in the Maltese Islands.

3.3. Grid Setup

The COMCOT tsunami modelling software uses a cascade of nested 'grids' constructed from bathymetric and topographic datasets to account for spatial resolution requirements when simulating a tsunami in different regions. In this study, four levels of numerical meshes at different spatial resolutions were used to simulate tsunami generation, propagation and coastal flooding (Figs. 1 and 2). We will now describe the bathymetric and topographic data sets that were used to generate the numerical grids, which were created internally in COMCOT for the simulation. The resolution of these grids (layers) are given in Table 3.

The data for the first level grid—layer 1—was taken from the GEBCO topographic and bathymetric database. The data set covers the entire Mediterranean at a spatial resolution of 1 arc-minutes and is used to simulate tsunami generation and propagation from distant sources. This grid was only active for the earthquake source simulation.

The data for the second level grid—layer 2—was derived from EMODnet bathymetry (http://www.emodnet-bathymetry.eu/) and has an original resolution of ~ 200 m. It covers the area between Pantelleria and the eastern edge of the Medina Escarpment.

For landslide simulations, the data for the first layer was also taken from the EMODnet bathymetry dataset. This was done because the landslide sources were closer to the Maltese Islands and a setup with a larger outer grid that covers the entire Mediterranean region was not required. To be consistent in the naming of the inner layers (3—6), the outer layer was called layer 2. The reader should note that this grid has a larger extent than the one used for the earthquake scenarios to be able to accommodate both landslide sources. The numerical grid resolution was increased to 0.25 arc-minutes in COMCOT for source A and to 0.1667 arc-minutes for source B to allow for appropriate resolution of the DEM in the landslide source region.

The dataset for the third level numerical grid—layer 3—covers the shallow waters around the Maltese Islands at a resolution of 10 m and was derived from bathymetric data from Micallef et al. (2013) and ERDF 156 project. The resolution for the numerical grids in the simulation was set to different resolutions in COMCOT for each scenario (Table 3). The COMCOT version we used does not allow the landslide numerical source domain to intersect with the next level inner numerical grid. As a result, layer 3 had to be reduced in size for the Gela Slide scenario (source B) (Fig. 1).

On the fourth level, three numerical grids (layers 4, 5, 6) were used to cover different areas of the Maltese Islands for both earthquake and landslide scenarios (Fig. 2). The original resolution of the bathymetric datasets vary between 1 m and 5 m (Micallef et al. 2013). The bathymetric datasets were combined with a 1 m resolution onshore grid for the entire Maltese Islands derived from LiDAR data from ERDF 156 project. A spatial resolution of about 0.01 and 0.02 arc-minutes (~ 17 to 34 m) was used for the numerical grids in COMCOT for earthquake simulations. For landslide simulations, the spatial resolution of the numerical grids was set to 0.0125 arc-minutes. Please refer to Table 3 for a summary of grid spacing used for the numerical grids.

The reader should note that the grid setup and different resolutions were chosen as a result of experimentation for the different scenarios. Higher resolution numerical grids come with an increased simulation runtime, which was initially kept short.

Table 3

Bathymetry data sources and numerical grid resolution

	Layer 1	Layer 2	Layer 3	Layer 4	Layer 5	Layer 6
Bathymetry data source	GEBCO	EMODnet	Micallef et al. (2013); ERDF 156			
Source A	–	0.25	0.0625	0.0125	0.0125	0.0125
Source B	–	0.25	0.0625	0.0125	0.0125	0.0125
Source C	2	0.5	0.1	0.02	0.02	0.02
Source D	1	0.25	0.05	0.01	0.01	0.01

Resolution given in arc-minutes

When numerical instabilities were observed, the resolution of the grid setup had to be adapted at the expense of longer overall simulation times.

4. Results

4.1. Source A: Largest Submarine Landslide on Outer Malta Plateau

A landslide on the outer Malta Plateau would generate significant inundation in many areas of Malta (Fig. 5). Due to its relief, the Malta Escarpment refracts the larger part of tsunami energy back into the comparably shallow waters to the west of the escarpment. Bathymetric features in this area create characteristic tsunami rays that channel energy towards Malta (Fig. 5a). Our inundation simulations of the inner grids show inundation with flow depths up to ~ 10 m in the low lying topographies of Mellieha Bay, Xemxija, Salini, Gzira, Msida, Marsaskala, St Thomas Bay, Marsaxlokk and Birzebbuga. We observe the common phenomenon of increased susceptibility to inundation in the funnel-shaped bays, whereas steeper coastal areas do not show any significant inundation. The first tsunami waves from this source arrive at the coast of Malta after ~ 40 min, and at the coast of Gozo ~ 5 min later (Table 4). The first arrivals will express as water retreating from the coast. We find the dominant oscillation periods of the arriving wave field to be quite different for the individual virtual tide gauges in Malta and Gozo (Fig. 5e–f). In Gozo we observe a dominant oscillation period of about 10 min, while in Valletta we measure 43 min. In other areas around the coast of Malta, oscillation periods span about 5–25 min. The reader should note that there is a large tsunami beam impacting the south of Sicily with offshore wave height larger than those observed in the vicinity of Malta. This demonstrates that a large submarine mass failure on the Malta Escarpment would also pose a significant threat to Sicily. We have not studied the inundation for this region, however, as no detailed grids had been set up.

Figure 5 ▶
Source A (submarine landslide on outer Malta Plateau). **a** Maximum tsunami wave height above mean sea level. **b–d** Inundation for different areas in Malta. **b–d** refer to the computational domains 4–6 of Fig. 2. **e** Oscillation periods and **f** frequency analyses of tsunami waves for virtual tide gauges in Malta and Gozo

4.2. Source B: Gela Slide

A tsunami generated by the Gela Slide source initially sends waves to the south-west and north-east, with the latter severely impacting the south-western Sicilian coastline (Fig. 6). While this source initially radiates most of its waves to the south-west of its location and towards Sicily, a significant part of this energy either reaches Malta directly or is reflected and scattered by the Malta Escarpment and southern Sicily. We estimate that the first wave will arrive in the northern part of Gozo after about 18 min, reaching Mellieha Bay 9 min later, and finally arriving in Valletta, Marsaskala and Marsaxlokk after a total travel time of only 30, 35 and 40 min respectively (Table 4). The tsunami would affect similar areas as the outer Malta Plateau landslide source, but would cause a much larger inundation with flow depths of 10 m and above in places (Fig. 6). The first arrivals of this source will express with rising waters and inundation as the start of the tsunami. The dominant oscillation periods of the arriving wave field are found to be around 20–25 min for all virtual tide gauges along the coast of Gozo and Malta (Fig. 6e–f).

4.3. Source C: 365 A.D. Earthquake in Western Hellenic Arc

The source simulated from the Hellenic Arc has an impact that is comparable to that of the submarine landslide on the outer Malta Plateau (source A) (Fig. 7). This is not surprising, considering that this is an earthquake of magnitude 8.4 that is highly directive towards Malta. We also observe similar tsunami rays as for the scenario for source A (Fig. 7a). This confirms that the Malta Escarpment serves as a potential amplifier of tsunamis generated along the escarpment or to its east. We observe

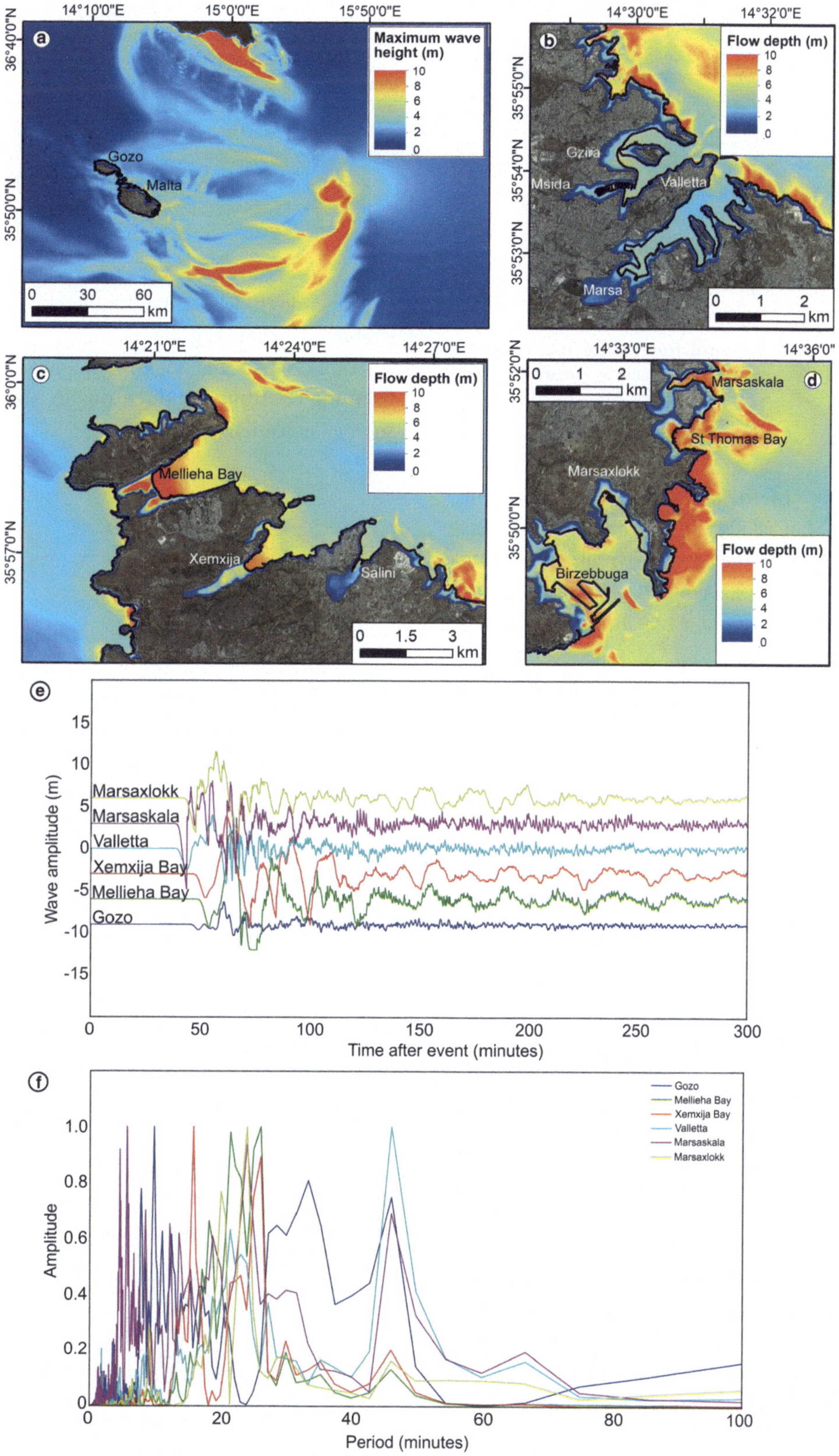
a
Maximum wave height (m)
Gozo
Malta
b
Flow depth (m)
Gzira
Msida
Valletta
Marsa
c
Mellieha Bay
Xemxija
Salini
d
Marsaskala
St Thomas Bay
Marsaxlokk
Birzebbuga
e
Wave amplitude (m)
Time after event (minutes)
Marsaxlokk
Marsaskala
Valletta
Xemxija Bay
Mellieha Bay
Gozo
f
Amplitude
Period (minutes)

Table 4

Tsunami travel time estimates in minutes (± 2 min)

	Source A	Source B	Source C	Source D
Gozo (14.28, 36.08)	45	18	82	50
Mellieha Bay (14.36, 35.97)	50	27	85	58
Valletta (14.52, 35.91)	40	30	78	42
Marsaskala (14.58, 35.86)	40	35	78	43
Marsaxlokk (14.55, 35.83)	43	40	82	50

These values have been estimated from virtual tide gauges in positions indicated in column one of the table as longitude/latitude pairs

inundation of up to ~ 10 m in the low lying topographies of Mellieha Bay, Xemxija, Salini, Gzira, Msida, Marsaskala, St Thomas Bay, Marsaxlokk and Delimara. However, the overall flow depths are lower than those for source A. The first tsunami waves from source C would arrive after ~ 78 min in Malta and ~ 4 min later in Gozo (Table 4). The tsunami generated by this source would start with inundation of the land and dominant oscillation periods are between 10 and 20 min for all virtual tide gauges (Fig. 7e–f).

4.4. Source D: 1693 A.D. Earthquake South-East of Sicily

The source chosen for this case radiates energy towards the north-west and south-east rather than directly towards Malta (Fig. 8). We observe that, once again, tsunami waves are refracted and enhanced by the Malta Escarpment and radiated towards Malta, with the escarpment serving as a 'secondary' source or an amplifier. This effect is seen as the reason for this source having a noticeable impact on Malta. However, inundation extents are lower than for source C, and flow depths are less than 1 m. The first tsunami waves from this source would arrive after ~ 42 min in Malta and ~ 8 min later in Gozo (Table 4). The onset of the tsunami would be observed as water rising and inundating low lying land. The dominant oscillation periods are between 30 and 35 min for all tide gauges, with the exception of Marsaxlokk, where the dominant oscillation period is about 22 min (Fig. 8e–f).

Figure 6 ►

Source B (Gela Slide). **a** Maximum tsunami wave height above mean sea level. Maximum tsunami elevation offshore in **a** is cut off above 90 m in post processing. **b–d** Inundation for different areas in Malta. In most areas, this scenario does causes significant inundation. **b–d** The computational domains 4–6 of Fig. 2. **e** Oscillation periods and **f** frequency analyses of tsunami waves for virtual tide gauges in Malta and Gozo

Table 5 lists the maximum inundation distances from the coastline for three rias in Malta for all four sources considered.

5. Discussion

5.1. Inundation Scenarios

All of the four scenarios simulated in this study cause inundation in densely populated and low-lying coastlines, rias and harbours on Malta. The land uses and infrastructure affected by the inundation include residential areas, agricultural areas, natural reserves, hotels, offices, restaurants, coastal amenities, main roads, yacht marinas, an international port and a power station. The largest inundation extents (Table 5) and flows depths (> 10 m) are produced by the two landslide sources and the western Hellenic Arc earthquake. The submarine landslide source simulations demonstrate how devastating the impact from landslide tsunamis can be. The Gela Slide (source B) produces by far the largest inundation extent and flow depth of all the sources. Times for the first waves to reach Malta from these sources range between 22 and 65 min.

Malta's coastal topography appears to influence the tsunami impact in two ways. Rias, bays and harbours tend to trap the tsunami energy, whereas the steep inclines along the eastern coastlines appear to limit inundation extent and increase flow depths.

5.2. Tsunami Source Recurrence Intervals

In the entire Mediterranean Basin, the earthquakes considered as sources C and D have recurrence

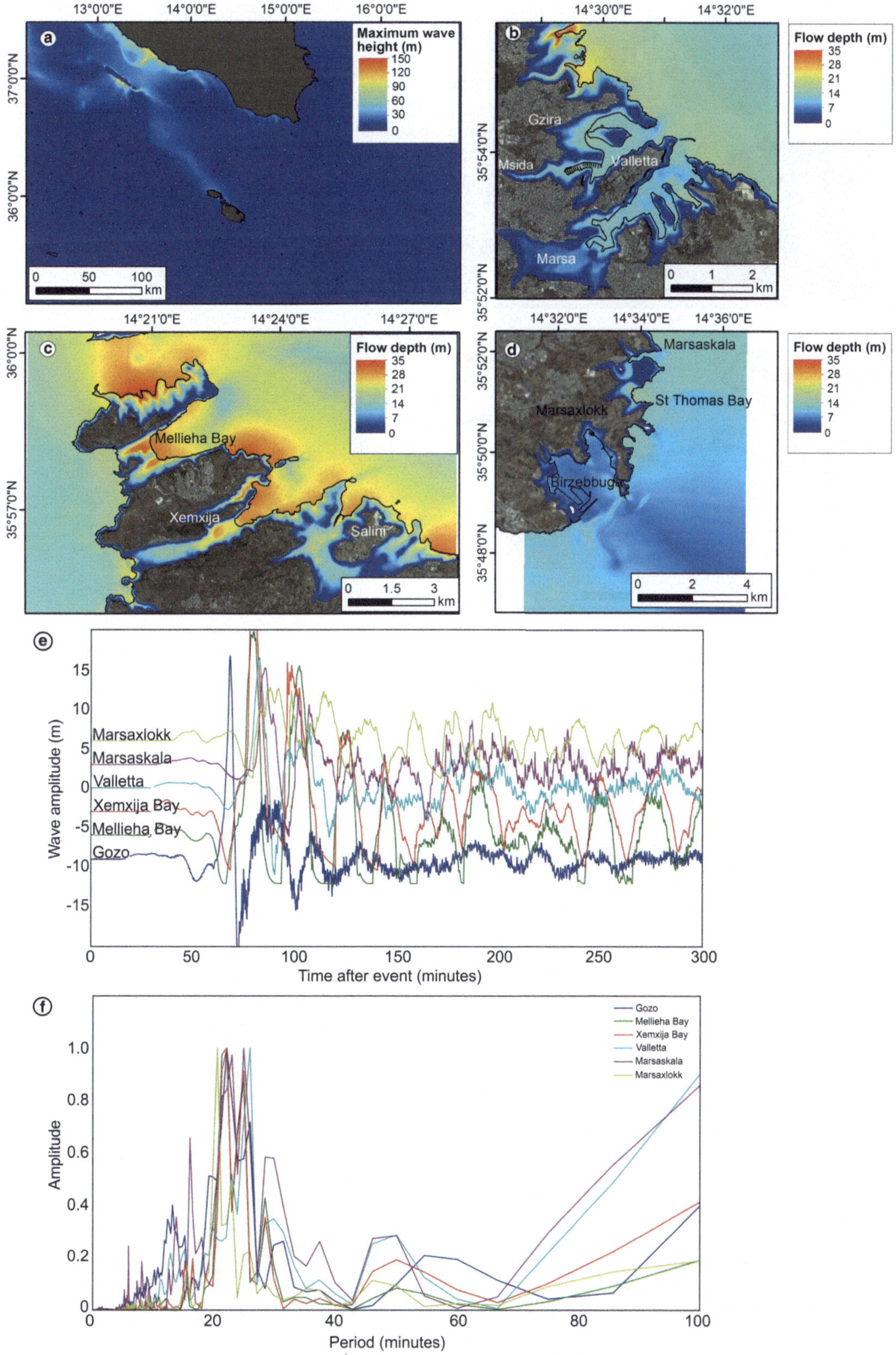

a
Maximum wave height (m)
150
120
90
60
30
0
13°0'0"E
14°0'0"E
15°0'0"E
16°0'0"E
37°0'0"N
36°0'0"N
0 50 100 km
b
Flow depth (m)
35
28
21
14
7
0
14°30'0"E
14°32'0"E
35°54'0"N
35°52'0"N
Gzira
Msida
Valletta
Marsa
0 1 2 km
c
14°21'0"E
14°24'0"E
14°27'0"E
36°0'0"N
35°57'0"N
Mellieha Bay
Xemxija
Salini
0 1.5 3 km
d
14°32'0"E
14°34'0"E
14°36'0"E
35°52'0"N
35°50'0"N
35°48'0"N
Marsaskala
St Thomas Bay
Marsaxlokk
Birzebbuga
0 2 4 km
e
Wave amplitude (m)
Time after event (minutes)
Marsaxlokk
Marsaskala
Valletta
Xemxija Bay
Mellieha Bay
Gozo
f
Amplitude
Period (minutes)
Gozo
Mellieha Bay
Xemxija Bay
Valletta
Marsaskala
Marsaxlokk

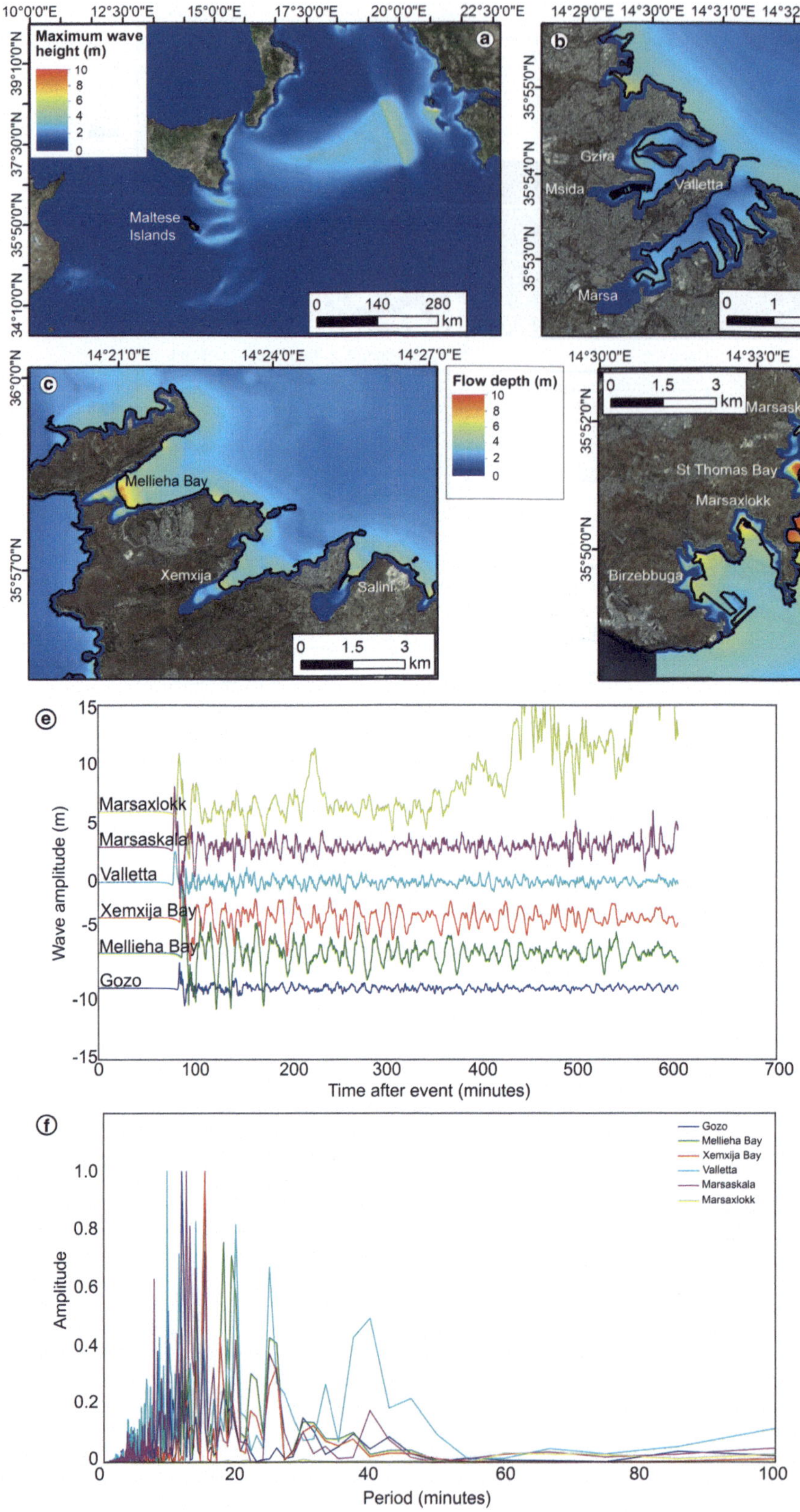

Maximum wave height (m)
10
8
6
4
2
0
Maltese Islands
0 140 280 km
10°0'0"E 12°30'0"E 15°0'0"E 17°30'0"E 20°0'0"E 22°30'0"E
39°10'0"N 37°30'0"N 35°50'0"N 34°10'0"N
Flow depth (m)
Gzira
Msida
Valletta
Marsa
0 1 2 km
14°29'0"E 14°30'0"E 14°31'0"E 14°32'0"E
35°55'0"N 35°54'0"N 35°53'0"N
Mellieha Bay
Xemxija
Salini
0 1.5 3 km
14°21'0"E 14°24'0"E 14°27'0"E
36°0'0"N 35°57'0"N
Marsaskala
St Thomas Bay
Marsaxlokk
Birzebbuga
14°30'0"E 14°33'0"E
35°52'0"N 35°50'0"N
Marsaxlokk
Marsaskala
Valletta
Xemxija Bay
Mellieha Bay
Gozo
Wave amplitude (m)
Time after event (minutes)
Gozo
Mellieha Bay
Xemxija Bay
Valletta
Marsaskala
Marsaxlokk
Amplitude
Period (minutes)

◀Figure 7
Source C (earthquake on western Hellenic Arc). **a** Maximum tsunami wave height above mean sea level. **b–d** Inundation for different areas in Malta. **b–d** The computational domains 4–6 of Fig. 2. **e** Oscillation periods and **f** frequency analyses of tsunami waves for virtual tide gauges in Malta and Gozo

intervals of 333 years and 10 years, respectively (Matias 2014). The landslide sources we simulated occurred in the past, so these are unlikely to be tsunami sources in the near future. We use landsliding rates estimated for the central Mediterranean basin for the Holocene by Urgeles and Camerlenghi (2013) to assess their recurrence interval. For source A, this is 5000 years, whereas for source B the recurrence interval is ~ 50,000 years.

According to Sørensen et al. (2012), estimated peak coastal (earthquake-generated) tsunami amplitudes of 2 m (± 1 m) and 5 m (± 1 m) in the eastern coast of the Maltese Islands have recurrence intervals of 500 years and 5000 years, respectively. The peak coastal tsunami amplitudes estimated for source C are in the order of 10 m, which categorises such an event as extremely rare (recurrence intervals of > 5000 years). However, as noted by Kamigaichi et al. (2009), the approach used by Sørensen et al. (2012) does not discriminate between direct and reflected waves, and it may have overestimated peak coastal tsunami amplitudes and underestimated recurrence intervals.

The study presented here is not intended to replace a full probabilistic tsunami inundation hazard assessment study (e.g. Goda et al. 2015; Mueller et al. 2015). Such a full probabilistic investigation would result in tsunami return intervals, which, in contrast to source recurrence intervals, describes how often we have to expect tsunami exceeding a given inundation area. In our study, we have conducted tsunami inundation studies to show potential worst case situations only.

5.3. The Impact of Urbanisation on Tsunami Inundation

Our simulations have not taken into consideration the impact of urbanisation on tsunami inundation. Previous studies in New Zealand have demonstrated that including building footprints in the DEM will lead to higher tsunami flow velocities and depths, but reduced inundation extent (Wang et al. 2017a, b). Urbanisation is traditionally considered in tsunami modelling by introducing a roughness parameter (Manning's roughness). No investigation into the Manning's Roughness for particular land use areas on the Maltese Islands has been conducted. From other studies we can conclude that a Manning's Roughness value of 0.013 is more appropriate for a smooth land cover than in densely built up areas, where the Manning's roughness can be assumed to be at least 5 times as high (see Griffin et al. 2015 and Shimozono et al. 2012 for further discussion). The Manning's roughness value assumed in our study was 0.013, which is conservatively low and will tend to overestimate the inundation extent and support our worst case assessment. In the built-up areas of Malta, the extent of inundation for such worst case situations will likely be smaller than described in this study, whereas in rural areas of Malta our estimates will be more realistic. No sensitivity study with respect to Manning's Roughness has been conducted.

5.4. The Influence of the Malta Escarpment on Tsunami Waves

The Malta Escarpment acts as a reflector and amplifier of tsunami waves. Such a considerable morphological step constitutes an impedance contrast for shallow water/tsunami waves. When entering from deeper water into a shallow shelf, wave energy that is not reflected will have to be accommodated in higher wave amplitudes and lower velocities. Refraction leads to a focussing of wave fronts from deep to shallow and a widening of wave fronts (and lower amplitudes) from shallow to deep. For these reasons, tsunami waves that are traveling from the east of the Malta Escarpment will be amplified by the escarpment when travelling onto the Malta Plateau. We also observe tsunami energy being reflected off the southern coast of Sicily and back again from Malta, getting trapped in the Sicily Channel. This creates a unique situation for Malta's tsunami exposure. The exposure of the north-eastern coastline to tsunamis is, to some degree, independent of the position of the

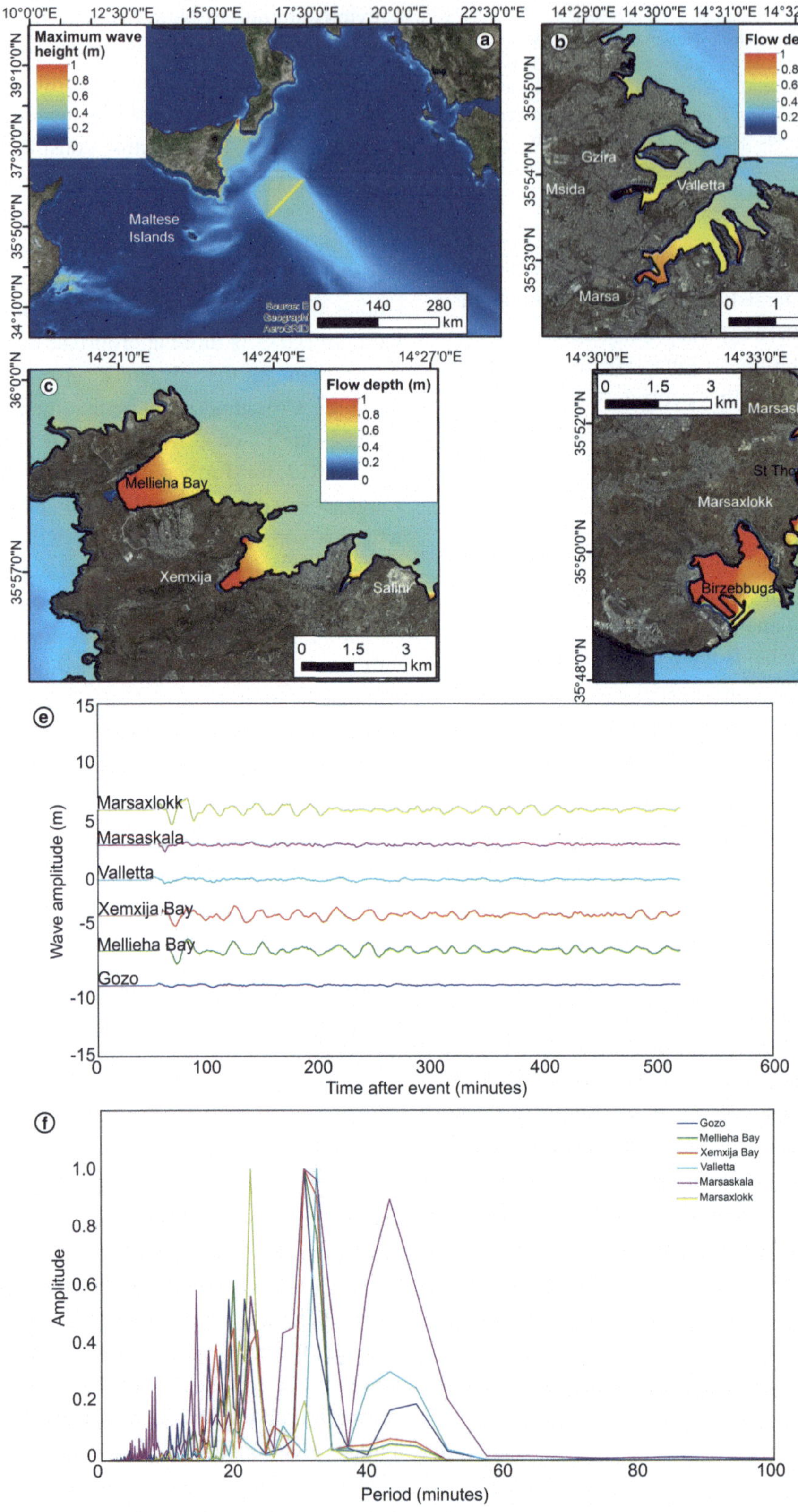

a
Maximum wave height (m)
1
0.8
0.6
0.4
0.2
0
Maltese Islands
0 140 280 km
10°0'0"E 12°30'0"E 15°0'0"E 17°30'0"E 20°0'0"E 22°30'0"E
39°10'0"N 37°30'0"N 35°50'0"N 34°10'0"N
b
Flow depth (m)
Gzira
Msida
Valletta
Marsa
0 1 2 km
14°29'0"E 14°30'0"E 14°31'0"E 14°32'0"E
35°55'0"N 35°54'0"N 35°53'0"N
c
Flow depth (m)
Mellieha Bay
Xemxija
Salini
0 1.5 3 km
14°21'0"E 14°24'0"E 14°27'0"E
36°0'0"N 35°57'0"N
d
Marsaskala
St Thomas Bay
Marsaxlokk
Birzebbuga
0 1.5 3 km
14°30'0"E 14°33'0"E
35°52'0"N 35°50'0"N 35°48'0"N
e
Marsaxlokk
Marsaskala
Valletta
Xemxija Bay
Mellieha Bay
Gozo
Wave amplitude (m)
Time after event (minutes)
15 10 5 0 -5 -10 -15
0 100 200 300 400 500 600
f
Gozo
Mellieha Bay
Xemxija Bay
Valletta
Marsaskala
Marsaxlokk
Amplitude
Period (minutes)
0 0.2 0.4 0.6 0.8 1.0
0 20 40 60 80 100

◀Figure 8
Source D (earthquake in south east of Sicily). **a** Maximum tsunami wave height above mean sea level. **b–d** Inundation for different areas in Malta. **b–d** The computational domains 4–6 of Fig. 2. **e** Oscillation periods and **f** frequency analyses of tsunami waves for virtual tide gauges in Malta and Gozo

primary source, with the Malta Escarpment and Sicily acting as secondary tsunami sources. This phenomenon, together with the low-lying nature of the north-eastern Maltese coastline, generates very consistent 'hot spot' areas, which makes it easier to identify predominantly exposed areas and focus emergency management efforts accordingly.

5.5. *Implications for Interpretation of Extreme Wave Morphological Indicators*

The majority of the extreme wave morphological indicators reported along the eastern Maltese coastline have a general freshness of form that suggests that they are not of great age (Mottershead et al. 2014). This was confirmed by radiocarbon dating of marine organisms encrusted on some coastal boulders, which suggests ages ranging between 800 AD and the 20th century (Biolchi et al. 2016). This eventually led to the interpretation that the majority of these deposits can be explained by moderately frequent storm waves. Figure 9 shows that the deposits could also be explained by tsunami waves generated by the 365 A.D. earthquake source for all sites, and the 1693 earthquake source for Xghajra and Water Park. It is interesting to note that only the Gela Slide can generate a tsunami wave that can mobilise the boulder at Ahrax Point (Fig. 9).

5.6. *Limitations and Future Research*

The scenario study presented here should be regarded as a starting point to fully assess the tsunami exposure of the Maltese Islands. It provides initial information to better understand the potential maximum impacts and documents potential end members of the tsunami hazard spectrum. Our study is based on conservative assumptions, including comparatively low Manning's roughness, no consideration of specific effects of urbanisation and a sliding solid block model for landslides, all of which may maximise inundation extent. Experimental and numerical investigations on landslide-generated tsunamis (Ma et al. 2013; Lindstrøm 2016) indicate that rigid landslides, e.g. solid blocks, tend to generate larger tsunami waves than deformable landslides such as granular flow or debris avalanche type of landslides. Yavari-Ramshe and Ataie-Ashtiani (2017, 2018) saw an increase in tsunami amplitudes by 16–18% when comparing rigid slides to loose material slides for tsunami generation in their studies. Rigid and deformable landslides can also generate tsunami waves with very different temporal and spatial distributions (Ma et al. 2013). It is conceivable that deformable landslides generate waves with longer tsunami wavelengths and as such can generate tsunami that have higher impact than the tsunami generated by the source approximations used here. To fully understand the tsunami hazard on the Maltese Islands, a much more detailed investigation is required. First, the source mechanisms in the scenarios investigated here are first-order approximations, so there is a need for further refinement. Further studies could also include the effects of non-uniform slip for the earthquake sources, the evolution of mass transport in the submarine landslide sources and investigation of the effects of different rheological models. Secondly, a tsunami forecasting database considering a representative number of potential tsunami sources should be established. Such a database would be useful in informing early tsunami evacuation planning and evacuation zoning. Once the full spectrum of the tsunami exposure is known, proactive tsunami mitigation efforts can be implemented.

Future studies should also include probabilistic inundation hazard and risk assessment. A detailed tsunami hazard assessment, initially only considering offshore wave heights and later potentially inundation, will help to put the tsunami hazard into perspective amongst other hazards, such as seismic hazard. Inundation hazard assessment will help to better understand tsunami risk and related potential losses. This will help the government to motivate prioritisation of tsunami hazard studies, mitigation studies and emergency management efforts.

The maximum inundation extent and run-up from the events considered in this study can be used as a first estimate to delineate the maximum extent of evacuation zones. However, such worst case estimates may create conservatively large evacuation zones, which should be avoided because over-evacuation also poses a hazard to property, human well-being and potentially life. More frequent events, which cause lower peak coastal tsunami amplitude and less inundation, can still pose a considerable hazard. Their impact should be investigated comprehensively to gain a complete picture of potential tsunami hazard. Given that the Maltese Islands are exposed to a fairly large set of local tsunami sources with travel times of only tens of minutes, public education on how to behave if a strong or long earthquake is felt is highly advisable. The potential of earthquakes to trigger submarine landslides, and whether the local population should be made aware of this additional hazard, should be investigated. The comparably small areas affected by tsunami waves, as demonstrated in our study, and the large number of escape routes (e.g. streets leading away from the shoreline) will support the effectiveness of self-evacuation.

Table 5

Approximated maximum inundation extent from the shore by example

	Source A (m)	Source B (m)	Source C (m)	Source D (m)
Marsa Harbour/ Albert Town	960	1600	850	200
Msida	1360	1980	1000	720
Il-Pwales/Xemxija	2200	4000[a]	1000	90

The "a" indicates that inundation spanned the full width of Malta

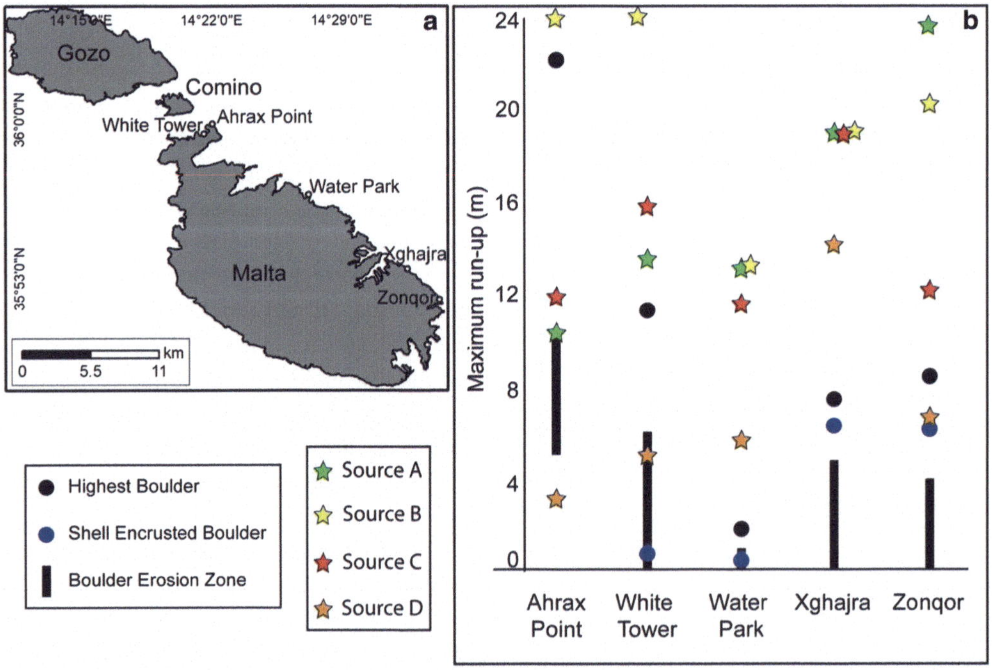

Figure 9
Plot of height distributions of extreme wave morphological indicators (adapted from Mottershead et al. 2014) and maximum run-up of the modelled tsunamis

6. Conclusions

We have investigated the tsunami inundation extent and flow depths of four tsunami scenarios that were designed to represent worst case situations for the island of Malta. These scenarios were modelled after historic and prehistoric large events and include two submarine landslide sources (outer Malta Plateau slide and Gela Slide) and two earthquake sources (365 A.D. western Hellenic Arc event and the 1693 south-east of Sicily event). All of the scenarios cause inundation in densely populated and low lying coastlines, rias and harbours that host high population densities and important infrastructure. The largest inundation extents and flow depths (> 10 m) are produced by the two landslide sources (A and B) and the western Hellenic Arc earthquake (source C). As we have not taken into account the impact of urbanisation, inundation is likely to be smaller than the modelled results for built-up areas, but similar in rural areas. Times for the first waves to reach Malta from these sources range between 22 and 65 min. Recurrence rates for the sources of the modelled events range between 5000 and 50,000 years (landslide sources), and 10–333 years (earthquake sources). The Malta Escarpment and Sicily play a role in amplifying and reflecting tsunami waves, and in generating consistent hot spots along the eastern coastline. Our study demonstrates that a potential tsunami hazard does exist for Malta, and serves as a first order and worst case investigation that is intended to help identify the upper limits of potential tsunami inundation for Malta. Future studies should focus on achieving a better understanding of the source mechanisms, develop a tsunami forecasting database, and carrying out a probabilistic inundation hazard and risk assessment. Information from such studies would be extremely valuable to inform civil protection and emergency management efforts.

Acknowledgements

We thank GNS Science in-house funding and the University of Malta Research Seed Fund for supporting this project financially. AM and DS are funded by EMODnet Bathymetry, which is financed by the European Union under Regulation 508/2014 of the European Parliament and of the Council of 15 May 2014 on the European Maritime and Fisheries Fund. We express our gratitude to Marco Taviani, Lorenzo Angeletti and the Planning Authority for providing access to data from the MEDCOR cruise (2009), DECORS cruise (2011) and ERDF project 156.

References

Argnani, A., & Bonazzi, C. (2005). Malta Escarpment fault zone offshore eastern Sicily: Pliocene-quaternary tectonic evolution based on new multichannel seismic data. *Tectonics*. https://doi.org/10.1029/2004TC001656.

Baratta, M. (1910). *La Catastrofe Sismica Calabro-Messinese (28 Dicembre 1908) [The Calabrio-Messinian seismic catastrophe (28 December 1908)]*. Rome: Presso la Societa Geografica Italiana.

Beavan, R. J., Wang, X., Holden, C., Wilson, K. J., Power, W. L., Prasetya, G., et al. (2010). Near-simultaneous great earthquakes at Tongan megathrust and outer rise in September 2009. *Nature*. https://doi.org/10.1038/nature09292.

Biolchi, S., Furlani, S., Antonioli, F., Baldassini, N., Causon Deguara, J., Devoto, S., Di Stefano, A., Evans, J., Gambin, T., Gauci, R., Mastronuzzi, G., Monaco, C., & Scicchitano, G. (2016). Boulder accumulations related to extreme wave events on the eastern coast of Malta. *Nature Hazards Earth Systems Science*. https://doi.org/10.5194/nhess-16-737-2016.

Borg, R. P., D'Amico, S., & Galea, P. (2016). Earthquake and people: The maltese experience of the 1908 Messina earthquake. In S. D'Amico (Ed.), *Earthquakes and Their Impact on Society. Earthquakes and their Impact on Society* (pp. 533–562). Cham, Switzerland: Springer.

Camilleri, D. H. (2006). Tsunami construction risks in the Mediterranean—Outlining Malta's scenario. *Disaster Prevention Management An International Journal*. https://doi.org/10.1108/09653560610654301.

Cho, Y. S. (1995). Numerical simulations of tsunami and runup. [PhD thesis]. Ithaca, (NY): Cornell University.

Civile, D., Lodolo, E., Accettella, D., et al. (2010). The Pantelleria graben (Sicily Channel, Central Mediterranean): An example of intraplate "passive" rift. *Tectonophysics*. https://doi.org/10.1016/j.tecto.2010.05.008.

D'Addezio, G., & Valensise, G. (1993). Metodologie per l'individuazione della struttura sismogenetica responsabile del terremoto del 13 dicembre 1990, in Contributi allo studio del terremoto della Sicilia Orientale del 13 dicembre 1990, Edited by E. Boschi and A. Basili, Int. Rep. 537, pp. 115–125, Ist. Naz. di Geofis., Rome.

De Martini, P. M., Barbano, M. S., & Pantosti, D., et al. (2012) Geological evidence for paleotsunamis along eastern Sicily

(Italy): An overview. *Nature Hazards Earth Systems Science*. https://doi.org/10.5194/nhess-12-2569-2012.

De Soldanis, G. P. F. A. (1746). Il Gozo Antico-Moderno e Sacro-Profano (Gozo, Ancient & Modern, Religious & Profane). Rabat, Gozo. (In Italian) Unpublished. In National Library of Malta, Libr.145.

Drago, A., Azzopardi, J., Gauci, A., Tarasova, R., & Bruschi, A. (2013) Assessing the offshore wave energy potential for the Maltese Islands. Institute of Sustainable Energy, University of Malta, Annual Conference, 21st March 2013, Qawra, Malta.

Enet, F., & Grilli, S. T. (2007). Experimental study of tsunami generation by three-dimensional rigid underwater landslides. *Journal of Waterway, Port, Coastal, and Ocean Engineering, 133*(6), 442–454.

Faccenna, C., Funiciello, F., Giardini, D., & Lucente, P. (2001). Episodic back-arc extension during restricted mantle convection in the Central Mediterranean. *Earth and Planetary Science Letters*. https://doi.org/10.1016/S0012-821X(01)00280-1.

Flemming, N. C. (1978). Holocene eustatic changes and coastal tectonics in the northeast mediterranean: Implications for models of crustal consumption. *Philosophical Transactions of the Royal Society of London. Series A, Mathematical and Physical Sciences*. https://doi.org/10.1098/rsta.1978.0065.

Foglini, F., Dalla Valle, G., Micallef, A., Berndt, C., Campiani, E., & Trincardi, F. (2011). Seafloor instability and mass wasting processes along the eastern Gela slope. In: Y. Yamada, K. Kawamura, K. Ikehara, Y. Ogawa, R. Urgeles, D. Mosher, J. D. Chaytor, M.C. Strasser (Eds.), *5th international symposium on submarine mass movements and their consequences*, Kyoto, Japan.

Foglini, F., Prampolini, M., Micallef, A., Angeletti, L., Vandelli, V., Deidun, A., Soldati, M., Taviani, M. (Eds.), (2016). Late Quaternary coastal landscape morphology and evolution of the Maltese Islands (Mediterranean Sea) reconstructed from high-resolution seafloor data. Geology and Archaeology: Submerged Landscapes of the Continental Shelf, 411. Geological Society of London.

Gailler, A., Schindelé, F., & Hébert, H. (2016). Impact of Hellenic Arc Tsunamis on Corsica (France). *Pure and Applied Geophysics*. https://doi.org/10.1007/s00024-016-1365-1.

Galea, P. (2007). Seismic history of the Maltese islands and considerations on seismic risk. *Annales Geophysicae*. https://doi.org/10.4401/ag-3053.

Galli, P. (2005). I terremoti del gennaio 1117. Ipotesi di un epicentro nel cremonese. *Quaterly, 22,* 207–234.

Gerardi, F., Smedile, A., Pirrotta, C., et al. (2012) Geological record of tsunami inundations in Pantano Morghella (south-eastern Sicily) both from near and far-field sources. *Nature Hazards Earth Systems Science*. https://doi.org/10.5194/nhess-12-1185-2012.

Goda, K., Li, S., Mori, N., & Yasuda, T. (2015). Probabilistic tsunami damage assessment considering stochastic source models: Application to the 2011 Tohoku earthquake. *Coastal Engineering Journal, 57*(03), 1550015.

Goes, S., Jenny, S., Hollenstein, C., et al. (2004). A recent tectonic reorganization in the south-central Mediterranean. *Earth and Planetary Science Letters*. https://doi.org/10.1016/j.epsl.2004.07.038.

Griffin, J., Latief, H., Kongko, W., et al. (2015). An evaluation of onshore digital elevation models for modeling tsunami inundation zones. *Frontiers in Earth Science, 3,* 32.

Gutscher, M. A., Dominguez, S., De Lepinay, B. M., et al. (2016). Tectonic expression of an active slab tear from high-resolution seismic and bathymetric data offshore Sicily (Ionian Sea). *Tectonics*. https://doi.org/10.1002/2015TC003898.

Gutscher, M. A., Roger, J., Baptista, M. A., et al. (2006). Source of the 1963 Catania earthquake and tsunami (southern Italy): New evidence from tsunami modeling of a locked subduction fault plane. *Geophysical Research Letters*. https://doi.org/10.1029/2005GL025442.

Illies, J. H. (1981). Graben formation—The Maltese Islands—A case history. *Tectonophysics*. https://doi.org/10.1016/0040-1951(81)90182-7.

Kamigaichi, O., Saito, M., Doi, K., et al. (2009). Earthquake early warning in Japan: Warning the general public and future prospects. *Seismological Research Letters*. https://doi.org/10.1785/gssrl.80.5.717.

Lindstrøm, E. K. (2016). Waves generated by subaerial slides with various porosities. *Coastal Engineering, 116,* 170–179. https://doi.org/10.1016/j.coastaleng.2016.07.001.

Liu, P. L. F., Cho, Y. S., Briggs, M. J., Synolakis, C. E., & Kanoglu, U. (1995). Run-up of solitary waves on circular island. *Journal of Fluid Mechanics, 302,* 259–285.

Liu, P. L. F., Cho, Y. S., & Fujima, K. (1994a). Numerical solutions of three-dimensional run-up on a circular island. In: Isaacson M, Quick M, editors. *Proceedings of international symposium: Waves—Physical and numerical modelling*; 1994 Aug 21–24; Vancouver, Canada. Vancouver (BC): University of British Columbia (pp. 1031–1040).

Liu, P. L. F., Cho, Y. S., Yoon, S. B., & Seo, S. N. (1994b). Numerical simulations of the 1960 Chilean tsunami propagation and inundation at Hilo, Hawaii. In M. I. El-Sabh (Ed.), *Recent development in tsunami research* (pp. 99–115). Dordrecht (NL): Kluwer Academic Publishers.

Liu, Y. R., Wang, X., Wu, Z., He, Z., & Yang, Q. (2018). Simulation of landslide-induced surges and analysis of impact on dam based on stability evaluation of reservoir bank slope. *Landslides*. https://doi.org/10.1007/s10346-018-1001-5.

Liu, P. L. F., Woo, S. B., & Cho, Y. S. (1998). *Computer programs for tsunami propagation and inundation*. Ithaca (NY): Cornell University (Cornell University Technical Report).

Ma, G., Kirby, J. T., & Shi, F. (2013). Numerical simulation of tsunami waves generated by deformable submarine landslides. *Ocean Modelling, 69,* 146–165. https://doi.org/10.1016/j.ocemod.2013.07.001.

Maesano, F. E., Tiberti, M. M., & Basili, R. (2017). The Calabrian Arc: Three-dimensional modelling of the subduction interface. *Scientific Report*. https://doi.org/10.1038/s41598-017-09074-8.

Magri, O. (2006). A geological and geomorphological review of the Maltese islands with special reference to the coastal zone. *Territoris, 6,* 7–26.

Matias, L. (2014). ASTARTE Assessment, STrategy And Risk Reduction for Tsunamis in Europe. http://www.astarte-project.eu/files/astarte/documents/deliverables/d2-9/D2.9-vs3-full.pdf. Accessed 12 Jan 2019.

Micallef, A., Berndt, C., & Debono, G. (2011). Fluid flow systems of the Malta Plateau, Central Mediterranean Sea. *Marine Geology*. https://doi.org/10.1016/j.margeo.2011.03.009.

Micallef, A., Camerlenghi, A., Georgiopoulou, A., Garcia-Castellanos, D., Gutscher, M. A., Lo Iacono, C., et al. (2019). Geomorphic evolution of the Malta Escarpment and implications for the Messinian evaporative drawdown in the eastern Mediterranean Sea. *Geomorphology, 327,* 264–283.

Micallef, A., Foglini, F., Le Bas, T., et al. (2013). The submerged paleolandscape of the Maltese Islands: Morphology, evolution and relation to quaternary environmental change. *Marine Geology*. https://doi.org/10.1016/j.margeo.2012.10.017.

Micallef, A., Georgiopoulou, A., Mountjoy, J., et al. (2016). Outer shelf seafloor geomorphology along a carbonate escarpment: The eastern Malta Plateau, Mediterranean Sea. *Continental Shelf Research, 22,* 22. https://doi.org/10.1016/j.csr.2016.11.002.

Minisini, D., Trincardi, F., Asioli, A., et al. (2007). Morphologic variability of exposed mass-transport deposits on the eastern slope of Gela Basin (Sicily channel). *Basin Research*. https://doi.org/10.1111/j.1365-2117.2007.00324.x.

Mottershead, D. N., Bray, M. J., & Soar, P. J. (2017). Tsunami landfalls in the Maltese archipelago: Reconciling the historical record with geomorphological evidence. In: Tsunamis: Geology, Hazards and Risks. Geological Society, London, Special Publications (p. 456).

Mottershead, D., Bray, M., Soar, P., & Farres, P. J. (2014). Extreme wave events in the central Mediterranean: Geomorphic evidence of tsunami on the Maltese Islands. *Zeitschrift für Geomorphol.* https://doi.org/10.1127/0372-8854/2014/0129.

Mottershead, D. N., Bray, M. J., Soar, P. J., & Farres, P. J. (2015). Characterisation of erosional features associated with tsunami terrains on rocky coasts of the Maltese islands. *Earth Surface Processing Landforms*. https://doi.org/10.1002/esp.3784.

Mountjoy, J. J., Wang, X., Woelz, S., Fitzsimons, S., Howarth, J. D., Orpin, A. R., et al. (2018). *Tsunami hazard from lacustrine mass wasting in Lake Tekapo, New Zealand. Subaqueous Mass Movements*. New Zealand: Geological Society, London, Special Publications. https://doi.org/10.1144/SP477.21.

Mueller, C., Power, W., Fraser, S., & Wang, X. (2015). Effects of rupture complexity on local tsunami inundation: Implication for probabilistic tsunami hazard assessment by example. *Journal of Geophysical Research, 120*(1), 488–502.

Okada, Y. (1985). Surface deformation due to shear and tensile faults in a half-space. *Bulletin of the Seismological Society of America, 75*(4), 1135–1154.

Papadimitriou, E. E., & Karakostas, V. G. (2008). Rupture model of the great AD 365 Crete earthquake in the southwestern part of the Hellenic Arc. *Acta Geophysica*. https://doi.org/10.2478/s11600-008-0001-6.

Papadopoulos, G. (2009). Tsunamis. In J. C. Woodward (Ed.), *The physical geography of the Mediterranean* (pp. 483–512). Oxford: Oxford University Press.

Papadopoulos, G. A., Gràcia, E., Urgeles, R., et al. (2014). Historical and pre-historical tsunamis in the Mediterranean and its connected seas: Geological signatures, generation mechanisms and coastal impacts. *Marine Geology*. https://doi.org/10.1016/j.margeo.2014.04.014.

Pedley, H. M., House, M. R., & Waugh, B. (1976). The geology of Malta and Gozo. *Proceedings of the Geologists' Association*. https://doi.org/10.1016/S0016-7878(76)80005-3.

Piatanesi, A., & Tinti, S. (1998). A revision of the 1693 Sicily earthquake and tsunami. *Geophysical Research Letters, 103,* 2749–2758.

Pirazzoli, P. A., Thommeret, J., Laborel, J., & Montaggioni, L. F. (1982). Crustal Block Movements from Holocene Shorelines: Crete and Antikythira (Greece). *Tectonophysics*. https://doi.org/10.1016/0040-1951(89)90105-4.

Polonia, A., Vaiani, S. C., & De Lange, G. J. (2016). Did the AD 365 Crete earthquake/tsunami trigger synchronous giant turbidity currents in the Mediterranean Sea? *Geology*. https://doi.org/10.1130/G37486.1.

Reuther, C. D., & Eisbacher, G. H. (1985). Pantelleria Rift—Crustal extension in a convergent intraplate setting. *Geology Rundschau*. https://doi.org/10.1007/BF01821214.

San Pedro, L., Babonneau, N., Gutscher, M. A., & Cattaneo, A. (2017). Origin and chronology of the Augias deposit in the Ionian Sea (Central Mediterranean Sea), based on new regional sedimentological data. *Marine Geology*. https://doi.org/10.1016/j.margeo.2016.05.005.

Shaw, B., Ambraseys, N. N., England, P. C., et al. (2008). Eastern Mediterranean tectonics and tsunami hazard inferred from the AD 365 earthquake. *Nature Geoscience*. https://doi.org/10.1038/ngeo151.

Shaw, B., & Jackson, J. (2010). Earthquake mechanisms and active tectonics of the Hellenic subduction zone. *Geophysical Journal International*. https://doi.org/10.1111/j.1365-246X.2010.04551.x.

Shimozono, T., Sato, S., Okayasu, A., Tajima, Y., Fritz, H. M., Liu, H., et al. (2012). Propagation and inundation characteristics of the 2011 Tohoku tsunami on the central Sanriku coast. *Coastal Engineering Journal, 54*(01), 1250004.

Sørensen, M. B., Spada, M., Babeyko, A., et al. (2012). Probabilistic tsunami hazard in the Mediterranean Sea. *Journal of Geophysics Research Solid Earth*. https://doi.org/10.1029/2010JB008169.

Stiros, S. C. (2010). The 8.5 + magnitude, AD365 earthquake in Crete: Coastal uplift, topography changes, archaeological and historical signature. *Quaternary International*. https://doi.org/10.1016/j.quaint.2009.05.005.

Stiros, S. C., & Drakos, A. (2006). A fault model for the tsunami-associated, magnitude $\geq$ 8.5 Eastern Mediterranean, AD 365 Earthquake. *Zeitschrift für Geomorphol Suppl, 216,* 54–63.

Tinti, S., Armigliato, A., Pagnoni, G., & Zaniboni, F. (2005). Scenarios of giant tsunamis of tectonic origin in the Mediterranean. *ISET Journal of Earthquake Technology, 42*(4), 171–188.

Trincardi, F., & Argnani, A. (1990). Gela submarine slide: A major basin-wide event in the plio-quaternary foredeep of Sicily. *Geo-Marine Letter*. https://doi.org/10.1007/BF02431017.

Urgeles, R., & Camerlenghi, A. (2013). Submarine landslides of the Mediterranean Sea: Trigger mechanisms, dynamics, and frequency-magnitude distribution. *J Geophysics Research Earth Surface*. https://doi.org/10.1002/2013JF002720.

Wang, X. (2008). Numerical modelling of surface and internal waves over shallow and intermediate water. [PhD thesis]. Ithaca, (NY): Cornell University. p. 245.

Wang, X., & Liu, P. L. F. (2006). An analysis of 2004 Sumatra earthquake fault plane mechanisms and Indian Ocean tsunami. *Journal of Hydraulic Research, 44*(2), 147–154.

Wang, X., & Liu, P. L. F. (2007). Numerical simulation of the 2004 Indian Ocean tsunami—Coastal effects. *Journal of Earthquake and Tsunami, 1*(3), 273–297.

Wang, X., Lukovic, B., Power, W. L., & Mueller, C. (2017a). High-resolution inundation modelling with explicit buildings. Lower Hutt (NZ): GNS Science. 27 p. (GNS Science report 2017/13), https://doi.org/10.21420/G2RW2N.

Wang, X., Mountjoy, J. J., Power, W. L., Lane, E. M., & Mueller, C. (2016). Coupled modelling of the failure and tsunami of a submarine debris avalanche offshore central New Zealand. In: G. Lamarche et al. (Eds.) *Submarine mass movements and their consequences: 7th international Symposium*. Advances in natural and technological hazards research, p. 41.

Wang, X., & Power, W. L. (2011). COMCOT: A tsunami generation propagation and run-up model. GNS Science Report 2011/43.

Wang, X., Power, W. L., Lukovic, B., & Mueller, C. (2017b). Effect of explicitly representing building on tsunami inundation: A pilot study of Wellington CBD. Paper O3C.3 in: Next generation of low damage and resilient structures: New Zealand Society for Earthquake Engineering Annual Conference. Wellington, N.Z.: New Zealand Society for Earthquake Engineering.

Watts, P., Grilli, S. T., Kirby, J. T., Fryer, G. J., & Tappin, D. R. (2003). Landslide tsunami case studies using a Boussinesq model and a fully nonlinear tsunami generation model. *National Hazard Earth System, 3,* 391–402.

Wijetunge, J. J., Wang, X., & Liu, P. L. F. (2008). Indian ocean tsunami on 26 December 2004: Numerical modelling of inundation in three cities on the south coast of Sri Lanka. *Journal of Earthquake and Tsunami, 2*(2), 133–155.

Yavari-Ramshe, S., & Ataie-Ashtiani, B. (2017). Subaerial landslide-generated waves: Numerical and laboratory simulations. In K. Sassa, M. Mikoš, & Y. Yin (Eds.), *Advancing culture of living with landslides. WLF 2017*. Cham: Springer.

Yavari-Ramshe, S., & Ataie-Ashtiani, B. (2018). On the effects of landslide deformability and initial submergence on landslide-generated waves. *Landslides, 16,* 37–53. https://doi.org/10.1007/s10346-018-1061-6.

(Received February 18, 2019, revised November 9, 2019, accepted December 5, 2019, Published online December 16, 2019)

Pure Appl. Geophys. 177 (2020), 1639–1657

https://doi.org/10.1007/s00024-019-02390-2

Pure and Applied Geophysics

The 1674 Ambon Tsunami: Extreme Run-Up Caused by an Earthquake-Triggered Landslide

Ignatius Ryan Pranantyo[1] and Phil R. Cummins[1]

Abstract—We present an analysis of the oldest detailed account of tsunami run-up in Indonesia, that of the 1674 Ambon tsunami (Rumphius in Waerachtigh Verhael van de Schuckelijcke Aerdbebinge, BATAVIA, Dutch East Indies, 1675). At 100 m this is the largest run-up height ever documented in Indonesia, and with over 2300 fatalities even in 1674, it ranks as one of Indonesia's most deadly tsunami disasters. We consider the plausible sources of earthquakes near Ambon that could generate a large, destructive tsunami, including the Seram Megathrust, the South Seram Thrust, and faults local to Ambon. We conclude that the only explanation for the extreme run-up observed on the north coast of Amon is a tsunami generated by an earthquake-triggered coastal landslide. We use a two-layer tsunami model to show that a submarine landslide, with an approximate volume of 1 km^3, offshore the area on Ambon's northern coast, between Seith and Hila, where dramatic changes in coastal landscape were observed can explain the observed tsunami run-up along the coast. Thus, the 1674 Ambon tsunami adds weight to the evidence from recent tsunamis, including the 1992 Flores, 2018 Palu and Sunda Strait tsunamis, that landslides are an important source of tsunami hazard in Indonesia.

Keywords: Eastern Indonesia, Ambon, tsunami hazard, landslide.

1. Introduction

Eastern Indonesia, and the Banda Sea in particular, is a region of very active and complex tectonics (Hamilton 1979; McCaffrey 1988; Spakman and Hall 2010; Pownall et al. 2013). Despite a historical record rich in major, destructive earthquakes and tsunamis, during the more recent era of instrumental seismology most of the major events have occurred in western Indonesia. The only way to better understand the tsunami threat in eastern Indonesia is therefore to glean as much information as we can from the historical record, which often consists of accounts that are sparse and difficult to interpret.

The oldest detailed tsunami account in Indonesia was documented by Rumphius (1675). A devastating earthquake rocked Ambon and its surrounding islands on 17 February 1674. The earthquake was followed by a massive tsunami about 100 m in run-up height which was only observed on the northern coast of Ambon Island while other areas experienced only minor tsunamis. The earthquake and tsunami caused more than 2300 fatalities, mostly on the northern shore of Ambon.

The source of the tsunami and earthquake is unknown. Løvholt et al. (2012) and Harris and Major (2017) speculated that it was triggered by an earthquake from south of Ambon and a landslide triggered by an earthquake from inside Ambon Bay, respectively. However, no attempt has been made to investigate this event further, particularly to answer why the extreme run-up was observed only on the northern coast of Ambon. Therefore, the primary source of the tsunami and earthquake remains open to question.

In the following sections of this paper, the tectonic setting around Ambon is discussed first, followed by our interpretation of the accounts of this event. The primary source of the earthquake and tsunami is investigated through analysis of the Rumphius document. Tsunami modelling is then performed to confirm the analysis. Lastly, the result of the analysis and the implications of the findings are discussed.

2. Tectonic Setting Around Ambon

Ambon is a small volcanic island that lies southwest of Seram Island (Fig. 1). It consists of two small

[1] Research School of Earth Sciences, Australian National University, Canberra, Australia. E-mail: ryan.pranantyo@anu.edu.au; phil.cummins@anu.edu.au

islands, Hitu and Leitimor, which are connected by a short isthmus (Fig. 2). Ambon is located to the southwest of Seram Island which is part of the outer Banda Arc. The islands are surrounded by major faults, namely the Seram Megathrust, Kawa Fault and the Banda Detachment (Fig. 1).

As summarised in Patria and Hall (2017), the Seram Megathrust stretches from Kai Islands in the east side to the northwest of Seram Island (Fig. 1). It is often described as a subduction zone (e.g. Hamilton 1979; Honthaas et al. 1998). However, others have argued that it is a foredeep produced by loading from a developing fold and thrust belt (Audley-Charles et al. 1979; Pairault et al. 2003; Spakman and Hall 2010). Through high-resolution bathymetry and seismic data, Patria and Hall (2017) confirmed the second hypothesis: that it is a result of oblique intraplate convergence.

The Kawa Fault is a prominent structure in central Seram Island (Fig. 1). The fault runs from Piru Bay, off the northern shore of Ambon, to the central south of the island on the northern side of the Banda Sea. The fault has a major left-lateral movement identified through geological observations (Pownall et al. 2013). A large earthquake which caused a catastrophic tsunami on Seram Island in 1899 was suspected to have ruptured the Kawa Fault (Soloviev and Go 1974).

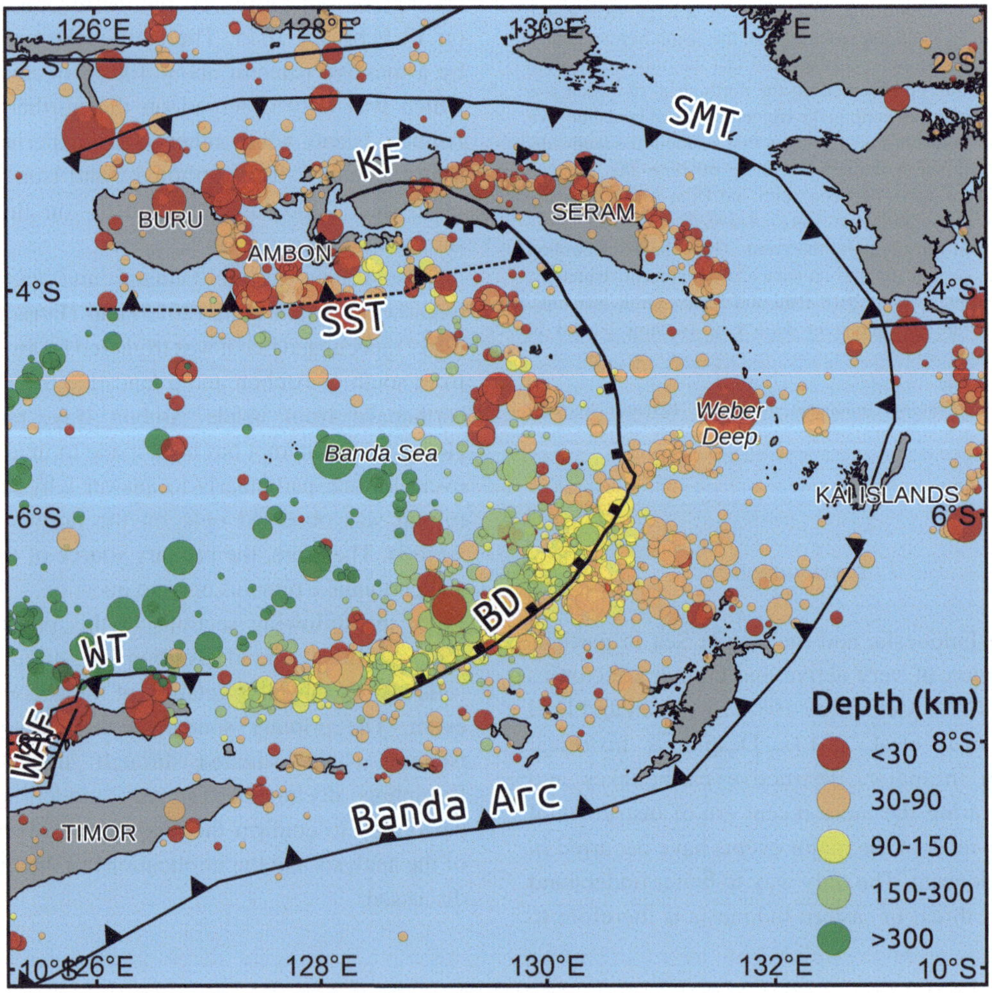

Figure 1
Tectonic setting of Ambon and its surrounding islands; *SMT* Seram Megathurst, *KF* Kawa fault, *SST* purported South Seram Thrust, *BD* Banda Detachment, *WT* Wetar thrust, *WAF* Wetar-Atauro fault

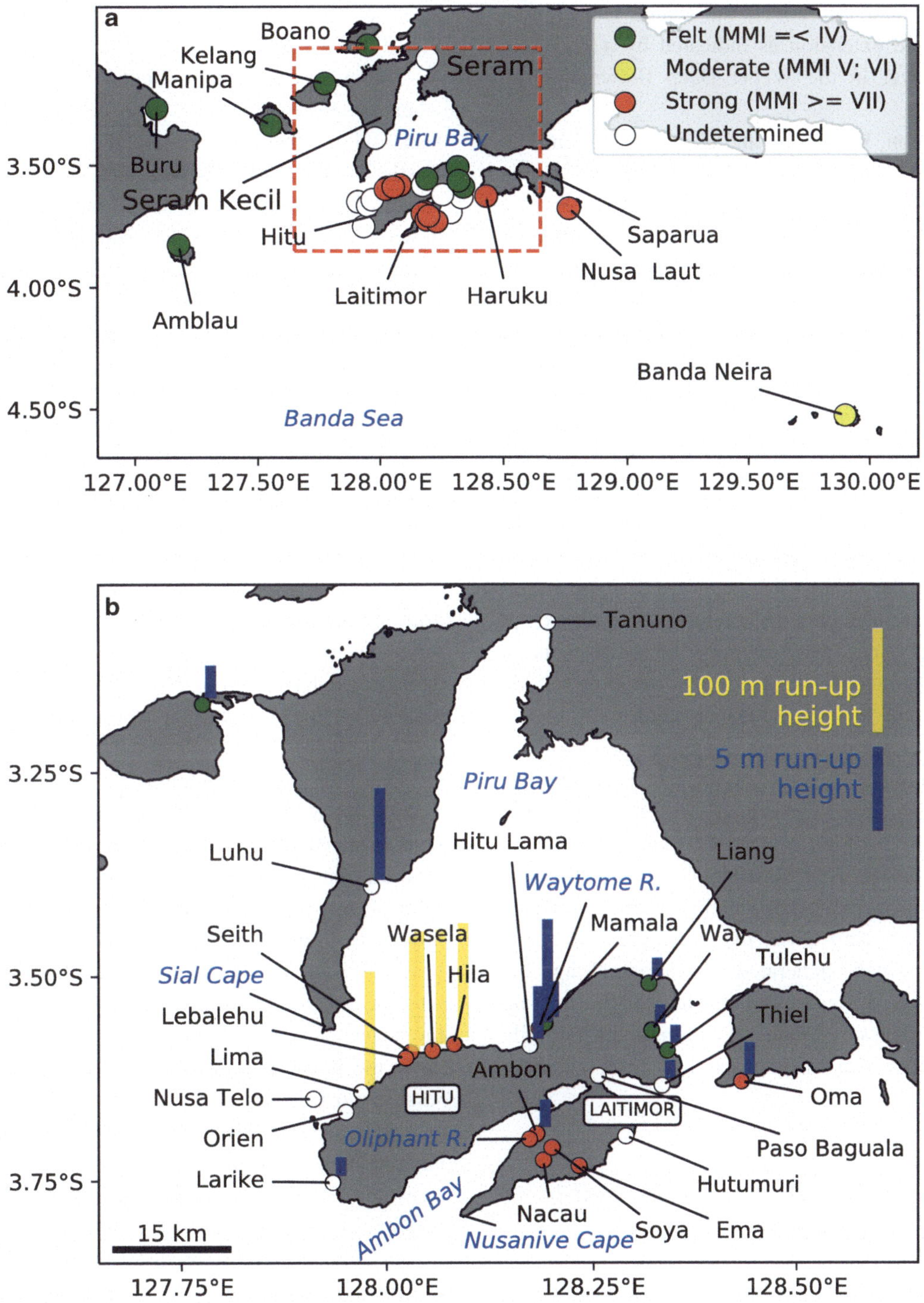

Figure 2

Historical accounts of the 1674 Ambon Island **a** earthquake and **b** tsunami. Location of map b is shown as the red dashed box. Colour circles represent interpreted earthquake intensity. Tsunami heights are indicated by the gold (100 m) and blue (5 m) color bars

At the southern end of Seram, Pownall et al. (2016) argue that the Kawa Fault continues to the Banda Sea and is related to the Banda Detachment (Fig. 1). The Banda Detachment is a recently discovered low-angle normal fault identified from geological observations and high-resolution bathymetry data. Moreover, the Banda Detachment exhibits rapid extension as indicated by a very thin sediment layer in the Weber Deep (Pownall et al. 2016).

On Ambon, Watkinson and Hall (2017) identified several faults via a digital elevation model of the Shuttle Radar Topography Mission (SRTM). First, a normal fault with a very steep angle was identified on the northern shore of the island. Then a lineament with NE–SW trend, crossing the city of Ambon was observed. Last, a suspected quaternary normal fault was observed on the southern side of Ambon Bay.

Ambon and its surrounding islands (Seram, Haruku, Saparua, Nusa Laut, and Buru) have experienced at least 14 tsunamis in the historical past (Table 1). There were two events each generated by volcano activities and landslides, earthquakes (seven events), and three events with an undetermined source. The majority of the events are categorised to be minor tsunamis with a maximum height up to 2 m, except for the 1674 and 1899 events. The 1674 tsunami event was started by strong ground motion felt on Ambon Island. An extreme tsunami run-up up to 100 m was observed only on the southern shore of the island with minor tsunamis in other areas (Rumphius 1675). The source of this tsunami will be discussed in the following sections. In the 1899 event, settlements on Seram Island were devastated by an earthquake that was suspected of originating from the Kawa Fault (Soloviev and Go 1974). It triggered multiple landslides which generated tsunamis along the northern shore of Seram Island. The tsunami rose up to 9 m at villages of Amahai, Paulohi, and Elpaputih, with no reports of tsunami on the north coast of Ambon.

3. *Historical Accounts of the Ambon Island 1674 Earthquake and Tsunami*

Historical accounts of the 17 February 1674 Ambon earthquake and tsunami were documented in the book "Waerachtigh Verhael Van de Schlickelijcke Aerdbebinge" written by Rumphius (1675). The book was translated into English with the title "The true history of the terrible earthquake" (Rumphius 1997). The historical accounts described below are based on the English translation and summarised in Fig. 2 and Table 2.

The event occurred at about 7:30 p.m. local time when people in Laitimor were celebrating Chinese Lunar New Year. In Ambon, the bells in Victoria Castle swung by themselves and people who were standing fell to the ground as the earth heaved up and down like the sea. Stone buildings collapsed and buried up to 80 people. Strong shaking was felt on the mountains in Laitimor, with rocks falling and ground cracking open. In Hutumuri near the coast, on the eastern side of Laitimor, seawater burst into the air like a fountain.

The earthquake was reported from Hitu as well. In Waytome River, on the northern side, the river water spurted to 6 m high. People in the north to northwest of Hitu heard a loud sound like canon-fire. They noticed two long, thin marks in the sky, extending from Luhu to Seith, shortly before the earthquake. Less than 15 min after the earthquake, villages between Lima and Hila were wiped away by a gigantic mountain of seawater. The seawater rose about 50 to 60 fathoms (approximately 90–110 m) to the top of the surrounding hills, and more than 2300 people perished.

This unusual phenomenon was reported at other places, but with much less intensity. Hitu Lama village, located approximately 15 km to the east of Hila, reported that the seawater rose only about 3 to 5 m, killing 35 people. A little further to the east, 40 houses in Mamala were swept away without fatalities. The settlement of Orien (present name; Ureng), which is located < 10 km west of Lima reported that the seawater rose and inundated land but it did not enter the houses. People in Larike, a village in the westernmost of Hitu, observed that the seawater rose < 1 m at the Rotterdam Redoubt. In the southern and eastern areas of Ambon Island, there was much less seawater oscillation reported, apart from some small boats being tossed over each other.

At Luhu in Seram Kecil, the seawater inundated trees and the dwellings of a company. The water rose to a height just over 5 m. At the northernmost tip of Piru Bay, half of the houses in Tanuno were engulfed by water, but without fatalities. Fishermen in Piru Bay said the sea remained calm with a noticeable ripple. A much lower seawater oscillation was noticed by people from Manipa, Salati, Haruku, Nusa

Table 1

Historical tsunamis on Ambon and its surrounding islands between 1659 and 1950

Date	Source	Description	References
9–11 November 1659	Volcano: Mt Teon	A tsunami was noticed at Ambon Bay (1–1.5 m)	Wichmann (1918), Soloviev and Go (1974)
17 February 1674	Landslide triggered by an earthquake	Strong ground motion devastated Laitimor region followed by an extreme tsunami only on the northern shore of Hitu (100 m). A prolonged sequences of aftershocks was reported, with the largest occurring on 6 May generating a weak tsunami at Ambon Bay	Wichmann (1918), Soloviev and Go (1974), Rumphius (1675)
28 November 1708	Undetermined	A strong "tidal wave" suddenly burst into Ambon Bay and oscillated throught the night until the following morning. An earthquake was felt a day before	Wichmann (1918), Soloviev and Go (1974)
5 September 1711	Earthquake	A strong earthquake was felt on Haruku, Saparua, Nusa Laut, and Banda Islands. At Paso Baguala, eastern shore of Ambon, seawater rose by 1.2 m and destroyed two houses	Wichmann (1918), Soloviev and Go (1974)
18 August 1754	Earthquake	Houses on Ambon were destroyed by an earthquake. It was followed by a tsunami observed at Hutumuri, Haruku, and Saparua	Wichmann (1918), Soloviev and Go (1974)
19 April 1775	Earthquake	In the morning, a strong ground motion felt at Ambon Bay made the seawater oscillate	Wichmann (1918), Soloviev and Go (1974)
25 August 1802	Earthquake	Seawater rose and damaged the coast of Ambon after a strong earthquake was felt on the island	Wichmann (1918), Soloviev and Go (1974)
16 December 1841	Earthquake	A moderate earthquake was felt on Ambon, Buru, and Amblau Islands that generated a tsunami. The tsunami rose up to 1.5 m in height and swept away houses at Ambon Bay and Amblau. The aftershocks were reported until 21 December	Wichmann (1918), Soloviev and Go (1974)
26 November 1852	Earthquake	The Banda Islands were devastated by strong ground motion that was felt on Ambon. A tsunami arrived 15 min later on the Banda Islands and was reported at Ambon Bay up to 2 m height	Wichmann (1918), Soloviev and Go (1974)
10 June 1891	Undertermined	In the evening, seawater at Saparua Bay retreated about 200 m and then rapidly returned, inundating the shore. An earthquake was felt a day before	Soloviev and Go (1974)
7 June 1892	Volcano: Mt Awu	Eruption of Awu Volcano. A tsunami was noticed at Ambon Bay (1.5 m)	Soloviev and Go (1974)
30 September 1899	Landslide triggered by an earthquake	A strong earthquake was suspected as originating from the Kawa Fault devastated the settlements of Seram Island. It generated a tsunami that swept away villages at Elpaputih and Taluti Bays on the southern shore of Seram up to 9 m height. The tsunami was reported in a village on the eastern shore of Piru Bay as well, and noticed on Ambon	Soloviev and Go (1974)
5 July 1904	Undertermined	In the morning, seawater suddenly retreated and then rapidly returned, inundating the shore. An earthquake was felt a day before	Soloviev and Go (1974)
8 October 1950	Earthquake	An M 7.3 earthquake at south of Ambon generated a tsunami that arrived about 15 min later at Ambon Bay	Latief et al. (2016)

Table 2

Tsunami and ground motion interpretation from Rumphius (1997)

Location (region: village)	Description	Estimated tsunami height (m)	Estimated earthquake intensity
Hitu: Larike, Rotterdam Redoubt	Seawater rose to 2 feet high around the redoubt. The water rose three times without causing major damage other than crashing a boat	< 1 (inundation depth)	Undetermined
Hitu: Nusa Telo	Seawater rose suddenly after first withdrawing in the direction of Orien. The water oscillated three times from two directions		Undetermined
Hitu: Orien	A loud roar was heard in the air. The seawater rose then was drawn in the direction of Nusa Telo. The water did not enter houses but only surrounded the barricades		Undetermined
Hitu: Lima, Haarlem Redoubt	Seawater came from the direction of Lebalehu and rose about 50 or 60 fathoms around the redoubt, carrying stones, mud, and sand. Large boulders were thrown on the first floor of the redoubt. A woman was swept away 20 fathoms behind the redoubt. Lime kilns were washed away, with at least 86 fatalities	90–110 (run-up height)	Undetermined
Hitu: Seith	Seawater rose about 50 to 60 fathoms, up to the windows of the fort. Villages between Seith, Lebalehu, and Wasela were washed away (but Hautunua village at a higher location was safe). Lime kilns were washed away, with more than 619 fatalities	90–110 (run-up height)	Strong
Hitu: Lebalehu	A 200 m-wide region around Lebalehu collapsed, making the beach very steep. This also happened between Seith and Hila (west of Fort Amsterdam), including Nukunali, Taela, and Wawani. An upwelling of seawater (to about 50 or 60 fathoms) originated from here, dividing into three: (i) east—Seith and Hila, (ii) west—Lima and Orien, (iii) to the sea. The water glowed like fire but was as black as coal and noisy. Lime kilns were destroyed	90–110 (run-up height)	Strong
Hitu: Hila, Fort Amsterdam	At least 1461 people were killed. Seawater rose between 50 and 60 fathoms high to the gallery of the fort and overtopped the roof. All the houses around the fort were swept away, particularly to the west and south. A grinding wheel, a drum, a large coral, fish cages, and cattle were swept away	90–110 (run-up height)	Strong
Hitu: Hitu Lama	Seawater rose 10 feet and swept away several houses. About 35 people were killed	3–5 (inundation depth)	Undetermined
Hitu: Mamala and Waytome River	40 houses were swept away, without fatalities. Limekilns were destroyed. Strong ground motion toppled equipment inside a warehouse. Eastern shore of Waytome River split open and water spurted 18–20 feet high	< 2 m (inundation depth)	Felt (Mamala)–Strong (Waytome River)
Hitu: Liang	The earthquake was felt with no damage to the houses		Felt
Hitu: Way	The earthquake was felt with no damage to the houses		Felt
Hitu: Tulehu	The earthquake was felt with no damage to the houses		Felt
Hitu: Thiel	A few houses were carried away by seawater		Undetermined
Laitimor: Paso Baguala, Fort Middleburg	Weapons and equipment were damaged. The earth around Baguala and Hutumuri cracked. Seawater came from the direction of Fort Victoria but did not overflow the isthmus		Strong
Laitimor: Hutumuri, near to the shore	Seawater burst upwards like a fountain		Strong
Laitimor: Cape Nusanive	Boats were tipped over and fishermen noticed the waves were a little higher than normal		Undetermined
Laitimor: Ambon, Victoria Castle	Bells in the castle swung by themselves. People around the castle fell over and saw the ground heaving up and down like the sea. Central walls of the town hall collapsed. A hospital was badly damaged. Boats at the mouth of Oliphant River were thrown into the stream. The bridge close by was almost shaken loose. Goods in houses shifted 3 to 4 feet without breaking. 79 people were killed and more than 35 injured when at least 75 buildings collapsed		Strong
Seram Kecil: Cape Sial	Seawater caused damaged		Undetermined

Table 2 *continued*

Location (region: village)	Description	Estimated tsunami height (m)	Estimated earthquake intensity
Seram Kecil: Luhu, Overburg Redoubt	Villages and boats were swept away. Seawater rose 3 fathoms above normal	5.5	Undetermined
Seram Kecil: Tanuno	Seawater rose once and half the houses were swept away, without fatalities. Fishermen claimed that the sea in Piru Bay was calm		Undetermined
Haruku: Oma	Earthquake described as violent, small rocks fell into the sea and a large boulder was moved. Seawater rose 6 feet above normal	2	Strong
Nusa Laut	Earthquake described as violent		Strong
Banda Islands: Neira	Earthquake described as moderate. Seawater rose a little, but no damage		Moderate
Buru	Earthquake was felt		Felt
Amblau	Earthquake was felt		Felt
Manipa	Earthquake was felt. Seawater rose at Fort Manipa. 40 houses swept away, but no fatalities	1–2	Felt
Kelang: Salati	Earthquake was felt. Seawater rose 6 feet above normal	2	Felt
Boano	Earthquake was felt		Felt
Laitimor: Nacau	Strong earthquake made seven houses collapse, large rocks fell		Strong
Laitimor: Ema -Soya	Road was cracked 2 to 3 feet wide in several places		Strong

Laut, and Banda Neira Island compared with the oscillation that was observed in Hitu and Laitimor. Aftershocks continued for at least 3 months. The two largest aftershocks occurred on 6 and 10 May.

What were the source of the ground motion and seawater phenomena described above, particularly on 17 February 1674? In the following sections, we will make use of the historical accounts of ground motion and tsunami observation to answer that question.

4. *Source Identification*

4.1. *Earthquake Source*

With regard to the tectonic setting around Ambon, there are five candidate faults that could have generated a large earthquake with the intensities and effects shown in Fig. 2: (1) the north Seram Megathrust, (2) the Kawa Fault, (3) an intraslab fault (4) the purported South Seram Thrust Fault, or (5) a local fault on Ambon. Each of these is qualitatively analysed below to identify the most credible source of this event.

In general, the strongest ground motions were felt on Laitimor, Oma (Haruku), and Nusa Laut (Fig. 2). The intensity of ground motion decreases toward Banda Neira in the south, and Boano in the north. Many buildings collapsed and ground cracked in various places on Laitimor. There was liquefaction at Hutumuri and Waytome Rivers according to the accounts of *'water spurt[ing] high [in]to the air'*. Recurring aftershocks were reported until at least 10 May. Judging by these observations, the source must have been a moderate to large earthquake with a local and shallow epicentre.

A shallow earthquake on the north Seram Megathrust would have been too far from Ambon to have these effects and if the earthquake had occurred on this fault, the islands of Boano, Kelang, and Manipa where moderate shaking was reported (Fig. 2) should have experienced stronger ground shaking than Ambon and Banda Neira. If the earthquake had originated from the Kawa Fault, the villages on Seram Island would have experienced more intense ground shaking as in the 1899 earthquake and tsunami event (Soloviev and Go 1974). A deep intraslab earthquake would generate ground motion felt over a broader region. However, deep intraslab earthquakes typically do not cause ground cracking and long aftershock sequences. The Benioff zone is over 100 km beneath Ambon (Spakman and Hall 2010), and an intraslab earthquake at this depth would likely generate strong ground motion distributed over a wider area than observed.

Therefore, credible sources for this event could have be the South Seram Thrust Fault or a local fault on Ambon. The South Seram Thrust was used in seismic and tsunami hazard maps of Indonesia (Irsyam et al. 2010; Horspool et al. 2014), where it is indicated as running from the south shore of Buru to Nusa Laut with northward dip (Fig. 1). Although there is apparently evidence for this fault in proprietary marine seismic survey results (J. Griffin, personal communication 2018), we note that it does not appear on the most recent revision of Indonesia's seismic hazard map (Irsyam et al. 2019). We therefore regard the existence of this fault as speculative, even though an earthquake on it could have caused the observed ground shaking. Brouwer (1921) and Watkinson and Hall (2017) have confidently identified quaternary faults on Ambon itself, one of which Harris and Major (2017) identified as the source of the 1674 earthquake without a clear explanation of why. We conclude that any of these faults on and near Ambon could have generated the observed ground shaking, but further investigation is required to determine exactly which one ruptured in 1674.

4.2. Tsunami Source

If we consider which of the five candidate faults previously mentioned might have been capable of directly generating the observed tsunami, it is immediately obvious that an intraslab event could not, since an earthquake deeper than 100 km in the Benioff zone beneath the Banda islands could generate only a weak tsunami at best. Okal and Reymond (2003) showed that the largest intraslab event ever recorded, the 1938 $M_w = 8.5$ Banda Sea event at 60 km depth, generated only a weak tsunami. A tsunami generated by the Seram Megathrust or South Seram Thrust would have to enter Piru Bay through narrow straits around Ambon Island, which would greatly attenuate the tsunami arriving on the north coast of Ambon. As we show in Sect. 6 below, instead of having highest run-up along the north coast of Ambon, the tsunami energy generated by these two scenarios would be concentrated between the western and southern coast of Hitu and Laitimor, respectively. The Kawa Fault has a strike-slip mechanism that might generate a significant local tsunami generation where it crosses the southern shore of Seram, such as occurred in the 1994 Mindoro (Imamura et al. 1995) and 2018 Palu (e.g. Jamelot et al. 2019) events. However, it seems unlikely this mechanism could generate high tsunami run-up only along the northern coast of Ambon.

The only fault that might generate a large tsunami on the northern coast of Ambon and nowhere else would be a local fault documented on the northern shore of Hitu (Brouwer 1921; Watkinson and Hall 2017). The fault has a normal mechanism that could generate vertical displacement of a water column for tsunami generation. However, the fault under consideration is only 16 km long, and therefore unlikely to generate an earthquake with magnitude < 7 and slip much higher than 2 m (Kanamori and Anderson 1975; Geller 1976; Wells and Coppersmith 1994), far too small to generate the observed tsunami run-up. Therefore, if none of the potential tsunamigenic earthquakes could have been capable of directly generating a tsunami with the observed run-up, the most plausible source of the tsunami is from an earthquake-triggered landslide in Piru Bay.

There is an indication in Rumphius' accounts 1997 that a massive coastal landslide occurred on the northern shore of Hitu, between Seith and Hila.

> The country around Lebalehu, a region once famed for its market and for being the most important Muslim meeting place, collapsed the width of a Musquet shot. There is no longer any beach there, but only a very steep precipice. Just the same is true between Ceyt (Seith) and Hila, even as far as the beach at the later place, along the west side of the Fort Amsterdam and beneath the Residence of Intche Tay. Including the Negeris Nukunali, Taela and Wawani, all this disappeared along with [the] roadstead where ships used to anchor. It seems likely that the aforementioned wall of water arose in the place just indicated, to wit directly below Lebalehu. It might even have come from Hitu because various people on board ships that were not far off shore, reported only a rippling of the waves. The mass of upwelling water divided into three parts. One went east to Ceyt (Seith) and Hila, the other west to the villages of Lima and Oried. The water stank so horribly that people on board ships close to the coast became ill, and it was so filthy that anyone who

had been immersed in it looked as if he had been hauled out of a mudbath.

In other words, there was a major change in the landscape around the coastal region: a gentle beach became very steep. Moreover, people saw that the water was dark and a roaring sound was heard from this area, which indicates that the seawater mixed with sediment and the tsunami source was near the people. The seawater colour and loud sound described in the accounts are similar to those described after the 1998 Aitape, Papua New Guinea (PNG) tsunami (Davies et al. 2003), which is thought to have been generated by a landslide. Unfortunately, there is no further information with regard to parameters of the landslide. According to the accounts, it was a partially subaerial landslide, with about 200 m of the previous shoreline collapsing into the sea. The lateral extent is thought to be at least 5 km along the coastline between the villages of Seith and Hila.

Further, the majority of the fatalities in this event were in this region and not many buildings collapsed in the earthquake. In addition, an extreme run-up laterally along the coastline can only be explained by a landslide. These observations are similar to the 1998 Aitape, PNG tsunami, which was generated by an underwater landslide (Okal and Synolakis 2004; Synolakis et al. 2002).

5. *Landslide-Generating Tsunami*

Tsunami generation is affected by both the vertical and length extent of the displace bathymetry (Heidarzadeh et al. 2014). Vertical displacement due to an earthquake normally reaches only a few metres at most, whereas it can easily reach hundreds of metres in a mass failure event. The dimensions of an earthquake rupture can extend to hundreds of kilometres but it is rare to see a huge landslide up to 100 km length. Therefore, a tsunami generated by an earthquake has a long wavelength as it travels across the ocean. On the other hand, a tsunami generated by a landslide loses its energy quickly as it travels because of its shorter wavelength, but it has a much larger amplitude in a local area. Therefore, tsunamis generated by earthquakes can be observed over a wide area, with a relatively constant run-up along the coastline. In contrast, a narrow region of extreme run-up is observed in a tsunami event generated by a landslide.

Okal and Synolakis (2004) investigated tsunami height profiles along a coastline from some past tsunami events. They found similar patterns in the 1946 Unimak (Alaska) and 1998 Aitape (PNG) tsunami profiles, which were both generated by landslides. The profiles had extreme run-up heights over a very narrow along-shore extent. Based on these findings, Okal and Synolakis (2004) developed a criterion for identifying landslide-generated tsunami based on the aspect ratio $I_2 = \frac{b}{a}$ between the maximum tsunami run-up height b and the distance a of the lateral extent of high tsunami run-up along the coastline. Thus, any event with I_2 larger than 0.0001 should be considered a tsunami generated by a landslide.

Accordingly, we calculate the I_2 of the tsunami observations on the northern coast of Hitu. It was difficult to determine the exact values of a and b because of the sparse data available (Fig. 3a). However, the data clearly shows a very rapid drop in run-up to the east and west of the maximum run-up height location between Lima and Hila. The estimated I_2 was > 0.006, which was 60 times larger than the limit suggested by Okal and Synolakis (2004). Therefore, this tsunami almost certainly was generated by a landslide.

6. *Tsunami Modelling*

Tsunami modelling was performed to confirm the analysis above using the JAGURS tsunami simulation code (Baba et al. 2015, 2017). The code numerically solves the shallow water wave equations in a spherical coordinate system with a finite-difference scheme. The digital elevation model (DEM) was built from a combination of nautical charts, a 90-m commercial bathymetry dataset provided by the TCarta Marine, the General Bathymetric Chart of the Oceans (GEBCO), and the SRTM-90m in a domain of nested grids (Fig. 4). The coarsest and finest grid resolution of the domain is approximately 1500 and

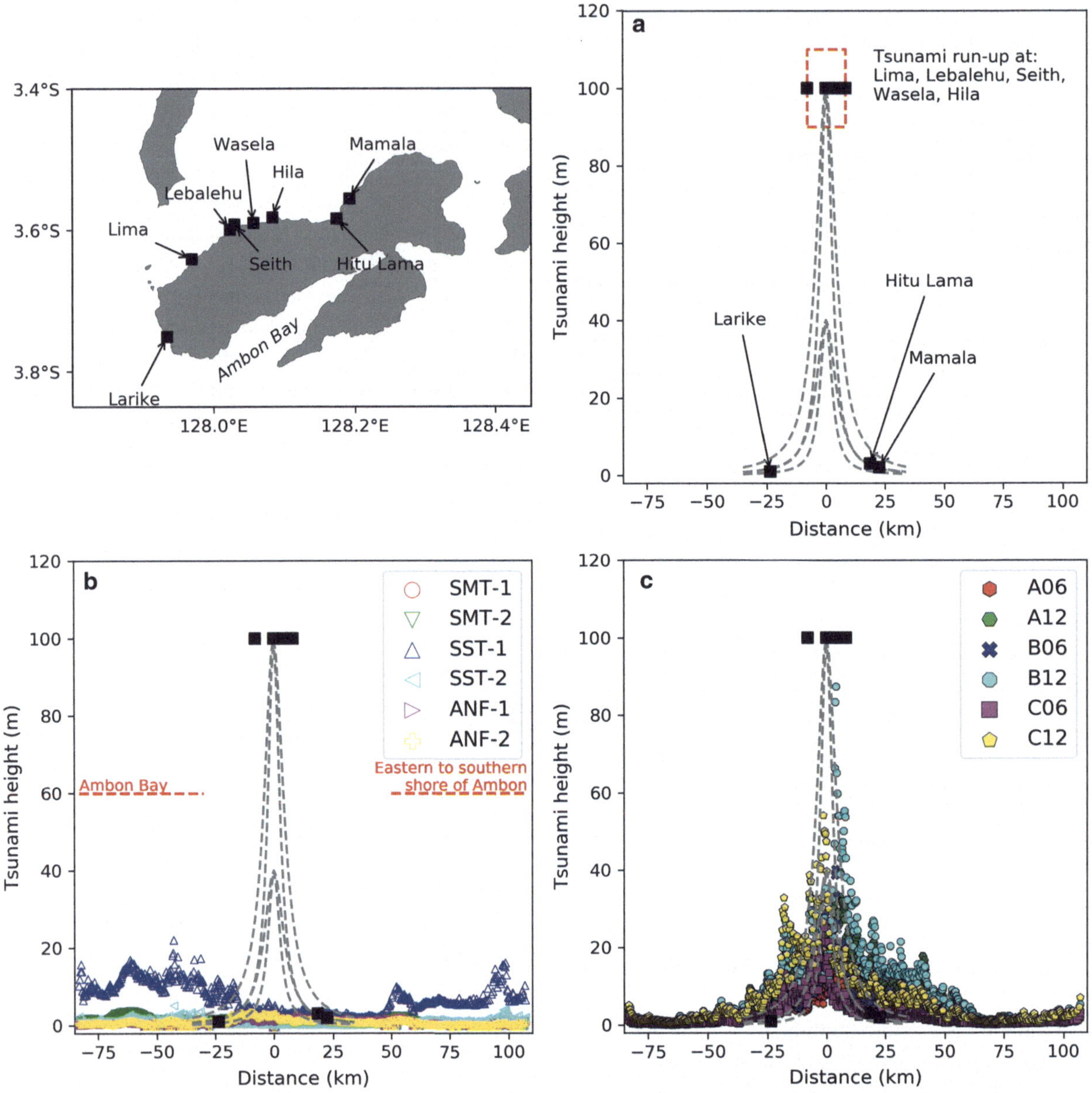

Figure 3

Tsunami height profiles along the coastline: **a** black squares are data; the dashed lines indicate I_2 of $a = 3$–5 km with $b = 40$–100 m; **b** only from tsunamigenic earthquake scenarios; **c** selected landslide-generating tsunami scenarios. Symbols with colour indicate the scenario codes as shown in Tables 3 and 4

167 m, respectively. A time-step of 0.5 s is set to satisfy the Currant stability condition.

Tsunami simulations were conducted for tsunamigenic earthquake (Table 3) and landslide-generated tsunami scenarios in this area (Table 4). The initial sea surface elevation from the earthquake scenarios were assumed to be equal to the earthquake deformation calculated from the Okada (1985) formula, using the parameters discussed below.

A two-layer model was utilised to simulate tsunami generation and propagation due to a landslide (Baba et al. 2019). As with the single-layer, earthquake-generated tsunami simulation, the two-layer simulation requires specification of ocean depth (i.e.,

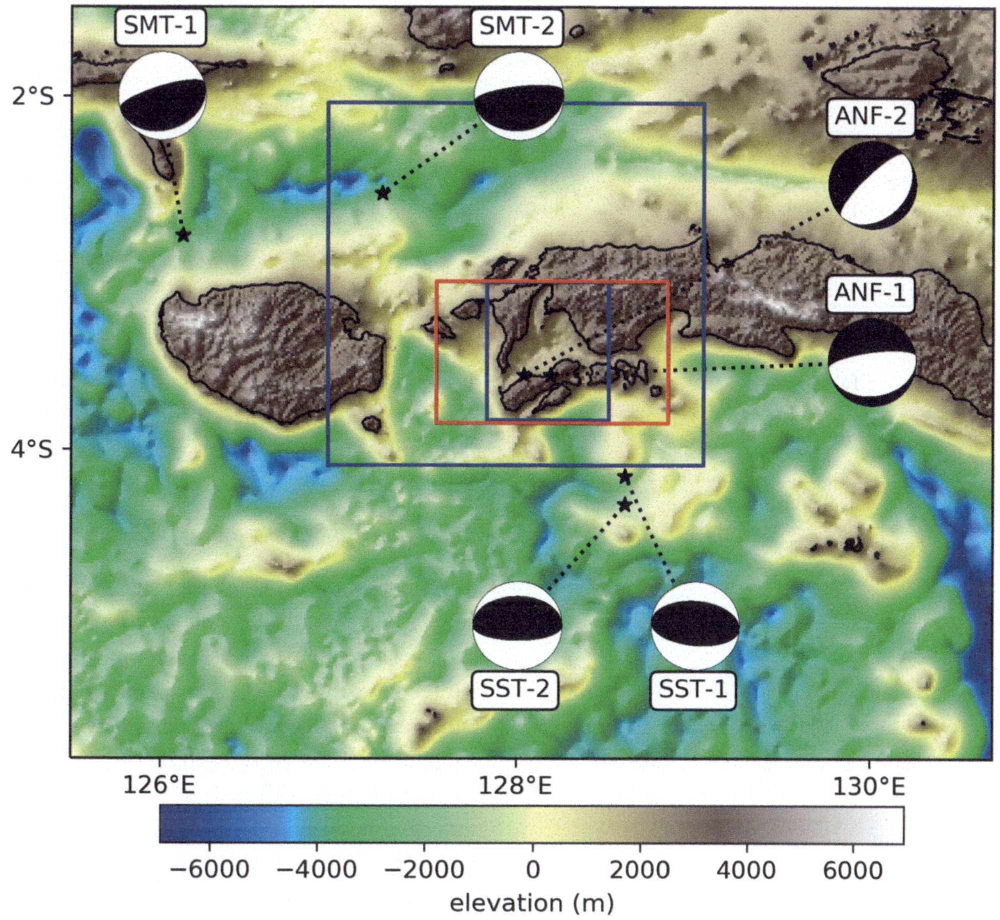

Figure 4
Domain model of tsunami modelling. The blue boxes are the nested-grids domain for the tsunamigenic earthquake model and red box shows the domain model for landslide-generating tsunami scenarios. Beach balls and black starts represent scenarios on Table 3

bathymetry). In addition, the two-layer simulation requires an initial condition for the depth to the top of a viscous fluid layer, which is allowed to flow in response to gravity, exciting a tsunami by displacing the ocean layer above it (see Baba et al. (2019) for further details of the two-layer algorithm). The two-layer simulation was performed in a single grid domain model, over a tsunami propagation time of 2 h with time-step of 0.25 s.

6.1. Tsunamigenic Earthquake Scenarios

SMT-1 and SMT-2 (Fig. 5) represent two earthquake scenarios from the Seram Megathrust zone. The M_w = 8.2 parameters of SMT-1 (Løvholt et al. 2012) were speculated to be the source of the events documented in 1657, 1708, 1876, and 1965 (note that here and in subsequent sections, we use M_w to denote an input parameter for tsunami modelling, not a magnitude estimated for a historical earthquake). According to Horspool et al. (2014) and Irsyam et al. (2010), the Seram Megathrust zone is capable of generating an earthquake with a maximum $M_w = 7.9$ to 8.2. Here, SMT-2 represents an earthquake with $M_w = 8.1$ with the fault parameters taken from Horspool et al. (2014).

SST-1 and SST-2 (Fig. 5) represent two scenarios from the South Seram Thrust Fault. Løvholt et al. (2012) assumed the 1674 Ambon tsunami was caused by an $M_w = 8.1$ earthquake on the South Seram Thrust Fault. The authors suggested that the same fault zone triggered the 1950 and 1983 events. The source parameters of the SST-1 scenario were taken from

Table 3

Tsunamigenic earthquake scenarios for the 1674 Ambon event

Scenario	Position[a]	Length (km)	Width (km)	Strike	Dip	Rake	Slip (m)	Mo (Nm)
2.2SMT-1	2.97° S, 126.13° E, 0 km	202	80	73°	20°	90°	8.0	2.24×10^{21}
SMT-2	2.55° S, 127.25° E, 5 km	180	90	83°	20°	90°	5.0	1.58×10^{21}
SST-1	4.15° S, 128.61° E, 0 km	179	80	− 85°	40°	90°	7.0	1.58×10^{21}
SST-2	4.31° S, 128.61° E, 5 km	70	40	− 90°	30°	90°	1.5	2.00×10^{20}
ANF-1	3.58° S, 128.19° E, 2 km	25	10	− 98°	70°	− 90°	5.0	1.12×10^{18}
ANF-2	3.58° S, 128.04° E, 2 km	20	10	− 135°	70°	− 90°	5.0	6.31×10^{18}

Position[a]: top-right corner coordinate of a fault plane

Table 4

Landslide-generating tsunami scenarios

Centre of the Gaussian		
Location	Longitude	Latitude
A	128.07° E	3.585° S
B	128.03° E	3.585° S
C	128.02° E	3.585° S
Gaussian parameters		
ID	Radius (m)	Thickness (m)
1	500	100
2	500	200
3	500	300
4	1000	100
5	1000	200
6	1000	300
7	2000	100
8	2000	200
9	2000	300
10	2500	100
11	2500	200
12	2500	300

their study. Another study conducted by Latief et al. (2016) indicated the source of the 1950 Ambon tsunami as being on the South Seram Thrust Fault, but it was located a little further south, with a smaller fault plane. The SST-2 scenario followed the parameters used in Latief et al. (2016) and is represented as an $M_w = 7.5$ earthquake.

Finally, ANF-1 and ANF-2 (Fig. 5) were designed to follow the normal fault, as indicated by Brouwer (1921) and Watkinson and Hall (2017), respectively. These two scenarios represent earthquakes with $M_w = 6.0$ to 6.5, respectively. They are only hypothetical scenarios because no detailed study has been conducted on these faults, other than categorisation as quaternary faults (Watkinson and Hall 2017). Here a dip of 70° was selected as the optimum angle of a normal fault earthquake to generate maximum vertical displacement.

6.2. *Landslide-Generating Tsunami Scenarios*

We tested 36 synthetic scenarios of landslide-generating tsunami models (Table 4). The landslide layer was assumed to have a Gaussian function on top of the recent DEM with a deformable material type. The radius and thickness of the Gaussian varied from 500 to 2500 m and between 100 and 300 m, respectively. It was located at three different places (A, B, C) near the shoreline between Seith and Hila. The locations were selected according to the accounts previously discussed. To satisfy the stability of the model, the landslide layer needed to be 'clipped', so that the landslide would always remain below the at-rest level of the sea surface. Therefore, the landslide layer looked like a Gaussian-shaped accumulation of sediment that is clipped where it emerges above the water surface (Fig. 6).

7. *Results and Discussion*

Piru Bay is a nearly closed sea surrounded by several narrow straits surrounding Ambon Island. It is difficult for much tsunami energy to propagate from outside the bay to Ambon's northern shore, particularly in the SMT-1, SMT-2, SST-1, and SST-2

Figure 5 ▶

Initial sea surface elevation from tsunamigenic earthquake scenarios shown in Table 3. SMT-1 and SMT-2 represent the Seram Megathrust; SST-1 and SST-2 are the South Serahm Thrust; and ANF-1 and ANF-2 are local faults on the Ambon scenarios. Each figure has a different colour scale

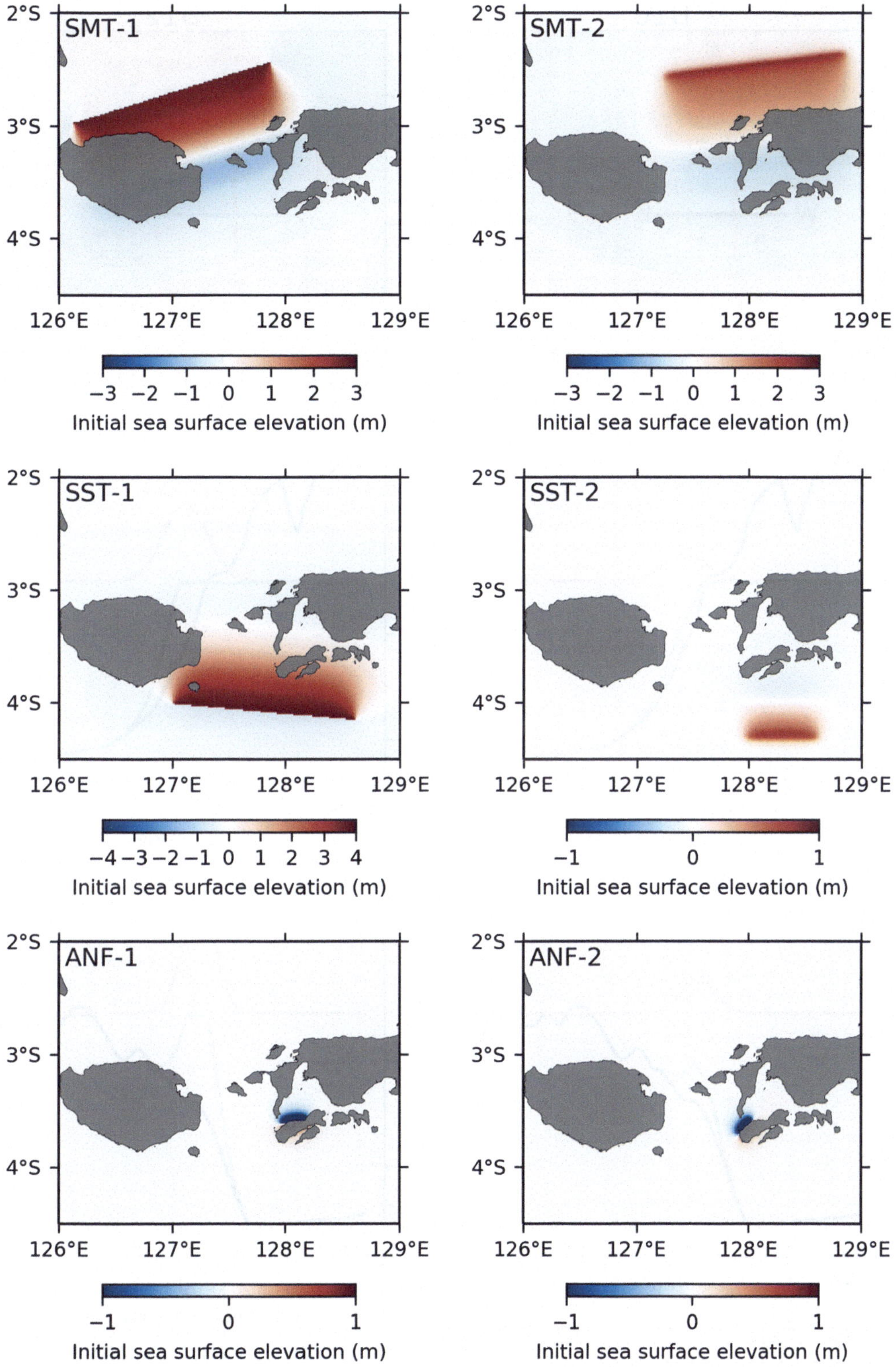
SMT-1
SMT-2
SST-1
SST-2
ANF-1
ANF-2
2°S
3°S
4°S
126°E
127°E
128°E
129°E
−3 −2 −1 0 1 2 3
−4 −3 −2 −1 0 1 2 3 4
−1 0 1
Initial sea surface elevation (m)

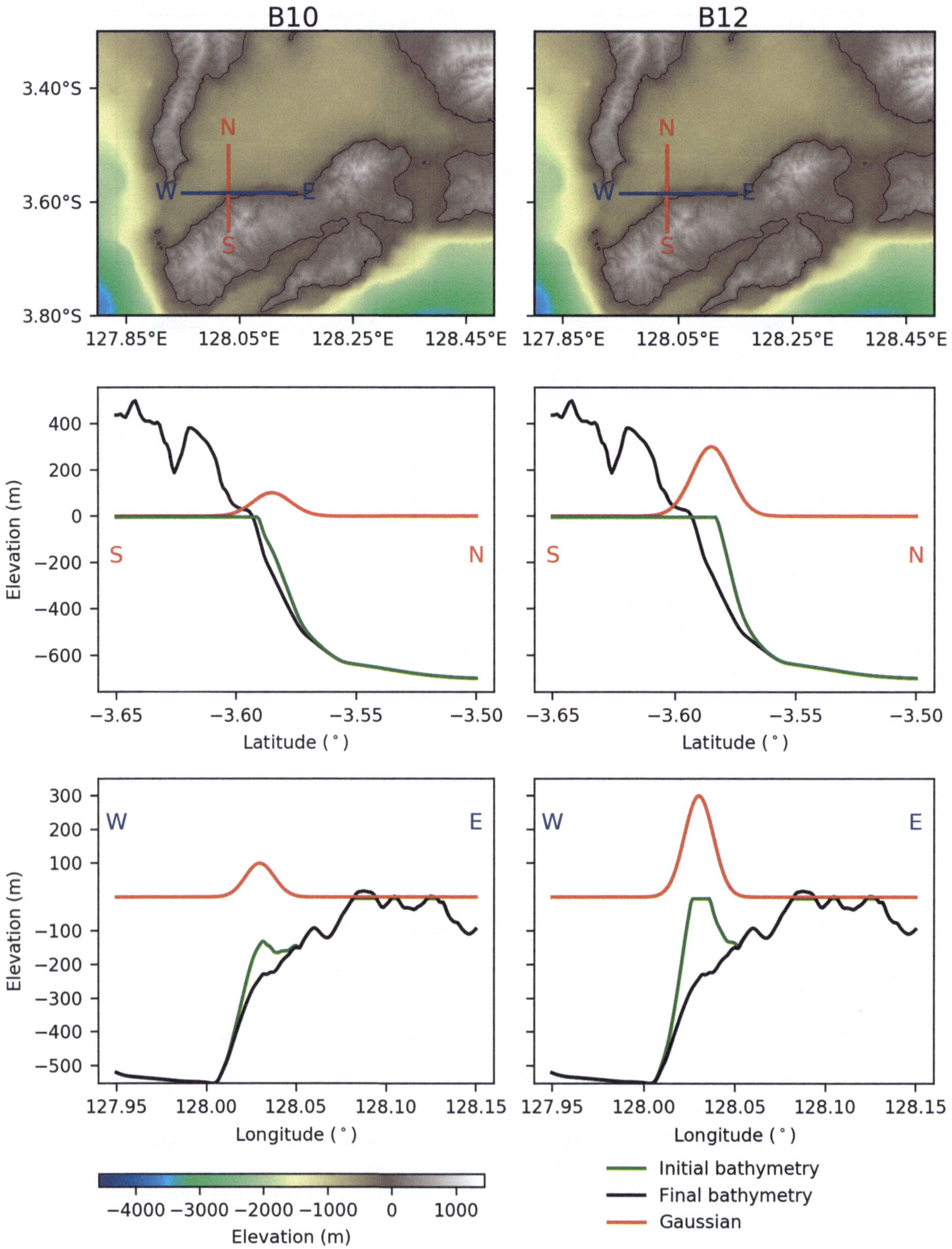
B10
B12
3.40°S
3.60°S
3.80°S
127.85°E
128.05°E
128.25°E
128.45°E
N
S
W
E
Elevation (m)
Latitude (°)
−3.65
−3.60
−3.55
−3.50
400
200
0
−200
−400
−600
Longitude (°)
127.95
128.00
128.05
128.10
128.15
300
100
−100
−300
−500
−4000 −3000 −2000 −1000 0 1000
Elevation (m)
Initial bathymetry
Final bathymetry
Gaussian

◀Figure 6
Illustration of selected tsunami landslide scenarios shown in Table 4. B10 and B12 are landslides with the centre of the Gaussian located at Point B (128.030° E and 3.583° S) with each radius of 2500 m and maximum thickness of 100 and 300 m, respectively. The middle and bottom figures are 'clipped' initial bathymetry (green), final bathymetry (black), and the Gaussian (red) profiles along the W–E and N–S direction

scenarios. Therefore, the maximum tsunami amplitude simulated was relatively small (Fig. 7). These results meant that a tsunami source inside the bay is required.

As suspected, no tsunamigenic earthquake model could generate a high tsunami inside Piru Bay only, with the maximum height located on the northern shore of Ambon, except the normal fault scenarios (Fig. 3b). However, the fault length limits the magnitude, requiring a huge slip to produce a very high tsunami. This would be an unrealistic model. The normal fault model scenarios in this study were already unrealistic, with a 5-m slip.

All the landslide-generating tsunami simulations result in a maximum tsunami height concentrated inside Piru Bay between Seith and Hila with minor tsunamis at the other places (Fig. 8). Through these simulations, we confirm the tsunami height distribution mentioned in the historical accounts (Rumphius 1675). The largest tsunami height near the shoreline is almost 80 m from the B12[1] landslide scenario, with an approximate volume of 1 km^3 (Fig. 3c).

The 100 m tsunami run-up between Seith and Hila villages could not be reconstructed for several reasons. First, there was no high-resolution DEM available to accommodate detailed inundation modelling. Therefore, the tsunami height profiles were extracted along the depth of 20 m and Green's law (Synolakis 1991) was used to estimate the run-up height on the shoreline. Second, any DEM would not represent the actual elevation at that time. Third, it was possible that the tsunami run-up reported had been exaggerated. For example, *'the water rose especially between those villages (Lima and Hila) and Seith to the top of the surrounding hills, estimated to be some 50 to 60 fathoms [90–110 m] high'*. This account could have meant a maximum run-up height due to water splash. According to the recent DEM, the closest hill with a height above 100 m was located about 500 to 1500 m from the shoreline, meaning the northern coast had a narrow backed strip coastal by steep topography. Moreover, some accounts noted seawater rising as high as the window of a redoubt in Seith and over-topping Fort Amsterdam in Hila. According to the online photos available, all the redoubts and forts are located near the coastline and are two to three floors high (approximately 20–30 m). Therefore, our simulated landslide scenario produced a reasonable tsunami height.

Although the primary source of the ground motion was still unclear, it was most likely from a local and shallow earthquake. From the reports of ground cracking and building damage in Laitimor region, the earthquake location was most likely south of Ambon or Nusa Laut, with the northern limit in the Laitimor region. Further investigation is highly recommended.

7.1. *Limitations*

Because our DEM cannot resolve evidence of a slump, confirmation of this finding through a high-resolution bathymetric survey, as well as a more sophisticated landslide-generating tsunami model is needed. The two-layer model in the JAGURS code considered only a submarine landslide type. Therefore, as noted earlier, the landslide layer had to be 'clipped' to keep it below sea level, for code stability, which may have resulted in underestimating the actual condition. In addition, the code was a frictionless model, so the slide layer did not stop moving during the simulation.

7.2. *Implications for the Other Historical Accounts*

The primary source of this earthquake and tsunami was confirmed from the sparse historical accounts (Rumphius 1997). While the precise mechanism of the earthquake remains unclear, the source of the tsunami could be confirmed to be a coastal landslide.

This study has shown a technique for optimising sparse and incomplete accounts. The historical accounts of tsunamis in Indonesia have been documented in several catalogues (e.g. Wichmann

[1] A landslide located at Point B from ID-12—Table 4.

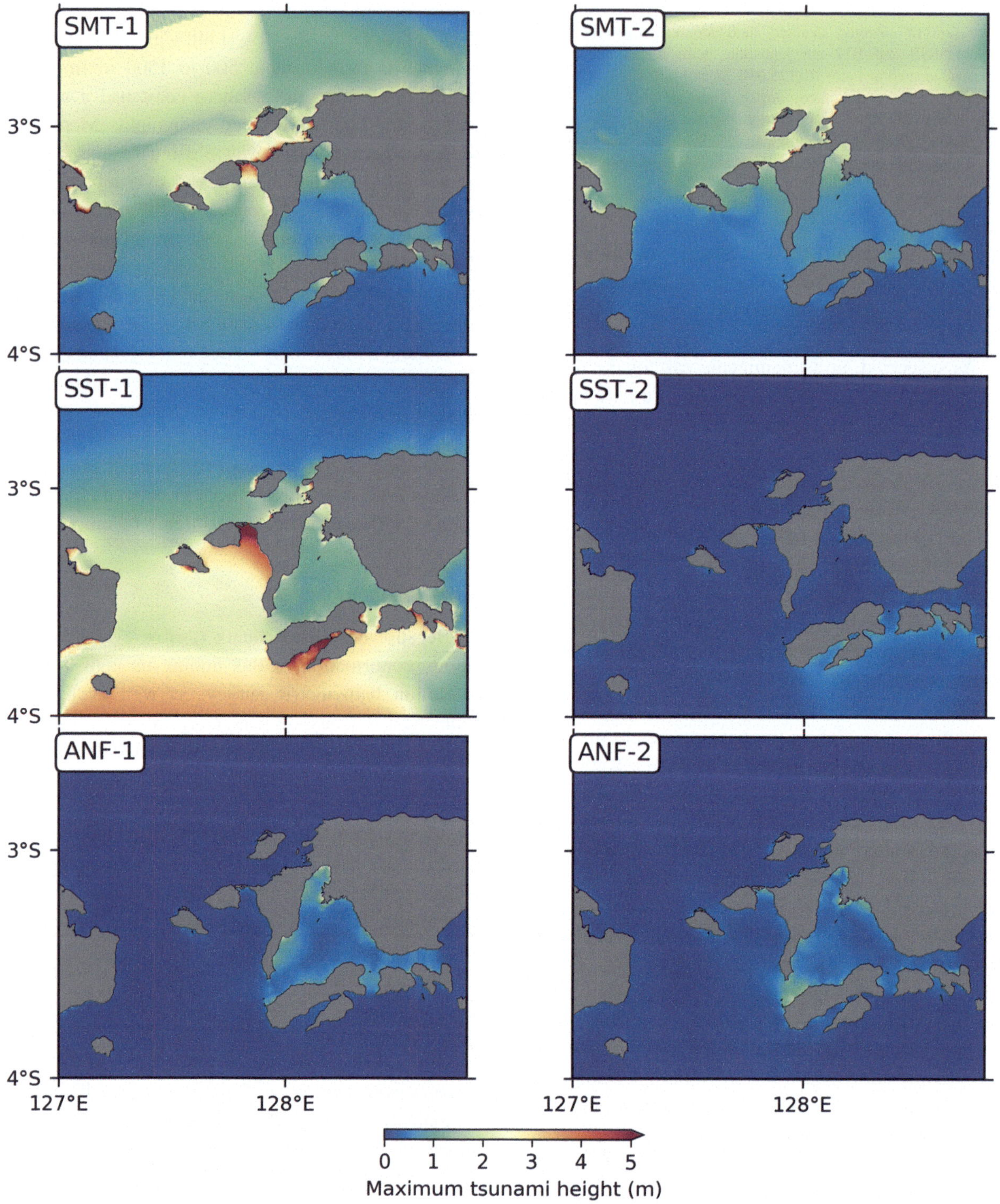

Figure 7
Maximum tsunami heights from the tsunamigenic earthquake scenarios shown in Fig. 5

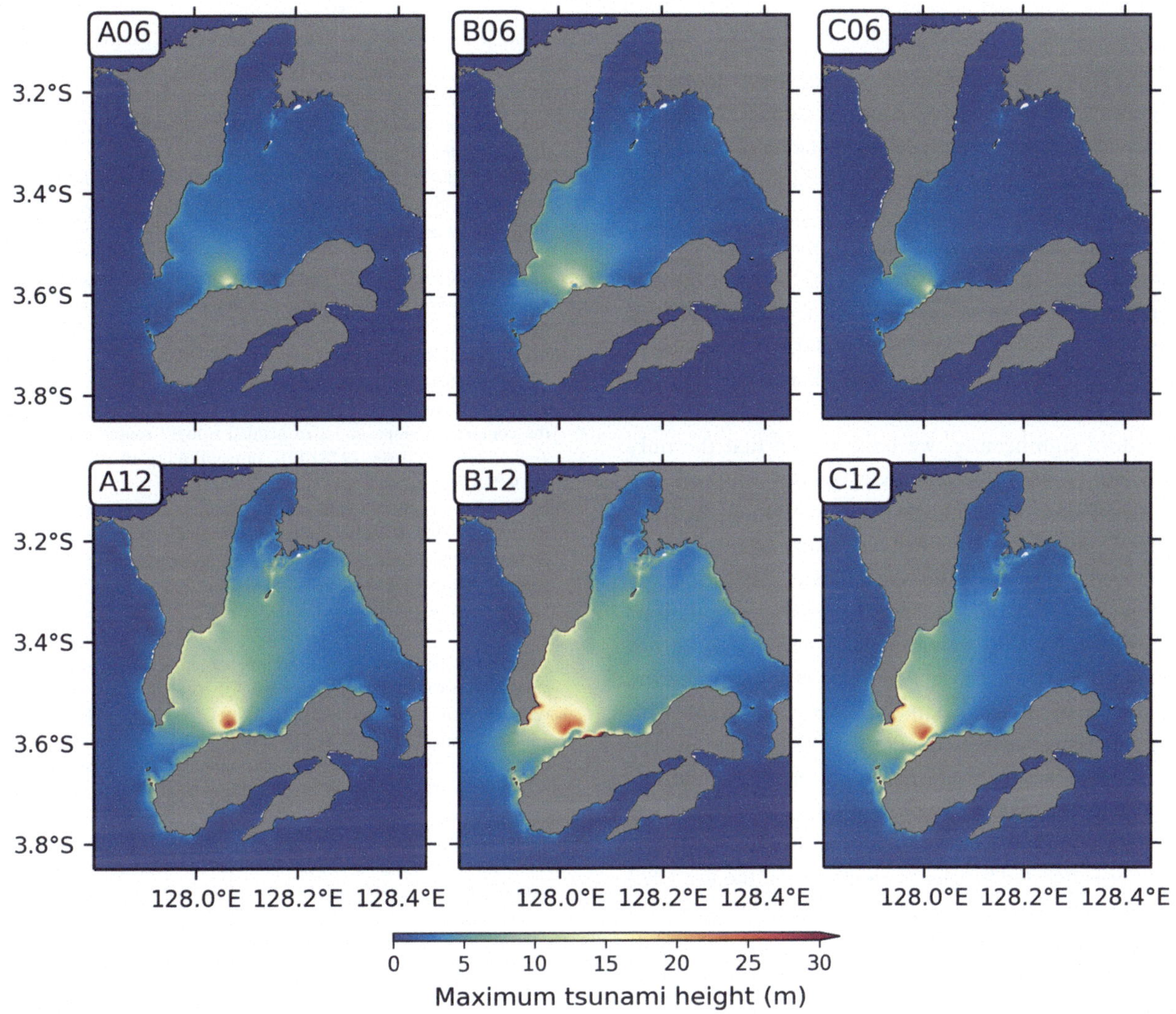

Figure 8
Selected maximum tsunami height from landslide scenarios shown in Table 4

1918, 1922; Soloviev and Go 1974; Soloviev et al. 1986). Generally, all accounts of tsunami events began by describing ground motion felt at various places with different intensities. These were followed by tsunami heights observation along the coastline. For the 1674 Ambon event, there is no detailed information regarding the source of the ground motion, nor the tsunami. By using similar technique that used in this study, the primary source of the tsunamis noted in the catalogue might be changed and/or updated in more detail. For example, the primary source of the devastating 1899 Seram tsunami was associated with an earthquake from the Kawa Fault. The kind of analysis we have employed here might help resolve whether the tsunami was generated by coseismic displacement where the strike-slip fault crosses the shore, such as occurred in the 1994 Mindoro (Imamura et al. 1995) and the 2018 Palu (Jamelot et al. 2019) events, or was mainly caused by the landslides observed along the southern coast of Seram (Soloviev and Go 1974).

Investigating other historical accounts will help to reveal more about the most likely primary source of the tsunami in each region. This will allow a more comprehensive tsunami hazard assessment.

8. Conclusion

In this study we have shown how historical accounts of the 1674 Ambon tsunami, one of Indonesia's largest and deadliest tsunami disasters, can be used to better understand its earthquake source and mechanism of generation. We have shown that, although the reports of ground motion intensity and damage fail to definitively identify the earthquake source, it is almost certain that it was local and shallow, probably either a crustal fault on Ambon itself or the putative South Seram Thrust Fault off its southern coast.

More significantly, we have shown that the only way to explain the extreme run-up only on the northern shore of Ambon, and in particular the very narrow lateral extent along the coast over which it occurred, is by attributing the tsunami generation to a submarine landslide. Tsunami scenario simulations showed that plausible earthquake sources could not generate such a run-up profile, but a submarine landslide of about 1 km^3 volume, consistent with eyewitness accounts of dramatic changes in the coastal landscape, could produce run-up commensurate with the historical observations.

Our analysis of the 1674 Ambon tsunami suggests that, as suggested by more recent events like the 1992 Flores and 2018 Palu and Sunda Straight tsunamis (Pranantyo and Cummins 2019; Sassa and Takagawa 2019; Giachetti et al. 2012; Patton et al. 2018, respectively), submarine landslides are an important component of the tsunami threat in Indonesia and should be considered in future tsunami hazard assessments. We believe further work on this and other historical events, especially when combined with paleotsunami and bathymetric surveys, can provide important constraints on the unique nature of the tsunami threat in regions of particularly complex and active tectonics like eastern Indonesia.

References

Audley-Charles, M., Carter, D., Barber, A., Norvick, M., & Tjokrosapoetro, S. (1979). Reinterpretation of the geology of Seram: Implications for the Banda Arcs and northern Australia. *Journal of the Geological Society*, *136*(5), 547–566.

Baba, T., Allgeyer, S., Hossen, J., Cummins, P. R., Tsushima, H., Imai, K., et al. (2017). Accurate numerical simulation of the far-field tsunami caused by the 2011 Tohoku earthquake, including the effects of Boussinesq dispersion, seawater density stratification, elastic loading, and gravitational potential change. *Ocean Modelling*, *111*, 46–54. https://doi.org/10.1016/j.ocemod.2017.01.002.

Baba, T., Gon, Y., Imai, K., Yamashita, K., Matsuno, T., Hayashi, M., & Ichihara, H. (2019). Modelling of a dispersive tsunami caused by a submarine landslide based on detailed bathymetry of the continental slope in Japan Nankai trough, southwest Japan. *Tectonophysics*, *768* (228182). https://doi.org/10.1016/j.tecto.2019.228182.

Baba, T., Takahashi, N., Kaneda, Y., Ando, K., Matsuoka, D., & Kato, T. (2015). Parallel implementation of dispersive tsunami wave modeling with a nesting algorithm for the 2011 Tohoku tsunami. *Pure and Applied Geophysics*, *172*(12), 3455–3472. https://doi.org/10.1007/s00024-015-1049-2.

Brouwer, H. (1921). Some relations of earthquakes to geologic structure in the East Indian Archipelago. *Bulletin of the Seismological Society of America*, *11*(3–4), 166–182.

Davies, H., Davies, J., Perembo, R., & Lus, W. (2003). The Aitape 1998 tsunami: Reconstructing the event from interviews and field mapping. In J. P. Bardet, F. Imamura, C. E. Synolakis, E. A. Okal, & H. L. Davies (Eds.), *Landslide tsunamis: Recent findings and research directions* (pp. 1895–1922). Basel: Springer. https://doi.org/10.1007/978-3-0348-7995-8_7.

Geller, R. J. (1976). Scaling relations for earthquake source parameters and magnitudes. *Bulletin of the Seismological Society of America*, *66*(5), 1501–1523.

Giachetti, T., Paris, R., Kelfoun, K., & Ontowirjo, B. (2012). Tsunami hazard related to a flank collapse of Anak Krakatau volcano, Sunda Strait, Indonesia. *Geological Society, London, Special Publications*, *361*(1), 79–90. https://doi.org/10.1144/SP361.7.

Hamilton, W. (1979). *Tectonics of the Indonesian region* (4th ed.). Washington: United States Government Printing Office.

Harris, R., & Major, J. (2017). Waves of destruction in the East Indies: The Wichmann catalogue of earthquakes and tsunami in the Indonesian region from 1538 to 1877. *Geological Society, London, Special Publications*, *441*(1), 9–46. https://doi.org/10.1144/SP441.2.

Heidarzadeh, M., Krastel, S., & Yalciner, A. C. (2014). The state-of-the-art numerical tools for modeling landslide tsunamis: A short review. In S. Krastel, et al. (Eds.), *Submarine mass movements and their consequences* (pp. 483–495). Cham: Springer.

Honthaas, C., Réhault, J. P., Maury, R. C., Bellon, H., Hémond, C., Malod, J. A., et al. (1998). A neogene back-arc origin for the Banda Sea basins: Geochemical and geochronological constraints from the Banda ridges (East Indonesia). *Tectonophysics*, *298*(4), 297–317. https://doi.org/10.1016/S0040-1951(98)00190-5.

Horspool, N., Pranantyo, I., Griffin, J., Latief, H., Natawidjaja, D., Kongko, W., et al. (2014). A probabilistic tsunami hazard assessment for Indonesia. *Natural Hazards and Earth System Sciences*, *14*(11), 3105–3122. https://doi.org/10.5194/nhess-14-3105-2014.

Imamura, F., Synolakis, C. E., Gica, E., Titov, V., Listanco, E., & Lee, H. J. (1995). Field survey of the 1994 Mindoro Island, Philippines tsunami. *Pure and Applied Geophysics*, *144*(3), 875–890. https://doi.org/10.1007/BF00874399.

Irsyam, M., Cummins, P., Asrurifak, M., Faisal, L., Natawidjaja, D., Widyantoro, S., et al. (2019). *Development of the 2017 national seismic hazard maps of Indonesia*. Earthquake Spectra. In press.

Irsyam, M., Sengara, I., Aldiamar, F., Widyantoro, S., Triyoso, W., Natawidjaja, D., Kertapati, E., Meilano, I., Suhardjono, Asrurifak, M., & Ridwan, M. (2010). Summary of study: Development of seismic hazard maps of Indonesia for revision of hazard map in SNI 03-1726-2002. Technical Report, Team for Revision of Seismic Hazard Maps of Indonesia, Bandung.

Jamelot, A., Gailler, A., Heinrich, P., Vallage, A., & Champenois, J. (2019). Tsunami simulations of the Sulawesi Mw 7.5 event: Comparison of seismic sources issued from a tsunami warning context versus post-event finite source. *Pure and Applied Geophysics*, *176*(8), 3351–3376. https://doi.org/10.1007/s00024-019-02274-5.

Kanamori, H., & Anderson, D. L. (1975). Theoretical basis of some empirical relations in seismology. *Bulletin of the Seismological Society of America*, *65*(5), 1073–1095.

Latief, H., Kodijat, A., Ismoyo, D., Bustamam, B., Adyasar, D., Nurbandika, N., & Rahayu, H. (2016). Air turun naik di tiga negeri. United Nations Educational, Scientific, and Cultural Organization, Office Jakarta - Indian Ocean Tsunami Information Centre, in Indonesian

Løvholt, F., Kühn, D., Bungum, H., Harbitz, C. B., & Glimsdal, S. (2012). Historical tsunamis and present tsunami hazard in eastern Indonesia and the southern Philippines. *Journal of Geophysical Research: Solid Earth*, *117*(B9), B09310. https://doi.org/10.1029/2012JB009425.

McCaffrey, R. (1988). Active tectonics of the eastern Sunda and Banda arcs. *Journal of Geophysical Research: Solid Earth*, *93*(B12), 15163–15182. https://doi.org/10.1029/JB093iB12p15163.

Okada, Y. (1985). Surface deformation due to shear and tensile faults in a half-space. *Bulletin of the Seismological Society of America*, *75*(4), 1135–1154.

Okal, E. A., & Reymond, D. (2003). The mechanism of great Banda Sea earthquake of 1 February 1938: Applying the method of preliminary determination of focal mechanism to a historical event. *Earth and Planetary Science Letters*, *216*(1–2), 1–15. https://doi.org/10.1016/S0012-821X(03)00475-8.

Okal, E. A., & Synolakis, C. E. (2004). Source discriminants for near-field tsunamis. *Geophysical Journal International*, *158*(3), 899–912. https://doi.org/10.1111/j.1365-246X.2004.02347.x.

Pairault, A. A., Hall, R., & Elders, C. F. (2003). Structural styles and tectonic evolution of the Seram Trough, Indonesia. *Marine and Petroleum Geology*, *20*(10), 1141–1160. https://doi.org/10.1016/j.marpetgeo.2003.10.001.

Patria, A., & Hall, R. (2017). The origin and significance of the seram trough, Indonesia.

Patton, J., Stein, R., & Sevilgen, V. (2018). Sunda Strait tsunami launched by sudden collapse of Krakatau volcano into the sea. *Temblor*. https://doi.org/10.32858/temblor.001.

Pownall, J., Hall, R., & Watkinson, I. (2013). Extreme extension across Seram and Ambon, eastern Indonesia: Evidence for Banda slab rollback. *Solid Earth*, *4*(2), 277–314. https://doi.org/10.5194/se-4-277-2013.

Pownall, J., Hall, R., & Lister, G. (2016). Rolling open Earth's deepest forearc basin. *Geology*, *44*(11), 947–950. https://doi.org/10.1130/G38051.1.

Pranantyo, I. R., & Cummins, P. R. (2019). Multi-data-type source estimation for the 1992 flores earthquake and tsunami. *Pure and Applied Geophysics*, *176*(7), 2696–2983. https://doi.org/10.1007/s00024-018-2078-4.

Rumphius, G. E. (1675). *Waerachtigh Verhael van de Schuckelijcke Aerdbebinge*. Dutch East Indies: BATAVIA.

Rumphius, G. E. (1997). True History of the terrible earthquake. (W. Buijze, translated by Beekman, E.M. and Foss, F. from (1675).

Sassa, S., & Takagawa, T. (2019). Liquefied gravity flow-induced tsunami: First evidence and comparison from the 2018 Indonesia Sulawesi earthquake and tsunami disasters. *Landslides*, *16*(1), 195–200. https://doi.org/10.1007/s10346-018-1114-x.

Soloviev, S., & Go, C. N. (1974). *Catalog of tsunamis in western coast of the Pacific Ocean* (pp. 1–30). Moscow: Academy of Sciences, USSR, Izdat Nauka.

Soloviev, S., Go, C., & Kim, X. (1986). *Catalog of tsunamis in the Pacific Ocean, 1969–1982* (p. 163). Moscow: MGC.

Spakman, W., & Hall, R. (2010). Surface deformation and slab-mantle interaction during Banda arc subduction rollback. *Nature Geoscience*, *3*(8), 562. https://doi.org/10.1038/NGEO917.

Synolakis, C. E. (1991). Green's law and the evolution of solitary waves. *Physics of Fluids A: Fluid Dynamics*, *3*(3), 490–491. https://doi.org/10.1063/1.858107.

Synolakis, C. E., Bardet, J. P., Borrero, J. C., Davies, H. L., Okal, E. A., Silver, E. A., et al. (2002). The slump origin of the 1998 Papua New Guinea tsunami. *Proceedings of the Royal Society of London Series A: Mathematical, Physical and Engineering Sciences*, *458*(2020), 763–789. https://doi.org/10.1098/rspa.2001.0915.

Watkinson, I. M., & Hall, R. (2017). Fault systems of the eastern Indonesian triple junction: Earth'svaluation of Quaternary activity and implications for seismic hazards. *Geological Society, London, Special Publications*, *441*(1), 71–120. https://doi.org/10.1144/SP441.8.

Wells, D. L., & Coppersmith, K. J. (1994). New empirical relationships among magnitude, rupture length, rupture width, rupture area, and surface displacement. *Bulletin of the Seismological Society of America*, *84*(4), 974–1002.

Wichmann, A. (1918). *Die Erdbeben Des Indischen Archipels Bis Zum Jahre 1857*. Amsterdam: Joahnnes Muller.

Wichmann, A. (1922). *Die Erdbeben Des Indischen Archipels von 1858 bis 1877*. Amsterdam: Uitgave van de Koninklijke Akademie van Wetenschappen.

(Received July 5, 2019, revised September 24, 2019, accepted December 5, 2019, Published online December 17, 2019)

Pure Appl. Geophys. 177 (2020), 1659–1674

https://doi.org/10.1007/s00024-019-02316-y

Pure and Applied Geophysics

Shoaling on Steep Continental Slopes: Relating Transmission and Reflection Coefficients to Green's Law

Jithin George,[1] David I. Ketcheson,[2] and Randall J. LeVeque[3]

Abstract—The propagation of long waves onto a continental shelf is of great interest in tsunami modeling and other applications where understanding the amplification of waves during shoaling is important. When the linearized shallow water equations are solved with the continental shelf modeled as a sharp discontinuity, the ratio of the amplitudes is given by the transmission coefficient. On the other hand, when the slope is very broad relative to the wavelength of the incoming wave, then amplification is governed by Green's Law, which predicts a larger amplification than the transmission coefficient, and a much smaller amplitude reflection than given by the reflection coefficient of a sharp interface. We explore the relation between these results and elucidate the behavior in the intermediate case of a very steep continental shelf.

Key words: Shoaling, tsunamis, Green's Law, reflection, transmission, continental shelf.

1. Introduction

This paper concerns the shoaling of long waves, such as tsunamis, as they pass from the deep ocean onto a shallower continental shelf. We seek to provide a better understanding of the degree to which waves are amplified during this transition. In particular we elucidate the connection between the reflection and transmission that would occur at a sharp interface and the prediction of Green's Law, which holds if the continental slope connecting the ocean and shelf is instead very gentle, as first derived by Green (1838). Green's law predicts a greater amplification than the transmission coefficient of the sharp interface, with little reflected energy. We will show that these two limiting cases can be naturally connected via a more general solution that holds for steep continental shelves, with intermediate degrees of transmission and reflection.

If the wavelength of an ocean wave is significantly greater than the ocean depth, then the one-dimensional shallow water equations can be used to model the propagation of a plane wave onto a planar bathymetry of the sort illustrated in Fig. 1. If we also assume that the amplitude of the wave is very small relative to the water depth everywhere, then the linearized shallow water equations can be used as a model. We consider only this case here. Green's Law is often used for nonlinear shoaling as a wave approaches a beach, but in this paper we are only concerned with shoaling onto the continental shelf, which typically has a depth of a few hundred meters, while the wave of interest has amplitude at most one or two meters. We also ignore possible effects of dispersion since our results are primarily of interest for waves that are long relative to the width of the continental slope and hence very long relative to the ocean depth. This work complements many other discussions in the literature of the applicability of Green's Law in nonlinear and dispersive cases, e.g. by Dzvonkovskaya et al. (2014), Grilli et al. (1994), Heron and Dzvonkovskaya (2015), Madsen et al. (2008), and Synolakis (1991).

DIK was supported by funding from King Abdullah University of Science and Technology. RJL was supported in part by a subcontract with Science and Technology Corporation (STC) under NASA Contract NNA10DF26C as part of the Asteroid Threat Assessment Project (ATAP).

[1] Department of Engineering Sciences and Applied Mathematics, Northwestern University, Evanston, USA. E-mail: jithindgeorge93@gmail.com

[2] Computer, Electrical, and Mathematical Sciences and Engineering Division, King Abdullah University of Science and Technology, 4700 KAUST, Thuwal 23955, Saudi Arabia. E-mail: david.ketcheson@kaust.edu.sa

[3] Department of Applied Mathematics, University of Washington, Seattle, WA 98195-3925, USA. E-mail: rjl@uw.edu

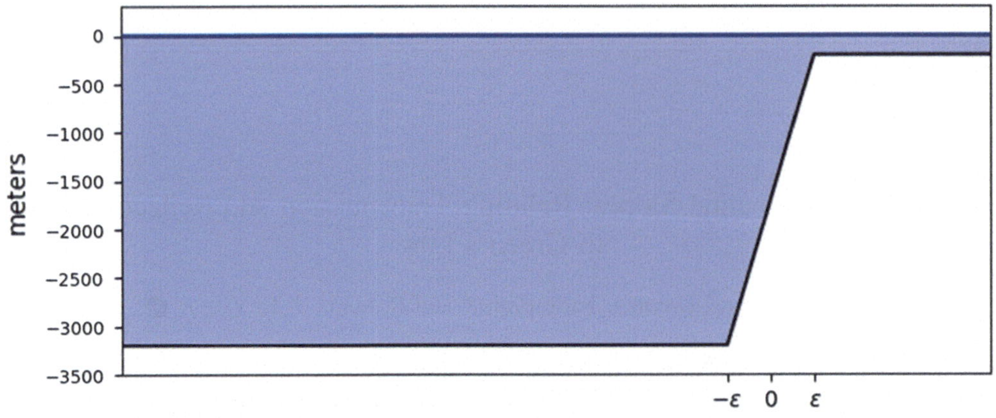

Figure 1
Bathymetry showing deep ocean for $x < -\epsilon$, a linear continental slope for $-\epsilon \leq x \leq \epsilon$, and a continental shelf for $x > \epsilon$

Our analysis makes use of approximations of a smooth continental slope by a piecewise-constant bathymetry separated by step discontinuities at which the classic transmission and reflection coefficients can be determined. The problem can then be viewed as a wave propagation problem in a layered medium. In Sect. 3 we explain how the multiple internal reflections in this approach combine to result in the structure shown in Sect. 4 when a simple hydraulic jump moves from the deep ocean onto the shelf on a linear slope. It is seen that first transmitted wave has an amplitude given by Green's Law, but behind this the wave then decays to an amplitude given by the net transmission coefficient for the depth difference between ocean and shelf.

Other classic papers have considered piecewise-constant bathymetry, for example Miles (1967) and Evans and Linton (1994), with extension to piecewise linear bathymetry by Kânoğlu and Synolakis (1998). These papers have generally considered sinusoidal wave trains, leading to the analysis of eigenfunctions in each layer and more complex coupling conditions at the interfaces, and have not related the results directly to Green's Law. We believe that our analysis and the figures presented give additional insight into the behavior of shoaling waves, and the manner in which the simple transmission and reflection coefficients for a hydraulic jump combine to yield Green's Law in the limit of a smooth slope.

The analysis presented here can be extended to other wave propagation problems to explore the transmission and reflection of waves passing through heterogeneous materials. The linearized shallow water equations turn out to be a special case in the sense that the "impedance" of the material is directly related to the wave speed, whereas for other wave propagation problems such as acoustics these two parameters describing the material can vary independently. We have extended the results presented in this paper to the more general case in George et al. (2019a), where we also provide additional mathematical results on the structure of the solution shown in this paper. Here we concentrate on providing a better physical understanding of the process of shoaling on continental shelves in the intermediate case between Green's Law and reflection/transmission coefficients.

In Sect. 5 we discuss possible implications of this work to realistic tsunami modeling and present results from a simulation of the 2011 tsunami originating from Japan as it shoals onto the continental slope of North America, as an illustration that amplification of a wave on the shelf may be smaller than suggested by Green's Law. Because of the fact that a real tsunami is not a plane wave approaching a uniform shelf in the normal direction, in order to accurately capture shoaling, and especially runup and inundation, it is generally necessary to perform a detailed simulation using suitable multi-dimensional equations. However, one-dimensional theory is still sometimes used to get an estimate of the amplitudes along a coastline, particularly in cases where many tsunamis must be simulated and nearshore amplitudes estimated over a large region, for example in probabilistic tsunami hazard assessment (PTHA), e.g. Davies et al. (2018), Grezio et al. (2017), Lorito et al. (2015), Løvholt et al. (2012), Løvholt et al. (2016), Glimsdal et al. (2019), Ward and Asphaug (2000). Green's Law is often considered a conservative over-estimate in these papers, and the analysis of this paper helps to confirm and illuminate this conclusion.

2. *Transmission, Reflection, And Green's Law*

The linearized shallow water equations in one dimension can be written in conservation form as

$$\begin{aligned} \eta_t(x,t) + (h(x)u(x,t))_x &= 0, \\ u_t + g\eta_x(x,t) &= 0, \end{aligned} \tag{1}$$

where $h(x)$ is the undisturbed fluid depth, $\eta(x,t) \ll h(x)$ is the linearized surface elevation (with $\eta = 0$ corresponding to the undisturbed sea level), $u(x, t)$ is the depth-averaged horizontal velocity, and $g = 9.81$ m/s^2 is the gravitational constant.

The linear hyperbolic Eq. (1) has the form $q_t(x,t) + (A(x)q(x,t))_x = 0$, where $q = [\eta, u]$ is the solution vector and the coefficient matrix A is given by

$$A(x) = \begin{bmatrix} 0 & h(x) \\ g & 0 \end{bmatrix}. \tag{2}$$

This matrix has eigenvalues and corresponding eigenvectors given by

$$\begin{aligned} \lambda_1 &= -\sqrt{gh(x)}, \quad r_1 = \begin{bmatrix} 1 \\ -\sqrt{g/h(x)} \end{bmatrix}, \\ \lambda_2 &= \sqrt{gh(x)}, \quad r_2 = \begin{bmatrix} 1 \\ \sqrt{g/h(x)} \end{bmatrix}. \end{aligned} \tag{3}$$

The eigenvalues are the wave speeds of left- and right-going waves. The eigenvectors reveal, for example, that a purely right-going wave on flat bathymetry with $h(x) \equiv h_0$ must have $u(x,t) = \sqrt{g/h_0}\eta(x,t)$ for all x and t [see, e.g. LeVeque (2002)].

We restrict our attention to piecewise linear bathymetry of the form

$$h(x) = \begin{cases} h_\ell, & x < -\epsilon, \\ h_\ell + \left(\frac{h_r - h_\ell}{2\epsilon}\right)(x+\epsilon), & -\epsilon \le x \le \epsilon, \\ h_r, & x > \epsilon. \end{cases} \tag{4}$$

as shown in Fig. 1. Here h_ℓ denotes the ocean depth and h_r the depth of the continental shelf. The region between $-\epsilon$ and ϵ is known as the continental slope, and has width 2ϵ with this notation.

As $\epsilon \to 0$, the slope in (4) becomes a step discontinuity. A wave of amplitude A hitting a sharp discontinuity in bathymetry is split into a transmitted wave of amplitude $C_T A$ and a reflected wave of amplitude $C_R A$, where the transmission and reflection coefficients (C_T and C_R, respectively) are given by

$$C_T = \frac{2\sqrt{h_\ell}}{\sqrt{h_\ell} + \sqrt{h_r}}, \qquad C_R = \frac{\sqrt{h_\ell} - \sqrt{h_r}}{\sqrt{h_\ell} + \sqrt{h_r}} = C_T - 1. \tag{5}$$

These expressions can be found in many references on linear wave propagation, e.g. LeVeque (2002), Levin and Nosov (2016), Mei (1992). They can be derived using the eigenvectors to the left and right of the discontinuity in coefficients in the equation, as outlined in the next section.

On the other hand, if $\epsilon > 0$ and if wavelength of the wave is significantly smaller than the width 2ϵ of the continental slope, then Green's Law predicts that the amplitude of the wave moving up the slowly varying slope will have been magnified by a factor of $(h_\ell/h(x))^{1/4}$ at each point on the slope, with a total magnification factor upon reaching the shelf of

$$C_G = (h_\ell/h_r)^{1/4}. \tag{6}$$

To illustrate the shoaling of waves governed by the shallow water equations, in Figs. 2 and 3 we show numerical solutions of the shallow water equations when the initial conditions consist of a square pulse wave approaching a continental slope with three different choices of the continental slope parameter ϵ. Although a square pulse is not a realistic tsunami, for the linear equations over flat bathymetry examining this case is very useful to help explain the behavior of the solution as the slope is made more steep. A square pulse can be viewed a pair of discontinuities (a positive jump followed by a negative jump), and we will first study the case of a single jump discontinuity

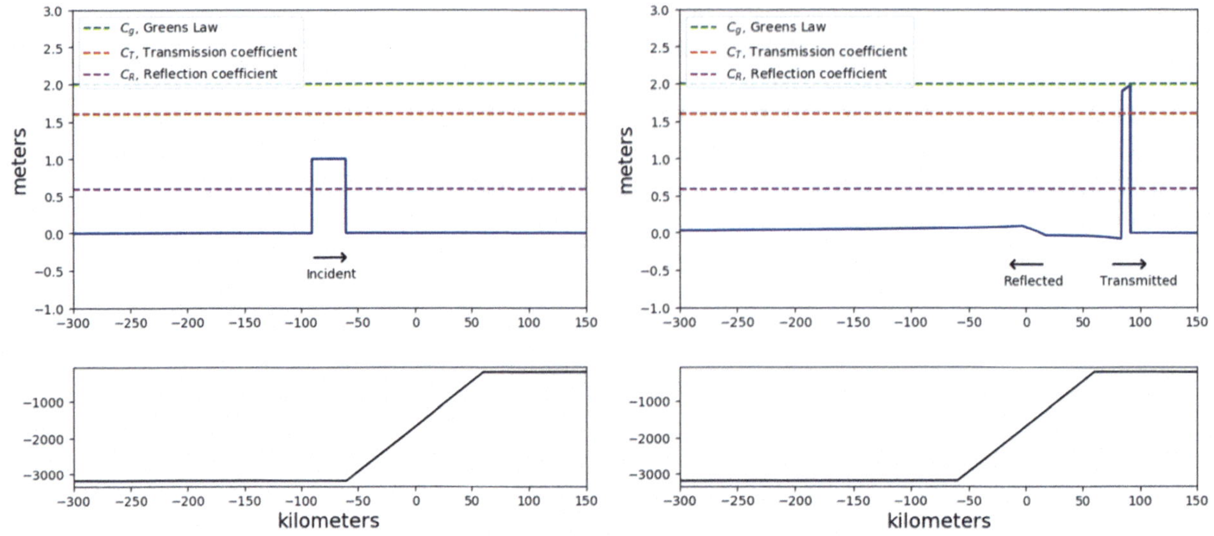

Figure 2
Left: The initial conditions consist of a square pulse of width 10 km and height 1 m propagating to the right toward a continental slope with $\epsilon = 60$ km. Right: Solution at a later time $t = 1800$ seconds, when the pulse has become narrower and taller on the shelf, with very little reflection. The amplitude is close to that predicted by Green's Law. The bottom plots show the bathymetry, with a different vertical scale

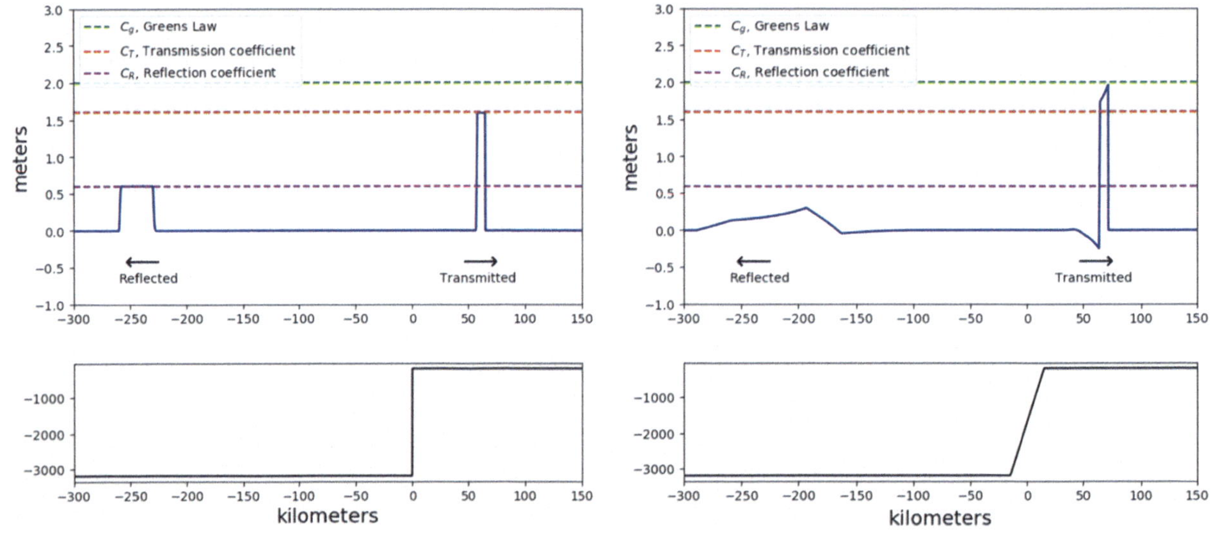

Figure 3
Left: solution when the smooth continental slope of Fig. 2 has been replaced with a sharp discontinuity. The solution at $t = 1800$ now consists of a right-going transmitted pulse on the shelf and a left-going reflected pulse in the ocean. The initial pulse width is again 10 km. Right: Solution with a steep slope ($\epsilon = 15$ km), showing a larger reflected wave than in the case shown in Fig. 2

propagating toward the shelf. We can then use the superposition of two such solutions to explain the behavior of the square pulse, giving insight also into the shoaling behavior of a more general wave.

For the tests shown in Figs. 2 and 3 we use $h_\ell = 3200$ m and $h_r = 200$ m, which are realistic values and were chosen to have a ratio of 16, so that $C_G = 2$, while $C_T = 1.6$ and $C_R = 0.6$. The solutions shown were generated using Clawpack (Clawpack Development Team 2017) on a very fine nonuniform grid with 10,000 grid cells distributed based on the wave speed in such a way that the Courant number is close to 1 in each cell for maximum accuracy, and we confirmed that the results are converged to plotting resolution. We solved the full nonlinear shallow water equations, but for these depths and amplitude the solution is visually identical to solutions of the linearized equations.

Figure 2 shows a case in which a square pulse moves up a slope that is broad relative to the width of the pulse, a case where Green's Law should approximately apply. The initial surface displacement is shown on the left, and consists of a pulse with amplitude $A = 1$ m, and with fluid velocity $u = \sqrt{g/h_\ell}$ in the pulse, to give a purely right-going wave (based on the eigenvectors). The plot on the right shows the solution at a later time when the wave has been amplified due to shoaling by roughly the expected factor of $C_G = 2$. The pulse also becomes narrower as the wave speed decreases. There is a small amplitude reflected wave in this case, and there is also a bit of a negative tail following the transmitted pulse that we will say more about later.

Figure 3 shows solutions to the equations with the same initial conditions as in Fig. 2 but on steeper slopes. In the left figure, the slope is replaced by a jump discontinuity at $x = 0$. In this case the incident pulse is partially transmitted as a narrower pulse with height $C_T A$ and partially reflected as a left-moving wave with height $C_R A$, as expected in this case.

Finally, the right side of Fig. 3 shows a case where the width of the continental slope is comparable to the width of the incoming pulse. In this case the solution is more complicated, consisting of a transmitted wave on the shelf that has a peak value close to that predicted by Green's Law but decreasing behind the peak and has a negative tail, along with a reflected wave with complicated structure. As the slope is made more broad, the amplitude of the reflected wave will decrease and eventually go to zero in the Green's Law limit. Our goal is to fully describe waves in this transitional regime of steep slopes and the manner in which this reflection disappears. In Sect. 5 we illustrate that realistic tsunamis can fall into this transitional region, suggesting that a better understanding of this case is important in practice.

3. Jump Discontinuities in the Wave And the Bathymetry

To begin with we first take as the approaching wave initial conditions with jump discontinuities in the elevation and velocity,

$$\eta(x,0) = \begin{cases} A, & x < -\epsilon, \\ 0, & x \geq -\epsilon \end{cases},$$
$$u(x,0) = \begin{cases} \sqrt{g/h_\ell}, & x < -\epsilon, \\ 0, & x \geq -\epsilon \end{cases}. \tag{7}$$

This can be viewed as a right-going jump discontinuity in surface elevation and velocity that is moving across the ocean, at the instant when it first encounters the continental slope, as shown in the $t = 0$ plot in Fig. 5. As shorthand we refer to this discontinuous wave as a hydraulic jump, or bore, although more properly this term is used in the nonlinear case, after shock formation.

If we also take $\epsilon = 0$ in (4) (a jump discontinuity in bathymetry, at the same point as the jump in the initial conditions), then the problem has the form of a "Riemann problem". Problems of this nature form the basis for much of the theory of hyperbolic PDEs, and of many numerical methods; see e.g. Chap. 9 of LeVeque (2002) for more discussion of this particular Riemann problem and its solution. The initial jump in (η, u) splits into a single left-going wave across which η jumps from η_ℓ to some state η_m, and a single right-going wave across which η jumps from η_m to η_r. There is a single intermediate state η_m since the solution must be continuous at $x = 0$. Similarly, the Riemann solution must have a single intermediate state u_m. Moreover, due to the form of the eigenvectors displayed in Eq. (3), across the left-going wave we must have $(u_m - u_\ell) = -\sqrt{g/h_\ell}(\eta_m - \eta_\ell)$, while across the right-going wave $(u_r - u_m) = \sqrt{g/h_r}(\eta_r - \eta_m)$. These conditions result in a system of two equations that, for the initial conditions (7), yield $\eta_m = C_T A$ and $\eta_m - \eta_\ell = C_R A$, with C_T and C_R given by (5).

For a continuous slope with $\epsilon > 0$, the solution behaves differently than in the singular case of a discontinuous continental slope. This will be elucidated in Sect. 4. First, however, note that a smooth continental slope can be viewed as the limiting case of a sequence of small jump discontinuities, and at each jump the transmission and reflection coefficients can be computed. Piecewise constant bathymetry of this nature defines a "layered medium" in the standard terminology of wave propagation problems; we provide a more general analysis of this case in George et al. (2019a). To understand the behavior of waves through such a medium, it is natural to first consider a layered medium with only a few intermediate layers. The single jump discontinuity between depths h_ℓ and h_r that we used to define the transmission and reflection coefficients in (5) can be partitioned into $N+1$ discontinuities separating N intermediate layers. Figure 4 shows the bathymetry if there are $N=0,1$ or 2 intermediate layers (top plots) and also shows how a single initial discontinuity of amplitude A propagates in the x–t plane in each case. With no intermediate layer (Fig. 4a, $N=0$), the wave splits into transmitted and reflected waves only once, with amplitudes C_TA and C_RA respectively. With $N=1$ interior layer the solution is already much more complicated, with internal reflections in the layer that result in an infinite sequence of waves eventually departing to both the right and to the left, as shown in Fig. 4b. The amplitude of later waves decays rapidly since each reflection coefficient is less than one. We will show in Appendix 1 that the first wave departing to the right (after transmission through each layer and no reflection) has amplitude greater than C_TA, but that subsequent departing waves (with 2, 4, 6, or more internal reflections) also affect the value of η and asymptotically they sum to C_TA. Similarly, the reflections that depart from the left (after 1, 3, 5, or more reflections) all sum to C_RA.

When the same jump from h_ℓ to h_r is split into three discontinuities, as shown in Fig. 4c, then there are many more internal reflections but the same asymptotic limits still hold. When additional layers are added, these asymptotic results continue to hold (see George et al. 2019a). Moreover, the first transmitted wave exiting to the right (with no internal reflections) asymptotes to magnitude C_GA as the number of layers increases, where C_G is the value predicted by Green's Law. This is shown analytically in Appendix 1.

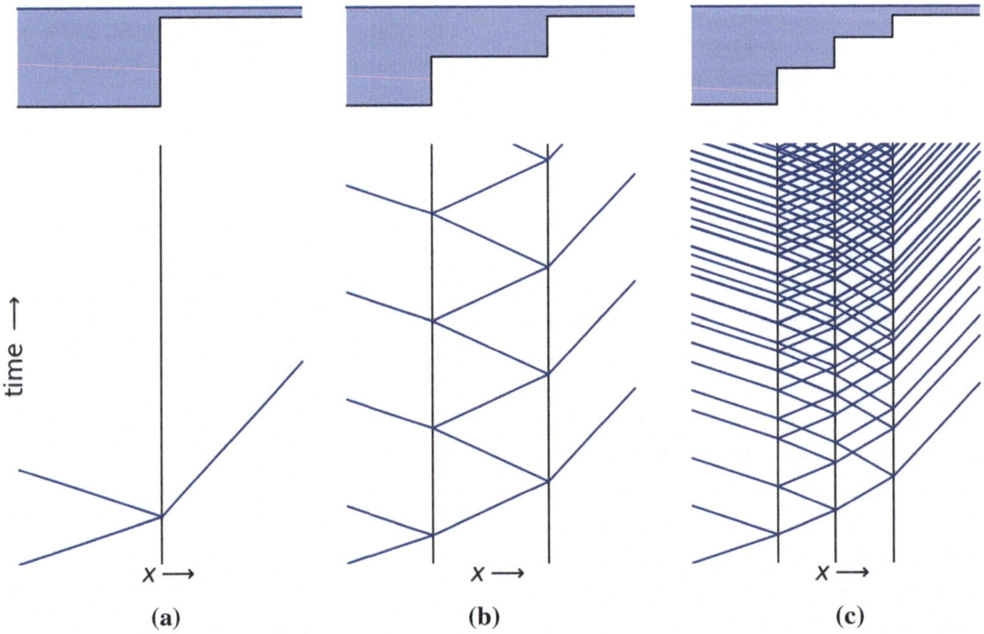

Figure 4
Plots in the x–t plane showing an initial right-going wave interacting with a layered medium (i.e. bathymetry with jump discontinuities, as shown at the top of each plot). **a** With a single interface, where only one transmitted and reflected wave are generated. **b** With two interfaces and a single intermediate layer. **c** With three interfaces and two intermediate layers

4. *The Bore on a Piecewise Linear Slope*

As the bore moves up a linear slope with bathymetry given by (4) for $\epsilon > 0$, it amplifies and also generates a reflected wave. The full wave form of the transmitted and reflected waves on a linear slope can be worked out from asymptotic limits of the layered medium case by considering all internal reflections and transmissions as $N \to \infty$. However, the number of waves that must be considered grows exponentially in N. We present this much more complicated analysis in George et al. (2019a).

Here we simply show a numerical approximation to this wave in Fig. 5, as computed with Clawpack on a very fine grid. Note that this is shown for arbitrary ϵ and we denote the solution to this problem by $(\eta^{[\epsilon]}(x,t), u^{[\epsilon]}(x,t))$ to indicate that this solution depends on the parameter ϵ defining the width of the continental slope. Solutions to (1) scale with ϵ in the following manner: if we calculate the solution $(\eta^{[1]}, u^{[1]})$ for $\epsilon = 1$ then for any other value $\epsilon > 0$ we have:

$$\eta^{[\epsilon]}(x,t) = \eta^{[1]}(x/\epsilon, t/\epsilon), \qquad u^{[\epsilon]}(x,t) = u^{[1]}(x/\epsilon, t/\epsilon). \tag{8}$$

This is easy to check by differentiating these with respect to x and t and inserting into (1).

As the bore moves up the continental slope to some point $x > -\epsilon$, its amplitude increases according to Green's Law to $(h_\ell/h(x))^{1/4}A$, based on the depth $h(x)$ at the point the bore has reached and the initial depth h_ℓ (this follows from the result in

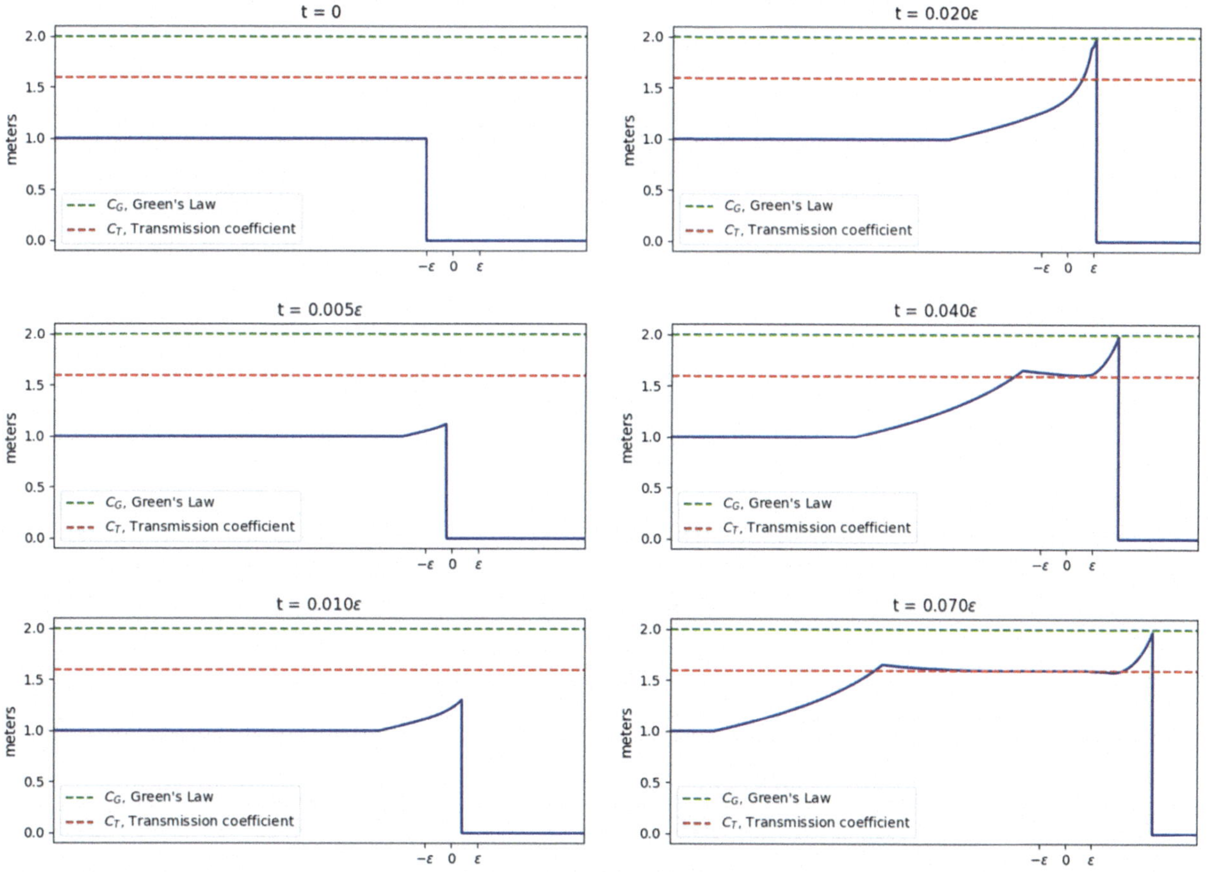

Figure 5
Left column: initial conditions $\eta^{[\epsilon]}(x,0)$ at $t=0$, and evolution to $t = 0.01\epsilon$, where ϵ is the half-width of the continental slope. Right column: Further evolution to $t = 0.07\epsilon$ seconds (when ϵ is in meters). At later times the left going and right-going waves propagate outward over constant bathymetry with no further change in shape

Appendix 1). Upon reaching the shelf at $x = \epsilon$, it has reached an amplitude $C_G A = (h_\ell/h_r)^{1/4} A$, and is followed by a decrease in amplitude down to the value $C_T A$ given by the transmission coefficient (5). As it propagates up the shelf, it also gives rise to a left-going wave that increases to just above $C_T A$ before settling down to the value $C_T A$. This reflected wave thus has overall amplitude $C_T A - A = C_R A$. This is seen in Fig. 5 and justified in George et al. (2019a).

Once we have observed the form of this solution, it is easy to understand the results shown in Figs. 2 and 3 and the transition between them. The square pulse initial data considered in Sect. 3 can be viewed as a jump discontinuity from $\eta = 0$ up to $\eta = A$ initially located at $x = \epsilon$, followed by another jump discontinuity from $\eta = A$ back down to $\eta = 0$ initially located at $x = \epsilon - w$, where w is the width of the square wave. The solution for each of these initial conditions has the form discussed above, and the full solution for the pulse is hence (by linearity) the sum of the solutions for each jump separately. Figure 6 shows each of these solutions separately on the top, and the sum of the two on the bottom. These are shown for two different choices of the width w of the initial pulse. As the width decreases, the two waves cancel out almost everywhere except near the initial peak location, where the amplitude jumps up to $C_G A$. Hence in the limit of a narrow pulse relative to the width ϵ of the continental slope ($w/\epsilon \to 0$), we recover the Green's Law limit of a pulse with amplitude $C_G A$. For larger values of w/ϵ there is less cancellation of the left-going waves and a larger apparent reflected wave.

Note that the results above concern the *amplitudes* of the waves. It is also interesting to consider the transmission of energy, along with the volume of water that is carried onto the shelf in the transmitted wave. These satisfy somewhat different relations that are summarized in Appendix 1.

5. *Implications for Real Tsunamis*

Both Green's Law and the more general theory presented here are based on the one-dimensional shallow water equations and would apply to a real tsunami only if it were a plane wave approaching a continental shelf in the normal direction, in a case where the shelf is invariant in the along-shore

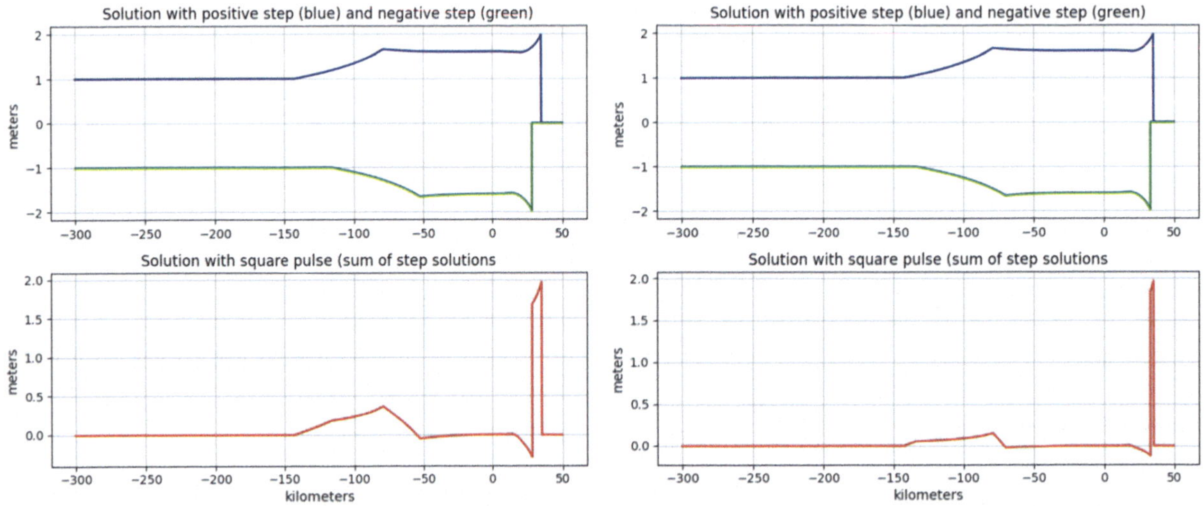

(a) **(b)**

Figure 6
The solution to the shoaling problem with square pulse initial conditions can found as the linear combination of the single bore solution and a shifted and negated version of the single bore solution. **a** A wider pulse has a larger shift between the two. **b** A narrower pulse has a smaller shift between the two, more cancellation of the reflected waves, and a transmitted pulse that more closely resembles the Green's Law prediction, with amplification by C_G and a smaller reflected wave

direction. A real tsunami is typically a complex wave train by the time it reaches a shelf, and is not necessarily approaching normal to the shore. Moreover, the continental shelf varies greatly over relatively short spatial scales, with undulations in width and sometimes deep valleys or outcroppings that will focus or defocus the approaching waves. Predicting the amplitude of the wave hitting the beach from the amplitude seaward of the shelf is generally not possible with any precision using one-dimensional theory alone, and simulations in two spatial dimensions with the appropriate shelf bathymetry and approaching wave train should be used for detailed inundation predictions. Nonetheless, as noted in the introduction, one-dimensional theory is still often used to get an estimate of the amplitudes along a coastline for some applications such as probabilistic tsunami hazard assessment.

Comparing this theory with a real tsunami is difficult because each tsunami and shelf bathymetry is unique. Here we simply show one example. We have attempted to identify a case where the leading tsunami wave is relatively planar and normal to the coast and the width of the continental slope is comparable to the width of the tsunami, where the results of this paper imply that amplification should be less than what Green's Law predicts.

Figure 7a shows a simulation of the tsunami arising from the 2011 Great East Japan earthquake (also known as the Tohoku event), performed with the GeoClaw software (Clawpack Development Team 2017) using an earthquake source model provided by the NOAA Center for Tsunami Research and recently used in a comparison study with tide gauge observations for this event by Adams and LeVeque (2017). This seafloor deformation file along with the GeoClaw code used to produce these figures is available in the code archive for this paper (George et al. 2019b).

Figure 7 also shows the leading wave of this tsunami as it is approaching the coast of Oregon. Black contours of bathymetry show the continental shelf and red contours show the wave elevation. Figure 7b shows the wave approaching the shelf, where it is roughly planar with a fairly uniform amplitude of about 0.15 m. Figure 7c shows the wave 25 min later, after passing onto the shelf. The wave has amplified by varying amounts. The broader shelf around latitude 44.2 causes a folding in of the wave along this latitude (due to the slower wave speed on the shallow shelf) and amplification along this latitude and lower amplitudes to the north and particularly to the south (e.g., along latitude 44.0).

In Fig. 8 we show transects of the bathymetry and of the solution along two transects, one at latitude 44.4, where the amplitude is roughly equal to the mean amplitude around this region, and one at latitude 44.2, where geometric focusing from the shelf geometry causes greater amplification. Even in the latter case we see that the wave amplification as it first passes onto the shelf is less than what is predicted by Green's Law, which is indicated by the dashed green curve. The solution is shown at three times along each transect, and the red solid line also shows the maximum sea surface at each point over the full time range from 9:00 to 9:50 hours after the earthquake, showing the amplification as the wave goes on to the shelf. The Green's Law value at each point x, based on the depth $h(x)$ at this point relative to the roughly uniform 3000 m depth seaward of the shelf. The blue dashed line at each x shows the amplification that would be predicted by the transmission coefficient from the deep ocean to depth $h(x)$, if the shelf were replaced by a step discontinuity up to the shelf depth at this location.

Based on the theory in this paper, we expect the wave to be amplified by more than the transmission coefficient defined above, but by less than Green's Law would suggest. Indeed we see that the red curve of maximum amplitude lies between the two dashed curves over most of the shelf at latitude 44.4 and at least initially as the wave moves onto the shelf at latitude 44.2. As the wave moves into much shallower water near the shore, Green's Law roughly applies to predict further amplification from the shelf depth to the nearshore depth, and we see that the actual wave amplitude grows even faster near shore due to local bathymetry effects and runup of the relatively broad wave. At latitude 44.2 additional amplification is observed due to the shelf geometry.

As an additional test, we also ran the one-dimensional shallow water equations on bathymetry given by a transect of the Oregon coast bathymetry taken at 44.2°N. This was run on a much finer grid

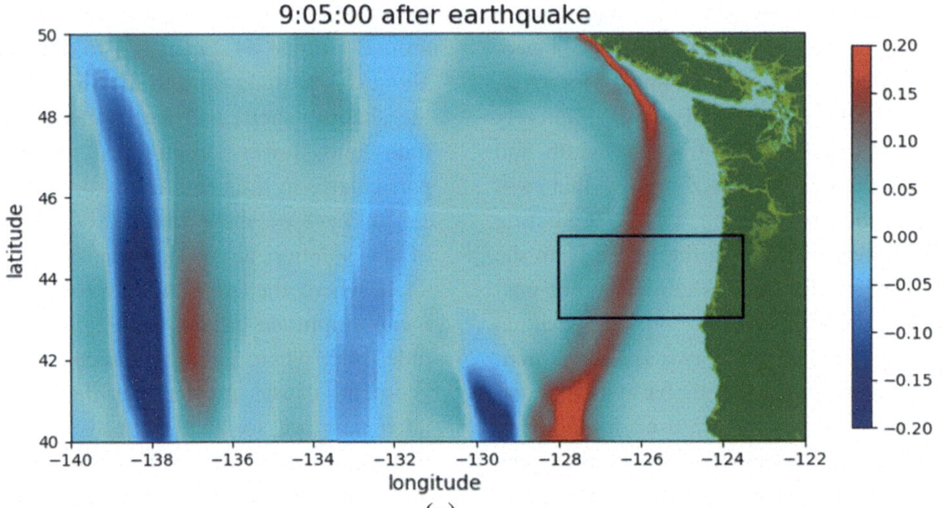
9:05:00 after earthquake
latitude
longitude
50
48
46
44
42
40
−140
−138
−136
−134
−132
−130
−128
−126
−124
−122
0.20
0.15
0.10
0.05
0.00
−0.05
−0.10
−0.15
−0.20

(a)

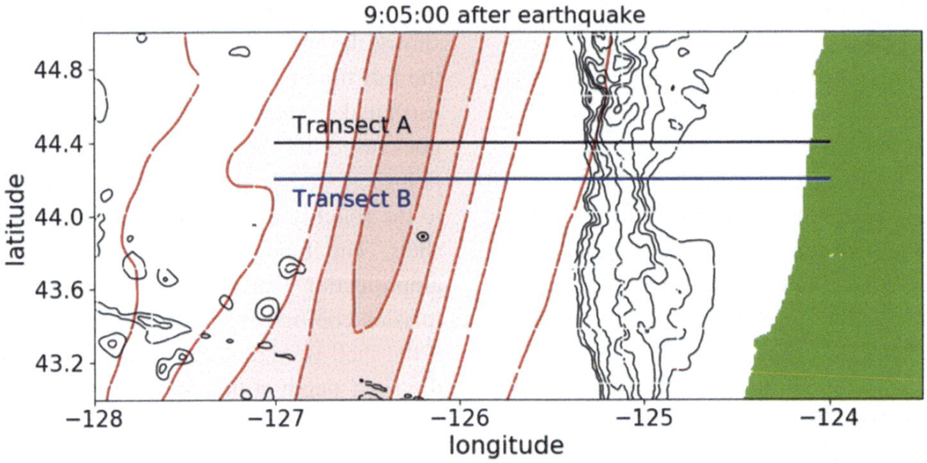
9:05:00 after earthquake
Transect A
Transect B
latitude
longitude
44.8
44.4
44.0
43.6
43.2
−128
−127
−126
−125
−124

(b)

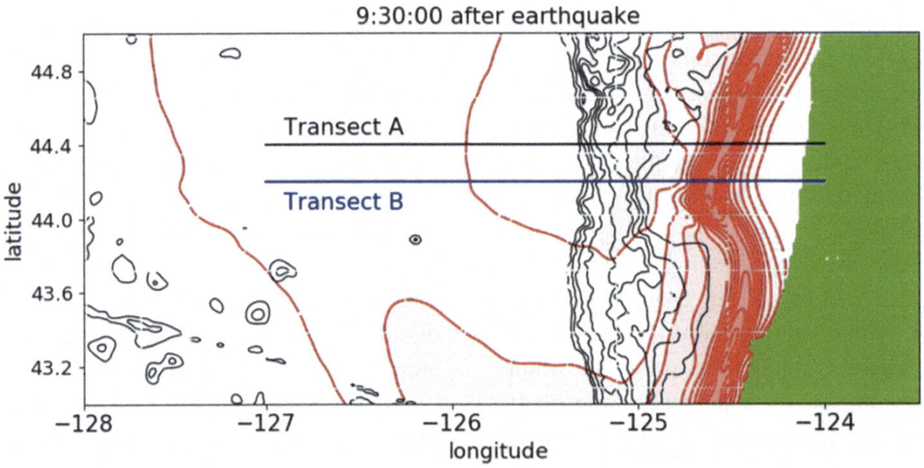
9:30:00 after earthquake
Transect A
Transect B
latitude
longitude
44.8
44.4
44.0
43.6
43.2
−128
−127
−126
−125
−124

(c)

◀Figure 7
a Portion of a simulation of the 2011 Tohoku Tsunami, 9:05 hours after the earthquake as the leading wave is approaching the west coast of North America. The black rectangle indicates the region shown in the other two plots. **b** Zoom on the wave approaching the central Oregon Coast. The black contours show the bathymetry of the continental slope (contours at depths of $300, 600, \ldots, 2700$ m. Red contours and shading show surface elevation above sea level, with contours at $0.025, 0.5, \ldots, 0.5$ m. **c** Zoom at a later time 9:30 h post-quake, after the wave has passed onto the continental shelf. Two transects are also indicated, along which vertical cross-sections of the bathymetry and tsunami are shown in Fig. 8

than the two-dimensional simulations shown above, using a transect of the etopo1 bathymetry with 1 arc-minute resolution, interpolated to the computational grid that again has 10,000 points and variable spacing, and is converged to plotting resolution. As initial conditions, a sech^2 wave is used as shown in Fig. 9, with a width chosen similar to the approaching wave in Fig. 7, and shown on a domain that has roughly the same length in x (250 km). Fig. 9 also shows the resulting wave at 1800 s. As in Fig. fig:japan1, the dashed green and blue lines show the values of C_G and C_T as x varies. The amplitude of the transmitted wave is higher than C_T but noticeably lower than C_G.

Because the bathymetry on this transect resembles a piecewise-constant medium with a relative flat region on the shelf slope with depth 1400 , it is also interesting to see how the amplitude compares to the theory for a layered medium with one intermediate layer, as developed in the appendices. That theory suggests that the amplitude might be better modeled by the product of the transmission coefficient from 3000 to 1400 m with the transmission coefficient from 1400 to 100 m (the approximate depth of the shelf on this transect).

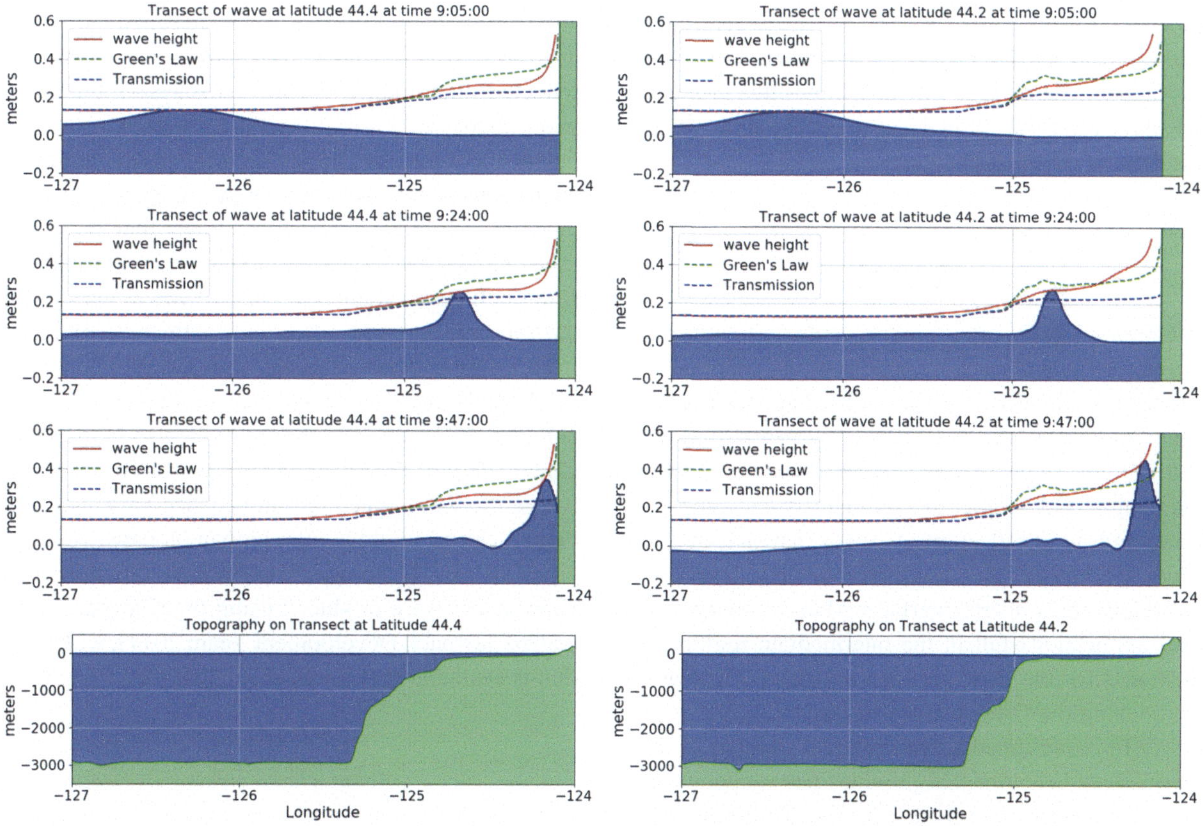

Figure 8
Vertical cross sections of the solution from Fig. 7 (and bathymetry in the bottom plot) along two different transects. Left column: transect A at latitude 44.4°N, Right column: transect B at latitude 44.2°N. See the text for discussion of the curves shown

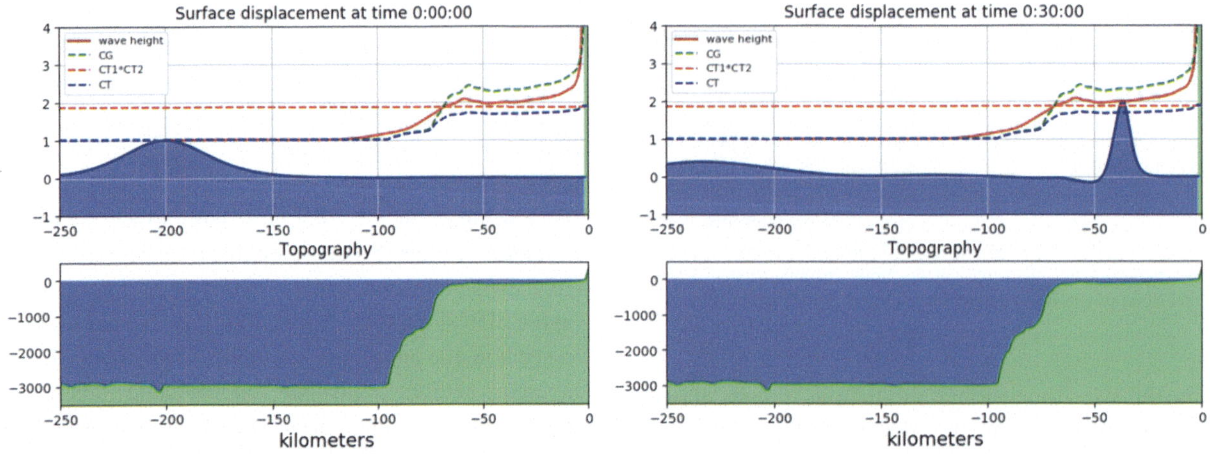

Figure 9
Left: the initial conditions consist of a sech^2 wave propagating to the right toward a continental slope, with bathymetry given by a transect at 44.2°N. Right: solution at a later time $t = 1800$ seconds. The red dashed curve shows the value $C_T^{\ell m} C_T^{mr} = 1.88$ discussed in the text. The bottom plots show the bathymetry, with a different vertical scale

This product, using the first two expressions in (11) of Appendix 1, works out to $C_T^{\ell m} C_T^{mr} = 1.88$, whereas C_T for these values is 1.69 while $C_G = 2.34$. In Fig. 9 we have added a dashed red line that shows this product, and indeed the transmitted wave matches this more closely than the other curves.

6. Conclusions

Green's Law is often used to estimate the amplification of tsunamis as they pass from the deep ocean onto a continental shelf. This is a good approximation if the continental slope is sufficiently gentle that the width of the slope region is large compared to the wavelength. For steeper slopes there is less amplification and more reflection of waves, asymptoting to values given by the transmission and reflection coefficient for a sharp interface. Many realistic tsunami applications fall in the intermediate region, as illustrated for the 2011 Tohoku earthquake tsunami approaching the Oregon coast in Sect. 5.

We have presented a mathematical analysis of the intermediate behavior that provides new understanding of the connection between the two limiting cases. In Sect. 3 we showed that for a wave consisting of a right-going jump discontinuity approaching a linear slope, there is a solution that varies in a self-similar manner as the width of the continental slope is varied. Viewing a square pulse as a linear superposition of two such waves shows the manner in which the reflected amplitude and energy decays as the slope width increases, even though the reflected volume is independent of the slope width. Approximating a continuous slope as a piecewise constant "layered medium", as done in Sect. 3, shows that the "first transmitted wave" satisfies Green's Law but that the sum of all transmitted and reflected waves agree with the transmission and reflection coefficients obtained from a single sharp interface.

Additional mathematical analysis that gives a more complete description of the solution, also in the more general case of the linear wave equation in a medium where the wave speed and impedance vary independently, can be found in our companion paper (George et al. 2019a). The numerical experiments shown in this paper paper were produced using the Clawpack software (Clawpack Development Team 2017), with code that is available at (George et al. 2019b).

Acknowledgements

The authors are grateful to Avi Schwarzschild for stimulating discussions in the early phase of this project.

Funding

The article was funded by National Aeronautics and Space Administration (NNA10DF26C) and King Abdullah University of Science and Technology.

A: Transmission of Volume and of Energy

Note that Green's Law and the analysis above applies to the amplitude of waves, not to their energy or the volume of water displaced by the wave. In this appendix we discuss how the results extend to these other quantities. Of course the total volume of water is conserved and so $\int_{-\infty}^{\infty}(h(x)+\eta(x,t))\,dx$ is constant in time. Moreover from this or directly from the first conservation law of (1) we see that the "wave volume" $\bar{\eta} \equiv \int_{-\infty}^{\infty}\eta(x,t)\,dx$ is constant in time and always equal to the initial wave volume. Similarly, the second conservation law of (1) shows that $\bar{u} \equiv \int_{-\infty}^{\infty}u(x,t)\,dx$ is constant in time. Suppose the initial elevation η and velocity u are identically 0 for $x > -\epsilon$ and consists of some purely right-going wave with finite volume $\bar{\eta}$ travelling over the deep ocean. Then $u(x,0) = \sqrt{g/h_\ell}\,\eta(x,0)$ for all x, and so $\bar{u} = \sqrt{g/h_\ell}\,\bar{\eta}$. For large times, the perturbed elevation approaches 0 over the continental slope region, and consists of a purely left-going reflected wave and a purely right-going transmitted wave. Thus for sufficiently large t we can decompose $\eta(x,t) = \eta^R(x,t) + \eta^T(x,t)$, and similarly for u, where we have $\eta^R(x,t) \equiv 0$ for $x > -\epsilon$ and $u^R(x,t) = -\sqrt{g/h_\ell}\,\eta^R(x,t)$ for all x, while $\eta^T(x,t) \equiv 0$ for $x < \epsilon$ and $u^T(x,t) = \sqrt{g/h_r}\,\eta^T(x,t)$ for all x. If we define $\bar{\eta}^R = \int_{-\infty}^{\infty}\eta^R(x,t)\,dx = \int_{-\infty}^{-\epsilon}\eta(x,t)\,dx$ and $\bar{\eta}^T = \int_{-\infty}^{\infty}\eta^T(x,t)\,dx = \int_{\epsilon}^{\infty}\eta(x,t)\,dx$, then by conservation of η and u, we obtain a linear system of two equations

$$\begin{aligned} \bar{\eta}^R + \bar{\eta}^T &= \bar{\eta}, \\ -\sqrt{g/h_\ell}\,\bar{\eta}^R + \sqrt{g/h_r}\,\bar{\eta}^T &= \sqrt{g/h_\ell}\,\bar{\eta}, \end{aligned} \tag{9}$$

which can be solved to find that the total reflected and transmitted volume are given respectively by

$$\begin{aligned} \bar{\eta}^R &= \left(\frac{\sqrt{h_\ell}-\sqrt{h_r}}{\sqrt{h_\ell}+\sqrt{h_r}}\right)\bar{\eta} \equiv \bar{C}_R\bar{\eta}, \\ \bar{\eta}^T &= \left(\frac{2\sqrt{h_r}}{\sqrt{h_\ell}+\sqrt{h_r}}\right)\bar{\eta} \equiv \bar{C}_T\bar{\eta}. \end{aligned} \tag{10}$$

Note that $\bar{C}_T = \sqrt{h_r/h_\ell}\,C_T$. As the pulse passes over a discontinuous jump in bathymetry, the amplitude increases by C_T but its width decreases by $\sqrt{h_r/h_\ell}$ due to the change in wave speed, and so its volume changes by the product of these, which is $\bar{C}_T$. The reflected wave, on the other hand, has the same width as the incident pulse, and so $\bar{C}_R = C_R$.

The transmission and reflection coefficients for volume from (10) apply regardless of the width of the continental slope, and in fact are the same for any specified variation of $h(x)$ provided it varies only over a finite region and is identically equal to h_ℓ and h_r away from the slope. They are also independent of the shape of the pulse $\eta(x,0)$ as long as the initial wave is purely right-going, $u(x,0) = \sqrt{g/h_\ell}\eta(x,0)$, and $\eta(x,0) = 0$ for $x > -\epsilon$.

Note that the square pulse shown in fig. 2 for a very broad continental slope also appears to be transmitted as a square pulse, and the change in wave speed again suggests the width of the pulse will be reduced by $\sqrt{h_r/h_\ell}$. But if the initial amplitude is A then the amplitude of the transmitted wave is close to C_GA rather than $\bar{C}_TA$, and so it seems that too much volume has been transmitted onto the shelf. However, recall that there is also a small negative trailing wave. It turns out this wave has amplitude that decreases as the width of the continental slope increases, while at the same time it spreads out farther, in such a way that its total volume is constant and roughly equal to $(\bar{C}_T - C_G)A < 0$, which cancels the excess volume that appears in the transmitted pulse. This follows from the more general results presented in George et al. (2019a).

It is interesting to consider the case of a localized variation in bathymetry, with $h_r = h_l$ and any variations restricted to $-\epsilon < x < \epsilon$, e.g., an underwater ridge or sill. Then $\bar{C}_R = 0$ and $\bar{C}_T = 1$, so the total volume reflected is zero while the transmitted volume is equal to the volume of the original wave. This does not mean, of course, that the transmitted wave has the same form as the incident wave, nor that there are no reflections (only that the integral of the reflected wave vanishes).

It is also interesting to note that the total energy propagated onto the shelf varies as the width of the slope or the initial pulse are varied. A wave that is purely left-going or right-going satisfies equipartition of energy between potential and kinetic energy (see e.g. Berger et al. 2011; Mei 1992), so it suffices to consider, for example, the potential energy given by $E(x,t) = \rho g \int_{-\infty}^{\infty} \eta^2(x,t)\,dx$. where ρ is the density of water (which is assumed to be constant in this calculation). Consider the case of a step function continental slope and an initial pulse of height A and width w, for which $E(x,0) = \rho g w A^2$. The reflected wave has height $C_R A$ and width w, while the transmitted wave has height $C_T A$ and width $w\sqrt{h_r/h_\ell}$. So the total potential energy is $\rho g(C_R^2 w + C_T^2 w\sqrt{h_r/h_\ell})A^2 = \rho g w A^2$, illustrating conservation of energy. For a very broad slope, for which Green's Law holds, the energy in the reflected wave goes to zero (even though its volume is constant, it becomes more spread out with smaller amplitude as the width of the continental slope increases, and the energy is quadratic in η, so this integral vanishes). The transmitted wave carries all of the energy; it has height $C_G A$ and width $w\sqrt{h_r/h_\ell}$, and hence energy $\rho g C_G^2 w\sqrt{h_r/h_\ell} A^2 = \rho g w A^2$.

B: Single Intermediate Layer

In this appendix and the next we derive two results mentioned in the paper on the propagation of a bore through a layered medium, starting with the case of a single intermediate layer where it is easy to show that the sum of all transmitted waves approaches $C_T A$ over time while the sum of all reflected waves approaches $C_R A$. The single-layer case is shown in fig. 4(b), where the intermediate step in bathymetry has depth h_m with $h_r < h_m < h_\ell$. Define the following transmission coefficients:

$$C_T^{\ell m} = \frac{2\sqrt{h_\ell}}{\sqrt{h_\ell}+\sqrt{h_m}}, \quad C_T^{mr} = \frac{2\sqrt{h_m}}{\sqrt{h_m}+\sqrt{h_r}}, \quad C_T^{m\ell} = \frac{2\sqrt{h_m}}{\sqrt{h_\ell}+\sqrt{h_m}}, \tag{11}$$

where for example the superscript mr indicates transmission of a right-going wave from the middle layer to the right layer, and $m\ell$ indicates transmission of a left-going wave from the middle layer to the left layer. Similarly define the reflection coefficients for internal reflections in the intermediate layer as

$$C_R^{mr} = \frac{\sqrt{h_m}-\sqrt{h_r}}{\sqrt{h_m}+\sqrt{h_r}}, \quad C_R^{m\ell} = \frac{\sqrt{h_m}-\sqrt{h_\ell}}{\sqrt{h_m}+\sqrt{h_\ell}} \tag{12}$$

Then an initial right-going wave consisting of a discontinuity in η of amplitude A leads to a first transmitted wave of amplitude $C_T^{\ell m} C_T^{mr} A$. This is larger than $C_T A$, where C_T is the transmission coefficient with no intermediate layer from (5), but not as large as $C_G A$, the value found in Appendix 1 for a smooth transition. This first wave is followed by transmitted waves that have undergone $2n$ internal reflections for $n = 1, 2, 3, \ldots$, each of which has amplitude

$$C_T^{\ell m}(C_R^{mr} C_R^{ml})^n C_T^{mr} A. \tag{13}$$

Note that $-1 < C_R^{mr} C_R^{ml} < 0$ and so the sum of all the transmitted waves, including the first wave with no reflections, is given by the geometric series

$$\sum_{n=0}^{\infty} C_T^{\ell m}(C_R^{mr} C_R^{ml})^n C_T^{mr} A = \left(\frac{C_T^{\ell m} C_T^{mr}}{1 - C_R^{mr} C_R^{ml}}\right) A = C_T A \tag{14}$$

As expected, the state behind the right-going waves decays to the same state $C_T A$ obtained behind the right-going transmitted wave in the case of no interior layers.

Similarly, the sum of all left-going waves that depart into the deep ocean after $2n+1$ internal reflections (for $n = 0, 1, 2, \ldots$) is given by

$$\sum_{n=0}^{\infty} C_T^{\ell m} C_R^{mr} (C_R^{ml} C_R^{mr})^n C_T^{m\ell} A = C_R A, \tag{15}$$

where C_R is the reflection coefficient in the case of no intermediate layers. But note that there is a time lag between each departing wave, so the composite reflected wave takes the form of a step function with infinitely many jumps that decay exponentially with n.

C Infinitely Many Layers and the First Transmitted Wave

We now consider the amplification of the wave as it is transmitted through each interface of a layered medium, ignoring all the reflected waves generated in the process. The final amplitude of this "first transmitted wave" will be denoted by $C_T^f A$, and we will show that $C_T < C_T^f < C_G$ and that this transmission coefficient asymptotes to the Green's Law coefficient C_G as the number of layers increases. A different approach to deriving this same result is taken in George et al. (2019a) and George (2018), where more general results are derived for the waves that also experience internal reflections.

The amplitude of the first transmitted wave in the case of a single intermediate layer can be represented by $A_{t_1 t_2}$, signifying transmission through the first and second interface:

$$A_{t_1 t_2} = C_T^{\ell m} C_T^{mr} A = \frac{2\sqrt{h_\ell}}{\sqrt{h_m} + \sqrt{h_\ell}} \frac{2\sqrt{h_m}}{\sqrt{h_m} + \sqrt{h_r}} A. \tag{16}$$

In the case of two intermediate layers, the amplitude of the first transmitted wave is given by

$$A_{t_1 t_2 t_3} = \frac{2\sqrt{h_\ell}}{\sqrt{h_{m_1}} + \sqrt{h_\ell}} \frac{2\sqrt{h_{m_1}}}{\sqrt{h_{m_1}} + \sqrt{h_{m_2}}} \frac{2\sqrt{h_{m_2}}}{\sqrt{h_{m_2}} + \sqrt{h_r}} A. \tag{17}$$

It can be seen that $A_{t_1 t_2 t_3}$ is larger than $A_{t_1 t_2}$ and that increasing the number of layers increases the amplitude of the first transmitted wave.

More generally, consider piecewise-constant bathymetry with N intermediate steps having depths h_i chosen with $h_i = h_\ell + i\Delta h$ where $\Delta h = (h_r - h_\ell)/N$ for $i = 1, 2, \ldots, N$. We also define $h_0 = h_\ell$ and $h_{N+1} = h_r$. Then the transmission coefficient at the interface between h_i and h_{i+1} is given by

$$C_T^i = \frac{2\sqrt{h_i}}{\sqrt{h_i} + \sqrt{h_i + \Delta h}} = \frac{2}{1 + \sqrt{1 + \frac{\Delta h}{h_i}}}.$$

This allows us to find the continuous limit of the leading amplitude by considering the scenario with infinitely many layers. The amplitude of the first transmitted wave would then be given by

$$\lim_{N\to\infty} \prod_{i=0}^{N} C_T^i A = \lim_{N\to\infty} \prod_{i=0}^{N} \left(\frac{2}{1 + \sqrt{1 + \frac{\Delta h}{h_i}}} \right) A \equiv C_T^f A, \tag{18}$$

Taking the log of C_T^f converts the infinite product into a sum,

$$\log(C_T^f) = \lim_{N\to\infty} \sum_{i=0}^{N} \log\left(\frac{2}{1 + \sqrt{1 + \frac{\Delta h}{h_i}}} \right). \tag{19}$$

Using the Taylor expansion then results in

$$\log(C_T^f) = -\lim_{N\to\infty} \sum_{i=0}^{N} \left(\frac{\Delta h}{4h_i} - \frac{\Delta h^2}{16h_i^2} + \cdots \right). \tag{20}$$

Since the sum of the Δh^2 and higher order terms vanish in the limit, we obtain

$$\log(C_T^f) = -\lim_{N\to\infty} \frac{h_r - h_\ell}{N} \sum_{i=0}^{N} \frac{1}{4h_i}. \tag{21}$$

This sum approaches an integral in the limit, giving

$$\log(C_T^f) = -\int_{h_\ell}^{h_r} \frac{1}{4x} dx = -\frac{1}{4} \log\left(\frac{h_r}{h_\ell} \right) \tag{22}$$

and hence

$$C_T^f = (h_\ell / h_r)^{1/4}. \tag{23}$$

This is exactly the Green's Law coefficient C_G from Eq. (6), showing that the first transmitted wave has amplitude $C_G A$ in the limit as the number of internal layers goes to infinity, i.e. as the step function bathymetry is smoothed out to a continuous continental slope. Note also that this argument is independent of the locations x_i corresponding to each jump discontinuity from h_i to h_{i+1}, and so the same result holds regardless of whether we discretize a narrow or wide slope, and regardless of the shape of the slope; it need not be a discretization of a linear slope of the sort shown in fig. 1.

References

Adams, L. M., & LeVeque, R. J. (2017). *GeoClaw Model Tsunamis Compared to Tide Gauge Results—Final Report.*http://hdl.handle.net/1773/41886.

Berger, M. J., George, D. L., & LeVeque, R. J. (2011). Adaptive mesh refinement techniques for tsunamis and other geophysical flows over topography. *Acta Numerica, 20,* 211–289.

Clawpack Development Team. (2017). *Clawpack software.*https://doi.org/10.5281/zenodo.1405834, http://www.clawpack.org, version 5.5.0.

Davies, G., Griffin, J., Løvholt, F., Glimsdal, S., Harbitz, C., Thio, H. K., et al. (2018). A global probabilistic tsunami hazard assessment from earthquake sources. *Geological Society, London, Special Publications, 456,* 219–244. https://doi.org/10.1144/SP456.5.

Dzvonkovskaya, A., Heron, M., Figueroa, D., & Gurgel, K. (2014). Observations and theory of a shoaling tsunami wave. In: *2014 Oceans—St. John's,* 1–5. https://doi.org/10.1109/OCEANS.2014.7003236.

Evans, D. V., & Linton, C. M. (1994). On step approximations for water-wave problems. *Journal of Fluid Mechanics, 278,* 229–249. https://doi.org/10.1017/S002211209400368X.

George, J.D. (2018) *Green's law and the Riemann problem in layered media.* Master's thesis, University of Washington.

George, J. D., Ketcheson, D. I., & LeVeque, R. J. (2019a). *A characteristics-based approximation for wave scattering from an arbitrary obstacle in one dimension.*, arXiv:1901.04158.

George, J. D., Ketcheson, D. I., & LeVeque, R. J. (2019b). *Code to accompany this paper.*, https://doi.org/10.5281/zenodo.3373993. https://github.com/rjleveque/shoaling_paper_figures.

Glimsdal, S., Løvholt, F., Harbitz, C. B., Romano, F., Lorito, S., Orefice, S., et al. (2019). A new approximate method for quantifying tsunami maximum inundation height probability. *Pure and Applied Geophysics.* https://doi.org/10.1007/s00024-019-02091-w.

Green, G. (1838). On the motion of waves in a variable canal of small depth and width. *Transactions of the Cambridge Philosophical Society, 6,* 457.

Grezio, A., Babeyko, A., Baptista, M. A., Behrens, J., Costa, A., Davies, G., et al. (2017). Probabilistic tsunami hazard analysis: Multiple sources and lobal applications. *Reviews of Geophysics, 55,* 1158–1198. https://doi.org/10.1002/2017RG000579.

Grilli, S. T., Subramanya, R., Svendsen, I. A., & Veeramony, J. (1994). Shoaling of solitary waves on plane beaches. *Journal of Waterway, Port, Coastal, and Ocean Engineering, 120*(6), 609–628. https://doi.org/10.1061/(ASCE)0733-950X(1994)120:6(609).

Heron, M., & Dzvonkovskaya, A. (2015). Conceptual view of reflection and transmission of a tsunami wave at a step in bathymetry. In *OCEANS 2015—MTS/IEEE Washington.* (pp 1–4), https://doi.org/10.23919/OCEANS.2015.7404520.

Kânoğlu, U., & Synolakis, C. E. (1998). Long wave runup on piecewise linear topographies. *Journal of Fluid Mechanics, 374,* 1–28.

LeVeque, R. J. (2002). Finite volume methods for hyperbolic problems. Cambridge University Press, http://amath.washington.edu/~claw/book.html.

Levin, B. W., & Nosov, M. A. (2016). *Physics of tsunamis* (2nd ed.). New York: Springer.

Lorito, S., Selva, J., Basili, R., Romano, F., Tiberti, M. M., & Piatanesi, A. (2015). Probabilistic hazard for seismically induced tsunamis: Accuracy and feasibility of inundation maps. *Geophysical Journal International, 200*(1), 574–588.

Løvholt, F., Glimsdal, S., Harbitz, C. B., Zamora, N., Nadim, F., Peduzzi, P., et al. (2012). Tsunami hazard and exposure on the global scale. *Earth-Science Reviews, 110*(1–4), 58–73. https://doi.org/10.1016/j.earscirev.2011.10.002.

Løvholt, F., Griffin, J., & Salgado-Gálvez, M. (2016). Tsunami hazard and risk assessment on the global scale. In R. A. Meyers (Ed.), *Encyclopedia of complexity and systems science* (pp. 1–34). Berlin: Springer. https://doi.org/10.1007/978-3-642-27737-5_642-1.

Madsen, P. A., Fuhrman, D. R., & Schäffer, H. A. (2008). On the solitary wave paradigm for tsunamis. *Journal of Geophysical Research: Oceans, 113*(C12), C12012. https://doi.org/10.1029/2008JC004932.

Mei, C. C. (1992). *The applied dynamics of ocean surface waves.* Singapore: World Scientific.

Miles, J. W. (1967). Surface-wave scattering matrix for a shelf. *Journal of Fluid Mechanics, 28,* 755–767. https://doi.org/10.1017/S0022112067002423.

Synolakis, C. E. (1991). Green's law and the evolution of solitary waves. *Physics of Fluids A: Fluid Dynamics, 3*(3), 490–491. https://doi.org/10.1063/1.858107.

Ward, S. N., & Asphaug, E. (2000). Asteroid impact tsunami: A probabilistic hazard assessment. *Icarus, 145*(1), 64–78.

(Received January 13, 2019, revised August 21, 2019, accepted August 24, 2019, Published online September 9, 2019)

Pure Appl. Geophys. 177 (2020), 1675–1693

https://doi.org/10.1007/s00024-019-02259-4

Pure and Applied Geophysics

Transformation of Long Surface and Tsunami-Like Waves in the Ocean with a Variable Bathymetry

A. M. Ermakov[1] and Y. A. Stepanyants[1,2]

Abstract—We consider a transformation of long linear waves in an ocean with a variable depth.We calculate the transformation coefficients (the coefficients of transmission and reflection) as the functions of frequency and the total depth drop for three typical models of bottom profile variation: (i) piecewise-linear, (ii) piecewise-quadratic, and (iii) hyperbolic tangent profiles.For all these cases, exact solutions are obtained, analysed and graphically illustrated. This allows us to derive the transformation coefficients in the analytic form for the subsequent comparison with the results of approximate, numerical, or experimental study. We show that the results obtained are in agreement with both the energy flux conservation and Lamb's formulae in the limiting case of zero frequency. We also study wave transformation on the underwater barriers and trenches of different shapes and compare the results obtained.

Key words: Shallow water, tsunami wave, variable bathymetry, wave transformation.

1. Introduction

The problem of linear wave transformation on the bottom unevenness has a long history. The development of rigorous mathematical methods for the description of wave transformation in the coastal zone represents an important and topical problem from both the academic and practical points of view, especially in application to the protection of marine engineering constructions (platforms, gas and oil pipelines, etc.) (Belibassakis et al. 2017) and coastlines against hazardous impacts of large oceanic waves including tsunami waves (Harris and Vila-Concejo 2013; Ferrario et al. 2014). As is well-known (see, for example, Pelinovsky 1996; Levin and Nosov 2009), tsunami waves in the open ocean can be treated in the same way as the linear waves in a shallow basin; their typical heights are about 0.5 m, and wavelengths are 300–500 km, whereas the average ocean depth is 4–5 km. Therefore, scattering of such waves on underwater mountain ranges or trenches can be studied within the framework of linear shallow water theory. Figure 1 from https://en.wikipedia.org/wiki/Continental_shelf illustrates schematically the continental slope located between the shelf where the depth varies linearly in average and continental foot, bordering the ocean bed. According to https://en.wikipedia.org/wiki/Continental_shelf, in the continental slope the depth varies greatly and can be approximates either by a linear function or quadratic function as shown in Fig. 1. The typical variation of the bottom angle is from about 3°–4° to 40°–45° and even more. The average bottom slope is about 3.5°, which is close to the slope of the mountain ranges. The continental slope covers a significant range of depths, in general, from 100–200 m to 3000–4000 m and has a typical width from 300 km up to 2000 km. This make the problem of wave transformation in the ocean with a variable bathymetry very topical.

The problem of wave propagation in the fluid with uneven bottom has a general solution in two limiting cases. When the bottom smoothly varies with the distance so that the characteristic scale of bottom variation is much greater than the wavelength, then the traditional JWKB method can be used to calculate wave parameters (amplitude, wavelength, and other parameters) (see, e.g., Massel 1989; Pelinovsky 1996; Dingemans 1997). In another limiting case when there is an underwater step-wise obstacle, the solution

[1] Faculty of Health, Engineering and Sciences, University of Southern Queensland, Toowoomba, QLD 4350, Australia. E-mail: ErmakovAndreiM@gmail.com; Yury.Stepanyants@usq.edu.au

[2] Department of Applied Mathematics, Nizhny Novgorod State Technical University, Nizhny Novgorod, Russia 603950.

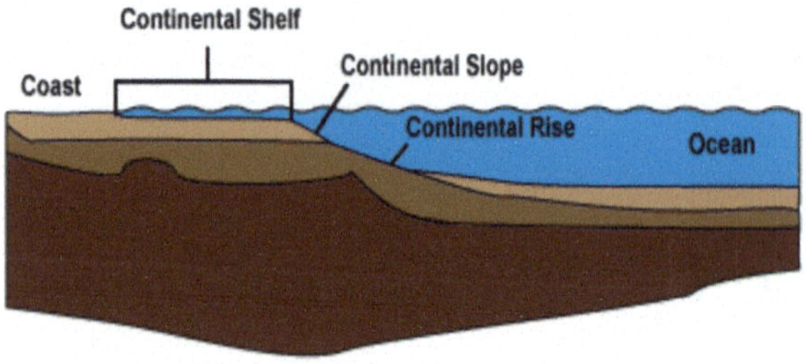

Figure 1
(Color figure online) A sketch of the typical oceanic profile (https://en.wikipedia.org/wiki/Continental_shelf)

for a wave field can be obtained on both sides from the step and then the solution can be matched at the boundary of two homogeneous regions. This procedure is not so trivial as may seem at first glance, because in general one should take into account not only travelling waves, but also the infinite series of evanescent modes (see Kurkin et al. 2015 and references therein). However, in the long-wave approximation contribution of evanescent modes into the wave scattering is negligible which significantly simplifies the problem (Dingemans 1997).

Besides these limiting cases the problem of long wave propagation can be solved analytically for some particular bottom profiles (Dean 1964; Mei 1990; Kânoğlu and Synolakis 1998; Jung et al. 2008; Xie et al. 2011) (see also Dingemans 1997), moreover for the linear bottom profile there is an exact solution of linearised hydrodynamic equations for waves of *arbitrary* wavelength (Stoker 1957; Sretenskii 1977). It has also been shown that in some particular cases exact solutions can be obtained for the reflectionless propagation of long linear and nonlinear waves in a fluid with a special bottom profiles (see Didenkulova and Pelinovsky 2013, 2016 and references therein).

Here we revise the problem of linear wave transformation in shallow water with three particular bottom profiles: (i) piecewise-linear, (ii) piecewise-quadratic, and (iii) hyperbolic tangent (tanh-) profiles. We show that in these particular cases the basic equations can be solved analytically and solutions can be presented in general in terms of hypergeometric functions, which can be reduced to elementary functions in particular cases. This allows us to derive the transformation coefficient and compare the results for all three profiles. Then we study wave transformation over underwater obstacles/trenches of finite width, present the results in the graphical forms, and analyse them. Our consideration is confined to the normal incidence of waves on the plane (one-dimensional) underwater obstacles. As is well-known the effect of wave transmission is the greatest in this case. Therefore, the wave impact on a coastal zone and marine engineering constructions is the most dangerous at the normal incidence.

The main goal of this publication is to draw readers' attention to the possibility of construction of exact analytical solutions to the problem of water wave transformation on rather typical bottom profiles and demonstrate what is common and what is the different in the transformation coefficients. In a separate publication, we plan to present a more detailed analysis of wave transformation in basins with the typical bottom configurations and compare the theoretical results with the data of laboratory and numerical experiments. The results obtained are applicable not only to long surface waves (swell, wind waves) in the coastal zones, but also to tsunami waves in the open oceans where such waves are essentially linear. Scattering of tsunami waves on underwater mountain ranges and canyons is a topical problem, and results obtained allow one to estimate the amplitude of transmitted wave without the detail and costly numerical modelling, but using simple analytic formulae.

2. *The Piecewise-Linear Bottom Profile*

Consider first long linear surface wave scattering when it propagates in water with the piecewise-linear bottom profile shown in Fig. 2. An exact analytical solution to this problem was obtained by Dean (1964) as early as the beginning of 1960's (see also the book by Dingemans 1997 where other approaches to this problem are described). We shall re-consider this problem and present the results obtained in a form suitable for the comparison with other results derived for the parabolic and tanh-profiles. In the long-wave approximation the basic hydrodynamic equations representing mass and momentum conservations are (Carrier and Greenspan 1958):

$$\eta_t^* + [U(\eta^* + h)]_x = 0, \tag{1}$$

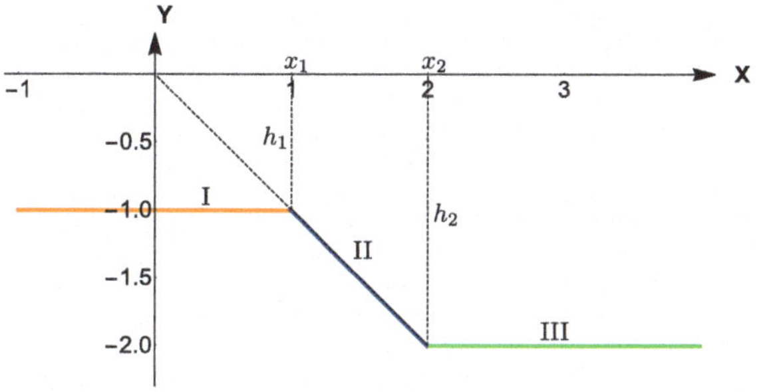

Figure 2
(Color figure online). A basin with the piecewise-linear bottom profile

$$g\eta^*_x + U_t + UU_x = 0, \tag{2}$$

where U is the depth averaged fluid velocity, η^* is the perturbation of a free surface, g is the acceleration due to gravity, and h is the water depth. We assume that the water depth is constant, $h = h_1$ when $x \le x_1$ and $h = h_2$ when $x \ge x_2$; in the interval $x_1 \le x \le x_2$ the depth changes linearly $h = \alpha x$ (see Fig. 2).

For the further analysis it is convenient to present the set of Eqs. (1) and (2) in the dimensionless form using the following change of variables:

$$V = \frac{U}{U_0}, \quad \eta = \frac{\eta^*}{\alpha l_0}, \quad \xi = \frac{x}{l_0}, \quad \hat{h} = \frac{h}{\alpha l_0} = \xi, \quad \hat{\omega} = \omega T, \quad \tau = \frac{t}{T}, \tag{3}$$

where $l_0 = x_2 - x_1$ (see Fig. 2), $T = l_0/\alpha g$, and $U_0 = (\alpha g l_0)^{1/2}$.

In the dimensionless variables the basic set of equations is:

$$\eta_\tau = -[V(\eta + h)]_\xi, \tag{4}$$

$$\eta_\xi = -V_\tau - VV_\xi. \tag{5}$$

Assuming that $\eta \ll h$, we obtain after linearisation and elimination of η the following wave equation:

$$V_{\tau\tau} - (Vh)_{\xi\xi} = 0. \tag{6}$$

In the intervals where the fluid depth is constant, i.e., for $x < x_1$ and $x > x_2$, an elementary solution to this equation can be sought in the form $V \sim e^{i(\hat{\omega}\tau \pm k\xi)}$. However in the interval where the bottom linearly varies, i.e., when $x_1 < x < x_2$ and the depth $h = \xi$ in the dimensionless variables, the solution should be sought in the form: $V = \Psi(\xi)e^{i\hat{\omega}\tau}$. After substitution of this solution into Eq. (6) we obtain:

$$\xi\Psi_{\xi\xi} + 2\Psi_\xi + \hat{\omega}^2\Psi = 0. \tag{7}$$

This is the standard Bessel equation whose solution is a linear combination of Bessel functions. Therefore for function $V(\xi, \tau)$ we obtain:

$$V(\xi) = \frac{1}{\hat{\omega}\sqrt{\xi}}\left[B_1 J_1(2\hat{\omega}\sqrt{\xi}) + B_2 Y_1(2\hat{\omega}\sqrt{\xi})\right]e^{i\hat{\omega}\tau}, \tag{8}$$

where $J_1(x)$ and $Y_1(x)$ are Bessel functions of the first and second kinds respectively. Then in the linear approximation we find from Eq. (4) for $\eta = \Phi(\xi)e^{i\hat{\omega}\tau}$, where

$$\Phi(\xi) = \frac{i}{\hat{\omega}}\left[B_1 J_0(2\hat{\omega}\sqrt{\xi}) + B_2 Y_0(2\hat{\omega}\sqrt{\xi})\right]. \tag{9}$$

Here $J_0(x)$ and $Y_0(x)$ are Bessel functions of the first and second kinds, but with zero indices.

Now we can write down the general solutions for the perturbation of a free surface in three different domains shown in Fig. 2:

$$\Phi_l = A_1 e^{-ik_1(\xi - \xi_1)} + A_2 e^{ik_1(\xi - \xi_1)}, \quad \xi \le \xi_1; \tag{10}$$

$$\Phi_c = \frac{i}{\hat{\omega}}\left[B_1 J_0(2\hat{\omega}\sqrt{\xi}) + B_2 Y_0(2\hat{\omega}\sqrt{\xi})\right], \quad \xi_1 \le \xi \le \xi_2; \tag{11}$$

$$\Phi_r = C_1 e^{-ik_2(\xi - \xi_2)} + C_2 e^{ik_2(\xi - \xi_2)}, \quad \xi \ge \xi_2. \tag{12}$$

where the wave numbers are

$$k_1 = \frac{\omega l_0}{\sqrt{gh_1}} = \frac{\hat{\omega}}{\sqrt{\xi_1}}, \quad k_2 = \frac{\omega l_0}{\sqrt{gh_2}} = \frac{\hat{\omega}}{\sqrt{\xi_2}}, \tag{13}$$

and $A_{1,2}, B_{1,2}, C_{1,2}$ are arbitrary constants. Note that in stationary inhomogeneous media a wave frequency is conserved in the process of wave transformation.

Solutions (10)–(12) can be matched at the boundaries of domains 1, 2 and 3 shown in Fig. 2 by means of the following boundary conditions reflecting the continuity of function $\Phi(\xi)$ and its first derivative:

$$\begin{gathered} \Phi_l(\xi_1) = \Phi_c(\xi_1), \quad \Phi_c(\xi_2) = \Phi_r(\xi_2), \\ \Phi'_l(\xi_1) = \Phi'_c(\xi_1), \quad \Phi'_c(\xi_2) = \Phi'_r(\xi_2). \end{gathered} \tag{14}$$

Using these boundary conditions, we consider below surface wave transformation on the bottom unevenness when the incident wave arrives from the left and from the right.

2.1. Wave Transformation in the Case of Decreasing Depth

This case corresponds to the situation when an incident wave arrives from the right in Fig. 2. Then we set the following coefficients in the solutions (10)–(12):

$$A_1 = 0, \quad A_2 = T_r, \quad C_1 = R, \quad C_2 = 1, \qquad (15)$$

where C_2 is the amplitude of an incident wave, R is the reflection coefficient (the amplitude of reflected wave), and T_r is the transmission coefficient (the amplitude of transmitted wave).

By substitution solutions (10)–(12) with the coefficients (15) into the boundary conditions (14), we obtain the transformation coefficients R and T, as well as the expressions for the wave amplitudes B_1 and B_2 in the transient region II (see Fig. 2):

$$\begin{aligned} R = \frac{1}{\Delta}\Big\{&\left[J_0\left(2\hat{\omega}\sqrt{\xi_2}\right) - \mathrm{i}J_1\left(2\hat{\omega}\sqrt{\xi_2}\right)\right]\\ &\left[Y_0\left(2\hat{\omega}\sqrt{\xi_1}\right) - \mathrm{i}Y_1\left(2\hat{\omega}\sqrt{\xi_1}\right)\right].\\ &-\left[J_0\left(2\hat{\omega}\sqrt{\xi_1}\right) - \mathrm{i}J_1\left(2\hat{\omega}\sqrt{\xi_1}\right)\right]\\ &\left[Y_0\left(2\hat{\omega}\sqrt{\xi_2}\right) - \mathrm{i}Y_1\left(2\hat{\omega}\sqrt{\xi_2}\right)\right]\Big\}, \end{aligned} \qquad (16)$$

$$\begin{aligned} T_r = \frac{2\,\mathrm{i}}{\Delta}\Big[&J_1\left(2\hat{\omega}\sqrt{\xi_1}\right)Y_0\left(2\hat{\omega}\sqrt{\xi_1}\right)\\ &-J_0\left(2\hat{\omega}\sqrt{\xi_1}\right)Y_1\left(2\hat{\omega}\sqrt{\xi_1}\right)\Big], \end{aligned} \qquad (17)$$

$$B_1 = -\frac{2\,\mathrm{i}\,\hat{\omega}}{\Delta}\left[Y_0\left(2\hat{\omega}\sqrt{\xi_1}\right) - \mathrm{i}Y_1\left(2\hat{\omega}\sqrt{\xi_1}\right)\right], \qquad (18)$$

$$B_2 = \frac{2\,\mathrm{i}\,\hat{\omega}}{\Delta}\left[J_0\left(2\hat{\omega}\sqrt{\xi_1}\right) - \mathrm{i}J_1\left(2\hat{\omega}\sqrt{\xi_1}\right)\right], \qquad (19)$$

where the determinant is:

$$\begin{aligned} \Delta = &\left[J_0\left(2\hat{\omega}\sqrt{\xi_2}\right) + \mathrm{i}J_1\left(2\hat{\omega}\sqrt{\xi_2}\right)\right]\\ &\left[Y_0\left(2\hat{\omega}\sqrt{\xi_1}\right) - \mathrm{i}Y_1\left(2\hat{\omega}\sqrt{\xi_1}\right)\right]\\ &-\left[J_0\left(2\hat{\omega}\sqrt{\xi_1}\right) - \mathrm{i}J_1\left(2\hat{\omega}\sqrt{\xi_1}\right)\right]\\ &\left[Y_0\left(2\hat{\omega}\sqrt{\xi_2}\right) + \mathrm{i}Y_1\left(2\hat{\omega}\sqrt{\xi_2}\right)\right]. \end{aligned} \qquad (20)$$

Solutions (10)–(12) with the coefficients (16)–(19) are shown in Fig. 3 for $|\Phi(\xi)|$ (see line a). A wavy dependence of function $|\Phi(\xi)|$ in the region III is caused by bits of incident and reflected waves,

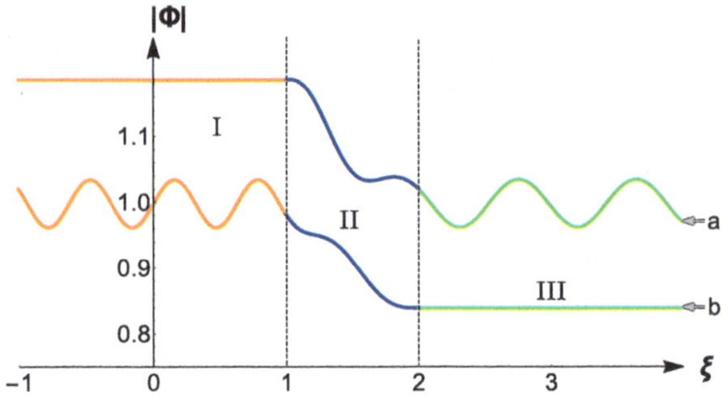

Figure 3
(Color figure online) Solutions for the free surface perturbation (10)–(12) with the coefficients (16)–(19) in terms of $|\Phi(\xi)|$. Line (**a**) pertains to the case when the incident wave arrives from the right where the depth $h_2 > h_1$, and line (**b**) pertains to the case when the incident wave arrives from the left where the depth $h_1 > h_2$. The Roman numerals correspond to the domains shown in Fig. 2

whereas in region I we have only one transmitted sinusoidal wave of a constant amplitude. Figure 4a shows the reflection coefficient and Fig. 4b the transmission coefficient as the functions of normalised frequency $\hat{\omega}$.

The plots shown in Figs. 3 and 4 were generated for the particular set of parameters with $h_1 = 1$ and $h_2 = 2$, and wave frequency $\hat{\omega} = 5$; for other parameters h_1, h_2, and $\hat{\omega}$ the plots are very similar to those shown in these figures.

In the limiting case $\hat{\omega} \to 0$ the formulae for the transformation coefficients reduce to the well-known expressions derived by Lamb (1932) for the step-wise bottom (see also Massel 1989; Kurkin et al. 2015 and references therein):

$$R = \frac{1-\sqrt{k}}{1+\sqrt{k}}, \quad T = \frac{2}{1+\sqrt{k}}, \qquad (21)$$

where $k = h_1/h_2$. Such reduction is quite natural, because in the limit $\hat{\omega} \to 0$ the wave lengths of scattered waves become much greater than the length of transient domain l_0.

The non-monotonic character of dependencies $|R|(\hat{\omega})$ and $|T|(\hat{\omega})$ is explained by the interference of waves within the transient zone $\xi_1 \le \xi \le \xi_2$ due to the reflections from the bottom edges at ξ_1 and ξ_2. However in the minima the reflection coefficient never vanishes (see Fig. 4), therefore the reflectionless propagation is impossible, although for some

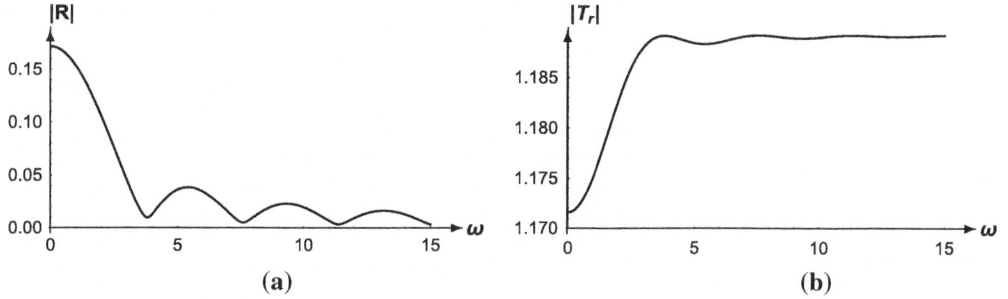

Figure 4
The reflection |R| (a) and transmission $|T_r|$ (b) coefficients as the functions of normalised frequency $\hat{\omega}$ for the fixed value of $h_2/h_1 = 2$. The plots pertain to the case when the incident wave arrives from the right where the depth $h_2 > h_1$

"quasi-resonant" frequencies the reflected wave can be very small.

As follows from Fig. 4, the amplitude of a transmitted wave can be greater than the amplitude of an incident wave ($T > 1$); this is a usual effect when a wave travels from a deeper region to a shallower region, herewith its wavelength decreases and amplitude increases. The energy flux conserves in this case and results in the relationship between R, T, and $h_{1,2}$. The energy flux can be determined as the density of wave energy (which is proportional to the squared wave amplitude) times the group speed (see, e.g., Mei 1990; Dingemans 1997), which in the dimensionless variables for the shallow-water case is $c(\xi) = \sqrt{h(\xi)}$. Therefore, the energy flux conservation can be presented as:

$$c_1 T^2 = c_2(1 - R^2), \quad \text{or} \quad \sqrt{\frac{h_1}{h_2}}T^2 + R^2 = 1. \qquad (22)$$

2.2. *Wave Transformation in the Case of Increasing Depth*

In this case when the incident wave arrives from the left we chose the coefficients of the system (10)–(12) as follows:

$$A_1 = 1, \quad A_2 = R, \quad C_1 = T_l, \quad C_2 = 0. \qquad (23)$$

By substituting solutions (10)–(12) with the coefficients (23) into the boundary conditions (14), we obtain again for the transformation coefficients R and T_l, as well as for the amplitudes of waves in the transient domain 2 (see Fig. 2) B_1 and B_2:

$$\begin{aligned} R = \frac{1}{\Delta}\Big\{&\Big[J_0(2\hat{\omega}\sqrt{\xi_2}) + \mathrm{i}J_1(2\hat{\omega}\sqrt{\xi_2})\Big] \\ &\Big[Y_0(2\hat{\omega}\sqrt{\xi_1}) + \mathrm{i}Y_1(2\hat{\omega}\sqrt{\xi_1})\Big]\cdot \\ &-\Big[J_0(2\hat{\omega}\sqrt{\xi_1}) + \mathrm{i}J_1(2\hat{\omega}\sqrt{\xi_1})\Big] \\ &\Big[Y_0(2\hat{\omega}\sqrt{\xi_2}) + \mathrm{i}Y_1(2\hat{\omega}\sqrt{\xi_2})\Big]\Big\}, \end{aligned} \qquad (24)$$

$$\begin{aligned} T_l = \frac{2\mathrm{i}}{\Delta}\Big[&J_1(2\hat{\omega}\sqrt{\xi_2})Y_0(2\hat{\omega}\sqrt{\xi_2}) \\ &-J_0(2\hat{\omega}\sqrt{\xi_2})Y_1(2\hat{\omega}\sqrt{\xi_2})\Big], \end{aligned} \qquad (25)$$

$$B_1 = \frac{2\hat{\omega}\mathrm{i}}{\Delta}\Big[Y_0(2\hat{\omega}\sqrt{\xi_2}) + \mathrm{i}Y_1(2\hat{\omega}\sqrt{\xi_2})\Big], \qquad (26)$$

$$B_2 = -\frac{2\hat{\omega}\mathrm{i}}{\Delta}\Big[J_0(2\hat{\omega}\sqrt{\xi_2}) + \mathrm{i}J_1(2\hat{\omega}\sqrt{\xi_2})\Big], \qquad (27)$$

where the determinant Δ is the same as in Eq. (20).

Solutions (10)–(12) with the coefficients (24)–(27) are shown in Fig. 3 for $|\Phi(\xi)|$ (see line b). Graphic of the reflection coefficient for the same depth ratio $h_2/h_1 = 2$ is the same as in Fig. 4a, and the graphic of transformation coefficient is similar to that shown in Fig. 4b, but with the additional multiplicative factor $\sqrt{h_2/h_1}$. Now the amplitude of transmitted wave is less than the amplitude of an incident wave ($T_l < 1$), but the energy flux still conserves and has the same form as in Eq. (22) with the depth interchange: $h_1 \longleftrightarrow h_2$. This is the direct consequence of the *reciprocity relationship* derived in (Mei 1990): $k_2 h_2 T_l = k_1 h_1 T_r$, where the wave numbers k_1 and k_2 are determined in Eq. (13).

3. The Piecewise-Quadratic Bottom Profile

In this section we consider the case when the the bottom profile in the transient domain between two semi-infinite intervals with a constant depth can be approximated by a smooth conjugation of two quadratic functions as shown in Fig. 5 by line 1. A similar problem was considered by Kajiura (1961) (see also Dingemans 1997) for the model when one quadratic function in the transient domain connects two bottom levels with the discontinuities of derivatives at the boundaries (see line 2 in Fig. 5). Kajiura derived the reflection coefficient by means of the JWKB method up to second order terms in the asymptotic expansion in small parameter $(kl_0)^{-1} \ll 1$, where k is the wave number and $l_0 = x_2 - x_1$ is the width of the transient domain. It has been shown that the reflection coefficient apart from an oscillatory structure of the dependence $R(\hat{\omega})$ has an asymptotic form $R \sim \hat{\omega}^{-1}$. This problem was reconsidered later by Mei (1990) who derived the exact solution for the transient domain with the quadratic bottom profile and obtained the transformation coefficients. Here we consider a similar problem, but with two quadratic functions representing smooth connection of two bottom levels as shown in Fig. 5. We show that the exact solutions can be constructed and the transformation coefficients can be derived in the analytic, although rather cumbersome form.

We assume that the bottom profile can be presented by two parabolic functions

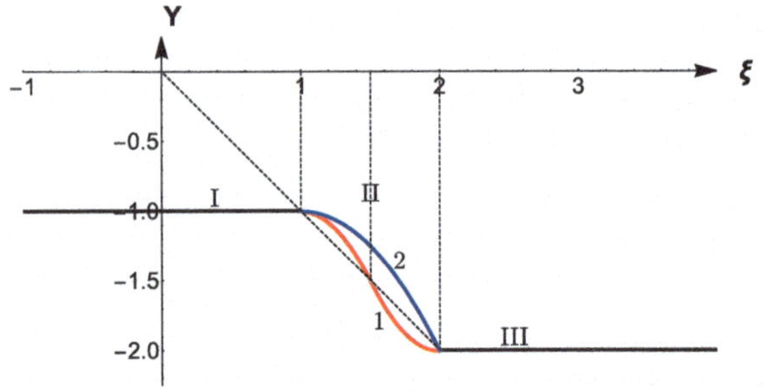

Figure 5
(Color figure online) A basin with the piecewise-quadratic bottom profile. Line 1 pertains to the case when two quadratic functions are conjugated at the middle of the transient zone ($\gamma = 0.5$—see the text) with smooth conjugation at the boundaries shown by dashed vertical lines. Line 2 pertains to the quadratic profile connecting two constant bottom levels as considered by Kajiura (1961) and Mei (1990)

$h(\xi) = a\xi^2 + b\xi + c$ with different coefficients smoothly matched with each other within the transient domain (at the point ξ_c) and with the constant bottom levels at the edges ξ_1 and ξ_2 of the transient domain. This can be achieved by the following choice of parabolic functions:

$$h_L(\xi) = \xi_1 + \frac{(\xi - \xi_1)^2}{\gamma(\xi_2 - \xi_1)}, \quad \xi_1 \le \xi \le \xi_c, \tag{28}$$

$$h_R(\xi) = \xi_2 - \frac{(\xi - \xi_2)^2}{(1-\gamma)(\xi_2 - \xi_1)}, \quad \xi_c \le \xi \le \xi_2, \tag{29}$$

where $\xi_c = \xi_2\gamma + (1-\gamma)\xi_1$ and $0<\gamma<1$. Varying the parameter γ one can obtain a class of smooth parabolic profiles connecting two constant bottom levels at different points; this can be treated as the spline approximation of a bottom profile. For simplicity we show in Fig. 5 only one such profile with $\gamma = 0.5$ (see line 1).

To construct a solution to the linearised set of Eqs. (4) and (5) it is convenient to reduce it to one equation for the perturbation of water surface $\eta(\xi, \tau)$:

$$\eta_{\tau\tau} - \left(\eta_\xi h\right)_\xi = 0 \tag{30}$$

Note that for the piecewise-quadratic bottom profile it is convenient to reduce the basic set of Eqs. (4) and (5) to one Eq. (30) for the surface elevation rather than to Eq. (6) for the fluid velocity, because the following Eq. (31) for function $\Phi \sim \eta$ reduces directly to the Legendre equation, whereas the governing Eq. (7) for function $\Psi \sim V$ reduces directly to the Bessel equation.

Substituting in Eq. (30) a solution in the form: $\eta(\xi, \tau) = e^{i\hat{\omega}\tau}\Phi(\xi)$ and the expressions for the bottom profile (28), (29), we obtain the Legendre equation for function $\Phi(\xi)$:

$$\left(a\xi^2 + b\xi + c\right)\Phi'' + (2a\xi + b)\Phi' + \hat{\omega}^2\Phi = 0, \tag{31}$$

where a, b, and c are some real constants. Solution to this equation can be expressed in general in terms of the linear combination of Legendre functions:

$$\Phi(\xi) = B_1 P_\nu^0\left(\frac{2a\xi + b}{\sqrt{b^2 - 4ac}}\right) + B_2 Q_\nu^0\left(\frac{2a\xi + b}{\sqrt{b^2 - 4ac}}\right), \tag{32}$$

where P_ν^0 and Q_ν^0 are the Legendre functions of the first and second kinds with $\nu = \left(\sqrt{1 - 4\hat{\omega}^2/a} - 1\right)/2$ (Polyanin and Zaitsev 2003), and B_1 and B_2 are arbitrary constants. In the particular case when $a = -\hat{\omega}^2/n(n+1)$, where n is a positive integer, one of these functions reduces to the Legendre polynomial $P_n(\xi)$.

Now solutions of Eq. (30) in all four domains shown in Fig. 5 can be presented as:

$$\Phi_l(\xi) = A_1 e^{-ik_1(\xi-\xi_1)} + A_2 e^{ik_1(\xi-\xi_1)}, \quad \xi \le \xi_1, \quad (33)$$

$$\Phi_{p_1}(\xi) = B_1 \bar{w}_1 + B_2 \bar{w}_2, \quad \xi_1 \le \xi \le \xi_c, \quad (34)$$

$$\Phi_{p_2}(\xi) = C_1 \tilde{w}_1 + C_2 \tilde{w}_2, \quad \xi_c \le \xi \le \xi_2, \quad (35)$$

$$\Phi_r(\xi) = D_1 e^{-ik_2(\xi-\xi_2)} + D_2 e^{ik_2(\xi-\xi_2)}, \quad \xi \ge \xi_2, \quad (36)$$

where $A_{1,2}, B_{1,2}, C_{1,2}$ and $D_{1,2}$ are arbitrary constants,

$$\bar{w}_1 = P_\nu^0\left[\frac{\xi - \xi_1}{\sqrt{\gamma\xi_1(\xi_1 - \xi_2)}}\right], \quad \bar{w}_2 = Q_\nu^0\left[\frac{\xi - \xi_1}{\sqrt{\gamma\xi_1(\xi_1 - \xi_2)}}\right], \quad (37)$$

$$\tilde{w}_1 = P_\mu^0\left[\frac{\xi - \xi_2}{\sqrt{(1-\gamma)\xi_2(\xi_2 - \xi_1)}}\right], \quad \tilde{w}_2 = Q_\mu^0\left[\frac{\xi - \xi_2}{\sqrt{(1-\gamma)\xi_2(\xi_2 - \xi_1)}}\right], \quad (38)$$

and the indices ν and μ are:

$$\nu = \frac{1}{2}\left[\sqrt{1 + 4\gamma\hat{\omega}^2(\xi_1 - \xi_2)} - 1\right], \quad \mu = \frac{1}{2}\left[\sqrt{1 + 4\hat{\omega}^2(1-\gamma)(\xi_2 - \xi_1)} - 1\right]. \quad (39)$$

Solutions (33)–(36) can be matched at the boundaries of domains in the points $\xi = \xi_1, \xi_c$ and ξ_2 with the help of boundary conditions reflecting the continuity of function $\Phi(\xi)$ and its first derivative:

$$\Phi_l(\xi_1) = \Phi_{p_1}(\xi_1), \quad \Phi_{p_1}(\xi_c) = \Phi_{p_2}(\xi_c), \quad \Phi_{p_2}(\xi_2) = \Phi_r(\xi_2), \quad (40)$$

$$\Phi_l'(\xi_1) = \Phi_{p_1}'(\xi_1), \quad \Phi_{p_1}'(\xi_c) = \Phi_{p_2}'(\xi_c), \quad \Phi_{p_2}'(\xi_2) = \Phi_r'(\xi_2). \quad (41)$$

This gives a set of six linear algebraic equations for the coefficients $A_{1,2}, B_{1,2}, C_{1,2}$ and $D_{1,2}$.

Using these boundary conditions, we consider below surface wave transformation on the bottom unevenness when the incident wave arrives from the left and from the right. In the former case we set the following coefficients in the system (33)–(36):

$$A_1 = 0, \quad A_2 = T_l, \quad D_1 = R, \quad D_2 = 1 \quad (42)$$

and after that determine the transformation coefficients R, T_l, as well as $B_{1,2}$ and $C_{1,2}$. In the latter case we set the following coefficients in the system (33)–(36):

$$A_1 = 1, \quad A_2 = R, \quad D_1 = T_r, \quad D_2 = 0 \quad (43)$$

and determine the transformation coefficients R, T_r, as well as $B_{1,2}$ and $C_{1,2}$. Omitting long and tedious calculations, we obtain the transformation coefficients $R(\hat{\omega})$ and $T_{l,r}(\hat{\omega})$ which are shown in Fig. 6 for three values of γ: $\gamma = 0.1$ (line 1), $\gamma = 0.5$ (line 2), and $\gamma = 0.9$ (line 3). The reflection coefficients are again the same for the left- and right-propagating incident waves. The transmission coefficients are different as in the previous model with the piecewise-linear bottom profile, but proportional to each other, $|T_l| = |T_r|\sqrt{h_2/h_1}$, in accordance with the reciprocity theorem. Figure 6b shows the coefficient T_r as the function of normalised wave frequency for the same three values of γ: $\gamma = 0.1$ (line 1), $\gamma = 0.5$ (line 5), and $\gamma = 0.9$ (line 3).

We do not present here formulae for the coefficients $B_{1,2}$ and $C_{1,2}$, as well as the explicit solutions (33)–(36), because they look very cumbersome. However in Fig. 7 we illustrate solutions obtained in terms of $|\Phi(\xi)|$ for the particular wave frequency and depths h_1 and h_2.

In the limiting case $\hat{\omega} \to 0$ the formulae for the transformation coefficients reduce again to the well-known Lamb formulae for the step-wise bottom profile (21). The dependencies of transformation coefficients on frequency, $|R|(\hat{\omega})$ and $|T|(\hat{\omega})$, are still non-monotonic, but much smoother than in the former case of quasi-linear bottom profile (cf. graphics of Figs. 6 and 4), because in this case the bottom profile is much smoother having discontinuity only in the second derivative at the edge points ξ_1 and ξ_2. The energy flux conservation in the form of Eq. (22) has been tested and confirmed on the derived solutions.

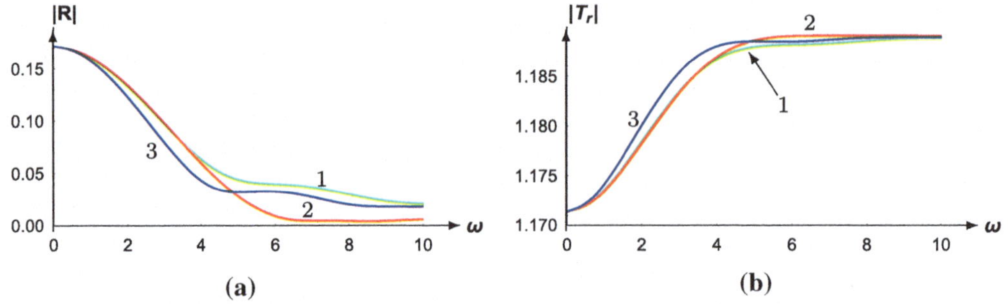

Figure 6
(Color figure online) The reflection R (frame a) and transmission T_r (frame b) coefficients as the functions of normalised frequency $\hat{\omega}$ for the fixed value of $h_2/h_1 = 2$ and $\hat{\omega} = 5$. Line 1 pertains to $\gamma = 0.5$, line 2—to $\gamma = 0.1$, and line 3—to $\gamma = 0.9$ (for the definition of γ see the text)

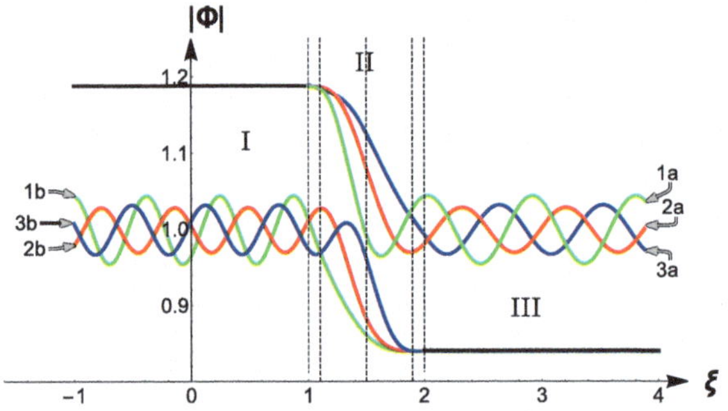

Figure 7
(Color figure online) Graphics of solutions for the free surface perturbation in terms of $|\Phi(\xi)|$. Lines 1a, 2a and 3a pertain to the case when the incident wave arrives from the right where the depth $h_2 > h_1$, and lines 1b, 2b and 3b pertain to the case when the incident wave arrives from the left where the depth $h_1 > h_2$. The numbers 1, 2, 3 correspond to the cases when $\gamma = 0.1, 0.5, 0.9$ respectively, and Roman numerals denote different spatial intervals as in Fig. 5. The plot was generated for the particular case when $h_1 = 1$ and $h_2 = 2$, and wave frequency $\hat{\omega} = 5$; for other parameters h_1, h_2, and $\hat{\omega}$ the plots are similar to these

4. Hyperbolic Tangent Bottom Profile

In this section we consider the bottom spatial variation that can be described by the tanh-function so that the fluid depth is: $h(\xi) = a\tanh(\xi) + b$, where $a = (h_2 - h_1)/2$ and $b = (h_2 + h_1)/2$. Figure 8 illustrates the tanh bottom profile (line 1) in comparison with the quadratic profile (line 2) and linear profile (dashed line) considered in the previous sections.

By substituting the depth profile into the equation for the velocity (6) we obtain:

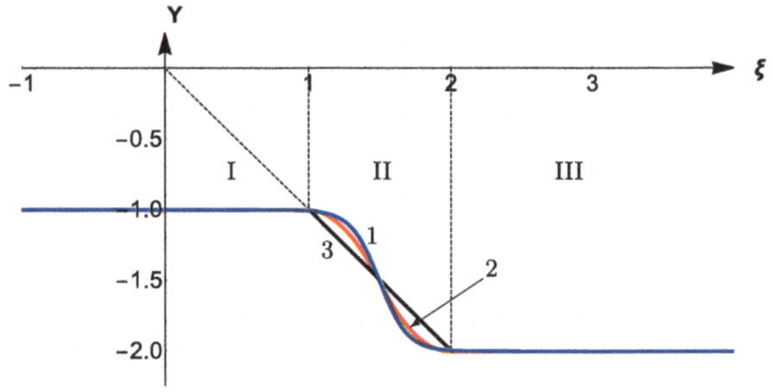

Figure 8
(Color figure online) The tanh bottom profile (line 1), combined quadratic profile (line 2) and linear profile (line 3). In the domains I and III the bottom quickly becomes constant, whereas in the domain II the bottom profile experiences a significant transition from one level to another level

$$\begin{aligned}&[a\tanh(\xi) + b]\Psi_{\xi\xi} + 2a\mathrm{sech}^2(\xi)\Psi_\xi \\ &\quad + \left[\hat{\omega}^2 - 2a\mathrm{sech}^2(\xi)\tanh(\xi)\right]\Psi = 0.\end{aligned} \tag{44}$$

The solution to this equation can be presented in terms of hyper-geometric functions:

$$\begin{aligned}\Psi = &\frac{(k^2-1)\left(1+\mathrm{e}^{2\xi}\right)}{2\left(1+\mathrm{e}^{2\xi}k^2\right)}\mathrm{e}^{\mathrm{i}\tilde{\omega}(\xi - \mathrm{i}\pi/2)} \\ &\times \left\{C_1 \cdot {}_2F_1\left[\frac{\mathrm{i}(k-1)\tilde{\omega}}{2}, -\frac{\mathrm{i}(k+1)\tilde{\omega}}{2}, 1-\mathrm{i}\tilde{\omega}, -\frac{\mathrm{e}^{-2\xi}}{k^2}\right]\right. \\ &+ C_2\left(\frac{\mathrm{e}^{-\xi+\mathrm{i}\pi/2}}{-k^2}\right)^{\mathrm{i}\tilde{\omega}} \\ &\left.\times\, {}_2F_1\left[-\frac{\mathrm{i}(k-1)\tilde{\omega}}{2}, \frac{\mathrm{i}(k+1)\tilde{\omega}}{2}, 1+\mathrm{i}\tilde{\omega}, -\frac{\mathrm{e}^{-2\xi}}{k^2}\right]\right\},\end{aligned} \tag{45}$$

where $k = \sqrt{h_2/h_1}$, $\tilde{\omega} = \hat{\omega}/\sqrt{h_2}$.

Substituting this solution into the Eq. (5), we obtain a solution for the perturbation of a free surface:

$$\Phi(\xi) = \frac{\sqrt{h_1}(k^2-1)e^{\pi/2}}{2} \Bigg(C_1 \Bigg\{ {}_2F_1\left[\frac{i(k-1)\tilde{\omega}}{2}, -\frac{i(k+1)\tilde{\omega}}{2}, 1-i\tilde{\omega}, -\frac{e^{-2\xi}}{k^2}\right] - \frac{k+1}{k}\,{}_2F_1\left[\frac{i(k-1)\tilde{\omega}}{2}, 1-\frac{i(k+1)\tilde{\omega}}{2}, 1-i\tilde{\omega}, -\frac{e^{-2\xi}}{k^2}\right]\Bigg\} e^{i\tilde{\omega}\xi} - C_2 k^{-2i\tilde{\omega}} \Bigg\{ {}_2F_1\left[-\frac{i(k-1)\tilde{\omega}}{2}, \frac{i(k+1)\tilde{\omega}}{2}, 1+i\tilde{\omega}, -\frac{e^{-2\xi}}{k^2}\right] - \frac{k+1}{k}\,{}_2F_1\left[-\frac{i(k-1)\tilde{\omega}}{2}, 1+\frac{i(k+1)\tilde{\omega}}{2}, 1+i\tilde{\omega}, -\frac{e^{-2\xi}}{k^2}\right]\Bigg\} e^{-i\tilde{\omega}\xi} \Bigg). \tag{46}$$

The asymptotics of this solution when $\xi \to \pm\infty$ are:

(1) $\xi \to +\infty$:

$$\Phi(\xi) = \frac{\sqrt{h_1}(k^2-1)}{2k}\left[-C_1 e^{i\tilde{\omega}(\xi - i\pi/2)} + C_2 k^{-2i\tilde{\omega}} e^{-i\tilde{\omega}(\xi - 3i\pi/2)}\right], \tag{47}$$

(2) $\xi \to -\infty$:

$$\Phi(\xi) = k^{i(k-1)\tilde{\omega}}\sqrt{h_1}\Gamma(-ik\tilde{\omega}) \left\{ C_2(k+1)\frac{\Gamma(i\tilde{\omega})}{\Gamma^2\left[-\frac{i(k-1)\tilde{\omega}}{2}\right]} - C_1(k-1)e^{2\pi\tilde{\omega}}\frac{\Gamma(-i\tilde{\omega})}{\Gamma^2\left[-\frac{i(k+1)\tilde{\omega}}{2}\right]} \right\} e^{i\tilde{\omega}(k\xi + 3i\pi/2)} + k^{-i(k+1)\tilde{\omega}}\sqrt{h_1} \left\{ C_2(k-1)\frac{\Gamma(i\tilde{\omega})\Gamma(ik\tilde{\omega})}{\Gamma^2\left[\frac{i(1+k)\tilde{\omega}}{2}\right]} - C_1 e^{2\pi\tilde{\omega}}\frac{k^2-1}{2k}\frac{\Gamma(-i\tilde{\omega})\Gamma(1+ik\tilde{\omega})}{\Gamma\left[\frac{i(k-1)\tilde{\omega}}{2}\right]\Gamma\left[1+\frac{i(k-1)\tilde{\omega}}{2}\right]} \right\} e^{-i\tilde{\omega}(k\xi - 3i\pi/2)}. \tag{48}$$

Thus, one can see that asymptotically solution (46) represents oppositely travelling waves which can be presented in the form:

$$\Phi(\xi) = A_1 e^{i\tilde{\omega}\xi} + A_2 e^{-i\tilde{\omega}\xi}, \qquad \xi \to +\infty \tag{49}$$

$$\Phi(\xi) = B_1 e^{ik\tilde{\omega}\xi} + B_2 e^{-ik\tilde{\omega}\xi}, \qquad \xi \to -\infty, \tag{50}$$

where $A_{1,2}$ and $B_{1,2}$ are the coefficients in front of corresponding exponential functions in Eqs. (47) and (48).

Then, we can obtain the transformation coefficients of surface waves passing over such bottom unevenness. If an incident wave arrives from the right (from the deeper region as shown in Fig. 8), then we set:

$$A_1 = 1, \quad A_2 = R_r, \quad B_1 = T_r, \quad B_2 = 0 \tag{51}$$

and readily derive from Eqs. (47) and (48) the transformation coefficients:

$$R_r = -k^{-2i\tilde{\omega}}\frac{k+1}{k-1}\frac{\Gamma(-i\tilde{\omega})\Gamma^2\left[\frac{1}{2}i(k+1)\tilde{\omega}\right]}{\Gamma(i\tilde{\omega})\Gamma^2\left[\frac{1}{2}i(k-1)\tilde{\omega}\right]}, \tag{52}$$

$$T_r = k^{i(k-1)\tilde{\omega}}\frac{2k}{k+1}\frac{\Gamma(-i\tilde{\omega})\Gamma(-ik\tilde{\omega})}{\Gamma^2\left[-\frac{1}{2}i(1+k)\tilde{\omega}\right]}\frac{\sinh(\pi\tilde{\omega})\sinh(k\pi\tilde{\omega})}{\mathrm{sech}^2\left[\frac{1}{2}(1+k)\pi\tilde{\omega}\right]}, \tag{53}$$

$$C_1 = \frac{-2e^{-\pi\tilde{\omega}/2}k}{\sqrt{h_1}(k^2-1)}, \quad C_2 = \frac{-2e^{3\pi\tilde{\omega}/2}k}{\sqrt{h_1}(k-1)^2}\frac{\Gamma(-i\tilde{\omega})\Gamma^2\left[\frac{1}{2}i(k+1)\tilde{\omega}\right]}{\Gamma(i\tilde{\omega})\Gamma^2\left[\frac{1}{2}i(k-1)\tilde{\omega}\right]}. \tag{54}$$

If an incident wave arrives from the left (from the shallower region), then we set:

$$A_1 = 0, \quad A_2 = T_l, \quad B_1 = R_l, \quad B_2 = 1 \tag{55}$$

and readily derive again from Eqs. (47) and (48) the transformation coefficients:

$$R_l = -k^{4i\tilde{\omega}}R_r, \quad T_l = k^{i(k-1)\tilde{\omega}}\frac{k+1}{2k}\frac{\Gamma^2\left[\frac{1}{2}i(1+k)\tilde{\omega}\right]}{\Gamma(i\tilde{\omega})\Gamma(ik\tilde{\omega})}, \tag{56}$$

$$C_1 = 0, \quad C_2 = \frac{e^{\frac{3\pi\tilde{\omega}}{2}}k^{i(1+k)\tilde{\omega}}\Gamma^2\left[\frac{1}{2}i(1+k)\tilde{\omega}\right]}{\sqrt{h_1}(-1+k)\Gamma(i\tilde{\omega})\Gamma(ik\tilde{\omega})}. \tag{57}$$

Figure 9 illustrates solution (46) for the incident wave arriving from the right (line 1a) and from the left (line 1b) in comparison with the solutions derived for the piecewise-quadratic bottom profile (lines 2a and 2b correspondingly). As one can see from this figure, the spatial periods of bits are shorter in the case of tanh bottom profile, whereas their amplitudes are slightly greater than in the case of piecewise-quadratic bottom profile.

The moduli of transformation coefficients $|R_r|$ and $|T_r|$ are shown in Fig. 10 for the same two bottom profiles and for the case when the incident wave arrives from the right. In the case when the incident wave arrives from the left the modulus of reflection coefficient is the same as in the former case: $|R_l| = |R_r|$, and the transmission coefficient $|T_l|$ is related with the transmission coefficient $|T_r|$ by the reciprocity relationship: $|T_l| = |T_r|\sqrt{h_2/h_1}$.

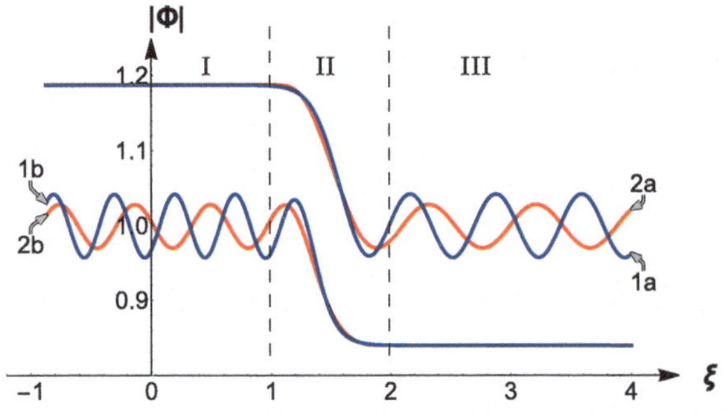

Figure 9
(Color figure online) Graphics of solutions for the free surface perturbation in terms of $|\Phi(\xi)|$ for the tanh bottom profile (lines 1a and 1b) and for the piecewise-quadratic bottom profile (lines 2a and 2b). Lines 1a and 2a pertain to incident waves arriving from the right, and lines 1b and 2b pertain to incident waves arrives from the left. The plot was generated for the particular case when $h_1 = 1$ and $h_2 = 2$, and wave frequency $\hat{\omega} = 5$; for other parameters h_1, h_2, and $\hat{\omega}$ the plots are similar to these

The general conclusion which follows from the comparison of transformation coefficients for all three cases of piecewise-linear, piecewise-quadratic, and tanh bottom profiles is that the smoother the profile, the smaller oscillations in the dependences of $|R(\hat{\omega})|$ and $|T(\hat{\omega})|$ (cf. lines 1 and 2 in Fig. 10 with the lines in Fig. 4). There are no oscillations in the dependencies $|R(\hat{\omega})|$ and $|T(\hat{\omega})|$ in the case of tanh-profile (see lines 1 in Fig. 10). In the case of piecewise-quadratic bottom profile the oscillations in such dependencies are very small and almost invisible (see lines 2 in Fig. 10), whereas in the case of piecewise-linear profile the oscillations are very well pronounced Fig. 4).

The conservation of energy flux in the form of Eq. (22) holds again, as expected.

Having solutions for the wave propagation over uneven bottom with the piecewise-linear, piecewise-quadratic, and tanh bottom profiles, we can solve problems with the more complicated bottom profiles, approximating them by sets of such functions for which exact solutions exist in the analytic forms. Below we present solutions for underwater trenches and barriers of different shapes. A similar problem was considered by Tokano (1960), Newman (1965), Kirby and Dalrymple (1983), Devillard et al. (1988) for a submerged rectangular bar and trench (see also Massel 1989 and Dingemans 1997 and references therein). In the paper by Rey et al. (1992) this problem was studied experimentally in the laboratory wave tank.

5. *Wave Scattering on an Underwater Trench and Barrier with the Linear Slopes*

Consider the bottom profiles shown in Fig. 11, which can be described by a set of linear functions $h(\xi) = a\xi + b$ with different parameters a and b. The left and right slopes of the trench/barrier can be described by the following parameters (see Fig. 11):

$$a_1 = \frac{h_2 - h_1}{\xi_2 - \xi_1}, \quad a_2 = \frac{h_2 - h_1}{\xi_3 - \xi_4}, \quad b_1 = \frac{h_1\xi_2 - h_2\xi_1}{\xi_2 - \xi_1}, \quad b_2 = \frac{h_1\xi_3 - h_2\xi_4}{\xi_3 - \xi_4}. \tag{58}$$

Solutions in the each interval of the piecewise-linear bottom profile is:

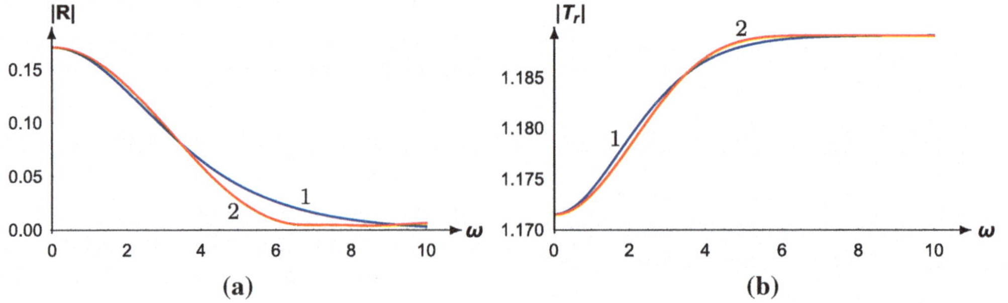

Figure 10
(Color figure online) The reflection (frame a) and transmission (frame b) coefficients for the tanh-profile (lines 1) and piecewise-quadratic (lines 2) bottom profile for the cases when the incident waves arrives from the right. The plots were generated for the particular case when $h_1 = 1$ and $h_2 = 2$, and wave frequency $\hat{\omega} = 5$

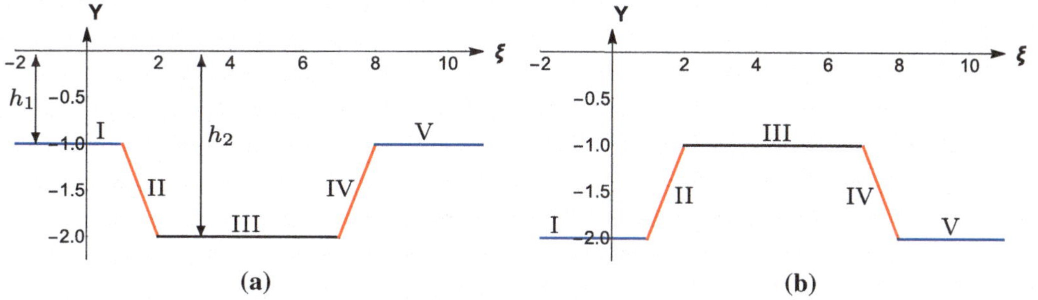

Figure 11
(Color figure online) Bottom profiles in the form of underwater trench (frame a) and barrier (frame b) with the linear slopes

$$\Phi_L = A_1 e^{-ik_1(\xi-\xi_1)} + A_2 e^{ik_1(\xi-\xi_1)}, \quad -\infty < \xi < \xi_1, \tag{59}$$

$$\Phi_{SL} = B_1 I_0\left(\frac{2i\omega}{a_1}\sqrt{a_1\xi + b_1}\right) + B_2 K_0\left(\frac{2i\omega}{a_1}\sqrt{a_1\xi + b_1}\right), \quad \xi_1 < \xi < \xi_2, \tag{60}$$

$$\Phi_C = C_1 e^{-ik_2\xi} + C_2 e^{i\xi_2 x}, \quad \xi_2 < \xi < \xi_3, \tag{61}$$

$$\Phi_{SR} = D_1 I_0\left(\frac{2i\omega}{a_2}\sqrt{a_2\xi + b_2}\right) + D_2 K_0\left(\frac{2i\omega}{a_2}\sqrt{a_2\xi + b_2}\right), \quad \xi_3 < \xi < \xi_4, \tag{62}$$

$$\Phi_R = E_1 e^{-ik_1(\xi-\xi_4)} + E_2 e^{i\xi_1(\xi-\xi_4)}, \quad \xi_4 < \xi < \infty. \tag{63}$$

When the incident wave propagates from the left to right in the geometry shown in Fig. 11, then we set the following coefficients $A_{1,2}$ and $E_{1,2}$:

$$A_1 = 1, \quad A_2 = R, \quad E_1 = T, \quad E_2 = 0. \tag{64}$$

To match solutions (59)–(63) at the points x_1, x_2, x_3, and x_4, we use eight boundary conditions which reflect continuations of function $\Phi(\xi)$ and its first derivative:

$$\Phi_L(\xi_1) = \Phi_{SL}(\xi_1), \quad \Phi'_L(\xi_1) = \Phi'_{SL}(\xi_1), \tag{65}$$

$$\Phi_{SL}(\xi_2) = \Phi_C(\xi_2), \quad \Phi'_{SL}(\xi_2) = \Phi'_C(\xi_2), \tag{66}$$

$$\Phi_C(\xi_3) = \Phi_{SR}(\xi_3), \quad \Phi'_C(\xi_3) = \Phi'_{SR}(\xi_3), \tag{67}$$

$$\Phi_{SR}(\xi_4) = \Phi_R(\xi_4), \quad \Phi'_{SR}(\xi_4) = \Phi'_R(\xi_4). \tag{68}$$

This leads to the set of eight linear algebraic equations for eight unknown coefficients R, T, $B_{1,2}$, $C_{1,2}$, and $E_{1,2}$. The set can be readily solved, but the solution looks very cumbersome, therefore we present below only a graphical illustration for the particular choices of parameters h_1, h_2, and $\hat{\omega} = 5$. Figure 12 shows wave filed amplitude $|\Phi(\xi)|$ for the trench with $h_2/h_1 = 2$ (frame a) and barrier with $h_1/h_2 = 2$ (frame b). In both cases it is supposed that the incident wave arrives from the left.

Figures 13 and 14 show the moduli of reflection and transmission coefficients respectively for the trench with $h_2/h_1 = 2$ (frames a) and barrier with $h_1/h_2 = 2$ (frames b). The oscillatory character of transformation coefficients as functions of wave frequency is the well-known phenomenon which has been noticed in many publications (see, e.g., Kajiura 1961; Massel 1989; Mei 1990; Dingemans 1997). The reason of this is the interference of standing waves within the trench or barrier.

In this section we have considered wave scattering on a symmetric trench or barrier; similarly one can consider a scattering on non-symmetric obstacles with the linear slopes or even more complicated piecewise-linear slopes.

6. *Wave Scattering on an Underwater Trench and Barrier with the Piecewise-Quadratic Slopes*

Consider now a wave scattering on the underwater obstacles (trench or barrier) when the left and right slopes can be described by a smooth conjugation of quadratic

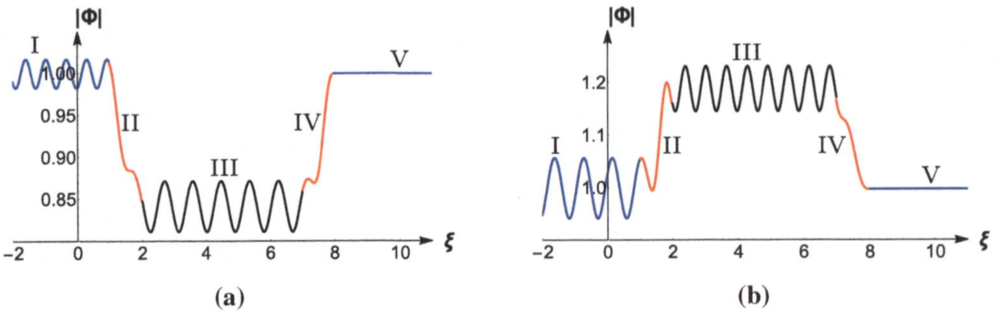

Figure 12
(Color figure online) Amplitude of surface wave field when an incident wave arrives from the left and scatters on the underwater trench $h_1 = 1$, $h_2 = 2$ (frame a) and barrier $h_1 = 2$, $h_2 = 1$ (frame b) with the linear slopes. In both cases the dimensionless frequency is $\hat{\omega} = 5$

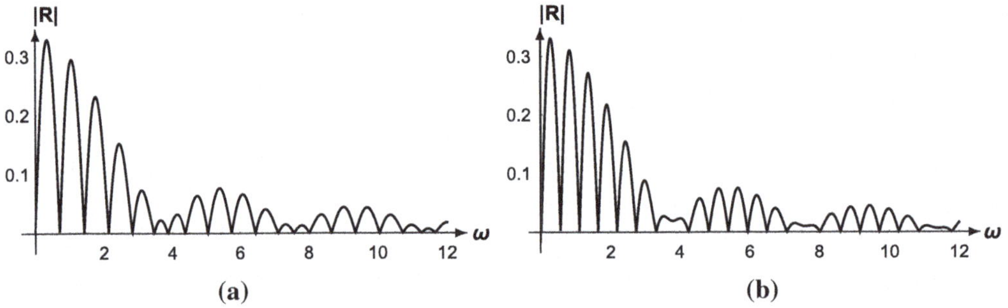

Figure 13
The moduli of reflection coefficients for a surface wave of dimensionless frequency $\hat{\omega} = 5$ scattering at the underwater trench (frames a) and barrier (frames b) with the linear slopes shown in Fig. 11

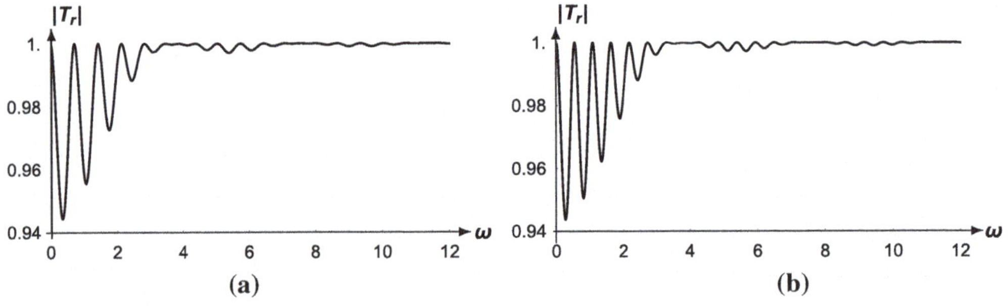

Figure 14
The moduli of transmission coefficients for a surface wave of dimensionless frequency $\hat{\omega} = 5$ scattering at the underwater trench (frames a) and barrier (frames b) with the linear slopes shown in Fig. 11

functions. The water depth over the uneven bottom in this case can be presented as $h(\xi) = a\xi^2 + b\xi + c$ (see Fig. 15). The basic differential equation for the free surface perturbation over the uneven bottom is:

$$\left(a\xi^2 + b\xi + c\right)\Phi''(\xi) + (2a\xi + b)\Phi'(\xi) + \omega^2\Phi(\xi) = 0. \tag{69}$$

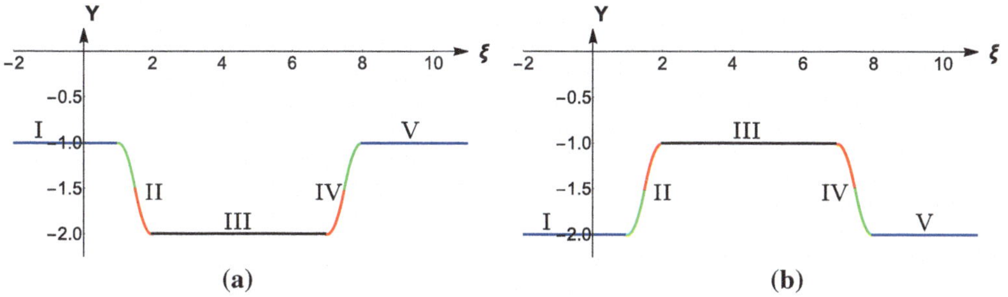

Figure 15
(Color figure online) Bottom profiles in the form of underwater trench (frame a) and barrier (frame b) with the piecewise-quadratic slopes

For the symmetric obstacles shown in Fig. 15 the coefficients of quadratic functions are:

$$a_1 = \frac{h_2 - h_1}{(\xi_2 - \xi_1)^2 \gamma}, \quad b_1 = \frac{2(h_1 - h_2)\xi_1}{(\xi_2 - \xi_1)^2 \gamma},$$
$$c_1 = h_1 + \frac{(h_2 - h_1)\xi_1^2}{(\xi_2 - \xi_1)^2 \gamma},$$
$$a_2 = \frac{h_2 - h_1}{(\xi_2 - \xi_1)^2 (\gamma - 1)}, \quad b_2 = \frac{2(h_1 - h_2)\xi_2}{(\xi_2 - \xi_1)^2 (\gamma - 1)},$$
$$c_2 = h_2 + \frac{(h_2 - h_1)\xi_2^2}{(\xi_2 - \xi_1)^2 (\gamma - 1)},$$
$$a_3 = \frac{h_1 - h_2}{(\xi_3 - \xi_4)^2 \gamma}, \quad b_3 = \frac{2(h_2 - h_1)\xi_3}{(\xi_3 - \xi_4)^2 \gamma},$$
$$c_3 = h_2 + \frac{(h_1 - h_2)\xi_3^2}{(\xi_3 - \xi_4)^2 \gamma},$$
$$a_4 = \frac{h_1 - h_2}{(\xi_3 - \xi_4)^2 (\gamma - 1)}, \quad b_4 = \frac{2(h_2 - h_1)\xi_4}{(\xi_3 - \xi_4)^2 (\gamma - 1)},$$
$$c_4 = h_1 + \frac{(h_1 - h_2)\xi_4^2}{(\xi_3 - \xi_4)^2 (\gamma - 1)}.$$

Here the parameter γ determines the positions of the conjugation of quadratic functions in the each slope:

$$\xi_{c1} = \xi_1 + \gamma(\xi_2 - \xi_1), \quad \xi_{c2} = \xi_3 + \gamma(\xi_4 - \xi_3). \tag{70}$$

Solutions to Eq. (69) can be presented in the dimensionless variables in terms of the Legendre polynomials $P_\nu(\xi)$ and $Q_\nu(\xi)$ [see Eqs. (34), (35)]:

$$\Phi(\xi) = B_1 P_\nu\left(\frac{2a\xi + b}{\sqrt{b^2 - 4ac}}\right) + B_2 Q_\nu\left(\frac{2a\xi + b}{\sqrt{b^2 - 4ac}}\right), \tag{71}$$

where $\nu = \left(\sqrt{1 - 4\hat{\omega}^2/a} - 1\right)/2$. Wave perturbations in each interval of ξ corresponding to different domains of bottom profile (see Fig. 15) can be presented as follows:

$$\Phi_L(\xi) = A_1 e^{-ik_1(\xi - \xi_1)} + A_2 e^{ik_1(\xi - \xi_1)}, \quad -\infty < \xi < \xi_1,$$
$$\Phi_{SL1}(\xi) = B_1 P(a_1, b_1, c_1) + B_2 Q(a_1, b_1, c_1), \quad \xi_1 < \xi < \xi_{c1},$$
$$\Phi_{SL2}(\xi) = C_1 P(a_2, b_2, c_2) + C_2 Q(a_2, b_2, c_2), \quad \xi_{c1} < \xi < \xi_2,$$
$$\Phi_C(\xi) = D_1 e^{-ik_2\xi} + D_2 e^{ik_2\xi}, \quad \xi_2 < \xi < \xi_3,$$
$$\Phi_{SR1}(\xi) = E_1 P(a_3, b_3, c_3) + E_2 Q(a_3, b_3, c_3), \quad \xi_3 < \xi < \xi_{c2},$$
$$\Phi_{SR2}(\xi) = F_1 P(a_4, b_4, c_4) + F_2 Q(a_4, b_4, c_4), \quad \xi_{c2} < \xi < \xi_4,$$
$$\Phi_R(\xi) = G_1 e^{-ik_1(\xi - \xi_4)} + G_2 e^{ik_1(\xi - \xi_4)}, \quad \xi_4 < \xi < \infty.$$

To smoothly match these functions we use the boundary conditions of continuity of function $\Phi(\xi)$ and its first derivatives at the boundaries of intervals like in Eq. (14) above. As a result, we obtain a set of sixteen linear algebraic equations for the coefficients $A_{1,2}, B_{1,2}, \ldots, G_{1,2}$ that can be readily solved.

Figure 16 shows wave field amplitudes $|\Phi(\xi)|$ for the trench with $h_2/h_1 = 2$ (frame a) and barrier with $h_1/h_2 = 2$ (frame b). In both cases, it is assumed that the incident wave arrives from the left.

Figures 17 and 18 show the moduli of reflection and transmission coefficients respectively for the trench with $h_2/h_1 = 2$ (frames a) and barrier with $h_1/h_2 = 2$ (frames b).

It is interesting that in the case of an obstacle with the piecewise-quadratic slopes the number of resonances in the low-frequency domain is greater than in the case of an obstacle of the same width, but with the linear slopes (cf. Figs. 13, 14, 17 and 18). The basic features of transmission coefficients including maxima of reflection coefficients are similar in both cases.

In this section we considered again wave scattering on a symmetric trench or barrier; similarly one can consider a scattering on non-symmetric obstacles with the piecewise-quadratic slopes or even more

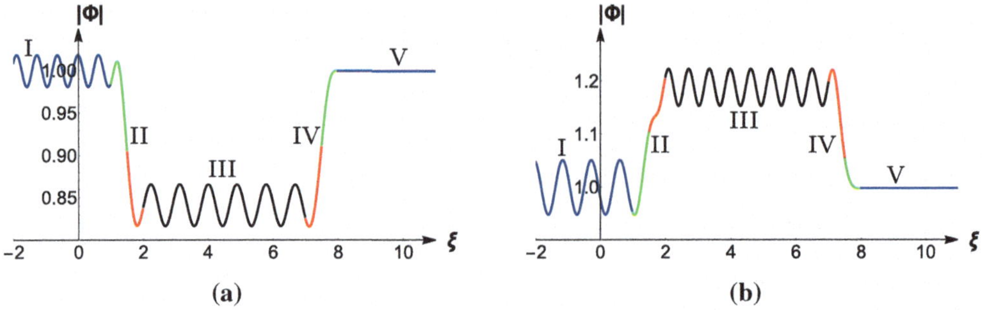

Figure 16
(Color figure online) Amplitude of surface wave field when an incident wave arrives from the left and scatters on the underwater trench $h_1 = 1$, $h_2 = 2$ (frame a) and barrier $h_1 = 2$, $h_2 = 1$ (frame b) with the piecewise-quadratic slopes (each segment consists of two different parabolic slopes). In both cases the dimensionless frequency is $\hat{\omega} = 5$

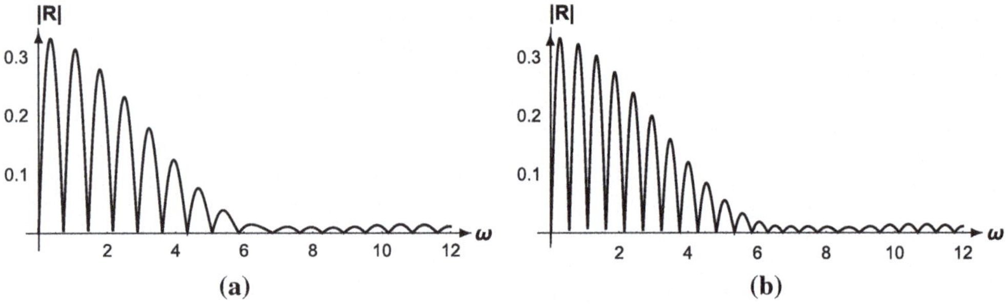

Figure 17
The moduli of reflection coefficients for a surface wave of dimensionless frequency $\hat{\omega} = 5$ scattering at the underwater trench (frames a) and barrier (frames b) with the piecewise-quadratic slopes shown in Fig. 15

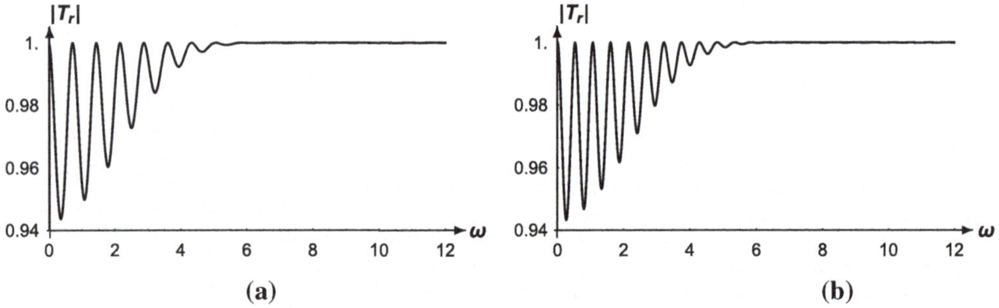

Figure 18
The moduli of transmission coefficients for a surface wave of dimensionless frequency $\hat{\omega} = 5$ scattering at the underwater trench (frames a) and barrier (frames b) with the piecewise-quadratic slopes shown in Fig. 15

complicated combination of piecewise-linear and piecewise-quadratic slopes. This opens a wide range of possibilities to approximate real bottom profiles with the set of such functions.

7. Wave Scattering on an Underwater Trench and Barrier with the Hyperbolic Tangent Profiles

In this section we consider wave transformation on the underwater obstacles with the slopes describing by tanh-functions. We assume that the characteristic width of each slope l_0 is much less than the distance L between them. The bottom profiles for the trench and barrier are very similar to those shown in Fig. 15; they are practically indistinguishable from their counterparts described by piecewise-quadratic functions.

When the obstacle slopes are far enough from each other so that $L \gg l_0$, then the wave fields at the big distances from the slopes can be considered in the form of travelling waves in a fluid of constant depth. (Note that in the considered case, the characteristic width of function $\tanh(\xi)$ is one—see the beginning of Sect. 4). Here we consider for the sake of simplicity the situation when the characteristic width of an obstacle is much greater than the wavelength of an incident wave $L \gg \lambda$. However, there are no principal limitations for the exact solution in the case when $L \sim \lambda$. However, the problem of matching solutions in the vicinity of a transient zone $-L/2 \le \xi \le L/2$ makes the problem much more complicated and cumbersome.

For the left slopes of obstacles centred around $\xi = 0$, a solution to the wave equation has the following asymptotics:

$$\Phi_0(\xi) = w_1 e^{ik\hat{\omega}\xi} + w_2 e^{-ik\hat{\omega}\xi}, \quad \xi \to -\infty,$$
$$\Phi_0(\xi) = \tilde{w}_1 e^{i\hat{\omega}\xi} + \tilde{w}_2 e^{-i\hat{\omega}\xi}, \quad \xi \to +\infty$$

where

$$w_1 = e^{-\pi\hat{\omega}/2}\sqrt{h_1}k^{i(k-1)\hat{\omega}}\Gamma(-ik\hat{\omega}) \times \left\{A_1 \frac{e^{\pi\hat{\omega}}(1-k)\Gamma(-i\hat{\omega})}{\Gamma^2\left[-\frac{1}{2}i(k+1)\hat{\omega}\right]} + A_2 \frac{e^{-\pi\hat{\omega}}(1+k)\Gamma(i\hat{\omega})}{\Gamma^2\left[-\frac{1}{2}i(k-1)\hat{\omega}\right]}\right\},$$

$$w_2 = e^{-\pi\hat{\omega}/2}\sqrt{h_1}k^{-i(k+1)\hat{\omega}}\Gamma(ik\hat{\omega}) \times \left\{-A_1 \frac{e^{\pi\hat{\omega}}(1+k)\Gamma(-i\hat{\omega})}{\Gamma^2\left[\frac{1}{2}i(k-1)\hat{\omega}\right]} + A_2 \frac{e^{-\pi\hat{\omega}}(k-1)\Gamma(i\hat{\omega})}{\Gamma^2\left[\frac{1}{2}i(1+k)\hat{\omega}\right]}\right\},$$

$$\tilde{w}_1 = -\frac{A_1}{2k}\sqrt{h_1}\left(k^2-1\right)e^{\pi\hat{\omega}/2},$$

$$\tilde{w}_2 = \frac{A_2}{2}\sqrt{h_1}k^{-1-2i\hat{\omega}}\left(k^2-1\right)e^{-3\pi\hat{\omega}/2}.$$

The asymptotics of a solution to the wave equation for the second slope centred around $\xi = L$ is:

$$\Phi_0(\xi) == \hat{w}_1 e^{i\hat{\omega}\xi} + \hat{w}_2 e^{-i\hat{\omega}\xi}, \quad \xi \to -\infty,$$
$$\Phi_0(\xi) == \breve{w}_1 e^{ik\hat{\omega}\xi} + \breve{w}_2 e^{-ik\hat{\omega}\xi}, \quad \xi \to +\infty$$

whereFor the incident wave arriving from the left we set the following coefficients:

$$w_2 = 1, \quad w_1 = R, \quad \tilde{w}_1 = \breve{w}_1, \quad \tilde{w}_2 = \breve{w}_2, \quad \hat{w}_1 = 0, \quad \hat{w}_2 = T. \tag{72}$$

Then we can solve the set of equations for R, T, as well as $\tilde{w}_1$, $\tilde{w}_1$ and obtain:

$$\breve{w}_1 = e^{-iL\hat{\omega}-k\pi\hat{\omega}/2}\sqrt{h_1}k^{i(k-1)\hat{\omega}}\Gamma(-i\hat{\omega}) \times \left\{B_1 \frac{e^{k\pi\hat{\omega}}(k-1)\Gamma(-ik\hat{\omega})}{\Gamma^2\left[-\frac{1}{2}i(k+1)\hat{\omega}\right]} + B_2 \frac{e^{-k\pi\hat{\omega}}(k+1)\Gamma(ik\hat{\omega})}{\Gamma^2\left[\frac{1}{2}i(k-1)\hat{\omega}\right]}\right\},$$

$$\breve{w}_2 = \frac{1}{2}e^{iL\hat{\omega}-k\pi\hat{\omega}/2}\sqrt{h_1}k^{i(k+1)\hat{\omega}}(k-1) \times \left\{B_1 \frac{e^{k\pi\hat{\omega}}(k+1)\Gamma(1+i\hat{\omega})\Gamma(-ik\hat{\omega})}{\Gamma\left[-\frac{1}{2}i(k-1)\hat{\omega}\right]\Gamma\left[1-\frac{1}{2}i(k-1)\hat{\omega}\right]} - B_2 \frac{2e^{-k\pi\hat{\omega}}\Gamma(i\hat{\omega})\Gamma(ik\hat{\omega})}{\Gamma^2\left[\frac{1}{2}i(k+1)\hat{\omega}\right]}\right\},$$

$$\hat{w}_1 = \frac{1}{2}B_1 e^{k(-iL+\pi/2)\hat{\omega}}\sqrt{h_1}\left(k^2-1\right),$$

$$\hat{w}_2 = -\frac{1}{2}B_2 e^{ikL\hat{\omega}-3k\pi\hat{\omega}/2}\sqrt{h_1}k^{2ik\hat{\omega}}\left(k^2-1\right).$$

$$R = \frac{4\pi^2}{\Delta_1} k^{2ik\hat{\omega}} \Gamma(-ik\hat{\omega}) \times \left\{ e^{2iL\hat{\omega}} (k-1)^2 k^{i(k+3)\hat{\omega}} \frac{\Gamma^2(i\hat{\omega})\Gamma^4[i(k-1)\hat{\omega}/2]}{\mathrm{sech}^2[(k+1)\pi\hat{\omega}/2]} - k^{i(k-1)\hat{\omega}} (k+1)^2 \frac{\Gamma^2(-i\hat{\omega})\Gamma^4[i(k+1)\hat{\omega}/2]}{\mathrm{sech}^2[(k-1)\pi\hat{\omega}/2]} \right\}, \tag{73}$$

$$T = \frac{(k^2-1)^2}{\Delta_2} e^{i(k+1)L\hat{\omega}} k^{2i(k+1)\hat{\omega}-1} \Gamma^4 \left[\frac{i}{2}(k-1)\hat{\omega}\right] \Gamma^4 \left[\frac{i}{2}(k+1)\hat{\omega}\right], \tag{74}$$

where

$$\Delta_1 = k^{i\hat{\omega}}(k^2-1) e^{iL\hat{\omega}} \hat{\omega}^2 \Gamma(ik\hat{\omega}) \Gamma^2\left[-\frac{i}{2}(k-1)\hat{\omega}\right] \Gamma^2\left[-\frac{i}{2}(k+1)\hat{\omega}\right] \times \left\{ k^{i(k+2)\hat{\omega}} (k-1)^2 e^{iL\hat{\omega}} \Gamma^2(i\hat{\omega}) \Gamma^4\left[\frac{i}{2}(k-1)\hat{\omega}\right] - k^{i(k-2)\hat{\omega}} (k+1)^2 e^{-iL\hat{\omega}} \Gamma^2(-i\hat{\omega}) \Gamma^4\left[\frac{i}{2}(k+1)\hat{\omega}\right] \right\}$$

$$\Delta_2 = 4k^{2i\hat{\omega}} e^{iL\hat{\omega}} \Gamma^2(ik\hat{\omega}) \left\{ k^{2i\hat{\omega}} e^{iL\hat{\omega}} (k-1)^2 \Gamma^2(i\hat{\omega}) \Gamma^4\left[\frac{i}{2}(k-1)\hat{\omega}\right] - k^{-2i\hat{\omega}} e^{-iL\hat{\omega}} (k+1)^2 \Gamma^2(-i\hat{\omega}) \Gamma^4\left[\frac{i}{2}(k+1)\hat{\omega}\right] \right\}. \tag{75}$$

The structures of the wave fields in terms of the dependencies $|\Phi(\xi)|$ for the trench with $h_2/h_1 = 2$ and barrier with $h_1/h_2 = 2$ are very similar to those shown in Fig. 16 for the piecewise-quadratic model. To minimise the number of illustrations, we do not present her the corresponding graphics. In Figs. 19 and 20 we show by lines 1 the moduli of reflection and transmission coefficients respectively for the trench with $h_2/h_1 = 2$ (frames a) and barrier with $h_1/h_2 = 2$ (frames b). For the comparison we show also by lines 2 similar coefficients derived in the previous Sect. 7 for the trench and barrier with the piecewise-quadratic slopes.

As one can see from these figures, the qualitative character of dependencies of transformation coefficients on frequency is similar. Moreover, even the maximum and minimum vales of transformation coefficients are the same. However, there are some differences. In particular, the number of significant maxima and minima in the low-frequency domain is greater in the case of obstacles with the tanh-slopes. And in this case the maxima of reflection coefficients decay slower with the frequency.

8. Conclusion

In this paper we have studied surface wave transformation on bottom unevenness in a shallow basin. We have obtained exact analytical solutions for three reference cases of (i) linearly increasing or decreasing bottom profiles on a finite spatial interval, (ii) piecewise-quadratic transient bottom profile between two constant values, and (iii) a transient bottom profile described by the tanh-function. We have studied also wave scattering on underwater barriers and trenches whose slopes can be described by the same functions (linear, piecewise-quadratic, and tanh-functions). The results obtained are in a good agreement with the results previously derived by different authors for the similar models (Kajiura

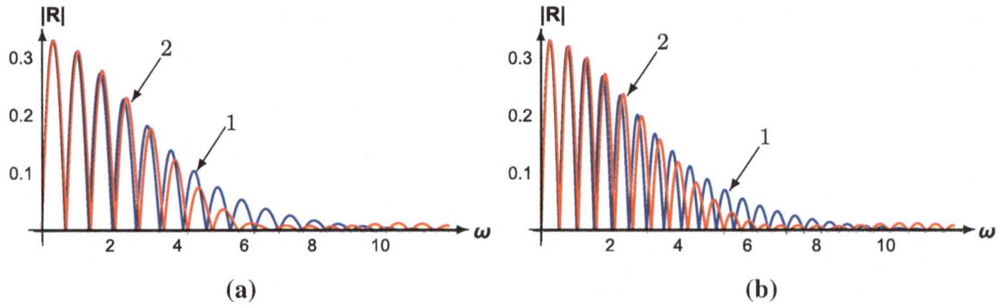

Figure 19
(Color figure online) The moduli of reflection coefficients for a surface wave of dimensionless frequency $\hat{\omega} = 5$ scattering at the underwater trench (frames **a**) and barrier (frames **b**). Lines 1 pertain to the obstacles with the tanh-slopes, and lines 2—to the obstacles with the piecewise-quadratic slopes shown in Fig. 15

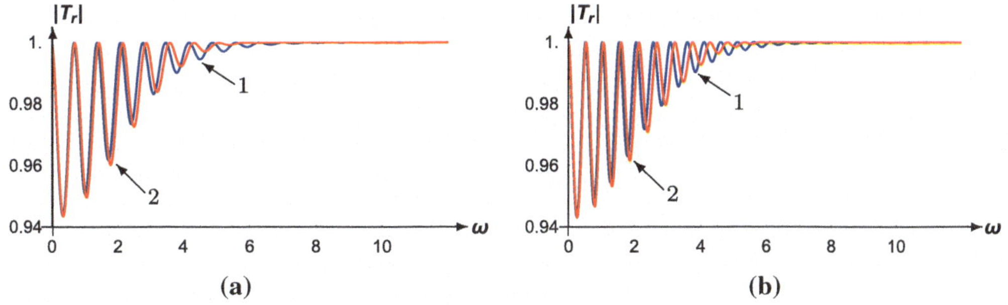

Figure 20
(Color figure online) The moduli of transmission coefficients for a surface wave of dimensionless frequency $\hat{\omega} = 5$ scattering at the underwater trench (frames **a**) and barrier (frames **b**). Lines 1 pertain to the obstacles with the tanh-slopes, and lines 2—to the obstacles with the piecewise-quadratic slopes shown in Fig. 15

1961; Dean 1964; Kirby and Dalrymple 1983; Massel 1989; Mei 1990; Dingemans 1997; Rey 1992; Rey et al. 1992; Kânoğlu and Synolakis 1998; Jung et al. 2008; Xie et al. 2011). In the limiting case when the wave frequency goes to zero we obtained the same transformation coefficients which are predicted by Lamb's theory (1932) for step-wise bottom.

In the meantime, there are some quantitative differences in the transformation coefficients for the different bottom profiles. In particular, the smoother the bottom profile is, the smoother the dependencies of transformation coefficients on frequency—cf. Figure 4 for the piecewise-linear profile with Fig. 6 for the piecewise-quadratic (line 2) and tanh (line 1) profiles. There are several well-pronounced oscillations in the dependence of $|R(\hat{\omega})|$ in Fig. 4. In particular, one can see that the reflection coefficient at some frequencies drops down to nearly zero. This provides almost reflectionless propagation (Didenkulova and Pelinovsky 2013, 2016) of an incident wave at the corresponding frequencies. We have tested the influence of basic parameters (the depth and frequency) on the structure of wave function $\Phi(\xi)$ and transformation coefficients. The results in all three model cases studied in the paper were qualitatively the same.

The situation is opposite for the wave scattering on underwater barriers or tranches. In particular, one can see fewer oscillations in the transformation coefficients shown in Figs. 13 and 14 for the barriers and tranches with the linear slopes in comparison with the barriers and tranches with the piecewise-quadratic or tanh-slopes (cf. with Figs. 19 and 20).

More complicated real bottom profiles can be approximated with higher accuracy by the combination of those considered here: linear, quadratic, and tanh-profiles; such an approach has been exploited in Refs. Rey (1992), Rey et al. (1992), Kânoğlu and Synolakis (1998), Jung et al. (2008), Xie et al. (2011). The results obtained can be used, in particular, for the protection of beaches against storm surges, swells, and tsunami waves, as well as for the validation of numerical codes. On the basis of analytical results for the rather realistic bottom profiles, it would be of interest to analyse records of real tsunami waves passing over underwater obstacles of known profiles.

In conclusion, it is necessary to mention a further possible development of the results obtained in this paper. One of the obvious generalisations should take into account a finite spectrum of oceanic waves, whereas results presented in this paper pertain to a monochromatic wave. Then, we have studied here the normal incidence of a wave to the underwater obstacle, whereas oceanic waves can approach the coastal zone at an arbitrary angle. Therefore, the two-dimensional generalisation of the problems considered here seems to be reasonable. For some cases, such a problem has been studied both theoretically (Kirby and Dalrymple 1983; Kânoğlu and Synolakis 1998; Dingemans 1997) end experimentally (Thomson et al. 2007). The third generalisation is related to the finiteness of amplitude of real oceanic waves. In the coastal zones where the depth decreases, the nonlinear effects become important. Therefore, it is topical to consider the nonlinear wave transformation on various bottom topography. Interesting exact solutions for the particular bottom profiles in the long-wave approximation were found in the papers (Mirchina and Pelinovski 1992; Pelinovsky 1996; Aksenov et al. 2018). Further efforts in the study of nonlinear effects in the coastal zones are highly desirable.

Acknowledgements

A.E. acknowledges the financial support obtained from the Australian Government Research Training Program Scholarship. Y.S. acknowledges the funding of this study from the State task program in the sphere of scientific activity of the Ministry of Education and Science of the Russian Federation (Project No. 5.1246.2017/4.6) and the grant of the President of the Russian Federation for state support of leading Scientific Schools of the Russian Federation (Grant No. NSH-2685.2018.5).

References

Aksenov, A. V., Dobrokhotov, S. Y., & Druzhkov, K. P. (2018). Exact step-like solutions of one-dimensional shallow-water equations over a sloping bottom. *Mathematical Notes, 104*, 915–921.

Belibassakis, K. A., Simon, B., Touboul, J., & Rey, V. (2017). A coupled-mode model for water wave scattering by vertically sheared currents in variable bathymetry regions. *Wave Motion, 74*, 73–92.

Carrier, G. F., & Greenspan, H. P. (1958). Water waves of finite amplitude on a sloping beach. *Journal of Fluid Mechanics, 4*, 97–109.

Dean, R. G. (1964). Long wave modification by linear transitions. *Journal of the Waterways, Harbors, and Coastal Engineering Division, 90*, 1–29.

Devillard, P., Dunlop, F., & Souillard, B. (1988). Localization of gravity waves on a channel with random bottom. *Journal of Fluid Mechanics, 186*, 521–538.

Didenkulova, I., & Pelinovsky, E. (2013). Analytical solutions for tsunami waves generated by submarine landslides in narrow bays and channels. *Pure and Applied Geophysics, 170*(9–10), 1661–1671.

Didenkulova, I., & Pelinovsky, E. (2016). On shallow water rogue wave formation in strongly inhomogeneous channels. *Journal of Physics A: Mathematical and Theoretical, 49*, 194001.

Dingemans, M. W. (1997). *Water wave propagation over uneven bottoms. Parts 1 and 2*. Singapore: World Scientific.

Ferrario, F., Beck, M. W., Storlazzi, C. D., Micheli, F., Shepard, C. C., & Airoldi, L. (2014). The effectiveness of coral reefs for coastal hazard risk reduction and adaptation. *Nature Communications, 5*, 37–94.

Harris, D. L., & Vila-Concejo, A. (2013). Wave transformation on a coral reef rubble platform. *Journal of Coastal Research, 65*, 506–510.

Jung, T.-H., Suh, K.-D., Lee, S. O., & Cho, Y.-S. (2008). Linear wave reflection by trench with various shapes. *Ocean Engineering, 35*, 1226–1234.

Kajiura, K. (1961). On the partial reflection of water waves passing over a bottom of variable depth. In *Proceedings of the tsunami meetings, 10th Pacific Science Congress, IUGG Monograph* vol. 24 (pp. 206–234). Honolulu, Hawaii: University of Hawaii.

Kânoğlu, U., & Synolakis, C. E. (1998). Long wave runup on piecewise linear topographies. *Journal of Fluid Mechanics, 374*, 1–28.

Kirby, J. T., & Dalrymple, R. A. (1983). Propagation of obliquely incident water waves over a trench. *Journal of Fluid Mechanics, 133*, 47–63.

Kurkin, A. A., Semin, S. V., & Stepanyants, Y. A. (2015). Transformation of surface waves over a bottom step. *Izvestiya, Atmospheric and Oceanic Physics, 51*(2), 214–223.

Lamb, H. (1932). *Hydrodynamics*. Cambridge: Cambridge University Press.

Levin, B. W., & Nosov, M. A. (2009). *Physics of tsunamis*. Berlin: Springer.

Massel, S. R. (1989). *Hydrodynamics of coastal zones*. Amsterdam: Elsevier.

Mei, C. C. (1990). Basic gravity wave theory. In J. B. Herbich (Ed.), *Handbook of coastal and ocean engineering vol. 1: Wave phenomena and coastal structures, chapter 1* (pp. 1–62). Houston: Gulf Publishing Company.

Mirchina, N. R., & Pelinovski, E. N. (1992). Nonlinear transformation of long waves at a bottom step. *Journal of Korean Society of Coastal and Ocean Engineers, 4*(3), 161–167.

Newman, J. N. (1965). Propagation of water waves past long two-dimensional obstacles. *Journal of Fluid Mechanics, 23*(1), 23–29.

Pelinovsky, E. N. (1996). *Tsunami wave hydrodynamics*. Nizhny Novgorod: Institute of Applied Physics of the Russian Academy of Sciences.

Polyanin, A. D., & Zaitsev, V. F. (2003). *Handbook of exact solutions for ordinary differential equations*. Boca Raton: Chapman and Hall/CRC.

Rey, V. (1992). Propagation and local behaviour of normally incident gravity waves over varying topography. *European Journal of Mechanics, B/Fluids, 11*, 213–232.

Rey, V., Belzons, M., & Guazzelli, E. (1992). Propagation of surface gravity waves over a rectangular submerged bar. *Journal of Fluid Mechanics, 235*, 453–479.

Sretenskii, L. N. (1977). *Theory of wave motions in a fluid*. Moscow: Nauka.

Stoker, J. J. (1957). *Water waves: The mathematical theory with applications*. Hoboken: Wiley.

Thomson, J., Elgar, S., Herbers, T. H. C., Raubenheimer, B., & Guza, R. T. (2007). Refraction and reflection of infragravity waves near submarine canyons. *Journal of Geophysical Research*, *112*, C10009.

Tokano, K. (1960). Effets d'un obstacle parallélépipedique sur la propagation de la houle. *La Houille Blanche*, *15*, 247–267.

Xie, J.-J., Liu, H.-W., & Li, P. (2011). Analytical solution for long-wave reflection by a rectangular obstacle with two scour trenches. *Journal of Engineering Mechanics*, *137*, 919–930.

(Received April 17, 2019, revised June 12, 2019, accepted June 14, 2019, Published online June 24, 2019)

The manufacturer's authorised representative in the EU is Springer Nature Customer Service Centre GmbH, Europaplatz 3, 69115 Heidelberg, Germany. If you have any concerns regarding our products, please contact ProductSafety@springernature.com

Printed and bound by CPI Group (UK) Ltd, Croydon, CR0 4YY
30/09/2026
02225492-0020